Fire and Life Safety Inspection Manual

Fire and Life Safety Inspection Manual

Eighth Edition

Robert E. Solomon, P.E.
Editor

National Fire Protection Association
Quincy, Massachusetts

Product Manager: Pam Powell
Developmental Editor: Robine Andrau
Editorial-Production Services: Techbooks
Composition: Techbooks
Cover Design: Groppi Advertising Design
Manufacturing Manager: Ellen Glisker
Printer: Courier Westford

Notice Concerning Liability: Publication of this work is for the purpose of circulating information and opinion among those concerned for fire and life safety and related subjects. While every effort has been made to achieve a work of high quality, neither the NFPA nor the authors and contributors to this work guarantee the accuracy or completeness of or assume any liability in connection with the information and opinions contained in this work. The NFPA and the authors and contributors shall in no event be liable for any personal injury, property, or other damages of any nature whatsoever, whether special, indirect, consequential, or compensatory, directly or indirectly resulting from the publication, use of or reliance upon this work.

This work is published with the understanding that the NFPA and the authors and contributors to this work are supplying information and opinion but are not attempting to render engineering or other professional services. If such services are required, the assistance of an appropriate professional should be sought.

NFPA No.: IM02
ISBN: 0-87765-472-7
Library of Congress Control No.: 2002108884

Printed in the United States of America
02 03 04 05 06 5 4 3 2 1

Contents

Preface vii

SECTION 1 GENERAL 1

Chapter 1 The Fire Inspector *Steven F. Sawyer* 3
Chapter 2 Inspection Procedures *Steven F. Sawyer* 7
Chapter 3 Housekeeping and Building Procedures *Jon Nisja* 19
Chapter 4 Report Writing and Record Keeping *Ronald R. Farr* 29

SECTION 2 BUILDING SYSTEMS AND FIRE PROTECTION SYSTEMS 37

Chapter 5 Building Construction Elements *Richard J. Davis* 39
Chapter 6 Classification of Construction Types *Richard J. Davis* 49
Chapter 7 Construction, Alteration, and Demolition Operations *Richard J. Davis* 59
Chapter 8 Protection of Openings in Fire Subdivisions *Richard J. Davis* 67
Chapter 9 Electrical Systems *Noel Williams* 79
Chapter 10 Heating Systems *Michael Earl Dillon* 95
Chapter 11 Air-Conditioning and Ventilating Systems *Michael Earl Dillon* 105
Chapter 12 Smoke-Control Systems *Michael Earl Dillon* 113
Chapter 13 Fire Alarm Systems *Lee Richardson* 123
Chapter 14 Water Supplies *Phillip A. Brown* 143
Chapter 15 Automatic Sprinkler and Other Water-Based Fire Protection Systems *Roland Huggins* 157
Chapter 16 Water Mist Systems *Jack R. Mawhinney* 177
Chapter 17 Special Agent Extinguishing Systems *Gerard G. Back III* 201
Chapter 18 Clean Agent Extinguishing Systems *Eric W. Forssell and Scott A. Hill* 209
Chapter 19 Portable Fire Extinguishers *Ralph J. Ouellette* 213
Chapter 20 Means of Egress *Ron Coté* 225
Chapter 21 Interior Finish, Contents, and Furnishings *Joseph M. Jardin* 231

SECTION 3 OCCUPANCIES 239

Chapter 22 Assembly Occupancies *Joseph Versteeg* 241
Chapter 23 Educational Occupancies *Joseph Versteeg* 249
Chapter 24 Day-Care Facilities *Joseph Versteeg* 255

Chapter 25 Health Care Facilities *Joseph M. Jardin* 259
Chapter 26 Ambulatory Health Care Facilities *Joseph M. Jardin* 267
Chapter 27 Detention and Correctional Occupancies *Joseph M. Jardin* 273
Chapter 28 Hotels *Joseph M. Jardin* 283
Chapter 29 Apartment Buildings *Joseph M. Jardin* 289
Chapter 30 Lodging or Rooming Houses *Joseph M. Jardin* 299
Chapter 31 Residential Board and Care Occupancies *Joseph M. Jardin* 305
Chapter 32 One- and Two-Family Dwellings *Joseph M. Jardin* 315
Chapter 33 Mercantile Occupancies *Joseph Versteeg* 321
Chapter 34 Business Occupancies *Joseph Versteeg* 327
Chapter 35 Industrial Occupancies *Joseph Versteeg* 335
Chapter 36 Storage Occupancies *Joseph Versteeg* 345
Chapter 37 Special Structures and High-Rise Buildings *Joseph Versteeg* 355

SECTION 4 PROCESS AND STORAGE HAZARDS 367

Chapter 38 Waste-Handling and Processing Systems *Sharon S. Gilyeat* 369
Chapter 39 Radioactive Materials *Wayne D. Holmes* 377
Chapter 40 Materials-Handling Systems *L. Jeffrey Mattern* 385
Chapter 41 General Storage *Stephen R. Hoover* 395
Chapter 42 Storage and Handling of Flammable and Combustible Liquids *Orville M. "Bud" Slye, Jr.* 415
Chapter 43 Gas Hazards *Theodore C. Lemoff* 431
Chapter 44 Combustible Dusts *William J. Bradford* 443
Chapter 45 Combustible Metals *Carl H. Rivkin* 453
Chapter 46 Chemicals *John A. Davenport* 463
Chapter 47 Plastics and Rubber *Steve Younis* 473
Chapter 48 Explosives and Blasting Agents *Lon Santis* 485
Chapter 49 Fireworks and Pyrotechnics *Ken Kosanke* 493
Chapter 50 Heat-Utilization Equipment *Richard A. Gallagher* 501
Chapter 51 Spray Painting and Powder Coating *Don R. Scarbrough* 517
Chapter 52 Welding, Cutting, and Other Hot Work *August F. Manz and Amy Beasley Spencer* 533
Chapter 53 Hazards of Manufacturing Processes *L. Jeffrey Mattern* 547
Chapter 54 Aerosol Manufacturing and Storage *Michael Madden* 559
Chapter 55 Protection of Commercial Cooking Equipment *R. T. "Whitey" Leicht* 573

Appendix Inspection Forms 587
Index 679
About the Editor 703

Preface

Building design and construction, the types and complexities of the systems that we now install in buildings, and the interaction between numerous systems that have to work in harmony to ensure our safety are more important than ever. The safety systems and related procedures that our building codes and occupancy standards demand cannot simply be installed never to be inspected again.

The inspection process is sometimes a long and tedious undertaking that does not appear to have immediate results or effects; it is generally not appreciated or even noticed by the public. It does, however, have major consequences for building owners, property managers, and insurance companies who assume some risk or liability for the use and purpose that a particular building or structure serves. In the worst-case scenario, equipment or building systems that malfunction because they were not inspected can result in serious injury or death to building occupants. First responders—fire fighters, EMS personnel, or on-site fire brigades—can also be unnecessarily jeopardized if they encounter a building that has an inadequate inspection plan.

The best fire inspection program is one nobody hears about; such a program identifies and corrects problems before they can have a negative impact on the building or the public. In some situations, fire inspection programs are nonexistent, inadequately staffed, or simply not supported with a strong enough level of commitment by the local government entity. In such cases, the consequences can lead to human suffering, devastation, financial loss, and business interruption.

At budget time, a commitment to an inspection program should be among the top priorities of a fire department or other authority having jurisdiction. Good inspection programs require expenditure of capital; investment in human resources to carry out the inspection program; and a commitment from the dedicated inspectors whose job it is to know the myriad code requirements, procedures, and processes. It is neither an easy nor a particularly glamorous job, but it is a crucial one that provides satisfaction when a potentially serious situation is uncovered before it becomes a problem. Fire loss statistics do not indicate every time something *didn't* happen, only when it *did* happen.

Although the protection of human life remains the paramount motivation to safely keep the building in operating order, the need to sustain continuance of an organization's mission is a close second. Office buildings, warehouses, and factories all serve some specific purpose for a community or a corporation. Specialized processes, such as cutting and welding, intentionally introduce ignition sources to and near some of these structures. The continuous process of following procedures and protocol, surveillance, common sense, and vigilance, coupled with ongoing inspection, testing, and maintenance of the structure's features and systems, is key to having a successful inspection program.

The purpose of the *Fire and Life Safety Inspection Manual* is to provide the most up-to-date information to those interested in fire protection, fire safety, and life safety inspections. This eighth edition of the *Inspection Manual* is a compilation of key inspection procedures, requirements, and regulations found in relevant resource documents. The manual identifies by system, occupancy, process, or some combination of these the dangerous and hazardous conditions that could be encountered in a structure and spells out the chief areas the inspector should be focused on during an inspection.

Inspectors should use the *Inspection Manual* to identify an existing deficiency, an imminently dangerous condition, or a fault in a procedure or a protocol that may result in a fire. It is one of the tools inspectors should have at their disposal. Inspectors should, however, verify all requirements by referring to the most recent edition of the relevant code or standard, as identified at the back of each chapter. The more than three hundred codes and standards maintained by the NFPA in the *National Fire Codes*® form the basis for the text, criteria, recommendations, and requirements that are found in the *Inspection Manual.*

Expanded and revised in this eighth edition are inspection checklist forms for selected chapters; these forms appear at the back of the book. The forms are not a recitation of every code requirement. Instead they can be used primarily as a memory jog. They are presented here in reproducible format for actual use in the field and are also available on NFPA's website at *http:///www.premiums.nfpa.org.*

In addition to the *Fire and Life Safety Inspection Manual,* NFPA's *Fire Protection Handbook* continues to be an excellent resource for both the rookie and the seasoned inspector. Also, new tools and resources that are more prevalent now than during previous editions of the *Inspection Manual* include personal computers, handheld computers, and the Internet. These tools allow inspectors to be more accurate in the hazards they identify, to follow up more efficiently, and to research a greater number of specialized problems or potential issues.

Much has changed since the previous edition of the *Inspection Manual,* but the philosophy it is based on remains constant—inspections can reveal deficiencies that must be corrected or abated, and doing so could prevent a potentially dangerous fire and life safety situation from developing.

ACKNOWLEDGMENTS

NFPA appreciates the time and effort the contributing authors and reviewers have made to the development of this eighth edition of the *Fire and Life Safety Inspection Manual.* As is true for all NFPA documents, a broadly based group of individuals provided commentary, resources, and, most importantly, their expertise to the subjects covered in the *Inspection Manual.* In terms of expertise, the authors and contributors represent the very best in their respective areas, as can be attested to by the brief biographical notes appearing at the beginning of each chapter. The editors and product management staff wish to thank all the individuals who revised, wrote, reviewed, or otherwise made a contribution, in one form or another, to this eighth edition of the *Fire and Life Safety Inspection Manual.*

Fire and Life Safety Inspection Manual

SECTION 1 General

CHAPTER 1 The Fire Inspector *Steven F. Sawyer*

CHAPTER 2 Inspection Procedures *Steven F. Sawyer*

CHAPTER 3 Housekeeping and Building Procedures *Jon Nisja*

CHAPTER 4 Report Writing and Record Keeping *Ronald R. Farr*

The Fire Inspector

CHAPTER 1

Steven F. Sawyer

The fire inspector is the key to the community's fire reduction program. Fire inspection is the means of discovering and eliminating or correcting deficiencies that pose a threat to life and property. The fire inspector may fall under the jurisdiction of either the public or the private sector, with most jurisdictions and some large industrial complexes having fire inspection personnel. In addition to performing fire inspection, today's fire inspectors may have other functions delegated to them, such as conducting fire investigations, providing public education, and examining plans.

Fire inspectors must possess excellent communications skills and be knowledgeable about property, occupancy and its contents, operations, and fire protection and life safety provisions of building and fire codes. They must have exceptional judgment and should understand their role in life safety and property conservation. Fire inspectors are part detective, part reporter, part technical consultant, part missionary, part educator, and part salesperson.

PHYSICAL CONDITION

While performing their duties, inspectors might have to climb ladders or stairs in tall buildings, walk long distances or for long durations, crawl into confined spaces, and lift or push heavy objects. Thus, they should have enough strength, agility, and stamina to perform the physical activities associated with these duties.

COMMUNICATION SKILLS

A vital part of the inspection process is discussing the problems or violations discovered and their potential solutions with owners, property managers, architects, engineers, lawyers, contractors, vendors, the fire service, and representatives from the insurance industry. Inspectors will also have to record the conditions found and actions taken. Therefore, they must be able to communicate clearly, both orally and in writing, and should use tact and discretion to maintain authority. This is critical to an effective, smoothly run fire prevention program.

AUTHORITY

Fire inspectors must know where their authority comes from and how to enforce the locally adopted fire and life safety regulations. If they are not able to gain compliance by using their communications skills, they may have to take legal action to force compliance, such as serving the property owner with a series of written notices of violation, each with a deadline for compliance. If, after serving the owner with such notices of violations, they are still not able to gain compliance, they will probably have to consult with a local or state government's attorney to start the appropriate legal action specified in the fire safety legislation.

In some situations policy may require immediate compliance, such as unlocking exit doors in an assembly occupancy. In other situations the inspector may have to issue a violation notice similar to a traffic ticket. In extreme cases, the inspector may have cause to order a building evacuated and closed.

Steven F. Sawyer is executive secretary of the International Fire Marshals Association and senior fire service specialist for NFPA. He has over 25 years of fire service experience as a deputy fire marshal and deputy fire chief.

The organization the inspector represents must support the inspector with clear authority and consistent policy enforcement. In addition, only enforceable and nationally recognized codes and standards, such as the *NFPA National Fire Codes®*, should be the basis for the jurisdiction's fire safety ordinances and laws.

KNOWLEDGE

The breadth of knowledge fire inspectors need is determined by the duties they are expected to perform and the types of facilities they will be inspecting, the materials contained in those facilities, and the operations the facilities house. Inspectors must also be familiar with construction practices, nationally recognized fire and life safety standards, and other agencies they can consult for advice, solutions to problems, or corrections for specific hazards.

Most important is for inspectors to be aware of their own limitations and to know when to ask questions. As is often said, there are no stupid questions, only stupid mistakes. If an owner asks a question to which the inspector does not know the answer, he or she should not try to fake it. Instead, telling the owner that the inspector will research the question and will get back to him or her with an answer is the best policy.

The interior layouts and other aspects of buildings are often altered with the introduction of new processes, new product lines, or new tenants. During periods of construction, renovation, or demolition, properties are especially vulnerable to fire and life safety concerns are increased. Inspectors should have sufficient knowledge of building construction and materials to recognize potentially hazardous conditions and to recommend temporary steps that can be taken during construction or renovation to provide for the fire and life safety of the structure.

Building Services

Building services can represent a fire hazard if they are not installed and maintained properly. Inspectors should be familiar with the fire hazards associated with electrical systems, heating systems, air conditioning and ventilating systems, waste-handling systems, and materials-handling systems.

Hazardous Materials

Inspectors should be familiar with the proper handling, storage, and protection of a wide variety of hazardous materials they might encounter during an inspection. Typical hazardous materials include flammable and combustible liquids, compressed or liquefied flammable gases, explosives, corrosives, reactive materials, unstable materials, toxic materials, oxidizers, radioactive materials, natural and synthetic fibers, combustible metals, and combustible dusts.

Process Hazards

Industrial processes can introduce unusual hazards. Inspectors should be able to recognize those hazards, know how to minimize them, and understand the fire protection methods appropriate for the hazards. The inspector may need to seek out industry experts or other written sources of hazardous processes.

Fire Protection Equipment

A variety of fire protection equipment might be installed or available on the premises. The most common are portable fire extinguishers, sprinkler systems, and standpipe and hose systems. However, some areas and processes may be protected by special fixed extinguishing systems. For example, a dipping or coating process might be protected by a system that uses carbon dioxide, dry chemical, or foam as the extinguishing agent. Inspectors should understand the application and operation of any extinguishing system equipment on the premises and should be able to evaluate the operational readiness of the private or public water supply systems.

A property might also be equipped with heat, smoke, and flame detection equipment. This equipment may provide early warning of a developing fire, and fire alarm systems and devices may alert the occupants and summon the fire department. Inspectors should be acquainted with the purpose and operation of such devices and systems.

SUMMARY

The main purpose of fire inspection is the discovery and correction of deficiencies that pose a threat to life and property. As this chapter points out, to perform their function effectively, fire inspectors must be in good physical condition; have excellent communication skills; have the authority to enforce locally adopted fire and life safety regulations; and possess a considerable breadth of knowledge of building services, hazardous materials, process hazards, fire protection equipment, construction practices, and nationally recognized fire and life safety standards.

BIBLIOGRAPHY

Cote, A. E., ed., *Fire Protection Handbook,* 18th ed., NFPA, Quincy, MA, 1997.

NFPA Codes, Standards and Recommended Practices

See the latest version of The NFPA Catalog for availability of current editions of the following documents.

NFPA 1031, *Standard for Professional Qualifications for Fire Inspector and Plan Examiner*
NFPA 1033, *Standard for Professional Qualifications for Fire Investigator*
NFPA 1035, *Standard for Professional Qualifications for Public Fire and Life Safety Educator*

CHAPTER 2

Inspection Procedures

Steven F. Sawyer

As pointed out in Chapter 1, fire inspectors are part detective, part reporter, part technical consultant, part missionary, part educator, and part salesperson. An inspection should inspire others to take action to reduce or eliminate fire and life safety hazards, encourage an improved attitude toward fire safety and building safety by management and employees, and provide a record of the findings and actions resulting from the inspection.

INSPECTOR'S EQUIPMENT

Personal Equipment

To conduct the fire inspection safely and efficiently, inspectors should have the proper equipment. Included in the equipment they will need is a visible means of identification, such as an identification card or badge; they should also wear a uniform or other appropriate attire to make them easily identifiable. Because inspectors usually get dirty during inspections, they may need to wear coveralls and perhaps overshoes in order to protect their uniform or street clothes and shoes. When conducting waterflow tests, they may need to wear boots.

Inspectors should be equipped with and use the same type of personal safety equipment as the jurisdiction requires the workers in the area being inspected to use. This could include a hard hat, safety shoes, safety glasses, gloves, and ear protection. In some environments inspectors may have to use respiratory protection devices.

Inspection Tools

The basic tools fire inspectors need are a flashlight, a notebook or clipboard on which to make sketches or record observations, report forms, and a pen or pencil. If a sketch is to be drawn, the accuracy of dimensions measured by pacing is often adequate. When greater accuracy is required, a 6-ft (182.8 cm) ruler or a 50-ft (1524 cm) measuring tape may be helpful. The inspector may also ask the owner for "as built" drawings, if they are available, from which accurate measurements can be taken.

More sophisticated equipment that inspectors might occasionally need includes gauges and connections for making waterflow measurements, a combustible gas detector for testing potentially hazardous environments, and other instruments for testing fire protection equipment and other safety items (Figure 2-1).

Steven F. Sawyer is executive secretary of the International Fire Marshals Association and senior fire service specialist for NFPA. He has over 25 years of fire service experience as a deputy fire marshal and deputy fire chief.

INSPECTION PREPARATION

If they are inspecting residential properties, inspectors will need little in the way of preparation after they have made a few inspections, except to remind themselves of chronic trouble areas that need to be checked carefully. If they are inspecting nonresidential properties, however, they should prepare themselves by

- Reviewing previous inspection reports, violation notices, surveys, and any construction plans

FIGURE 2-1 Pitot Tube Assembly Used for Making Waterflow Measurements

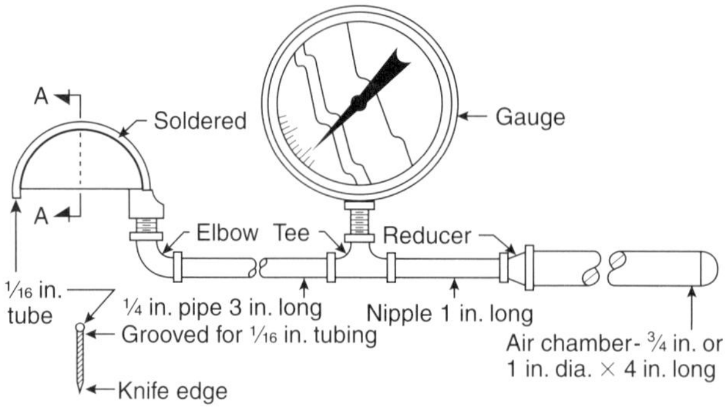

- Refreshing themselves about the operations and activities carried out on the premises
- Preparing a list of the more important points to be investigated before starting the inspection (The appropriate occupancy chapter in this book is a good place to start.)

If they are inspecting a property for the first time, they can add items to the inspection list from NFPA's current editions of *Fire Protection Handbook;* NFPA 1, *Uniform Fire Code;* NFPA *101*®, *Life Safety Code*®*;* local building codes, if applicable; and other NFPA codes and standards on specific occupancies, hazards, or fire protection features, such as sprinklers and standpipes. Inspectors may not have to do much preparation before inspecting small properties, such as one-story mercantile establishments or business occupancies. They should, however, review information on larger, more complex properties and the processes they contain before beginning the inspection.

Inspections are usually conducted during normal business hours, although advance arrangements can be made for inspections at other hours. For example, the inspector may visit a property at night to observe conditions during the night shift or at other times to check on special operations or as part of a permit. Normal business hours for many occupancies such as nightclubs or theaters may be at night.

The initial inspection should be performed by appointment. If an inspection time is likely to harm a good relationship between the inspector and the owner or manager, it might be prudent to reschedule the inspection, provided there is no evidence of an immediate fire or life hazard. For subsequent follow-up inspections, the element of surprise can be effective in determining true operating conditions. However, checking a restaurant's hood and duct system during peak mealtime hours is a good example of when *not* to conduct an inspection. Briefly walking through the facility to make sure exits are not blocked in the dining area during this busy time is appropriate, however. Follow-up inspections can be done with no appointment.

INTRODUCTIONS

Inspectors can make some general mental observations as they enter the property. They can observe the general occupancy, the condition of exterior housekeeping and maintenance, some building construction features, and the height of the facility. However, inspectors should not begin an exterior inspection without first introducing themselves to management and seeking permission to inspect the property.

If not part of the facility's staff, the inspector should make an effort to create a favorable impression in order to ensure cooperation and courteous treatment. Inspectors should enter the premises by the main entrance, seek out someone with authority, introduce themselves, and state the nature of their business. Inspectors should ask for permission to inspect the premises, not demand it. Inspectors have no reason to be irritated if they have to wait before receiving attention, especially if they have arrived without an appointment. The person the inspector needs to see may have other important matters to attend to first.

It would be wise to spend a reasonable amount of time making sure that whoever is in charge of the property understands why the facility is being inspected and answering any questions the property owner or manager might have, particularly during the first inspection of a property. Most properties have been inspected at some time, and records of such inspections can usually be obtained from the inspection agency's files. The records often contain plans that could save the inspector much time or work.

Inspections should always be conducted in the company of the property owner or a designated representative. This representative will help the inspector gain access to all parts of the property and will obtain answers to necessary questions. Of course, if the inspector is an employee of the occupant, there will be no need for a guide once the initial inspection has been made. Inspectors should work in pairs when inspecting residential properties to eliminate any potential complaints of impropriety.

INSPECTION OBSERVATIONS

The inspector either has or will be developing his or her own technique and methodology for inspecting a facility, and this methodology must be flexible to allow for variations and unexpected observations during the inspection. Either before or during the inspection, the inspector must ascertain both general and specific information in order to review and generate the appropriate recommendations or code compliance requirements. In addition to the customary information inspectors will obtain about the specific occupancy classification, they must determine several general facts before they can completely evaluate the occupancy and determine which code requirements it must meet.

Occupancy Classification

Inspectors should evaluate how the facility is used and determine which specific occupancy classification it falls under. Doing so will enable the inspector to choose the appropriate checklist and code requirements in order to accurately conduct the inspection and make the appropriate evaluations based on the *Life Safety Code* or the local building code as appropriate to will help in classifying the occupancy.

Sequence

The inspector should start by touring the outside of the facility to observe how the building or buildings relate to one another and to adjacent properties. A site plan of the property will help to visualize the layout of the premises. Obtaining an overall view of the property from the top of the tallest building might also be helpful.

Whether a building is inspected from top to bottom or from bottom to top is of little consequence; it is the inspector's choice. It is important, however, to conduct the inspection systematically and thoroughly. No area should be omitted. Every room, closet, attic, concealed space, basement, or other place where fire could start should be inspected. If the inspector is barred from an area for security reasons, he or she should note it on the inspection report. It may be advisable to obtain U.S. Department of Transportation (DOT), U.S. Department of Defense (DOD), Federal Aviation Administration (FAA), or other appropriate clearances to gain access to these particular areas.

The following gives a general indication of what the inspector should look for while going through the property. More specific information is contained in subsequent chapters and from the list the inspector prepared in the preliminary research.

Exterior

While touring the exterior areas of the property, the inspector should record the address, the names and types of occupancies, exterior housekeeping and maintenance, exterior evidence of building-construction type, any construction problems, and the building height. The inspector should note the location and character of potential exposures and the arrangement and condition of outdoor storage. He or she should also note the conditions affecting fire department response and fireground operations, including the location of public and private fire hydrants.

Accessibility is an important factor. Fire lanes should be well marked, unobstructed, and wide enough to allow fire apparatus to pass. Vehicular activity should be limited to the pickup and discharge of passengers, and parking should be prohibited in these areas. Hydrants and other sources of water must be accessible. Sprinkler valves must be open, and sprinkler and standpipe connections must be capped, free of debris, and accessible. The inspector should determine in which direction flammable liquids will flow if they are spilled and what sort of drainage facilities are provided.

Construction Classifications

An important point of all inspections is to determine accurately the construction classification of the building or structure. Such a determination will normally be based on NFPA 220, *Standard on Types of Building Construction,* or on the local building code. The inspector will need to be familiar with the definitions of the individual construction types defined in NFPA 220 or the local building code in order to accurately classify the different types of construction being inspected.

Once the inspector is familiar with the exact definitions of the different construction types, he or she will need to determine the similarities and distinguishing features of each category. A facility will often be composed of multiple construction types, and it is common for construction classifications to change as the building undergoes renovations, including alterations and additions. These factors can make the overall classification process complicated, and it may be impossible to determine one overall construction classification if the building is composed of multiple construction types. However, inspectors can simplify the classification process by dividing the structure during the inspection into sections based on building configurations and construction, renovations, alterations, and additions. Classifying the structure accurately is essential because the construction classification(s) will significantly affect the code requirements for the overall level of life safety and property conservation that can be provided inherently within the structure.

The type of construction and the materials used will influence the ease of ignition and the rate of fire spread. The integrity of fire-resistive walls and floor/ceiling assemblies must be assured. Openings in fire-rated walls must be protected to retard or prevent the spread of fire to other areas. Doors in fire-rated walls must be kept closed or close automatically to ensure a reasonably safe avenue of escape for the occupants and to restrict fire spread. If holes are made in these assemblies for the passage of services and utilities and the voids are not sealed, they could allow fire to spread horizontally and vertically throughout the facility.

Inspecting the integrity of exit enclosures is very important. Inspectors should check the door, penetrations, and other openings into each exit enclosure while inspecting each floor. Then they should inspect each exit stair enclosure for its full length. In taller buildings it is recommended that this be done from the top down for ease. Inspectors should use a different elevator to go back to the top each time, taking this opportunity to note if Phase II fire fighter service is provided for that elevator car. (See Chapter 20, "Means of Egress," for more information on exit enclosures.)

Many of these items are not readily obvious, and inspectors may find it necessary to examine concealed spaces, such as the voids above suspended ceilings, the interiors of shafts, and stair enclosures, to make sure that the integrity of these fire protection features has not been breached.

Building Facilities

Water distribution systems, heating systems, air-conditioning and -ventilating systems, electrical distribution systems, gas distribution systems, refuse-handling equipment, and conveyor systems all play an important role in the fire hazard potential of the premises. They must be properly installed, used, and maintained in order to minimize the hazard. Although inspectors are not responsible for maintaining such systems, they should be able to determine whether the equipment is being properly used and maintained. This may mean reviewing the equipment's maintenance records as part of the inspection process.

Hazards of Contents

The level of hazard of the contents of a building are categorized as low, ordinary, and high in of the *Life Safety Code.* The evaluation of the hazard level of the building contents will have a significant impact on the fire safety evaluation and the resulting recommendations. Therefore, it is critical for inspectors to be familiar with the following definitions of each category. (See of the *Life Safety Code,* 2000, for the exact definitions.)

Low Hazard. Low hazard contents are classified as those of such low combustibility that no self-propagating fire therein can occur. Storage of noncombustible materials is classified as low hazard. In occupancies not otherwise defined as low hazard, it is assumed that, even where the actual contents hazard is normally low, there is sufficient likelihood that some combustible materials or hazardous operations will be introduced in connection with building repair or maintenance or some psychological factor might create conditions conducive to panic. Because of this likelihood, life safety features cannot safely be reduced below those specified for ordinary hazard contents.

Ordinary Hazard. Ordinary hazard contents are classified as those that are likely to burn with moderate rapidity or to give off a considerable volume of smoke. Ordinary hazard classification represents the conditions found in most buildings and is the basis for the general requirements of the *Life Safety Code.* The fear of poisonous fumes or explosions is necessarily a relative matter to be determined on a judgment basis. All smoke contains some toxic fire gases but, under conditions of ordinary hazard, there should be no unduly dangerous exposure during the period necessary to escape from the fire area, assuming there are ample and properly arranged exits.

High Hazard. High hazard contents shall be classified as those that are likely to burn with extreme rapidity or from which explosions are likely. High hazard contents include occupancies where flammable liquids are handled or used or are stored under conditions involving possible release of flammable vapors; where grain dust, wood flour or plastic dust, aluminum or magnesium dust, or other explosive dusts are produced; where hazardous chemicals or explosives are manufactured, stored, or handled; where cotton or other combustible fibers are processed or handled under conditions producing flammable flyings; and other situations of similar hazard.

Inspectors should be aware that the classifications used by NFPA 13, *Standard for the Installation of Sprinkler Systems,* may be different. For example, a business occupancy will have "ordinary hazard contents," as defined by the *Life Safety Code.* For purposes of selecting a sprinkler design density, however, NFPA 13 will define an office as a "light hazard occupancy."

At the time of the inspection, inspectors must determine the hazard level of the building's contents based on their observations of the actual contents of the building or structure. Controlling the hazards of materials depends on storing, handling, using, and disposing of them properly. In this regard, inspectors should pay particular attention to housekeeping and storage practices. They should also be familiar with any process that might cause a fire hazard or any special features of the property that might present special problems.

During the inspection, inspectors might want to use the process of elimination to accurately determine the hazard level of the contents. They should begin by asking the question "could a self-propagating fire occur within that space?" This question should be based on the type and burning characteristics of the fuel located in the building and its specific arrangement in relation to other fuel stations. The low hazard level of contents category does not imply that no fire can occur; it implies only that fire will not spread from one combustible item to another. Low hazard contents are rarely found in occupancies; thus this condition normally would not be a major classification during most inspections. Because the vast majority of structures have contents classified as ordinary hazard, it is normally best to skip this category and determine if the contents fall into the high hazard category.

When making this determination inspectors must use a great deal of judgment based on their experience in the field and their ability to make observations and assess burning characteristics of various fuels. To classify contents as having a high hazard, the contents would have to burn at a very fast rate and have dramatic burning characteristics that could render the occupied space unsafe at a faster rate than the occupants could evacuate. High hazard contents could explode in the occupied area and also produce significant and unusual amounts of poisonous fumes, thus exposing the occupants to a high level of personal hazard. Flammable liquids, gases, dusts, or solid combustibles with a very high rate of heat release are included in this category.

Inspectors will often be able to easily eliminate categories of low and high hazard contents, which leaves only contents of ordinary hazard. To ascertain if contents fall under this classification, inspectors need to determine if the contents in the building are liable to burn with moderate rapidity or give off a considerable volume of smoke, but would not necessarily produce poisonous fumes or explosions. This classification includes typical combustion products such as carbon monoxide and hydrogen cyanide.

It is commonly believed that the most hazardous classification will prevail as the overall classification for the building, but this is not usually the case. Normally, when some contents are of a high hazard, the area is protected as a subcategory, but it will not be the determining overall hazard level of contents classification. Inspectors must make sure that the provisions of special protection are provided and are adequate before areas of high hazard contents can be segregated from the overall classification.

To make this concept tangible, consider the following example of an educational facility. In a college, typical classrooms and office areas normally would have a sufficient amount of fuel in a configuration that would allow a self-propagating fire to occur, but neither poisonous fumes nor explosions would be produced. As a result, this area would be classified as having ordinary hazard contents.

Restrooms probably have some amount of fuel that could allow a fire to begin; but in the appropriate type of construction, the fire most likely would burn without significantly affecting the structure or the egress time of the occupants. As a result, these areas can be classified appropriately as having low hazard contents. There may also be laboratories that utilize high-pressure reactors for research purposes, that store considerable amounts of flammable liquids or flammable gases, or that store a host of other hazardous materials that would classify that part of the occupancy as having high hazard contents.

If the following requirements are met, then the appropriate hazard level of contents classification would be ordinary, but inspectors must consider all three classifications when making their evaluation:

1. The high hazard items are appropriately protected and segregated from other areas.
2. The low hazard items do not contribute significantly to the overall square footage of the facility.
3. The ordinary hazard items predominate.

It is very common for a structure to have either two or all three hazard categories because the hazard level of contents may change as one moves through the facility. Determining the hazard level of contents will allow inspectors to make a more precise assessment of the facility, and the correct occupancy classification will direct inspectors to the use of the appropriate code requirements.

Fire Detection and Alarm Systems

Often, a property will be equipped with fire detection and alarm devices and systems. The purpose of such equipment is to detect the presence of fire, alert the occupants, notify the fire department, activate fire suppression systems, and perform other functions such as close doors, turn off air handling equipment, or a combination of these functions. Inspectors should understand the function of, and be able to identify, the major components of these systems. Routine inspections should ensure that manually operated fire-alarm devices are clearly marked, accessible to occupants, and properly maintained. Tests should be performed by a representative of the owner and witnessed by the inspector to confirm that the systems are in operating condition. (For further information on fire detection alarm systems, see Chapter 13.)

Fire Suppression Equipment

Inspectors should carefully check the fire suppression equipment on the premises. Typical equipment includes sprinklers and standpipe systems and portable fire extinguishers. Routine inspections should determine that sprinkler valves are open, sprinklers are unobstructed, the system has not been altered, and the sprinkler system has been extended to cover building additions. Standpipes should be checked for proper operation and to ensure that caps are in place and hose valves are closed.

Inspectors should determine that portable fire extinguishers of the proper size and type are provided for any given hazard and that they are serviceable, clearly identified, and accessible to the occupants. Inspectors should also check special extinguishing systems for special hazards to ensure that they have been maintained and are serviceable, and inspectors should conduct or witness periodic operational tests of fire-extinguishing equipment. (See Chapters 15, 17, and 18 for further information on these systems.)

Surveying and Mapping

INSPECTION SUMMARY

During the initial inspection, inspectors should gather information that will be used to prepare a site plan if one does not already exist. Such information will include construction features, occupancy data, fire protection features, and exposures.

The site plan is a scaled drawing that indicates the locations and dimensions of the buildings and fire protection equipment (including water-distribution systems) and the specific hazards and hazardous processes in each building (Figure 2-2). To

FIGURE 2-2 A Typical Site Plan

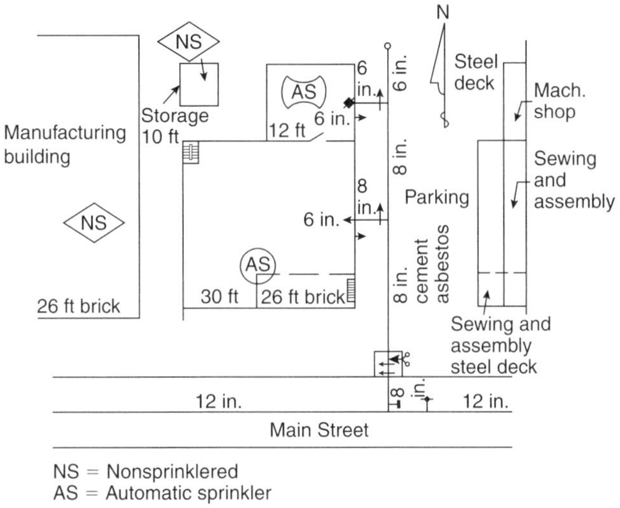

show details of the fire protection features, the inspector may have to draw a series of side sketches, which need not be drawn to scale. These maps should be incorporated into the fire department prefire plan.

Closing Interview

At the conclusion of the facility tour, the inspector should discuss the results with the owner or the owner's representative. Conditions that seriously jeopardize the safety of the occupants and the property itself may have been found and should be corrected immediately. In the case of an in-house fire inspector, the inspector or his or her supervisor often has the authority to remedy hazardous situations. Imminent safety to life issue should be addressed immediately; others may be corrected over a predetermined time. However, all inspectors will have to rely on their regulatory authority to convince the owner or the representative that corrective action should be taken at once.

Reports

All violations noted during the inspection should be included in the written report to the owner and filed in the building file. Items corrected during the inspection should also be added to the report for future reference. There will be times when items that do not present an immediate threat to life safety will have to be corrected at a substantial cost to the owner. In such cases, inspectors should go back to their office, research the adopted codes and standards to ensure that they have accurately noted the code violation, and produce a typewritten correction order. The inspector should call the owner or the owner's representative to schedule a meeting, should deliver the correction order in person, and should fully explain the reasoning behind the requirement.

During the inspection process, inspectors may be asked to clarify a provision of a code or standard or be asked a technical question. In a field of knowledge as complex and diversified as fire protection, it is impossible for anyone to know all the answers. For example, NFPA develops more than 300 codes and standards. It is better for inspectors to admit they do not know the answer than to try to bluff their way through. Remember, the inspector must have the trust of those responsible for making and financing corrections to violations. To gain trust, the inspector must always be completely honest.

The inspector should write down the question and the name and telephone number of the individual and should tell the person that he or she will provide an answer. The inspector now has the time needed to research the question, consult with other enforcement officials, talk to his or her supervisor, or call the organization responsible for the requirement. As soon as possible, the inspector should call back with an answer.

For less urgent conditions or conditions that will take time to correct, inspectors should explain their recommendations clearly so that the owner fully understands the problem and the options available to correct them. Inspectors should express their view in easy-to-understand terms. They should not engage in arguments, technicalities, or petty faultfinding, any of which will antagonize the people inspectors most want to influence. In all cases, inspectors should explain any appeal process or procedures for granting equivalencies.

A written report should be prepared for each inspection. The amount of detail required will depend on the character and purpose of the inspection and the local requirements. In general, every report should include the following information:

1. Date of inspection
2. Name of inspector
3. Name and address of property, noting the name and title of the person(s) interviewed, and phone numbers
4. Name and address of owner (or agent if a different location), and phone number
5. Names of tenants of a multiple occupancy building (but not necessarily the name of every tenant in an apartment building or office building)
6. Type of occupancy (if mixed occupancy, each principal occupancy and its location; if an industrial plant, the principal raw materials and finished product)
7. Dimensions of buildings, including height and construction type
8. Factors that could contribute to fire spread inside buildings, such as open stairways, elevator and utility shafts, and lack of vertical and horizontal cutoffs
9. Common fire hazards, such as open flames, heaters, and inadequate wiring
10. Special fire hazards, such as hazardous materials and their storage, handling, use, and processes
11. Extinguishing, detection, and alarm equipment
12. Employee fire safety organization
13. Adequacy and accessibility of exits
14. Exposures, including factors making fire spread possible between buildings
15. Recommendations or notations of violations

The purpose of this report is to describe the property and its use, hazards, and fire protection without going into unnecessary detail. An inspection report should give the reader a clear understanding of the conditions found and the corrections needed.

A checklist might be adequate for routine procedures, such as determining whether a sprinkler valve is open. When a measurement, such as water or air pressure, is to be checked, however, provision should be made for entering the actual measurement.

Hazardous practices and conditions are best treated in the narrative form. Inspectors who are required to describe the conditions they have observed are likely to do a more thorough job than those who merely complete a checklist. A checklist cannot be devised to take into account every situation that could conceivably arise, and an inspector could easily miss some hazard that a checklist does not include. Some agencies computerize data from the inspections they conduct.

The inspector's recommendations or correction orders for reducing hazards and improving protection constitute an important part of the reporting process. Recommendations or correction orders can be prepared as a separate document and submitted to the property owner or manager for consideration. A copy should be filed with the inspection report.

If the purpose of the inspection is code enforcement, the inspector should identify the code violations and give a date by which compliance is expected. Follow-up inspections should then be conducted to ensure proper compliance with the requirements.

DAILY INSPECTIONS

In many facilities, there are items that should be checked daily or at some more frequent interval. These items should be compiled into a list that inspectors can use to guide them and to ensure that each item is checked. The following is a partial list that might apply to an individual facility:

1. Check that exit doors are not improperly locked.
2. Check the control valves, fire department connections, and gauges on sprinkler and standpipe systems.
3. Check the status lamps on fire pump control panels to be certain the equipment is energized.
4. Check the status lamps and trouble lights on the fire alarm panels.
5. Check that all fire doors are closed.

In addition, there could be other items in the various departments of the facility that require a daily or periodic check. One convenient routine is to provide a card for each item to be checked. These cards should be kept at the location to be checked, and the employee responsible for the inspection should be required to initial, check, and record the necessary observations on the appropriate card. Entries should show the date, the time, and the name of the person making the observation. Bar-code readers also can be used for this purpose in many cases.

It is not enough for management to specify that daily checks must be conducted. The individuals assigned to make the checks must feel that if the matter is important enough to be recorded, it must be done correctly. The property manager or on-site fire inspector should review the cards or records weekly, and the results should be summarized in the weekly report of loss prevention activities.

SUMMARY

Fire inspectors need to observe, report, educate, and inform the people they come in contact with about fire safety and fire and life safety hazards within their buildings. The inspector must be prepared to perform inspections by having the proper equipment; by being familiar with the applicable fire and life safety regulations, codes, and standards; and by having reviewed previous inspection reports.

During inspections, inspectors should note various features of the building such as its construction type, its occupancy classification, the hazards of its contents, the processes performed within it, its fire protection systems, and any additional fire and life safety items. The information collected should be developed into a written report to the owner on deficiencies noted and corrective action to be taken and into prefire plans for the fire department. Records of each inspection should be maintained with the inspection agency.

BIBLIOGRAPHY

Cote, A. E., ed., *Fire Protection Handbook,* 18th ed., NFPA, Quincy, MA, 1997.

NFPA Codes, Standards, and Recommended Practices

See the latest version of The NFPA Catalog for availability of current editions of the following documents.

NFPA 1, *Uniform Fire Code*

NFPA 13, *Standard for the Installation of Sprinkler Systems*
NFPA *101*®, *Life Safety Code*®
NFPA 170, *Standard for Fire Safety Symbols*
NFPA 220, *Standard on Types of Building Construction*
NFPA 1031, *Standard for Professional Qualifications for Fire Inspector and Plan Examiner*
NFPA 1201, *Recommendations for Developing Fire Protection Services for the Public*
NFPA 1452, *Guide for Training Fire Service Personnel to Conduct Dwelling Fire Safety Surveys*

Housekeeping and Building Procedures

CHAPTER 3

Jon Nisja

Practicing good housekeeping, which can be described as plain common sense, is an effective fire prevention measure. People do not need an extensive background in fire protection to recognize poor housekeeping practices that pose a fire safety risk. Poor housekeeping practices should serve as a caution to the inspector; places that have poor housekeeping practices often have other fire safety deficiencies.

Effective indoor and outdoor housekeeping practices accomplish the following major objectives:

1. Eliminating unwanted fuels
2. Removing obstructions
3. Controlling sources of ignition
4. Improving safety for fire-fighting and emergency response personnel

Certain aspects of housekeeping are common to almost all types of occupancies; other aspects are unique to certain occupancies. To describe every aspect of housekeeping for all of the various occupancies would be impossible. Inspection personnel must be able to recognize housekeeping problems and take actions to eliminate them and educate the property owners on the importance of reducing these types of hazards. Inspectors who take the time to educate property owners will find that the incidence of housekeeping problems will decrease.

PRINCIPLES OF GOOD HOUSEKEEPING

Three basic requirements of good housekeeping are equipment arrangement and layout; material storage and handling; and operational neatness, cleanliness, and orderliness.

Equipment Arrangement and Layout

Looking at the equipment being used and related storage and workspaces can reveal housekeeping problems. Production processes can cause aisles to be blocked or obstructed with storage. These problems can often be corrected with relatively simple modifications to the working area, equipment, or procedures.

Material Storage and Handling

Adequate storage provisions are lacking in many buildings. This lack of space encourages people to store materials wherever it is convenient for them to do so. Many occupancies use carts or powered equipment to move materials from one area in a building to another. Often, the carts or moving equipment are left in undesirable locations.

These housekeeping practices can cause egress aisles and exit doors to be blocked. Another egress concern is storage under stairways (Figure 3-1). The area under stairs

Jon Nisja, a supervisor of inspections for the Minnesota State Fire Marshal Division, has been involved in fire prevention inspections and investigations since 1982. His areas of interest include building construction, means of egress, and fire protection systems (sprinklers and fire alarms).

FIGURE 3-1 Combustible Storage Under Wooden Stairway

should not be used for storage unless it is separated from the stairs by fire-rated construction. Badly stored material can obstruct access to fire-protection equipment, such as fire extinguishers (Figure 3-2), control valves, and fire-alarm pull stations; it can also impair the proper operation of passive fire-protection equipment, such as fire doors.

FIGURE 3-2 Access to Fire Extinguisher Blocked by Storage

In some cases the storage arrangement—for example, high-piled combustible storage with very narrow aisles and unstable piles—may pose a risk to fire-fighting personnel. Such an arrangement could hamper fire-fighting operations, and the piles could collapse on fire fighters, especially when the boxes became saturated with water or are weakened by fire damage.

Operational Neatness, Cleanliness, and Orderliness

Good housekeeping practices can be as simple as keeping all areas as neat and clean as possible. Emptying trash and waste on frequent-enough intervals to prevent accumulation is one example of an effective housekeeping practice. The frequency of trash removal may vary substantially depending on the amount of trash or waste generated. Some occupancies, such as large retail stores, have very complex and involved trash- and waste-handling operations to remove, compact, and/or bale materials generated in their operation. Figure 3-3 shows a production area in a scene shop that needs to be cleaned and organized.

Many states and jurisdictions are required to participate in environmental recycling programs. Although these materials are not considered trash or waste, they represent the same types of housekeeping problems—they are often combustible and they are frequently stored in undesirable arrangements or locations. The number and size of containers is also much greater. Individual containers are located at each work area, with larger collection points in key areas.

Inspection personnel should check for general cleanliness in the building. Obviously, the level of cleanliness is relative to the operation or type of business. Some businesses, such as wood shops, repair garages, and agricultural mills, will be messier than a typical restaurant, school, or office occupancy. A business that keeps its operation neat and clean will have less risk of fire. Fire inspectors can check whether wastebaskets or trash cans are available so that the building's occupants will find tidiness convenient and whether the facility has a regularly scheduled waste-removal program.

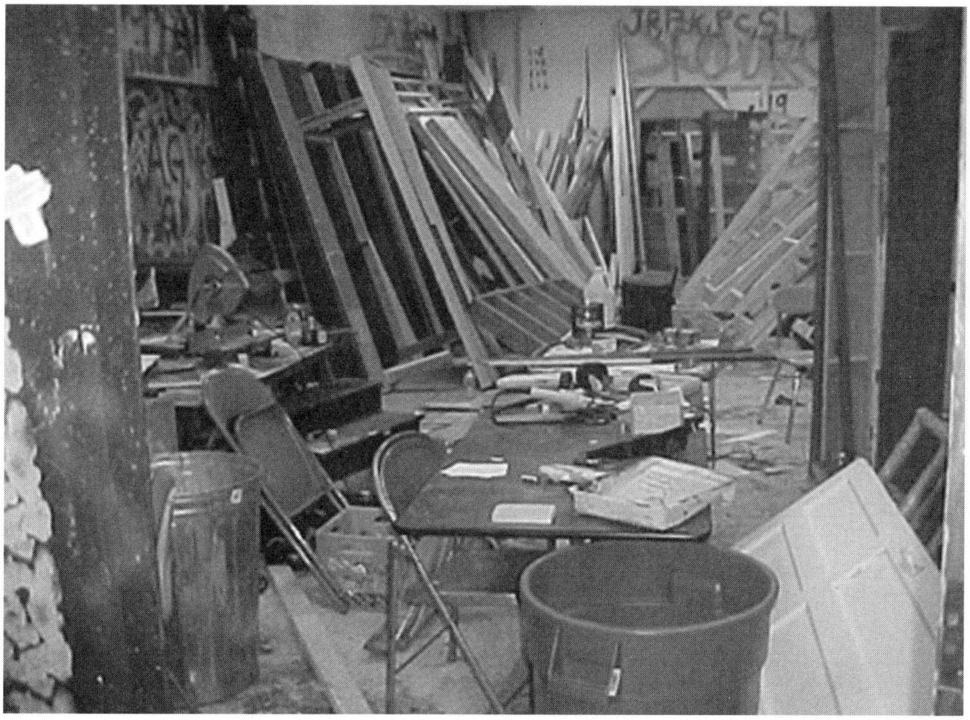

FIGURE 3-3 Woodworking Shop with Blocked Aisles and Debris Everywhere

HOUSEKEEPING PROBLEMS

Different types of occupancies or businesses have different housekeeping problems depending on the nature and type of processes taking place in them.

Flammable/Combustible Liquid Spills

Spills can happen whenever flammable or combustible liquids are handled or used. Businesses handling flammable or combustible liquids should have an adequate supply of absorptive materials and tools to control, contain, or clean up spills. The most common type of absorptive material is a granular product similar to cat litter; it is available from automotive parts stores, hardware stores, and safety supply stores.

Flammable/Combustible Liquid Waste Disposal

Environmental concerns have made many types of flammable and combustible liquids expensive and somewhat difficult to dispose of. Flammable liquid wastes should never be drained into sewers or dumped onto the ground. Fire inspectors should review the flammable liquid disposal procedures with the facility's management. The plan may involve contracting with a hazardous waste disposal contractor.

Paintings, Coatings, Finishes, and Lubricants

Paints, grease, oil, and similar materials are sources of readily combustible residues. Spray finish booths, exhaust ducts, fans, and motors need to be cleaned frequently to prevent dangerous accumulation of residues. Filters in spray finish operations should be installed and replaced frequently to minimize residue accumulation. Figure 3-4 shows a paint spray booth where the filters have deteriorated and need to be replaced.

FIGURE 3-4 Spray Finish Booth with Deteriorated Filters

Because of the high number of fires associated with spray-finishing operations, the area or spray booth is required to have an automatic fire suppression system. Fire inspectors need to note how sprinklers are protected against residue accumulation in spray finish booths and ducts. One method is to cover the sprinkler with a light plastic or paper bag that is changed on a regular basis to avoid excessive accumulation or residue.

Many motors, engines, compressors, and similar equipment require frequent lubrication for proper operation. Although some lubrication is necessary, an excessive amount can become a source of fuel. Excessive lubrication can also attract dirt and dust and cause overheating of the equipment.

Oily Waste and Rags

Oily waste and rags are commonly found in restaurants for cleaning cooking equipment, vehicle repair garages, industrial occupancies, paint-spraying operations, and building maintenance areas. Because they can spontaneously ignite, oily waste and rags are a fire safety hazard. Clean cotton waste or wiping rags represent less of a fire problem but still pose a risk due to contamination with certain types of oils. Oily waste and rags should be stored in metal containers with tight-fitting covers. Commercially made containers are available for this purpose (Figure 3-5).

Packing and Shipping Materials

Almost all packing and shipping materials are combustible and represent a severe fire risk. Cardboard, paper, Styrofoam, expanded plastics, excelsior, straw, and similar materials are used for packing and shipping. These materials are all relatively easy to ignite and have relatively high rates of heat release. They are treated as clean waste; large quantities should be kept in special vaults or storerooms.

FIGURE 3-5 Container for Oily Waste and Rags

Fire inspectors should note the condition of shipping and receiving rooms. They should pay particular attention to large quantities of accumulated waste around packaging or unpackaging operations. A regularly scheduled program for cleaning and waste removal should be instituted immediately.

BUILDING CARE AND MAINTENANCE

Maintenance and upkeep procedures inside the building can reduce the probability of ignition and limit the available fuel supply. Some maintenance procedures, however, can pose fire risks.

Floor Cleaning and Treatment

Floor cleaning, treatment, or refinishing can be a fire hazard if flammable solvents or finishes are used. In addition, the removal, sanding, and refinishing of floor surfaces—especially wood floors—can generate combustible dusts and residues. This risk has been reduced recently because the finishing industry has developed newer products and finishes that are not classified as flammable liquids.

If a floor-cleaning or -refinishing operation is conducted, there should be adequate ventilation and only materials having a flash point above the highest room temperature should be used. In addition, nonsparking equipment should be used and there should be no open flames in the area. The containers for the solvents or cleaning compounds should be labeled as to their flammability or combustibility. If no labels are present, material safety data sheets should be reviewed.

Some floor treatments and dressings contain oils or compounds that can spontaneously ignite. Oily mops, towels, or rags used to apply these treatments or dressings should be stored in the same manner as oily rags, that is, in metal containers with tight-fitting lids.

Flammable Cleaning Solvents

The use of flammable cleaning solvents is becoming fairly rare today as many nonflammable solvents have been developed that have high flash points, are very stable, and have limited toxicity problems. Flammable liquids are still used for some cleaning purposes; examples are alcohols and paint thinners. Flammable liquids used for cleaning should be stored in safety cans with tight-fitting lids and used only for dispensing small quantities. Flammable liquids should not be stored in open pails, buckets, dip tanks, or in containers that may be degraded by the liquid.

Kitchen Cooking Hoods, Exhaust Ducts, and Equipment

Grease accumulation on cooking hoods in kitchens, on the hood's grease filters, or inside the exhaust duct represents a serious fire safety risk. This accumulated grease can be ignited from the sparks or heat from the cooking operation or from a small fire on the cooking surface.

The kitchen hood and filters should be inspected and cleaned on a very regular basis. Inspections will reveal how often cleaning will need to occur; it is recommended that the owner conduct inspections at least weekly. It may have to be inspected and cleaned daily depending on the amount of food cooked and the method of cooking.

The entire hood, grease-removal appliances, exhaust ducts, fans, and related equipment need to undergo a thorough cleaning on a regular basis. This can be a very messy and difficult job, especially in the ductwork itself. Commercial firms that specialize in this type of cleaning should be utilized. Once again, the amount of food cooked and the method of cooking will determine the necessary cleaning frequency. If any grease is observed dripping from the kitchen hood, filters, exhaust duct, or the exterior of the building, an immediate cleaning is needed. It is important that filters be in place and clean to ensure proper operation.

Dust and Lint Removal

Dust and lint can accumulate in combustible or even explosive quantities on walls, ceilings, motors, heating equipment, structural members, or inside ducts and conveying equipment. Removing it can be dangerous because an explosion can occur if it is not done correctly. In most cases, the dust or lint accumulation can be removed using vacuum-cleaning equipment with an explosion-proof motor.

Compressed Gas Cylinders

Many facilities need to use compressed gas cylinders as part of their operation. There are restrictions on the amounts of certain types of compressed gases, such as flammable gases, permitted in an area. Compressed gas cylinders should also be secured in a manner to prevent them from being knocked over. Figure 3-6 provides an example of an unsecured compressed gas cylinder. If the cylinder tips over and the neck of the cylinder ruptures, the cylinder can be propelled like a rocket and do considerable damage.

FIGURE 3-6 Unsecured Gas Cylinder

CONTROL OF SMOKING

By controlling smoking, a common ignition source can be eliminated. Smoking introduces a flame (lighter or matches) and a smoldering heat source (the cigarette or cigar). Cigarette smoking is generally on the decline in the United States. In addition, many states and communities have enacted smoking regulations that limit smoking in public buildings and workplaces. These factors have led to a decrease in the emphasis that fire inspectors need to place on the control of smoking. It has, however, also led to an increase in people smoking in areas where smoking represents a fire hazard. The fire inspector should look for evidence of smoking—ashtrays, ashes, or cigarette butts—in prohibited areas.

The are situations where smoking should definitely be prohibited and "No Smoking" signs should be prominently displayed. Smoking should always be forbidden near flammable liquids, both indoors and outdoors; near flammable gases, such as LP-Gas and acetylene; and in areas where there are large quantities of combustibles, such as retail and mercantile occupancies. Smoking should also be prohibited in areas where dust accumulations are present—such as woodworking plants—and in areas where there are combustible decorations.

In areas where smoking is allowed, approved smoking receptacles should be provided. This type of receptacle keeps the cigarette inside the receptacle rather than allowing it to fall out as it burns down. These smoking receptacles can also be filled with sand to assist in extinguishing the smoking materials.

Control of smoking materials is a topic that fire inspectors can cover when giving public fire safety education presentations to community, civic, or workplace groups. Emphasis must be placed on not smoking in bed and on having working smoke alarms.

OUTDOOR HOUSEKEEPING

Outdoor housekeeping issues are typically easy to correct. One example might be premise identification—that is, the posting of an address—to aid emergency responders in finding the building. Often the identification numbers are missing or the posted address has deteriorated. Larger complexes may have separate streets or roads and a private building numbering system.

Poor housekeeping outside of a facility can threaten the fire safety of accessory structures, products stored outside, building utility equipment, and the structure itself. Accumulations of rubbish, tall grass, weeds, and waste materials adjacent to structures, exterior storage, and utility equipment are common problems.

Wildland Interface

Buildings and facilities constructed near large wildland areas without an intervening firebreak are of growing concern. Several businesses and apartments have been lost as the result of a fast-moving wildland or forest fires.

Weeds, Grass, and Vegetative Materials

Weeds, grass, shrubs, and trees can cause various concerns for the fire inspector. These materials can block the view of or impair access to fire protection equipment, such as hydrants and fire department connections. Trees or bushes can become overgrown and block exit doors.

Tall grass and weeds should never be allowed to grow near flammable or combustible liquid storage or liquefied petroleum gas (LP-Gas) installations. A fire involving grass or weeds could have a disastrous outcome if it reaches the tanks containing the product. These situations can be corrected by cutting the weeds or grass, trimming the shrubs and trees, or applying herbicides or weed killers.

Many chemicals used as herbicides are chlorate compounds. Chlorate compounds are oxidizing agents and can contribute to fire conditions, especially during protracted dry periods when the vegetative materials have dried out.

Waste, Rubbish, and Refuse Disposal

Goods, production materials, and waste products stored in piles outdoors should be separated from one another and kept away from combustible buildings. The separation will help prevent a fire from spreading from one pile to another and will provide a means of fire apparatus access.

Waste, rubbish, and similar combustibles should never be stored or allowed to accumulate next to flammable or combustible liquid or flammable gas storage tanks or containers. Figure 3-7 shows an example of combustible debris stored next to an LP-Gas tank.

It is common to see idle or damaged wood pallets stored outside next to a building. Another common fire safety problem is the storage of waste rubber tires outside a building. Both of these are dangerous practices because these materials burn very intensely and can cause severe damage to the building or its contents.

Dumpsters and similar waste receptacles should be located at least 10 ft from combustible buildings and should not be placed under roof eaves or overhangs. Combustible waste materials from industrial and manufacturing operations are commonly stored on-site before being hauled away. These accumulations should be at least 20 ft from buildings; it may be preferable to increase this distance to 50 ft if there are relatively large quantities of combustible materials. Having these outdoor storage areas fenced to prevent unwanted persons from getting access is a good idea. Ignition sources, such as from smoking or from cutting, welding, and other hot work operating tools should be kept at a safe distance of 20 ft or more.

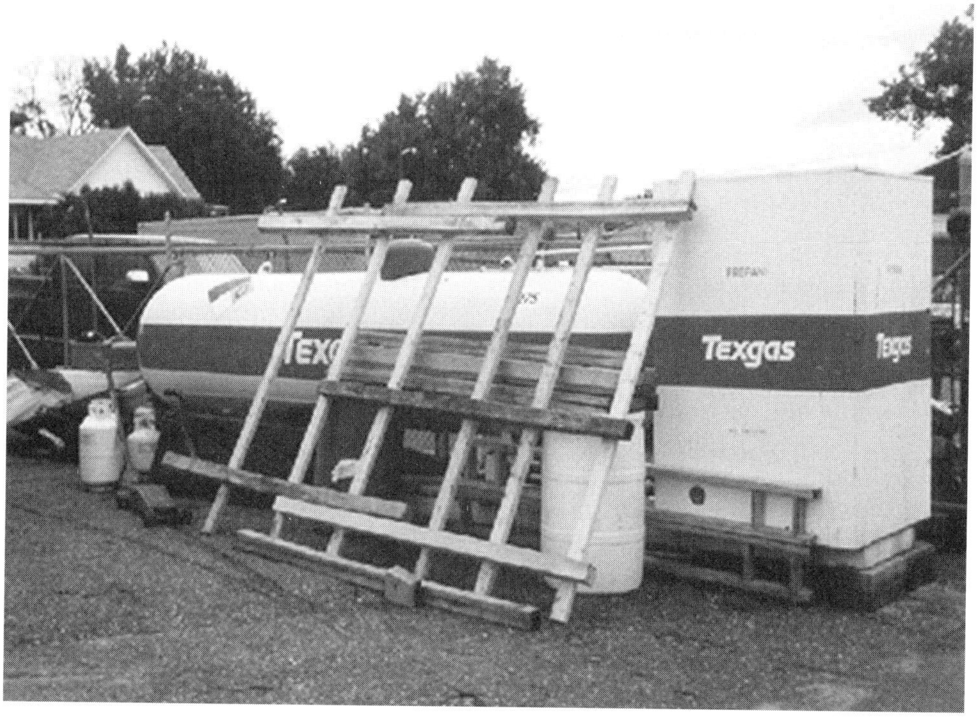

FIGURE 3-7 Excessive Storage and Debris Under and Adjacent to LP-Gas Storage Tank

Snow Removal

In some areas of the United States, snow accumulation can pose additional fire problems. Snow and ice must be removed from around fire hydrants and fire protection equipment. For larger buildings it may be necessary to plow snow from fire lanes and fire apparatus access roads. Ice accumulations on gas meters can cause the meter or regulator to overpressure and fail. One of the most common problems, however, is the failure to keep exits and outside egress paths shoveled or cleared of snow.

SUMMARY

Fire safety problems related to housekeeping practices can be very easy to recognize and often involve minimal expense to correct. Although recognizing and correcting these deficiencies may involve minimal effort, the importance of good housekeeping practices in preventing ignition and minimizing a fire's impact cannot be overstated. Many devastating fires have been caused by a failure to provide or maintain good housekeeping. The correction of housekeeping-related problems improves fire safety by eliminating excessive fuels, removing obstructions to fire safety and egress features, controlling sources of ignition, and increasing safety for responding personnel.

Storage and operational issues often lead to housekeeping problems. The fire safety solution may be something simple, such as removing waste materials more often or finding an alternate storage arrangement. Working with property owners to educate them on the dangers of these unsafe practices can assist in preventing future housekeeping problems.

BIBLIOGRAPHY

Cote, A. E., ed., *Fire Protection Handbook,* 18th ed., NFPA, Quincy, MA, 1997.

Cote, A. E., and Linville, J. L., eds., *Industrial Fire Hazards Handbook,* 3rd ed., NFPA, Quincy, MA, 1990.

"Good Housekeeping as a Means of Protection," in *The Handbook of Property Conservation,* Factory Mutual System, Norwood, MA, 1983, pp.169–172.

Overview Manual, 3rd ed., Industrial Risk Insurers, Hartford, CT, 1989.

NFPA Codes, Standards, and Recommended Practices

See the latest version of The NFPA Catalog for availability of current editions of the following documents.

NFPA 1, *Uniform Fire Code*

NFPA 13, *Standard for the Installation of Sprinkler Systems*

NFPA 25, *Standard for the Inspection, Testing, and Maintenance of Water-Based Fire Protection Systems*

NFPA 58, *Standard for the Storage and Handling of Liquefied Petroleum Gases*

NFPA *101*®, *Life Safety Code*®

NFPA 241, *Standard for Safeguarding Construction, Alteration, and Demolition Operations*

NFPA 914, *Code for Fire Protection of Historic Structures*

CHAPTER 4

Report Writing and Record Keeping

Ronald R. Farr

Fire safety inspections are a valuable component of a complete fire prevention program. In the opinion of many fire safety experts, fire safety inspections are one of the most important non-fire-fighting functions the fire service provides. Important components in the inspection process are identifying deficiencies that are observed during the inspection, communicating those deficiencies to the person responsible for corrective action, and maintaining a record of the inspection. Solutions are discussed with the owner, architect, engineer, and others who may be a part of the corrective action process. Documenting the conditions found during the inspection and the actions taken and communicating these issues both orally and in a written form are necessary to the procedure.

DOCUMENTING THE INSPECTION

The findings of fire safety inspections and all information pertaining to contacts with the facility must be documented. Providing a formal inspection report each time is not always necessary; furnishing details of the contact for the files, however, is important. Fire inspectors must develop good writing skills to communicate important issues they have uncovered during the inspection. A written overview of the issues—whether in the form of a memo, a written inspection report, or a formal letter—is important for appropriate documentation. Policies within the fire-prevention bureau may dictate whether an inspection report or formal letter is to be used.

If the inspector decides to use the written inspection report format, he or she should style the document like a "who, what, where, why, and how" investigative report, indicating the following:

- *Who* are the responsible parties (i.e., who has been made aware of the deficiency and will be taking corrective action?)
- *What* issue needs to be corrected (i.e., what has been identified as the deficiency or violation?)
- *Where* the deficiency was located
- *Why* this issue was identified (i.e., what section of the fire code regulates this issue?)
- *How* this issue can be corrected (i.e., what needs to be done to correct the problem?)

Rather than a written inspection report, the fire inspector may choose to send a letter that identifies the deficiencies and the required corrective action. Such a letter is generally prepared in a formal style. In some cases, it may be appropriate for the inspector to use both formats, that is, to write an inspection report and to follow it up with a formal letter.

Ronald R. Farr, currently fire marshal with the Kalamazoo Township Fire Department in Kalamazoo, Michigan, has over 30 years' experience in fire safety. He is a member of NFPA and the International Fire Marshals Association.

WRITTEN INSPECTION REPORT/FORMAL LETTER

The purpose of providing a written inspection report is to relay to the reader information regarding code enforcement that has been obtained during the inspection. Without going into detail, the inspection report describes the property and its use, the hazards, and the fire protection systems. In a written inspection report or formal letter, the inspector is generally concerned with presenting facts and evidence to prove a point, draw a conclusion, or justify a recommendation. In some instances the inspection report will be used only to document the inspection and the action to be taken to correct the problem.

The inspector may use the inspection report to develop the formal inspection letter that is sent to the owner or manager of the property. This information—whether as a written report or formal letter—must be presented in a professional and businesslike manner. It should not be opinionated, biased, emotional, or unfair. Code violations must be identified and a date must be given by which compliance is expected or a plan of action to abate the violations is established. Follow-up inspections should then be conducted to ensure proper compliance with the requirements.

Written documentation should be prepared for each inspection. The amount of detail required will depend on the character and purpose of the inspection. In general, every letter or report should include the following information:

1. Name, address, and phone number of the property inspected
2. Name, address, and phone number of the owner (The inspector should obtain the name of a local contact person or representative if the owner is not on site or is not readily available.)
3. Name of the person present during the inspection if not the owner
4. Date of the inspection
5. Type of occupancy (If the building has multiple or mixed occupancies, the inspector should gather information for all of the occupancies within it or treat each occupant as an independent facility.)
6. Fire protection features such as automatic fire suppression systems, portable fire extinguishers, and fire alarm and detection systems
7. Deficiencies or violations found and corrective actions required or taken
8. Date for a reinspection
9. Name of the inspector

The inspector may also wish to include additional information on an inspection report that pertains to the means of egress; special hazards or processes; dimensions of the building, including the height, area, and construction type; fire brigades or internal fire safety teams; and exposure problems with adjacent buildings.

When the inspector provides recommendations (for issues not required by code), the inspector should group them and place them at the end of the report or letter. The inspector must make sure the owner understands that recommendations are just that and not code requirements. When a deficiency is identified and the inspector is responding with a formal letter, he or she should list the section number from the fire code relating to the issue in the narrative section of the letter. In the final portion of the written report or formal letter, the inspector should offer to answer any questions regarding the inspection the owner might have.

If the inspector has prepared a handwritten report in the field, he or she may leave a copy of it at the facility. Field written reports are generally checklists and may also include an area to be used to identify minor deficiencies. If the inspector returns to the office to prepare the written report or formal letter, he or she must write it in a timely fashion. The written report or formal letter can then be mailed. If necessary, depending on legal requirements for due notice, it can be sent by certified or registered mail so there is a receipt of delivery.

For less urgent conditions or conditions that will take time to correct, the inspector should explain the recommendations clearly so that the owner understands the problem and the options available to correct the problem. Inspectors should express their views in easy-to-understand terms and should not engage in arguments, technicalities, or petty fault finding, any of which will antagonize the people they most desire to influence. In all cases, inspectors should explain the appeal processes or the procedures for granting equivalencies.

An inspection report should give the owner a clear understanding of the conditions found and the corrective action necessary. A checklist may be adequate for routine procedures, such as determining whether a sprinkler valve is open or whether portable fire extinguishers are present. When a measurement, such as water or air pressure, is to be checked, however, the report should provide a place for entering the actual measurement data. Checklists can be a good guide to be used as a basis for information collection.

Hazardous practices and conditions are best recorded in the narrative form rather than on a checklist. Inspectors who are required to describe the conditions they have observed are likely to do a more thorough job than those who merely complete a checklist. A checklist cannot be devised to take into account every situation that could conceivably arise, and an inspector could easily miss some hazard that a checklist does not include. Some agencies computerize data from the inspections they conduct. In these situations the terminology and data classifications contained in NFPA 901, *Standard Classification for Incident Reporting and Fire Protection Data,* can be helpful.

Recommendations or correction orders given for reducing hazards and improving protection constitute an important part of the reporting process. Recommendations or correction orders (in the form of a formal inspection letter) can be prepared as a separate document and submitted to the property owner or manager for consideration. A copy should be filed with the inspection report.

Follow-up Inspection Letter

FOLLOW-UP INSPECTIONS

Follow-up inspections are made to ensure that the deficiencies noted in the inspection report have been corrected. Fire inspectors should confirm the time and date of the follow-up inspection with the owner prior to their arrival. In other cases, the inspector may not want to make an appointment for a follow-up inspection.

When performing a follow-up inspection, the inspector does not need to inspect the entire occupancy again. Instead, he or she should inspect only the problem areas included in the inspection report to verify that the hazards have been corrected. If, however, a violation is discovered during a follow-up inspection that was not found during the initial inspection or is a new violation, the inspector should identify it. All of this information should then be noted in the follow-up inspection report.

If all deficiencies have been corrected, the inspector should commend the owner for taking the appropriate actions. Then, to close out the file, the inspector should send a follow-up letter stating that the follow-up inspection found the violations to have been corrected. Inspectors should exercise caution, however, when issuing such a letter so they do not give the owner the impression that the facility is hazard free. The inspector could have missed something that might cause a problem later. This letter also gives the inspector a second opportunity to thank the owner/occupant for his or her cooperation.

If the owner/occupant is making a conscientious effort to comply, but some deficiencies remain to be corrected, the inspector should commend the owner/occupant on the progress made to that point. The inspector should then set a date and time for yet another follow-up inspection and should add a written update of the findings to the inspection files, with the original copy going to the owner/occupant.

Final Notice

If the hazards have not been corrected and it is apparent that the owner/occupant has made no effort to correct them, the inspector should issue a final notice with a date for another inspection. The final notice should inform the owner/occupant exactly what legal action will be taken if full compliance is not attained by the date specified. The fire inspector must follow through with appropriate legal action. If the inspector uncovers a hazard during an inspection that requires immediate corrective action and was corrected at that time, he or she still needs to record the hazard on the report and make a note that the issue was corrected at the time of the inspection.

REPORT WRITING

The majority of report writing consists of documenting serious or numerous fire code violations and the conditions found in the building. Reports are designed to provide vital and useful information to the building owner/occupant and the fire department or inspection agency. The information in the report should be factual and should be delivered in a manner that is easily understood and not misleading. Although perhaps difficult to do at times, inspectors must not inject their personal feelings into the document. Doing so could reduce their credibility.

Producing a well-written report requires the use of complete sentences, proper grammar, and an appropriate choice of words. These basic writing techniques apply to any type of report writing. A report may appear to contain the necessary information, but improper grammar or incomplete sentences may render it incomprehensible to the reader. Two methods to remedy report-writing problems are to practice writing skills and to enlist someone else's assistance in proofreading the report.

Misspelled words reflect poorly on the writer and will make the reader question the report's technical content. Questions regarding spelling should be resolved by consulting a dictionary. Most word-processing programs have built-in spelling-correction features. Caution, however, should be exercised when spell-check programs are used because such programs identify only misspelled words, not misused words, for example, words that sound the same but are spelled differently. Run-on sentences or sentence fragments make understanding the writer's meaning difficult. Good report writing calls for short, clear sentences. For correct word usage and punctuation, a good, easy-to-use manual should be consulted.

Taking extra time to prepare and write a report will help in the catching of common writing mistakes. Once the report is grammatically correct and the proper use of words and the appropriate punctuation have been assured, effort should be devoted to making the report neat in appearance and legible. In general, typewritten or computer-generated reports are preferable.

Following these basic report-writing guidelines will enable fire inspectors to present themselves and their agency in a positive and professional manner. Written and verbal communications are an integral part of a fire inspector's everyday activity, so it is crucial that clear and effective communications become second nature. It is beyond the scope of this manual to provide extensive instructions in business communications; therefore, individuals in a fire inspector positions are urged to complete a business communications course.

INSPECTION INFORMATION NOTE TAKING

During the inspection fire inspectors should record sufficient information pertaining to the identified issues so they can prepare accurate reports. Each inspector will develop his or her own method or style of note taking. If the inspector writes incomplete notes, he or she may not be able to remember what the deficiency or hazardous issue was. This is especially true if the inspector conducts several inspections during a shift and then returns to the office to prepare written reports or responses for each. To avoid such a situation, notes should be easy to understand and should provide an orderly approach to inspection information. Specific locations should be identified to assist the inspector in remembering the details of the issue. Once inspectors find a note-taking style they prefer, they should follow that style throughout all inspections to provide consistency.

At times the inspector will be unable to give an immediate answer to a question. The inspector should not feel embarrassed to inform the questioning person that a period of time will be needed to obtain an answer. If that happens, the inspector should write down what information is needed and should inform the person that he or she will receive a response shortly. The inspector will then have time to research the question and provide a correct answer. The inspector must remember to respond to the query. The inspector should never guess at an answer to a question.

WRITING REPORTS AND LETTERS

Draft Letter

In today's business world a large number of issues are handled orally. Writing, however, is still the preferred and most accurate method of communication. A well-written report or letter that documents the issues can help in demonstrating a high level of professionalism. Most people can, with a little practice, write a report or letter that intelligently describes an issue or hazard.

Inspectors who have never written letters or inspection reports before should familiarize themselves with basic letter-writing techniques and should practice writing letters and reports. Inspectors need to have a basic understanding of the mechanics of letter writing and the required content necessary for a letter. They can obtain several samples of well-written letters to use as a guide and can follow a basic format or style the department may have instituted. Samples can be found in IFSTA's *Fire Inspection and Code Enforcement* or from other inspecting agencies in the fire inspector's region.

Once the inspector settles down to write the letter, he or she may want to first develop a draft letter and review the draft to make sure all the information is included and necessary corrections have been made. The inspector also needs to make sure the content of the letter is understandable. Having another inspector read the letter will provide a check on the letter's accuracy and clarity. Another method of ensuring the owner will understand the document is to have someone not familiar with inspection issues read the letter and give the inspector feedback. These checks are ways of ensuring that the inspector has not written something inappropriate, misleading, or inaccurate.

Mechanics of Letter Writing

The mechanics of letter writing identifies issues such as what type of stationery to use, how to set up the letter on the page, and whether to use a memo format or a formal letter style. Department policies may dictate what style of written communication to use and when to use it. Generally, fire prevention bureau letterhead stationery and the use of the formal letter style of written communication to describe fire safety issues are preferred.

When developing the content of a formal letter, the inspector needs to understand the principles of good letter writing, correct grammar, and appropriate word usage. The contents of the letter must be reviewed to ensure that unnecessary words and phrases are eliminated and that all necessary information is included. The inspector must remember that a piece of communication will become an official document and part of the property's file and could also become evidence in legal proceedings.

Fire inspectors must also remember that a piece of correspondence is a reflection of them as well as of the department or agency that employs them. The appearance of the letter can lead the reader to form an opinion regarding the inspector's level of professionalism and attention to details.

Elements of Inspection Letters

Fire inspection letters will contain most of the following: return address (if not written on department letterhead stationery), date, inside address, reference/subject, salutation, body of the letter, closing/signature/name/title, enclosure(s), and copies.

A basic letter-writing format, in order of appearance, is as follows:

1. *Letterhead or return address:* If fire prevention bureua letterhead stationery is not used, the inspector should include the complete return address of the agency.
2. *Date:* The date the letter is written. The preferred method is to spell out the month with numerical identification for the day and year (May 5, 2001) rather than a total numerical identification (5/5/01). Using the numerical identification, however, is acceptable for a memo.
3. *Inside address:* The person's name with the title; if the title is short (e.g., Dr.), it generally appears on the first line; if the title is long or has multiple words (e.g., Senior Project Manager), it should be placed on the next line. The address, including the business name, follows. If the inside address is long, it should also be separated and placed on two lines to provide a balanced appearance.
4. *Subject/reference:* This line states in a few words the reason or purpose of the letter. It may also contain an inspection reference number or case number.
5. *Salutation:* Contains the person's name to whom the letter is being written. The salutation can also include the person's title.
6. *Body of the letter:* In general it contains the issues or deficiencies that are being addressed. When listing violations, the inspector may do so by following the route taken during the inspection. This can also assist the recipient in identifying the location of the needed corrective action. The inspector should remember to include the section from the fire code that identifies the issue as a deficiency. At the conclusion of the body of the letter, the inspector needs to identify what type of response is expected and when a reinspection will be made to verify corrective action.
7. *Closing/signature/name/title:* A complimentary closure, such as "Sincerely," followed by four spaces and then the inspector's name and title.
8. *Enclosure(s):* Lists the items included with the letter or indicates the number of items enclosed with the letter after the word "enclosure(s)."
9. *Copies:* Indicates who have been sent copies of the letter.

RECORD KEEPING

Cross-Referenced Filing System

Another important and valuable part of the fire prevention program involves maintaining a record-keeping or -filing system and accounting for all activities that may pertain to a particular business or address. Maintaining accurate inspection information of past and current activities about an occupancy is a critical component. These records can provide a historical overview of what has happened or what is happening with the building or address. Such issues as alterations to the building, violations and corrective actions that have taken place, and permits that may have been issued can provide the fire department with valuable information.

Generally, a file will be maintained on all occupancies within the jurisdiction and maintained in the fire prevention bureau offices. These files will contain materials such as the following:

- Inspection reports, forms, and written communications
- Permits or licenses
- Complaint information or investigations
- Court notices or notices of violations
- Plan reviews, approvals, or denials
- Modifications or variances
- Fire inspection reports
- Fire investigations

Files will also contain information relative to fire suppression or detection systems in the occupancy, hazardous materials or operations that are present, and information gathered during a prefire survey. All of this information is valuable when operational procedures for a fire department response is being planned.

The preferred method of filing information is by the street address of the facility. A building will undergo many changes during its life in its use, its owner, and the building name, but the address will stay constant. When a filing or record-keeping system is established, it should be a cross-referenced filing system that allows inspectors to locate facilities by either their business name or the address of the occupancy.

Departmental policy and any legal requirements of the jurisdiction determine what is to be included in the departmental files and permit filing to be done on a timely basis. The inspector must remember that departmental files are official documents that are open to the public for review and must be properly maintained.

Electronic Records

Many fire service agencies document fire safety inspection information and other important departmental information electronically. The use of computerized inspection forms and record-keeping systems has been credited with improving efficiency in the overall record-keeping process. Computerized programs have helped inspectors to better track and schedule required inspections, including follow-up inspections. Most fire service agencies find it easier to use prepared programs rather than putting the required resources into developing one of their own programs.

As electronic technology improves, inspectors will probably be using handheld computers in the field to record inspection information, and then, when they return to their office, downloading this information into the main computer system for further processing.

Inspectors should take care to back up electronic information and have paper copies to guard against the possibility of electronic failure. Departmental policy should determine who has access to computer files and what the limits of such access are. The agency must also provide all employees with proper training to ensure the program's efficiency.

SUMMARY

This chapter has provided information that is necessary to document fire code violation during a fire safety inspection. An overview has been provided on how to write a fire inspection report or formal letter in reference to an inspection and how to maintain files relative to occupancies within a jurisdiction.

BIBLIOGRAPHY

Cote, A. E., ed., *Fire Protection Handbook,* 18th ed., NFPA, Quincy, MA, 1997.
IFSTA, *Fire Inspection and Code Enforcement,* 6th ed., Fire Protection Publications, Stillwater, OK, 1998.

NFPA Codes, Standards, and Recommended Practices

See the latest version of The NFPA Catalog for availability of current editions of the following documents.

NFPA 901, *Standard Classification for Incident Reporting and Fire Protection Data*
NFPA 1031, *Standard for Professional Qualifications for Fire Inspector and Plan Examine.*

SECTION 2 Building Systems and Fire Protection Systems

CHAPTER 5 Building Construction Elements *Richard J. Davis*

CHAPTER 6 Classification of Construction Types *Richard J. Davis*

CHAPTER 7 Construction, Alteration, and Demolition Operations *Richard J. Davis*

CHAPTER 8 Protection of Openings in Fire Subdivisions *Richard J. Davis*

CHAPTER 9 Electrical Systems *Noel Williams*

CHAPTER 10 Heating Systems *Michael Earl Dillon*

CHAPTER 11 Air-Conditioning and Ventilating Systems *Michael Earl Dillon*

CHAPTER 12 Smoke-Control Systems *Michael Earl Dillon*

CHAPTER 13 Fire Alarm Systems *Lee Richardson*

CHAPTER 14 Water Supplies *Phillip A. Brown*

CHAPTER 15 Automatic Sprinkler and Other Water-Based Fire Protection Systems *Roland Huggins*

CHAPTER 16 Water Mist Systems *Jack R. Mawhinney*

CHAPTER 17 Special Agent Extinguishing Systems *Gerard G. Back III*

CHAPTER 18 Clean Agent Extinguishing Systems *Eric W. Forssell and Scott A. Hill*

CHAPTER 19 Portable Fire Extinguishers *Ralph J. Ouellette*

CHAPTER 20 Means of Egress *Ron Coté*

CHAPTER 21 Interior Finish, Contents, and Furnishings *Joseph M. Jardin*

Building Construction Elements

CHAPTER 5

Richard J. Davis

The type of construction and the materials used in a building influence the building's life-safety and property-protection requirements. Inspectors have a major responsibility in determining that those requirements are met throughout the life of the building. To discharge that responsibility, they must know the functions of the various structural elements of a building, which is the topic of this chapter, and understand the significant characteristics of the various construction types, which is the topic of Chapter 6, "Classification of Construction Types." Because space in this manual is limited, inspectors should refer to NFPA's *Fire Protection Handbook* and Frank Brannigan's *Building Construction for the Fire Service,* both of which contain significant additional information.

DEFINITIONS

Knowing the meaning of key terms is essential to an understanding of how the different building construction elements interact.

- *Bearing wall:* A bearing wall is any wall meeting either of the following two classifications: (1) any metal or wood stud wall that supports more than 100 lb/linear ft (1400 N/m) of vertical load in addition to its own weight or (2) any concrete or masonry wall that supports more than 200 lb/linear ft (2900 N/m) of vertical load in addition to its own weight.
- *Dead loads:* Dead loads are loads consisting of the weight of all materials of construction incorporated into the building, including, but not limited to, walls, floors, roofs, ceilings, stairways, built-in partitions, finishes, cladding and other similarly incorporated architectural and structural items, and fixed service equipment, including the weight of cranes.
- *Environmental loads:* Environmental loads are loads caused by the environment, including wind load, snow load, ice load, rain load, earthquake load, and flood load.
- *Foundation systems:* Foundation systems include foundation walls, footings, posts, piers, piles, caissons, or slabs-on-grade.
- *Live loads:* Live loads are loads produced by the use and occupancy of the building.
- *Nonbearing wall:* A nonbearing wall is any wall that is not a bearing wall.

Richard J. Davis, a senior engineering specialist with FM Global, is chair of the NFPA Technical Committee on Construction, Alteration, and Demolition Operations and a member of the NFPA Technical Committee on Building Construction and the NFPA Technical Committee on Structures and Construction.

FRAMING MEMBERS

In general, the structural components of a building can be divided into two groups: those elements that support the structure, or its framing members, and those that enclose the working, storage, and living spaces, that is, its nonbearing walls, floors, ceilings, and roofs. The framing members form the skeleton of a building, which supports the building and everything attached to it and is part of the *dead load.* The frame also supports the *live load,* or the building's contents and its occupants, as well as the *environmental loads.*

Foundation Systems, Columns, and Bearing Walls

The structural frame is supported on foundation systems, which transfer the loads imposed on the building to the earth below. Columns or bearing walls are located on top of these footings and support the floor or floors above and the roof. The failure of a column or columns from fire exposure is critical because it can result in the collapse of a floor or, in extraordinary cases, of the entire building.

Horizontal Structural Members

Trusses, beams, girders, joists, and rafters are all structural members that support a ceiling, floor, or roof. Live loads and environmental loads must follow a load path and are typically applied to the cladding or decking. These loads are transferred through various roof or floor structural members to columns or bearing walls, into foundation systems, and into the earth.

Trusses are composed of steel, wood, combinations of steel and wood, or concrete. They include elements of upper and lower chord members connected to each other with vertical and/or diagonal web members to span large distances and to transmit the load directly or indirectly to the building's columns or bearing walls. Because of their relatively light weight and the type of connections, wood truss members are more commonly used in residential construction or in small commercial projects.

A *beam* is a relatively large horizontal structural member [e.g., 4 × 6 in. to 6 × 10 in. (100 × 150 mm to 150 × 250 mm) wood beam] into which other members, such as joists [e.g., 2 × 8 in. or 2 × 10 in. (50 × 200 mm or 50 × 250 mm) wood members], can be framed. If a beam is supported only at its ends, it usually is called a *simple span beam.* If it spans three or more supports, it is a *continuous span beam.*

A *girder* is a deep beam, such as a glued and laminated (glu-lam) wood beam, into which other beams are generally framed.

A *joist* is one of a series of smaller, parallel members used to directly support either the floor, in which case they are called floor joists, or the ceiling, where they are called ceiling joists. Joists generally are framed into beams or bearing walls (Figure 5-1). Wood joists are usually spaced 16 in. (400 mm) on center.

A rafter is similar to a joist except that it supports the roof. Rafters are closely spaced and are usually framed into beams or bearing walls.

For industrial and larger commercial construction, steel framing as well as concrete construction is often used. For construction such as an insulated steel deck, the deck is supported by secondary structural framing such as open-web steel joists. The depth of the joists varies considerably, depending primarily on its span but also on the design load and its spacing on center. The first number in a steel joist designation is its depth in inches. For example, a 24K6 joist is 24 in. (600 mm) deep. The "K" indicates that the steel has a yield stress of 50,000 lb/in.^{2} (35.2 kg/mm^2). The last number in this example, "6," is the chord designation, which is the relative ranking of the joist's strength. Everything else being equal, the higher the chord designation, the stronger the joist.

Measuring the depth of the joist and its spacing on center is relatively easy. For new construction, this measurement should be taken and the results checked against the design drawings.

Steel joists are supported by primary structural members, such as wide-flange, solid-web steel beams, or open-web joist girders. Steel joists are usually 6 ft (1.8 m) or more on center and often span 40 to 50 ft (12.2 to 15.2 m). For wide-flange beams, the "W" indicates that it is a wide flange. The first number in its designation is the *approximate* depth in inches; the second number is the exact weight in

FIGURE 5-1 Joists

pounds per linear foot. For example, a W10 × 49 is approximately 10 in. (250 mm) deep and weighs exactly 49 lb/linear ft (73 kg/m). For deeper heavier beams, however, the actual depth may vary considerably from the first number in the designation. Exact dimensions can be found in the American Institute of Steel Construction's *Steel Design Manual,* 2001.

Fire-Protective Coating

Fire-resistance ratings for steel structural framing are usually achieved by the application of a spray-applied fire-protective coating. Particularly for important buildings, such as high-rise buildings, the building owner's representative should provide field test data for spray-applied fire-protective coatings. Two tests that should be used are ASTM E-605 for thickness and density and ASTM E-736 for cohesion and adhesion.

Verifying that the thickness of the coating meets the minimum required for the specified fire rating and size of the structural member is important and relatively easy. To measure the thickness, special gauges (Figure 5-2) are available, or a piece of wire or a straightened paper clip can be used and pushed into the coating up to the steel, marked, and then measured after it is taken out. Measured thicknesses should be checked against listing requirements, such as those in the UL *Fire Resistance Directory.*

Verifying that the density of the coating meets specified criteria is also important. If the listed coating is modified, it may still meet the thickness criteria but may not be completely effective because it is too light or lacks cohesive strength. Likewise, if the surface is not compatible with the coating, the adhesive strength will not be adequate and the coating may peel off prematurely. It is recommended that an independent laboratory perform these density, adhesion, and cohesion tests, that it check the results against listing requirements and that it give the report and the conclusions to the inspector for review.

FIGURE 5-2 Gauge to Measure Thickness of Fire Protection Coating

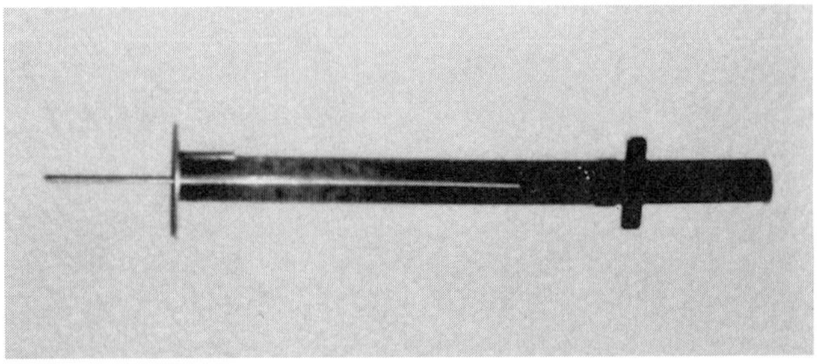

In many cases, it may be practical for the same procedure to be followed for the coating thickness. In some cases, the coating thickness needed may be adjusted up or down because the actual steel member differs in size from the member shown in the listing. This situation is acceptable provided that the design professional specifies that the assembly will still attain the intended rating, that is, 1 hour, 2 hour, or 3 hour.

Typically, structural steel framing is not required to be prime painted prior to the application of spray-applied coatings. The surface should be free of dirt, oil, or loose mill scale. If a primer paint is used, it must be compatible with the coating. If an improper paint is used or the surface is not properly cleaned, the coating may fall off prematurely.

Inspectors should be particularly aware of any building renovation/rehabilitation projects that could alter the applied coatings. Coatings may be inadvertently removed when ductwork is added or modified, when new communication wiring is added to above ceiling areas, and even when new pipe for automatic sprinkler systems is installed.

WALLS, FLOOR/CEILING ASSEMBLIES, AND ROOFS

Walls

Walls serve a variety of functions, including security, privacy, weather resistance, fire resistance, and structural resistance. They can be classified as bearing or nonbearing walls, common (or party) walls, shear walls, or fire walls or fire barrier walls. Often times, a single wall will have two or more of these functions and may have multiple classifications.

A bearing wall supports more than its own weight, such as a floor or roof; a nonbearing wall typically supports only its own weight. More specifically, a metal or wood stud wall that supports more than 100 lb/linear ft (1400 N/m) of vertical load in addition to its own weight is considered a bearing wall. A concrete or masonry wall that supports more than 200 lb/linear ft (2900 N/m) in addition to its own weight is also considered a bearing wall.

A common, or party, wall is a single wall that is common to two separate areas or buildings. In some cases, the wood floor joists of both buildings that share a common wall are placed in the same opening, providing a hidden path for fire to spread.

A shear wall acts to brace a portion of a building against the lateral forces of wind, earthquake, or similar loads. It resists, by its stiffness, the forces applied parallel to its length.

Finally, a fire wall is used to subdivide a building or separate buildings from each other. It has both a fire resistance rating and structural stability. Fire walls may be freestanding (cantilevered), double walls, or tied walls, as defined by NFPA 221, *Standard for Fire Walls and Fire Barrier Walls.* When a fire wall is used in a single-story building or on the top floor of a building, it typically penetrates the roof and forms a parapet. A fire barrier wall is a fire-resistant wall that does not have the structural stability of a fire wall and is often used to isolate a special hazard within a building, such as a flammable liquids storage area.

Openings in these fire walls and fire barrier walls are protected with automatic-closing or self-closing devices. Fire walls usually have a fire resistance rating that ranges from 1 to 4 hours, but their differentiation from fire barrier walls is not with regard to fire resistance. Required fire resistance ratings are determined from code requirements and are typically based on the occupancy use.

Fire walls should remain structurally sound and should not permit the spread of fire through, under, over, or around the wall, even if the structure on one side of it burns out and collapses. The efficiency of a fire wall depends on its own integrity and on the reliability of its closing devices. Only certain types of fire wall design will provide such stability. For more information the inspector should refer to NFPA 221.

Pipe and conduit penetrations should be checked to assure the piping penetrates the wall at points permitted in the listed assembly. In walls rated for 3 hours and greater, the penetrations must be not more than 3.0 feet (0.9 m) above the floor. Ductwork should be checked to assure that slip joints are provided near the wall and dampers are provided within the wall. To assure that a true fire wall has been provided, the inspector should check for a parapet, which is a vertical extension of the fire wall that divides the roof and extends at least 30 in. (762 mm) above the top surface of the roof cover.

Fire barrier walls have fire resistance ratings ranging from 20 minutes to 4 hours, but they are usually not structurally independent. Fire barrier walls rely on the building frame for support, and so their structural integrity is dependent on passive protection of the building frame, automatic sprinkler protection, or some combination of the two. "Smoke barriers," which may or may not have a fire resistance rating, are primarily intended to limit the passage of smoke. Other terms commonly used are "building separation walls," "area separation walls," "fire partitions," and "occupancy separation walls." Criteria for fire barrier walls can be found in NFPA 221.

A veneered wall consists of a single wythe, or thickness, of brick facing that is laterally supported by wood or steel studs, by concrete, or by masonry. Fire performance is based on the construction of this "backing system." Fire resistance for concrete and masonry will depend on the thickness, type of aggregate used, and cover distance for reinforcing steel. For additional information on fire resistance, see ASCE/SFPE 29-99.

Postoccupancy modifications to fire walls and fire barrier walls may include installation of pipe, wiring, ductwork, or similar components that may breech the wall and jeopardize its integrity. Inspectors should verify that any such penetrations have been properly sealed, provided with a damper, or otherwise configured to maintain the rating that is intended for the wall.

Floor/Ceiling Assemblies

Fire safety in buildings is also influenced by the floor/ceiling assemblies that must support the dead and live loads placed on them. Some assemblies have been tested in accordance with nationally recognized standards and have received a specific fire resistance rating. Such assemblies can be found in the UL *Fire Resistance Directory* (Volume 1) or in the Gypsum Association's *Fire Resistance Design Manual.* This rating does not mean that they are impervious to fire for the full rating period, however. For example, rated steel floor/ceiling assemblies can easily be made ineffective by the removal of ceiling tiles. The installed assembly must be identical to the assembly listed. Rated assemblies are tested only against fire from below when, in fact, fire can enter an assembly without hindrance either laterally or from above. When a floor/ceiling or roof/ceiling assembly is given a fire resistance rating, it is a rating for the total assembly, not the ceiling itself.

Other assemblies may be combustible in themselves and may contain concealed spaces that are difficult to access and that allow for rapid fire spread. The space between a ceiling and the floor or roof above may sometimes be used as part of the building's air-handling system or a convenient location to conceal electrical wiring, piping for domestic water supply, automatic sprinkler systems, or communication wiring (Figure 5-3). Control of combustible materials in these spaces is tightly regulated by NFPA 90A, *Standard for the Installation of Air-Conditioning and Ventilating Systems.*

Roofs

Roofing assemblies are constructed of a combination of materials and in a variety of configurations. Basically, a roof consists of supports, such as beams or rafters; a deck; insulation; and a covering. Environmental loads on the roof are carried by the roof deck and are transmitted along rafters and beams to the columns, foundation systems, and soil below. Typical roof deck materials are concrete in industrial and commercial construction and steel and wood sheathing in residential construction and smaller commercial projects.

Insulation is generally adhered or mechanically fastened to the top of the roof deck and is covered on top by a roof covering that provides weatherproofing. Roof coverings can be prepared, built-up, or single-ply membranes. Examples of fire-retardant Class A coverings are brick, exposed concrete, concrete paver blocks, and concrete or clay tile; fiber-reinforced cement tile; slate; and copper or steel panels or shingles. Large stone used on ballasted roofs and a minimum of 4 lb/ft^2 (20 kg/m^2) of pea gravel embedded in a flood coat of asphalt or coal tar are also considered to be Class A coverings. Fire-retardant-treated wood shingles are recommended over

FIGURE 5-3 Space Above Ceiling Used to Hide Piping, Electrical Wiring, and Communication Wiring

untreated wood shingles. Also, fire retardancy may be lost over time due to weathering; replacement is recommended at the end of their life span. Untreated wood shingles have been used in the past but are now prohibited in many communities because they ignite readily and can produce fire brands, which may ignite surrounding combustibles and buildings.

A built-up roof membrane typically consists of three to five layers, known as "plies," of roof membrane attached to a wood or concrete roof deck or insulation and to each other at the job site, with hot asphalt or some other adhesive securing the layers. Built-up roofs laid directly on metal decking can allow fire spread on the deck underside and should not be used in this way. The use of gravel surfacing or special coatings, such as asphalt emulsions or fibrated aluminum coatings, will increase the resistance of a built-up roof to exterior fire exposure.

Single-ply roof membranes generally consist of flexible, water-resistant sheets of a variety of plastic- or rubber-based products. These membranes are typically applied over rigid insulation. Reinforced membranes are usually attached to the deck with mechanical fasteners. Unreinforced membranes are usually adhered or are ballasted with large gravel or concrete paving blocks.

The spacing of fasteners is critical to the wind-uplift resistance of the roof. With single-ply roof covers that are thin, the outline of the top of the fastener plate can usually be seen under the membrane or be felt by rubbing the hand or foot over the top of the cover. Also, with steel deck roofs, viewing the underside of the deck where fasteners penetrate allows for a check of the fastener spacing. Differentiating between the various types of fasteners used is important. With mechanically attached roof covers, some fasteners are used to secure the roof cover to the deck; others are used to fasten rigid insulation below the cover to the deck. Also, fasteners found at the overlaps of steel deck sections, usually about every 3.0 ft (0.9 m), are called side-lay fasteners and are used to connect adjoining deck section but do not add any direct securement for the insulation or roof cover.

Elastomeric coatings, such as acrylic, silicone, polyurethane, and so on, are used as spray-applied roof coverings over urethane foam insulation, which is sprayed in place on top of the roof deck, existing roof cover, or thermal barrier.

Torch-applied roof systems are bituminous roofing systems that are heated with a torch (as it is rolled onto its substrate), which melts the asphalt that saturates the membrane. The membrane is then immediately secured to the substrate, usually a base ply, below. Guidance on safety in the use of torch-applied roof systems can be found in NFPA 241, *Standard for Safeguarding Construction, Alteration, and Demolition Operations.* Nationally recognized standards exist to evaluate roof assemblies for interior combustibility, surface burning over the top surface, wind resistance, and hail resistance.

Roof systems such as insulated steel deck are rated for interior fire exposure based on test standards such as FM 4450 or 4470. Insulated steel decks may contain combustible above-deck components but may be considered Class 1, or limited combustible, if they are listed based on FM 4450 or 4470. Some insulated steel deck assemblies are Class 2, or combustible, due to the type and quantity of combustible components above the deck. One classic example is any insulation that is secured directly to steel deck with a full mop of asphalt.

Class 1 insulated steel deck assemblies are suitable for Type I and Type II construction (see Chapter 6, "Classification of Construction Types"); Class 2 steel deck assemblies are suitable for Type III, Type IV, or Type V construction. Thermoplastic insulation used over a thermal barrier on top of steel deck can be considered limited combustible and is acceptable for Type I or Type II construction.

NFPA 256, *Standard Methods of Fire Tests of Roof Coverings,* FM 4470, and UL 790 are tests used to evaluate fire performance of roof coverings exposed to exterior fire and rated as A, B, or C. A Class A rating is the best rating and is preferred; however, codes allow use of a Class B or a Class C rating in many applications. Some smooth-surfaced, bituminous roofs use coatings such as fibrated aluminum and asphalt emulsion to improve exterior fire resistance.

Tests to determine hail resistance include FM 4470 (moderate or severe hail), FM 4473 (Class 1, 2, 3, or 4 hail), or UL 2218 (Class 1, 2, 3, or 4 hail). A Class 2 hail rating, or "moderate" rating, is based on resistance to 1½-in. (38-mm) diameter hail. A Class 4 hail rating is based on exposure to 2-in. (50-mm) diameter hail.

FM 4470 is also used to evaluate wind-uplift resistance of roof covers. Ratings are 60 psf (2.86 kN/m^2), 75 psf (3.58 kN/m^2), 90 psf (4.30 kN/m^2), 105 psf (5.02 kN/m^2), 120 psf (5.74 kN/m^2), and so on. (Note that psf stands for pounds per square foot.) The required rating (psf) is based on the roof weight, geographic wind speed, and ground roughness.

See *Building Construction for the Fire Service* and the *Fire Protection Handbook* for additional information on structural elements.

SUMMARY

A variety of different components can be used in the construction of the building framing and envelope. It is important that the size, spacing, span, and method of connection for various members comply with design drawings. Although it is not practical to check every detail of construction, the inspector can easily measure and check some important items in the field. Postoccupancy changes or modifications made to the building during its life can alter or degrade the expected performance of construction systems. A careful inspection of penetrations and applied coatings on structural members can be valuable to ensure that the building's performance is as intended.

BIBILIOGRAPHY

AISC *Steel Design Manual,* 3rd ed., American Institute of Steel Construction, Chicago, IL, 2001.

ASCE/SFPE 29-99, *Standard Calculation Methods for Structural Fire Protection,* American Society of Civil Engineers/Society of Fire Protection Engineers, 1999.

ASTM E-605-96, *Standard Test Method for Thickness and Density of Sprayed Fire-Resistive Material (SFRM) Applied to Structural Members,* American Society for Testing and Materials, Conshohocken, PA, 1996.

ASTM E-736-00, *Standard Test Method for Cohesion/Adhesion of Sprayed Fire-Resistive Materials Applied to Structural Members,* American Society for Testing and Materials, Conshohocken, PA, 2000.

Brannigan, F. L., *Building Construction for the Fire Service,* 3rd ed., NFPA, Quincy, MA, 1992.

Building Materials Directory, Underwriters Laboratories, Inc., Northbrook, IL, issued annually.

Cote, A. E., ed., *Fire Protection Handbook,* 18th ed., NFPA, Quincy, MA, 1997.

Factory Mutual Research Approval Guide, Building Materials Volume, Norwood, MA, issued annually.

FM 4450, *Test Standard for Class 1 Insulated Steel Roof Decks,* FM Global, Norwood, MA, 1989.

FM 4470, *Approved Standard for Class 1 Roof Covers,* FM Global, Norwood, MA, 1986.

FM 4473, *Specification Test Protocol for Impact Resistance Testing of Rigid Roofing Materials by Impacting with Freezer Ice Balls,* FM Global, Norwood, MA, 2000.

Fire Resistance Directory, Vol. 1, Underwriters Laboratories, Inc., Northbrook, IL, issued annually.

GA-600, *Fire Resistance Design Manual,* Vol. 16, Gypsum Association, Washington, DC, 2000.

UL 2218, *Standard for Impact Resistance of Prepared Roof Covering Materials,* Underwriters Laboratories, Inc., Northbrook, IL, 1998.

NFPA Codes, Standards, and Recommended Practices

See the latest version of The NFPA Catalog for availability of current editions of the following documents.

NFPA 90A, *Standard for the Installation of Air-Conditioning and Ventilating Systems*
NFPA 203, *Guide on Roof Coverings and Roof Deck Constructions*
NFPA 220, *Standard on Types of Building Construction*
NFPA 221, *Standard for Fire Walls and Fire Barrier Walls*
NFPA 241, *Standard for Safeguarding Construction, Alteration, and Demolition Operations*
NFPA 251, *Standard Methods of Fire Tests of Building Construction and Materials*
NFPA 255, *Standard Method of Test of Surface Burning Characteristics of Building Materials*
NFPA 256, *Standard Methods of Fire Tests of Roof Coverings*
NFPA 703, *Standard for Fire Retardant Impregnated Wood and Fire Retardant Coatings for Building Materials*

Classification of Construction Types

CHAPTER 6

Richard J. Davis

To determine a building's life safety and fire protection requirements, inspectors must be familiar with the basic construction elements of a building (which is the topic of Chapter 5, "Building Construction Elements"), but they must also know and understand the fire resistance rating characteristics of the various construction types. A well-established means of codifying fire protection and fire safety requirements for buildings is to classify them by types of construction based on the materials used for the structural system and on the degree of fire resistance afforded by each structural element. This chapter reviews the five basic types of building construction that are defined in NFPA 220, *Standard on Types of Building Construction,* and used throughout the built environment.

BACKGROUND OF BUILDING CLASSIFICATION TYPES

Terminology

In early codes, only two classifications of construction were identified: (1) "fireproof" and (2) "nonfireproof." When it was recognized that no material or building is totally fireproof and that the building contents can produce a significant fire without involving the structure, the term "fireproof" was replaced by the term "fire resistive." Designing buildings that will resist fire without suffering serious structural damage is possible. Appropriate fire-resistive design, balanced against anticipated fire severity, is the objective of structural fire protection requirements in current codes.

Other significant terms related to building classification types include the following:

- *Noncombustible:* A material that, in the form in which it is used and under the conditions anticipated, will not ignite, burn, support combustion, or release flammable vapors when subjected to fire or heat.
- *Limited combustible:* As applied to a building construction, a material not complying with the definition of noncombustible material, which, in the form in which it is used, has a potential heat value not exceeding 3500 Btu/lb (8141 kJ/kg).
- *Low hazard:* Contents of such low combustibility that no self-propagating fire therein can occur.
- *Ordinary hazard:* Contents that are likely to burn with moderate rapidity or to give off a considerable volume of smoke.
- *High hazard:* Contents that are likely to burn with extreme rapidity or from which explosions are likely.

Basic Types of Construction

Several distinct types of construction that use combustible framing were originally classified based on the materials used in the exterior wall construction (i.e., masonry or wood) and the type and size of the structural members (i.e., heavy timber versus conventional framing). As fire resistance ratings for construction assemblies began to be recognized in building codes, classifications of building types were added for

Richard J. Davis, a senior engineering specialist with FM Global, is chair of the NFPA Technical Committee on Construction, Alteration, and Demolition Operations and a member of the NFPA Technical Committee on Building Construction and the NFPA Technical Committee on Structures and Construction.

both noncombustible and combustible types of construction, based on the degree of fire resistance provided. Currently five basic types of construction are recognized. Two—Type I and Type II—are identified as noncombustible construction types and three—Type III, Type IV, and Type V—as combustible construction types.

Table 6-1 gives the fire resistance rating requirements for structural elements, interior bearing walls, floor construction, and roof construction of the five basic types of construction. Each type of construction has anywhere from one to three different subcategories, which are indicated by a three-digit Arabic notation. This notation, which accompanies the construction type, identifies the fire resistance rating required for three basic elements of the building: (1) the first digit corresponds to the exterior bearing wall, (2) the second digit corresponds to the primary structural elements, and (3) the third digit corresponds to the floor construction.

The table lists the building components that are considered essential to the stability of the building as a whole and that compose the building's structural system. The members of the floor or roof panels that have no connection to the columns are considered part of the floor construction or part of the roof construction and are not classified as a part of the structural system.

TABLE 6-1 Fire Resistance Ratings (in Hours) for Type I through Type V Constructions

	Type I		Type II			Type III		Type IV	Type V	
	443	332	222	111	000	211	200	2HH	111	000
Exterior Bearing Walls										
Supporting more than one floor, columns, or other bearing walls	4	3	2	1	0[1]	2	2	2	1	0[1]
Supporting one floor only	4	3	2	1	0[1]	2	2	2	1	0[1]
Supporting a roof only	4	3	1	1	0[1]	2	2	2	1	0[1]
Interior Bearing Walls										
Supporting more than one floor, columns, or other bearing walls	4	3	2	1	0	1	0	2	1	0
Supporting one floor only	3	2	2	1	0	1	0	1	1	0
Supporting roofs only	3	2	1	1	0	1	0	1	1	0
Columns										
Supporting more than one floor, columns, or other bearing walls	4	3	2	1	0	1	0	H[2]	1	0
Supporting one floor only	3	2	2	1	0	1	0	H[2]	1	0
Supporting roofs only	3	2	1	1	0	1	0	H[2]	1	0
Beams, Girders, Trusses, and Arches										
Supporting more than one floor, columns, or other bearing walls	4	3	2	1	0	1	0	H[2]	1	0
Supporting one floor only	3	2	2	1	0	1	0	H[2]	1	0
Supporting roofs only	3	2	1	1	0	1	0	H[2]	1	0
Floor Construction	3	2	2	1	0	1	0	H[2]	1	0
Roof Construction	2	1½	1	1	0	1	0	H[2]	1	0
Exterior Nonbearing Walls[3]	0[1]	0[1]	0[1]	0[1]	0[1]	0[1]	0[1]	0[1]	0[1]	0[1]

▭ Those members that shall be permitted to be of approved combustible material.

[1]See A-3-1 (table) in NFPA 220.

[2]"H" indicates heavy timber members.

[3]Exterior nonbearing walls meeting the conditions of acceptance of NFPA 285, *Standard Method of Test for the Evaluation of Flammability Characteristics of Exterior Non-Load-Bearing Wall Assemblies Containing Combustible Components Using the Intermediate-Scale, Multistory Test Apparatus,* shall be permitted to be used.

Source: NFPA 220, 1999, Table 3-1.

TYPE I CONSTRUCTION

Type I construction is construction in which the structural members are designated noncombustible or limited combustible and have a fire resistance rating as specified in Table 6-1. Monolithic reinforced concrete construction, precast concrete construction, and protected steel-frame construction are all examples of Type I noncombustible construction. This classification is divided into two subcategories: Type I (443) and Type I (332). (Currently, there is discussion of modifying Type I (443) to Type I (442).) The basic difference between the subcategories is in the level of hourly fire resistance rating specified for the structural system.

The fire resistance rating requirements for Type I (443 and 332) construction were selected because they ensure reasonable fire safety for the structure in occupancies with low to ordinary hazard contents. In occupancies with higher fire loads and high hazard contents or in order to limit fire spread within the occupancy, codes may require that the fire resistance–rated construction be supplemented by additional protection, usually including an automatic fire-extinguishing system. Even in occupancies with moderate fire loads, such as in mercantile and in some factory industrial and storage uses, supplementary fire safety precautions are required. These include either restrictions on the building's allowable height and area or requirements for automatic fire-extinguishing equipment.

In Type I construction, only noncombustible materials are permitted for the structural elements of the building. This is an accepted regulation that appears in practically every model building code. If combustible structural materials were allowed in noncombustible building types, the whole concept of their allowable use (i.e., the building's height and area) would be rendered meaningless. However, for practical reasons, the use of some combustible materials in Type I and Type II buildings are permitted for nonstructural components. Roof coverings, some types of insulating materials, and limited amounts of interior finish and flooring have been traditionally recognized as not adding significantly to the fire hazard or fire load, if these materials are properly regulated and qualified by fire tests.

Some codes have attempted to regulate combustible materials by using a slightly different definition of noncombustible materials, which includes two or three alternatives that allow for the acceptance of materials having relatively low fuel content and surface-burning characteristics. The purpose of this definition was to recognize certain materials or nonhomogenous assemblies containing limited amounts of combustible materials, for example, gypsum wallboard, which is used as a fire-resistive material even though it is covered with paper. These alternate definitions include limits on surface flame spread rating (per NFPA 255, *Standard Method of Test of Surface Burning Characteristics of Building Materials*) and on the heat content (per NFPA 259, *Standard Test Method for Potential Heat of Building Materials*). Rather than complicate the definition for the accommodation of certain materials, a more fundamental approach is to define limited uses and combustibility characteristics of materials that may be acceptable in buildings of noncombustible construction.

TYPE II CONSTRUCTION

Type II construction is a construction type in which the structural elements are entirely of noncombustible or limited combustible materials permitted by the code and protected to have some degree of fire resistance, 2 hour [Type II (222)], 1 hour [Type II (111)], or completely unprotected except for exterior walls in Type II (000) construction. Typical of Type II noncombustible construction are metal-framed, metal-clad buildings, and concrete-block buildings with Class 1 insulated steel deck roofs supported by unprotected open-web steel joists.

The fire resistance rating required in Type II (222 or 111) construction will afford adequate fire safety for residential, educational, institutional, business, and assembly occupancies, without supplementary restrictions. Building height limits, however, are commonly prescribed for this type of construction. In occupancies with low-hazard

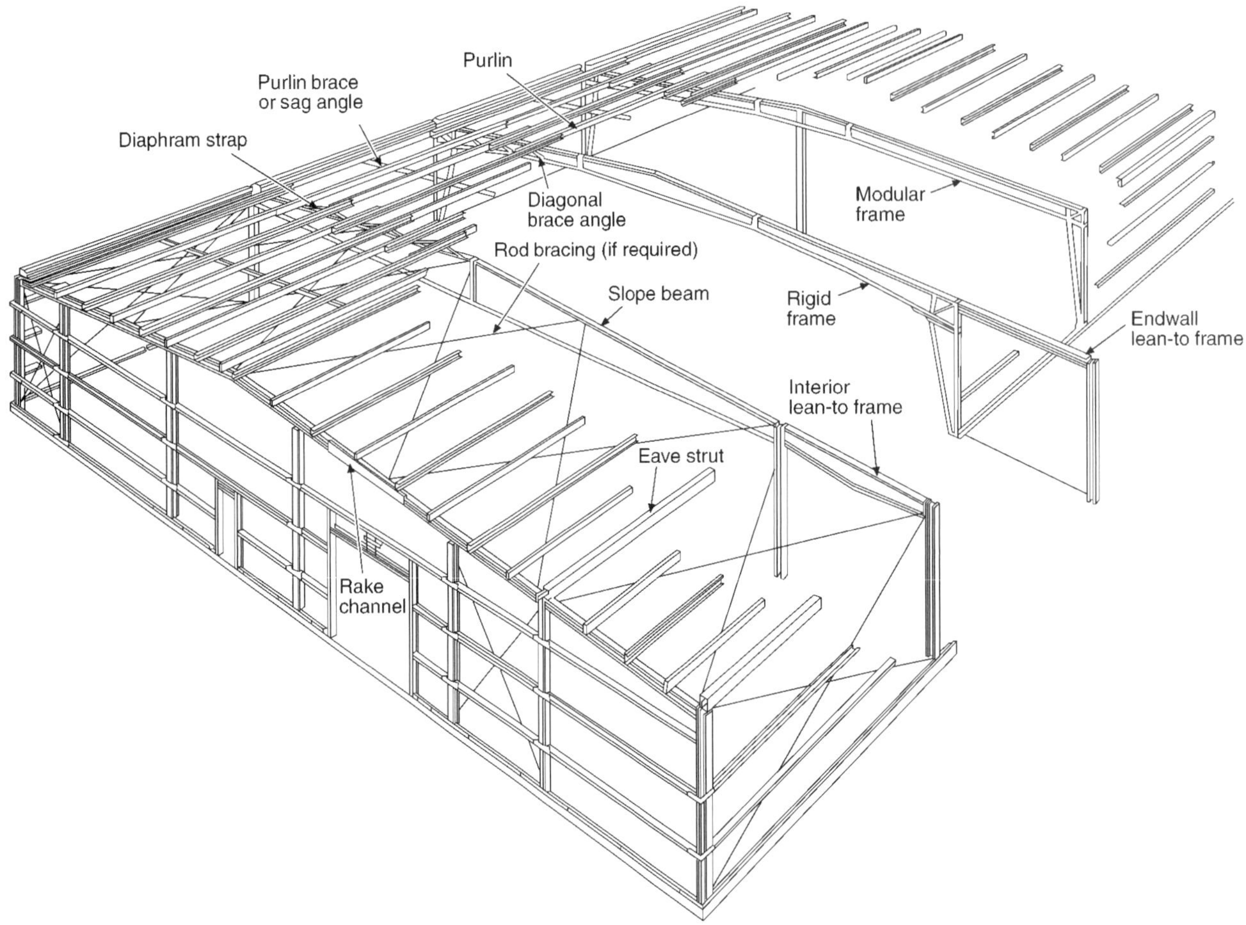

FIGURE 6-1 Framing System of a Representative Building of Type II Noncombustible Construction
Source: *Fire Protection Handbook,* NFPA, 1997, Figure 7-2A.

contents, the absence of fuel in noncombustible construction reduces potential risk of a fire starting within the structural members and helps prevent the spread of fire if the contents are involved. When Type II construction is used for occupancies with a greater fire loading, the code may require additional fire safety precautions, such as more stringent building area limitations and automatic fire-extinguishing equipment.

No matter the hazard content level, the noncombustible structural system is invaluable because it prevents fire from spreading through concealed spaces or involving the structure itself. Because of this attribute, a fire in a building of noncombustible construction can often times be controlled more readily. Figure 6-1 illustrates a structural system of a representative building of Type II noncombustible construction. Shown is a preengineered pitched roof and lean-to framing with structural elements of unprotected steel.

TYPE III CONSTRUCTION

Type III construction is a construction type in which all or part of the interior structural elements may be of combustible materials or any other material permitted by the applicable building code. Type III construction is widely used for mercantile buildings, schools, churches, motels, and apartment houses. The exterior bearing walls of Type III construction are required (1) to be of approved noncombustible or limited-combustible materials, such as brick, concrete, or reinforced concrete; and (2) to have a degree of fire resistance depending on the horizontal separation and the fire load.

Type III construction is further divided into protected and unprotected subcategories. Protected construction, Type III (211), has a 1-hour fire resistance rating for the floors and structural elements. Type III (200) construction has no fire resistance rating for the floors or structural elements. Whether or not fire resistance is provided, it is essential that all concealed spaces be properly fire blocked in buildings of combustible construction. Codes require that this be done in all furred spaces, partitions, ceiling spaces, and attics. Codes are very specific as to the materials to be used for a fire block and the locations where a fire block is required. To be effective, a fire block must completely close off and subdivide the combustible construction into limited areas, thereby restricting the spread of fire and hot gases and allowing additional time for detection and evacuation of the building or area involved.

The 1-hour fire resistance rating provided in Type III (211) construction offers a measure of safety for fire fighting and occupant evacuation before the construction itself becomes involved. Combustible parts of any fire-rated assembly will, however, be burning actively before the end of the rated time period. For this reason, that portion of the fire load represented by combustible structural members must be considered as part of the total potential fire load, whether or not the construction is protected.

TYPE IV CONSTRUCTION

Type IV construction is a construction type in which structural members—that is, columns, beams, arches, floors, and roofs—are composed of unprotected wood (solid or laminated) with large cross-sectional areas. No concealed spaces are permitted in the floors and roofs or other structural members, with minor exceptions. NFPA 220, *Standard on Types of Building Construction,* and most model building codes are specific in the minimum dimensions permitted for the various wood structural members and minimum fire-resistive ratings required for interior columns, arches, beams, girders, and trusses of materials other than wood that may be permitted as acceptable alternatives to wood members (Table 6-2).

TABLE 6-2 Recommended Minimum Nominal Dimensional Requirements for Type IV (2HH) Construction

	Supporting Floors	Supporting Roofs
Columns	8 in. × 8 in.	6 in. × 8 in.
Beams and girders	6 in. × 10 in.	4 in. × 6 in.
Arches*	8 in. × 8 in.	6 in. × 8 in., 6 in. × 6 in., 4 in. × 6 in.
Trusses	8 in. × 8 in.	4 in. × 6 in.
Floors	3-in. T&G† covered with 1 in. T&G flooring or ½-in. wood structural panel; or 4-in. laminated planks on edge with 1-in. T&G flooring or ½-in. wood structural panel	
Roof assemblies		2-in. T&G; or 3-in. laminated planks on edge; or 1⅛-in. wood structural panels

For SI units: 1 in. = 25.4 mm.

*See NFPA 220, *Standard on Types of Building Construction,* for applications of minimum size of members.

†T&G is tongue and groove.

FIGURE 6-2 Elements of a Building of Type IV Construction. Note the large size of the columns and beams and the absence of concealed spaces. The exterior wall at far left is of lightweight corrugated steel.
Source: *Fire Protection Handbook*, NFPA, 1997, Figure 7-2B.

FIGURE 6-3 Variation of Type IV Construction. Shown are haunched arches made of laminated wood (glue-laminated construction). Beams are anchored to the arches by steel hangers.
Source: *Fire Protection Handbook*, NFPA, 1997, Figure 7-2C.

Walls, both interior and exterior, including structural members framed into them, can be of approved noncombustible or limited-combustible materials. Brick and stone were the traditional materials used in early heavy timber, or "mill" construction. Exterior walls may be of heavy timber construction if they are over 30 ft (9.1 m) from the property line and provided that the 2-hour fire resistance rating is maintained.

During a fire, heavy timber construction performs better than do conventional wood frame structures because the structural members are larger, have a smaller surface-to-mass ratio, and take longer to burn. As the wood member burns, a layer of char develops which acts like insulation and slows down the rate of burning. The large wood members, therefore, can continue to carry their structural loads for longer durations due to the mass of unburned wood.

Heavy timber construction is more appropriately considered a building system, not just a construction type using large-size structural members. It was developed during the mid-1800s by insurance interests for the purpose of reducing fire losses in the many textile factories, paper mills, and storage buildings in the New England states. Through the intelligent use of combustible materials of sufficient mass, the absence of concealed spaces, and by paying attention to details to avoid sharp corners and ignitable projections, the chance of rapid spread of fire are lessened and the probability of serious structural damage is reduced. Examples of heavy timber construction are shown in Figures 6-2 and 6-3.

TYPE V CONSTRUCTION

Type V construction is a type of construction in which the structural members are entirely of wood or any other approved material (Figure 6-4). Depending on the exterior horizontal separation, the exterior walls may or may not be required to be fire resistive.

In Type V construction, walls and partitions are typically framed with 2 × 4-in. wood studs attached to wood sills and plates. Wood boards, plywood sheets, various composition boards, or foamed plastic are then nailed to the studs. Over this underlayment is placed a layer of building paper and then the finished material. The exterior wall covering can be any one of a variety of materials, including wood

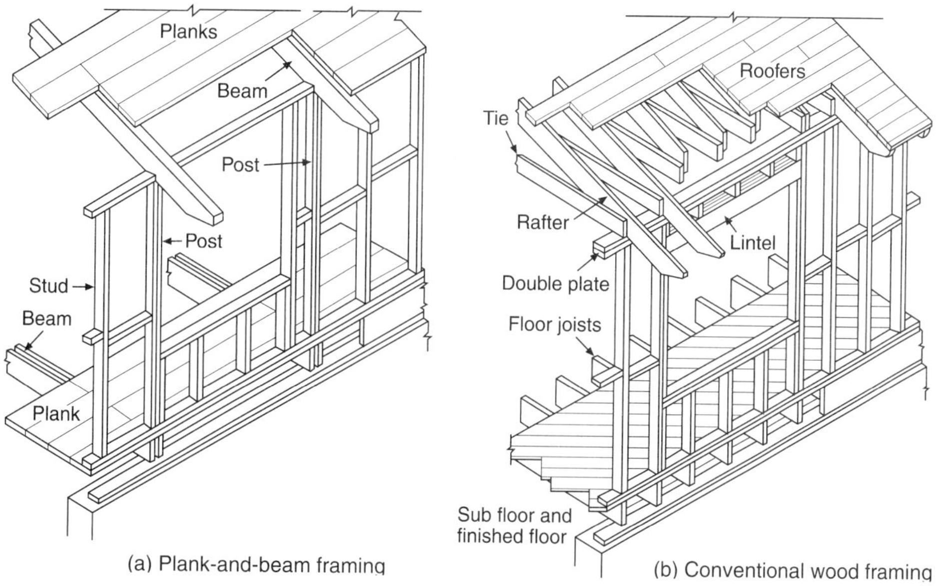

FIGURE 6-4 Two Variations on Basic Type V Construction: (a) Plank-and-Beam Framing with a Few Large Members Replacing Many Small Members Used in Typical Wood Framing and (b) Conventional Wood Framing (Western or Platform Construction)
Source: *Fire Protection Handbook*, NFPA, 1997, Figure 7-2D.

shingles; wood, plastic, or metal clapboards; matched boards; brick veneer; sheet metal cladding; stucco; or fiber-reinforced corrugated cement panels.

Type V construction is more vulnerable to fire, both internally and externally, than any other building type. Accordingly, it is essential that greater attention be given to the details of construction of this basically light wood-frame building. The use of a fireblocking in exterior and interior walls at ceiling and floor levels, in furred spaces, and other concealed spaces can retard the spread of fire and hot gases in these vulnerable areas. Type V construction is subdivided into two subcategories—Type V (111) construction, which has 1-hour fire resistance rating throughout, including the exterior bearing walls; and Type V (000) construction, which has no fire protection or fire resistance rating requirements, except for the exterior walls when horizontal separation is less than 10 ft (3 m).

MIXED TYPES OF CONSTRUCTION

Where two or more types of construction are used in the same building, it is generally recognized that the requirements for occupancy or height and area would apply for the least fire-resistive type of construction. However, in cases where each building type is separated by adequate fire walls (see NFPA 221, *Standard for Fire Walls and Fire Barrier Walls*) having appropriate fire resistance, each portion may be considered as a separate building.

Another general limitation included in some model building codes prohibits construction types of lesser fire resistance to support construction types having higher required fire resistance. In the event of a fire, the risks of a major structural collapse are generally too great to permit this type of design. This limitation does not necessarily apply where construction supports nonbearing separating partitions that provide protection for exit corridors or tenant spaces.

SUMMARY

Fire resistance ratings for construction assemblies are recognized in building codes based on the degree of fire resistance provided. Currently five basic types of construction are recognized. Type I and Type II are identified as noncombustible construction types. Type III, Type IV, and Type V are identified as combustible construction types.

Examples of Type I construction are monolithic reinforced concrete construction, precast concrete construction, and protected steel frame construction. Typical of Type II noncombustible construction are metal-framed, metal-clad buildings, and concrete-block buildings with Class 1 insulated steel deck roofs supported by unprotected open-web steel joists. The exterior walls of Type III construction are required to be of noncombustible or limited-combustible materials, such as brick, concrete, or reinforced concrete. Type III construction is widely used for mercantile buildings, schools, churches, motels, and apartment houses. Type IV construction, also called heavy timber, or "mill," construction, is characterized by large wood columns and beams and an absence of concealed spaces. In Type V construction the structural members are entirely of wood or any other approved material.

BIBLIOGRAPHY

Brannigan, F. L., *Building Construction for the Fire Service,* 3rd ed., NFPA, Quincy, MA, 1992.

Building Materials Directory, Underwriters Laboratories, Inc., Northbrook, IL, issued annually.

Cote, A. E., ed., *Fire Protection Handbook,* 18th ed., NFPA, Quincy, MA, 1997.

Factory Mutual Research Approval Guide, Building Materials Volume, Norwood, MA, issued annually.

Fire Resistance Directory, Vol. 1, Underwriters Laboratories, Inc., Northbrook, IL, issued annually.

NFPA Codes, Standards, and Recommended Practices

See the latest version of The NFPA Catalog for availability of current editions of the following documents.

NFPA 220, *Standard on Types of Building Construction*
NFPA 221, *Standard for Fire Walls and Fire Barrier Walls*
NFPA 241, *Standard for Safeguarding Construction, Alteration, and Demolition Operations*
NFPA 255, *Standard Method of Test of Surface Burning Characteristics of Building Materials*
NFPA 259, *Standard Test Method for Potential Heat of Building Materials*

Construction, Alteration, and Demolition Operations

CHAPTER 7

Richard J. Davis

CASE STUDY

Cutting Torch Heats Metal Bolts, Igniting Wooden Structure

In Georgia a fire damaged an unoccupied university office building under renovation when a worker using a cutting torch to remove metal stairs unintentionally started a chain of events that led to a concealed fire in a wall space. The fire burned undetected, spreading to the third floor and attic.

Built in 1901, the three-story structure was constructed of heavy timber with a brick veneer measuring 9000 ft (836 m) per floor. It had been used for office space until the renovation project began. All the fire detection and suppression systems had been disabled or removed during construction, and only temporary power remained in part of the building.

The fire department received a 911 call at 6:28 P.M. from a passerby reporting the fire using their cell phone. Fire fighters responded to find heavy smoke and flames coming from the third floor and attic of the building. As fire fighters from two departments began a defensive attack, a second alarm was sounded. It took eight engines, two truck companies, two squads, and numerous support staff to extinguish the blaze.

The fire started in a wall void as a result of a worker's removal of metal bolts from a third-floor stairwell by use of a cutting torch. A small fire occurred during his work, but he said that he had extinguished it with a bucket of water, then left for the day about an hour later. There was no evidence of fire or smoke when he left.

Investigators determined that the cutting torch had heated the metal bolts, which then conducted heat to wooden structural members inside the wall space. The resulting fire then smoldered for a while. Oxygen supplied primarily through the roof vents, appears to have helped the flames spread from the concealed spaces to the attic.

The building, which was valued at more than $1 million, was a total loss. Contents losses and injuries were not reported.

Source: "Fire Watch," *NFPA Journal*, May/June, 2001.

This unoccupied historic university building, which was under renovation, was destroyed in a fire that began in a wall void as a worker removed bolts with a cutting torch. (Photo courtesy of Bob Morris/*Savannah Morning News*)

Richard J. Davis, a senior engineering specialist with FM Global, is chair of the NFPA Technical Committee on Construction, Alteration, and Demolition Operations and a member of the NFPA Technical Committee on Building Construction and the NFPA Technical Committee on Structures and Construction.

Fire inspectors are frequently faced with hazards introduced during construction, alteration, and demolition operations. Most buildings are more vulnerable to fire at these times than at any other because the amount of combustibles and hazardous materials and the number of potential ignition sources present are often greater than usual, and the facility's fire protection systems may be impaired or even inoperative. Many of the comments to follow could apply to two or even all three of these factors.

CONSTRUCTION

Construction projects progress more rapidly in areas important to fire protection once bvasic foundation walls are completed and the building begins to take shape. Thus, they should be inspected more frequently after framing is underway. However, planning and scheduling water supplies for fire control during construction must be reviewed during the planning and permit stages so that the various installations needed, such as access roads, water mains, temporary water storage, sprinkler systems, hydrants, standpipes for multistory buildings, portable extinguishers for contractors and fire watch services, and fire walls, are properly planned and physically in place when needed. Making sure that fire protection during construction is clearly laid out in the plans and specifications or included in the permit conditions can prevent many problems, ranging from disagreements to major fire loss during construction. The building owner or contractor should appoint one individual to oversee all fire protection duties.

Site Preparation

In large projects, site preparation should include not only the removal of vegetation and combustible debris from the site, but also the appropriate layout of the contractors' temporary buildings, trailers, and material storage yards so that they will be neither a fire exposure to new construction nor obstruct access routes for fire fighters and their equipment. Most codes contain provisions that can be invoked to cover problem areas, but such after-the-fact solutions are seldom as satisfactory as planned layouts.

Roadways with an all-weather driving surface should be provided for fire apparatus. They should be at least 20 ft (6.1 m) wide and have at least 13 ft 6 in. (4.1 m) of clearance.

If permanent water mains and hydrants cannot be installed during the site preparation and foundation phases of the project, temporary water supplies, such as on-site tanks or tank trailers, temporary or surface mains, and pumps, may have to be included in the project. Aboveground swimming pools make excellent temporary water storage and can be moved easily as needed. It is important that these items, which could involve substantial cost, be included in the contracts. Temporary water supplies must be properly located to protect against damage from construction equipment or activities, and they should be designed with protection against freezing or other site-specific perils.

If special equipment, such as pumps or normally closed water supply lines, are present, guards or other persons on site 24 hours per day must know how to operate them. This means knowing not only how to start a pump but how to prime it, if necessary, and understanding which valves need to be kept open or closed.

Temporary Structures

Nearly all temporary structures associated with construction projects are made of combustible construction, whether they have a metal or plywood skin or ride on trailer wheels or skids. They will burn rapidly and can be the source of a major loss.

In the worst situation, a temporary structure, such as a job office, a tool or supply shed, a warming or locker room, or a carpenter or paint shop, is set up within the structure under construction. A fire in such a temporary structure can quickly involve the major structure. Because this approach is favored by contractors for ease of travel, security, economy, and weather protection, it can become the norm unless the regulatory agencies resist it. The situation is more difficult to handle at an urban site, where space in the streets is limited or nearby buildings must be rented at significant cost. If space is available, the temporary units should be separated by 30 ft (9.1 m) or more (depending on the size of the exposing unit) from the main building and from each other to minimize the loss potential from a "shack" fire. However, this amount of space can easily require many acres, so a compromise is generally necessary, depending on the hazard, value, protection, and construction features of both the temporary and permanent structures.

Temporary enclosures of coated fabric or plastic are often used to protect workers and construction operations from the weather until the building is enclosed. If fabric is used, the inspector should make sure that it is fire-retardant-treated tarpaulin. If plastic is used, it should be flame resistant and pass Test No. 2 noted in NFPA 701, *Standard Methods of Fire Tests for Flame Propagation of Textiles and Films.* In either case, the material should be fastened securely to a rigid steel or wood frame to prevent it from coming in contact with an ignition source, such as a temporary heater. The tarpaulin or plastic sheet should be anchored properly so that it cannot be torn by the wind.

Process Hazards

Although some process hazard is inherent in such job-site shops as carpenter, welding, pipe, and paint facilities, the inspector can encourage segregation of incompatible uses, such as carpentry and welding, and discourage the accumulation of flammable liquid in paint, fuel, and lubrication areas. Substandard heating appliances, stovepipes, bonfires, and substandard liquid- or gas-fuel-handling systems are proven fire hazards and should not be tolerated.

If local fire codes do not provide the authority to cope with these problems, the adoption of stronger codes should be encouraged. NFPA codes and standards that provide specific fire safety criteria addressing specific hazards are available (see Bibliography).

Housekeeping

Prompt removal of trash from a construction or remodeling site is critical to fire safety. Because it is an overhead cost to the contractor, however, the inspector must often take a firm stand to ensure reasonable compliance. If trash chutes are used, they should be located on the outside of the building. They should be of noncombustible construction and as straight as possible to prevent debris from piling up inside.

Storage of new material on site can present both a fire exposure and an obstruction to emergency access. To prevent this from happening, large amounts of combustible materials should be well separated from the building. An inspector can have considerable influence on material storage and delivery practices.

Theft and Vandalism

Although there is no sure prevention measure, a clean job site that is fenced, secured, lighted where needed, and attended by guards is much less likely to experience a fire started by a thief or vandal. Again, a realistic balance of cost and exposure is necessary.

Other Hazards

The number and size of engine-driven forklifts, crew lifts, excavators, and so on, all of which must be refueled on site and all of which contain hot exhaust systems capable of igniting trash, spilled fuel, or plastic weather enclosures, are significant. The person handling job-site safety for the general contractor should enforce and require in writing job-site safety rules on fuel storage and handling, equipment shutdown during fueling, and fire extinguisher availability.

A fuel-dump arrangement and fuel-handling rules also must be designed for the site. Fuel should be stored separately from the building under construction and from major temporary structures, and indoor fueling should be restricted to devices that cannot be moved readily. Only those fueling systems that minimize accidental spills, such as safety cans, automatic shut-off nozzles, and approved pump systems, should be used, and extinguishers should be provided according to NFPA 10, *Standard for Portable Fire Extinguishers.* Further discussion of fuel storage and handling can be found in Chapter 42 of this book and in NFPA 30, *Flammable and Combustible Liquids Code.*

Open-flame and spark-producing equipment—such as cutting torches, arc welders, soldering and grinding tools, and roofing fusion machines—must be strictly controlled. The basics of fire prevention are simple: Combustibles in or below work areas must be monitored, the equipment used must have the proper safety controls, a fire watch must be used, and extinguishers and hoses must be readily available (Figure 7-1). Still, hot work fires in construction projects are frequent. Further discussion of hot work hazards and controls can be found in NFPA 51, *Standard for the Design and Installation of Oxygen-Fuel Gas Systems for Welding, Cutting, and Allied Processes;* NFPA 51B, *Standard for Fire Prevention During Welding, Cutting, and Other Hot Work;* and NFPA 241, *Standard for Safeguarding Construction, Alteration, and Demolition Operations.*

Roofing materials, whether the older hot-mopped felt, tar, and gravel system; the torch-applied modified bitumen; or the cold-applied cut back asphalt system,

FIGURE 7-1 Fire Watch Monitoring a Hot Work Area
Source: *Introduction to Employee Fire and Life Safety*, NFPA, 2001, Figure 10-5.

all have a common problem: Much of the work is done with flammable materials that are heated near or above their flash points. Workers who are careless with torches, inaccurate or nonexistent controls for asphalt pots, punklike action of fiber insulation, or cigarette smoking provide potential sources of ignition. The inspector should not allow torches to be used near areas in which combustible dusts or oils may accumulate, such as exhaust hoods. Propane tanks on which frost has built up should not be heated with the torch flame. Instead, a larger tank should be recommended. Strict controls are necessary on the location and temperature control for asphalt pots. The pots should never be placed on the roof or under roofs or canopies, and their temperature controls must be automatic and working properly.

NFPA 241 contains guidelines for fire-safe roofing operations. For additional information refer to FM Global's *Property Loss Prevention Data Sheet 1-33, Safeguarding Torch-Applied Roof Installations.*

Needless to say, application and kettle areas must have a sizable (2-A:20-B:C) portable extinguisher within 25 ft (7.6 m)–20 ft (6.1 m) for torch-applied equipment—of the immediate work area because a roofing fire can be just as hot and rapid as any flammable liquid fire. Where practical, charged hose lines should be available. If attempts to extinguish a kettle fire with hand extinguishers are unsuccessful, water from a hose line must be applied in a fine spray due to the potential frothing action of hot asphalt. Roofing mops soaked with tar have been known to ignite spontaneously and cause fires. Used mops should not be left indoors or near ignition sources or combustible materials. Rather, they should be "spun" or cleaned thoroughly and safely stored.

Fire Protection

When NFPA *101*®, *Life Safety Code*®, or the building code requires that a building have standpipes, they should be installed on a floor-by-floor basis and in accordance with NFPA 14, *Standard for the Installation of Standpipe, Private Hydrant, and Hose Systems.* The standpipe can be either temporary or permanent (and rigged as a dry standpipe in cold climates) until the building is enclosed. In most cases, the fire department connection must be temporary because the permanent location will not be completed and might not be accessible during much of the construction. Regardless of the type of building under construction, portable extinguishers must be provided in rating and spacing suitable to the construction activity and in accordance with NFPA 10, *Standard for Portable Fire Extinguishers.*

Sprinklers should be put into service and fire walls or fire barrier walls should be built as soon as possible after the building shell is finished. Only steel; fire-retardant-treated wood; or limited amounts of combustible formwork, scaffolding, or shoring are acceptable in unsprinklered areas.

ALTERATIONS AND ADDITIONS

More and more older buildings are being altered or renovated, some to preserve the architecture of an age gone by. As older buildings are rehabilitated, every effort should be made to bring them into compliance with present-day building and fire codes. However, this is not always possible. In such cases, equivalent protective measures might be acceptable in lieu of meeting certain code requirements. These equivalencies might involve the installation of automatic sprinkler protection, smoke detection systems, and smoke control features. Fire inspectors are responsible for policing those protection systems as they would those in a new building.

Sometimes renovations have to be made while a building is partially occupied. It is extremely important for life safety that exits for occupants are properly maintained. It is the inspector's job to make sure that the exits are accessible and have been properly identified. In addition to exits for occupants, sufficient means of escape should be provided for construction workers in the area under renovation, in accordance with the *Life Safety Code*. Inspectors should be just as diligent in inspecting these means of escape as they are in inspecting the public exits.

If a building addition blocks an existing exit, an alternate means of egress must be provided. The inspector must inspect the alternate route. The exit should be free of debris and stored materials. It should be properly identified and lighted. The exit discharge should be clear of parked vehicles and other obstructions.

If the alterations involve hot work, such as cutting, welding, and heating, a permit system should be used; for further information see NFPA 51B. Whenever practical, combustibles should be removed from at least 35 ft (10.7 m) in all directions. If this cannot be done, the inspector should recommend covering combustibles with a fire-resistant tarp, sweeping up combustible dusts, and wetting down wood floors. A fire watch provided with extinguishers or hoses should be assigned to the area for at least 30 minutes after the work has been finished; this time period should be extended to 60 minutes in torch-applied roof installations.

DEMOLITION

Demolition operations have many of the previously noted attendant hazards of construction operations, as well as a few others. The hazards of cutting torches, flammable liquids, and trash accumulations are as common in demolition operations as they are in construction operations.

Before demolition begins, gas pipes should be turned off and capped outside the building. Explosives should be stored and used in accordance with NFPA 495, *Explosive Materials Code.*

Early in a demolition project, flammable liquids and combustible oils should be drained from tanks and machinery and immediately removed from the building. The removal of residue and sludge deposits is also important, especially in areas where cutting torches are used. Torches should not be used to cut through walls, floors, ceilings, or roofs containing combustible materials.

Fixed fire protection systems and fire walls should be maintained for as long as possible. Sprinkler and standpipe systems should be modified so they can be dismantled floor by floor as demolition progresses downward from the top floor or system by system in large one-story buildings. This will preserve protection in the adjacent floors or areas. Either type of protection can be converted readily from a wet-pipe system to a dry-pipe system, if minimal heat (40°F, 4.4°C) cannot be maintained at remote parts of the building.

Generally, chutes are provided to carry demolition rubble and debris from the upper floors to trucks or mobile trash receptacles below. These should be constructed as described in the earlier section on housekeeping. The use of inside chutes, which would necessitate cutting holes in the floor, thereby creating an unprotected vertical opening through which fire could spread rapidly from floor to floor, should be discouraged.

UNDERGROUND OPERATIONS

General

A number of the guidelines mentioned earlier for construction, alteration, and demolition operations apply to underground operations as well. However, other concerns also apply to these operations. Structures and equipment should be constructed of noncombustible materials as much as is practical. Because of possible changes in underground operations, it is important to maintain current written procedures for evacuation, fire prevention inspections, and other emergency procedures.

Fire Protection

A water supply should be provided wherever combustibles are present. Outlets with standard fittings compatible with local fire department equipment should be provided so that the maximum travel distance does not exceed 150 ft (46 m). Drainage systems should be provided to remove sprinkler and fire hose discharge.

Extinguishing equipment should be provided for belt conveyors at their head, tail, drive mechanism, and take-up pulley areas and at maximum intervals of 300 ft (91 m) along their length and along the tunnel, such that the maximum travel distance on a horizontal plane does not exceed 300 ft (91 m). Areas near and under belt conveyors should be kept free of accumulating combustibles. Interlocks should be provided to shut down the conveyor drive on activation of any fire protection system or on stoppage or slowdown of the conveyor belt.

Flammable and Combustible Liquids

Class I flammable liquids should not be permitted underground or within 100 ft (30.5 m) of a tunnel portal or shaft opening. Storage of other flammable liquids should be diked or otherwise positioned so that spills will not flow away from the storage area.

Security and Compartmentation

A check-in/check-out system should be provided at each underground entrance for each person in transit to and from the underground operation. Sections of underground that are no longer in use should be barricaded and marked. Fire and smoke barriers should be used to limit fire spread and provide areas of refuge for occupants where egress is not practical.

Electrical Equipment

Electrical equipment should be protected against physical damage and suitable for damp locations. Oil-filled transformers should not be used underground unless they are enclosed by fire-resistant materials, vented to the outside, and diked to contain the entire contents of the transformer.

SUMMARY

Construction, alteration, and demolition operations can result in sever fires occurring frequently. Control of combustibles and hot work operations will reduce the probability of a fire occurring. Proper fire protection management, prefire planning, and reasonable fire protection will help limit the extent of damage from a fire.

BIBLIOGRAPHY

Cote, A. E., ed., *Fire Protection Handbook,* 18th ed., NFPA, Quincy, MA, 1997.

Fisher, G. L. "Getting Through Construction," *Fire Prevention,* No. 248, April 1992, pp. 20–22.

FM Global Property Loss Prevention Data Sheet 1-33, Safeguarding Torch-Applied Roof Installations, September 2000.

Gray, J. A., and Arditi, D., "Fire Prevention and Protection During Construction," *Journal of Applied Fire Science* Vol. 4, No. 1, 1994–1995, pp. 53–68.

Tremblay, K. J., "Sparks from Saw Ignite Church Balcony," *NFPA Journal,* November/December, 1999, p. 17.

Wolf, A., "Seventeen Die in Dusseldorf Airport Terminal Fire," *NFPA Journal,* July/August, 1996.

NFPA Codes, Standards, and Recommended Practices

See the latest version of The NFPA Catalog for availability of current editions of the following documents.

NFPA 10, *Standard for Portable Fire Extinguishers*
NFPA 14, *Standard for the Installation of Standpipe, Private Hydrant, and Hose Systems*
NFPA 30, *Flammable and Combustible Liquids Code*
NFPA 51, *Standard for the Design and Installation of Oxygen-Fuel Gas Systems for Welding, Cutting, and Allied Processes*
NFPA 51B, *Standard for Fire Prevention during Welding, Cutting, and Other Hot Work*
NFPA 54, *National Fuel Gas Code*
NFPA 58, *Liquefied Petroleum Gas Code*
NFPA *101*®, *Life Safety Code*®
NFPA 241, *Standard for Safeguarding Construction, Alteration, and Demolition Operations*
NFPA 495, *Explosive Materials Code*
NFPA 701, *Standard Methods of Fire Tests for Flame Propagation of Textiles and Films*

Protection of Openings in Fire Subdivisions

CHAPTER 8

Richard J. Davis

CASE STUDY

Four Die in Apartment Blaze

Two 35-year-old adults and their two children, ages 5 and 2, died when fire engulfed their four-story apartment building in Utah. Several other occupants were injured when they jumped from upper floor windows.

The 16-unit building was constructed of unprotected wood framing with brick veneer and a flat, built-up composition roof. Unenclosed stairwells were located at each end of the building. Corridor fire doors divided each floor in half, but they were found blocked open. Each apartment contained smoke detectors, which operated, but the hallways and stairwells had no detectors. There were no sprinklers.

A second-floor tenant called 911 at 11:03 A.M. to report fire and smoke. While fire fighters were en route, other calls came in reporting that conditions were deteriorating and occupants were trapped and jumping from windows. Fire fighters arrived six minutes after the call and immediately began rescue operations using ground ladders. Fire in the stairwells made it difficult to enter the building.

Crews eventually made some headway and searched the ground floor, but heavy fire conditions prevented them from reaching the upper two floors. After 45 minutes, fire broke throught the roof, forcing the incident commander to switch to a defensive attack, as paramedics treated and transported injured civilians.

The fire started in the basement under the stairs when an undetermined source ignited an upholstered recliner. The fire spread up the north stairwell, down the first-floor hallway, and through the open fire doors to the other stairwell and upper floors.

Fire fighters found all four victims in the hallway, where they'd apparently gone to try to escape. The tenants who remained in their apartments and went to windows survived. Two occupants were injured when they jumped from windows, and two sustained smoke inhalation injuries.

The building, valued at $600,000, was a total loss. Damage to its contents, valued at $200,000, was estimated at $150,000.

Source: "Fire Watch," *NFPA Journal*, September/October, 1997.

Richard J. Davis, a senior engineering specialist with FM Global, is chair of the NFPA Technical Committee on Construction, Alteration, and Demolition Operations and a member of the NFPA Technical Committee on Building Construction and the NFPA Technical Committee on Structures and Construction.

One method of limiting the spread of fire and smoke in a structure is to divide the interior into compartments using fire walls and fire barrier walls and rated floor/ceiling assemblies. However, fire subdivisions can be expected to delay the spread of fire from the room or area of origin to other parts of the structure only if they are constructed and maintained properly and if the openings in them are properly protected.

FIGURE 8-1(a) Double Fire Wall with No Connection
Source: NFPA 221, 2000, Figure A.2.5(a).

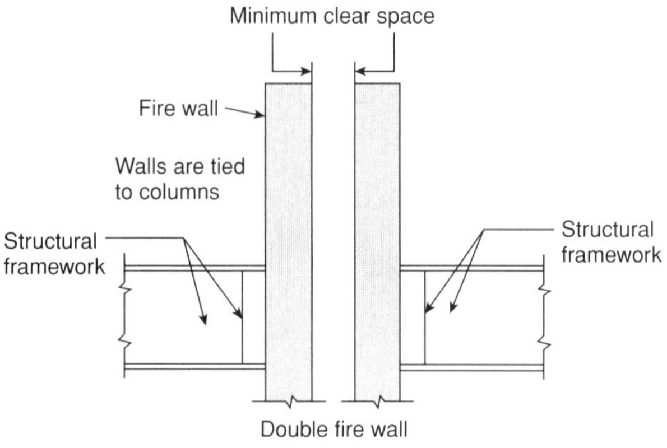

FIGURE 8-1(b) Double Fire Wall with Separate Horizontal and Vertical Flashing Sections
Source: NFPA 221, 2000, Figure A.2.5(b).

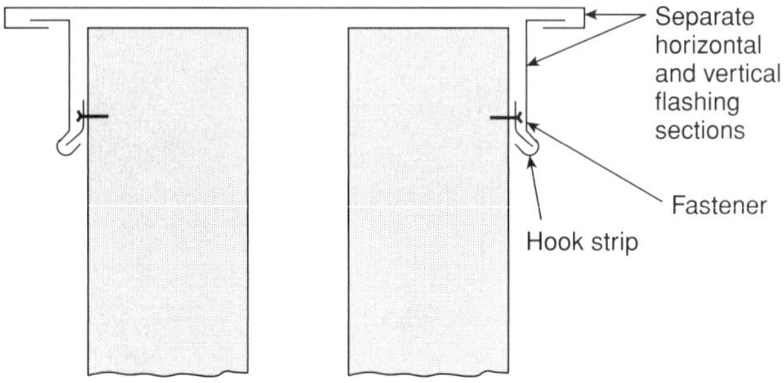

THE HAZARD AND FORMS OF PROTECTION

It is not uncommon for heated, unburned pyrolysis products to flow out of the area of initial involvement, mix with air, and ignite. Such flame extension can even occur over noncombustible surfaces. Flames can heat interior finish materials to the point at which they release pyrolysis products of their own, which also ignite and contribute to the intensity of the extending flame.

Properly maintained opening protectives in fire walls and fire barrier walls (see NFPA 221, *Standard for Fire Walls and Fire Barrier Walls*) is essential to contain a fire until automatic or manual fire suppression is effective (Figure 8-1). Part of the inspector's responsibility, therefore, is to determine that these openings are properly protected and that nothing has been done to nullify or defeat their protection.

A variety of opening protectives in fire walls and fire barrier walls is available. The method selected will depend on the type, function, and configuration of the opening. Typical protection measures include firestopping, fire-resistive construction, fire doors, and fire resistance–rated glazing materials, although special problems may require other forms of protection.

Fire walls are intended to retain their structural integrity and remain stable in the event of building collapse as the result of an uncontrolled fire on either side of the wall. Fire barrier walls rely on the building framing for stability.

VERTICAL OPENINGS

Unprotected openings in floors and ceilings, referred to as vertical openings, may permit the extension of fire from one floor to another.

Floor/Ceiling Penetrations

Unsealed gaps, created when holes are made through floor/ceiling assemblies for routing cables, conduits, or pipes, permit the passage of fire and smoke from floor to floor. One method used to seal these gaps involves modular devices that are sized for the pipe, conduit, or cable and that contain an organic compound that expands when heated to seal the penetration. Other methods of sealing such penetrations include the use of foamed-in-place fire-resistant elastomers, various caulking materials, and poured- or troweled-in-place compounds. In addition, there are bags of fire-resistant material that can be placed around penetrating pipes, cables, or conduits. When exposed to fire, these bags expand and fuse to prevent the passage of the fire products.

When plastic pipes or conduits penetrate a fire subdivision, they will melt and create a large opening. In some cases noncombustible piping can be used where the piping penetrates the subdivision. In other cases special penetration seals, such as intumescent types, are needed that will expand as the result of fire exposure and will fill the void. Many of these materials have been tested and listed or approved, and all can provide the required protection if they are properly installed and maintained.

The penetrating objects should be supported well enough to keep them from placing any mechanical stress on the seal that could pull the sealant from the opening. Where "temporary" routing of utilities or control cables is a fairly common occurrence, workers tend to neglect to seal the gaps (Figure 8-2). The inspector should be alert for such conditions. Often, utility lines are hidden in closets or above drop ceilings and are not obvious during a casual visual inspection. The locations of these concealed, but accessible, fire barrier penetrations should be noted so that they are not overlooked during the inspection process.

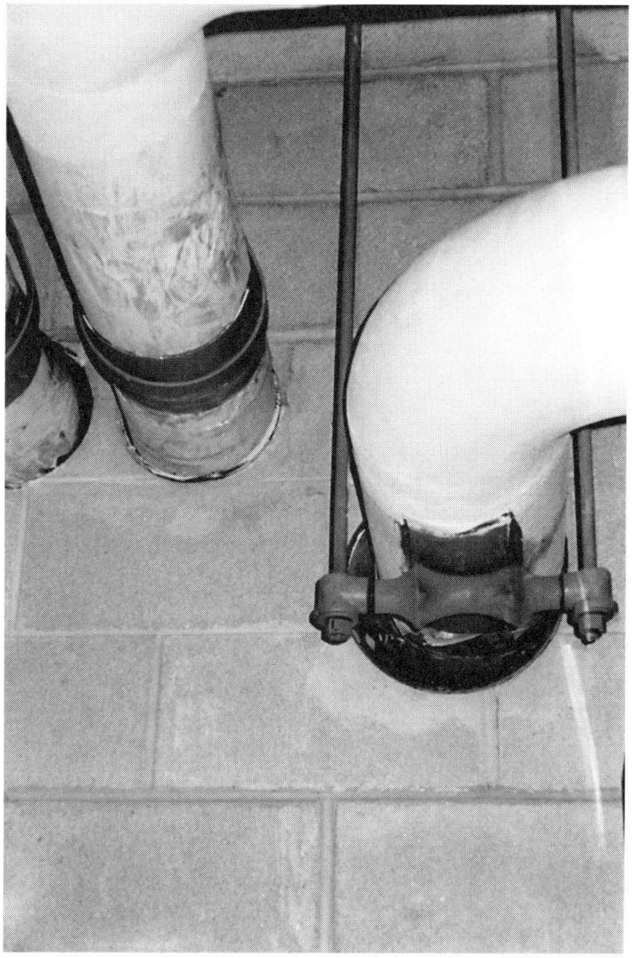

FIGURE 8-2 Pipe Penetration Through Fire Wall with No Penetration Sealant

Stair Enclosures, Shafts, and Chutes

Certain vertical openings cannot be sealed because their functions require them to communicate between floors. Examples include stair enclosures, elevator shafts, utility shafts, and chutes for mail, laundry, or trash. Such openings should be enclosed in fire-resistive construction. Openings in the walls of utility shafts should be protected with self-closing fire-rated doors approved or listed for the purpose. Openings in the walls of stair enclosures and elevator shafts must be protected by rated self- or automatic-closing fire door assemblies.

It is not uncommon for occupants to prop open stair doors for the sake of convenience. Wood blocks, wedges, or pieces of wire or rope near a fire door indicate that the occupants are blocking or holding the fire door open. This, of course, defeats the purpose of a fire door and is a condition that must be corrected immediately. If this situation becomes common and impossible to enforce, you may consider requiring the installation of magnetic hold-open devices with appropriate actuation devices.

Escalators

Openings made in floor/ceiling assemblies to accommodate escalators present a unique protection problem because enclosing them in fire-resistive construction is not practical (Figure 8-3).

FIGURE 8-3(a) Escalators

FIGURE 8-3(b) Sprinklers Around Escalators
Source: NFPA 13, 1999, A.5.13.4.

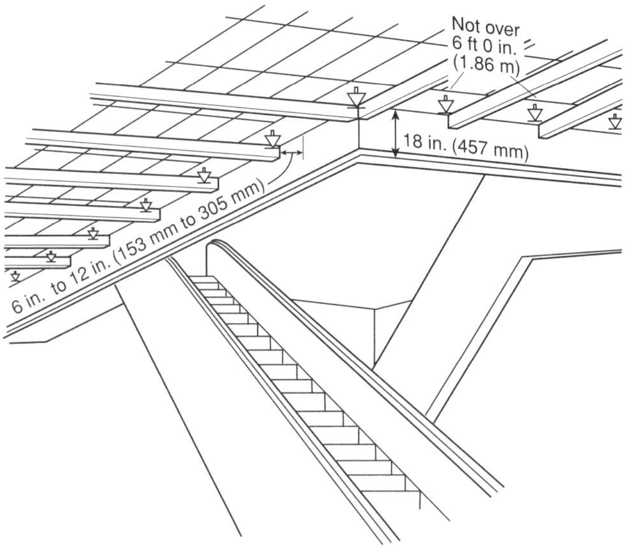

However, there are alternative forms of protection. In a fully sprinklered building, NFPA 13, *Standard for the Installation of Sprinkler Systems,* provides a protection of escalators known as the sprinkler-draft curtain method.

Another method relies on a combination of automatic fire or smoke detection equipment, an automatic exhaust system, and an automatic deluge water curtain. Another method involves filling the opening with a dense water spray pattern from open, high-velocity water-spray nozzles. The water-spray system is operated automatically by heat or smoke detection and is equipped with manual control valves to minimize water damage. (Details on these systems can be found in the *Life Safety Code® Handbook* and the *Automatic Sprinkler Systems Handbook.*) Inspectors should examine the control valves in these systems as they would those in other water-based fire protection systems to make sure that they are open.

Another method is to protect the opening with a partial enclosure of fire-resistive construction in a "kiosk" configuration. This enclosure is equipped with self-closing doors. The inspector should check that the doors are in operating condition and that the self-closing feature has not been circumvented in any way.

HORIZONTAL OPENINGS

If left unprotected, openings in fire walls, called "horizontal openings," will permit fire to spread in the horizontal plane throughout the floor of origin. Some corridors, in particular, must be protected, not only because they are a path for the horizontal spread of fire, smoke, and toxic gases, but also because they are a part of the means of egress through which the occupants must pass in order to exit the building.

Fire Door Assemblies

One of the most widely used opening protectives in fire-resistive walls is the fire door assembly. Tests conducted by independent testing laboratories in accordance with accepted test methods determine the fire resistance rating of a fire door assembly. Fire door assemblies are given an hourly rating, as are the openings in which they are placed, as illustrated in Table 8-1.

TABLE 8-1 Fire Protection Ratings for Doors and Associated Support Construction

Component	Walls & Partitions* (Hours)	Fire Doors* (Hours)
Elevator hoistways	2	1½
	1	1
Vertical shafts	2	1½
(including stairways, exit, and refuse chutes)	1	1
Fire walls	4	3
	3	3
Fire barrier	4	3
	3	3
	2	1½
	1	¾
Horizontal exit	2	1½
Corridors, exit access	1	¾
	½	20 min
Smoke barrier	1	20 min
Smoke partition	½	20 min
Exterior Walls	4	3
	3	1½
	2	1½
	1	¾

*Consult NFPA *101,* for applicable requirements.

Listed fire doors must be identified by a label, a listing mark, or a classification mark that is readily visible. Labels or classification marks may be of metal, paper, or plastic, or they may be stamped or diecast into the item. Very large fire doors may not have a listing mark if they exceed the size of the door the testing laboratory can physically test. However, the laboratory may furnish the door with a certificate of inspection that states that it conforms to the same requirements of design, materials, and construction as a rated fire door, even though it has not been subjected to an actual test.

Ratings. Each fire door classification has specific applications. Where a fire wall is provided to separate two buildings or divide a building into two fire areas, the use of a 3-hour fire door assembly is required. In industrial and storage buildings that have vehicle openings, two fire doors are normally required for reliability of closure and are mandated by NFPA 221 on 4-hour–rated fire walls. Openings in double fire walls should have one fire door in each separate wall or two fire doors in a freestanding fire-resistive vestibule (Figure 8-4). If the corridor is provided as a smoke barrier only, a 20-minute-rated fire door is acceptable.

Openings in walls enclosing hazardous areas also can be protected with fire door assemblies. Depending on the local codes or ordinances in effect, the fire-resistance rating required for a specific application may vary from those given here.

Doors in openings in exterior walls that might be subjected to severe fire exposure from outside the building and doors protecting openings in 2-hour enclosures

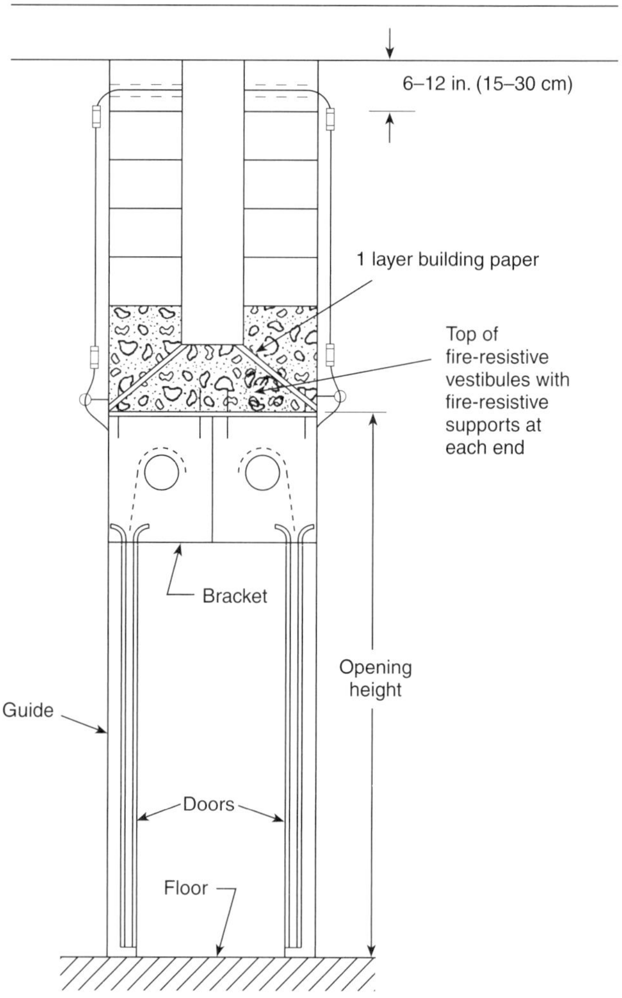

FIGURE 8-4 Double Doors on a Freestanding Vestibule
Source: NFPA 221, 2000, Figure A.5.2.

of vertical building openings each carry a 1½-hour fire-resistance rating. One-hour stair enclosures are protected by 1-hour fire doors. Rated ¾-hour fire door assemblies are used to protect openings in the exterior walls of buildings that might be subjected to a light or moderate fire exposure from outside the building. Rated ¾-hour fire door assemblies are used in some room-to-corridor openings, especially to isolate a hazardous area from the corridor.

Fire door assemblies with ½-hour and 20-minute fire-resistance ratings are intended primarily for smoke control. These doors are used across corridors in which a smoke barrier is required and to protect openings in walls with fire-resistance ratings of up to 1 hour that are installed between a habitable room and a corridor.

Construction. Several types of construction are used in the manufacture of fire doors. Composite doors are flush doors made of a manufactured core material with chemically impregnated wood edge banding. They are faced with untreated wood veneer or laminated plastic, or they are encased in steel.

Hollow metal doors are made in flush and panel designs of 20-gauge (0.036-in., 0.9-mm) or heavier steel. Metal-clad doors are flush or panel-design swinging doors with metal-covered wood cores or stiles and rails and insulated panels covered with 24-gauge (0.024-in., 0.6-mm) or lighter steel. Sheet metal doors are made in corrugated, flush, or panel designs of 22-gauge (0.028-in., 0.7-mm) steel or lighter, while rolling steel doors are fabricated of interlocking steel slats or plate steel (Figure 8-5).

Metal- or tin-clad doors are of two- or three-ply wood core construction. They are covered with 30-gauge galvanized steel or terne plate, with a maximum size of 14 × 20 in. (356 × 508 mm) or with 24-gauge (0.024-in., 0.6-mm) galvanized steel sheets with a maximum width of 48 in. (1220 mm). Curtain-type doors consist of interlocking steel blades or a continuous formed-spring steel curtain installed in a steel frame. Wood core doors consist of wood, hardboard, or plastic face sheets bonded to a wood block or a wood particle board core material with untreated wood edges.

Special-purpose fire door assemblies called horizontal sliding accordion or folding doors also are available. They are self- or automatic-closing doors, and some of them are power-operated. Under some codes, folding doors are permitted within a means of egress as horizontal exits or in smoke barriers under certain restrictions, such as the provision of backup power for power-operated doors. Materials used in these types of doors vary.

Door Closing. Fire doors must be self-closing or close automatically in the event of fire (Figure 8-6). A suitable door holder/release device can be used, provided the automatic-release feature is actuated by automatic fire detection devices. Generally, automatic release that is accomplished with fusible links is permitted only in limited areas. If there is any significant distance in elevation between the top of the opening and the underside of the ceiling or roof, fire detection devices used to actuate the door should be located near the top of the opening *and* the underside of the ceiling or roof.

Automatic-closing fire doors are sometimes required to begin closing not more than 10 seconds after the release device has actuated. Where applicable, this should be verified because a door holder/release device with an excessive time-delay feature could allow a large volume of smoke to pass through the opening before the door closes.

Maintenance and Inspection. Fire doors that are normally open during working hours should be closed during nonworking hours. Combustible material should not be stored near an opening in a fire wall because the fire might spread through the opening before the protective device can operate. Also, most fire doors do not limit heat transmission. Heat from a fire can be conducted through the door and can

FIGURE 8-5 Surface-Mounted Rolling Steel Doors
Source: NFPA 80, 1999, Figure B-48.

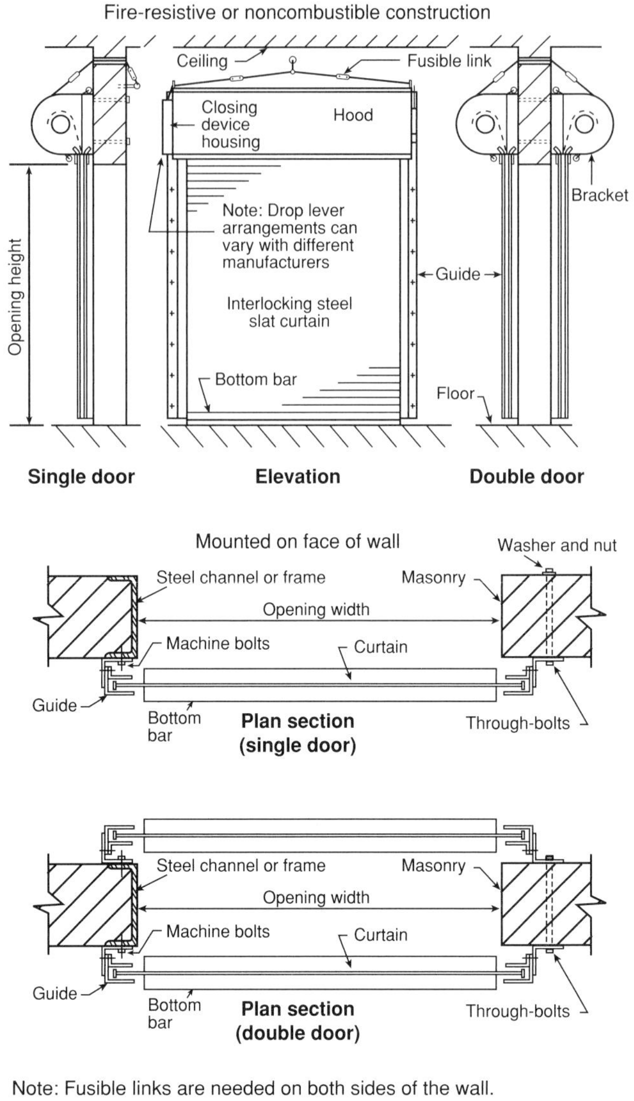

FIGURE 8-6 Closing Devices for Center-Parting, Horizontally Sliding Doors (Inclined Track)
Source: NFPA 80, 1999, Figure B-38.

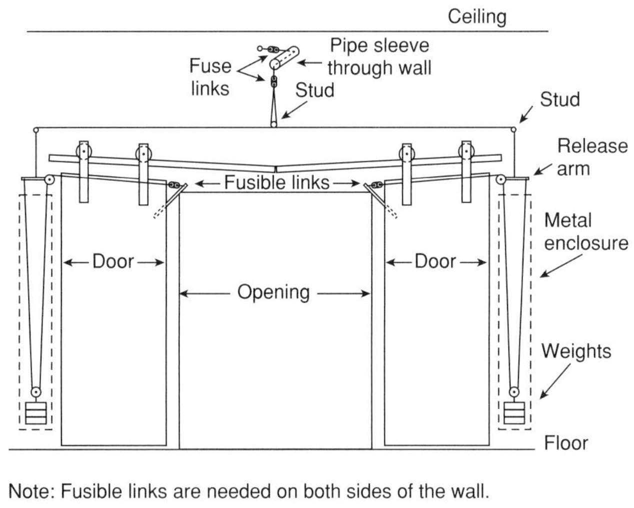

FIGURE 8-7 Fire Door Wedged Open

radiate from the opposite side of the door, igniting nearby combustibles. To prevent this situation, combustibles should be kept away from the opening for a distance of at least one times the larger dimension of the opening.

Inspectors should make sure that fire doors are not obstructed or blocked in any way or intentionally wedged open so that self-closing is not possible (Figure 8-7). Should they find that fire doors have been intentionally blocked or wedged, they should determine the reason for it and take appropriate action. Where doors have been fastened open to improve ventilation, other ventilating means should be provided. Guards or railings should be provided where necessary to prevent damage from materials-handling equipment. Doing so also discourages employees from piling stock against or near the door. Inspectors should ensure that the movement of balance weights is free and unobstructed and that fusible links are of the proper temperature rating. They should also verify that fusible links have not been made inoperative by paint, corrosion, or other external conditions. For example, they may have wires that would prevent closure, even if the links did activate, rendering the elements inoperative. Self-closing and automatic-closing fire doors should be tested to ensure that they are operational and that swinging doors latch when closed. Inspectors should be aware that heat-actuated closing devices are suitable only when property protection is desired and life safety is not being considered.

Metal- or tin-clad doors have some special features that require attention. Inspectors should make sure that the door has proper lap over the opening. The binders are sometimes filled with blocking to make the door easier to open. These blocks must be removed. The door should have chafing strips, which help maintain its fire resistance. Inspectors should note the condition of the door. Is the metal covering corroded, torn, or battered? Is there evidence of dry rot? Indications of dry rot include edges caving in and screws pulling out. Tapping the door with a weighted object such as a hammer can give some indication of the extent of the rot damage.

When inspecting fire doors, inspectors should perform an operating test to make sure they do not jam or stick and that the hardware is complete and undamaged. For sliding doors, the inspector should check the automatic-closing mechanism by lifting the counterbalance weight or dropping the suspended weight. When door closers are equipped with a fusible link, the test may be limited to general observation of the device. However, if spare links are available, a preferred test method is to fuse it using an electric heat gun. It is also important to ensure that the link is exposed so heat can reach it.

Rolling steel doors should be tested when an opening is not in use and reset only by qualified personnel so that any malfunction that might occur will not interfere with the normal activities on the premises. Vertical guides for rolling steel fire doors should be bolted to the framing through slotted holes to allow for expansion during fire exposure. Likewise, a gap should be provided between the bottom of the guides and the floor.

Fire Shutters

Fire shutters are used to protect openings in exterior walls. If the potential fire exposure from outside the building is severe, 1½-hour shutters are used. Where the potential fire exposure is moderate or light, ¾-hour shutters are used. If fire shutters are installed on the outside of the opening, they should be protected against the weather to ensure proper operation. Shutters must be equipped to close automatically in case of fire.

These devices should not be overlooked on the inspection tour. Although fire shutters are no longer used to any great extent in new construction, they are still found on older structures.

Fire Windows and Vision Panels

Wired glass, glass blocks, ceramic glass, and other glazing materials are used in ¾-hour fire-resistance-rated window assemblies. Fire window assemblies are designed to protect openings in corridors, rooms, and exterior walls where the potential exposure is moderate or light. Some fire window assemblies are equipped with automatic closing devices that are actuated by automatic fire detection equipment. When inspecting fire window assemblies, inspectors should be sure the closing devices are in operating condition.

Wired glass is frequently used as a vision panel in smoke-stop barriers and in fire door assemblies that protect stairway enclosures (Figure 8-8). While conducting the inspection, inspectors should be on the lookout for situations in which plain window glass might have been used to replace broken wired glass. They should also be aware that glazing that does not contain wire but that has been tested successfully for specific hourly fire rating is available. One fire-rated glazing product is made of a ceramic material, looks like single-thickness plate glass, but does not provide significant resistance to heat transfer.

Also available on the commercial market are other glazing materials that meet the performance for use in fire resistance–rated assemblies. For example, one type has two or three layers of glazing sandwiching a gel-like material, making the

FIGURE 8-8 Vision Panel in a Fire Door

glass considerably thicker than other types of conventional glazing. That type of assembly is intended to insulate in a manner similar to that of a fire barrier wall.

Sills

In buildings with noncombustible floor construction, sills should also be constructed of noncombustible materials. Special sill construction is not required, provided that the floor structure is extended through the door opening. Table 8-2 lists the sill clearance requirements for fire door assemblies. Criteria prohibiting the extension of combustible material through an opening protected by a fire door assembly have been revised. Combustible floors may extend through door openings required to be protected by ½-hour or 20-minute fire-rated assemblies. Combustible floor coverings can now extend through openings protected by 1½-, 1-, or ¾-hour fire door

TABLE 8-2 Clearances Under Bottoms of Doors

	Swinging Doors with Builders Hardware		Swinging Doors with Fire Door Hardware		Horizontally Sliding Doors		Vertically Sliding Doors		Special Purpose Horizontally Sliding Accordion or Folding Doors	
Clearance Between	**in.**	**mm**	**in.**	**mm**	**in.**	**mm**	**in.**	**mm**	**in.**	**mm**
Bottom of door and raised noncombustible sills	⅜	9.5	⅜	9.5	⅜	9.5	⅜	9.5	⅜	9.5
Floor where no sill exists	¾	19.1	¾	19.1	¾	19.1			¾	19.1
Rigid floor tile	⅝	15.9								
Floor coverings	½	12.7	½	12.7	½	12.7			½	12.7

Source: NFPA 80, 1999, Table 1.11.4.

assemblies, as long as the floor covering has a minimum critical radiant heat flux of 22 kW/m^2. The clearance between the covering and the bottom of the door should not exceed ½ in. (12 mm).

SPECIAL PROBLEMS

Duct and materials-handling systems that penetrate walls, partitions, floors, and ceilings contribute to both the horizontal and vertical spread of fire. Duct systems are discussed in Chapter 11, "Air-Conditioning and Ventilating Systems," and in Chapter 55, "Protection of Commercial Cooking Equipment," and conveyors are discussed in Chapter 40, "Materials-Handling Systems." The protection of openings for various materials-handling systems is also covered in *FM Global Property Loss Prevention Data Sheet 1-23.*

SUMMARY

The protection of openings in a fire subdivision is critical in order to maintain the integrity of the subdivision. Fire doors must be inspected to assure they are of the proper rating, that they are not blocked open, and that they are adequately maintained. A reliable method is needed to assure that the opening is clear where material-handling systems pass through an opening.

BIBLIOGRAPHY

Cote, A. E., ed., *Fire Protection Handbook,* 18th ed., NFPA, Quincy, MA, 1997.
Coté, R., *Life Safety Code® Handbook,* 8th ed., NFPA, Quincy, MA, 2000.
FM Global Property Loss Prevention Data Sheet 1-23, Protection of Openings, January 2000.

NFPA Codes, Standards, and Recommended Practices

See the latest version of The NFPA Catalog for availability of current editions of the following documents.

NFPA 13, *Standard for the Installation of Sprinkler Systems*
NFPA 80, *Standard for Fire Doors and Fire Windows*
NFPA 80A, *Recommended Practice for Protection of Buildings from Exterior Fire Exposures*
NFPA 82, *Standard on Incinerators, Waste, and Linen Handling Systems and Equipment*
NFPA 90A, *Standard for the Installation of Air Conditioning and Ventilating Systems*
NFPA 90B, *Standard for the Installation of Warm Air Heating and Air Conditioning Systems*
NFPA 91, *Standard for Exhaust Systems for Air Conveying of Vapors, Gases, Mists, and Noncombustible Particulate Solids*
NFPA 92A, *Recommended Practice for Smoke-Control Systems*
NFPA *101®*, *Life Safety Code®*
NFPA 105, *Recommended Practice for the Installation of Smoke-Control Door Assemblies*
NFPA 204, *Standard for Smoke and Heat Venting*
NFPA 221, *Standard for Fire Walls and Fire Barrier Walls*
NFPA 252, *Standard Methods of Fire Tests of Door Assemblies*
NFPA 257, *Standard for Fire Tests of Window Assemblies*

Electrical Systems

CHAPTER 9

Noel Williams

A fire started by an electrical malfunction spread rapidly along combustible decorative materials in a California furniture store, destroying the store before it moved to adjacent businesses in the building. It was finally stopped by fire fighters.

The single-story building contained five separate occupancies joined by common walls. Neither smoke detection nor fire suppression systems had been installed in the building, which was closed for the night when the fire broke out.

The blaze was discovered by a store employee, who was preparing to leave shortly after 8:00 P.M., when he heard a buzzing in the electrical panel. Seeing flames in the front display area, he tried to called 911 from two phones, but both lines were disabled. He then left the store and called 911 from a restaurant across the street.

The fire department arrived 7 minutes later and found the store fully involved in fire. Fire fighters initially used hose lines to protect exposures, which included a restaurant and palm trees across the street, the latter already on fire. While en route, the incident commander had ordered additional resources when subsequent calls confirmed the incident was a structure fire. Arriving fire fighters were asked to protect exposures and then to cut a trench on the north side of the building to stop the flames from spreading. These actions eventually allowed fire fighters to contain the fire and control the blaze, although off-duty personnel had to be called to the scene early on. Fire attack continued for 4 hours before the fire was declared under control, with 6 hours of overhaul after that.

Investigators determined that the fire started in the front display area, where an electrical outlet showed signs of arcing. Numerous electrical cords, including chains of extension cords and multiple outlet boxes, had evidently been used with the one primary outlet. The panel buzzing heard just before the fire was discovered was the circuit breaker trying to trip at a time when power to a large part of the store lighting had already been interrupted.

Apparently, an electrical arc ignited combustible decorations, and the fire quickly spread along other combustibles to engulf the store.

Damage to the structure and its contents was estimated at $1.2 million and $230,000, respectively. Several exposures, including a vehicle, trees, a mailbox, and signal lights, were also damaged, for an additional $16,000. There were no injuries.

Source: "Fire Watch," *NFPA Journal,* September/October, 1999.

CASE STUDY

Extension Cords Cause Electrical Arc

Noel Williams spent almost 25 years supervising electrical projects as a licensed master electrician and electrical inspector. He has taught a class on NFPA 70, National Electrical Code®, *for over 10 years and has coauthored two books for NFPA.*

Electrical systems are best examined and tested by qualified electrical inspectors because they have the special skills needed to correct electrical system deficiencies that may cause fires. Fire inspections are carried out more frequently than electrical inspections, however, it is likely that a fire inspector, rather than an electrical inspector or a qualified electrician, will detect potential problems. The fire inspector must be aware of the signs and symptoms of the potential fire hazards presented by electrical systems.

The causes of electrical fires can be placed in four broad categories: damaged electrical equipment, improper use of electrical equipment, accidents, or defective installations. By learning to recognize the signs of potential trouble, the inspector can eliminate many sources of electrical failure as a cause of fire.

WIRING AND APPARATUS

Electrical fires are due principally to arcing and overheating. Arcing occurs when electrical current flows across gaps in otherwise conductive pathways. These gaps may be created in the normal operation of equipment, such as in switches or in motors with brushes. They may also be created at loose splices and terminals or where wire or other conductor insulation has been damaged and is in close proximity to other damaged conductors or grounded metal enclosures or surfaces. Arcing produces enough heat to ignite nearby combustible materials, such as insulation, and can throw off particles of hot metal that can cause ignition. Arcing can also melt metal conductors and produce sparks.

The conditions that create an arc usually cause overcurrent protective devices, such as fuses and circuit breakers, to operate, making the duration and resultant heat exposure very brief. However, intermittent arcing, such as might occur due to accidental damage, can sometimes happen without tripping overcurrent devices. Special devices, called arc-fault circuit-interrupters (AFCIs) are designed to detect such arcing, but they are currently required in only certain residential applications.

Overheating is more subtle, harder to detect, and slower to cause ignition, but it is equally capable of causing a fire. Conductors and other electrical equipment may generate a dangerous level of heat when they carry a current in excess of rated capacity. Overloading may cause conductors to overheat to the point at which the temperature is sufficient to ignite nearby combustible materials. Insulation failure caused by overheating also can lead to arcing between conductors or between conductors and adjacent grounded objects.

Common Faults

Conduits, Raceways, and Cables. Among the obvious faults are badly deteriorated and improperly supported conduits, raceways, and cables. Where these items enter boxes, cabinets, and other equipment, they should be terminated in proper fittings that hold them securely in place without damaging the conductor insulation. Conduits that are not supported properly may pull apart and expose conductors and insulation to damage.

Cables should be protected from mechanical damage where they pass through walls or floors. They also should be protected from overload, which is not as immediately obvious as mechanical damage. One way to determine whether a cable or the conductors in a conduit are overloaded is to touch the cable or conduit. Depending on the load, the cable or conduit may feel abnormally warm or even hot. However, this is a very subjective test. Many common conductors and equipment in normal use may operate at temperatures up to 167°F (75°C). This is much too hot to touch comfortably. Where overheating is suspected, it should be further investigated by a qualified person. Infrared scanning equipment is available to make a more objective assessment of actual and relative temperatures in a system without requiring any physical contact.

Circuit Conductors. Single conductors usually are installed in raceways, but they may be installed on insulators in free air or in cable trays. Open conductors are more common in industrial occupancies and in older buildings. Like cables, branch circuit conductors must be supported properly along their length and at the point at which they terminate in junction, switch, and outlet boxes. Conductors should not be exposed to excessive external heat, which will hasten the deterioration of their insulation. Circuit conductors also may be subject to electrical overload where fuses or circuit breakers are of the incorrect value. To detect overloaded conductors, look for discoloration of the terminals or of the surfaces of conduits and boxes.

Flexible Cords. Several unsafe practices involving flexible cords may result in fires. One of the most common unsafe practices is using flexible cords or extension cords in place of what should be the fixed wiring of a building. Extension cords should be used only to connect temporary portable equipment, not as part of permanent wiring. Nor should they be used to supply equipment that will load them beyond their rated capacity. Extension cords are not required to have the same current carrying capacity as the branch circuit wiring supplying wall receptacle outlet. Therefore it is extremely important to ensure that the equipment supplied by the cord has a lower wattage or amperage rating than that of the extension cord. The terminations of the conductors in a flexible cord should not be relied on to provide mechanical support. Rather, cords should be clamped in a connector or knotted in an approved manner to keep stresses from being transferred to the conductor terminals. Flexible cords should never be run where they can be damaged by vehicles, carts, or pedestrian traffic. Nor should they be left coiled or hanked or run under rugs or carpets. Extension cords and other flexible cords should not be attached to building surfaces, woodwork, pipes or other equipment, or run through doors or windows or through holes in walls, floors, or ceilings. Damaged cords should be replaced, not repaired or spliced. However, broken or damaged cord ends may be replaced using listed devices.

There are many legitimate uses for flexible cords. Flexible cords can be used for pendants, for connecting portable appliances or lamps, and for some permanent equipment such as submersible pumps or equipment that must be frequently moved or interchanged. Some devices such as portable power taps may be attached to building surfaces or furniture, but the cords themselves should not be attached to the building.

Boxes and Cabinets. Outlet, switch, and junction boxes and cabinets are used to protect the equipment and connections they house and to contain the sparks, arcs, or hot metal that may be produced in the equipment. All such boxes should be equipped with the proper cover. Boxes and cabinets are made with prepunched "knockouts" that can be removed to allow the installation of cable connectors and the entrance of a cable. Only those "knockouts" that are necessary to accommodate the conductors entering the box should be removed. All other openings must be closed, including any knockouts that may have been removed in error or result from modification to an existing installation. The number of wires in a box or cabinet must not exceed the number for which it was designed.

When observing outlet and switch boxes, inspectors should look for cracked or broken switch and outlet assemblies, discolored devices, or covers that indicate overheating. If any are found, they should be replaced promptly.

Switchboards and Panelboards. On some switchboards and panelboards (Figure 9-1a and b), there are exposed live parts from which occupants must be protected. This can be done by placing a cage or barrier around these open switchboards. Such boards usually also employ bus bars, which should be adequately supported. During

FIGURE 9-1(a) Generator Panelboard

the inspection, switchboards and panelboards should be checked for deterioration, dirt, moisture, tracking, and poor maintenance. Inspectors should also make sure that the surrounding area is kept clear to allow quick and ready access. Nothing should be stored in this working space or on top of switchboards or panelboards. The working space is intended to provide a safe area for persons who need access to or have to service the equipment, so even noncombustible materials cannot be stored in the working space.

Lamps and Light Fixtures. Light fixtures are subject to deterioration and poor maintenance. With age, the insulation on fixture wires can dry, crack, and fall away, leaving bare conductors. Sockets may become worn and defective, and the fixtures themselves may loosen in the mountings. Fixtures should not be mounted directly on combustible ceilings unless specifically listed for the purpose. Since lamps often

FIGURE 9-1(b) Panelboard with MCC and Smaller Panelboards Along Same Wall

operate at temperatures high enough to ignite combustible material, they should be mounted far enough away from materials such as paper or cloth, which may be used as shades or placed nearby, so that their continuous operation does not ignite them. Oversized lamps cause excessive temperatures in fixtures, and these temperatures can damage the supply conductors or ignite nearby combustibles. The inspector should take care to ensure that lamps are of the proper size and type and that the fiberglass thermal barrier is in the fixture canopy. Discolored globes or lenses can indicate improper lamp size. Most fixtures are marked with the appropriate lamp types and maximum ratings. Newer, recessed fixtures have thermal protectors that will turn the fixture off if incorrect lamp size results in high temperatures. Unguarded portable lamps may ignite ordinary combustibles if placed in contact with them, and a broken lamp may ignite combustible dust in suspension or flammable vapors in the atmosphere. Although halogen lamp elements now come in lower ratings, combustible building materials should be kept well away from such fixtures.

Grounding

Lightning, accidental contact with a high voltage source, surface leakage due to conductive dirt or moisture, and breakdown of insulation on conductors can cause hazardous voltages in electrical distribution systems and equipment. If the affected equipment is permitted to "float" at a dangerous voltage, anyone who comes in contact with it and a point of different potential, such as ground, will receive a serious, if not fatal, shock. Grounding electrical equipment helps to eliminate this shock hazard.

Grounding facilitates the operation of the overcurrent devices installed in ungrounded conductors. One of the system's conductors is grounded, and all the metal parts that could be energized are connected to it through equipment-grounding conductors and bonding jumpers. If a ground fault occurs, this will provide a path to the grounded system conductor, which will cause the overcurrent device to operate. Grounded conductors are usually identified by a white or gray coloring. Equipment grounding conductors are usually bare or identified by a green color.

Metal cable armor, raceways, boxes, and fittings, as well as the frames and housings of electrical machinery, must be grounded. Certain electrical tools and cord-and-plug-connected appliances, such as washers, dryers, air conditioners, pumps, and so on, must be grounded through a third contact in the line plug.

A grounding electrode is connected to the system to stablize the voltage to ground and to limit voltages due to lightning, line surges, or unintentional contact with higher voltage lines. A metallic underground water-piping system must be used as the grounding electrode where it is available and where the buried portion of the pipe is more than 10 ft (3.05 m) long. If a metal underground water pipe is the only grounding electrode, it must be supplemented by an additional electrode to ensure the integrity of the grounding electrode system. Grounded steel building frames, concrete-encased electrodes installed in footings, grounding rings or grids, and driven grounding rods are other electrodes that may be used. These may be used either as the main electrode, or to supplement the water pipe electrode (Figure 9-2).

Fire sprinkler system piping is prohibited by NFPA 13, *Standard for the Installation of Sprinkler Systems,* and NFPA 24, *Standard for the Installation of Private Fire Service Mains and Their Appurtenances,* from being used as an electrode for grounding of electrical systems in buildings. However, this prohibition does not relieve the NEC requirement for bonding of metal piping systems. Although sprinkler piping cannot be used for grounding the system, the piping must be connected (bonded) to the system to prevent the piping from becoming a fire or shock hazard if it inadvertently becomes energized. Bonding ensures that a pathway is available that will allow enough current to flow to operate the fuse or circuit breaker of a circuit that could energize the piping.

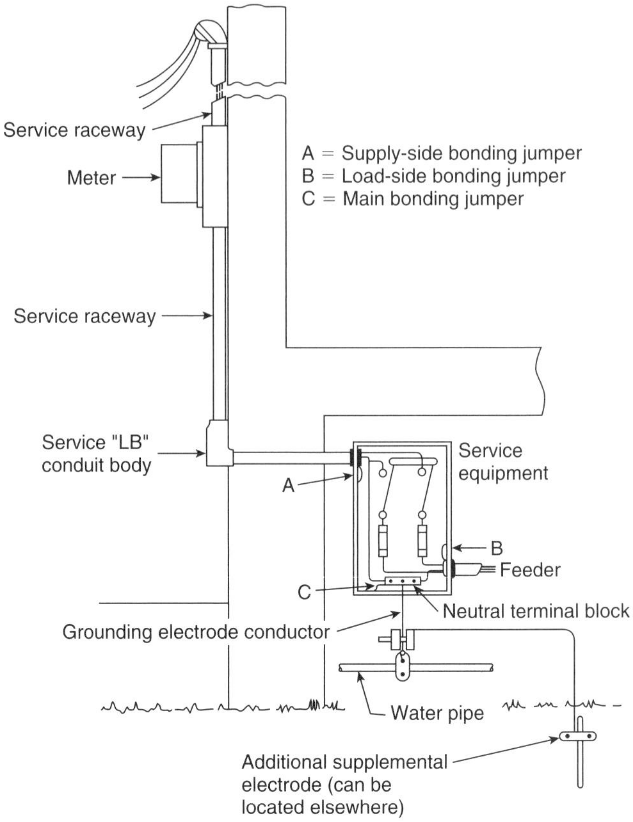

FIGURE 9-2 Grounding at a Typical Small Service (AC, single-phase, three-wire, 120/240 Y)

Ground clamps and connectors should be checked periodically to ensure that they are tight and that the ground connection is being maintained. When new electrical machinery or equipment is installed on the premises, it should be inspected to see that it has been connected properly to the grounding system.

Overcurrent Protection

Conductors and equipment are provided with overcurrent protection that opens a circuit if the current reaches a value that will cause an excessive or dangerous temperature in the conductor or the conductor insulation. Fuses and circuit breakers are the most commonly used overcurrent devices for the protection of feeders and circuits and of equipment. Thermal overload devices or electronic sensors may be used in conjunction with circuit breakers and fuses to protect equipment such as motors.

Plug Fuses. There are two types of plug fuses, the Edison base and the type S. Both can be either of the quick-acting or the time-delay type. The Edison base plug fuse is familiar to most people. Edison base fuses were widely used in older installations, but their use is now restricted to replacements in existing installations where there is no evidence of overfusing or tampering. Type S fuses may be used in new installations. The type S plug fuse is designed to prevent the use of pennies or other bridging schemes to bypass the fuse and to prevent the use of incorrectly sized fuses.

Cartridge Fuses. Cartridge fuses are made in quick-acting and time-delay types. They are also made for one-time use or with renewable links. However, the renewable link cartridge fuses have two drawbacks: Two or three links can be installed at one time, thereby increasing the fusing current and defeating the purpose of the fuse, and the fuse can be left with loose connections when a link is replaced, which results in overheating. Generally, the two or three fuses installed in a switch will all have the same rating. If one fuse has a different rating from the others, this may be an indication of overfusing and should be investigated further. Fuses are sometimes replaced by the most available fuse rather than by the proper size.

Circuit Breakers. Circuit breakers are available in a number of styles. The most common has two nonadjustable trips, one of which is thermal to detect overloads and the other magnetic to detect short circuits. These are called inverse-time or thermal-magnetic circuit breakers. Another type has adjustable trip units, which may have either conventional or solid-state sensing units. Some circuit breakers have shunt-trip features that allow them to be operated from remote locations. An example is the type of circuit breaker used to shut down equipment under kitchen hoods in restaurants. Among the special types of circuit breakers used are motor-operated breakers, ground-fault sensing breakers, and motor-circuit protectors. Circuit breakers may also include ground-fault circuit-interrupter (GFCI) protection and arc-fault circuit-interrupter (AFCI) protection. Any of these devices may feel warm under normal loads, but none should be too hot to touch.

Thermal Overload Devices. These devices are not intended to protect against short circuits. Rather, they protect against overload. Examples are the thermal overload devices included in many small motors, the thermal protectors in recessed fixtures, and the thermal protectors in fluorescent ballasts.

Ground-Fault Circuit-Interrupters. These devices sense when the current passes to ground through any path other than the proper path. When this occurs, the ground-fault circuit-interrupter (GFCI) trips almost instantly, stopping all current flow in the circuit. GFCIs are extremely important for life protection in wet locations. GFCI protection is provided by a special circuit breaker located in the panelboard or by a GFCI receptacle installed in the outlet box.

GFCIs are primarily intended to protect people from shock hazards. However, GFCIs are also used to prevent fires due to faults in "heat tape" used for freeze protection of water piping in mobile homes and manufactured housing. Similar devices, known as Equipment Ground Fault Protective Devices (EGFPD) are used for the same purpose in other occupancies where pipeline heating or deicing and snowmelting equipment are installed. EGFPDs do not provide acceptable shock protection for people and are intended only to protect equipment.

Arc-Fault Circuit-Interrupters. Arc-fault circuit-interrupters (AFCIs) are devices that can detect the presence of arcing faults on circuits even where the current does not rise to a value that will trip the overcurrent device. They are intended to help prevent fires by disconnecting the circuits where such damage occurs. The first requirement in NFPA 70, *National Electrical Code*® for these devices was effective in 2002 and applied only to circuits supplying receptacle outlets in, bedrooms of dwelling units. The requirements may have had different effective dates or may have been required in other locations by certain jurisdictions.

Transformers

INDUSTRIAL EQUIPMENT

Dry-type and fluid-filled transformers are used in both industrial and large commercial occupancies. Dry-type transformers are the most common in newer commercial constructions although other types may be encountered, while fluid-filled transformers are more common in industrial plants or older buildings. In most cases, dry-type transformers do not require a separate room or vault, but they must be separated from combustible materials and the area in which they are located must be adequately ventilated. Oil-filled transformers usually are required to be installed in a vault with 3-hour fire-rated floor, walls, doors, ceilings, and sills to contain the contents of the transformer should they spill.

New transformer fluids, classified as less flammable or nonflammable, are available. When these new fluids are used, the requirements for vaults are reduced or eliminated. Some older transformers might contain askarel and will have to be marked and eventually replaced due to environmental concerns.

Under conditions of full load, transformers operate at elevated temperatures. Many will be too hot to touch for more than a few seconds. All transformers should be provided with adequate ventilation, and the clearance requirements marked on the transformer should be maintained. Materials should not be stored on top of transformer enclosures.

Outdoor transformers should be located in such a way that leaking fluids will drain away from buildings or be contained in place. They should be placed in such a way that they will not expose exits or windows to fire in the event of transformer failure.

FIGURE 9-3(a) Cooling Tower Motors

FIGURE 9-3(b) Industrial Air Compressors

FIGURE 9-3(c) Pump Motors with Panelboard in Background

Motors

Motors and rotating machines can cause mechanical injury as well as a shock hazard and should be treated with caution [Figure 9-3(a), 9-3(b), and 9-3(c)]. Many motors start automatically, so even a motor at rest should be treated as though it were running.

The sparks or arcs that result when a motor short circuits can ignite nearby combustibles. Bearings can overheat if they are not properly lubricated. And dust deposits or accumulations of textile fibers can prevent heat from dissipating from the motor.

The inspection of the motors should indicate that there are no combustibles in the immediate vicinity of the motor or its controls, that the equipment is cleaned properly and maintained, and that it has the proper overcurrent protection. Motors are designed to operate without overheating under normal conditions, but they also are designed to operate with a temperature rise well above ambient under normal full-load conditions. A hot casing may indicate a potential problem and should be examined closely.

Medium Voltage Equipment

Most of the equipment discussed above can operate on medium-voltage, as well as low-voltage, systems. Medium voltage usually is considered to be in excess of 1000 V. Equipment rated as high as 15,000 V is common in large buildings and industrial complexes. Because of the severe shock hazard associated with this equipment, the inspector must use extreme caution when inspecting it.

HAZARDOUS AREAS

Electrical hazardous areas are those in which flammable liquids, gases, combustible dusts, or readily ignitible fibers or flyings are present in sufficient quantities to represent a fire or explosion hazard. Special electrical equipment is necessary in these areas. The special equipment is intended to keep the electrical system from becoming a source of ignition for the flammable or combustible atmosphere. This equipment is usually specifically listed or identified for the Class, Division, and, in Class I and II, the group of materials or chemicals that creates the classified area. Portable equipment should be similarly listed or identified for use. Complete definitions of the classes and divisions of hazardous locations and of the wiring methods and types of electrical equipment to be used in each are covered in Article 500 of NFPA 70, *National Electrical Code®*.

Class I, Division 1

Class I, Division 1 locations include areas in which ignitible concentrations of flammable gases or vapors exist under normal conditions; areas in which ignitible concentrations of flammable gases or vapors may exist frequently because of repair or maintenance operations or leakage; and areas in which the breakdown or faulty operation of equipment or processes may cause the simultaneous failure of electrical equipment. Electrical equipment used in these locations must be the explosion-proof type or the purged-and-pressurized type approved for Class I locations.

Class I, Division 2

These locations include areas in which volatile flammable liquids or flammable gases, which normally are confined to closed containers or systems that allow them to escape only during accidental rupture, breakdown, or abnormal operation of equipment, are handled, processed, or used. They also include areas in which positive mechanical ventilation normally prevents the development of ignitible concentrations of gases or vapors that could become hazardous should the ventilating equipment fail or operate abnormally, as well as areas adjacent to, but not cut off from, Class I, Division 1 locations to which ignitible concentrations of gases or vapors could be communicated.

Class II, Division 1

These classified locations include areas in which combustible dust is, or may be, in suspension in the air continuously, intermittently, or periodically under normal operating conditions in large-enough quantities to produce explosive or ignitible mixtures. Areas in which mechanical failure or abnormal operation of equipment might result in explosive or ignitible mixtures and provide a source of ignition through simultaneous failure of electrical equipment are also classified as Class II, Division 1, as are areas in which combustible dusts of an electrically conductive nature might be present. Class II, Division 1 locations also include areas in which a buildup of combustible dust on horizontal surfaces over a 24-hour period exceeds ⅛ in. (3.1 mm).

Class II, Division 2

These classified locations include areas in which combustible dust normally is not suspended in the air in quantities sufficient to produce explosive or ignitible mixtures and in which dust accumulations normally are not sufficient to interfere with the normal operation of electrical equipment or other apparatus. They also include areas in which the infrequent malfunction of handling or processing equipment might result in dust in suspension in the air and in which these dust accumulations could be ignited by the abnormal operation or failure of electrical equipment or other apparatus. Areas in which a buildup of combustible dust on horizontal surfaces is ⅛ in. (3.1 mm) or less, obscuring the surface color of the equipment, are also classified as Class II, Division 2 locations.

Class III, Division 1

This classified location includes areas in which easily ignitible fibers or materials producing combustible flyings are handled, manufactured, or used.

Class III, Division 2

These locations include areas in which easily ignitible fibers are stored or handled, except during the manufacturing process.

Class I, Zones 0, 1, and 2

The zone classification system provides an alternative to the Division 1 and 2 classifications in Class I areas. This classification system provides for some alternate protection methods. The classification of areas must be done by a qualified registered professional engineer. The area classification should be well documented and the documentation should be readily available. Zones 0 and 1 correspond roughly to Division 1, and Zone 2 corresponds roughly to Division 2.

STATIC ELECTRICITY

Precautions against sparks from static electricity should be taken in locations in which flammable vapors, gases, or dusts or easily ignited materials are present. Only qualified persons should be allowed to test for static charges in these locations, since unintended discharges can ignite the hazardous atmosphere.

Measures that will bring the hazard of static electricity under reasonable control are humidification, bonding, grounding, ionization, conductive floors, or a combination of these methods.

Humidification

Humidity alone is not a completely reliable means of eliminating static charges. To reduce the danger of static, however, the relative humidity should be high. If practical, relative humidities should be as high as possible, even up to 75%, as long as this does not create undue hardship.

Some industrial operations cannot be performed at humidities high enough to mitigate the danger of static.

Bonding and Grounding

Bonding minimizes the differences in electrical potential between metallic objects. There is practically no difference between two metal objects connected by a bond wire, because the bond wire is used to dissipate charges that might otherwise accumulate on a piece of equipment.

The term *grounding* describes connections made to minimize the difference in electrical potential between objects and the ground and to provide a low-resistance path to actuate the overcurrent device. In such cases, the ground wire can carry a current from power circuits that is much larger than the static current. You should check the ground connections when they are installed and frequently thereafter to detect corrosion, loose connections, or damage.

Flowing gases, liquids, or granular solids, such as sand, generate static. Thus, their containers should be bonded or grounded (Figure 9-4). When gasoline is transferred from a drum to a can, for example, the drum and the can should be bonded

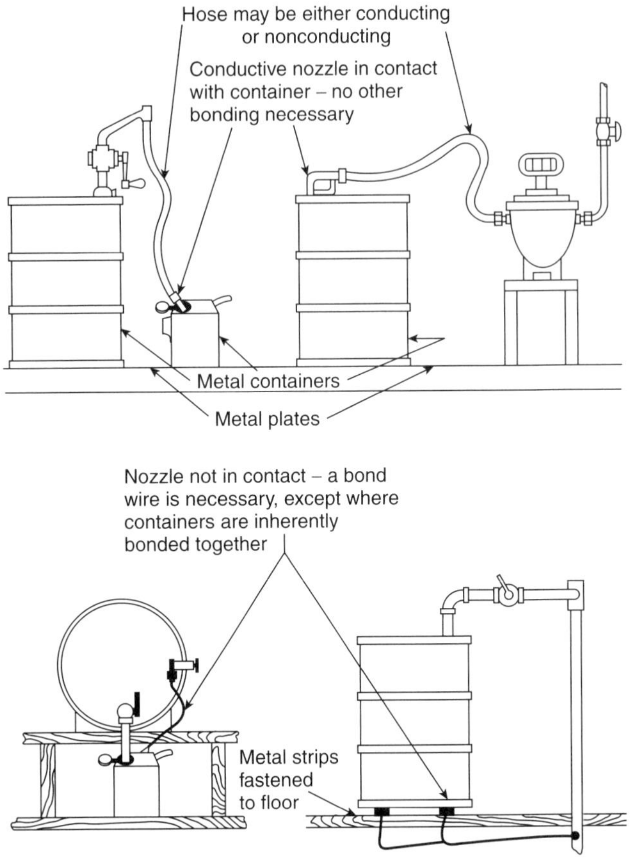

FIGURE 9-4 Recommended Methods of Bonding Flammable Liquid Containers During Container Filling
Source: NFPA 77, 1993, Figure 4-8.4.

together by an electrically conductive tube that is firmly in contact at both ends or securely attached to a grounding wire before the liquid is transferred (see Figure 9-4).

The currents involved in static discharge are quite small, even though the voltage may be quite high, so small conductors are adequate to dissipate the charges. In some cases, very small conductors are adequate where the connection is fixed and protected, such as at a small flammable storage cabinet. However, generally the bonding and grounding conductors and connections should be of substantial construction where they are exposed to damage or frequent interchange. The connections should be installed in such a way that they can be easily verified and inspected.

Ionization

Ionization is the process of increasing the conductivity of air so that it will conduct static charges away from the area.

One ionization technique employs the tendency of static to concentrate on the surface of least radius of curvature, such as a sharp point. A metal bar with needle points (static comb) or with metallic tinsel removes static from moving sheet materials. Another technique uses a so-called electrical neutralizer, which produces an alternating electrical field through which the electrified sheet material passes. Yet another technique, used on printing presses, uses a flame to ionize the surrounding air. Static may also be ionized by alpha radiation from a radioactive surface.

The hazards introduced by these various techniques, as well as their respective effectiveness in removing static charges, must be considered.

LIGHTNING PROTECTION

The risk of damage due to lightning is very high in some areas and low in others. The incidence of lightning varies greatly by geographic location. For example, lightning is common in Florida and rare in Alaska. Lightning risks also vary by the type of construction or the use of a building, as some systems and processes are more vulnerable to lightning that others. For these reasons, lightning protection is usually not required by local codes. However, where required or used, lightning protection must be properly installed and maintained for fire safety.

Lightning protection systems are installed to provide an alternate, nondestructive path for lightning to follow to earth. When lightning follows this path, building materials are spared the heat and mechanical forces that result when the energy of the lightning stroke passes through a structure. Any part of a building that is likely to be struck, such as chimneys, ventilators, steeples, dormers, and other projections, should be protected. This is done by installing a series of air terminals, down conductors, and secondary conductors and ground terminals. Surge arrestors also may be installed to protect the building's electrical system.

Air terminals are installed on the edges of a building's roof and on its vertical projections and connected by conductors. Down conductors are used to provide at

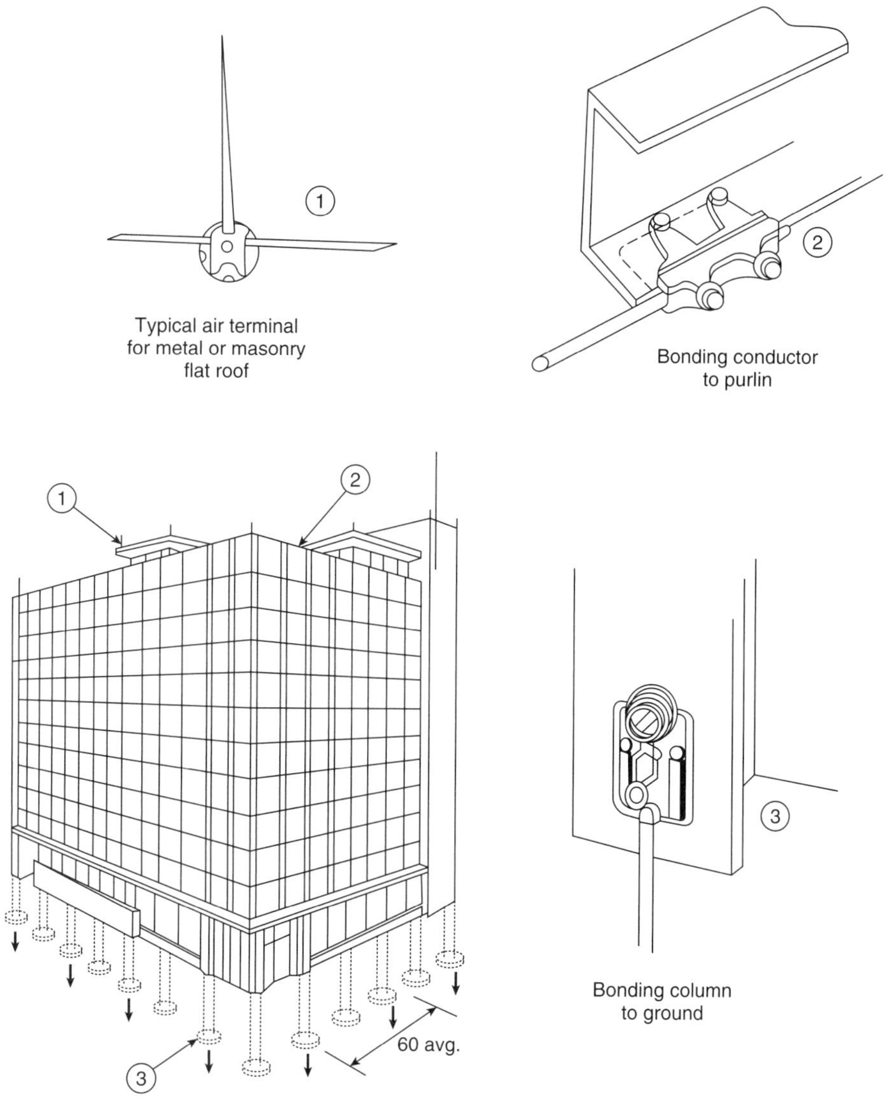

FIGURE 9-5 Grounding and Bonding of Lightning Down Conductors. Water pipe grounds (if pipes are metallic) can be made at 1, 2, or 3.

least two paths to the ground terminals. Metal objects nearby are bonded to the system with secondary conductors, which prevent damage from sideflashing. Bonding also helps to reduce the risk of induced voltages in materials near lightning conductors. In some cases, the metal frame of a building may be used as the down conductors. In all cases, the conductors must be resistant to corrosion and have the mechanical strength necessary to function as intended. Most components, including air terminals, down conductors, and connectors must be listed for the purpose (Figure 9-5). All of these conductors are constructed of heavy-gauge copper or aluminum to resist corrosion.

Because the materials used for lightning protection systems are strong, they require very little maintenance, so inspectors may not have to inspect these systems as frequently as they do other systems. In fact, NFPA 780, *Standard for the Installation of Lightning Protection Systems*, recommends inspecting them every 5 years. When they are inspected, conductors should be checked for excessive corrosion or mechanical damage. Down conductors near the ground may be damaged by vehicles, and connections to ground terminals may be disturbed by mowers or other groundskeeping tools. Connections should be checked throughout for tightness, and air terminals should be inspected to make sure they are secure and in place.

SUMMARY

Fires can occur in electrical equipment or be ignited by electrical equipment. The design of safe electrical systems includes compliance with NFPA 70, *National Electrical Code®*, and other codes and standards. Electrical inspectors verify compliance with these codes and standards when buildings are initially constructed. Over the course of time, additions, alterations, use or abuse, and maintenance or lack of maintenance may alter the level of safety that was originally present. Fire inspectors are the ones most likely to be able to observe these changes, so they must be able to recognize common signs and symptoms of electrical failures.

Most fire inspectors may not be qualified to inspect the internal wiring and parts of electrical equipment. Therefore, the goal of a fire inspection should be to ensure that the electrical system does not provide an ignition source for other materials, including exceptionally flammable or combustible materials. This chapter has provided some guidelines for inspections of existing electrical wiring and equipment. Generally, the fire inspector should verify that enclosures and wiring remain intact, that equipment is properly located and maintained, that overcurrent devices are installed, and that grounding systems are in place and undamaged. These steps will help to ensure the continued safe operation of an electrical system.

This chapter has also covered some systems that are not part of the normal power wiring in a building. Lightning and static both present a source of ignition that may be significant in certain areas or certain facilities. The fire inspector should verify that these systems also remain functional and in good repair.

BIBLIOGRAPHY

Cote, A. E., ed., *Fire Protection Handbook,* 18th ed., NFPA, Quincy, MA, 1997.

Electrical Appliance and Utilization Equipment Directory, Underwriters Laboratories, Inc., Northbrook, IL. Issued annually. This is a listing of the electrical appliances and devices that have been tested and found to be safe for use.

Electrical Construction Materials Directory, Underwriters Laboratories, Inc., Northbrook, IL. Issued annually. A directory of tested and listed construction materials such as circuit breakers, wires, transformers, industrial control equipment, electrical service equipment, and fixtures and fittings.

Hazardous Location Equipment Directory, Underwriters Laboratories, Inc., Northbrook, IL. Issued annually. A listing of electrical components and equipment that have been tested and listed for use in hazardous atmospheres.

Sargent, J. S., and Williams, N., *NFPA Electrical Inspection Manual with Checklists,* NFPA, Quincy, MA, 1999.

NFPA Codes, Standards, and Recommended Practices

See the latest version of The NFPA Catalog for availability of current editions of the following documents.

NFPA 13, *Standard for the Installation of Sprinkler Systems*
NFPA 24, *Standard for the Installation of Private Fire Service Mains and Their Appurtenances*
NFPA 70, *National Electrical Code®*
NFPA 70B, *Recommended Practice for Electrical Equipment Maintenance*
NFPA 70E, *Standard for Electrical Safety Requirements for Employee Workplaces*
NFPA 77, *Recommended Practice on Static Electricity*
NFPA 79, *Electrical Standard for Industrial Machinery*
NFPA 496, *Standard for Purged and Pressurized Enclosures for Electrical Equipment*
NFPA 497, *Recommended Practice for the Classification of Flammable Liquids, Gases, or Vapors and of Hazardous (Classified) Locations for Electrical Installations in Chemical Process Areas*
NFPA 499, *Recommended Practice for the Classification of Combustible Dusts and of Hazardous (Classified) Locations for Electrical Installations in Chemical Process Areas*
NFPA 780, *Standard for the Installation of Lightning Protection Systems*

Heating Systems

CHAPTER 10

Michael Earl Dillon

Most large buildings use oil- or gas-fired hot water or steam boilers to generate building heat. Hot water or steam piping systems distribute this heat to air-conditioning units, unit heaters, finned tube radiation units, and, in some cases, ice- and snow-melting systems. Many smaller buildings use packaged equipment, often referred to as unitary equipment, which contain an extended combustion chamber called a heat exchanger to heat the air passing over it without the use of an intermediate fluid such as steam or hot water.

As a result of both the energy crisis and stricter environmental regulations for flue gas exhaust emissions, there have been major changes in building heating systems since the mid-1970s. This chapter examines various heating systems, different heating fuels, and the installation of heat-producing equipment, as well as the inspector's role in inspecting heating systems. Not discussed here are systems used for industrial processes and power generation. These systems are discussed in Chapter 50, "Heat-Utilization Equipment."

HEATING SYSTEMS IN LARGE AND SMALL BUILDINGS

Large Buildings

Many large buildings have abandoned their heavy oil- and coal-fired boiler plants because installing and operating the necessary pollution abatement devices cost too much. In many cases, the savings in the cost of fuels simply could not offset the higher operating costs. Nor could space always be found to store the quantities of coal necessary to run such plants. And in large cities, removing the coal ash became a major cost and environmental problem.

Building owners became acutely aware of energy costs as they saw their utility bills skyrocket. They added insulation to exterior walls, replaced single-pane windows with insulating glass, and covered minimally insulated or noninsulated pipe with energy-saving insulation. They repaired and replaced leaking and noninsulated chimneys and exhaust flues. Boilers and heating units were replaced with more efficient units so that less energy escaped up the stacks.

Many of these changes supported greater fire safety and contributed to building safety. For example, the added insulation on pipes and other hot surfaces reduced the transmission of heat to adjacent surfaces, which reduced the likelihood of building fires. Efficient boilers and heating units reduced the temperatures of the stacks, which lessened the chances of chimney, chase, and roof fires.

Also, changes to codes have lessened fire risks. Fuel-fired heating equipment above a certain energy input capacity is often required to be located in rooms separated from other parts of the building by fire-resistive barriers. Thus, in new buildings the fire hazard from heating systems has been reduced; in a large number of existing buildings, however, the problem remains.

Michael Earl Dillon, P.E., president and CEO of Dillon Consulting Engineers, headquartered in Long Beach, California, is one of the leading experts on smoke control and heating, ventilating, and air-conditioning systems. He currently serves on five NFPA technical committees, including smoke management and air conditioning.

Residential and Small Buildings

Most small buildings use a warm-air furnace and air-conditioning unit to control temperature, and these systems have been engineered so that they are very energy efficient. In fact, the flue gas temperatures of some units have become so low that condensation within the flue has become a major problem. When conducting an inspection, the inspector should check the flue pipes for any sign of acid corrosion. Doing so is especially important for the newer combustion-efficient systems. Furnaces and ductwork have been insulated to increase operating efficiencies, but proper clearance between such devices and surrounding materials should still be maintained.

Despite these generally positive conditions, the energy crisis, with its high fuel costs, has created a new hazard within buildings. Occupants increasingly use portable devices to provide individual or area heating. These devices are not built into the building systems, and the building owners may not even be aware they are being used. Where they are permitted, the cords and extensions should be inspected. Portable electrical heaters with undersized cords and extensions overheat and cause fires. Unvented portable kerosene heaters, which have caused numerous deaths and fires, should not be permitted in places of public assembly. NFPA *101*®, *Life Safety Code*®, prohibits such devices in educational occupancies, day-care facilities, health-care occupancies, detention and correctional occupancies, and in all residential occupancies except one- and two-family dwellings.

BURNER CONTROLS AND BOILERS

Fuel permitted to collect in the combustion chamber of a furnace in the absence of an ignition source could explode if it is ignited. Therefore, safety considerations require that fuel burners be equipped with controls to cut off the fuel supply in the event of a malfunction.

Primary safety controls shut off the fuel supply in the event of flame or ignition failure. Interlock circuits shut off the fuel supply if an induced or forced draft fails, if atomization fails, if dangerous fluctuations in fuel pressure occur, or if the oil temperature in burners requiring heated oil falls below the required minimum. The inspector should verify that all of these controls operate satisfactorily by checking the operators' logs to confirm that periodic tests are being performed.

Boiler rooms often improperly become storage areas for building materials and chemicals. Such materials should be removed from these rooms. Good housekeeping practices should be enforced. In addition, building detection and fire suppression systems should be installed in such equipment rooms. The inspector should confirm that these systems are tested periodically.

Oil

Today there are a number of grades of fuel oil used by oil-fired furnaces and steam boilers (Figure 10-1), No. 1 and No. 2 being the most common. Numbers 5 and 6 fuel oil are known as "heavy" oils and must be heated if they are to flow. As a result, systems using Nos. 5 and 6 fuels have a complex system for oil heating that usually requires extensive pollution-abatement equipment to deal with the impurities in the oils. The inspector should verify that this equipment is tested periodically.

Oil leaks and spills should be kept to a minimum and cleaned when necessary. Sawdust should not be used on spills because it adds to the risk of fire.

Gas

LP-Gas, an LP-Gas-and-air mixture, and natural gas are also used to fuel heat-producing devices. Because LP-Gas vapors are heavier than air, the inspection of

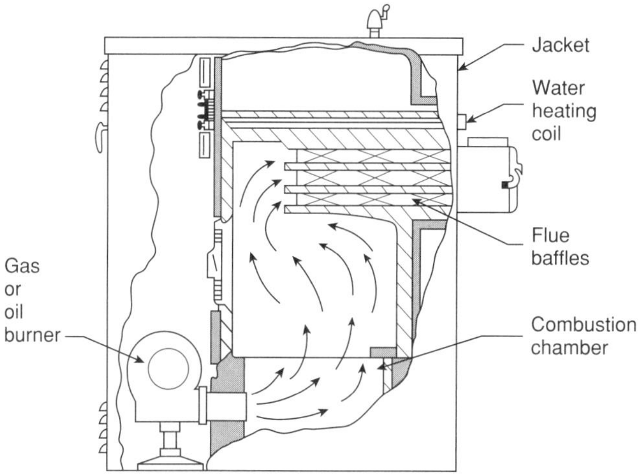

FIGURE 10-1 Oil-Fired Steam Boiler

such equipment in below-grade indoor locations is especially critical. Rooms containing LP-Gas piping and equipment should be well ventilated in such a manner that no vapors can accumulate in low areas.

Natural gas usually is preferred as a fuel because it is clean burning, generally available, and requires no storage facilities. Unlike LP-Gas, it is lighter than air, so the vapors will rise. Leaks are usually detected easily because of the distinctive odor added by gas companies. Many burners are arranged to use either gas or oil. Similarly, burners can be arranged to use LP-Gas or natural gas. Burners that are not properly equipped to use multiple fuels cannot be safely used with other than their intended fuel.

Solid Fuel

Wood-burning and coal-burning stoves are occasionally used as decorative or backup heat sources. Coal requires no special handling or storage. Wood stored outdoors should be up off the ground and not in direct contact with buildings, so as to allow air to circulate on all sides. Preferably, it should be protected from wetting. These fuels generate very high temperatures, so solid fuel appliances require special attention. Clearances to combustibles below and around the appliances and their chimneys and chimney connectors are critical.

In addition to stoves, the inspector may occasionally encounter solid-fuel central warm-air furnaces. Some of these are manually stoked; others are automatic. Manufacturer's literature should be carefully reviewed.

Furnaces

Central warm-air furnaces are either of the gravity type or the forced-air type. Gravity furnaces are mounted on the floor and heat only the spaces above them. Aside from the lack of a blower and motor, gravity type furnaces are generally distinguishable from their forced-air cousins by their larger supply ducts. They should be equipped with high-temperature-limit controls that shut off the fuel supply when the temperature of the discharge air reaches a predetermined level.

Forced warm-air furnaces (Figure 10-2) are equipped with plenums, which can become hot enough to ignite adjacent combustibles. Such furnaces should be equipped with a limit control to shut down the fuel supply when the temperature in the plenum or at the entrance to the supply duct reaches a predetermined level. As with all warm-air furnaces, maintaining adequate clearances from combustibles

FIGURE 10-2 Oil-Fired Forced-Air Furnace

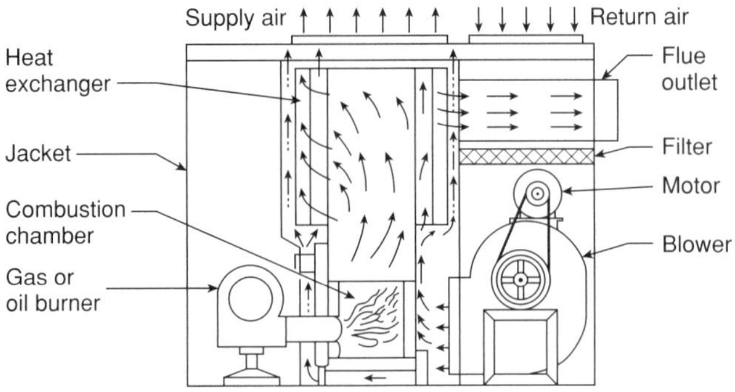

is important. The appropriate clearances will typically be identified on a permanent tag attached to the furnace.

All oil- or gas-fired furnaces should be provided with fuel shutoff valves located in the piping system just ahead of the appliance fuel connector. Additionally, most modern equipment is equipped with pilot safety devices that will not allow fuel to flow to the main burner or pilot if the pilot is extinguished. Some older equipment safeties will allow fuel to flow to the pilot, and very old equipment lacks such safety devices.

Wall Furnaces. Self-contained indirect-fired gas or oil heaters installed in or on a wall are called wall furnaces (Figure 10-3). They supply heated air by gravity or with the aid of a fan, and they are either directly vented or connected to a vent or chimney. Wall furnaces generally do not have high-temperature-limit controls.

Floor Furnaces. Floor furnaces are gravity heating appliances suspended beneath a floor and drawing their combustion air from the space below (Figure 10-4). Venting is accomplished by a horizontal vent connector beneath the floor connecting to a

FIGURE 10-3 Typical Gas-Fired Wall Furnace

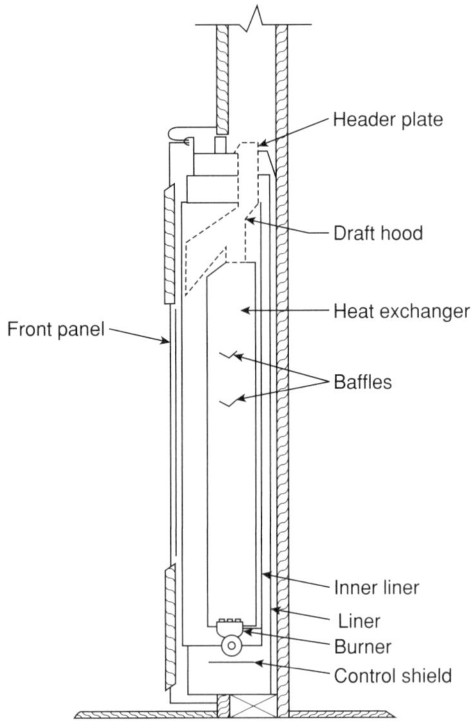

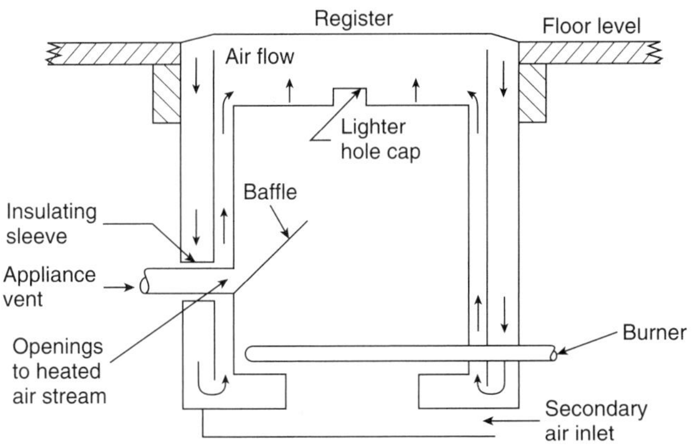

FIGURE 10-4 Floor Furnace

vertical vent or chimney. Since the vent connector is usually in an unconditioned crawl space, these connectors should be of Type B or Type L construction. Maintaining proper clearances to combustibles, including to movable objects such as doors, drapes, drawers, and so on, is extremely important because these appliances generally do not have temperature-limiting controls.

Type B and Type L vent systems are multiwall metal vents (chimneys) that are tested and listed by recognized laboratories such as UL. They are double-wall vent systems with an aluminum (Type B vent) or stainless steel (Type L vent) inner wall and a galvanized steel outer wall. The double-wall construction minimizes the surface temperature of the vent and retains heat in the vent to maintain buoyancy of the vent gases. Type B vents are used for most gas-fueled appliances, and Type L vents are used for the higher vent-temperature-oil and solid-fuel appliances.

Unit Heaters. Unit heaters are self-contained, automatically controlled, chimney- or vent-connected air-heating appliances equipped with a fan for circulating air (Figure 10-5). They can be mounted on the floor or suspended and are generally not equipped with limit controls. Unit heaters connected to a duct system can be

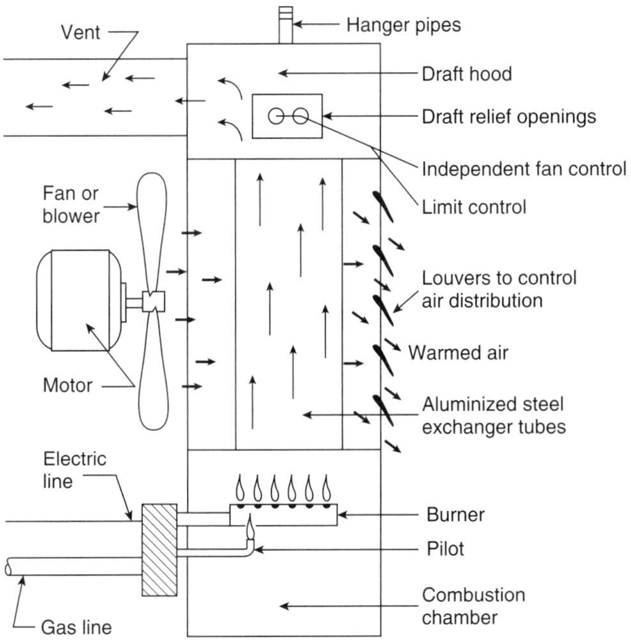

FIGURE 10-5 Typical Gas-Fired Unit Heater

considered central-heating furnaces and should have the same safeguards. Ducted unit heaters should be equipped with temperature limit controls to assure the duct temperature does not exceed acceptable limits.

Heat Pumps. A heat pump is a forced-air heating system that uses refrigeration equipment so the hazards of such an arrangement are similar to those of electrical, refrigeration, and heating equipment. When heat is wanted, it is taken from a heat source and given up to the conditioned space. When cooling is desired, the heat is removed from the space. Heat pumps that use supplemental heating units are equipped with an interlock that prevents the compressor from operating when the indoor air-circulating fan is not operating. Heat pumps are usually equipped with temperature-limit controls.

Solid Fuel Stoves

Solid fuel stoves are wood- or coal-burning, freestanding radiant-heat appliances connected to a chimney. Most draw combustion air from the space in which they are located, but some newer ones have outdoor combustion air ducted directly to the firebox. They all generate very high temperatures and have not temperature-limiting controls. Therefore, substantial clearances from combustible materials must be maintained, including from floors and materials concealed within the building construction, such as wood studs, or heat-dissipating shielding must be installed to protect combustibles from ignition.

HEAT DISTRIBUTION

Warm-Air Systems

Horizontal supply ducts, vertical ducts, risers, boots, and register boxes can reach hazardous temperatures if the high-temperature-limit control malfunctions. Thus, keeping combustibles away from them is essential. NFPA 90B, *Standard for the Installation of Warm Air Heating and Air-Conditioning Systems,* contains required clearances for many warm-air ductwork configurations.

Steam and Hot-Water Systems

Low-temperature hot-water systems are those operating at a temperature no greater than 250°F (121°C). Hot-water pipes and radiators in systems operating with a maximum temperature of 150°F (66°C) require no installation clearances. Those supplied with hot water up to 250°F (121°C) or with low-pressure steam [no more than 15 psig (103 kPa) gauge] require a clearance of 1 in. (25 mm) from combustibles. Where these pipes pass through a floor, wall, or ceiling, the clearance at the penetration must be at least ½ in. (12.5 mm), which is provided by means of a thimble, or sleeve, of noncombustible material.

Hot-water systems operating at temperatures no higher than 250°F (121°C) are considered *low-temperature hot-water systems.* Those operating between 250°F (121°C) and 350°F (177°C) are considered *medium-temperature hot-water systems.* Systems having temperatures of 350°F (177°C) and above are considered *high-temperature hot-water systems.*

Despite the common practice of dividing steam systems into multiple pressure categories along similar lines to hot-water systems, there are really only two that are germane to inspections under U.S. codes and standards. These are *low-pressure steam systems,* which operate at pressures no greater than 15 psig (103 kPa) gauge, and *high-pressure steam systems,* which operate at pressures greater than 15 psig (103 kPa).

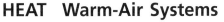

Medium- and high-temperature hot-water and high-pressure steam systems are occasionally used in hospitals, very large buildings, and some district or campus heating systems. The high temperatures and pressures encountered in these systems pose very serious hazards if released from containment. Leaks must never be approached by persons who have no specialized training in these systems. Surface temperatures of equipment and piping can be high enough to ignite adjacent combustibles.

Clearances

INSTALLATION

A major consideration in the installation of any heat-producing appliance is its effect on nearby combustibles. Wood and other combustibles can ignite at temperatures well below their usual ignition temperatures if they are continually exposed to moderate heat over long periods of time. For this reason, installation clearances of equipment, ducts, piping, vents, and connectors are of the utmost importance. Extensive information on clearances is given in the *Fire Protection Handbook;* in NFPA 211, *Standard for Chimneys, Fireplaces,Vents, and Solid Fuel-Burning Appliances;* and in NFPA 54, *National Fuel Gas Code.* Inspectors should always check these distances during inspections.

Listings of tested heating equipment indicate the materials upon which the equipment can be mounted, such as combustible floors, fire-resistive floors extending specific distances beyond the equipment, masonry floors, or metal-over-wood floors. The *Fire Protection Handbook* covers these materials thoroughly by type of appliance; laboratory listings cover them by manufacturer's model. For listed appliances, the manufacturer's installation instructions contain this information and should be with the equipment. Check the materials and consult the references for specific products.

Combustion Air

As a result of energy conservation measures, buildings are being insulated more thoroughly, and cracks and crevices are being sealed. However, there must be enough air available for combustion, for ventilation, and for replacing the volume lost in venting the combustion products to the outdoors An oxygen-starved fire causes an incomplete combustion reaction, which results in the vent clogging from soot accumulations and the production of higher than usual quantities of toxic by-products such as carbon monoxide. Additional air for ventilation also helps carry away the heat that develops on the surface of the equipment and within the space that encloses it.

Equipment rooms that contain combustion equipment should be inspected to ensure a positive means of supplying combustion air. This is especially important if there are any exhaust fans (including clothes dryers) operating in the area because they could draw a reverse flow down the stack or flue. NFPA 54 contains specific recommendations on how to supply the air required for combustion and ventilation.

Pilot and Burner Safeties. Most modern fuel-fired appliances are equipped with pilot and burner safeties, which will cut off the flow of fuel to the pilot and the main burner if the pilot is extinguished or its flame size has been diminished to the point where it cannot effectively assure main burner ignition. Some older equipment might be equipped with burner safeties that only cut off fuel to the main burner when the pilot is extinguished or diminished. Very old equipment is likely to have no pilot or burner safeties unless it has been retrofitted with them. All fuel-fired equipment should have such safeties.

Chimney and Vent Connectors. Chimney and vent connectors are those lengths of pipe or conduit that connect the heat-producing appliance to the chimney or vent. Connectors are made of noncombustible, corrosion-resistant material, such as steel or refractory masonry, that can withstand flue gas temperatures and resist physical damage.

Metal connectors may be of single-wall or double-wall construction. Connectors must be short, well fitted and supported, and continuously pitched upward toward the chimney or vent. They also should have adequate clearance from combustibles.

Vents. Vents are listed equipment used with specific types of heat-producing equipment. In buildings in which vertical openings must be protected, vents should be enclosed in fire-resistive construction. For buildings less than four stories high, the construction may have a 1-hour fire resistance rating. For buildings of four stories or more, the shaft construction must have a 2-hour rating. Vents must also have specific clearance to combustibles, usually 1 in. (25 mm), and where not enclosed in a fire-resistive shaft should be fireblocked with metal as they pass through floors or ceilings. They are to terminate sufficiently far above roofs or away from adjacent walls that they will draft properly. A way to check if the appliance is drafting properly is to hold a flame near the draft hood opening and see if it is drawn toward the draft hood (if the draft hood is internal, hold the flame to the upper air intakes to the appliance). Condensation on the inside of windows is a good indication there is a serious problem with appliance venting at the premises.

Chimneys

There are three major types of chimneys: masonry, factory built, and metal.

Masonry Chimneys. A masonry chimney should be inspected along its entire length, so far as it is accessible. The inspector can examine the inside of the chimney by placing a mirror in a connector opening or the cleanout opening and using captured sunlight. A sufficiently bright flashlight can also be used when sunlight is absent or offsets render it useless. On the roof, the inspector should note the condition of the mortar, the chimney lining, and the flashing, and should look for evidence of cracking or settling. If a solid-fuel appliance is connected to the chimney, the inspector should check to see that a spark-arrestor cap is installed and should note the number of flues. In the attic and the basement, the inspector should check for cracks and loose mortar. All chimney connections should also be checked and matched to the appropriate flues and liners. If a solid-fuel appliance is connected to the chimney, no gas- or oil-fired equipment should be connected to it. If mortar has begun crumbling from between the bricks, openings can be expected to develop all the way through the chimney wall.

Factory-Built Chimneys. Factory-built chimneys (Figure 10-6) are lightweight assemblies and good draft producers. Some types resemble Type B gas vents but are larger and heavier. The materials used in their construction meet certain requirements for heat and corrosion resistance. Factory-built chimneys are available as listed assemblies for low- and medium-heat appliance service. They should be inspected in accordance with the manufacturer's installation instructions, which should be located with the equipment.

Metal Chimneys. Metal chimneys are suitable for all classes of appliances, but they are not subjected to safety testing of any kind. The major hazard to look for when inspecting these chimneys is inadequate clearance from combustibles where they penetrate floors, ceilings, and roofs. The conditions under which metal chimneys can be used are quite limited and are spelled out in detail in NFPA 211, *Standard for Chimneys, Fireplaces, Vents, and Solid Fuel-Burning Appliances.*

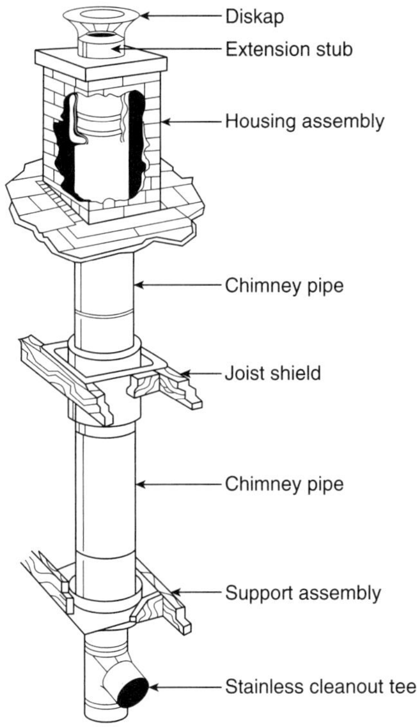

FIGURE 10-6 Typical Factory-Built (Type L) Chimney

CHECKLIST OF CHIMNEY DEFECTS

The following is a list of common and important chimney defects that, individually or in combination, constitute sufficient reason for requiring that a masonry chimney be repaired or rebuilt.

1. The design or proportionate dimensions of the chimney are structurally unsound.
2. There is evidence of settling or cracking due to inadequate footings or other causes.
3. The chimney rests upon, or is wholly or partly carried by, wooden floors, beams, or brackets, or it is hung by metal stirrups from wooden construction. The chimney is used to support any wooden floor or roof beams.
4. The chimney increases in size, has projecting masonry, or is set back within 6 in. above or below the rafters or roof joists.
5. The chimney is unlined, and its walls are not as thick as required.
6. The masonry is unbonded or improperly bonded, or the sections are not properly anchored or reinforced.
7. The mortar is weak.
8. Old mortar is decayed, due to the action of the flue gases, or it is weathering. The chimney is not properly finished at top.
9. The brickwork is not laid up around the lining. In other words, the lining was dropped into place after the walls were constructed.
10. Linings are cracked or broken.
11. There is no fire clay or metal thimbles at the openings for connectors.
12. Connector openings are found in more than one story for a single flue, and no provision has been made effectively closing unused openings.
13. The flues show leakage in a smoke test.
14. Flue linings are not complete from 8 in. (200 mm) below the connector openings to the top of the chimney.

15. There is a reduction in the cross-sectional area of a flue at any point.
16. The flue is positioned at a greater than 30° angle with vertical.
17. The chimney does not extend at least 3 ft (900 mm) above a flat roof or 2 ft (600 mm) above the edge of a gable or hipped roof.
18. Woodwork, particularly beams and joists, is within 2 in. (50 mm) of the outside surface of the entire chimney.

SUMMARY

Since the mid-1970s, as the result of both the energy crisis and stricter environmental regulations for flue gas exhaust emission, major changes have taken place in building heating systems This chapter examined the effects of these changes and provided information on boiler controls and boilers, on various heating fuels, on types of furnaces and on the installation of nonindustrial heat-producing equipment Throughout emphasis was placed on what the inspector should be looking for when inspecting heating systems.

BIBLIOGRAPHY

Boiler and Pressure Vessel Code, Section IV, American Society of Mechanical Engineers, New York, NY.

Chimneys, Factory-Built Residential Type and Building Heating Appliance, ANSI/UL 103, Underwriters Laboratories, Inc., Northbrook, IL.

Cote, A. E., ed., *Fire Protection Handbook,* 18th ed., NFPA, Quincy, MA, 1997.

Gas and Oil Equipment Directory, Underwriters Laboratories, Inc., Northbrook, IL. Issued annually.

Medium Heat Appliance Factory Built Chimneys, ANSI/UL 959, Underwriters Laboratories, Inc., Northbrook, IL.

NFPA Codes, Standards, and Recommended Practices

See the latest version of The NFPA Catalog for availability of current editions of the following documents.

NFPA 30, *Flammable and Combustible Liquids Code*
NFPA 31, *Standard for the Installation of Oil-Burning Equipment*
NFPA 54, *National Fuel Gas Code*
NFPA 58, *Liquefied Petroleum Gas Code*
NFPA 70, *National Electrical Code®*
NFPA 85, *Boiler and Combustion Systems Hazards Code*
NFPA 86C, *Standard for Industrial Furnaces Using a Special Processing Atmosphere*
NFPA 86D, *Standard for Industrial Furnaces Using Vacuum as an Atmosphere*
NFPA 90B, *Standard for the Installation of Warm Air Heating and Air-Conditioning Systems*
NFPA 97, *Standard Glossary of Terms Relating to Chimneys, Vents, and Heat-Producing Appliances*
NFPA *101®*, *Life Safety Code®*
NFPA 211, *Standard for Chimneys, Fireplaces, Vents, and Solid Fuel-Burning Appliances*

CHAPTER 11

Air-Conditioning and Ventilating Systems

Michael Earl Dillon

Air-conditioning systems control the temperature and humidity of air, clean it, and distribute it to meet the requirements of a conditioned space. There are many types of air-conditioning systems, such as those that provide filtered, cooled, and dehumidified air in summer and heated, humidified air in winter.

The mechanical equipment associated with air-conditioning systems can cause significant damage to building components or the system itself and cause injury to persons if improperly started, stopped, or operated. The status of a piece of equipment or a system should never be changed if its operation is not completely understood. When inspecting air-conditioning and ventilating systems, inspectors must understand what the system consists of and how it functions. Air-conditioning systems have three major components: the air intake system, the conditioning equipment, and the distribution system (Figure 11-1).

AIR INTAKE SYSTEM

Some systems mix fresh air with recirculated air; others use fresh air exclusively. In either case, there must be an air intake duct to introduce fresh air into the system. The opening of this duct should be protected with a grille or screen to prevent foreign materials from entering the system. The most common screening is wire mesh with an opening dimension no smaller than ¼ in. (6.25 mm) and no larger than ½ in. (12.5 mm) because smaller dimensions tend to become clogged too easily and larger dimensions fail to exclude many common vermin. The inspector should make sure that the duct is not broken, clogged, or missing and that it is free of rubbish, mold, and debris.

CONDITIONING EQUIPMENT

Unless the equipment consists of simple rooftop or ground-mounted package units, fans, air-heating and -cooling units, and filters should be installed in a room that is separated from the rest of the building by construction with a 1-hour fire resistance rating. Doors to these enclosures are to be 1-hour labeled and automatic closing.

Most building codes usually require large units above a certain energy input capacity to be in separated rooms. Also, units incorporating combustible filtration beds, such as activated carbon, can pose a serious fire hazard. Automatic sprinklers not otherwise required to be present in the building may, with the approval of the authority having jurisdiction, be substituted for such fire-resistive enclosures, but are of little effectiveness if installed only within the room. The fuel and the ignition source are within the equipment, which is designed to exclude water and any particulates present in the room.

Some systems use smoke detectors to stop the fan and close dampers during a fire. In accordance with *NFPA 72®*, *National Fire Alarm Code®*, it is important to use this type of detector in order to provide effective detection and avoid nuisance alarms. These detectors cannot function properly if there is too little or too much airflow over them. These devices should be inspected and tested periodically.

Michael Earl Dillon, P.E., president and CEO of Dillon Consulting Engineers, headquartered in Long Beach, California, is one of the leading experts on smoke control and on heating, ventilating, and air-conditioning systems. He currently serves on five NFPA technical committees, including smoke management and air conditioning.

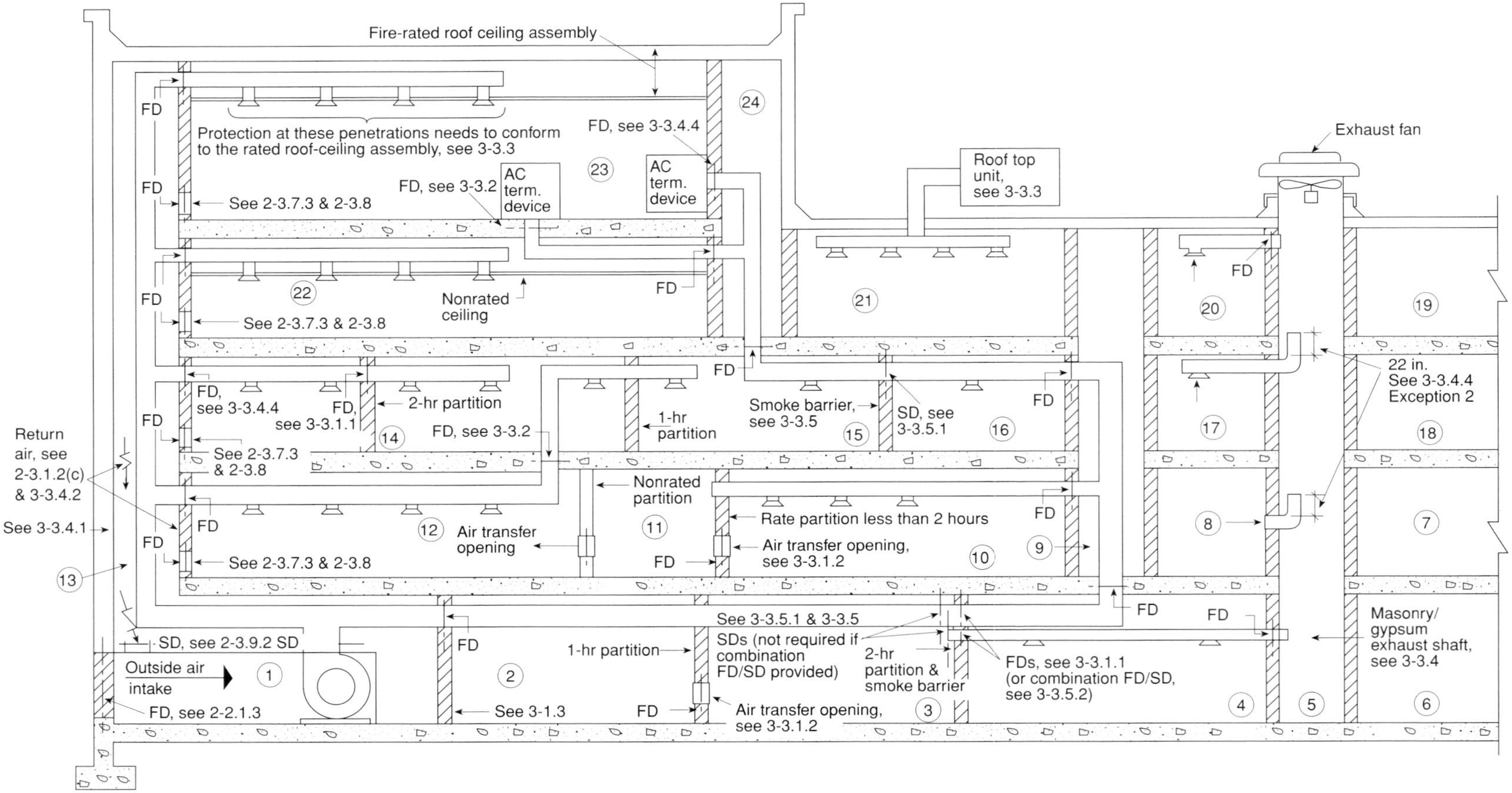

Notes:

Return/Exhaust duct requirements are similar

SD = Smoke damper

FD = Fire damper

(1) (2) (3) . . . . = Space identification

FIGURE 11-1 Typical Installation of an Air-Conditioning System Showing Penetration Protection in a Building

Note: Paragraph numbers refer to NFPA 90A, *Standard for the Installation of Air-Conditioning and Ventilating Systems.* Explanations can be found in that standard.

Source: NFPA 90A, 1999, Figure A-3-3.

The inspector should also inspect equipment rooms to make sure they are clean and that nothing is being stored in them. Such enclosures are often improperly used to store air filters, paper products, and light bulbs; they should not be used to store anything.

Fans. Lack of lubrication and accumulation of dust are two of the greatest enemies of fans and motors. Both can cause the equipment to overheat so much that it actually becomes an ignition source.

The inspector should check the fan belts for wear and proper tensioning and should ensure that the proper number, size, and configuration of belts are present. They should be adjusted or replaced as necessary. Although fans are often located in places that are difficult to reach, they should nonetheless be included in the inspection program.

Heating and Cooling Equipment. The hazards of cooling equipment are related to the hazards of electrical installations and to those of the refrigerant itself. (Proper wiring and grounding are discussed in Chapter 9, "Electrical Systems.") All common refrigerants are toxic to some degree, so a leak in the system can be hazardous to health and life safety. In some instances, exposure to high concentrations of halogenated refrigerants can manifest itself in the victim as cardiac arrhythmia and be mistakenly assumed to be a heart attack. If the victim is subsequently administered epinephrine by the emergency responders, death of the victim is a likely and unnecessary outcome.

Some refrigerants present a combustibility hazard as well, and all common systems incorporate combustible liquids as internal lubricants continuously circulated throughout the entire system. An often unrecognized problem associated with refrigeration units is the sudden rupture hazard of the pressurized refrigerant.

The fire experience of air-conditioning refrigeration units is generally good as long as the cooling equipment is properly installed and maintained. The inspector should be sure to check the frequency and quality of maintenance performed on the equipment and the housekeeping in the vicinity of the equipment. Again, these rooms frequently end up being used to store everything from office supplies to chemicals to lawnmowers. Refrigeration machinery rooms are not to be used as storerooms. Recommendations for the installation of mechanical refrigeration equipment are contained in the ASHRAE/ANSI 15, *Safety Code for Mechanical Refrigeration* (see Bibliography).

The hazards of air-heating equipment depend on the method of heating used. Heating equipment is discussed in Chapter 10, "Heating Systems."

Air-Cleaning Equipment. The purpose of most filters and air cleaners is to remove entrained dust and other particulate matter from the air stream. The filtered particles accumulate in the filter or on the air cleaner collector plates and, if ignited, could burn and produce a large volume of smoke. The products of combustion could be circulated throughout the building by the air distribution system, posing a threat to life and potential for significant property damage.

Many systems have pressure difference gauges that visibly or audibly indicate an excessive pressure drop across the filters, letting the occupants know that the filters should be cleaned or replaced. Filters should have either a Class 1 or Class 2 rating in accordance with UL 900, *Test Performance of Air Filter Units.* Most systems have disposable filters, which should be discarded when dirty and replaced with new clean filters. Some systems have a washable medium, which should be cleaned in accordance with the manufacturer's instructions and recoated with adhesive. This adhesive must have a flash point not lower than 325°F (163°C) as measured in the Pensky–Martens closed tester (ASTM D-93, *Standard Test for Flash Point by Pensky–Martens Closed Tester*).

Electronic air cleaners use electrostatic precipitation to remove particulate matter. Entrained particles pass through electrostatic fields and are collected either on a filter or on charged plates. Because electronic air cleaners use potentially lethal voltage and current combinations, they are equipped with interlocks that shut down the unit if a door or access panel is opened. The inspector should check that the interlocks are intact and have not been bypassed. Some systems have an automatic wash cycle for proper plate operation. These systems should be examined to ensure they are operating properly and to ascertain that the correct cleaning solvents are being used. Other systems use disposable filters, which are simply discarded when dirty and replaced with new clean ones.

Gas absorption systems are commonly used to remove volatile organic compounds that can have unhealthful effects or cause offensive odors. These filtration units use materials that can be fire, health, or reactive hazards. The units should be carefully examined for compliance with the manufacturer's installation and maintenance instructions. Their contents should be properly and clearly identified, with the appropriate placarding per NFPA 704, *Standard System for the Identification of the Hazards of Materials for Emergency Response,* in place.

DISTRIBUTION EQUIPMENT

Conditioned air is distributed throughout the building by means of the duct system. During a fire, this same duct system could disperse smoke and toxic gases instead of breathable air throughout the building.

Generally, ducts are of metal, masonry, fiberglass, or other approved materials. ASHRAE and the Sheet Metal and Air-Conditioning Contractors National Association (SMACNA) publish information about the construction of ducts. NFPA 90A, *Standard for the Installation of Air-Conditioning and Ventilation Systems,* and NFPA 90B, *Standard for the Installation of Warm Air Heating and Air-Conditioning Systems,* contain information about construction and installation practices.

UL 181, *Factory-Made Air Ducts and Connectors,* classifies duct materials according to flame spread, smoke developed, and flame penetration. Class 0 materials have a flame spread and smoke developed rating of 0. Class 1 materials have a flame spread rating of 25 or less, with no evidence of continued progressive combustion, and a smoke developed rating of not more than 50. Class 2 materials have a flame spread rating greater than 25 but not more than 50, with no evidence of continued progressive combustion, and a smoke developed rating of not more than 50 for the inside surface of the duct and not more than 100 for the outside surface. Class 0 and Class 1 materials must pass a 30-minute flame penetration test and Class 2 materials a 15-minute flame penetration test.

UL 181 also characterizes duct material's resistance to fungal growth because ducts can also harbor health hazards in the form of debris and biological. All duct systems should be regularly inspected and, if necessary, cleaned to mitigate the health and fire hazards.

Ducts can create both vertical and horizontal openings in fire barriers. Where the ducts pass through fire barriers or fire walls, adequate firestopping must be provided to seal the space between the duct walls and the edges of the opening. If properly installed and firestopped, steel sheet metal ducts in the thicknesses commonly used can protect an opening in a fire barrier wall for up to 1 hour.

Openings in a fire wall, ceiling, or floor may have to be protected with a fire damper, or combination fire/smoke damper. These dampers are mounted in heavy-gauge metal sleeves, which require clearances around them to permit free expansion of the damper and its sleeve in order to not bind the mechanism and interfere with its proper operation. The perimeter is fitted with metal angles, which keep the damper in place during the fire and help to seal off the required gap around them.

Dampers should be inspected, cleaned, and tested for proper operation at least once every 4 years. Fire dampers are listed in accordance with UL 555, *Fire Dampers.*

Penetrations of ceilings that are part of fire-resistive floor/ceiling or roof/ceiling assemblies require ceiling dampers. Ceiling dampers are different from fire dampers in that they also retard the passage of heat and not just flame. Ceiling dampers are listed in accordance with UL 555C, *Ceiling Dampers.*

Fire dampers are effective in limiting the passage of flame, and ceiling dampers are effective in limiting the passage of both flame and heat, but neither is very effective at limiting the passage of smoke. Smoke dampers or combination fire/smoke dampers are necessary to protect smoke barrier penetrations. These dampers must be arranged to close on the detection of smoke. *NFPA 72* provides guidance for the proper application of smoke detectors for both early detection and proper control of these dampers. This approach will limit smoke migration within buildings through the duct systems. Smoke dampers are listed in accordance with UL 555S, *Leakage Rated Dampers for Use in Smoke Control Systems.* Combination dampers are listed under both of the applicable categories.

Fire and ceiling dampers can fail to close properly if the airflow through them is not stopped prior to their operation. In such systems, these dampers should be specifically listed for dynamic operation. Smoke and combination fire/smoke dampers will always be dynamic operation dampers because they are opened and closed with electric motors actuated by smoke detection. Dampers listed for dynamic operation will state on their label the maximum velocity of air through them. If more than one damper is ganged together to protect an opening, then the velocity of air must be calculated on the assumption that only the smallest area damper is receiving the entire airflow of the duct.

CAUTION: The inspector must exercise extreme caution when testing any damper for operation because doing so can be very dangerous. A sudden powerful movement of the mechanism can sever fingers or hands.

Smoke Control

Two recognized approaches to controlling smoke in buildings make use of the air-conditioning system. The passive approach requires that smoke dampers be closed in coordination with doors and other such assemblies protecting the smoke barriers that define the smoke zone involved. In the active approach, the air-conditioning system is sometimes used to exhaust the products of combustion to the outdoors to prevent smoke migration from the fire area.

Smoke-control systems are necessarily complex to test and inspect since they must involve architectural elements such as floors, walls, doors, and windows. They also involve the electrical power system, the fire alarm system, the building management and control system, the fire sprinkler system, and the building heating ventilation, and air-conditioning systems. This subject is treated in greater detail in Chapter 12, "Smoke-Control Systems."

Ventilating Systems

Special ventilation systems are often needed to remove flammable vapors, corrosive vapors or fumes, grease-laden air from cooking equipment, or combustible dusts from an occupancy. Among the hazards such systems present is the possibility that sparks generated by fans, foreign materials in the air stream, or overheated bearings will ignite the flammable materials or vapors.

To reduce the hazard of fire, fans should be of noncombustible construction, accessible for maintenance, and structurally sound enough to resist wear. Occupants also should be able to shut the fans down from a remote location. In systems used to exhaust flammable solids or vapors, fan blades and housings should be constructed of nonsparking material to minimize the possibility of spark generation. Electrical wiring and motors are covered in NFPA 70, *National Electrical Code*®. In systems used to exhaust corrosive vapors, it is often necessary to use nonmetallic materials such as fiber-reinforced plastic, commonly but imprecisely referred to as "fiberglass."

CAUTION: The inspector must exercise caution because some laboratory exhaust systems can contain deposits of extremely hazardous materials such as crystallized perchloric acid, which can produce violent explosions when suddenly or sharply disturbed. Laboratory exhaust systems serving fume hoods and the like are not permitted by NFPA 45, *Standard on Fire Protection for Laboratories Using Chemicals,* to have automatic fire dampers incorporated into the ductwork. Common sense extrapolation would imply that smoke dampers should also be excluded.

These special exhaust systems should be independent of other ventilating systems and of one another. They should be vented directly outdoors or, in some very special cases, to approved containment or incineration systems by the shortest route and should not pass through fire walls. For specific hazards, these systems might contain special extinguishing systems. A schedule for inspecting, testing, and cleaning the system should be developed if one is not already in use. (Commercial kitchen exhaust systems are discussed in Chapter 55, "Protection of Commercial Cooking Equipment.")

MAINTENANCE

A maintenance and cleaning schedule is the key to safely operating air-conditioning and ventilating systems. When inspecting the equipment, the inspector should look for signs of rust and corrosion, especially on moving parts; check the condition of the filters and the electrical wiring; examine air ducts for accumulations of combustible dust and lint; and recommend cleaning if necessary.

The fire protection devices associated with the system—that is, the fire suppression and smoke-control equipment, the alarms, the fire and smoke dampers, and so on—should be tested periodically as part of the maintenance program. If inspectors do not witness these tests or conduct them themselves, they should ask to see the records of the tests that were performed. Such records must be specific as to the devices, both individually and the systems as an integrated whole, that were tested; and they should specify by whom and on what dates these tests were conducted. The records should contain the deficiencies discovered and the remedies implemented to remove and, if possible, avoid such deficiencies.

SUMMARY

Air-conditioning systems control the temperature and humidity of air, clean it, and distribute it to meet the rquirements of a conditioned space. In the summer these systems provide filtered, cooled, and dehumidified air; in the winter they provide heated, humidified air. There are three main components of air-conditioning systems: the air-intake system, which introduces fresh air into the system; the conditioning equipment, which consists of fans, air heating and cooling units, and filters; and the distribution system, which consists of the duct system. The chapter indicated what inspectors should be looking for when inspecting the various components of air-conditioning and ventilating systems.

BIBLIOGRAPHY

ASTM D-93, *Standard Test for Flash Point by Pensky–Martens Closed Tester,* American Society for Testing and Materials, Conshohocken, PA.

Cote, A. E., ed., *Fire Protection Handbook,* 18th ed., NFPA, Quincy, MA, 1997.

Cote, A. E., and Linville, J. L., eds., *Industrial Fire Hazards Handbook,* 2nd ed., NFPA, Quincy, MA, 1984.

Safety Code for Mechanical Refrigeration, ASHRAE/ANSI 15, American National Standards Institute, New York, NY.
UL 181, *Factory-Made Air Ducts and Connectors,* Underwriters Laboratories, Inc., Northbrook, IL.
UL 555, *Fire Dampers and Ceiling Dampers,* 3rd ed., Underwriters Laboratories, Inc., Northbrook, IL.
UL 555C, *Ceiling Dampers,* Underwriters Laboratories, Inc., Northbrook, IL.
UL 555S, *Leakage Rated Dampers for Use in Smoke Control Systems,* Underwriters Laboratories, Inc., Northbrook, IL.
UL 900, *Test Performance of Air Filter Units,* Underwriters Laboratories, Inc., Northbrook, IL.

NFPA Codes, Standards, and Recommended Practices

See the latest version of The NFPA Catalog for availability of current editions of the following documents.

NFPA 45, *Standard on Fire Protection for Laboratories Using Chemicals*
NFPA 70, *National Electrical Code®*
NFPA 72®, National Fire Alarm Code®
NFPA 90A, *Standard for the Installation of Air-Conditioning and Ventilating Systems*
NFPA 90B, *Standard for the Installation of Warm Air Heating and Air-Conditioning Systems*
NFPA 91, *Standard for Exhaust Systems for Air Conveying of Vapors, Gases, Mists, and Noncombustible Particulate Solids*
NFPA 92A, *Recommended Practice for Smoke-Control Systems*
NFPA 92B, *Guide for Smoke Management Systems in Malls, Atria, and Large Areas*
NFPA 96, *Standard for Ventilation Control and Fire Protection of Commercial Cooking Operations*
NFPA *101®, Life Safety Code®*
NFPA 704, *Standard System for the Identification of the Hazards of Materials for Emergency Response*

Smoke-Control Systems

CHAPTER 12

Michael Earl Dillon

Most meaningful smoke-control systems are complex. The systems are much more than just an arrangement of mechanical items. Although it is true that they have fans ducts and dampers associated with them, they also necessarily involve the fire detection and alarm systems; the HVAC control system; and the building's architectural features, which include floors, walls, doors, the exterior envelope, openings between floors around convenience stairs and escalators, and even elevator systems.

Before any successful inspection of a smoke-control system can begin, it is important to first determine the exact design intent of the smoke-control system. This information should be fully documented and kept in the building at a location acceptable to the authority having jurisdiction. Since these documents are of critical importance in the event of a fire, this location should be secure, protected from fire, and immediately accessible to the incident commander.

If such documentation is not readily available at the time of inspection, an agreement should be arrived at regarding its compilation and location in the future. Once the appropriate documentation regarding the smoke-control system design has been received, it must be carefully reviewed to determine what portions of the system require what type of inspection.

EARLY SMOKE-CONTROL SYSTEMS

Many older systems, which were constructed under requirements of earlier codes and design documents, consisted of relatively simple components to inspect. Some of these older systems required nothing more than openable or breakable windows or panels of a certain minimum dimension at a specified interval around the perimeter of the building on each level. If these were breakable windows, the earlier code usually required that the breakable pane be identified in some manner for location by the fire department during a fire. This identification was usually in the form of a small sticker in the lower corner of the window. If this type of system is present in the building to be inspected, then it is a simple matter to walk the perimeter of the building on each floor and look for the openable windows or panels of breakable glass and determine whether the original requirement is still being met.

From a practical point of view, this inspection is valuable only for verification purposes because such a system does not actually accomplish any meaningful smoke control during a real fire event. This type of system necessarily relies on favorable winds and manual intervention of the fire department on the fire floor. The manual intervention of the fire department consists of one or more responders finding their way through the smoke-filled floor to the perimeter wall and manually opening or breaking the appropriate designated panels. This procedure is, of course, unduly hazardous to both personnel on the fire floor and those on the ground below. If the prefire plan for the building anticipates using these panels or windows, it is of extraordinary importance that their location be well documented and readily available to the incident commander.

Michael Earl Dillon, P.E., president and CEO of Dillon Consulting Engineers headquartered in Long Beach, California, is one of the leading experts on smoke control and on heating, ventilating, and air-conditioning systems. He currently serves on five NFPA technical committees, including smoke management and air conditioning.

FIGURE 12-1 Smoke Concentration (*C*) versus Time (*t*) Illustrating Exponential Decay

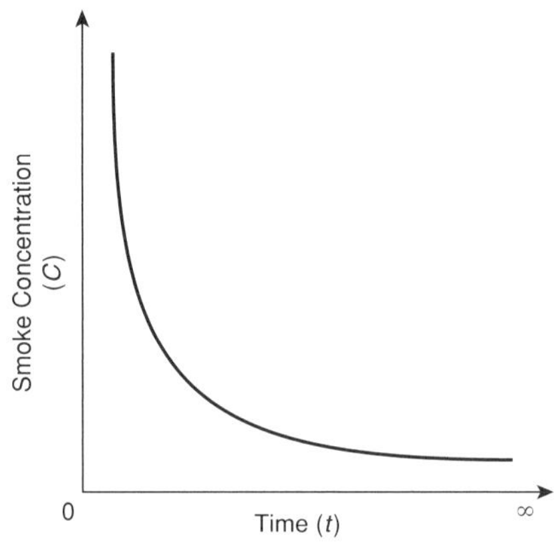

Other early systems simply specified a number of air changes per hour. One common number was six air changes per hour (6 ACH). Such systems are significantly more difficult to inspect. The common misconception is that the term "air changes" refers to complete removal or replacement of the air and its contaminants within the space of a relatively short period of time. For instance, six air changes per hour is the equivalent of one air change in 10 minutes. This requirement does not, however, mean that all the smoke and air within the space has been removed and replaced with smoke-free air in 10 minutes.

In a perfect world, the normal ventilation system is designed to fully mix the air and will, therefore, accomplish the actual turnover in an exponential decay manner (Figure 12-1). In simple terms, this means that for a perfectly arranged ventilation system accomplishing six air changes per hour, 37 percent of the original smoke contamination will still remain at the end of 10 minutes if no other smoke is introduced during that period of time.

From a practical point of view, what this situation means to an inspector is that no amount of "testing" through the use of smoke candles, theatrical foggers, or smoke cannons will provide meaningful information regarding compliance. The only method that can accurately determine compliance is the taking of flow measurements made at appropriate locations within the air extraction ducts themselves. The measurements require the proper use of special instruments by trained technicians. Therefore, the inspector should not undertake such an activity because it requires tools, instruments, training, and access to control over portions of the system, all of which are not generally available.

MORE RECENT DEVELOPMENTS

More recent smoke-control code requirements rely on one of four types of engineered designs: (1) the pressurization method, (2) the exhaust method, (3) the airflow method, and (4) the passive system.

Where such designs are mechanical, they are calculated to produce a minimum pressure difference across barriers to assist in confining smoke to its zone of origin (the pressurization method), a minimum airflow counter to the natural migratory path of the smoke (the airflow method), or a minimum exhaust rate to prevent the smoke layer from descending below a preestablished height above the floor along the egress path (the exhaust method). Where such designs are passive, they are an arrangement of smoke-tight construction that sufficiently inhibits migration of smoke for a considerable period of time to allow for safe egress (the passive system).

None of these systems can properly be called a "purge" system. In fact, there is no such commonly recognized system. Each of these and the unique inspection requirements of each are discussed later in this chapter.

SMOKEPROOF ENCLOSURES

Another early aspect of smoke control was the so-called smokeproof enclosure. In its various incarnations, it consisted of stairs positively pressurized by the introduction of supply air and a variety of vestibule arrangements.

Common naturally ventilated systems use vestibules ventilated directly to the outdoors or into a designated smoke shaft. When used, exterior vestibules require some attention regarding their configurations. If the opening to the vestibule is geometrically configured such that it can be compromised by an exposure hazard from the building itself, then it is of little practical use during a fire.

The impact of climate is also an issue. In many parts of the country, weather conditions can cause ice formation along the egress path through the vestibule, freeze the doors of the vestibule and stair closing, or result in obstructive accumulations of snow. Inopportune winds can also create problems by adversely affecting door operation through the imposing of force against the door that either makes opening it difficult or prevents it from closing.

The interior vestibules connected to smoke shafts suffer from many of these same problems to one degree or another. They also present an open line of air communication between all the levels of the enclosure system. The open line gives smoke that has contaminated one level the ready opportunity to spread and contaminate all other levels. When inspecting these types of systems, inspectors must examine the door arrangements to ensure that they will not be blocked by snow or ice or adversely affected by wind. Inspectors must also determine that appropriate geometry, shielding, or shaft construction prevents fire exposures from other parts of the building.

Common early mechanically ventilated systems used vestibules that were arranged with very high air-exchange rates and configured to be at a necessarily negative pressure relative to both the fire floor itself and the stair to which they were attached. These systems suffer from the following serious physical flaws:

- Negative pressure draws smoke into vestibule.
- Negative pressure in vestibule increases force required to open stair door.
- Negative pressure in vestibule increases force required to close tenant space door into vestibule.

First, the specified exchange rates are impossible to achieve when the construction and doors serving the vestibules need to meet all of the other requirements typically imposed by various codes and when doors are in a closed position. This is because the excess exhaust quantity, typically 150 percent of supply, cannot be made up by the leakage rate it would be allowed by the prescribed construction and door clearances.

It is also obvious that there will be a negative pressure within the vestibule as a result of the design specification of the early codes. For a typical vestibule, the negative pressure achieved within it as a result of these prescriptions is approximately 0.03 in. (7.46 Pa) water gauge. This level of negative pressure is sufficient to assure a steady flow of smoke from the fire floor in the exit vestibule.

Second, this negative pressure, when added to the positive pressure within the stair, necessarily results in a minimum stair door opening force of 23 lb (102 N), measured at the doorknob or lever. And, third, it also results in a requirement for the door closer on the vestibule to floor door to apply a closing force of at least 32 lb (142 N), measured at the doorknob or lever.

Some more recent codes omitted the requirements for the vestibule and simply required the stair to be pressurized. A typical pressure in this instance was 0.15 in. (37 Pa) water gauge. This configuration also has problems with effectively keeping smoke out of the stair tower and requiring excess opening forces to be exerted to gain entry. This latter situation is especially true if there is a mechanical smoke-control system on the floor. Systems based on minimum air change rates do not adequately account for the negative pressure that may be created. Attempts to mitigate this pressure problem by providing makeup air to the floor, simply mixes the smoke more thoroughly on the floor and distributes it more widely in addition to providing an additional source of oxygen for the fire. Attempts to mitigate the entrainment of smoke into the stair tower by adding excess air to provide a minimum velocity through one or more open doors simply exacerbates the pressure problem since door opening and closing times are significantly shorter than the air flow modulating times of control systems.

Among the more recent code requirements is the provision for a positively pressurized stair tower and a positively pressurized vestibule for entering the stair. In this scheme, the stair tower is pressurized to 0.05 in. (12.4 Pa) water gauge, positive, relative to the vestibule, and the vestibule is pressurized to 0.05 in. (12.4 Pa) water gauge, positive, relative to the fire floor. Under this scenario, door-opening forces can easily be held to less than 15 lb (67 N), measured at the doorknob or lever, assuring the proper closure of the door.

No matter which type of mechanically ventilated smoke-proof enclosure the building has, the proper method for inspecting its performance is to, in Step 1, activate the pressurization system for the stair and vestibules and turn on the smoke-control system, if any, for each floor in turn. With the doors closed and the systems on, the inspector then reads the pressure difference created across the doors and measures the door-opening force to assure compliance with the maximum allowable. The pressure difference can be measured with a manometer or a high-quality, calibrated, electronic pressure gauge. The measuring device is connected to a length of flexible tubing, which is inserted under the door to a distance of 1 ft (0.3 m) or more and carried 90 degrees in the direction of airflow. The pressure difference is then read directly. If it meets the minimum pressure difference required by the design and prevailing code requirement, Step 1 is finished.

For Step 2, the inspector measures the door-opening force, using either a belt tension tester to push the door open at the doorknob or lever or a fish scale to pull the door open at the doorknob or lever. In either instance, if the door closes properly and the opening force is equal to or less than the maximum allowable, Step 2 is finished.

ELEVATOR SMOKE CONTROL

If a code requirement affecting the building or one of the design requirements called for pressurized elevator hoistways, then the procedures to inspect these features would be the same as those for pressurized stairways discussed previously. The inclusion of elevator lobbies or vestibules at car landings is analogous to the vestibules for the smokeproof enclosures. In either event, vertical exit enclosures containing stairs, or shafts containing elevator hoistways, are nothing more than very tall smoke-control zones that just happen to have a number of smoke-control zones adjacent to them.

If smoke control is used in conjunction with elevator systems, it is imperative that the pressure differences across the doors be accurately measured and maintained. Very small pressure differences across elevator door systems can cause these systems to bind, prevent proper operation, and prevent the cars from moving. Such a situation can render the elevator system useless, strand occupants within the cars, interfere with Phase I elevator recall, and affect Phase II elevator fire-fighter service.

In modern smoke-control system design, the analysis required will generally lead the designer to the conclusion that the elevator system is a critical element in the overall smoke-control scheme. It is obviously important that any occupants within the elevators be allowed to safely leave the elevator system as soon as possible, since the hoistway shaft is a natural path for smoke movement in the building. Leaving individuals trapped within the elevator system could expose them to unsafe concentrations of smoke. The full coordination of the elevator recall system, automatic electrical power shutoff, and sprinkler operation within the elevator system is often an integral part of the overall smoke-control design.

Pressurization Method

MECHANICAL SMOKE-CONTROL SYSTEMS

The pressurization method of mechanical smoke-control relies on the existence of reasonably airtight construction completely surrounding the smoke zone. In a typical multistory office building, the zones may very well correspond to the individual floors. In detention and correctional occupancies, there will typically be two or more zones per floor, with self-closing or automatic-closing doors and automatic-closing smoke dampers arranged to isolate the zone. The function of other dampers is to open and allow exhaust air to be taken from the zone.

Either by means of this exhaust or in conjunction with supply air to areas outside the zone, a negative pressure difference is produced within the zone relative to the surrounding zones. For sprinklered building, this pressure difference is generally on the order of 0.05 in. (12.4 Pa) water gauge. Measuring this pressure difference is simple, as it was for the stair tower tests.

The reliability of the system, however, is dependent on many other factors. The construction of the barriers to form the zone must be properly assembled, and the penetrations through them properly protected. At the perimeter of the floor slab where the slab meets the exterior wall, a proper smoke-tight perimeter fire barrier is needed. Door operation at the perimeter of the smoke zone is critical both for isolating the zone and ensuring that no undue force is required to open the doors in order to exit the fire area.

Each of the smoke dampers must be inspected in order to determine whether they are properly labeled and listed. This labeling and listing includes such things as

- Degradation temperature for which the dampers and their operators are rated
- Fusing temperature at which the dampers will close if they also serve as fire dampers
- Characteristics of the sleeve in which they are to be installed
- Any clearances required around them
- Maximum velocity of the air traveling through them

The inspector must examine each of these to ensure that the appropriate device has been properly installed and to determine whether the controls and the monitoring means are operative. The initial inspection of these devices for both installation and function is time consuming and labor intensive. For this reason, much of this type of inspection is often assigned to third-party agents acting as special inspectors under contract to the owner, the architect, or the engineer (but not the contractor) and responsible to the authority having jurisdiction.

The fire detection and alarm systems, which ultimately provide the signs/output on which the system depends, must also be inspected at a level not usually encountered with simple fire alarm systems. When the alarm system activates the smoke-control system, detailed program instructions—as opposed to audio or visual signals to occupants—are being transmitted as electronic signals. Inspecting these systems requires an additional level of sophistication on the part of the inspector and the test instruments required.

Beyond the complex and time-consuming original testing and acceptance of the smoke-control system, there is the ongoing testing that is required to ensure continued reliability of the system. Because it is unreasonable to expect the inspector to have to have either the time or the personnel to do comprehensive ongoing testing and inspection of smoke-control systems, it is often more appropriate to simply check randomly selected zones and components of the system for occasional testing.

UL 864, *Control Units for Fire-Protective Signaling Systems,* is the test standard for fire alarm control units. It has been expanded to include certain types of building automation systems (BAS) and in particular has a category (UUKL) that qualifies devices for use in smoke-control systems. It is possible to review the logs created by a printer connected to a UL 864 system that has been configured under UUKL to treat all elements of the smoke-control system as "dedicated." Such logs will reveal any deficiencies that existed and whether they have been remedied.

For very large and complex systems, it may be appropriate to require third-party, special inspectors to conduct the bulk of the inspections and to produce a written report for filing with the authority having jurisdiction. This approach ensures that thorough inspections are in fact conducted while at the same time relieving the burden on the inspector and allowing the personnel to simply observe a small portion of the inspection process carried out by the special inspection agency.

Exhaust Method

The exhaust method is another form of mechanical smoke-control often encountered in large enclosed spaces such as atria. This approach relies on exhausting the smoke from high in the space as it accumulates in a layer above the highest exit pathway. For this approach to be successful, the rate of smoke extraction must be at least equal to the rate at which the smoke is being produced and transported into the overhead layer. In this instance, "smoke" is more than just the products of combustion. It is also the air that has become entrained with the products of combustion and that has been rendered inseparable from them.

Inspecting the system is relatively straightforward. Each of the means by which the smoke-control system is activated must be tested to demonstrate that it functions appropriately. This may be by smoke detection, sprinkler flow indication, or a manual override switch on the fire fighter's control panel.

The exhaust volume produced by the exhaust fans must be verified to be in accordance with the design exhaust rate. This testing is most appropriately accomplished by the use of a Pitot tube traverse. The inspector must also determine that the exhaust has in fact been taken from well up into the intended smoke layer in order to ensure that the smoke will in fact be extracted from above the occupants' heads and also not distributed to other spaces.

The inspector must also determine that no adverse crosscurrents are being caused within the space at points that are likely to coincide with the location of the fire itself. If air is drawn toward the fire or the rising plume at too high velocity, the plume can be deflected and additional air entrained, causing the smoke volume delivered to the overhead smoke layer to be greater than designed for.

Airflow Method

The airflow method of mechanical smoke control is rarely used. To work properly, it requires a very special architectural geometry, which is seldom encountered in practice. The theory behind this method is that a linear airflow pattern directed horizontally along a straight path can prevent smoke flow in the direction opposite of the airflow. The precise velocity necessary to accomplish this smoke flow prevention is a function of the fire size, its heat output, and the temperature difference. The method also requires that there be a method of exhaust relief of sufficient capacity to accept the significant volumes of air used to produce the necessary velocity.

Testing and inspecting the system require detailed measurements of airflow quantities, velocities, and directions. Of course, as is the case with all systems, the inspector must inspect the installation of each component and functionally test it for appropriate operation and for proper configuration by the control system.

PASSIVE SMOKE-CONTROL SYSTEMS

Passive smoke control has been around for almost as long as buildings have been built. It simply was not given any title or recognition. More recently, passive smoke barriers have been incorporated in many buildings, particularly those housing institutional uses such as hospitals and detention centers.

As the term implies, passive smoke-control systems rely on passive barriers to prevent or restrict the movement of smoke from the place where it is being generated to the areas from which it is to be excluded. Typically these barriers consist of walls or floors. Walls and floors can be effective smoke barriers only if openings and penetrations through them are properly sealed or otherwise protected. It is important that doors and dampers be installed in openings and that they be inspected for proper installation and operation.

The numerous penetrations of walls and floors by pipes, conduits, cabling, and the like must be properly sealed to prevent smoke flow through them. Additionally, the joints at the top of the wall and at the intersection of perimeter of walls, including the floor, must be sealed to prevent the leakage of smoke through these elements. Many listed penetration sealant systems and joint systems are available to accomplish this smoke tightening of the construction.

Unfortunately, the inspection of such systems is time consuming and labor intensive. Therefore, delegating this type of inspection and testing to an independent third-party organization is most appropriate. At the completion of the installation of the joint treatments, the penetration treatments, the dampers, the doors, and other protective means to prevent smoke flow, a pressure test of the space can be arranged to determine the effectiveness of the smoke-tightening measures and establish the equivalent leakage area.

Once the equivalent leakage area has been established through calculations based on the pressure difference developed for a given flow rate, this value can be used as a benchmark in subsequent years to determine whether the system has deteriorated to the point where it needs remedial action. Reconducting the same pressure test and recalculating the data can help make this determination.

In certain large area buildings like warehouses and factories, passive style smoke and heat vents may be present. These vents are normally actuated by a fusible link. Basic inspection of such vents should ensure that the link has no residue or buildup on it. In addition, no roof-mounted equipment or appurtenances should obstruct the free opening of the vent.

SMOKE-CONTROL SYSTEM ACCEPTANCE TESTING

When smoke-control systems are first installed, they require thorough testing prior to acceptance. This testing is necessary to verify that the intent of all of the design elements is being accomplished. Three methods by which this testing can be accomplished are as follows:

- Employing accepted test and balance procedures to verify that airflows and pressure differences are being achieved
- Employing tracer gas analysis to verify the migratory path of an airborne material
- Employing an actual fire of the design fire size and observing the results

		STAND-BY POWER AVAILABLE	BMS NORMAL ON	BMS NORMAL OFF	MANUAL FIRE ALARM BOX	FIREFIGHTER'S VOICE NOTIFICATION TO OCCUPANTS SYSTEM	FIREFIGHTER'S CONTROL PANEL STAIR 1 PRESSURIZATION FAN "ON"	FIREFIGHTER'S CONTROL PANEL STAIR 1 PRESSURIZATION FAN "OFF"	FIREFIGHTER'S CONTROL PANEL STAIR 2 PRESSURIZATION FAN "ON"	FIREFIGHTER'S CONTROL PANEL STAIR 2 PRESSURIZATION FAN "OFF"	FIREFIGHTER'S CONTROL PANEL STAIR DOORS "UNLOCK"	FIREFIGHTER'S CONTROL PANEL ELEVATOR LOBBY DOORS "CLOSE"	FIREFIGHTER'S CONTROL PANEL LEVEL 1 SWITCH "ON"	FIREFIGHTER'S CONTROL PANEL LEVEL 1 SWITCH "OFF"	SPRINKLER FLOW LEVEL 1	FIREFIGHTER'S CONTROL PANEL LEVEL 2 SWITCH "ON"
CENTRAL STATION NOTIFICATION	ALARM/SUPERVISORY/TROUBLE	YES	N/A	N/A	ON								ON		ON	
FIRE CONTROL ROOM ANNOUCIATION	AUDIBLE AND VISUAL SUPERVISORY SIGNALS (ALARM/SUPERVISORY/TROUBLE)		N/A	N/A	ON								ON		ON	
OCCUPANT NOTIFICATION ALL LEVELS (ALL CALL)	AUDIBLE (VOICE CAPABLE) AND VISUAL ALARM SIGNALS					YES										
OCCUPANT NOTIFICATION LEVEL 1	AUDIBLE (PRERCORDED MESSAGE/VOICE CAPABLE) AND VISUAL ALARM SIGNALS	YES	N/A	N/A		YES							ON		ON	
OCCUPANT NOTIFICATION LEVEL 2	AUDIBLE (PRERCORDED MESSAGE/VOICE CAPABLE) AND VISUAL ALARM SIGNALS	YES	N/A	N/A		YES							ON		ON	
OCCUPANT NOTIFICATION LEVEL 3	AUDIBLE (PRERCORDED MESSAGE/VOICE CAPABLE) AND VISUAL ALARM SIGNALS	YES	N/A	N/A		YES										
OCCUPANT NOTIFICATION LEVEL 4	AUDIBLE (PRERCORDED MESSAGE/VOICE CAPABLE) AND VISUAL ALARM SIGNALS	YES	N/A	N/A		YES										
OCCUPANT NOTIFICATION LEVEL 5	AUDIBLE (PRERCORDED MESSAGE/VOICE CAPABLE) AND VISUAL ALARM SIGNALS	YES	N/A	N/A		YES										
OCCUPANT NOTIFICATION LEVEL 6	AUDIBLE (PRERCORDED MESSAGE/VOICE CAPABLE) AND VISUAL ALARM SIGNALS	YES	N/A	N/A		YES										
OCCUPANT NOTIFICATION LEVEL 7	AUDIBLE (PRERCORDED MESSAGE/VOICE CAPABLE) AND VISUAL ALARM SIGNALS	YES	N/A	N/A		YES										
OCCUPANT NOTIFICATION LEVEL 8	AUDIBLE (PRERCORDED MESSAGE/VOICE CAPABLE) AND VISUAL ALARM SIGNALS	YES	N/A	N/A		YES										
OCCUPANT NOTIFICATION LEVEL 9	AUDIBLE (PRERCORDED MESSAGE/VOICE CAPABLE) AND VISUAL ALARM SIGNALS	YES	N/A	N/A		YES										
OCCUPANT NOTIFICATION LEVEL 10	AUDIBLE (PRERCORDED MESSAGE/VOICE CAPABLE) AND VISUAL ALARM SIGNALS	YES	N/A	N/A		YES										
OCCUPANT NOTIFICATION LEVEL 11	AUDIBLE (PRERCORDED MESSAGE/VOICE CAPABLE) AND VISUAL ALARM SIGNALS	YES	N/A	N/A		YES										
OCCUPANT NOTIFICATION LEVEL 12	AUDIBLE (PRERCORDED MESSAGE/VOICE CAPABLE) AND VISUAL ALARM SIGNALS	YES	N/A	N/A		YES										
OCCUPANT NOTIFICATION LEVEL 13	AUDIBLE (PRERCORDED MESSAGE/VOICE CAPABLE) AND VISUAL ALARM SIGNALS	YES	N/A	N/A		YES										
OCCUPANT NOTIFICATION LEVEL 14	AUDIBLE (PRERCORDED MESSAGE/VOICE CAPABLE) AND VISUAL ALARM SIGNALS	YES	N/A	N/A		YES										
OCCUPANT NOTIFICATION LEVEL 15	AUDIBLE (PRERCORDED MESSAGE/VOICE CAPABLE) AND VISUAL ALARM SIGNALS	YES	N/A	N/A		YES										
OCCUPANT NOTIFICATION LEVEL 16	AUDIBLE (PRERCORDED MESSAGE/VOICE CAPABLE) AND VISUAL ALARM SIGNALS	YES	N/A	N/A		YES										
OCCUPANT NOTIFICATION LEVEL 17	AUDIBLE (PRERCORDED MESSAGE/VOICE CAPABLE) AND VISUAL ALARM SIGNALS	YES	N/A	N/A		YES										
OCCUPANT NOTIFICATION LEVEL 18	AUDIBLE (PRERCORDED MESSAGE/VOICE CAPABLE) AND VISUAL ALARM SIGNALS	YES	N/A	N/A		YES										
OCCUPANT NOTIFICATION LEVEL 19	AUDIBLE (PRERCORDED MESSAGE/VOICE CAPABLE) AND VISUAL ALARM SIGNALS	YES	N/A	N/A		YES										
OCCUPANT NOTIFICATION LEVEL 20	AUDIBLE (PRERCORDED MESSAGE/VOICE CAPABLE) AND VISUAL ALARM SIGNALS	YES	N/A	N/A		YES										
OCCUPANT NOTIFICATION LEVEL 21	AUDIBLE (PRERCORDED MESSAGE/VOICE CAPABLE) AND VISUAL ALARM SIGNALS	YES	N/A	N/A		YES										
OCCUPANT NOTIFICATION LEVEL 22	AUDIBLE (PRERCORDED MESSAGE/VOICE CAPABLE) AND VISUAL ALARM SIGNALS	YES	N/A	N/A		YES										
OCCUPANT NOTIFICATION LEVEL 23	AUDIBLE (PRERCORDED MESSAGE/VOICE CAPABLE) AND VISUAL ALARM SIGNALS	YES	N/A	N/A		YES										
RECALL LOW RISE PE-1		YES	N/A	N/A												
RECALL LOW RISE PE-2		YES	N/A	N/A												
RECALL LOW RISE PE-3		YES	N/A	N/A												
RECALL LOW RISE PE-4		YES	N/A	N/A												
RECALL LOW RISE PE-5		YES	N/A	N/A												
LOW RISE ELEVATOR MACHINE ROOM POWER SHUNT	DISCONNECT POWER ONLY AFTER TIME DELAY ADEQUATE TO COMPLETE RECALL	YES	N/A	N/A												
LOW RISE PE SPRINK PREACTION VALVE	OPEN VALVE ONLY AFTER POWER DISCONNECTED	YES	N/A	N/A												
RECALL HIGH RISE PE-6		YES	N/A	N/A												
RECALL HIGH RISE PE-7		YES	N/A	N/A												
RECALL HIGH RISE PE-8		YES	N/A	N/A												
RECALL HIGH RISE PE-9		YES	N/A	N/A												
RECALL SE-10		YES	N/A	N/A												
HIGH RISE ELEVATOR MACHINE ROOM POWER SHUNT	DISCONNECT POWER ONLY AFTER TIME DELAY ADEQUATE TO COMPLETE RECALL	YES	N/A	N/A												
HIGH RISE ELEVATOR MACHINE ROOM PREACTION VALVE	OPEN VALVE ONLY AFTER POWER DISCONNECTED	YES	N/A	N/A												
SF-1	OSA VENTILATION SUPPLY FAN	YES	ON	OFF									ON		ON	ON
REMOVED SF-2																
REMOVED SF-3																
REPLACED SF-4	STAIR 1 SUPPLY FAN	YES	N/A	N/A	ON		ON	OFF					ON		ON	ON
SF-5	STAIR 2 SUPPLY FAN	YES	N/A	N/A	ON				ON	OFF			ON		ON	ON
REPLACED EF-1	SMOKE CONTROL SYSTEM FLOOR DEPRESSURIZATION EXHAUST FAN	YES	ON	OFF									ON		ON	ON
REMOVED EF-2																
ADDED TO STAND-BY POWER EF-3	RESTROOM AND GENERAL EXHAUST	YES	ON	OFF												
REMOVED EF-4																
REMOVED EF-5																
EF-26	UPS EQUIPMENT ROOM EXHAUST	YES	ON	OFF												
AC-1	AIR CONDITIONING UNIT SUPPLY FAN LEVEL 1	NO	ON	OFF									OFF		OFF	
AC-2	AIR CONDITIONING UNIT SUPPLY FAN LEVEL 2	NO	ON	OFF												OFF
AC-3	AIR CONDITIONING UNIT SUPPLY FAN LEVEL 3	NO	ON	OFF												
AC-4	AIR CONDITIONING UNIT SUPPLY FAN LEVEL 4	NO	ON	OFF												
AC-5	AIR CONDITIONING UNIT SUPPLY FAN LEVEL 5	NO	ON	OFF												
AC-6	AIR CONDITIONING UNIT SUPPLY FAN LEVEL 6	NO	ON	OFF												
AC-7	AIR CONDITIONING UNIT SUPPLY FAN LEVEL 7	NO	ON	OFF												
AC-8	AIR CONDITIONING UNIT SUPPLY FAN LEVEL 8	NO	ON	OFF												
AC-9	AIR CONDITIONING UNIT SUPPLY FAN LEVEL 9	NO	ON	OFF												
AC-10	AIR CONDITIONING UNIT SUPPLY FAN LEVEL 10	NO	ON	OFF												
AC-11	AIR CONDITIONING UNIT SUPPLY FAN LEVEL 11	NO	ON	OFF												
AC-12	AIR CONDITIONING UNIT SUPPLY FAN LEVEL 12	NO	ON	OFF												
AC-13	AIR CONDITIONING UNIT SUPPLY FAN LEVEL 13	NO	ON	OFF												
AC-14	AIR CONDITIONING UNIT SUPPLY FAN LEVEL 14	NO	ON	OFF												
AC-15	AIR CONDITIONING UNIT SUPPLY FAN LEVEL 15	NO	ON	OFF												
AC-16	AIR CONDITIONING UNIT SUPPLY FAN LEVEL 16	NO	ON	OFF												
AC-17	AIR CONDITIONING UNIT SUPPLY FAN LEVEL 17	NO	ON	OFF												
AC-18	AIR CONDITIONING UNIT SUPPLY FAN LEVEL 18	NO	ON	OFF												
AC-19	AIR CONDITIONING UNIT SUPPLY FAN LEVEL 19	NO	ON	OFF												
AC-20	AIR CONDITIONING UNIT SUPPLY FAN LEVEL 20	NO	ON	OFF												
AC-21	AIR CONDITIONING UNIT SUPPLY FAN LEVEL 21	NO	ON	OFF												
AC-22	AIR CONDITIONING UNIT SUPPLY FAN LEVEL 22	NO	ON	OFF												
AC-23	AIR CONDITIONING UNIT SUPPLY FAN LEVEL 13 ELEVATOR MACHINERY ROOM	NO	ON	OFF												
AC-24	AIR CONDITIONING UNIT SUPPLY FAN LEVEL 13 ELEVATOR MACHINERY ROOM	NO	ON	OFF												
AC-25	AIR CONDITIONING UNIT SUPPLY FAN LEVEL 13 ELEVATOR MACHINERY ROOM	NO	ON	OFF												
SFD.SA.01.01	OSA INTAKE AT LEVEL 1 MER 16X36 w/MVD	YES	OPEN	CLOSE									CLOSE	CLOSE	CLOSE	OPEN
SFD.SA.01.02	SA TO SERVICE ELEVATOR LOBBY LEVEL 1	YES	OPEN	CLOSE									CLOSE	CLOSE	CLOSE	
SFD.SA.01.03	SA TO FIRE CONTROL ROOM FROM VV1-7	YES	OPEN	CLOSE									CLOSE	CLOSE	CLOSE	
SFD.RA.01.01	RA FROM FIRE CONTROL ROOM TO CEILING PLENUM	YES	OPEN	CLOSE									CLOSE	CLOSE	CLOSE	
SFD.EA.01.01	R-E/3 CONNECTION TO LEVEL 1 CEILING PLENUM 46X22	YES	CLOSE	CLOSE									OPEN	CLOSE	OPEN	CLOSE
SFD.SA.02.01	R-OSA/3 INTAKE CONNECTION TO LEVEL 2 MER 16x36 w/MVD	YES	OPEN	CLOSE									OPEN		OPEN	CLOSE

FIGURE 12-2 Partial Sample of an Event Matrix

Appropriate guidance and details on the selection of procedures for testing a smoke management system can be found in *ASHRAE Guideline 5—1994, Commissioning of Smoke Management Systems;* NFPA 92A, *Recommended Practice for Smoke-Control Systems;* and NFPA 92B, *Smoke Management Systems in Malls, Atria, and Large Areas.*

EVENT MATRIX

An important part of the original acceptance testing of the system is the thorough documentation on its as-built condition. In addition to the information discussed earlier in this chapter regarding dampers, doors, walls, floors, fans, and the like, the inspector needs detailed information regarding control input and output functions. This information has been variously presented in a cause-and-effect chart, a sequence of operations matrix, and an event matrix. The term *event matrix* is gaining in popularity. No matter what it is called, the use of such a document is of vital importance.

An event matrix, as used in this discussion, is a chart in which input is displayed on one axis and output is displayed on the other axis. Figure 12-2 shows a partial sample of an event matrix. When properly constructed, such a chart has each input device in a fire alarm system, such as initiating devices, sprinkler flow indicators, smoke detectors, heat detectors, and manual pull stations, listed on one axis. Some devices could be grouped by the zone in which they are located and the circuit they are on.

All the alarm-initiating devices should be listed whether or not they have a control function over the smoke-control system. In an event matrix, a nonfunction is also an important piece of information. Manual switches used by the fire department to control devices should be listed as if they were initiating devices.

Each device that needs to be controlled is listed on the other axis. These control devices include components such as automatic-closing doors, smoke dampers, fans, and elevator recall systems. Even devices that are not part of the smoke-control system should be listed because, as noted, a nonfunction remains important.

The event matrix serves as the fundamental document in determining what status each device in the system must assume, given the input of some other device. By organizing the information in such a manner, conflicts created by devices, which might otherwise go undetected, become apparent. It also serves to determine which devices in the system can be safely ignored. When completed, it serves as the checklist for tracking completion of the acceptance testing.

In addition, the final event matrix, which emerges from the completion of the acceptance testing, becomes the record by which all future verification of continued proper operation is conducted. In short, without a complete and comprehensive event matrix, subsequent reinspections of the smoke-control system cannot be properly conducted.

SUMMARY

Smoke-control design, including the physical implementation of the design and its ultimate acceptance, maintenance, and utility is a complex multidisciplinary effort. To be successful, it requires a coordinated approach involving technically competent designers and craftspeople.

Recent smoke-control code requirements rely on one of four types of engineered designs. In the pressurization method of mechanical smoke-control, a reasonably airtight construction completely surrounds the smoke zone. The second type, the exhaust method, relies on exhausting the smoke as it accumulates in a layer above the highest exit pathway. To work properly, the airflow method of mechanical smoke control requires a seldom found architectural geometry; this third type, therefore, is rarely used. The fourth type, the passive smoke-control system, relies on passive barriers such as walls and floors to prevent smoke movement from place of origin to other areas.

An event matrix, which is a cause-and-effect chart documenting the as-built condition of the facility, is a useful record for the inspector to use when conducting a reinspection of the smoke-control system.

BIBLIOGRAPHY

ASHRAE Guideline 5—1994, Commissioning of Smoke Management Systems. American Society of Heating, Refrigeration and Air Conditioning Engineers, Inc., Atlanta, GA.

UL 864, *Control Units for Fire-Protective Signaling Systems,* Underwriters Laboratories, Inc., Northbrook, IL.

NFPA Codes, Standards, and Recommended Practices

See the latest version of The NFPA Catalog for availability of current editions of the following documents.

NFPA 92A, *Recommended Practice for Smoke-Control Systems*

NFPA 92B, *Smoke Management Systems in Malls, Atria, and Large Areas*

Fire Alarm Systems

CHAPTER 13

Lee Richardson

CASE STUDY

Safety Systems Limit Hospital Fire Damage

A large Illinois hospital undergoing a sprinkler retrofit quickly discovered the benefits of upgrading its system when sprinklers confined a fire caused by careless smoking to the basement storage room in which it began.

The five-story hospital, which was of unprotected, noncombustible construction, was occupied and operating at the time of the fire. Smoke and heat detectors providing full coverage were monitored by a municipal fire-alarm system. The system also covered the new wet-pipe sprinkler system, 80 percent of which had been installed at the time the fire occurred. When the sprinklers activated, the smoke doors operated, the building's air-handling units shut down, and the staff initiated its emergency response plan.

The fire department received an automatic alarm, followed by a call from hospital staff reporting a working fire, at 6:00 P.M. Arriving fire fighters found a single sprinkler operating in a basement storage room that had recently been stocked with palletized cardboard boxes of toilet paper. The sprinkler had confined the fire to its area of origin, so fire fighters extended a hose line from a nearby standpipe to the area of origin and extinguished it.

Fire damage was limited to the object of origin, while smoke damage extended to the room and adjacent hallway. Some smoke also migrated to the third floor through a loose fitting on a pneumatic tube delivery system. The hospital staff quickly moved patients out of the affected area.

The fire was started by a carelessly discarded cigarette, which ignited the wooden pallets, allowing flames to spread to the boxes of toilet paper. Officials estimated the loss at $5000. There were no injuries.

The fire department stated that, "as a result of all life safety systems operating as designed, damage was held at $5000, and there was no fire extension."

Officials further noted that, had it not been for the sprinklers, "heavy smoke and fire would have filled the entire...basement and would have caused significant disruption to their operation."

Source: "Fire Watch," *NFPA Journal*, January/February, 1998.

Lee Richardson, a senior electrical engineer in NFPA's Electrical Engineering Department, is the NFPA staff liaison for NFPA 72®, Nation Fire Alarm Code®. *He has been with NFPA for 7 years and has a background of more than 20 years in the area of signaling and control systems.*

Fire alarm systems are vital to minimizing life and property losses during fires. They provide early fire detection, warn occupants to evacuate or relocate, initiate fire safety functions, and notify the fire department to respond.

This chapter provides fire inspection personnel with an overview of fire alarm system features as well as an understanding of the key inspection points needed to assure fire alarm system installations are in compliance with applicable codes and standards. Although this chapter highlights how to determine the applicable regulations that apply and may, therefore, be helpful to plans examiners, it assumes that these regulations have already been identified and that the inspector has been provided with approved plans.

This chapter assumes that the inspector's role consists of (1) making a visual inspection of the installation to ensure that it complies with both the plans approved by the plans examiner and the installation requirements of *NFPA 72®*, *National Fire Alarm Code®*, and (2) verifying that the installation records are complete and that required testing has been done and documented in accordance with *NFPA 72*. It is not expected that extensive functional testing will be included in the inspection. Qualified inspection and test personnel must perform any limited functional tests in accordance with *NFPA 72*. References and information in this chapter concerning *NFPA 72* are based on the 1999 edition.

FIRE ALARM SYSTEM OVERVIEW

System Components

In a simplified view, fire alarm systems are typically composed of initiating devices, notification appliances, and control units (i.e., panels). Initiating devices include manual fire alarm boxes (pull stations), smoke detectors, smoke alarms, heat detectors, waterflow switches, tamper switches, and other types of detection devices that provide input signals to the system. Notification appliances include horns, speakers, strobes, text displays, and other types of appliances that provide audible, visible, or tactile outputs. Control units, which are used to process input and output signals, can be a single unit or an interconnected combination of several units. In addition, these units may also control ancillary equipment such as automatic door closers.

Notification Signals

Fire alarm systems, which typically provide occupant notification at the building, are known as protected premises, or local, fire alarm systems. Fire alarm system signals notify occupants of the potential danger and of the need for complete evacuation of the premises. The signals are transmitted via audible and visual notification appliances located throughout the premises. Audible appliances are required to provide a special three-pulse evacuation signal.

In some occupancies, such as hospitals and high-rise buildings, the building fire protection plan provides for selective evacuation or directed relocation rather than complete evacuation. In such cases speakers are included as a part of the system design for notification to provide emergency voice/alarm communications service. For this type of service, NFPA *101®*, *Life Safety Code®*, and *NFPA 72* provide special rules that address the design and the required operation of the system.

Protected premises fire alarm systems can also provide for the actuation of fire suppression systems or the operation of protected premises fire safety functions, such as the following:

- Elevator recall
- Elevator shutdown
- Operation of components of the heating, ventilating, and air-conditioning system
- Door release
- Door unlocking

Supervising Stations

Some fire alarm systems alert off-site emergency forces by transmitting signals to a supervising station or to a public fire service communications center. The three different types of supervising stations addressed in *NFPA 72* are as follows:

- Central supervising station
- Proprietary supervising station
- Remote supervising station

Separate rules are provided for each of these supervising stations. In central station service, a structured contractual arrangement provides the subscriber with several required elements of service, including runner service. Proprietary supervising stations are those that are provided by the owner of the protected premises. Remote supervising stations involve the use of a monitoring service that is not provided under the more structured contractual arrangements required for central station service.

Fire alarm systems that are monitored by a supervising station are required to retransmit fire alarm signals to the fire department or public fire service communications center. Some communities use municipal fire alarm systems to provide public access to fire boxes located throughout the community. These systems send signals directly to the public fire service communications center. In some cases the local authority will permit a protected premises to use the municipal fire alarm system to transmit fire alarm signals to the public fire service communications system. An auxiliary fire alarm system provides the interface between the protected premises and municipal fire alarm system.

Categories of Signals

Fire alarm systems use three categories of signals: fire alarm signals, supervisory signals, and trouble signals. These three categories are generally required to be kept independent of each other so that signals are not mixed. Fire alarm signals, such as a signal from a manual fire alarm box or smoke detector, indicate the presence of a fire. Supervisory signals are signals that may indicate the inability of a fire protection system to do its job. A valve tamper switch of a sprinkler system is an example of a supervisory signal initiating device. Trouble signals indicate the presence of a fault condition in a monitored circuit or component.

System Design Requirements

SYSTEM DESIGN REQUIREMENTS, APPROVAL, AND DOCUMENTATION

The requirements of *NFPA 72* apply to the performance, installation, inspection, testing, and maintenance of all fire alarm systems, including those that are required and those that are not required. Local codes may require a building to include a fire alarm system as a part of the features provided for the building fire protection, depending on the type of occupancy. When a fire alarm system is required, certain basic attributes of the system are usually specified as a part of the requirement for the system. Typically included are the following:

1. Requirements of type of initiation to actuate the system (manual initiation, automatic detection, extinguishing system operation)
2. Requirements for occupant notification
3. Requirements (if needed) for emergency forces notification
4. Requirements for the actuation of building fire safety functions

5. Requirements related to the location of operator controls and annunciation
6. Requirements for the supervision of certain features by on-site or off-site personnel

When local codes do not require a fire alarm system but the building owner or occupant decides to install a fire alarm system, the system must still be installed and maintained in accordance with *NFPA 72.* In this case the system designer must specify the basic attributes of the system in accordance with the performance needs intended for the system. The authority having jurisdiction (AHJ) must approve the performance standards planned for the design of the system.

Once the basic attributes of the fire alarm system are known, the installation rules of *NFPA 72* are applied. In a number of places, *NFPA 72* introduces a requirement with the conditional phrase "if required" or "when required." The basic attributes of the system specified by local codes or by the system designer provide the answers to the questions raised by these conditional phrases.

Because the terms used in *NFPA 72* often have special meaning, understanding the meaning of the terms used is important to properly interpret the rules of the code. Many terms are common to most NFPA codes and standards, for example, the terms "approved," "authority having jurisdiction," and "listed." Other terms, however, are unique to *NFPA 72,* for example, the terms "operating mode, public" and "operating mode, private" refer to the way in which notification appliances are used and not to public and private buildings. Clarity of the meaning of these terms is essential to proper application of the requirements in the chapter on notification appliances for public and private mode operation.

Approval and Documentation

NFPA 72 requires system designers to be experienced and qualified in the proper design, application, installation, inspection, testing, and maintenance of fire alarm systems. The code addresses requirements for approval and documentation of fire alarm system design and installation plans and requires that the AHJ be notified prior to the installation or alteration of equipment or wiring.

NFPA 72 also requires the preparation of a record of completion for each system to document the system design and installation. The information that must be provided in the record of completion includes written confirmation that the system has been properly installed, inspected, and tested. The installation, inspection, and testing must be done by installation and service personnel qualified to perform this work. A permanent record of inspections, testing, and maintenance of the system is also required.

Inspection personnel using this manual should confirm that the preliminary approvals and the installation, inspection, and testing records required by *NFPA 72* are complete up to the point of his or her inspection. Once this has been verified, the inspector should verify the system installation against these records and check to assure compliance with the installation rules in *NFPA 72.* The following sections highlight some of the key inspection points that should be checked. Inspectors should consult *NFPA 72* for additional information on specific requirements.

GENERAL SYSTEM REQUIREMENTS

Equipment Listing

One of the fundamental requirements of *NFPA 72* is the requirement that all fire alarm products must be listed for the specific fire alarm system application for which they are used. Furthermore, equipment must be installed and used in accordance with the provisions of the listing and with the manufacturer's instructions, which are usually included as the basis of equipment listing. The manufacturer's instructions should be consulted as a starting point for any listing issues that may arise. The listing agency and associated product listing directories published by listing agencies are additional sources to be consulted.

Fire Alarm Control Units (Panels) and Annunciation

The fire alarm system may include one or more fire alarm control units. *NFPA 72* does not specify the location required for fire alarm control units except as may be required where annunciation functions are integral to the control unit. Local codes would determine whether system annunciation is required. The rules in *NFPA 72* covering annunciation include a requirement that the annunciation means be readily accessible to personnel responding to a fire and be located as required by the AHJ. Aside from annunciation, local codes might dictate the location of controls or require the location to be acceptable to the AHJ. The inspector should confirm that the fire alarm control unit or units are located in accordance with these rules and with the approved installation plans.

Where more than one fire alarm control unit is provided as a part of the system design, these control units must be arranged to function as a single system. The arrangement and interconnection of master and local fire alarm control units must comply with the rules in Chapter 3 of *NFPA 72,* 1999 edition.

NFPA 72 permits the use of combination systems that join non–fire alarm functions and fire alarm functions. Where combination systems are used, the fire alarm functions must always take precedence. All fire alarm control units, including those that are part of combination systems must be listed for this purpose.

When fire alarm control units are installed in areas that are not continuously occupied, the areas must be protected by an automatic smoke detector installed at the location of the fire alarm control unit. In the context of this requirement, the term "continuously occupied" means occupied 24 hours per day, 7 days per week, 365 days per year. The installation of the smoke detector must conform to the requirements of *NFPA 72,* Chapter 2.

Power Sources

In general, each fire alarm system must be provided with a primary and secondary power source. The primary supply is usually the normal electrical service for the building. The secondary supply is often a battery system integral with the fire alarm control unit. Where the normal building service is used for the primary supply, the connection to the fire alarm system must be to a dedicated branch circuit. The disconnecting means for the branch circuit must have a red marking, must be identified with the label "FIRE ALARM CIRCUIT CONTROL," and must be accessible only to authorized personnel. The location of the disconnecting means must be permanently identified at the fire alarm control unit.

Manual Fire Alarm Boxes

INITIATING DEVICES

Manual fire alarm boxes, often called manual pull stations, are used to manually initiate an alarm signal. Operation of a manual fire alarm box (Figure 13-1) may require one action, such as pulling a lever, or two actions, such as lifting a cover and then pulling a lever. In some institutional occupancies, local codes may permit the use of key-operated manual alarm fire alarm boxes (Figure 13-2).

Anytime a fire alarm system includes an automatic fire detector or waterflow detection device, at least one manual fire alarm box is required. The manual fire

FIGURE 13-1 Typical Single-Action Manual Fire Alarm Box
Source: *National Fire Alarm Code® Handbook,* NFPA,1999, Exhibit 1.19; Protectowire; photo courtesy of Mammoth Fire Alarms Inc., Lowell, MA.

alarm box must be located where required by the AHJ. When local codes require a fire alarm system to have manual fire alarm initiation, manual fire alarm boxes must be installed throughout the protected area. They must be unobstructed, accessible, and installed in accordance with *NFPA 72.* Manual fire alarm boxes must be located within 5 ft (1.5 m) of the exit doorway opening at each exit on each floor. Where there is a group opening exceeding 40 ft (12.2 m), a manual fire alarm box must be located within 5 ft (1.5 m) of each side of the opening. Additional manual fire alarm boxes must be provided so that the travel distance from any location does not exceed 200 ft (61 m) to a manual fire alarm box.

Manual fire alarm boxes must be positioned within easy reach. The operable part of the manual fire alarm box must be at least 3½ ft (1.1 m) but not more than 4½ ft (1.37 m) above the floor.

FIGURE 13-2 Typical Key-Operated Manual Fire Alarm Box
Source: *National Fire Alarm Code® Handbook,* NFPA,1999, Exhibit 1.20; Fire Control Instruments; photo courtesy of Mammoth Fire Alarms, Inc., Lowell, MA.

Smoke Detectors

In many types of fire scenarios, detectable levels of smoke will precede detectable levels of heat and for these situations the use of smoke detectors is prominent. In addition, local codes may include specific requirements to use smoke detection in particular occupancies.

Types of Smoke Detectors. Devices for the detection of smoke can include several types of smoke detectors. The most common type is a spot detector. Other types of smoke detectors include projected beam-type smoke detectors and air-sampling smoke detectors.

- Spot-type smoke detectors are individual devices that usually detect smoke by means of a photoelectric sensor or ionization chamber within the device enclosure.
- Projected beam detectors are made up of two units arranged so that one unit sends a beam of light across a space to a separate photoelectric receiving unit.
- Air-sampling smoke detectors draw air through an air-sampling line or lines to the smoke detection unit.

Location of Smoke Detectors. The following information regarding the location of smoke detectors are primarily for spot-type smoke detectors. For other types of smoke detectors, the code should be consulted for additional rules that apply.

In general, spot-type smoke detectors must be located on the ceiling at least 4 in. (100 mm) from a side wall to the near edge of the detector, or they must be located on a side wall at least 4 in. (100 mm) from the ceiling to the top edge of the detector and not more the 12 in. (300 mm) from the ceiling to the top edge of the detector (Figure 13-3). Additional rules apply where the ceiling surface includes solid joists or beams. Note that the smoke detector listing must indicate its suitability for wall or ceiling mounting as appropriate for the installation.

Unless the manufacturer specifies otherwise or other considerations such as high air movement or stratification dictate differently, the spacing that is most often used as a guide for spot-type smoke detectors is 30 ft (9.1 m) where the ceiling is flat (nonsloping) and smooth. The arrangement of detectors must be such that no point on the

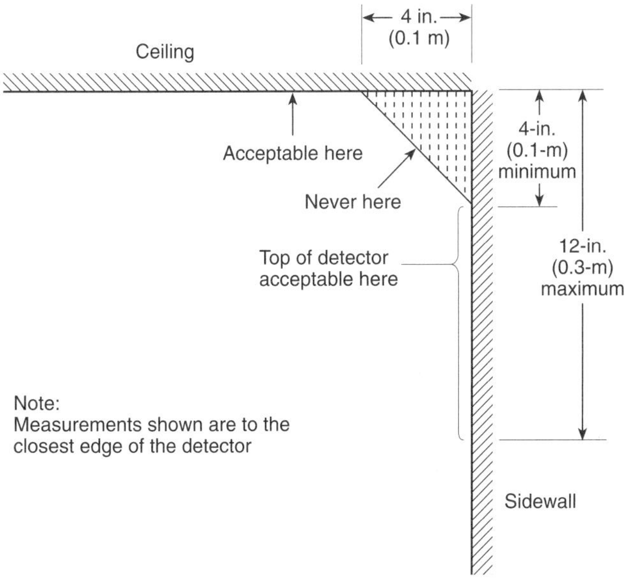

FIGURE 13-3 Proper Mounting for Detectors
Source: *National Fire Alarm Code*®, NFPA, 1999, Figure A-2-2.2.1.

ceiling is more than 0.7 times the selected spacing. If the selected spacing used is 30 ft (9.1 m), all points on the ceiling must fall within 21 ft (6.37 m) of a detector.

Where the ceiling surface involves solid joists or beams, special rules are used to modify the rules for smooth ceiling spacing. The code defines solid joist construction as a ceiling that has solid structural or solid nonstructural members projecting down from the ceiling surface for a distance of more than 4 in. (100 mm) and spaced at intervals of 3 ft (0.9 m) or less. Beam construction is defined as a ceiling that has solid structural or solid nonstructural members projecting down from the ceiling surface for a distance of more than 4 in. (100 mm) and spaced at intervals of more than 3 ft (0.9 m). These terms apply to the rules for heat detectors as well as smoke detectors. A solid joist over 1 ft (0.3 m) in depth is treated as a beam for smoke detector spacing purposes. The specific rules for smoke detectors regarding this type of construction are included in *NFPA 72,* Chapter 2.

If the ceiling is flat (nonsloping), is 12 ft (3.66 m) or less in height, and has beams or solid joists of 12 in. (0.3 m) or less in depth, the spacing for smooth ceilings is used with an exception. The exception is that in the direction perpendicular to the run of the beams or solid joists, the maximum spacing for detectors must be reduced to one-half of the smooth ceiling spacing. For beam ceilings, spot-type detectors are permitted to be located on the ceiling or on the bottom of the beam. For solid joist ceilings, the detector must be mounted on the bottom of the joist.

If the ceiling is flat, is more than 12 ft (3.66 m) high, or has beams whose depth exceeds 12 in. (0.3 m), spot-type detectors must be located on the ceiling in every beam pocket. The spacing for detectors would be based on each beam pocket being treated as a separate ceiling space, with detectors spaced in accordance with the rules for smooth ceiling spacing in each space. For beam ceilings, spot-type detectors are required to be located on the ceiling in the beam pocket. For solid joist ceilings, the detector must be mounted on the bottom of the joist.

If the ceiling is sloped, additional rules are specified in the code to modify the rules for flat ceilings. The code should be consulted for these additional rules.

The location and installation of smoke detectors must take into account the various other factors that can influence their performance. *NFPA 72* provides specific rules for various conditions, including the following:

1. Raised floors and suspended ceilings
2. Partitions
3. Heating, ventilating, and air-conditioning system effects
4. Plenums
5. Environmental limits
6. Stratification
7. High-rack storage
8. High air movement areas

Heat Detectors

Heat Detector Types. In some types of fire scenarios, heat detection will provide the best response to the fire situation. Designs for the detection of heat can include several types of heat detectors, including the following:

1. Fixed-temperature heat detectors
2. Rate-compensated fixed-temperature heat detectors
3. Rate-of-rise heat detectors
4. Combination detectors
5. Line-type heat detectors

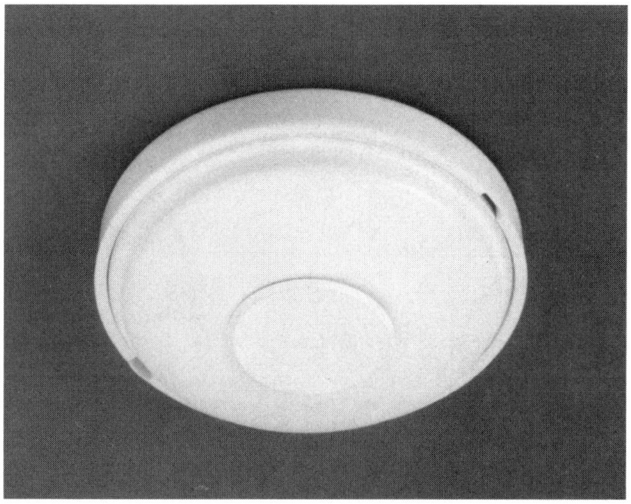

FIGURE 13-4 Typical Fixed-Temperature Only or Fixed and Rate-of-Rise (Nonrestorable) Heat Detector
Source: Chemetronics®, Ashland, MA.

Fixed-temperature heat detectors initiate an alarm when the detecting element reaches a predetermined fixed temperature (Figure 13-4). During a rapid temperature rise, the operation of a fixed-temperature heat detector can lag behind the actual air temperature because of the time needed for the heat to penetrate the device enclosure.

Rate-compensated fixed-temperature heat detectors are designed to anticipate the temperature lag and provide a response closer to that of the actual air temperature (Figure 13-5). Rate-of-rise heat detectors operate at a predetermined rate of temperature change.

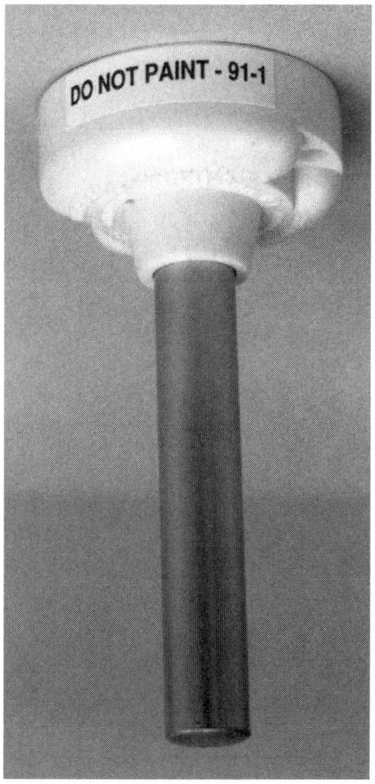

FIGURE 13-5 Typical Rate-Compensated Heat Detector
Source: *National Fire Alarm Code® Handbook,* NFPA,1999, Exhibit 1.31; Thermotech; photo courtesy of Mammoth Fire Alarms, Inc., Lowell, MA.

FIGURE 13-6 Combination Rate-of-Rise and Fixed-Temperature Heat Detectors
Source: *National Fire Alarm Code® Handbook,* NFPA, 1999, Exhibit 1.6; courtesy of Mammoth Fire Alarms, Inc., Lowell, MA.

FIGURE 13-7 Typical Combination Smoke-and-Heat Detector
Source: *National Fire Alarm Code® Handbook,* NFPA, 1999, Exhibit 1.7; System Sensor: photo courtesy of Mammoth Fire Alarms, Inc., Lowell, MA.

Combination detectors can contain more than one element to respond to a fire. Examples of such detectors include a combination rate-of-rise and fixed-temperature heat detector and a combined smoke-and-heat detector. Figures 13-6 and 13-7 illustrate typical combination detectors. Line-type heat detectors are typically made of heat-sensitive cable.

Location of Heat Detectors. In general, spot-type heat detectors must be located on the ceiling at least 4 in. (100 mm) from a side wall, or they must be located on a side wall at least 4 in. (100 mm) but not more the 12 in. (300 mm) from the ceiling. In applications where the ceiling falls under the definition of "solid joist construction," the detectors must be mounted at the bottom of the joist. In applications where the ceiling falls under the definition of "beam construction," where the beams are less than 12 in. (300 mm) in depth and are spaced less than 8 ft (2.4 m) on center, the detector is permitted to be installed on the bottom of the beam.

The temperature classification for fixed-temperature heat detectors and rate-compensated fixed-temperature heat detectors must be selected with consideration for the maximum expected ambient ceiling temperature. The temperature rating of the detector must be at least 20°F (11°C) above the maximum expected temperature at the ceiling. Table 13-1 provides temperature-classification and color-coding requirements based on maximum ceiling temperature.

The performance of heat detectors includes a "listed spacing," which is based on the spacing that the listing agency has used in the testing of the heat detector to verify its operation and temperature rating. When heat detectors are installed, the installed spacing must take in to account this listed spacing. Modifications to the spacing must reflect adjustments to account for ceiling height and construction. Where the ceiling is smooth and flat (nonsloping), heat detectors must be installed such that all points on the ceiling are within a distance of 0.7 times the listed spacing of a heat detector (Figure 13-8).

TABLE 13-1 Temperature Classification and Color Coding Requirements Based on Maximum Ceiling Temperature

Temperature Classification	Temperature Rating Range °F	Temperature Rating Range °C	Maximum Ceiling Temperature °F	Maximum Ceiling Temperature °C	Color Code
Low*	100–134	39–57	20 below	11 below	Uncolored
Ordinary	135–174	58–79	100	38	Uncolored
Intermediate	175–249	80–121	150	66	White
High	250–324	122–162	225	107	Blue
Extra high	325–399	163–204	300	149	Red
Very extra high	400–499	205–259	375	191	Green
Ultra high	500–575	260–302	475	246	Orange

*Intended only for installation in controlled ambient areas. Units shall be marked to indicate maximum ambient installation temperature.

Source: *National Fire Alarm Code*®, NFPA,1999, Table 2-2.1.1.1.

Where the ceiling surface involves solid joists or beams, special rules are used to modify the rules for smooth ceiling spacing (as described above under "Smoke Detectors"). In applications where the ceiling falls under the definition of "solid joist construction," the spacing for smooth ceilings is used except that in the direction perpendicular to the run of the joists, the maximum spacing for detectors must be reduced to one-half of the smooth ceiling spacing.

In applications where the ceiling falls under the definition of "beam construction" [unless the beam is more than 18 in. (460 mm) deep and the beam spacing is greater than 8 ft (2.4 m)], the spacing for smooth ceilings is used except that in the direction perpendicular to the run of the beams the maximum spacing for detectors must be reduced to two-thirds of the smooth ceiling spacing. Where the beam is more than 18 in. (460 mm) deep and the beams spacing is greater than 8 ft (2.4 m), each bay formed by the beam must be treated as a separate area. If the ceiling is sloped, additional rules are specified in the code to modify the rules for flat ceilings.

In addition, if the ceiling is more than 10 ft (3.05 m) high, heat detector spacing must be reduced by the factors that range from 0.91 times the listed spacing for

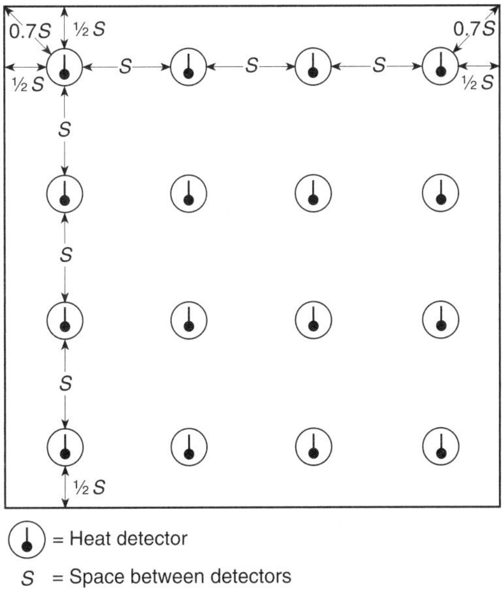

FIGURE 13-8 Arrangement for Spot-Type Heat Detectors for Square or Rectangular Spaces

Source: *National Fire Alarm Code*®, NFPA,1999, Figure A-2-2.4.1(a).

ceiling heights up to 12 ft (3.66 m) to 0.34 times the listed spacing for ceiling heights up to 30 ft (9.14 m). It should be noted that the reduction in listed spacing required due to ceiling height, solid joists, beams, or slopes must all be factored into the final spacing. As an example, if the listed spacing is 30 ft (9.14 m) and the ceiling height is 30 ft (9.14 m) and the ceiling construction is solid joist construction, the spacing at right angles to the joists would be adjusted by two factors: 0.34 for the ceiling height and 0.5 for the joists. The resulting spacing at right angles to the joist would be $30 \times 0.34 \times 0.5$ ft or 5.1 ft (1.55 m).

Radiant-Energy Detectors

Requirements for the selection, location, and spacing for fire detectors that sense the radiant energy produced by burning substances are contained in Chapter 2 of *NFPA 72*. Flame detectors and spark/ember detectors fall into this category. The type and quantity of radiant-energy-sensing fire detectors must be determined and the performance characteristic of the detector and an analysis of the hazard must all be considered before the system is installed.

Alarms from Suppression System Operation

Where an automatic sprinkler system is used, the fire alarm system is often required by other codes and standards to transmit a waterflow alarm to an approved supervising station or public fire service communications center. When the automatic sprinkler system operates, the waterflow-actuated fire alarm initiating device will initiate a fire alarm signal. The number of waterflow switches permitted on an initiating device circuit is limited to five.

The operation of automatic fire suppressions systems other than waterflow is also required to initiate an alarm signal at the protected premises. This would include wet chemical systems over commercial cooking equipment or clean-agent-extinguishing systems.

The fire alarm system is required to monitor the normal standby condition of these extinguishing or suppression systems by means of listed supervisory initiating devices. If someone closes a sprinkler system control valve or otherwise impairs the fire protection system, the supervisory initiating device will cause the fire alarm system control unit to indicate a "supervisory off-normal condition." When the off-normal condition is cleared, the supervisory initiating device will cause the fire alarm system control unit to indicate a "supervisory restoration to normal" signal. The number of supervisory devices permitted on an initiating device circuit is limited to 20.

Special consideration must be given to the design and installation of automatic fire suppression system alarm initiating devices and supervisory signal initiating devices and their circuits so they are not subject to tampering, opening, or removal without initiating a trouble signal.

Concealed Detectors

NFPA 72 requires the use of remote alarm indicators for in-duct smoke detectors in specified conditions. The location and labeling information required for these remote alarm indicators is contained in Chapters 2 and 3 of *NFPA 72*.

Coverage and Cross Zoning

Coverage. When local codes require a fire alarm system to have total (complete) coverage, partial coverage, or selective coverage for smoke or heat detection, the requirements of Chapter 2 of *NFPA 72* apply. The locations that require detection are addressed for each type of coverage.

Cross Zoning. In some situations, only where permitted by the AHJ cross zoning of initiation devices is effective in minimizing false activations of special hazard extinguishing systems. The rules in *NFPA 72,* Chapter 3, address the use of two automatic detectors to initiate an alarm response or the actuation of an extinguishing system.

NOTIFICATION APPLIANCES

General

When local codes require a fire alarm system to have occupant notification, notification appliances must be installed in accordance with the requirements of *NFPA 72.* Local codes will usually specify a framework of requirements based on the type of occupancy. Typically, occupant notification will involve the installation of audible and visible notification appliances (horns and strobes) throughout the protected premises to provide for total evacuation of the premises. Exceptions and special conditions may also be specified as a part of these framework rules, depending on the occupancy, for example, permitting partial or selective evacuation or directed relocation of building occupants. Emergency voice/alarm communications notification equipment, including the use of fire alarm speaker appliances, provides a means to achieve these functions.

In addition to local codes, *NFPA 72* requires fire alarm systems that are provided for the purpose of evacuation or relocation of occupants to have one or more notification appliances listed for the purpose on each floor of the building and located so that they have the characteristics described in Chapter 4 of *NFPA 72,* for public mode or private mode operation, as required.

Different rules apply, depending on whether public or private mode operation is used. Public mode operation applies when signaling to occupants for total, partial, or selective evacuation or directed relocation. When permitted by local codes, private mode operation applies for circumstances such as occupancies where occupants are not capable of evacuating themselves. Private mode operation involves signaling only to personnel concerned with implementation of emergency action initiation, such as attendants or other personnel required to evacuate or relocate occupants.

Audible Notification Appliances

The requirements for audible notification appliances are found in *NFPA 72,* Chapter 4. The basic rule for audible public mode signals is that they must have a sound level that meets the greater of the two following conditions:

- A sound level at least 15 dBA above the average ambient sound level
- A sound level at least 5 dBA above the maximum sound level having a duration of at least 60 seconds.

The term *average ambient sound level* involves an A-weighted measurement over a 24-hour period. A-weighted measurements are made using instruments that account for the normal attenuations that occur in human hearing. The notification appliance sound levels specified above, measured 5 ft (1.5 m) above the floor, are required to be provided throughout the occupiable area. When audible notification appliances are used to signal sleeping areas, the sound levels must also be measured at the pillow level and a third condition—a sound level of at least 70 dBA—must also be satisfied. For sleeping areas all three conditions must be satisfied.

The performance levels specified here depend on ambient sound levels. The code also requires a minimum appliance sound level (appliance rating) of 75 dBA, measured at 10 ft (3 m) from the appliance, irrespective of the ambient level. The code limits the appliance sound level, measured at minimum hearing distance, to not more than 120 dBA. Whenever the average ambient sound level is greater than 105 dBA, the code also requires the use of visible notification appliances.

The requirements for the performance of audible private mode signals are found in Chapter 4 of *NFPA 72*. They must have a sound level that meets the greater of the two following conditions: a sound level at least 10 dBA above the average ambient sound level or a sound level at least 5 dBA above the maximum sound level having a duration of at least 60 seconds. These levels are measured 5 ft (1.5 m) above the floor in the occupiable area. The minimum rating for private mode audible appliances is 45 dBA at 10 ft (3 m), and the maximum sound level is again 120 dBA at minimum hearing distance.

The code permits the use of listed, wall-mounted audible notification appliances or listed, ceiling-mounted audible notification appliances. Wall-mounted audible notification appliances that are not part of a combination audible/visible appliance must be mounted such that their top is at least 90 in. (2.3 m) above the finished floor and not closer than 6 in. (152 mm) to the finished ceiling. Different mounting heights are permitted by exception if the sound pressure level requirements are met for the operating mode used. Where combination audible/visible appliances are installed, the mounting requirements for visible appliances must be followed.

The rules in *NFPA 72* do not specify the specific location for audible notification appliances. However, audible notification appliances must provide the required sound levels throughout the occupiable area. After considering the ambient sound levels, the designer of the system must decide in advance where to place the appliance based on the area to be covered and the rating of the appliance. Attenuation through doors, walls, partitions, and furnishings must be taken into account when determining appliance locations. At the completion of the installation, the system must provide the sound levels required by the *NFPA 72*.

Where notification appliances (speakers) are installed for emergency voice/alarm communications, they must be capable of reproducing voice announcements with voice intelligibility. Tone signals from speakers must comply with the sound level requirements noted above for audible appliances.

Visible Notification Appliances

Signaling by means of visible notification appliances is by the illumination of the area surrounding the appliance or by direct viewing of the appliance. The rules specified in *NFPA 72* for the spacing of visible notification appliances in rooms are based on the appliances providing a minimum illumination of 0.0375 lm/ft^2 (0.4037 lm/m^2). The rules for spacing visible notifications appliances in corridors is based on the direct viewing of the appliances.

The code permits the use of listed wall-mounted visible notification appliances or listed ceiling-mounted visible notification appliances. Table 13-2 provides the spacing requirements for wall-mounted visible notification appliances in rooms and Table 13-3 provides the spacing requirements for ceiling-mounted visible notification appliances in rooms. In locating visible appliances in rooms it is important that the appliance be located to provide complete coverage for the room size and strobe intensity. For wall-mounted appliances this would normally be at the center of the longest wall.

In general, whenever more than two visible notification appliances are located in any field of view in a large room, they must be spaced a minimum of at least

TABLE 13-2 Room Spacing for Wall-Mounted Visible Notification Appliances

Maximum Room Size		Minimum Required Light Output (Effective Intensity) (cd)		
ft	m	One Light per Room	Two Lights per Room (Located on Opposite Wall)	Four Lights per Room; One Light per Wall
20 × 20	6.1 × 6.1	15	NA	NA
30 × 30	9.14 × 9.14	30	15	NA
40 × 40	12.2 × 12.2	60	30	15
50 × 50	15.2 × 15.2	95	60	30
60 × 60	18.3 × 18.3	135	95	30
70 × 70	21.3 × 21.3	185	95	60
80 × 80	24.4 × 24.4	240	135	60
90 × 90	27.4 × 27.4	305	185	95
100 × 100	30.5 × 30.5	375	240	95
110 × 110	33.5 × 33.5	455	240	135
120 × 120	36.6 × 36.6	540	305	135
130 × 130	39.6 × 39.6	635	375	185

NA: Not allowable.

Source: *National Fire Alarm Code*®, NFPA, 1999, Table 4-4.4.1.1(a).

55 ft (17.7 m) apart or they must be synchronized. The mounting height for wall-mounted visible notification appliances must be such that the entire lens is at least 80 in. (2.03 m) but not greater than 96 in. (2.43 m) above the finished floor. The height permitted for ceiling-mounted visible notification appliances is limited to 30 ft (9.14 m) in accordance with Table 13-3.

TABLE 13-3 Room Spacing for Ceiling-Mounted Visible Notification Appliances

Maximum Room Size		Maximum Ceiling Height		Minimum Required Light Output (Effective Intensity); One Light (cd)
ft	m	ft	m	
20 × 20	6.1 × 6.1	10	3.05	15
30 × 30	9.14 × 9.14	10	3.05	30
40 × 40	12.2 × 12.2	10	3.05	60
50 × 50	15.2 × 15.2	10	3.05	95
20 × 20	6.1 × 6.1	20	6.1	30
30 × 30	9.14 × 9.14	20	6.1	45
40 × 40	12.2 × 12.2	20	6.1	80
50 × 50	15.2 × 15.2	20	6.1	115
20 × 20	6.1 × 6.1	30	9.14	55
30 × 30	9.14 × 9.14	30	9.14	75
40 × 40	12.2 × 12.2	30	9.14	115
50 × 50	15.2 × 15.2	30	9.14	150

Source: *National Fire Alarm Code*®, NFPA, 1999, Table 4-4.4.1.1(b).

Special rules apply to visible notification appliances used for signaling in sleeping areas. In general visible notification appliances must be located within 16 ft (4.87 m) of the pillow. The intensity of the appliance must be at least 110 cd unless the appliance is mounted less than 24 in. (610 mm) from the ceiling, in which case it must be at least 177 cd.

OTHER FIRE ALARM SYSTEM FUNCTIONS

Building Fire Safety Functions

Fire safety functions are those that are intended to increase the level of life safety for occupants or to control the spread of the harmful effects of the fire. As listed above, these functions include the following: elevator recall; elevator shutdown; control of smoke dampers, fire dampers, fans, smoke doors, and fire doors; door release; and door unlocking. The basic requirements to include building fire safety functions originate primarily from local codes.

Suppression System Actuation (Interface)

The requirements for suppression system operation are contained in other codes and standards. However, if automatic or manual activation of a fire suppression system is to be performed through a fire alarm control unit, the control unit must be listed for the releasing service. The inspector should refer to the definitions of "fire alarm system" and "fire alarm control unit."

Each space protected by an automatic fire suppression system actuated by the fire alarm system must contain one or more automatic fire detectors. These fire detectors must be installed to sense the fire condition and subsequently initiate the sequence to allow the suppression system to discharge its agent.

Suppression systems or groups of systems must be controlled by a single control unit that monitors the associated initiating device(s), actuates the associated releasing device(s), and controls the associated agent release notification appliances. If the releasing panel is located in a protected premises having a separate fire alarm system, it must be monitored for alarm, supervisory, and trouble signals by—but not be dependent on or affected by—the operation or failure of the protected premises' fire alarm system.

Offsite Notification

When offsite notification is required or elected there are four different systems that are available for this purpose: central station fire alarm systems, proprietary supervising station fire alarm systems, remote supervising station fire alarm systems, and auxiliary fire alarm systems. These were described briefly under "Fire Alarm System Overview." Chapter 5 of *NFPA 72* provides the requirements for the three supervising stations and Chapter 6 of *NFPA 72* provides the requirements for auxiliary fire alarm systems.

Chapter 5 in *NFPA 72* also contains specific requirements for six different communications methods. The most common communications method used is the digital alarm communicator system. This system involves the use of a digital alarm communicator transmitter (DACT) located at the protected premises and a digital alarm communicator receiver (DACR) usually located at the supervising station.

When a digital alarm communicator system is used, redundancy must be provided in the transmission channels. Eight different combinations of transmission channels are permitted. These involve the use of a telephone line (number) in combination with an additional communications channel, such as an additional telephone line or other means.

A DACT must be connected to the public switched telephone network upstream of any private telephone system at the protected premises. These connections must be under the control of the subscriber. The connection must be made to a loop start, not a ground start, telephone circuit. This telephone line is not required to be a dedicated line.

In accordance with the testing methods specified in *NFPA 72*, the response time from the point of actuation of the initiating device at the protected premises to the receipt of the signal at the supervising station must not exceed 90 seconds. However, the rules for DACT operation permit a sequence of attempts for successful transmission that can greatly exceed 90 seconds. Despite these allowances for DACT transmission, the statistical probability of a successful DACT transmission occurring well within 90 seconds (on the first attempt) is very high.

FIRE ALARM SYSTEMS IN RESIDENTIAL OCCUPANCIES

Smoke Alarms versus Smoke Detectors

Chapter 8 of *NFPA 72* addresses fire-warning equipment for dwelling units, including single- and multiple-station smoke alarms and household fire alarm systems. A smoke alarm is distinguished from a smoke detector by the fact that the smoke alarm is a single unit that provides detection and a warning signal all in one unit. A smoke detector involves detection only and requires the use of a fire alarm control unit and separate notification appliances as a part of a fire alarm system to complete the warning function.

Requirements

Local codes specify requirements for smoke alarms in one- and two-family dwellings and other residential type of occupancies. Often these codes will also recognize the use of a fire alarm system to fulfill the functional requirements specified for the smoke alarms.

For some residential types of occupancies such as multifamily dwellings, local codes may specify requirements for a building fire alarm system with additional provisions for smoke alarms (or smoke detectors of a household fire alarm system) in specific locations in the dwelling unit. These additional provisions may specify that the smoke alarms (or smoke detectors of a household fire alarm system) sound only within the individual dwelling or living unit and not actuate the building fire alarm system unless otherwise permitted by the AHJ. Note that requirements for the building fire alarm system occupant notification also include signaling to the dwelling or living units. Sound-level requirements for notification appliances must also be met in these spaces, especially in sleeping areas. It is important to consult local codes for clarification about the specific requirements for the occupancy.

Household fire alarm systems are required to have two independent power sources, consisting of a primary source that uses a normal building service and a secondary source that consists of a rechargeable battery or standby generator that can operate the system for at least 24 hours in the normal condition followed by 4 minutes of alarm.

Smoke alarms are required to be powered from a normal building service along with a secondary battery source that is capable of operating the device for a least 7 days in the normal condition followed by 4 minutes of alarm. Alternately, smoke alarms can be powered by a nonreplaceable primary battery that is capable of operating the device for at least 10 years followed by 4 minutes of alarm, followed by 7 days of trouble.

Under certain conditions local codes may allow the use of smoke alarms powered by only a replaceable 1-year battery. *NFPA 72* allows the use of these types of smoke alarm only when specifically permitted by a statute or a Code Provision as the *Life Safety Code.*

Installation

Smoke alarms (or system smoke detectors) must be installed in locations that comply with the locations specified in local codes. The location of these devices must be selected to consider conditions such as high or low temperature, humidity, or sources of smoke that can lead to nuisance alarms. Smoke alarms or smoke detectors must not be closer than 3 ft (1 m) from the door to bathroom or kitchen. Smoke alarms or smoke detectors that are located within 20 ft (6.1 m) of a cooking appliance and are equipped with an alarm silencing means or are of the photoelectric type are considered acceptable. Additional guidance concerning clearance from dead air spaces and other concerns was previously discussed.

Chapter 8 of *NFPA 72* requires the replacement of smoke alarms located in one- and two-family dwellings and apartment buildings within 10 years of the date of installation. Note that this rule does not apply to smoke detectors.

Inspection, Testing, and Maintenance

NFPA 72 requires the inspection, testing, and maintenance of smoke alarms in one- and two-family dwellings to be in accordance with the manufacturer's instructions. This includes cleaning, testing, and battery replacement. Some smoke alarm manufacturers require weekly testing; others may require only monthly testing. The inspector should consult the manufacturer's instructions for specific requirements. Testing and maintenance for smoke alarms in other than one- and two-family dwellings and for household fire alarm systems must comply with the requirements of Chapter 7 of *NFPA 72,* as well as the manufacture's requirements.

SYSTEM WIRING

The installation of fire alarm equipment, including wiring, must comply with NFPA *70, National Electrical Code®(NEC®).* The *NEC* specifies requirements for non-power-limited fire alarm (NPLFA) and power-limited fire alarm (PLFA) circuits. PLFA circuits are distinguished from NPLFA circuits by the equipment listing. The equipment marking or listing must indicate the power-limited feature or source.

The wiring methods permitted on the supply side of NPLFA circuits must comply with the requirements of *NEC* 2002 Chapters 1 through 4. The wiring methods for NPLFA circuits that are outlined include the use of the wiring methods of *NEC* 2002 Chapter 3. The NEC also permits the use of multiconductor NPLFA cables.

The wiring methods permitted on the supply side of PLFA power sources must also comply with *NEC* 2002 Chapters 1 through 4. The *NEC* permits the use of either NPLFA wiring method and materials or the use of PLFA methods and materials.

The *NEC* places limitations on the mixing of cables and conductors used for fire alarm systems. Specific rules are provided for both NPLFA circuits and for PLFA circuits. These rules preclude the mixing of NPLFA circuits with PLFA circuits. They also restrict other combinations, especially with non–fire alarm applications.

There are also rules in *NFPA 72* that have an impact on fire alarm system wiring. For example, *NFPA 72* requires fire alarm circuits to be designated by class, style, or both. This designation is based on performance needs of the system. The class designation of a circuit is based on a circuit's ability to perform under the conditions of a single open or nonsimultaneous single ground fault on a circuit conductor. Class A circuits require a redundant circuit path to each device or appliance. The type

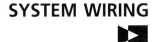

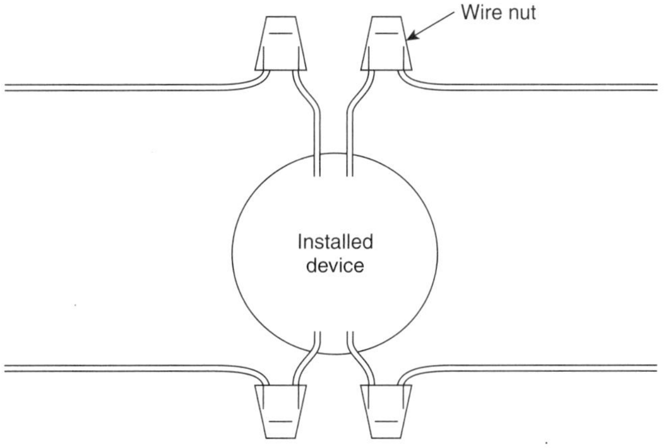

FIGURE 13-9 Correct Wiring Method for Pigtail Connections

circuit designation used may have an impact on how the wiring can be run. As an example, Class A circuits must be run so that the outgoing and return conductors of the circuit are routed separately.

In addition, *NFPA 72* requires circuits to be monitored for integrity. The interconnecting conductors of the circuit must be monitored so that the occurrence of a single open or single ground-fault condition is indicated. As a result, connections for each device or appliance must be made so that the opening of any connection causes a trouble signal. Looping the conductor around the device or appliance terminal and then continuing to the next device or appliance can result in an unsatisfactory connection. Listed devices and appliances that are provided with duplicate terminals for each circuit connection must be terminated by cutting the wire and making each connection separately. Figure 13-9 demonstrates the correct wiring method for pigtail connections; Figure 13-10 shows the incorrect method.

A wiring technique called "T-tapping," the parallel connection of one or more devices or appliances at the middle of a circuit, is also forbidden for all initiating device and notification appliance circuits. It is permitted for signaling line circuits if permitted by the designer and equipment listing. As a point of understanding, initiating device circuits are those that have no means of identifying the initiating device that operated it. Signaling line circuits are able to carry multiple signals and

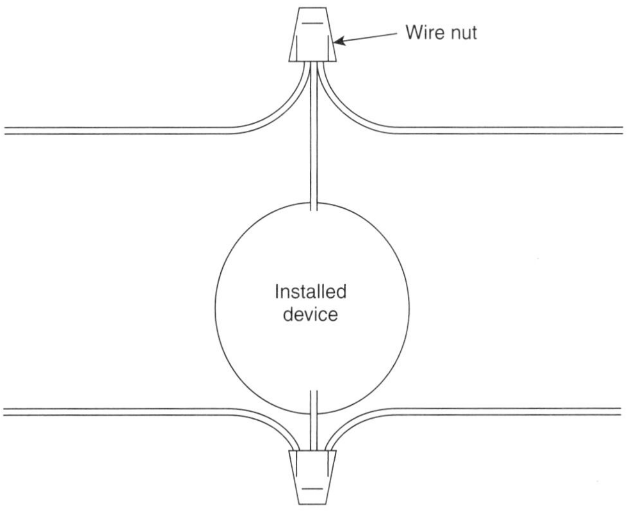

FIGURE 13-10 Incorrect Wiring Method for Pigtail Connections

can identify the device or other equipment sending the signal. Addressable systems use signaling line circuits.

Additional *NFPA 72* requirements that can have an impact on wiring are the requirements for system survivability. These requirements involve the design, protection, and layout of circuits to support continued operation during a fire. These circuits are associated with emergency voice/communications systems and fire command centers.

SUMMARY

Fire alarm systems provide early detection, signal occupants to evacuate or relocate, initiate fire safety functions, and notify the fire department to respond. These systems are composed of (1) initiating devices, such as manual fire alarm boxes, smoke detectors, smoke alarms, heat detectors, waterflow switches, and tamper switches; (2) notification appliances, for example, horns, speakers, and strobes, that provide audible, visible, or tactile outputs; and (3) control units, which process input and output signals and may control ancillary equipment, such as automatic door closers.

Some fire alarm systems alert off-site emergency forces by transmitting signals to a supervising station or to a public fire service communications center. Still other alarm systems actuate automatic fire suppression systems. The inspector should become familiar with the requirements of *NFPA 72,* which covers the performance, installation, inspection, testing, and maintenance of all fire alarm systems, both those that are required and those that are not required.

BIBLIOGRAPHY

Bunker, M. W., and Moore, W. D., eds., *National Fire Alarm Code® Handbook,* NFPA, Quincy, MA, 1999.

Cote, A. E., ed., *Fire Protection Handbook,* 18th ed., NFPA, Quincy, MA, 1997.

Bukowski, R. W., and O'Laughlin, R. J., *Fire Alarm Signaling Systems Handbook,* 2nd ed., NFPA, Quincy, MA, 1994.

Carson, W. G., and Klinker, R. L., *Fire Protection Systems: Inspection, Test, and Maintenance Manual,* 3rd ed., NFPA, Quincy, MA, 2000.

NFPA Codes, Standards, and Recommended Practices

See the latest version of The NFPA Catalog for availability of current editions of the following documents.

NFPA 70, *National Electrical Code®*
NFPA 72®, National Fire Alarm Code®
NFPA *101®*, *Life Safety Code®*

Water Supplies

CHAPTER 14

Phillip A. Brown

A general understanding of a facility's piping layout will help the inspector evaluate the level of protection provided by the individual sprinkler, standpipe, water spray, or foam-water systems. The inspector should, therefore, inspect the facility's water supply arrangements before going on to inspect its water-based fire-extinguishing systems. After providing information on water supply sources—including both public and private water supply systems—the chapter examines valves, the metering of water supply sources, private fire service mains, yard hydrants, and monitor nozzles. It then looks at the types and inspection requirements for fire pumps and water storage tanks and concludes with information to help the inspector conduct the relevant flow tests.

USING NFPA 25

In 1992 the National Fire Protection Association (NFPA) published the first edition of NFPA 25, *Standard for the Inspection, Testing, and Maintenance of Water-Based Fire Protection Systems,* which established, for the first time, a set of definitive minimum requirements relating to the inspection of water-based fire protection systems, including their water supply components. Because it is a standard, NFPA 25 can be made a mandatory reference by another standard or code or can be directly adopted into law.

Inspectors should become familiar with the requirements of NFPA 25. They should also be aware that standardized forms are now available for use in inspecting automatic sprinkler systems, standpipe systems, private fire service mains, fire pumps, water storage tanks, water spray fixed systems, and foam-water sprinkler systems in accordance with the requirements of NFPA 25.

Where NFPA 25 is enforced, the inspector should check to make sure that the property owner is keeping the records required by the standard to demonstrate compliance with all inspection, testing, and maintenance requirements. It is understood, however, that the inspector may not be the party who actually performs the inspection, testing, and maintenance procedures in accordance with NFPA 25.

Phillip Brown is a certified fire protection specialist with the American Fire Sprinkler Association and a NICET Level IV designer. He has a degree in fire science and has been an active participant in automatic fire sprinkler design and installation since 1963.

WATER SUPPLY SOURCES AND ARRANGEMENTS

The first step in inspecting a facility's water supply is to identify the supply sources. Inspectors can begin this step during the initial exterior inspection as they note the locations of hydrants and exterior control valves and determine whether elevated water storage tanks, groundwater storage reservoirs, or other sources of water exist. The inspector should also note the location of exterior fire department connections because these devices are considered supplemental sources of water supply. Visible external attachments of sprinkler systems, such as water motor alarm gongs and drains, can help the inspector locate the position of system control equipment within a building complex.

When identifying the water supply sources, the inspector may find building site plans and previous inspection reports extremely helpful because they may point out unlikely sources of supply, such as swimming pools or decorative ponds. Property owners or their representatives can also be of assistance in this regard. For large facilities or industrial complexes, maps and other information regarding the water supplies should be available at the facility, the local fire department, or from an insurance rating bureau.

In a city or developed suburban area, the water supply for fire protection commonly comes from a public water system. In other areas, it may come from private water systems, which can consist of a combination of tanks, reservoirs, and pumps. In many locations, the fire protection supplies are a combination of public and private supplies. It is common to augment the pressure of a public water supply with a private fire pump that supplies the water pressures needed to effectively operate the water-based fire-extinguishing system.

Sources of water supply may include surface lakes, rivers, and impounded supplies, but these are considered acceptable for fire protection use only if they are available on a year-round basis. For this reason, the inspector should make a special effort to inspect these sources during periods of freezing weather and drought.

VALVES AND METERING OF WATER SUPPLY SOURCES

Water supplied to private fire systems from a public water supply is sometimes metered. To allow accurate metering of small flows while avoiding high pressure losses for large flows, special check valve assemblies are equipped with a bypass that contains a meter. The normal water flow is through the small metered bypass. In the event of a fire and a large flow demand, the large flow will lift a weighted clapper in the main waterway, allowing water to flow unobstructed through the system. Obviously, the fire flow is not metered in such a case.

Backflow Protection for Public Water Supplies

Whatever the arrangements, the water used for fire protection may have to be segregated by check valves from the potable water that is delivered through the public water supply system. This is necessary to prevent backflow, a reverse flow of water that could take place if the pressure of the public water supply system dropped below that of the private fire protection systems it serves.

A number of studies have shown that potable water from a public supply stored in the piping of a fire-extinguishing system over a long period can become aesthetically objectionable. Although the water does not pose a health risk, its backflow into a public supply is undesirable. For this reason, private fire protection systems have traditionally been separated from public water supplies by means of check valves or alarm check valves, which permit only one-way flow into the fire protection systems. Control valves are generally located at the check valves to isolate the systems and to permit repair of the check valves.

Special backflow prevention devices are generally required where fire protection systems are served by additional nonpotable water supply sources or where additives, such as antifreeze or corrosion inhibitors, are used. These devices are usually either double check valve assembly backflow preventers or reduced-pressure principle backflow prevention assembly (RPBA) devices. The more elaborate RPBA-type devices are generally required only where contamination of the public water supply is a possibility. This would be the case where the additional water supply source was an open reservoir or where certain antifreeze or other additives were mixed into the water supply.

The fire department connection to a sprinkler or standpipe system does not warrant the use of a backflow prevention device unless the fire department's prefire planning includes drafting provisions for supplying the connection from an open pond or other potential source of contamination.

All backflow preventers must be tested annually. A forward flow test must be performed flowing the system demand, including hose stream demand, if hydrants or inside hose stations are installed downstream of the backflow preventer.

The appendix of NFPA 24, *Standard for the Installation of Private Fire Service Mains and Their Appurtenances,* addresses the conditions under which backflow prevention equipment on fire protection systems may be recommended. Because backflow prevention devices have inherent pressure losses, they should not be installed on existing fire protection water supply mains unless the available water supply pressures and system demands have been analyzed by a qualified fire protection contractor or fire protection engineer. Otherwise, the system's ability to perform its intended fire suppression function could be impaired.

Inspecting Valves

The single most important feature the inspector can check is the position of the control valves. Water supply system control valves must be in the "open" position. Post indicator valves used in yards indicate their position by means of an "open" or "shut" sign visible through a window in the valve face. Outside screw and yoke (OS & Y) valves indicate their position by means of the valve screw or stem. Because a visible screw or stem means the valve is open, the valve is sometimes referred to as an "outside screw and yoke" valve. Figure 14-1 illustrates an OS & Y valve in the closed position as indicated by the screw stem, which does not extend past the handwheel. Other types of valves use indicating markers. Since the key-operated gate valves frequently used by public water utilities are not indicating valves, water must often be flowed through the fire protection system to verify whether these valves are open.

Each control valve must be identified and have a sign indicating the system or portion of the system it controls. NFPA 25 requires that each valve be secured in its normal open or closed position by means of a seal or a lock or that it be electrically supervised in accordance with the applicable NFPA standards. For valves controlling flow to wall and roof outlets, an exception waives this requirement.

NFPA 25 requires that all types of valves be inspected weekly, including hose valves, pressure-regulating valves, and the valves that isolate backflow prevention devices. An exception permits monthly inspections for valves secured with locks or valves electrically supervised in accordance with applicable NFPA standards.

When inspectors inspect valves, they must verify that the valves are in the normal open or closed position and that they are properly sealed, locked, or supervised. The

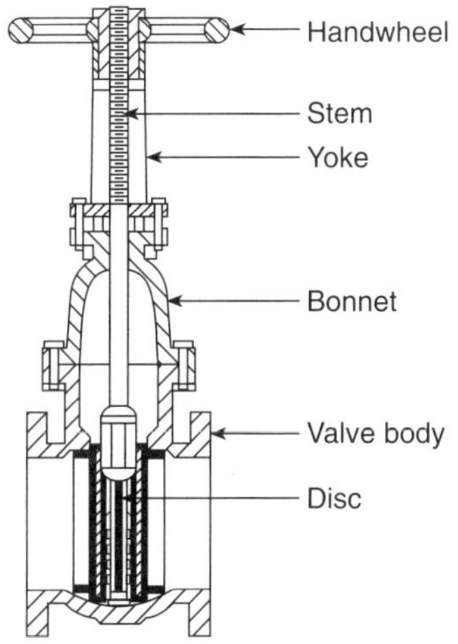

FIGURE 14-1 Typical Outside Screw and Yoke (OS & Y) Valve in Closed Position
Source: NFPA 25, 2002, Figure A.12.1(h).

valves must also be accessible, free from external leaks, and provided with appropriate identification and the appropriate wrenches.

Control valves must be tested annually, with the valve opened until spring or torsion is felt in the rod, indicating that the rod has not become detached from the valve. Each control valve must annually be operated through its full range and then returned to its normal position.

Inspecting Fire Department Connections

Fire department connections must be inspected quarterly to verify that they are visible and accessible, that couplings and swivels have not been damaged and rotate smoothly, and that plugs or caps are in place and undamaged. Gaskets should be in place and in good condition, and the automatic drain valve (when provided) should be in place and operating properly. Identification signs also should be in place. Finally, the check valve should be examined to make sure it is not leaking.

If the plugs or caps are not in place, the inspector must check the interior of the fire department connection for possible obstructions and verify that the valve clapper is operational over its full range. The inspector must remove the plugs or caps and check the interior of the fire department connection for debris. Caution is required when this task is performed. Fire department connections provide a convenient hiding place for such items as used drug needles.

PRIVATE FIRE SERVICE MAINS, YARD HYDRANTS, AND MONITOR NOZZLES

Private Fire Service Mains

A private fire service main is piping on private property located between a source of water and a sprinkler system, standpipe system, water spray system, foam-water system, hydrant, or monitor nozzle. When it is connected to a public water system, the private fire service main begins at a point designated by the public water utility, usually at a manually operated valve near the property line.

In some cases—for example, as with residential sprinkler systems—there may be no exclusive private fire service main. Instead, a combined service main might serve both domestic and fire protection demands. Special care must be taken in these situations to ensure that the water needed for fire protection purposes will be available even in times of peak domestic demand. NFPA 13R, *Standard for the Installation of Sprinkler Systems in Residential Occupancies Up to and Including Four Stories in Height,* provides a means of estimating simultaneous domestic demand through combined supply piping. A similar situation may also exist in industrial facilities, if a combined service main serves both fire protection and process demands.

In other cases, a private fire service main might serve a system of yard hydrants or monitor nozzles that protect a factory complex, tank farm, or other special hazard. This system could include hose or hydrant houses, located over or next to hydrants, to provide ready access to hose nozzles, hose wrenches, gaskets, and spanners.

Yard Hydrants

The authority having jurisdiction determines the needed fire flow and hydrant locations for a yard hydrant system. The standard also specifies that, for average conditions, hydrants must be placed at least 40 ft (12.2 m) from buildings. Hydrant outlets must have the NH (or American National Fire Hose Connection Screw Threads) standard external threads for the size outlet supplied, as specified in NFPA 1963, *Standard for Fire Hose Connections.* Where local fire departments do not use the NH threads, the authority having jurisdiction must designate the threads or connection to be used.

NFPA 25 calls for quarterly inspections of any hose or hydrant houses to check accessibility, repair physical damage, and replace missing equipment. Fire hose and required equipment in the hose houses should be maintained annually to ensure that they are in a usable condition.

Dry barrel and wall hydrants must be inspected annually; wet hydrants must be inspected at least annually. All hydrants should also be checked after each operation. In addition to ensuring that hydrants are accessible and wrenches available, the inspector should confirm that the outlets are tight and that the nozzle threads and hydrant-operating nut are not worn. Leaks in outlets or at the top of the hydrant should be repaired.

The inspector must check drainage from the barrel of dry barrel and wall hydrants. The presence of water or ice may indicate a faulty drain, a leaky hydrant valve, or a high groundwater table. The inspector could use the following checklist when inspecting hydrants:

1. Check that the hydrant is set up plumb with outlets at least 18 in. (457 mm) above the ground. The hydrant should be unobstructed, easily accessible, and clear of snow in the winter.
2. Open and close the hydrant to verify that it is working properly. Note the direction of turn and the number of turns needed to open it fully. Posting a sign on the hydrant showing this information is useful but not required. Caution is required when opening or shutting a hydrant. It should be opened and closed very slowly to prevent damage to the hydrant and to prevent water hammer.
3. Check that the hydrant drains properly. If the drain is working properly, a suction can be felt at the outlets immediately after the valve is closed. A small drain in the base of the barrel of frostproof types is closed when the main valve is open. It is, however, arranged in such a way as to permit water to drain out of the barrel when the main valve is shut. If the hydrant has been installed properly (with about a barrel of small stones under it), water will drain away. If the drain is working properly and the main valve is tight, any problems due to water freezing in the barrel will be avoided.
4. Check for leaks. The main valve should close tightly. When the main valve is open wide, with the hydrant outlets capped, there should be no flow from the drain valve. Look for leaks in mains near the hydrant; stethoscopelike listening devices can be used for this purpose.
5. Check the hydrants for freezing during cold weather by "sounding," or striking one of the open outlets with the hand. Water or ice in the barrel shortens the length of the "organ tube" and raises the pitch of the sound. With experience, the inspector can detect the presence of ice or water. Sometimes, lowering a weight on a stout cord into the barrel can help determine the presence of water or ice in the barrel. If water is found in the barrel, it must be pumped out and the defective drains or valves repaired. Using salt or antifreeze in the barrels is of limited value in preventing freezing, and the corrosive effect can impair the operation of the hydrant. If the hydrant is only slightly bound by ice, tapping the arm of a wrench on the nut can release the stem. Only moderate blows should be used to prevent breaking the valve rod.
6. Flush hydrants annually to remove debris. (Flow for approximately 60 seconds.)

Monitor Nozzles

Monitor nozzles must be inspected semiannually. Leakage and damage must be repaired, corroded parts must be cleaned or replaced, and the nozzle must be lubricated or otherwise protected. Monitor nozzles should be lubricated annually.

If the private fire service mains contain a mainline strainer, the inspector must check the strainer annually to make sure it is not plugged or fouled. If it is, it must be cleaned. The strainer must also be cleaned after each significant system flow. If corrosion is evident, the strainer must be repaired or replaced.

Because most private fire service main piping is located underground, it is not easy to inspect. However, the results of required flow testing at 5-year intervals can indicate whether the available water flow has deteriorated. Exposed portions of private fire service mains must be inspected annually; and leaks, damage, or weaknesses in restraint methods must be repaired. Corroded piping must be cleaned or replaced and coated.

The inspector should make sure that the property owner is keeping records on the items needing inspection and on the annual maintenance of hydrants, hose/hydrant houses, monitor nozzles, and mainline strainers. These records should include evidence of the required annual tests of hydrants, the semiannual tests of monitor nozzle range and operation, and the 5-year piping flow tests.

FIRE PUMPS

Types of Fire Pumps

Centrifugal Pump. The centrifugal-type pump is the pump of choice for providing water under pressure to fire protection systems. These are single- or multistage pumps, depending on the number and arrangement of the impellers. Horizontal and vertical centrifugal pumps are available in capacities up to 5000 gpm (18,925 L/min), with a net pressure ranging from 40 to 400 psi (2.8 to 27.6 bar). Pump capacity is rated on the discharge of one stage in gallons per minute (liters per minute); pressure rating is the sum of the pressures of the individual stages minus head loss.

Horizontal shaft centrifugal fire pumps should be operated under positive suction head, especially with automatic or remote-control starting. If the location requires suction lift, a vertical turbine-type pump should be used.

Vertical Turbine Pump. The vertical turbine pump is used in streams, ponds, and pits. Such a pump consists of a motor of right-angle gear drive, a column pipe and discharge fitting, a drive shaft, a bowl assembly housing the impellers, and a suction strainer. Its operation is similar to a multistage horizontal pump.

Fire Pump Assembly. A fire pump assembly includes the water supply suction and discharge piping and valving; the pump; the driver, which may be an electric motor, a diesel engine, or a steam turbine; the controller; and auxiliary equipment. The auxiliary equipment includes the shaft coupling; the automatic air release valve; pressure gauges; the circulation relief valve, which is not used for diesel drive, with heat exchanger; pump test devices; the pump relief valve and piping; alarm sensors and indicators; right-angle gear sets for engine-driven vertical shaft turbine pumps; and the pressure maintenance, or jockey pump, and accessories.

Inspecting Fire Pumps

A fire pump assembly is usually inspected weekly to verify that it is in proper condition (Table 14-1). This inspection is performed in conjunction with a weekly test of an electric, diesel, or steam-driven pump. The test for electric motors should last for at least 10 minutes; the test for diesel engines should last for a minimum of 30 minutes. These tests are conducted with an automatic start, but without flowing water.

Fire pumps are also tested annually by flowing water at "no load," "rated load," and "peak load" flow conditions. These three flow points relate to the three points

TABLE 14-1 Summary of Fire Pump Inspection, Testing, and Maintenance

Item	Activity	Frequency
Pump house, heating ventilating louvers	Inspection	Weekly
Fire pump system	Inspection	Weekly
Pump operation		
No-flow condition	Test	Weekly
Flow condition	Test	Annually
Hydraulic	Maintenance	Annually
Mechanical transmission	Maintenance	Annually
Electrical system	Maintenance	Varies
Controller, various components	Maintenance	Varies
Motor	Maintenance	Annually
Diesel engine system, various components	Maintenance	Varies

Source: NFPA 25, 2002, Table 8.1.

on a pump-discharge curve that any fire pump is required to meet (Figure 14-2) as follows:

- The first point of the pump-discharge curve is churn or shut-off, with the pump operating at rated speed with the discharge valve closed; the total head at shut-off must not exceed 140 percent of the rated head at 0 percent capacity.
- The second point is rating; the curve should pass through or above the rated capacity and head, providing at least 100 percent of the rated head at 100 percent capacity.
- The third point is overload; at 150 percent of rated capacity, the total head should not be less than 65 percent of rated total head.

When inspecting a fire pump, the inspector should first read the manufacturer's rating data and compare the data to the records of recent tests. Records should be maintained on suction pressure, discharge pressure, pump RPM, flow (based on Pitot

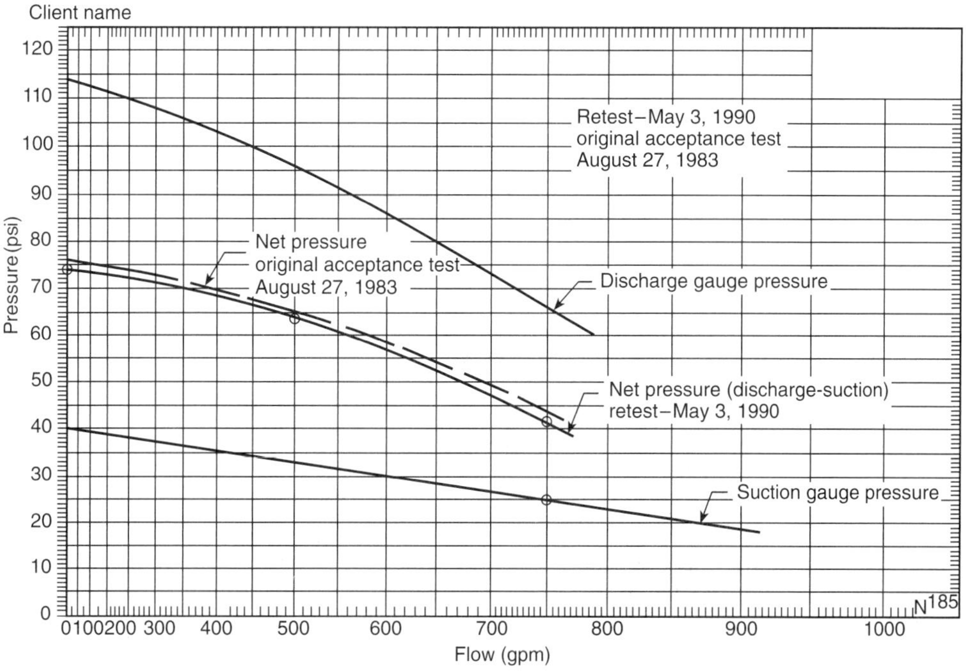

FIGURE 14-2 Fire Pump Retest
Source: NFPA 25, 2002, Figure A.8.3.5.3(I).

pressure readings from flowing nozzles or on a test meter), and current and voltage of the electric driving motor, using a number of different flow points, including the churn and overload conditions.

When inspecting a fire pump that is supposed to start automatically, inspectors should test it by opening a test connection. With the pump started, inspectors should watch for signs of leakage, overheating, and irregular performance. They should make sure all alarms and relief valves are operating satisfactorily. They should note whether the pump is aligned correctly with the driving motor or turbine and whether the stuffing box glands are leaking. Inspectors should watch pressure gauges for erratic performance, which could indicate poor suction, obstructions, inadequate water supply, or insufficient immersion of the suction pipe. Finally, they should close all outlets, including the relief valve, and note whether the pump shuts off at the correct pressure.

The purpose of the annual test is to compare the performance of the pump assembly to the performance recorded in earlier tests and at the time of initial field acceptance. Thus, to verify that all inspections and test results are being properly recorded and retained, interpreting the results correctly is important. Reduced pumping capacity and adjustments for changes in pump speed must be evaluated against the fire protection system demand.

The inspector must make sure that the fire pump room is kept clean, dry, orderly, and free of miscellaneous stored materials and that proper temperatures are maintained. Diesel-driven pumps must be ventilated adequately to supply air to the engine and to remove hazardous vapors.

The inspector should check the condition and reliability of any storage, of the lubrication systems, and of the oil and fuel supplies. If the pump is taking suction from a public water supply, the annual flow test should verify that the operation of the pump does not reduce the suction pressure at the pump below the minimum pressure permitted by local authorities, which is generally 10 to 20 psi (0.7 to 1.4 bar) In areas subject to earthquakes, the inspector should also make sure that the fire pump assembly and its associated piping and valving are properly braced and supported to withstand the possible horizontal and vertical forces.

WATER STORAGE TANKS

Types of Water Storage Tanks

NFPA 22, *Standard for Water Tanks for Private Fire Protection,* addresses three types of tanks that are used to store fire protection water: elevated tanks on towers or building structures (gravity tanks), water storage tanks that are at grade or below grade level, and pressure tanks.

Gravity Tanks

Gravity tanks are generally made of steel or wood. Steel tanks range in capacity from 5000 to 500,000 gal (18.93 to 1892.5 m^3); wood tanks range in capacity from 5000 to 100,000 gal (18.93 to 378.5 m^3). Depending on the requirements, these tanks may be located on the roof of a building or raised on an independent steel tower so that the tank's bottom capacity line is 75 to 150 ft (22.9 m to 30.5) above ground. Reinforced concrete towers may also be used. In some cases, concrete tanks have been constructed within the buildings or other structures they supply.

For a complete inspection of a gravity tank, the inspector must usually climb the tower and descend into the tank itself to check the condition of the interior. Appropriate and legally required precautions should be taken prior to entering confined spaces such as tanks or valve pits. Interior inspections are required every 5 years. Steel tanks without corrosion protection and pressure tanks require an interior inspection every 3 years. Inspectors who are inexperienced, however, should not attempt to descend into a tank or climb a tower until they have received instruction from an appropriately experienced person and have practiced these actions.

Ground Suction Tanks

Ground suction tanks are made of steel, wood, fiberglass, or concrete and are set on a foundation of concrete, crushed stone, or sand. A concrete ring wall usually surrounds the foundation.

Embankment-supported rubberized-fabric tanks are also used as ground suction tanks. They are available in 20,000- and 50,000-gal (75.5 and 189.3 m^3) sizes, and in 100,000-gal (378.5 m^3) increments up to 1 million gal (3785 m^3). These tanks have a reservoir liner with an integral flexible roof and are designed to be supported by earth on the bottom and on four sides.

For both elevated and ground suction tanks, a valve pit or house is usually built to contain valves, tank heaters, and other fittings. The inspector should check these houses or pits for appropriate waterproofing and drainage and for a water-level indicator or a high- and low-water electrical alarm. The gauge is normally installed in a heated room where it is readily accessible. There should be an overflow pipe at least 3 in. (76 mm) in diameter.

Pressure Tanks

Pressure tanks, whose capacities range from 2000 to 15,000 gal (7.5 to 56.6 m^3) are used to supply sprinkler and standpipe systems, hose lines, and water spray systems. Sometimes they are connected to fire pumps and gravity tanks. The tank is normally kept about two-thirds full of water, with an air pressure of at least 75 psi (5.2 bar). Tanks should be housed in noncombustible structures unless they are installed in a heated room within a building.

When inspecting a pressure tank, the inspector should read the water level in the sight gauge, check the pressure gauge, and compare these readings to previously recorded readings and to the fire protection system demand criteria. The inspector should also make sure that the pump is operating correctly to fill the tank and should examine the air compressor for capacity and maintenance condition.

Inspecting Water Storage Tanks

NFPA 25 requires that the water level and the condition of the water in storage tanks be checked monthly or daily during cold weather. The water temperature must never be permitted to drop below 40°F (4°C). Temperature can be maintained with a heat exchanger and water-circulation system.

The exterior of a tank should be checked at least quarterly for signs of damage or weakening. The area that surrounds the tank and its support structure must also be checked quarterly to ensure that it is free of combustible storage, debris, and other material that could present a fire exposure; accumulated material that could accelerate corrosion or rot; and ice buildup. The supporting embankments of embankment-supported tanks should be inspected for signs of erosion.

Exterior surfaces that have been painted, coated, or insulated should be checked at least annually for signs of degradation. Provided expansion joints should be inspected for leaks and cracks annually. The hoops and grillage of wooden tanks should be checked at least every other year.

The interior of most tanks must be inspected at least every 5 years. However, the interior of steel tanks that have no corrosion protection must be checked every 3 years. The interior of pressure tanks should be inspected at 3-year intervals at the most, and air pressure should be checked monthly.

The inspector must make sure that the required testing and maintenance is being carried out and recorded. During cold weather, the testing includes monthly tests of the low-water temperature alarms and the high-water temperature limit switches on tank heating systems. High- and low-water-level alarms must be tested at least semiannually, and pressure gauges and water level indicators must be tested for accuracy at 5-year intervals.

Tank drain valves should be fully opened and closed at least annually and the tank vents cleaned. Cathodic protection, if provided, must be maintained in accordance with manufacturer's instructions. Sediment is to be drained or flushed from tanks at least semiannually.

In areas subject to earthquakes, the inspector should make sure that the tank and its support structure are braced to withstand the possible horizontal and vertical forces. Water supply piping from the tank to the system should be braced at appropriate intervals, but the piping should also be flexible where needed to accommodate differential building movement.

The inspector can use the following checklist when inspecting water storage tanks:

1. Note the name of the tank manufacturer and the installing company and features of structural design, foundations, and wind and earthquake loadings that were addressed during initial installation.
2. Note whether the tank site is clear of weeds, brush, rubbish, and piles of combustible material, which might cause the steelwork to fail through fire or corrosion. Determine whether the tank would be safe should a fire occur in nearby buildings.
3. Note whether the tank is used for any purpose other than fire protection.
4. Read the mercury gauge or other water-level indicator and consult the records kept of such readings throughout the year.
5. Find out whether the tank has an overflow (this item does not apply to pressure tanks). Ask that the tank be filled to overflow to test the water-level indicators and the tank-filling arrangements.
6. Note the general maintenance of the tank structure, the tank itself, and its accessories.
7. Ask how recently the owner has had the tank completely inspected by an experienced tank contractor. Where water conditions make frequent cleaning necessary, ask when the tank was last cleaned.
8. Consult the records of weekly valve inspections. Note whether the valves were found to be wide open and properly supervised. Be sure each valve has been given an operating test.
9. Inspect every valve pit. Their construction and arrangement should be satisfactory, with adequate clearance around the pipes. The valves in the pit, the manhole, and the ladder should be in good repair. The pit should be waterproof and drain properly.
10. Consult the daily records of tank temperature readings kept during cold weather. Ask how the heating system is checked during any plant closing and during freezing weather.
11. Ask about any experiences with ice on any part of an outdoor tank structure.
12. Inspect the tank heating system, particularly any separate above-grade heater house. Note the construction of the heater house, and determine whether the roof will properly support frostproof casing and any other loads imposed on it.
13. Review the records of tank painting and estimate the condition of the paint. Note whether the paint surface inside the tank has been checked within 2 years.

14. If a steel tank has cathodic protection, ask when the supplier last inspected the equipment.
15. For pressure tanks, make sure that the pressure is at least 75 psi (5.2 bar).

MINIMUM WATER SUPPLIES FOR FIRE FIGHTING

A number of formulas have been used to determine needed fire flow in municipalities, but many variables influence the calculations. For facilities protected by water-based fire protection systems, the system design generally dictates the minimum water supply requirements. NFPA 1142, *Standard on Water Supplies for Suburban and Rural Fire Fighting,* contains the minimum requirements for fire flow for facilities located in areas without public water supply systems.

Some plumbing and health codes in North America require that public water supply pressures not be reduced below a minimum positive pressure of 10 to 20 psi (0.7 to 1.4 bar). For this reason, it is common to evaluate the capacity of both public and private water mains and hydrants as the flow available at a minimum net positive pressure of 20 psi (1.4 bar). Some large municipal water supply systems, such as that of Chicago, have fairly large flows available, but only at pressures of 30 to 40 psi (2.08 to 2.78 bar). Other municipal water supplies are kept at pressures exceeding 100 psi (7 bar). The combination of available flows and pressures determines whether a particular water supply can meet the demand of a fire protection system or whether pumps or on-site tanks will be needed.

Flow Testing Water Main

Chapter 4 of NFPA 25, 1998 edition, requires a flow test of underground and exposed pipe. This flow test should be conducted at 5-year intervals, at the minimum.

Using two hydrants nearest the point at which the building fire service connects to the main, the inspector can use the following test procedure to flow test hydrants on a water main (Figure 14-3). When public mains are involved, inspectors must be sure to secure permission for the test from the water utility.

1. Attach the gauge to Hydrant A and obtain the static pressure.
2. Attach a second gauge to Hydrant B and remove the cap from the other 2½-in. (63.5 mm) outlet.
3. While Hydrant B is uncapped, measure the diameter of the outlet to check its size. Although the inside diameter of the hydrant opening is usually close to 2½ in. (63.5 mm), some butts are sufficiently different to require a fairly close measurement. For convenience, this should be taken to the nearest hundredth of an inch (0.254 mm).
4. Feel inside the outlet to determine its shape—that is, either smooth, right angle, or projecting.
5. Slowly open Hydrant B and flush thoroughly, making sure a full flow is established for accurate measurement.

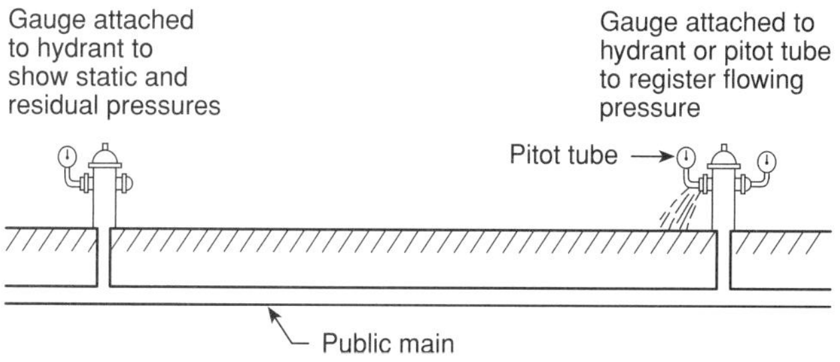

FIGURE 14-3 Method of Conducting Flow Tests
Source: NFPA 13, 1999, Figure A-9-2.2.

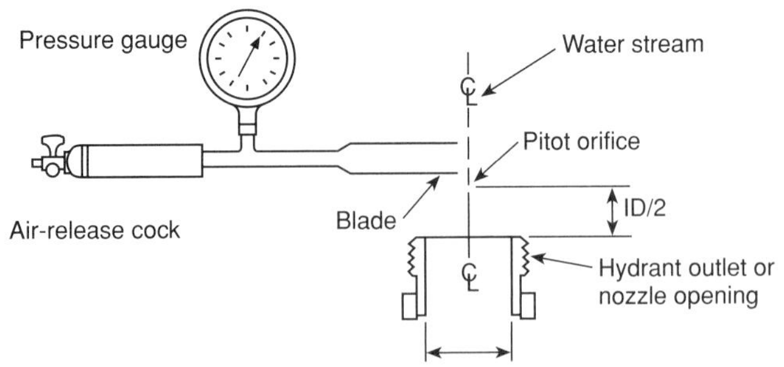

FIGURE 14-4 Pitot Tube Position
Source: NFPA 291, 1995, Figure 2-6.

The accuracy of the test is slightly better if the inspector reads the pressure on a gauge on a Pitot tube held in the flowing stream instead of on a gauge on Hydrant B. The position of the tube in the stream should be varied to get the most representative reading. In most openings, the Pitot reading will be best with the tube near the middle of the stream, held about half the outlet diameter away from the face of the outlet (Figure 14-4).

Discharge is computed using a formula. The inspector can compute the static pressure difference at the top line of sprinklers by deducting, from the minimum reading of the gauge on Hydrant A, 0.434 psi/ft (0.1 bar/m) the highest level of sprinklers are above the gauge. The inspector can compute discharge from the hydrant opening or nozzle by using the following formula:

$$Q = 29.84\, cd^2 \sqrt{p},$$

where Q is the *gallons* per minute, c is a coefficient of discharge for the opening, d is the opening's diameter in *inches,* and p is the pressure read at the gauge on the hydrant flowing or on the Pitot tube gauge in *pounds per square inch.*

The values in Table 14-2 are computed using the preceding formula, with c equaling 1.00. If the outlet meets the hydrant barrel with a smooth and rounded edge a c value of 0.9 should be used. If the outlet meets the hydrant barrel with a sharp right angle edge, a c value of 0.8 should be used. If the outlet edge projects into the hydrant barrel, a c value of 0.7 should be used. The inspector should use these factors to modify the flow calculations appropriately (Figure 14-5).

Flow Testing a Yard System

Figure 14-6 illustrates a yard system fed by a city connection, which provides an example of how water flow in such a system is checked.

In Figure 14-6, Gauge No. 1 illustrates street pressure. Gauge No. 2 is located as close to the street connection and the meter as possible. The amount by which

TABLE 14-2 Theoretical Discharge Through Circular Orifices

		Diameter of Orifice	
Pressure (psi)	Head (ft^2)	2½ in. (Flow, gpm)	4 in. (Flow, gpm)
15	34.61	722	1849
30	69.21	1021	2614
46	106.12	1264	3237
60	138.42	1444	3697
90	207.63	1769	4527
120	276.84	2042	5228

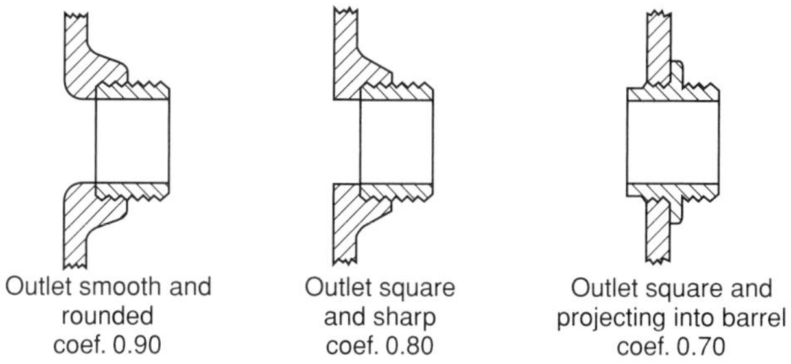

FIGURE 14-5 Three General Types of Hydrant Outlets and Their Coefficients
Source: NFPA 14, 2000, Figure B.5.

the reading of Gauge No. 2 differs from that of Gauge No. 1 shows the loss between the two points. Meters frequently cause large pressure drops.

A Pitot gauge reading is taken at the point at which water is flowing during a test. The difference in pressure should be explained by the expected friction losses in the piping system between the Pitot gauge and Gauge No. 2.

The pressure for a specific flow at the end of the system at the Pitot gauge should be the same as it is for the same flow at Gauge No. 1 minus the calculated losses in the piping between Gauges No. 1 and the Pitot gauge. If the observed pressure is less than calculated, the inspector should take readings at other outlets between Gauge No. 1 and the Pitot gauge and calculate the pressure losses to be expected for a given flow until the obstruction is located.

With tests such as these, the inspector can plot a hydraulic-grade line for a given run of piping that will often show otherwise undetectable sources of pressure drop. The following list of defects found in water supply systems point to the value of carefully checking private systems of underground piping:

1. The system was never connected to the city main.
2. The system was connected to the city main, but the gate valve was never opened or was only partly open.
3. The meters were broken or clogged.
4. The mains were smaller than indicated on plans.
5. Flow areas in the mains were seriously reduced because of sedimentation, mud, or hard deposits on the inside.
6. There was serious leakage in the underground systems.
7. The mains or valves were frozen.
8. The hydrants were inoperative.
9. The valves were entirely or partly closed due to improper or careless operation.
10. The check valves were installed in the wrong direction.
11. The check valves leaked.
12. Unknown valves and meters existed.
13. Suction on the fire pump was blocked.
14. The drop pipe on a gravity tank was frozen, as were other tank connections.

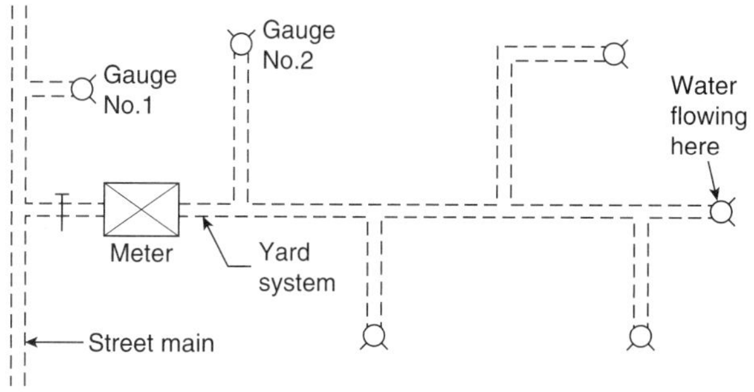

FIGURE 14-6 Method of Testing City Water Supply to Private (Yard) System
Source: *Fire Protection Handbook*, NFPA, 1997, Figure 6-7G.

SUMMARY

A water-based fire protection system is only as reliable as its water supply source. The first step in inspecting a fire protection system is determining how and from where the fire protection system will get its water supply. A water supply source with closed or partially closed valves will drastically affect the ability of the fire protection system to function properly. Improperly maintained fire pumps and water storage tanks could likewise severely affect the fire protection system's performance. Through proper testing and maintenance of the fire protection water supplies, the fire protection system connected to it will be able to function as it was designed to.

BIBLIOGRAPHY

Backflow Protection for Fire Sprinkler Systems, National Fire Sprinkler Association, Patterson, NY, 1990.

Carson, W. G., and Klinker, R. L., *Fire Protection Systems: Inspection, Test & Maintenance Manual,* 3rd ed., NFPA, Quincy, MA, 2000.

Cote, A. E., ed., *Fire Protection Handbook,* 18th ed., NFPA, Quincy, MA, 1997.

NFPA Codes, Standards, and Recommended Practices

NFPA 13, *Standard for the Installation of Sprinkler Systems*

NFPA 13D, *Standard for the Installation of Sprinkler Systems in One- and Two-Family Dwellings and Manufactured Homes*

NFPA 13E, *Recommended Practice for Fire Department Operations in Properties Protected by Sprinkler and Standpipe Systems*

NFPA 13R, *Standard for the Installation of Sprinkler Systems in Residential Occupancies Up to and Including Four Stories in Height*

NFPA 14, *Standard for the Installation of Standpipe, Private Hydrant, and Hose Systems*

NFPA 15, *Standard for Water Spray Fixed Systems for Fire Protection*

NFPA 16, *Standard for the Installation of Foam-Water Sprinkler and Foam-Water Spray Systems*

NFPA 20, *Standard for the Installation of Stationary Pumps for Fire Protection*

NFPA 22, *Standard for Water Tanks for Private Fire Protection*

NFPA 24, *Standard for the Installation of Private Fire Service Mains and Their Appurtenances*

NFPA 25, *Standard for the Inspection, Testing, and Maintenance of Water-Based Fire Protection Systems*

NFPA 110, *Standard for Emergency and Standby Power Systems*

NFPA 291, *Recommended Practice for Fire Flow Testing and Marking of Hydrants*

NFPA 1142, *Standard on Water Supplies for Suburban and Rural Fire Fighting*

NFPA 1962, *Standard for the Care, Use, and Service Testing of Fire Hose Including Couplings and Nozzles*

NFPA 1963, *Standard for Screw Threads and Gaskets for Fire Hose Connections*

Automatic Sprinkler and Other Water-Based Fire Protection Systems

CHAPTER 15

Roland Huggins

CASE STUDY

Sprinkler Prevents Loss of Shopping Center

One sprinkler extinguished a small fire in a department store in North Carolina before it spread to the eight other tenants of a shopping mall. The fire occurred at night when the store was closed.

The single-story department store, which measured 50 × 20 ft (15 × 6 m), had concrete block walls with a brick veneer and a flat, steel-supported roof. A monitored, wet-pipe sprinkler system that provided full coverage protected the property. There were no smoke detectors.

The fire department received the alarm at 11:23 P.M. from a municipal system alarm connected to the water flow. When fire fighters arrived, they connected and supported the sprinkler system, forced the front door, and advanced a hose line into the retail display area.

When they discovered that the sprinkler had extinguished the fire, the crews overhauled the scene and began investigating its cause. Apparently, a large display cabinet had damaged an electrical cord under it, causing resistance that led to heat buildup. The wooden cabinet and carpet ignited, and the fire spread to material on the wall before the sprinkler activated.

The unit of origin, which had a value of $72,000, suffered $5000 in structural damage. Damage to its contents, valued at $50,000, was estimated at $2500. There were no injuries, and adjoining tenants were not affected by the incident.

Source: "Fire Watch," *NFPA Journal*, May/June, 2000.

Because water is an efficient and almost universally available and acceptable agent, the use of water-based fire protection systems is continually increasing. Residential sprinkler systems are becoming more practical; and newer, more effective systems are being designed for warehouses and other large occupancies. In fact, the hazards of a particular building are an important factor in determining what kind of fire protection is needed for that building. This chapter describes some of these systems and provides information on inspecting them properly.

Sprinkler systems are found much more often now than they used to be, due both to code requirements and as an equivalent alternative to those code requirements. Therefore, it is important for the inspector to verify that any sprinkler system in the building is in service and fully operational. More and more often, communities are allowing construction modifications for some of the code requirements if the building is protected throughout with a sprinkler system. With this in mind, the inspector should recognize that some items required by the code might not be present when a sprinkler system has been installed in lieu of those items. This "voluntary" sprinkler system then becomes a required system because of its use as an equivalent

Roland Huggins, director of technical services for the American Fire Sprinkler Association, is a graduate of the University of Maryland and is registered in fire-protection engineering. He is a member of several NFPA technical committees, including NFPA 13 Technical Correlating Committee; NFPA 13 Discharge Criteria Committee; and NFPA 5000™, Building Construction and Safety Code™, *Technical Correlating Committee. He is also a member of SFPE and participates on SFPE committees addressing performance-based design.*

alternative. It should meet all the standards required by NFPA 13, *Standard for the Installation of Sprinkler Systems*, or NFPA 13R, *Standard for the Installation of Sprinkler Systems in Residential Occupancies Up to and Including Four Stories in Height*, as appropriate.

NFPA 25 INSPECTIONS

In 1992 NFPA adopted NFPA 25, *Standard for the Inspection, Testing, and Maintenance of Water-Based Fire Protection Systems*, which discusses water supply systems, sprinkler systems, standpipes, fire pumps, water storage tanks, water spray systems, and foam water sprinkler systems. With the widespread adoption of NFPA 25 across the country, most water-based fire protection systems are likely to already have been thoroughly inspected. If this is the case, a review of the inspection records should be made to confirm that a complete inspection was conducted and to identify any deficiencies. The inspector could also perform a spot check of some of the facilities to confirm the thoroughness and accuracy of the inspections.

An NFPA 25 inspection confirms only the mechanical condition of the fire protection system; it does not confirm the adequacy of a system to control the fire hazard. As stated within NFPA 25, "The owner or occupants shall not make any changes in the occupancy, use or process, or the materials used or stored in the building without evaluation of the fire protection systems for their capability to protect the new occupancy, use, or materials." This change includes relocating or adding new walls because doing so often creates unprotected areas. Unprotected areas are of particular concern in storage facilities because the adequacy of the system is affected not only by a change of contents but also by how the contents are stored (see Chapter 36, "Business Occupancies"). An NFPA 25 inspection should include a question to the owner on whether any changes have occurred. If a system evaluation is performed, it should be based on the requirements of NFPA 13.

CLASSIFYING THE OCCUPANCY

Hazard Classes

An important step in inspecting fire protection systems is to know the fire hazard presented by the occupancy of the facility that is protected by a system. Note that the term *occupancy* (when used to determine the design hazard) is used differently from how it is used in the building codes. The codes apply the term to a facility's use, such as a business occupancy. This chapter defines the term as it is used in NFPA 13, that is, by hazard. For simplicity's sake, occupancies are classified into three basic hazard categories when the installation of automatic sprinkler systems are contemplated. These are light hazard, ordinary hazard, and extra hazard. Except for the protection of storage occupancies, these categories are used when sprinkler systems are designed and installed in accordance with the requirements of NFPA 13.

Light Hazard Classification. The light hazard classification includes occupancies such as dwellings, apartments, churches, hotels, schools, offices, and public buildings, where the quantity of combustible materials is relatively low and fires are expected to have a low rate of heat release.

Ordinary Hazard Classification. The ordinary hazard classification includes ordinary mercantile, manufacturing, and industrial occupancies, and these fall into two groups. Ordinary Hazard Group 1 occupancies are those in which the combustibility of the contents is relatively low, the quantity of materials is moderate, stockpiles are no higher than 8 ft, and fires are expected to release heat at a moderate rate. Ordinary Hazard Group 2 occupancies include properties in which the quantity and combustibility of the contents is moderate to high, stockpiles are no higher than 12 ft, and fires are expected to have moderate to high rates of heat release. This group includes cereal mills, textile and printing plants, shoe factories, feed mills, piers and wharves, and paper-manufacturing plants.

Extra Hazard Classification. The extra hazard classification includes two main groups. Extra Hazard Group 1 includes occupancies that contain little or no flammable or combustible liquids but that may nonetheless have severe fires. These include die-casting and metal-extruding plants, sawmills, rubber-production facilities, and upholstering operations using plastic foams. Extra Hazard Group 2 includes occupancies that contain moderate to substantial amounts of flammable or combustible liquids or extensive shielding of combustibles. These include buildings that house asphalt saturating, flammable-liquid spraying, open-oil quenching, solvent cleaning, plastics processing, varnish and paint-dipping processes, or manufactured home or modular building assemblies (where enclosures have combustible interiors).

In addition to these hazard groupings, there are special occupancy conditions that require consideration. Such conditions include high-piled combustibles, a variety of flammable and combustible liquids, combustible dusts, chemicals, aerosols, and explosives.

In occupancies containing such high-hazard items, water alone may not be an effective extinguishing or fire control agent unless it is combined with an additive. For most occupancies, however, water is the primary media for controlling or suppressing fires. Water-based fire protection systems can be designed to meet a variety of situations, but they must have an automatic and reliable water supply source. As with any fire protection system, a sprinkler system must be operable at all times.

Critical Inspection Times

There are two occasions when inspectors must be at their most observant and critical: when a new sprinkler or some other water-based fire protection system is being installed and when the installed system is shut down for repairs, inspection, or modification. During these periods, the occupancy will lack its principal means of automatic fire protection.

When systems must be shut down for any reason, the proper authorities—that is, the fire department, the insurance company, and the corporate safety office—should be notified. This will allow the fire department to alter the response and the type of equipment that is dispatched to the fire. If the impairment is planned, all the necessary replacement parts and equipment should be in place to minimize the amount of downtime for the system. NFPA 25 provides a regimented protocol for taking a fire protection system out of service and a procedure for placing the system back into service once repairs have been completed.

Inspectors should determine whether the building they are inspecting is covered by NFPA 13. They also should determine whether it is a light, ordinary, or extra hazard occupancy or storage. It may also be a structure that may be covered by some other occupancy-based code or standard, many of which are referenced in NFPA 13. This information will help inspectors assess the capability of the automatic sprinkler system. Most of the flaws or errors will be fundamental and easy to find; others may be discovered during routine tests. Inspections, testing, and maintenance should be documented to ensure compliance with NFPA 25.

Sprinkler System Water Supply

When sprinkler systems fail to control a fire, there are generally two primary reasons: the water supply was insufficient for the particular hazard or a valve on the supply line was closed. Therefore, one of the inspector's first acts during an inspection is to verify that there is enough water available to the sprinkler system. The main drain flow test required by NFPA 25 can be used as an indicator to determine whether there has been a reduction in the available water supply. This procedure is discussed later in this chapter. The system plans should indicate the piping layout and the location of the valves. Inspectors should arrange to observe the operation of each valve in order to verify that it is performing well. They should also note if there is a written sign or some other indication that the valve should remain open or closed.

The fire department connection, if there is one, merits the inspector's attention. This connection is considered an auxiliary water supply source. The inspector should remove the outlet caps and examine the threads, then examine the interior for rags and other debris as well as hazardous items such as needles. If no caps are present, the potential for debris increases. The inspector should note whether the threads are compatible with those of the local fire department or whether adapters are needed.

The condition of any yard hydrants and indicator post valves outside the facility should be checked to make sure they are easy to operate. The inspector should determine whether outside stem and yoke (OS & Y) valves are supposed to be locked in the open position or whether they are supervised open by some other means. The inspector should make sure the valve pit is kept clean and is well maintained.

The inspector should ensure that the appropriate keys or wrenches to any underground valves are readily accessible. If the property contains municipal hydrants, the inspector should check with the local fire department to determine how the hydrants will be operated during a fire. The fire department should plan to support the system through the fire department connections.

Central station, proprietary, and remote supervisory services are available to provide continual electrical surveillance of the valves in water systems. The inspector should determine whether the building or plant being inspected has such services and, if they do, the extent and effectiveness of the supervision.

If the valves are not electrically supervised by means of a signaling service to a remote location or some other constantly attended point, they should either be locked open or sealed. Sealed valves should be located within fenced enclosures under the control of the owner and inspected weekly. Electrically supervised valves are inspected monthly.

Inspection reports on valve conditions should identify the valve number and note whether it is normally open or closed. The report should also note whether the valve is sealed properly or locked and whether it is in good operating condition and free of leaks. In addition, this report should discuss the valve's accessibility and note whether the valve key or wrench is in place.

THE SPRINKLER SYSTEM

All automatic fire sprinklers, regardless of the type of system, should be in good working condition or should be replaced. They must not be caked with dust, grease, or paint, particularly on the heat-responsive element. They should not be bent or otherwise damaged, and guards should be provided for sprinklers located in areas prone to damage or abuse. The sprinklers should not be obstructed by light fixtures; heating, ventilation, and air-conditioning equipment; cables; stored materials; or the movement of overhead doors or windows. When looking for obstructions, the inspector should focus primarily on the areas in the facility that have changed. The allowances for obstructions vary depending on the type of sprinkler and are provided in NFPA 13. NFPA 25 requires that this type of evaluation be performed once a year. These inspections are expected to be conducted from the floor and can impact the level of detail for high ceilings. Additionally, if the sprinklers are installed in an inaccessible area, the yearly inspection is not required.

Sprinklers should be free of corrosion. Listed corrosion-resistant or special coated sprinklers should be used in areas in which enough chemicals, moisture, or corrosive vapors exist to cause corrosion. Such coatings must be applied only by the sprinkler manufacturer; field application of any type of plating or coating is not acceptable.

When inspecting the sprinkler system, the inspector should be sure to check the temperature ratings of any sprinklers near such heat sources as ovens, unit heaters, and skylights. NFPA 13 provides detailed information on the color-coded intermediate- or higher rated sprinklers needed in such areas.

The orientation of sprinklers should be checked to ensure that upright sprinklers have been installed in the upright position; pendent sprinklers, in the pendent position; and sidewall sprinklers, in the proper position with respect to their coverage area. This orientation is of particular concern when a sprinkler that has activated is being replaced. To improve the likelihood of proper replacement, each unique model of sprinkler manufactured after January 2001 is permanently marked with a one- or two-character manufacturer symbol, followed by up to four numbers. This identification number is intended to identify sprinkler operating characteristics. Each change in orifice size, response characteristics, or deflector (distribution) characteristics results in a new model number.

All sprinklers made before 1920 should be replaced, and representative samples of sprinklers in all systems should be submitted for operational testing after a certain period of service. For most sprinklers, this period is 50 years. However, there are some exceptions. Sprinklers with temperature ratings of 325°F (163°C) and higher that have been exposed on a regular basis to the maximum allowable ambient temperatures should be tested at 5-year intervals. Quick-response sprinklers should initially be tested after 20 years of service and at 10-year intervals thereafter. Dry sprinklers, which are typically used to extend protection into a cold area from a wet-pipe system, should be tested at 10-year intervals.

The inspector should verify that a sufficient supply of spare sprinklers of each type and rating used on the system is stored on the premises in a cabinet where it will not be exposed to corrosion or high temperatures. This requirement does not apply to dry sprinklers, since each typically has a unique length for each specific location. The cabinet should also contain any special wrenches needed to replace the sprinklers.

Piping should be checked once a year when the sprinklers are being inspected to make sure that it is in good condition, free from mechanical damage, and not being used to support fixtures, ladders, or any other loads. Hangers and seismic bracing should be tight and intact. If the atmosphere is corrosive, the piping should be protected by an appropriate covering. Where mechanical fittings are used, the protection should extend to the bolts connecting them by the use of stainless steel or a galvanized coating.

The inspector should examine any records available to determine whether the interior of the piping has been inspected recently. Piping systems should be examined internally when abnormal conditions are detected during routine system tests. These conditions include, but are not limited to, defects on intake screens from a raw water supply source; foreign objects in the pump suction lines, dry-pipe valves, or deluge valves; obstructions in the sprinkler pipe due to system alterations; and false trips in a dry-pipe system.

Systems whose piping or sprinklers are plugged, systems that discharge foreign material during tests, and systems that show other signs of obstruction should be flushed and reexamined at intervals of not more than 5 years. Flushing connections are provided at the ends of the mains on all newer systems, and flushing procedures for various configurations of sprinkler systems are described in NFPA 25.

The inspector should record pressure gauge readings of the system and compare them with the information provided during the acceptance test and with previously recorded readings. There are ways to test alarm valves and water flow devices, and the inspector should arrange for such tests, as described later in this chapter. NFPA 25 requires that the pressure gauges of wet-pipe systems (Figure 15-1) be read

FIGURE 15-1 Tests and Drain Connection for a Wet Pipe Sprinkler System Riser
Source: NFPA 13, 1999, Figure 5-14.2.4.

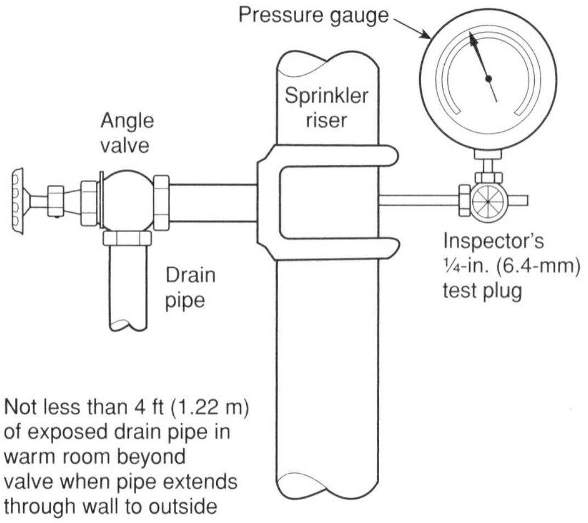

monthly and that those of dry-pipe, preaction, and deluge systems be read weekly. If the air pressure in dry-pipe and preaction systems is supervised at a constantly attended location, however, monthly inspections are permitted. The pressure gauges that monitor the air pressure in systems of this type must also be inspected and the air pressure recorded.

TYPES OF SPRINKLER SYSTEMS

Wet-Pipe Sprinkler Systems

A wet-pipe sprinkler system is a system that employs automatic sprinklers attached to a piping system containing water and connected to a water supply so that water discharges immediately from sprinklers when they are opened by heat from a fire. An antifreeze sprinkler system is effectively a wet-pipe system except that the water has been mixed with an antifreeze liquid. Often, a small antifreeze system is connected to a wet-pipe system to protect localized areas exposed to freezing temperatures, such as a loading dock. The minimum temperature for an antifreeze system depends on both the liquid and the concentration. (Typical liquids are shown in Table 15-1.)

TABLE 15-1 Antifreeze Solutions to Be Used If Potable Water Is Connected to Sprinklers

Material	Solution (by volume)	Specific Gravity at 60°F (15.6°C)	Freezing Point °F	Freezing Point °C
Glycerine	50% water	1.133	−15	−26.1
C.P. or U.S.P. grade*	40% water	1.151	−22	−30.0
	30% water	1.165	−40	−40.0
Hydrometer scale 1.000 to 1.200				
Propylene glycol	70% water	1.027	+9	−12.8
	60% water	1.034	−6	−21.1
	50% water	1.041	−26	−32.2
	40% water	1.045	−60	−51.1
Hydrometer scale 1.000 to 1.200 (subdivisions 0.002)				

*C.P.—chemically pure; U.S.P.—United States pharmacopoeia 96.5%.
Source: NFPA 13, 1999, Table 4-5.2.1.

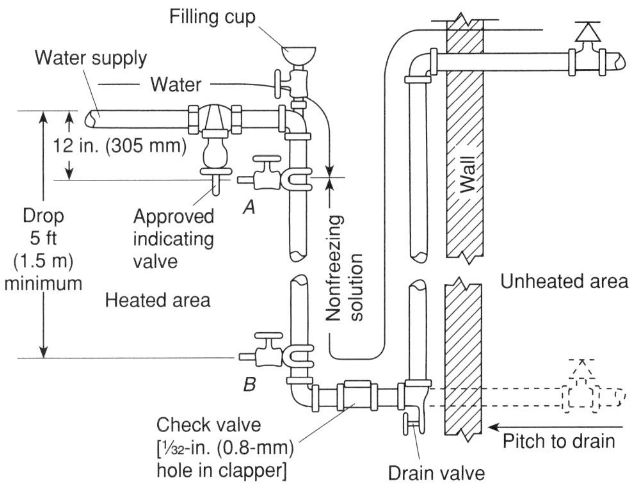

Notes:
1. Check valve shall be may be omitted where sprinklers are below the level of valve *A*.
2. The $\frac{1}{32}$-in. (0.8-mm) hole in the check valve clapper is needed to allow for expansion of the solution during a temperature rise, thus preventing damage to sprinklers.

FIGURE 15-2 Arrangement of Supply Piping and Valves
Source: NFPA 13, 1999, Figure 4-5.3.1.

More is not always better since at higher concentrations the freeze point increases. The antifreeze concentration across the system can also vary over time. For this reason, it is important that the system be drained and refilled when the concentration is being checked.

The only real difference between an antifreeze system and a wet-pipe system is that additional devices are located at the point of connection to the untreated water. On older systems, the interface consists of a U-shaped pipe arrangement (Figure 15-2).

This arrangement isolates most of the antifreeze solution (although some can migrate through the hole in the isolation check valve) and allows draining, refilling, and testing of the antifreeze solution. The hole in the check valve is needed to account for changes in pressure caused by thermal expansion from seasonal changes in the temperature. New installations generally require the use of a backflow preventor to isolate the antifreeze system (Figure 15-3). To account for the thermal expansion, an expansion chamber is required with the backflow preventor.

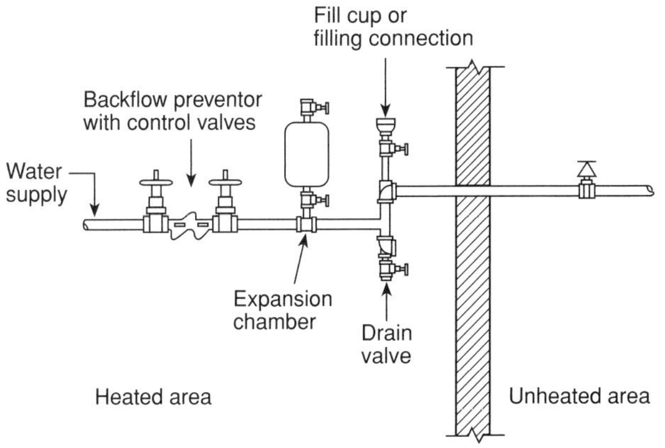

FIGURE 15-3 Arrangement of Supply Piping with Backflow Device
Source: NFPA 13, 1999, Figure 4-5.3.2.

Dry-Pipe Sprinkler Systems

Dry-pipe sprinkler systems differ from basic wet-pipe sprinkler systems in several ways. First, the water is kept from reaching the sprinklers by the pressure of compressed air or nitrogen acting on the system side of the valve. Water does not enter the system until the system air pressure drops below a predetermined point. When one or more sprinklers operate in the area of a fire, the air pressure drops, the dry-pipe valve automatically opens, and water enters the system. Water is then discharged from any open sprinklers.

Another difference between dry- and wet-pipe sprinkler systems involves the restricted use of pendent sprinklers. In areas subject to freezing, the dry pendent-type sprinkler can be used; standard pendent sprinklers are allowed only on return bends in heated areas. Typically, dry-pipe systems use upright sprinklers and horizontal sidewall sprinklers when installed so no water is trapped.

Check valves are sometimes installed in the branches of older dry-pipe systems to improve the water delivery time to open sprinklers. In such cases, drain valves are connected by a bypass around each check valve to allow the system to be drained. This practice is no longer permitted on newer systems because the reliability of the valve as the result of missed maintenance or the threat of ice at the valve is a concern.

Dry-pipe systems with a capacity of more than 500 gal (1893 L) must have quick-opening devices. Gridded dry-pipe systems, no longer permitted for new installations, are required to have a quick-opening device when the system's volume exceeds 350 gal (1325 L). This quick-opening device is usually an accelerator placed close to the dry-pipe valve. There will also be a soft disk globe or angle valve in the connection between the dry-pipe sprinkler riser and the accelerator. Quick-opening devices are inspected and tested at the same frequency as the dry-pipe valve. The maintenance of the quick-opening device is critical to proper system operation and its ability to protect the hazard. The dry-pipe valve and the supply pipe must be protected against cold temperatures and mechanical injury. A heated enclosure is required for this purpose.

The air pressure in the dry-pipe system may come from a shop system or an automatic compressor. The inspector should check that the air pressure in the system corresponds to that listed on the instruction sheet for the dry-pipe valve or is 20 psi (1.4 bar) above the valve's calculated trip pressure. The same requirement applies when nitrogen is used as the source for the system pressure.

Pressure gauges are installed on the water and air sides of the dry-pipe valve, at the source of air pressure, in each independent pipe from the air supply to the system, and at the exhausters and accelerators. The inspector can identify these gauges and determine their purpose by consulting the system installer's plans or the manufacturer's literature.

The inspection procedures for dry-pipe systems are much like those for wet-pipe systems except that the inspector will have to evaluate the source of the air pressure and examine the ancillary equipment unique to this type of system.

Preaction and Deluge Systems

Preaction and deluge systems use a supplemental detection system to initiate the flow of water. In a preaction system, only one or a few sprinklers will open; in a deluge system, all the sprinklers are open type and will discharge simultaneously.

When checking the location and spacing of fire detection devices, as well as the functions of the system, inspectors should refer to the manufacturer's literature, the listing criteria, and the appropriate installation standards such as *NFPA 72®, National Fire Alarm Code®*. If there are more than 20 sprinklers on a preaction system, the sprinkler piping and fire detection devices must be supervised automatically. Because a preaction system cannot have more than 1000 closed sprinklers, very large areas will require more than one preaction valve. The fire detection devices and systems of a deluge system must also be supervised automatically, regardless of the number of sprinklers on the system. The supervisory air on noninterlock and double interlock preaction systems is required to be at least 7 psi (0.5 bar). This minimum level is required because the system can operate like a dry-pipe system for which a minimum pressure is needed to ensure the sprinkler cap lifts off its seat when the sprinkler activates.

If the fire detection devices in the circuits are not accessible, additional testing devices for each circuit must be accessible, and they must be connected to the circuit at a point that will ensure an adequate test. The system should include testing apparatus that will produce enough heat or impulse to operate any normal fire detection device. In hazardous locations where explosive materials or vapors are present, a hot water, steam, or other nonignition method may be used.

Water Spray Systems

Water spray systems are designed to control and extinguish fires and to protect exposures in special situations; they are not meant to replace sprinkler systems, although they are similar to sprinkler systems except for the pattern of water spray discharge. Water spray can be designed for a variety of discharge rates and patterns, including ultra-high-speed response in milliseconds. The type of system used depends on the extent of the hazard and the required water discharge.

Water spray systems are commonly used to protect vessels in which flammable liquids and gases are stored, as well as to protect electrical transformers, oil switches, rotating electrical machinery, electrical cable trays and runs, conveyor systems, wall openings, and similar fire problem elements. They consist of fixed piping and water spray nozzles designed specifically to discharge and distribute water over the area to be protected (Figure 15-4).

Water flow is started either manually or automatically, usually by the actuation of separate detection equipment. Water spray systems have a heavy demand for water because simultaneous, high-density discharge from many nozzles is often needed. As with sprinkler systems, it is important that the water supplies for water spray systems be adequately designed and reliably maintained.

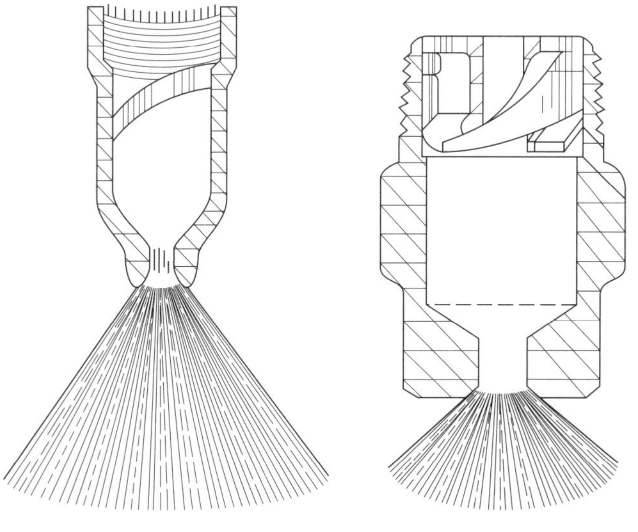

FIGURE 15-4 Water Spray Nozzles with Internal Spiral Passages

Steam Smothering Systems

Steam smothering systems were used many years ago, but they are rarely employed today and are not recommended for fire protection. They can be used to protect ovens or cargo spaces on ships, aircraft engine test facilities, and coal pulverizer systems. An accepted application is 8 lb (3.6 kg) of steam per minute for each 100 ft^3 (2.8 m^3) of volume.

TESTING SPRINKLER SYSTEMS

Wet-Pipe and Dry-Pipe Sprinkler Systems

Sprinkler systems must be maintained in accordance with NFPA 25. The water supply pressure, indicated by a gauge located above the indicator gate valve controlling the system but below the system alarm or dry valve, should be observed.

If the system riser contains an alarm check valve, as it does in a wet system, or contains a dry-pipe valve, as it does in a dry system, the inspector should note the reading of the system pressure gauge, which is usually located at or immediately above the alarm or dry valve. On a wet-pipe system, the system gauge should read the same or, in some cases, slightly higher than the supply gauge. This situation is the result of the system check or the alarm valve capturing pressure surges from the water supply. On a dry-pipe system, the system pressure is the pressure of the air or of the nitrogen, and it is usually considerably less than the water supply pressure.

A test known as the "main drain test" should be conducted annually and should follow the procedure established in NFPA 25:

1. Record the pressure on the water supply gauge.
2. If the system is equipped with an alarm check valve, close the alarm line.
3. Open the main drain valve.
4. When the flow has stabilized, record the residual pressure from the water supply gauge.
5. Close the main drain valve and note the time it takes the gauge to reach its original static reading.
6. Place the valve on the alarm line back in its normal open position.

No flow measurements are taken during this test, only readings from the pressure gauges. The pressures noted should be compared to the results from previous tests. This test simply provides a comparison with the original main drain test pressure values and can be used to detect deterioration in the system water supply. It is possible for the water supply to the system to degrade either by increased water flow to other nearby facilities, by mineral buildup within the underground supply main, or by a partially closed valve.

The gauges used on a sprinkler system must be replaced or tested every 5 years for a verification of their accuracy. A calibrator or inspector's gauge should be used to determine whether the system gauge is accurate to within 3 percent of the full-scale reading of that gauge. If the system gauge is not accurate to within 3 percent, it must be recalibrated or replaced.

Preaction and Deluge Systems

The same basic procedures used to test wet-pipe systems may be applied to preaction and deluge systems. The interior of the deluge valve housing should be inspected annually to determine whether there has been any corrosion or other deterioration that might impair operation of the device. The water level in preaction valves that use priming water must be checked quarterly. Too much priming water could interfere with the ability of the valve to open.

TABLE 15-2 Summary of Minimum Inspection, Testing, and Maintenance

Item	Activity	Frequency
Gauges (dry, preaction, and deluge systems)	Inspection	Weekly/monthly
Control valves	Inspection	Weekly/monthly
Alarm devices	Inspection	Quarterly
Gauges (wet pipe systems)	Inspection	Monthly
Hydraulic nameplate	Inspection	Quarterly
Buildings	Inspection	Annually (prior to freezing weather)
Hanger/seismic bracing	Inspection	Annually
Pipe and fittings	Inspection	Annually
Sprinklers	Inspection	Annually
Spare sprinklers	Inspection	Annually
Fire department connections	Inspection	Quarterly
Valves (all types)	Inspection	
Alarm devices	Test	Quarterly/semiannually
Main drain	Test	Annually
Antifreeze solution	Test	Annually
Gauges	Test	5 yrs
Sprinklers—extra-high temp.	Test	5 yrs
Sprinklers—fast response	Test	At 20 yrs and every 10 yrs thereafter
Sprinklers	Test	At 50 yrs and every 10 yrs thereafter
Valves (all types)	Maintenance	Annually or as needed
Obstruction investigation	Maintenance	5 yrs or as needed

Source: NFPA 25, 2002, Table 5.1.

Preaction and deluge systems must be flow tested annually to ensure that all the valves and devices perform as intended. Special arrangements must be made to flow test the deluge system. Equipment that is subject to damage must be removed or protected so that the discharge of water from the system will not affect it. Under special circumstances, this test may be postponed for as long as 3 years. For example, if the system protects sensitive equipment that can be damaged during a test, NFPA 25 permits a longer period between such tests. If the test must be postponed, the inspector can complete a trip test without discharging water into the system.

Double interlock preaction systems have the attributes of both dry-pipe and preaction systems. Thus, they are inspected, tested, and maintained in the same manner as a dry-pipe or a preaction system. Because double interlock systems are somewhat more complex than other types of sprinkler systems, the inspector will need to take extra time to ensure that they are operating normally. Combined dry-pipe and preaction systems must be constructed so that either system will still operate if the fire detection system fails and the fire detection system will operate if either of the sprinkler systems fail.

Chapters 5 and 12 of NFPA 25, 2002 edition, provide additional information on inspecting, testing, and maintaining sprinkler systems. Table 15-2 outlines the basic elements of those items associated with the inspection, testing, and maintenance of sprinkler systems.

Water Spray Systems

Water spray systems must be inspected and maintained regularly. When inspecting a water spray system, the inspector must make sure that the operating and maintenance

TABLE 15-3 Inspection, Test, and Maintenance Frequencies

Item	Activity	Frequency
Backflow preventer	Inspection	
Check valves	Inspection	
Control valves	Inspection	Weekly (sealed)
Control valves	Inspection	Monthly (locked, supervised)
Deluge valve	Inspection	
Detection systems	Inspection	
Detector check valves	Inspection	
Drainage	Inspection	Quarterly
Electric motor	Inspection	
Engine drive	Inspection	
Fire pump	Inspection	
Fittings	Inspection	Quarterly
Fittings (rubber-gasketed)	Inspection	Quarterly
Gravity tanks	Inspection	
Hangers	Inspection	Quarterly
Heat (deluge valve house)	Inspection	Daily/weekly
Nozzles	Inspection	Monthly
Pipe	Inspection	Quarterly
Pressure tank	Inspection	
Steam driver	Inspection	
Strainers	Inspection	Mfg. instruction
Suction tanks	Inspection	
Supports	Inspection	Quarterly
Water supply piping	Inspection	
UHSWSS*—detectors	Inspection	Monthly
UHSWSS*—controllers	Inspection	Each shift
UHSWSS*—valves	Inspection	Each shift
Backflow preventer	Operational test	
Check valves	Operational test	
Control valves	Operational test	Quarterly
Deluge valve	Operational test	
Detection systems	Operational test	
Detector check valve	Operational test	
Electric motor	Operational test	
Engine drive	Operational test	
Fire pump	Operational test	
Flushing	Operational test	Annually
Gravity tanks	Operational test	
Main drain test	Operational test	Quarterly
Manual release	Operational test	Annually
Nozzles	Operational test	Annually
Pressure tank	Operational test	
Steam driver	Operational test	
Strainers	Operational test	Annually
Suction tanks	Operational test	
Waterflow alarm	Operational test	Quarterly
Water spray system test	Operational test	Annually
Water supply flow test	Operational test	
UHSWSS*	Operational test	Annually
Backflow preventer	Maintenance	
Check valves	Maintenance	
Control valves	Maintenance	Annually
Deluge valve	Maintenance	
Detection systems	Maintenance	

TABLE 15-3 *Continued*

Item	Activity	Frequency
Detector check valve	Maintenance	
Electric motor	Maintenance	
Engine drive	Maintenance	
Fire pump	Maintenance	
Gravity tanks	Maintenance	
Pressure tank	Maintenance	
Steam driver	Maintenance	
Strainers	Maintenance	Annually
Strainers (baskets/screen)	Maintenance	5 yrs
Suction tanks	Maintenance	
Water spray system	Maintenance	Annually

*Ultra high speed water spray system

Source: NFPA 25, 2002, Table 10.1.

instructions and layouts are available or are posted at the control equipment and at the plant's fire headquarters. Among the items to be inspected are strainers, piping, control valves, heat-actuated devices, detectors, and spray nozzles, especially those with strainers. Because the nozzles are directional-type spray nozzles, the inspector should verify their position and alignment; making sure that the small water passages are kept clear is important. If spray nozzles are subject to paint vapors and other coatings, blowoff caps should be used to protect the nozzles against foreign matter and corrosion.

Operational flow tests of water spray systems are required annually to ensure that the spray nozzles are not obstructed and that they are positioned properly to protect the intended hazard. During the test, the inspector should record the time it takes the detection system to actuate the deluge valve and the time it takes the water to reach the nozzle system. The inspector should also record the pressure readings at the remote spray nozzles to establish that they have enough pressure to project the necessary pattern.

Chapter 10 of NFPA 25 provides information on inspecting, testing, and maintaining water spray systems. Table 15-3 outlines which elements of the system must be inspected, tested, and maintained at what intervals.

FOAM EXTINGUISHING SYSTEMS

Fire-fighting foam is a combination of water and concentrated liquid foaming agent. The foam floats on the surface of flammable and combustible liquids and forms a covering that excludes air, cools the liquid, and seals the layer of vapor. It can also form a blanket over transformers and other irregularly shaped items to smother flames. There are several kinds of foaming agents, and their effectiveness varies with the type of application and the properties of the fire being considered.

Foam can be applied by portable devices or fixed extinguishing systems. In either type of application, the resultant solution must be at the right proportion, and the application must be continuous and consistent. Foam breaks down and its water content vaporizes when it is directly exposed to heat and flame. If it is applied in sufficient volume, however, it can overcome this loss and can control and eventually extinguish the fire. The smothering layer can also be broken and dispersed by mechanical or chemical action or by turbulence from air or fire gases. Nevertheless, automatic extinguishing systems can still apply foam efficiently.

Types of foaming agents include aqueous film-forming, fluoroprotein foaming, film-forming fluoroprotein, protein foaming, high-expansion foaming, synthetic hydrocarbon surfactant foaming, low-temperature foaming, "alcohol-type," chemical, and powder. These last two are practically obsolete and are not used in newly designed systems.

TABLE 15-4 Inspection, Test, and Maintenance Frequency

System/Component	Activity	Frequency
Discharge device location (sprinkler)	Inspection	Annually
Discharge device location (spray nozzle)	Inspection	Monthly
Discharge device position (sprinkler)	Inspection	Annually
Discharge device position (spray nozzle)	Inspection	Monthly
Foam concentrate strainer(s)	Inspection	Quarterly
Drainage in system area	Inspection	Quarterly
Proportioning system(s)—all	Inspection	Monthly
Pipe corrosion	Inspection	Quarterly
Pipe damage	Inspection	Quarterly
Fittings corrosion	Inspection	Quarterly
Fittings damage	Inspection	Quarterly
Hangers/supports	Inspection	Quarterly
Water supply tank(s)	Inspection	
Fire pump(s)	Inspection	
Water supply piping	Inspection	
Control valve(s)	Inspection	Weekly/monthly
Deluge/preaction valve(s)	Inspection	
Detection system	Inspection	See *NFPA 72*
Discharge device location	Test	Annually
Discharge device position	Test	Annually
Discharge device obstruction	Test	Annually
Foam concentrate strainer(s)	Test	Annually
Proportioning system(s)—all	Test	Annually
Complete foam-water system(s)	Test	Annually
Foam-water solution	Test	Annually
Manual actuation device(s)	Test	Annually
Backflow preventer(s)	Test	Annually
Fire pump(s)	Test	See Chapter 8
Water supply piping	Test	Annually
Control valve(s)	Test	See Chapter 12
Strainer(s)—mainline	Test	See Chapter 10
Deluge/preaction valve(s)	Test	See Chapter 12
Detection system	Test	See *NFPA 72*
Backflow preventer(s)	Test	See Chapter 12
Water supply tank(s)	Test	See Chapter 9
Water supply flow test	Test	See Chapter 4
Foam concentrate pump operation	Maintenance	Monthly
Foam concentrate strainer(s)	Maintenance	Quarterly
Foam concentrate samples	Maintenance	Annually
Proportioning System(s) Standard Pressure Type		
Ball drip (automatic type) drain valves	Maintenance	5 yrs
Foam concentrate tank—drain and flush	Maintenance	10 yrs
Corrosion and hydro. test	Maintenance	10 yrs
Bladder Tank Type		
Sight glass	Maintenance	10 yrs
Foam concentrate tank—hydro. test	Maintenance	10 yrs
Line Type		
Foam concentrate tank—corrosion and pickup pipes	Maintenance	10 yrs
Foam concentrate tank—drain and flush	Maintenance	10 yrs
Standard Balanced Pressure Type		
Foam concentrate pump(s)	Maintenance	5 yrs *(see Note)*
Balancing valve diaphragm	Maintenance	5 yrs
Foam concentrate tank	Maintenance	10 yrs

TABLE 15-4 *Continued*

System/Component	Activity	Frequency
In-Line Balanced Pressure Type		
Foam concentrate pump(s)	Maintenance	5 yrs *(see Note)*
Balancing valve diaphragm	Maintenance	5 yrs
Foam concentrate tank	Maintenance	10 yrs
Pressure vacuum vents	Maintenance	5 yrs
Water supply tank(s)	Maintenance	See Chapter 9
Fire pump(s)	Maintenance	See Chapter 8
Water supply	Maintenance	Annually
Backflow preventer(s)	Maintenance	See Chapter 12
Detector check valve(s)	Maintenance	See Chapter 12
Check valve(s)	Maintenance	See Chapter 12
Control valve(s)	Maintenance	See Chapter 12
Deluge/preaction valves	Maintenance	See Chapter 12
Strainer(s)—mainline	Maintenance	See Chapter 10
Detection system	Maintenance	See *NFPA 72*

Note: Also refer to manufacturer's instructions and frequency. Maintenance intervals other than preventive maintenance are not provided, as they depend on the results of the visual inspections and operational tests. For foam-water systems in aircraft hangars, refer to the inspection, test, and maintenance requirements of NFPA 409, *Standard on Aircraft Hangars,* Table 6-1.1.

Source: NFPA 25, 2002, Table 11.1.

Foam-water sprinkler and spray systems are effective in protecting areas in which flammable and combustible liquids are processed, stored, and handled. These include aircraft hangars, oil–water separators, pump areas and oil-piping manifolds, petroleum piers, warehouses containing large quantities of combustible and flammable liquids, and similar installations. The foam is discharged in essentially the same pattern as water is discharged from a nozzle designed to discharge water.

When inspecting such equipment, inspectors should watch for indications of corrosion. They should also make sure that orifices are not clogged, valves do not stick, and electrical parts do not malfunction. In addition, they should review test results.

The foam system should meet the requirements of fire protection standards and should otherwise show evidence of regular inspections and maintenance. Regular tests should confirm the dimensions and configuration of the discharge pattern, the percentage of foam concentrate in the finished solution, the degree of foam expansion in the finished compound, the rate at which water drains from the foam after discharge, and the film-forming ability of the foam concentrate. Inspectors should inspect the system's mechanical equipment as they would that of a sprinkler system. Chapter 11 of NFPA 25 provides the requirements for properly inspecting and testing foam-water sprinkler systems. Table 15-4 outlines the basic elements associated with inspecting, testing, and maintaining foam-water sprinkler systems.

STANDPIPE AND HOSE SYSTEMS

Standpipe and hose systems provide a means of manually applying water to fires in buildings. However, they do not take the place of automatic fire protection systems. They are usually needed where automatic protection is not provided and in areas to which hose lines from outside hydrants cannot easily reach.

Types of Standpipe Systems

There are five types of standpipe systems as follows:

1. *Automatic wet system:* The most common standpipe system is an automatic wet system, which is charged at all times.
2. *Semiautomatic dry system:* A semiautomatic dry system is equipped with remote control devices at each hose station that admit water into the system.
3. *Automatic dry-pipe system:* A third type of system is an automatic dry pipe system for unheated buildings. In this system, which is similar to a dry-pipe sprinkler system, a dry-pipe valve prevents water from entering the system until the stored air pressure in the discharge side falls below the water supply pressure.
4. *Manual dry system:* A manual dry system has no permanent water supply. The system is composed of a pipe that contains air at atmospheric pressure and receives its water supply from a fire department pumper.
5. *Manual wet system:* A manual wet system, which has a permanent water supply, is composed of a small-diameter water supply pipe that is connected to the system to keep it filled at all times. The water supply for both flow and pressure is provided by a fire department pumper.

In most buildings, a combined sprinkler/standpipe system riser can be used. For some combined riser arrangements, the water supply and the pressure for the sprinkler system must be adequate, but the flow and pressure for the standpipe system might have to be augmented through the fire department connection. This type of arrangement is considered a manual wet standpipe system because it uses a fire department pumper to maintain the appropriate standpipe pressures.

Inspecting and Testing Standpipes

Periodic inspection of all portions of standpipe systems is essential. Inspectors should make sure that the valves in the automatic sources of water are open and should test the supervisory means of such valves. They should examine the threads at the fire department connection and verify that the waterway is not clogged with any foreign material.

The inspector should check the valve at each discharge outlet or hose station for leakage and should examine the hose threads. Where hose is provided for occupant use, the inspector should check its condition and the condition of the nozzle and sees to it that the hose is stored properly. Any pressure regulating valves provided must be tested once every 5 years to ensure that they are properly set and adjusted. Chapter 6 of NFPA 25 provides information on inspecting, testing, and maintaining standpipes (Table 15-5).

Inspecting and Testing Hose Threads and Fire Hose

Fire hose found in commercial or industrial environments is generally intended either for occupant use in dealing with incipient fires or for use by trained fire fighters or industrial fire brigade members in attacking a fire. The former is known as a 1½-in. (40 mm) diameter occupant use hose and the latter as a 2½-in. (65 mm) diameter attack hose. The inspector must understand the differences between them because their testing requirements differ.

Inspection Requirements for Fire Hose. When inspecting fire hose, the inspector should first determine whether the proper length of hose is available, where it will

TABLE 15-5 Standpipe and Hose System Inspection, Testing, and Maintenance

Components	Activity	Frequency
Control valves	Inspection	Weekly/monthly
Pressure regulating devices	Inspection	Quarterly
Piping	Inspection	Quarterly
Hose connections	Inspection	Quarterly
Cabinet	Inspection	Annually
Hose	Inspection	Annually
Hose storage device	Inspection	Annually
Alarm device	Test	Quarterly
Hose nozzle	Test	Annually
Hose storage device	Test	Annually
Hose	Test	5 yrs/3 yrs
Pressure control valve	Test	5 yrs
Pressure reducing valve	Test	5 yrs
Hydrostatic test	Test	5 yrs
Flow test	Test	5 yrs
Main drain test	Test	Annually
Hose connections	Maintenance	Annually
Valves (all types)	Maintenance	Annually/as needed

Source: NFPA 25, 2002, Table 6.1.

be used, and whether it is stored in hose houses, on racks or reels, or on some type of cart or vehicle (Figure 15-5). The inspector should also make certain that it is the proper type of hose. Attack hose can be used in an occupant use environment, but occupant use hose should not be used where attack hose is required.

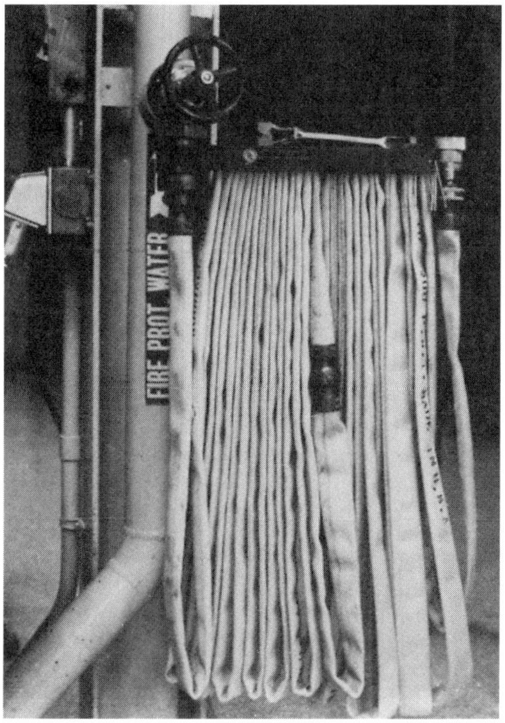

FIGURE 15-5 Conventional Pin Rack
Source: NFPA 25, 2002, Figure A.3.3.17.1.

Occupant use hose manufactured after July 1987 has a service test pressure of 150 psi (10.3 bar) stenciled on the jacket. Occupant use hose manufactured before July 1987 is either single-jacket hose stenciled with a proof pressure of 300 psi (20.7 bar) or is unlined standpipe hose.

Next, the inspector should make sure that the hose has no mechanical damage, such as an abrasion or a cut, and that it has not been damaged by heat, mildew or mold, acid, gasoline or oil, rodents, or any other environmental condition. If there is evidence that the hose is subject to any of these conditions, the inspector should examine the hose carefully and test it to determine whether it can still be used. The inspector should also make sure that the hose is stored in such a way as to protect it from these conditions.

Then the inspector should examine all couplings to make sure they have not been damaged. The inspector should physically connect or disconnect all the hose stored at a particular location to ensure that the couplings turn freely and are not out of round; that the threads have not been damaged; and that all the threads, including those on the nozzles and hydrant outlets, are compatible. The inspector should make sure there are no signs of corrosion, that the coupling has not slipped on the hose, and that there is no other sign of damage from use or misuse. Only those lubricants specified by the coupling manufacturer should be used on the couplings.

Next, the inspector should check the gaskets in the couplings to make sure that they fit properly, are flexible, have not deteriorated or been cut or torn, and do not protrude into the waterway. The inspector should also determine whether the hose has been service tested at the required frequency. In addition, the hose should be stretched out periodically for inspection and then repacked or refolded so that folds are not in the same place.

After hose has been used, it must be dried. The inspector should make sure that suitable facilities, such as hose racks, hose towers, or drying cabinets, are available for this procedure. Hose should not be laid out to dry on a hot surface such as concrete or asphalt.

Finally, the inspector should make sure that the hose threads used throughout the facility are the same. If they are not compatible with the threads used by the local public fire department or with the threads on municipal hydrants, the inspector should ensure that an adequate supply of adaptors are available where they will be needed.

Testing Requirements for Fire Hose. Attack hose must be tested annually; occupant use hose must be tested 5 years after the date of purchase and every 3 years after that. Hose manufactured after July 1987 will be stenciled with the service test pressure. For occupant use hose, this will be a minimum of 150 psi (10.3 bar). Attack hose has a minimum service test pressure of 300 psi (20.7 bar).

If the hose was manufactured before July 1987, a table in NFPA 1962, *Standard for the Care, Use, and Service Testing of Fire Hose Including Couplings and Nozzles,* specifies the test pressure. For older hose, the pressure stenciled on the hose is the proof pressure—that is, the pressure to which new hose is tested after being coupled. The hose should not be service tested to the proof test pressure.

The procedure for service testing fire hose is outlined in NFPA 1962. Essentially, it involves stretching the hose out in lengths no longer than 300 ft (91.4 m), connecting it to a water pressure source, and flooding it and expelling all the air. The pressure then should be partially raised while the hose is checked for leaks and marked at the couplings. Next, the pressure is raised to the test pressure and held there for 5 minutes. The hose is then checked for slippage at the couplings, and the test is concluded.

SUMMARY

Water provides an excellent means for controlling unwanted fires. Many different types of sprinkler and standpipe systems, utilizing water, have been developed. In order for these systems to function properly, they must be adequately maintained. Such maintenance consists of checking the condition or status of many different components, which vary depending on the type of system. In addition, to ensure that the systems are mechanically sound, the inspector must check that the water supply is still adequate and that there have been no changes to the facility or its contents without a fire protection evaluation. Almost as important as the inspections themselves are the records documenting the inspections. These not only show that the inspections were performed but also provide the data needed to determine if water supplies are degrading. The mere presence of a sprinkler system or a standpipe system does not mean that it is adequate. A thorough inspection program is necessary for such assurance.

BIBLIOGRAPHY

Cote, A. E., ed., *Fire Protection Handbook,* 18th ed., NFPA, Quincy, MA, 1997.
Carson, W. G., and Klinker, R. L., *Fire Protection Systems: Inspection, Test & Maintenance Manual,* 3rd ed., NFPA, Quincy, MA, 2000.

NFPA Codes, Standards, and Recommended Practices

See the latest version of The NFPA Catalog for availability of current editions of the following documents.

NFPA 11, *Standard for Low-Expansion Foam*
NFPA 11A, *Standard for Medium- and High-Expansion Foam Systems*
NFPA 13, *Standard for the Installation of Sprinkler Systems*
NFPA 13D, *Standard for the Installation of Sprinkler Systems in One- and Two-Family Dwellings and Manufactured Homes*
NFPA 13E, *Recommended Practice for Fire Department Operations in Properties Protected by Sprinkler and Standpipe Systems*
NFPA 13R, *Standard for the Installation of Sprinkler Systems in Residential Occupancies Up to and Including Four Stories in Height*
NFPA 14, *Standard for the Installation of Standpipe, Private Hydrant, and Hose Systems*
NFPA 15, *Standard for Water Spray Fixed Systems for Fire Protection*
NFPA 16, *Standard for the Installation of Foam-Water Sprinkler and Foam-Water Spray Systems*
NFPA 24, *Standard for the Installation of Private Fire Service Mains and Their Appurtenances*
NFPA 25, *Standard for the Inspection, Testing, and Maintenance of Water-Based Fire Protection Systems*
NFPA 72®, National Fire Alarm Code®
NFPA 1961, *Standard on Fire Hose*
NFPA 1962, *Standard for the Care, Use, and Service Testing of Fire Hose Including Couplings and Nozzles*
NFPA 1963, *Standard for Fire Hose Connections*
NFPA 1964, *Standard for Spray Nozzles (Shutoff and Tip)*

Water Mist Systems

CHAPTER 16

Jack R. Mawhinney

The first edition of NFPA 750, *Standard on Water Mist Fire Protection Systems,* came out in 1996. Prior to that year, water mist system installations were almost entirely in the marine sector, particularly in marine machinery spaces and as equivalents to marine sprinklers on passenger ships. Currently, the marine sector is still the largest market for water mist systems. That is beginning to change, however.

Since the first issue of NFPA 750, water mist systems have become increasingly acceptable forms of fire protection in land-based industrial applications, such as petroleum industry machinery spaces, turbine and diesel engine enclosures, a limited number of "control rooms," and wet benches. Water mist systems are now being installed in a growing number of nonindustrial land-based applications as well. Systems have been installed in historic churches in remote rural areas in Norway and Finland, in historic hotels and heritage buildings in Europe and the United Kingdom, in records storage archives and art galleries, and as smoke removal systems in computer room subfloors. These "new" systems have advanced to the stage where regular maintenance involving periodic inspection and testing must be performed.

This chapter focuses on the inspection and testing of water mist systems necessary to ensure that a system will work as intended for many years after installation. It describes the inspection, testing, and maintenance requirements as they apply to systems with common features and provides guidance on what must be done to maintain the system in operating condition.

OVERVIEW

There are a variety of types of water mist systems. Some systems are similar to sprinkler systems, with wet-pipe and thermally activated nozzles. There are also preaction systems, deluge systems, twin-fluid systems, total compartment systems that depend on enclosure integrity to function, and local application systems that require no enclosure. Some systems rely on compressed gases stored in high-pressure cylinders to force stored water out of a tank in a short duration discharge. Other systems, connected to pumps, provide extended discharge for 30 minutes or more. Each type involves its own special hardware and controls, which in turn have specific inspection and testing requirements.

Water mist systems involve features not traditionally used with water-based fire protection systems. These include the use of high-pressure stored gas cylinders; low-pressure water storage cylinders and tanks; filters and strainers; assemblies of positive displacement pumps; pneumatic pressure maintenance pumps; customized electrically or pneumatically activated valves; and nontraditional piping materials, including stainless steel tubing with compression fittings.

Water mist systems introduced the use of very small orifice nozzles, banks of water storage cylinders interconnected with compressed gas cylinders, nontraditional pump systems, electrically activated primary control valves, and complex control software. Compared to past practices, water mist fire protection systems permit the use of hardware and control features that in the past were viewed as failure prone.

Jack R. Mawhinney, P. Eng., is a senior engineer at Hughes Associates, Inc., based in Baltimore, Maryland. Prior to joining Hughes Associates in 1996, he was a senior research officer at the Institute for Research in Construction, National Research Council Canada.

In this light, diligence in carrying out periodic inspection, testing, and maintenance is essential to ensuring long-term viability of these systems.

The newness of water mist systems means that there has not been a lot of time to collect data on long-term reliability and maintainability of their component parts. The primary focus has been on developing design criteria for different hazard applications and on inspecting and testing brand new systems as part of acceptance testing. Factors that affect long-term performance, such as deterioration of quality of stored water, clogged screens and filters, loss of atomizing medium through slow leakage, deterioration of O-ring seals in nozzles or valves, and internal pipe corrosion have not had a chance to show up in the service record.

According to NFPA 750 and standard practice in the fire protection community, the owner or occupant (henceforth referred to as the owner) of the premises where the water mist system is installed is responsible for its inspection and maintenance. In industrial facilities, the owner's own employees may have the expertise necessary to conduct weekly, monthly, and in some cases routine semiannual inspections. A representative of the manufacturer of the water mist system (who is assumed to have technical mastery of the technology) should be involved in semiannual or annual inspection and testing that will involve actually operating the equipment and restoring the system to operating condition.

ACCEPTANCE TESTING OF WATER MIST SYSTEMS

The record of the acceptance tests for a system is the baseline to which future inspection data must be compared. Chapter 9 of NFPA 750, 2000 edition, states the requirements for acceptance testing of newly installed water mist systems. The inspection and maintenance program for a system assumes that all tasks described in NFPA 750 have been performed and the system functioned satisfactorily. Because subsequent routine inspections are intended to ensure that the conditions confirmed by the acceptance test are maintained, reviewing what is involved in the acceptance inspections and testing is worthwhile. Annual flow testing is intended to repeat the most important tests from the acceptance test procedures.

Flushing and Cleaning

Thorough flushing of system piping is necessary to remove welding slag, "purge" paper, metal shavings or filings, thread sealant compounds, or bits of Teflon tape (although forbidden by most manufacturers of water mist systems, the latter appears nonetheless). Every dead-end portion of piping should have been opened and flushed. The acceptance flushing tests will also have revealed unexpected problems with the system, such as water hammer phenomena, transport of debris into the piping, plugging of screens or filters, or too-low inlet pressure on the suction side of a pump.

Hydrostatic Test

The hydrostatic test ensures that the piping system is intact and will not leak under operating conditions. Low-pressure systems are hydrostatically tested at 200 psi (13.8 bar) for 2 hours or 50 psi (3.5 bar) over the working pressure if working pressure is greater than 150 psi (10.4 bar). NFPA 750 requires that intermediate- or high-pressure piping systems be tested to 1.5 times the normal operating pressure for a period of 2 hours. The system documentation will confirm that the piping system passed the appropriate hydrostatic test. If any changes have been made to the piping since the acceptance tests, the entire piping system must be retested to the appropriate pressure.

In water mist systems with open nozzles, it is not easy to repeat a hydrostatic test on a periodic basis. Each nozzle must be removed and replaced with a plug. After the test is conducted, the plugs must be removed and the nozzles reinstalled. It is important to confirm that a piping system is still intact after months or years have passed. Even annual visual inspection of every meter of system piping may not reveal a loose fitting or failed gasket. Where conditions permit, an annual trip test of a system should be performed to demonstrate that water reaches each nozzle and that system pressure is maintained. If there is doubt about the condition of the piping system, there is no alternative but to periodically repeat the hydrostatic test.

Review of Mechanical Components

The review of mechanical components of a system conducted as part of the acceptance test is confirmation that the piping system complies with the design and installation documents and hydraulic calculations. Nozzles may have been added to an area because of obstructions, without regard for the hydraulic performance of the system. Improperly positioned nozzles; inadequate pipe or tube capacity; and absence of hangers, supports, or cylinder restraints should be identified and repaired at the time of the original acceptance testing. From that point forward, periodic maintenance inspections of the mechanical components of the system should be used to confirm that important control valves are open, the detection/control system works, nozzles have not been blocked or obstructed, filters and strainers are in place, and supervisory switches are in working order.

Review of Electrical Components

The importance of the electrical interface to the performance of a water mist system cannot be overemphasized. The requirements of NFPA 70, *National Electrical Code®*; *NFPA 72®*, *National Fire Alarm Code®*; and NFPA 20, *Standard for the Installation of Stationary Pumps for Fire Protection,* that are intended to ensure the reliability of electrical power for fire alarms and fire pumps, among other things, should have been observed in the system design. Some types of water mist systems depend on a separate detection system with control panel to release banks of cylinders, open electrical (solenoid) valves, and start fire pumps. The water mist system may be interlocked electrically with other mechanical systems to shutdown compartment ventilation. Some specialized applications may also interrupt fuel and lubricating oil flow to a turbine. To avoid damaging the turbine, however, the lube-oil supply must be maintained to allow the turbine time to come to rest (coast-down time). The interface between the detection system, the water mist system and related electrically controlled systems are therefore essential features of the system. The system documentation, including acceptance test reports, identifies the characteristics of a particular system that must be maintained.

Preliminary Functional Test

At the time of acceptance testing of a system, the people with specialized knowledge of each associated system are available to confirm the functionality of electrical and mechanical components of the system. During periodic maintenance inspections, however, it is difficult to reunite the disciplines to carry out the same thorough check on performance. Difficult or not, it is imperative that the annual maintenance inspections confirm that interlocks and interdependent electrical systems are still functioning properly. A specialist service company may have to send several people to cover the range of expertise needed to properly inspect and test to verify its intended performance.

System Operational Tests

NFPA 750 requires that each water mist system be put through a full discharge test as the culmination of the acceptance test procedures. The discharge test confirms that the alarm and detection system works, valves open or close, nozzle layout is correct, discharge patterns are correct, screens are not plugged, and that hydraulic performance of the system is as per the design. On marine systems installed as equivalent to sprinklers on cruise ships, water is actually flowed through the system by releasing a mist nozzle on each sectional system and capturing the water in a drum. In new machinery space installations, water-sensitive equipment can be temporarily covered to protect it, so that a full discharge test can be conducted. After the installation phase is over, however, it becomes more difficult to conduct a full discharge test. In many heritage installations, concern about water damage overrides the recommendation for a discharge test. The alternative allowed by NFPA 750 is to confirm the water supply by flowing water through test connections. Where conditions in a facility allow such a test to be conducted as part of a scheduled annual inspection, that opportunity should be taken.

The acceptance test procedures described in Chapter 9 of NFPA 750 capture the elements of water mist systems in general. Whether a system is installed on a ship, on an off-shore drilling platform, or in a land-based application, the general steps to verify functionality and performance are the same. A record of all acceptance tests must be kept as part of the system documentation required by Chapter 8 of NFPA 750.

System Documentation and Test Record

The system documentation and test record is the baseline to which ongoing maintenance inspections and tests are compared. It should contain at least the following information:

1. The name, date, and details of the fire test protocol used as the basis for design.
2. A statement of what the water mist system is expected to achieve, for example, (control), (suppress), or (extinguish) fires larger than _____ kilowatts, in Class fuel for a period of (or within) ______ minutes in a compartment of _____ dimensions.
3. As-built drawings of the water mist system.
4. A copy of the hydraulic calculation documents.
5. As-built drawings of the detection system, alarm/releasing panel, and supervisory circuits.
6. A "cause-and-effects matrix" for the system, showing how the detection system interacts with the water mist system and all ancillary devices or events.
7. Written sequence of operation of the detection/releasing/supervisory system.
8. Written sequence of operation of the water mist system
9. Detailed data sheets on all mechanical components.
10. Detailed data sheets on all electrical components.
11. A copy of the manufacturer's design and installation manual.
12. Manufacturer's written inspection and maintenance instructions
13. Copies of all inspection, testing, and maintenance reports.

Design criteria for water mist systems are determined through full-scale fire testing conducted by each manufacturer. Fire test laboratories or other agencies, such as the International Maritime Organization, establish the fire test protocols. To "pass" one of the test protocols means that the prototype system achieved specific objectives, such as "extinguished concealed test fire," or "controlled temperature at a certain location." At the time of installation all parties should understand the performance objectives of the water mist system. It is important to include a clear statement of the performance objectives in the system documentation for future reference. A complete record of the original installation is the foundation upon which future maintenance practice is built.

SYSTEM INSPECTION, TESTING, AND MAINTENANCE

Chapter 10 of NFPA 750 states that "all components and systems shall be inspected and tested to verify that they function as intended." The owner is responsible for meeting that requirement.

Inspection

Inspecting a component means that a person who is knowledgeable about its intended function examines it, usually visually, to ascertain whether it appears to be in its normal state and that there is no obvious condition that could immediately or at a future time prevent it from performing as it was intended to perform. If anything is noted that is out of the ordinary, a closer examination is warranted and the condition should be corrected if necessary.

Testing

Testing involves the actual operation of devices and is more involved than inspection of the components. There are different levels of testing. Some components can be tested without involving other components in the system. For example, a fire pump can be started weekly to confirm that electric power or diesel fuel is available to the electric or diesel motor and that starting circuits work. There are other tests that cannot be performed without involving most elements of the system. Testing a pressure regulating valve or an automatic sectional control valve may require releasing compressed gas cylinders or flowing water into overhead piping. In that case significant expertise is required to drain the system, replace gas cylinders, and put the system back in operating order.

The same people who do the weekly or monthly inspections can perform some of the periodic tests. For example, the owner's employee can usually perform a weekly pump or compressor startup. The person who performs the more involved annual testing must have the expertise necessary to work with all system elements.

NFPA 750 provides a table that specifies inspection and testing frequencies for generalized equipment likely to be used in water mist systems. Due to the minimal field experience with water mist systems at the time the first edition of NFPA 750 was drafted, this table of testing frequencies in NFPA 750 was generated from established practice with similar components in traditional fire protection systems, such as sprinkler, water spray, carbon dioxide, and halon systems. The result was a set of requirements for inspection and testing of generic equipment that addresses some but not all of the details specific to water mist systems. The manufacturers' inspection, testing, and maintenance advice provided as part of the design and installation manual unique to their product was intended to supplement to supplement the NFPA 750 table.

Table 16-1 combines the recommendations provided in the table in NFPA 750 with manufacturers' recommended inspection and testing frequencies for the variety of equipment encountered in water mist systems. It also incorporates recent field experience gained from acceptance testing and commissioning of water mist systems designed by a

TABLE 16-1 Recommended Frequencies for Inspection and Testing of Water Mist Systems

Item	Task	Weekly	Monthly	Quarterly	Semi-annual	Annual	Other
Water supply (general)	Check source pressure			X			
	Check source quality (*first year)				X*	X	
	Test source pressure, flow, quantity, duration					X	
Water storage tanks (see Figures 16-3 and 16-4)	Check water level (unsupervised)	X					
	Check water level (supervised)		X				
	Check sight glass valves are open		X				
	Check tank gauges, pressure			X			
	Check all valves, appurtenances				X		
	Drain tank, inspect interior, and refill					X	
	Inspect tank condition (corrosion)					X	
	Check water quality					X	
	Check water temperature						Extreme weather
Water storage cylinder (high pressure) (see Figures 16-1 and 16-2)	Check water level (load cells)				X		
	Check water level (unsupervised)			X			
	Check support frame/ restraints					X	
	Check vent plugs at refilling					X	
	Check cylinder pressure on discharge					X	
	Inspect filters on refill connection					X	
Additive storage cylinders	Inspect general condition, corrosion			X			
	Check quantity of additive agent				X		
	Test quality of additive agent					X	
	Test additive injection, full discharge test					X	
Water recirculation tank ("break tank," see Figure 16-7	Check water level (unsupervised)		X				
	Check water level (supervised)			X			
	Inspect supports, attachments				X	X	
	Test low water level alarm					X	
	Check water quality, Drain, flush and re-fill					X	
	Test operation of float operated valve					X	

TABLE 16-1 *Continued*

Item	Task	Weekly	Monthly	Quarterly	Semi-annual	Annual	Other
	Test pressure at outlet during discharge					X	
	Test backflow prevention device (if present)					X	
	Inspect & clean filters, strainers, cyclone separator					X	
Compressed gas cylinders (see Figures 16-1 to 16-6)	Inspect support frame and cylinder restraints			X			
	Check cylinder pressure (unsupervised)		X				
	Check cylinder pressure (supervised)			X			
	Check cylinder control valve is open		X				
	Check cylinder capacity & pressure rating					X	
	Check cylinders meet DOT specifications					X	
	Confirm compressed gas meets specifications (moisture, cylinder pressure)					X	
	Hydrostatic test cylinders						5 to 12 yrs
Plant air, compressors, & receivers	Check air pressure (unsupervised)	X					
	Check air pressure (supervised)		X				
	Start compressor	X					
	Check compressor/receiver capacity, changes				X		
	Check compressed air moisture content					X	
	Clean filters, moisture traps				X		
	Test full capacity, duration, and any changes in other demands					X	
Pumps and drivers (see Figures 5 and 6)	Start pump driver, no flow	X					
	Start pump, full flow test through flow meter or system discharge					X	
	Inspect pump skid, mechanical components				X		
	Check suction supply valves are open			X			
	Check suction NPSH					X	
	Check pump oil levels, lubricated packings					X	
	Check amp draw, volts, RPM					X	

(continues)

TABLE 16-1 *Continued*

Item	Task	Weekly	Monthly	Quarterly	Semi-annual	Annual	Other
	Inspect filters, strainers			X			
	Check seawater transfer valve			X			
	Test pressure relief valve					X	
	Test pressure maintenance valves (unloaders)					X	
	Test programmable controller performance					X	
	Test all pressure/flow switches					X	
	Inspect pressure gauges					X	
Standby pump	Inspect moisture trap, oil injection (pneumatic)			X			
	Check compressed gas supply, inlet air pressure		X				
	Check outlet water (standby) pressure		X				
	Test start/stop pressure settings for standby pressure			X			
Pneumatic valves (see Figure 16-2)	Check cylinder valves, master release valves		X				
	Inspect all tubing associated with release valves			X			
	Test solenoid release of master release valve					X	
	Test manual release of master release valve					X	
	Test operation of slave valves					X	
	Reset all pneumatic cylinder release valves					X	
	Test on–off cycling of valves intended to cycle					X	
System control valves	Check position (open, closed) unsupervised		X				
	Check position (open, closed) supervised			X			
	Check pressure conditions at all valves			X			
	Test operation of solenoid valves					X	
	Test setting on Unloader Valves					X	
	Test setting of Pressure Relief Valves					X	
	Test back-flow prevention devices					X	
Control equipment	Test functionality of releasing panel					X	
	Test programmable logic controller					X	

TABLE 16-1 *Continued*

Item	Task	Weekly	Monthly	Quarterly	Semi-annual	Annual	Other
	Test automatic releasing mechanisms					X	
	Test manual releasing mechanisms					X	
	Inspect detector circuits, detector locations, pull stations			X			
	Test detector system zoning and functionality					X	
	Test pressure & flow switches					X	
	Test trouble (supervisory) signals					X	
	Test remote and local alarms					X	
Water mist system piping & nozzles	Inspect nozzle locations, potential obstructions				X		
	Inspect piping, supports					X	
	Inspect nozzle screens, strainers					X	After discharge test
Enclosure features, interlocks ventilation	Inspect enclosure integrity				X		
	Test interlocked systems (e.g., ventilation shutdown)					X	
	Test shutdown of fuel/ lubrication systems					X	

number of manufacturers. Table 16-1 groups water mist system features under the general categories of water supply, water storage tanks and cylinders, compressed gas cylinders, compressors and receivers, pumps, control valves, control equipment, piping, and enclosure features.

The following discussion identifies important features of the different types of water mist systems and provides commentary on aspects of the systems that require special attention.

High-Pressure, Single-Fluid, Compressed Gas-Driven Systems with Stored Water

COMPRESSED GAS-DRIVEN WATER MIST SYSTEMS

Figures 16-1 and 16-2 are illustrations of high-pressure, compressed gas-driven systems with stored water. The numbers appearing on these illustrations identify key components. High-pressure, gas-driven systems utilize water stored in pressure-rated cylinders (No. 15), which are connected by pneumatic lines to unregulated compressed gas cylinders (No. 2). Although the compressed gas cylinders are at between 2250 and 2900 psi (155 and 200 bar), the water storage cylinders and the piping system are typically pressurized to about 1450 psi (100 bar) when the compressed gas is released, due to pressure losses in the cylinders headers.

These systems have a "decaying pressure" discharge pattern. The system pressure decays over a 10-minute period from about 1450 psi (100 bar) to 290 psi (20 bar) or less. A signal from the releasing panel opens a solenoid valve (No. 6a) that

FIGURE 16-1 High-Pressure, Gas-Driven System with Stored Water (Preengineered System)

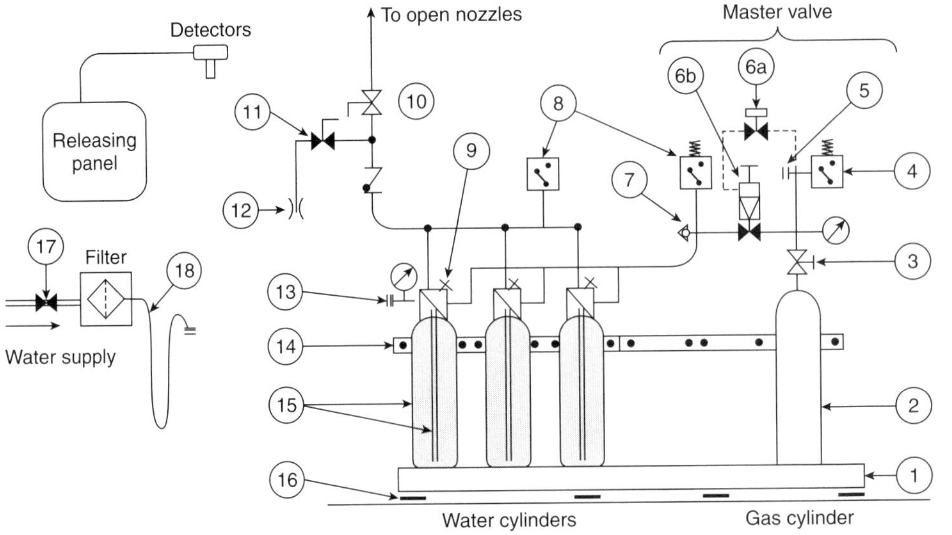

Key

1. Steel base and frame
2. Compressed gas cylinder (driving medium)
3. Cylinder control valve
4. Pressure switch, supervise cylinder pressure
5. Burst disc
6(a) Solenoid operated master release valve
6(b) Pneumatic release valve
7. Micro-leakage valve
8. Pressure switches, alarm if system trips
9. Vent port, for filling water cylinders
10. Primary system or sectional control valve
11. Test connection and drain
12. Test orifice (alternative to full discharge)
13. Cylinder discharge header with filling port
14. Cylinder rack wuth restraints
15. Pressure-rated water cylinders with dip tube
16. Optional load cells to monitor water cylinders
17. Water supply valve, normally closed
18. Filter and hose with adaptor fitting for filling cylinders.

FIGURE 16-2 High-Pressure, Gas-Driven System with Multiple Units Connected for Sequential or Simultaneous Discharge

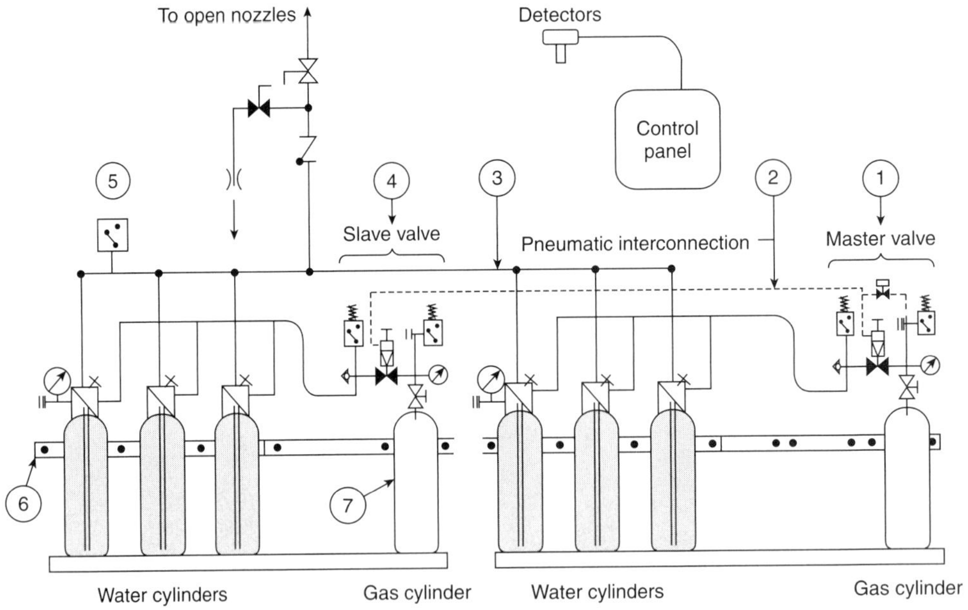

Key

1. Solenoid operated master release valve (with local manual release).
2. Pneumatic tubing interconnecting "master" to "slave" valve
3. Discharge header
4. Pneumatically activated slave valve (no local manual release)
5. Pressure switch, alarm when system trips
6. Cylinder rack with restraints
7. Compressed gas cylinders (driving medium)

directs compressed gas into the pneumatic valve (No. 6b) to cause it to open. The water cylinders must meet U.S. Department of Transportation (DOT) requirements for high pressure.

Compressed Gas Cylinders. Compressed gas cylinders are available in a variety of heights and diameters. It is possible for a purchaser to specify the wrong volume of compressed gas cylinder. It should be patently obvious if the wrong size cylinder has been installed because the cylinder head valve will not be at the correct height to connect to the flexible tubing. However, an inexperienced maintenance person could set the cylinder on a brick to correct for the height difference. It is very important to ensure that correct size compressed gas cylinders are provided each time cylinders are replaced.

Compressed gas cylinders (No. 2 in Figure 16-1) are purchased locally from commercial suppliers. Different countries have different pressure ratings for cylinders. In Europe, Asia, and Australia, 2900-psi (200 bar) cylinders are standard. In North America, commercially available cylinders are typically 2250 psi (155 bar). It is important to check the actual cylinder pressure whenever a new cylinder is installed. It is not uncommon to find a commercially supplied cylinder at, say, 2177 psi (150 bar), slightly less than expected. The system documentation will state what the minimum acceptable cylinder pressure is. The pressure switch (No. 4 in Figure 16-1) is intended to supervise cylinder pressure.

Water Cylinders. Water cylinders are modified gas cylinders, with a special head assembly. Refilling is done (slowly) via a vent port (No. 9 in Figure 16-1). The cylinder is known to be full when water starts to flow out of the open vent ports in adjacent cylinders. Only an experienced technician should perform the filling operation. A permanently mounted water supply (No. 17), with filter and hose valve (No. 18) should be mounted near the unit to facilitate refilling with filtered water.

Once the cylinders are filled and the vent ports sealed, it is not possible to visually confirm that water storage cylinders (No. 15) are full of water. Some end users have mandated the use of load cells to monitor the weight of the cylinders. This is the best way to supervise water level. Regardless of whether the unit is inspected weekly or annually, it is not possible to confirm the presence of water in the cylinders without either opening the vent plugs or using load cells to monitor weight.

Preengineered Systems. Systems similar to the one shown in Figure 16-1 are referred to as preengineered systems. Each set of water cylinders and compressed gas cylinders can supply only a fixed number of nozzles, with a maximum permitted length for each size of tubing. These systems are sometimes assembled as shown in Figure 16-2, in which several sets of water and compressed gas cylinders are connected in a "master/slave" arrangement.

If discharged simultaneously, multiple units increase the number of nozzles that can be supplied. If discharged sequentially, multiple units extend the duration of protection, for example, from 10 to 20 minutes. The integrity of pneumatic tubing connecting master and slave valves (No. 2 in Figure 16-2) (for simultaneous release) is seldom or never supervised. For that reason, pneumatic valves should be inspected carefully for possible problems with the control tubing. Generally it can only be determined to function properly at the time of an annual trip test of the system.

End of Discharge. During discharge testing, it is difficult to identify the end of discharge because the noise and apparent mist generation does not stop abruptly, as it does when a pump is shut off. The *end of discharge* can be identified as the time at which the pressure at the remote nozzle falls below the minimum specified operating pressure for the nozzle. The minimum operating pressure for a high-pressure, single-fluid, gas-driven system, however, is not the same as the minimum operating pressure for a nozzle on a pumped system. The system documentation should specify the minimum pressure that indicates the end of a 10-minute discharge. In annual testing of decaying pressure systems, it is advisable to install a pressure gauge on the most hydraulically remote nozzle during a full-scale discharge.

Low-Pressure, Twin-Fluid, Compressed Gas-Driven Systems with Stored Water

Twin-fluid, compressed gas-driven systems utilize water and compressed gas in two separate piping streams, which are joined at the nozzle to generate water mist. Figure 16-3 shows a low- or intermediate-pressure, gas-driven system utilizing tanks for storage of water, combined with high-pressure compressed gas cylinders and a pressure regulating valve. The water storage tank is connected to a bank of compressed gas cylinders, with the compressed gas used to drive the water out of the tank and

FIGURE 16-3 Low-Pressure, Twin-Fluid, Compressed Gas-Driven System

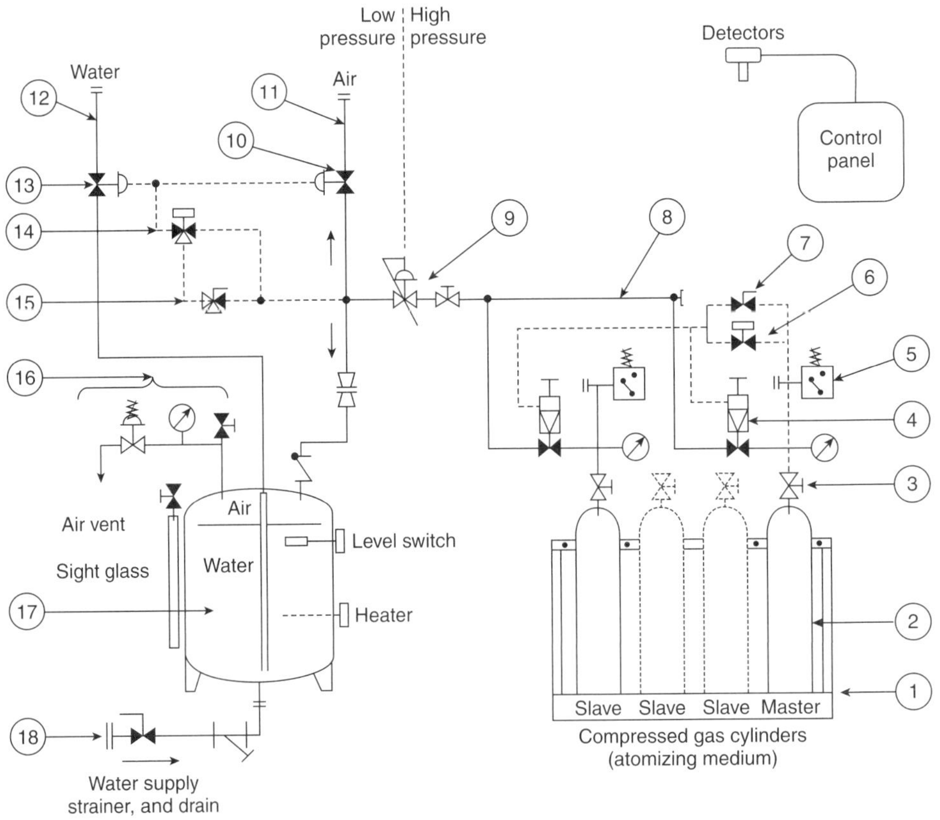

Key

1. Steel base and frame
2. Compressed gas cylinders (atomizing medium)
3. Cylinder control valve
4. Pneumatic cylinder release valve
5. Pressure supervisory switch with burst disc
6. Solenoid operated master release valve
7. Manually operated master release valve
8. ½" high pressure tubing manifold
9. Air pressure control valve (high to low pressure)
10. Air-actuated globe valve (cycle air line)
11. Airline to twin-fluid nozzles (low pressure)
12. Waterline to twin-fluid nozzles (low pressure)
13. Air-actuated globe valve (cycle water line)
14. Low pressure solenoid valves (for operating air-actuated globe valves)
15. Manual release valve (opens globe valves)
16. Pressure gauge, pressure relief valve and vent valve
17. Low pressure rated water tank
18. Drain and re-fill connection with strainer.

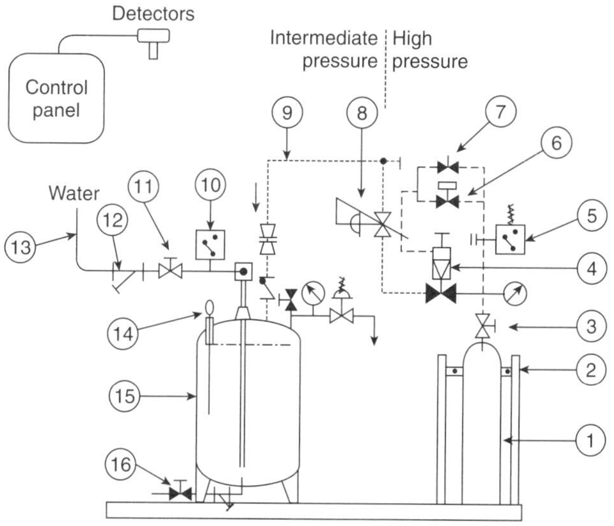

Key

1. Compressed gas cylinder
2. Steel frame and cylinder restraints
3. Cylinder control valve
4. Pneumatic cylinder release valve
5. Pressure switch with burst disc
6. Solenoid operated master release valve
7. Manually operated master release valve
8. Pressure regulating valve, high to low
9. Air-line tubing
10. Pressure switch, alarm on discharge
11. Primary system control valve
12. Strainer on discharge line
13. Water line to nozzles
14. Water level indicator (dipstick)
15. Water tank rated for intermediate-pressure
16. Drain and refill connection with strainer

FIGURE 16-4 Low- or Intermediate-Pressure, Single-Fluid, Compressed Gas-Driven System

into the distribution piping. The pressure at the entry to the water storage tank is regulated to provide a sustained, low to intermediate pressure on the system nozzles for as long as the gas supply lasts. Pressure in the water storage tank and the distribution piping is typically less than 290 psi (20 bar).

Low- and Intermediate-Pressure, Single-Fluid, Compressed Gas-Driven Systems with Stored Water

In low- and intermediate-pressure, single-fluid, compressed gas-driven systems the compressed gas is used solely to drive the water out of the tank and into the distribution piping. Unlike in the high-pressure systems, the water and air pressure that enters the tank and piping in a low- or intermediate-pressure system is regulated (No. 8 in Figure 16-4) so that tank pressure does not exceed a certain value, generally less than 500 psi (34 bar). The discharge pressure at the nozzle is maintained at an approximately constant level until the supply of water or compressed gas is depleted. There are therefore both high- and low-pressure components in the hardware. The system controls include air cylinder valves, regulating valves, solenoid valves, and a fire alarm and control panel (FACP).

PUMP-DRIVEN WATER MIST SYSTEMS

Water mist systems utilizing a pumped water supply represent a third distinct type of hardware. The pump-driven system may be low, intermediate, or high pressure. Some of the intermediate-pressure nozzles operate at a pressure of 191 psi (13 bar). A pump is, therefore, required to raise normal fire protection water pressure to the intermediate range of 191 to 500 psi (13 to 34 bar). Other nozzles in the high-pressure range require the use of pumps to provide system pressure of (2029 psi)

FIGURE 16-5 Pump-Driven High-Pressure Water Mist System

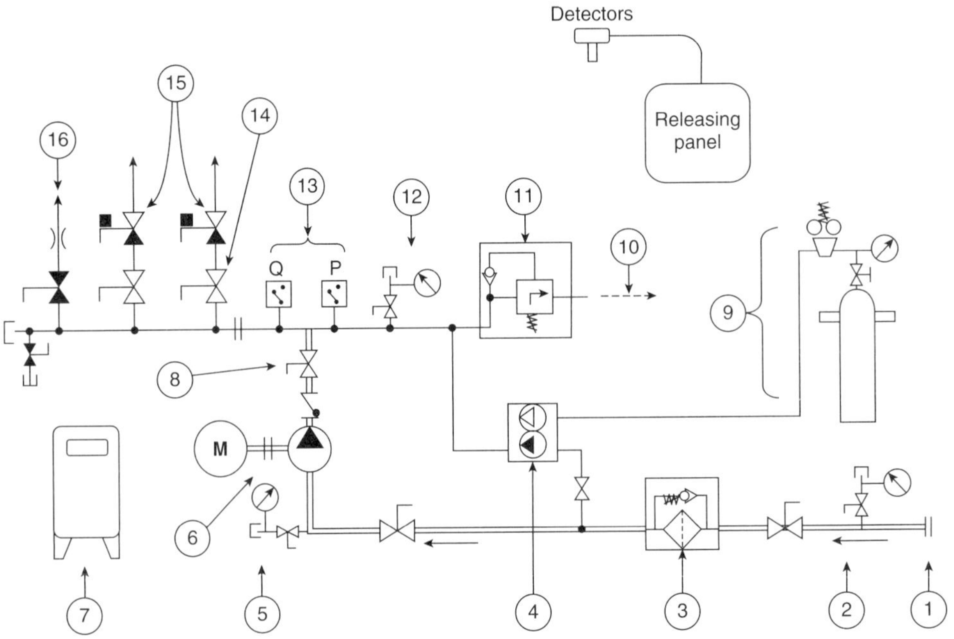

Key

1. Connection to water supply
2. Gauge, low pressure
3. Filters or screens with bypass
4. Stand-by pressure maintenance pump (pneumatic or electric)
5. NPSH gauge (+/–)
6. Pump and driver (electric or diesel)
7. Programmable Controller (PLC)
8. Primary system control valve
9. Air supply and regulator for pneumatic pressure maintenance pump (4)
10. Unloader valve discharge line to drain or break tank.
11. Large capacity unloader valve
12. Pressure gauge
13. Pressure (P) and flow (Q) switches connected to PLC (7)
14. Manual isolation valve
15. Sectional control valves to mist systems (solenoid release)
16. Test connection with flow meter

140 bar and higher. Despite the difference in operating range, both systems involve pumps, power supplies, and controllers. Maintenance and inspection requirements relating to pumps and controllers differ from compressed gas-driven systems.

Figures 16-5 and 16-6 show water mist systems that utilize pumps. They illustrate features typical of both high- and low-pressure water mist systems that make use of pumps. Although the figures represent a particular high-pressure system for a marine application, the system layout could equally well apply to an intermediate-pressure system.

On large, engineered systems protecting many different compartments, pumps are more practical to use than are self-contained cylinders. In large machinery spaces, the minimum flow duration may be 30 minutes or more. It is impractical to provide a dedicated local storage reservoir water for 30-minute duration of flow, so a pump will be used to draw from the plant or ship water supply.

Both figures show pumps supplemented by gas cylinder-driven subsystems. The subsystems are typical of marine installations and may not be present in land-based installations. The pumped system consists of one or more water supplies (fresh water and seawater), a break tank, recirculation lines, screens and filters, pumps and drivers, a controller, and a detection and control panel. The piping system may supply any open or closed nozzles, and there will likely be electrically controlled sectional control valves that must be opened before water can flow to the individual zones. A detection system and programmable logic controller are needed to complete the system controls.

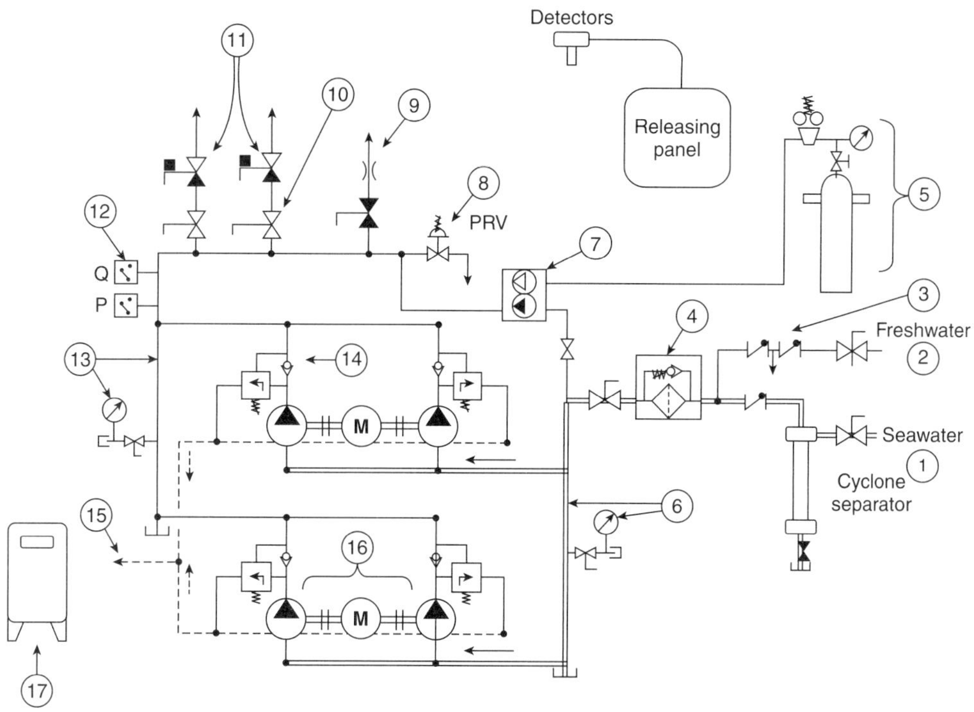

FIGURE 16-6 Positive Displacement (PD) Pump Assembly with Unloader Valves on Each Pump and Pressure Relief Valve on Discharge Manifold

Key

1. Seawater supply with cyclone separator
2. Fresh water supply
3. Backflow device (optional)
4. Filters or screens with bypass
5. Air-supply and regulator for stand-by pressure maintenance pump
6. Suction manifold with NPSH gauge (+/–)
7. Pneumatic stand-by pressure pump
8. Pressure relief valve
9. Test connection with flow meter
10. Isolation valve for sectional valves
11. Solenoid actuated sectional control valves
12. Control sensors, Pressure (P) and Flow (Q)
13. Discharge manifold with pressure gauge
14. Unloader valve (one per pump)
15. Unloader valve discharge by-pass line (to drain or break tank)
16. Positive displacement pumps (2 per motor)
17. Programmable pump controller

The systems shown in the figures are high-pressure systems. The pumps shown are positive displacement (PD) type, which generate much higher pressures than do centrifugal pumps. With the exception of the compressed gas-driven subsystems, the basic diagrams could apply equally well to low- or intermediate-pressure water mist systems. Intermediate-pressure water mist systems can sometimes use a centrifugal pump to boost a strong plant water supply, but PD pumps are also used.

NFPA 25, *Standard for the Inspection, Testing, and Maintenance of Water-Based Fire Protection Systems,* provides general requirements for maintenance of centrifugal fire pump systems. NFPA 20, *Standard for the Installation of Stationary Pumps for Fire Protection,* now has selected requirements for positive displacement pumps. The provisions of NFPA 25 should be followed to the extent they apply to a particular water mist system.

Controllers

Positive displacement pump assemblies are controlled by a programmable controller. For example, the pump system may need to provide different flows to different zones in a multizone system. In a centrifugal pump system, a single pump is designed to meet both the largest and the smallest demand. With PD pumps, the control system must be programmed to start the correct number of pumps for each precalculated demand.

Furthermore, in a standard (centrifugal) fire pump installation, once the pump is started, it must run for a minimum period of time or until it is manually stopped.

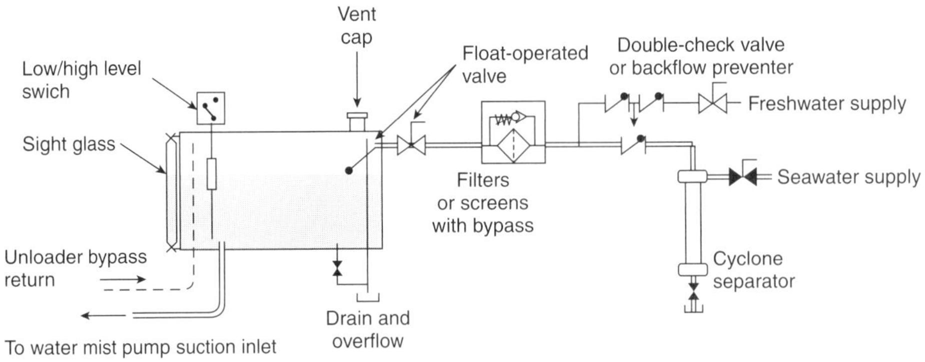

FIGURE 16-7 Break Tank Connection in Supply to Positive Displacement Pumps (Marine Systems)

In contrast, the control logic for PD pumps may automatically stop a motor at any time, if system pressure exceeds a certain limit. This type of shut-off feature must be in accordance with the manufacturer's instructions. Because of a number of such special aspects to pump control, the procedure for conducting the annual flow test on water mist pumps will be more complex than standard fire pump practice.

Break Tanks and Recirculation Lines

Positive displacement pump systems involve unloader valves that direct unused flow either to waste or to a recirculation line. In marine systems especially, unused flow is circulated back to an atmospheric break tank. A break tank is illustrated in Figures 16-7. A float switch in the break tank opens an automatic inlet valve. Fresh water enters the break tank from a filtered water supply. The break tank connects directly to the suction supply line to the pumps. The break tank separates the water mist system from the potable water supply.

It is difficult to check the integrity of the recirculation lines because the opening of any particular unloader valve cannot be predetermined. If the recirculation piping needs to be checked hydrostatically at any time after the original acceptance testing, water must be deliberately added into the system at a convenient point.

Filters

Positive displacement pumps are more sensitive to net positive suction head (NPSH) problems than are centrifugal pumps. The need to filter all water going to the system adds hydraulic resistance to the suction inlet line. If additional filters are needed on the suction line, friction losses in the suction supply may become too high. A suction booster pump may be used to increase the suction line pressure to overcome friction losses through the filters. Starting of the PD pumps must then be synchronized with starting of the suction booster pump.

It is important that filters be able to pass the maximum system demand flow rate for an extended period of time, without clogging. Unlike with manual-fill systems, which only need to pass the flow needed to refill a reservoir in a reasonable period of time, the filters on the suction supply to a pumped system have to be sized for the maximum system flow rate. Frequent examination of filters and the possible addition of instrumentation to identify plugged filters during a flow test may be needed.

Some high-pressure pump unit suppliers install a gauge on the suction side of the pumps that is not suitable for reading in the range of suction line pressures. This can be corrected by the installation of a new gauge with the appropriate scale. At the other end of the spectrum, another pump system manufacturer uses two banks of filters in parallel, with a pressure transducer connected to the pump controller. If the head loss across the first bank of filters becomes too high (as the filters become clogged), a solenoid valve opens and reroutes the flow through the redundant bank of filters. As an owner becomes more familiar with the type of problems caused by the filters, modifications to the filter lines should be implemented if necessary.

The maintenance plan for a water mist pump system must be designed to evaluate the specific details of the suction supply to the pumps. Installing temporary supplemental instrumentation to measure performance factors during the annual discharge test is recommended.

Freshwater/Seawater Transfer

On marine systems, water supply to the fire pumps may be taken from freshwater or seawater sources. Hardware to accomplish the transitions must be maintained for reliability, which means that inspection and testing of the transfer valves must be included in the annual testing. Concern about corrosion caused by seawater in the piping system inhibits the desire to conduct full-scale flow test from the seawater source. If one is not already present, installing a permanent test connection line that can be used to test the seawater source is recommended. Afterward it can easily be flushed with freshwater without involving overhead piping.

Pneumatic Jockey Pump

Water mist systems may utilize a pneumatic jockey pump to maintain the base system standby pressure. A clean air supply to run the pneumatic pump must be maintained. This source is sometimes provided as a cylinder of compressed gas (nitrogen) with a regulator. Air can also be taken from a dedicated compressor or plant air supply.

TESTING OF WATER STORAGE TANKS

Water Storage Tank Components

As illustrated earlier in Figures 16-3 and 16-4, the following typical components are associated with water storage tanks:

1. Tank rated for at least 500 psig (34 bar) pressure and lined to prevent internal corrosion
2. Top-mounted water discharge line with internal dip tube
3. Top-mounted air inlet line with orifice disk and check valve
4. Sight glass, dip-stick or weigh-scale for indication of water level
5. Air vent valves
6. Pressure gauges
7. Relief valve
8. Level switch (optional)
9. Immersion heater (optional)
10. External insulation (optional)
11. Bottom-mounted combined drain and refill line
12. Strainer on the refill line; in some cases a strainer on the outlet line
13. Electrical connection to ground

The owner's primary maintenance concerns with the water storage tank should be to confirm that (1) it contains the right quantity of water, (2) the water quality is acceptable, and (3) the control valves are set correctly. It is also important to confirm that the ambient temperatures at the tank location are within the acceptable range of 40° to 130°F (4° to 54°C) and that there are no signs of leakage or corrosion.

Tank Sight Glass

In tanks equipped with a tank sight glass, the tank fill level can be checked by inspecting the sight glass. The tank sight glass will be calibrated to show, as a minimum, the full, refill, and low-level status of the water in the tank. In addition to being used to check the capacity of the tank, discoloration or presence of particulates visible in the sight glass may be a sign of potential contamination in the water. While the level in the water storage tank should be checked weekly (unsupervised arrangement), manufacturer's practice or recommendations may also suggest that the valves on the sight glass be verified to be in the open position.

The system shown in Figure 16-3 has both a sight glass and a level switch that indicate that the water tank is filled to the proper level. Because of the level switch, the tank qualifies as "supervised"; therefore a monthly visual check of the sight glass is all that is required. The level switch will be monitored by the fire alarm control panel, which will indicate a trouble signal if the water level is too low. Presumably a person will monitor the panel and notice the trouble signal.

A low-level trouble signal will require that a maintenance person identify the reason for the drop in water level, fix the problem, and refill the tank to the proper level. Some manufacturers have installed a dipstick device to permit regular confirmation of water level in the vessel. The maintenance personnel must remove a cap and check the level indicator on the dipstick. Such devices do not qualify as supervised, so must be checked weekly.

The recommended frequency for testing the water level switch is annually, at the time the tank is scheduled to be drained and refilled. In fact the operation of the level switch should be checked any time that the tank is emptied and refilled.

Pressure Relief Valve

Most of the storage tank designs incorporate a pressure relief valve on the tank discharge line, which is intended to protect against overpressurizing the tank if the air pressure control valve malfunctions. Ordinarily, the tank is at atmospheric pressure and is only pressurized if the gas cylinders are released or perhaps during filling if the water pressure is high enough and the air vent valves are not opened.

The pressure relief valve should be manually operated semiannually. Even during the annual discharge test, the air pressure control regulator valve setting should not be changed to purposely overpressurize the tank. The manufacturer should be consulted about how to confirm the proper performance of this safety feature.

The relief valves sometimes installed on the discharge side of a pressure regulating valve are provided to ensure that system components are not subjected to pressure that may damage any downstream components. Pressure relief valves should be manually operated on a semiannual basis to confirm their functionality.

Water Quality

NFPA 750 requires the water source for filling the tank to be of potable water quality with respect to dissolved minerals and particulates. The quality of the refill water source should be checked before refilling. Early experience with similar storage tanks on Alaska's North Slope found that lake water used to fill the tank had bacteria content that resulted in bacteriological growth (algae) inside the tank. The strainer on the fill line was of no use for stopping bacteria entering the tank. Algae growth cannot be detected from outside the tank unless it happens to discolor the sight glass. If it does occur, and there are no strainers or filters on the discharge line from the tank, there is nothing to stop algae from being pushed into the piping system and plugging nozzles and special pressure regulating valves.

A combined refill and drain line with a strainer at the bottom of the tank is utilized to ensure that no large particulate matter enters the tank during filling. The drain line should not have a strainer, so that debris or algae growth can be washed out of the tank without restriction. Some manufacturers install a strainer on the outlet line from the water storage vessel. In that case, the strainer must pass the full discharge flow without excessive hydraulic head loss.

To identify whether bacteriological growth is occurring, a water sample should be taken from the tank via the drain valve. Over the first year, sampling and inspecting the water quality for bacteriological activity at quarterly intervals are recommended. It may be necessary to send the samples to a laboratory for analysis. If algae growth is identified during a quarterly inspection, the following changes to maintenance procedures for systems that utilize water stored in closed tanks or cylinders should be considered:

- Using bromination or chlorination during the filling operation to kill bacteria. (Biocides other than those used in potable water systems should not be used because of potential health safety concerns. If the distribution system is comprised of stainless steel pipe or tube, ozone treatment should be considered instead. Stainless steel is susceptible to chlorine corrosion; bromine, being a similar halogen, may have similar adverse effects.)
- Installing additional strainers or filters on the outlet of the tank. This could have a negative impact on the hydraulics of the system. The flow capacity and hydraulic resistance of the additional strainers must be accounted for so that minimum nozzle pressures are still achieved.
- Ensuring that all strainers are easily accessible for checking (i.e., not located under the tank) and that the screen can be removed for examination without having to empty the tank.
- Inspecting water quality quarterly for the first year, then semiannually thereafter only if water quality is consistently acceptable throughout the first year.
- Analyzing water quality (laboratory) semiannually the first year, then annually thereafter if the water quality is consistently acceptable over the course of the first year.

DISCHARGE TESTING

Full Discharge Test

The annual full discharge testing of the water mist system is intended to verify the operation of all components, from the detection system to the releasing valves and regulators. Only the manufacturer's representative should undertake this testing. A full discharge test is the best way to confirm the performance of all components, including duration of discharge and timing cycles. Only by flowing water through the actual distribution system and nozzles can one be sure that both the water supply and the compressed gas pressure are sufficient to sustain the desired duration and that water is delivered to the nozzles at the required pressure. The owner needs to provide the opportunity to conduct this disruptive test.

Modified Discharge Test

Although water-sensitive equipment can be covered by plastic, for various reasons it is not always possible to conduct a full compartment discharge test of every water mist system. Provided that a full discharge test was conducted at the time of the acceptance testing of the new system, it could be acceptable to conduct a modified discharge test during the annual test period.

A modified discharge test could be conducted by means of a test connection designed to simulate the hydraulic demands of the system. Orifice disks installed on separate water and air lines (in a twin-fluid system, for example) and piped to a suitable location could be used to simulate the actual flow rates. Test connections for such purposes are not always a standard feature. A number of the first water mist systems to be installed did not include test connections. If the difficulty of conducting effective annual tests creates problems for the ongoing maintenance of the system, test connections should be retrofitted to piping.

Cycling of Systems

The cycling feature of a particular system is usually reserved for use in water mist applications for marine service. This feature allows the flow rate and discharge pressure to vary under a range of parameters. The cycling modes discussed in NFPA 750 include constant, decaying, uniform, and nonuniform modes. Cycling permits the system to replenish after a predetermined discharge time or to utilize a variable flow and pressure discharge rate to provide compartment cooling following a fire. Inspecting the boundaries of the system performance under the conditions of cycling requires the use of fairly skilled technicians to determine if the system performs at or near its design level.

Discharge Time

The specific hazard being protected against, the size of the area or compartment being protected, and the configuration of the hazard(s) in the space can all have an influence on the design discharge time expected for a particular system. A 30-minute minimum discharge time is specified by NFPA 750. This fluid system or systems that utilize a compressed gas source must maintain a large enough quantity of each fluid (water plus compressed gas source) for the anticipated duration. Any evaluation of the system in future years must be able to show that any system discharge still meets the original time parameters for mist application.

TESTING OF OVERHEAD PIPING, VALVES, NOZZLES, AND DETECTION EQUIPMENT

Table 16-1 recommends periodic inspection and testing of all functional elements of the water mist system. These include the sectional control valves, drains, flow switches and alarm devices, piping, hangers, and nozzles. Procedures for such inspection or testing are similar to those for other types of fire protection systems. Practices described in *NFPA 72®, National Fire Alarm Code®*, and NFPA 25 should be followed. There are several unique aspects of water mist systems that should be highlighted.

Sectional Control Valves

It is particularly important to test the sectional control valves (see No. 15 in Figure 16-5 and No. 11 in Figure 16-6) used in large water mist systems. These valves are electrically activated. Compared to standard fire protection valves, they are of a nonstandard valve design. They may be nonindicating; knowing how to open them manually may not be clear. Even during a trip test, it is not possible to tell by visual inspection whether they are fully or partially open. On high-pressure systems, friction loss through sectional control valves may be very high. In order to keep friction losses within reasonable bounds, the total flow to a system may be divided among two or more sectional valves. This situation introduces the possibility that one of the valves in the group could fail to open and cause the hydraulic performance of the system to be adversely affected. The only way to test sectional control valves is to open them under full pressure and flow water.

In facilities where a full flow test can be conducted annually, the performance of sectional valves can be tested. If it is not possible to do a full system discharge, an alternative means of testing the functionality of the valves must be found. One possibility is to select nozzles in a limited area where water can be discharged without damage, and then to flow water through a limited portion of the overall system. Alternatively, it may be necessary to design the system so that there is a test line on each individual zone in a water mist system on the system (downstream) side of each sectional control valve.

If a sectional control valve does not "seat" properly, water will leak past the valve and begin to fill the overhead piping. On a large system with high ceilings, the leakage will not be noticed until water starts to "dribble" out of the nozzle. One way to supervise the system for slow leakage is to install leakage traps on the discharge side of each sectional control valve. Any leakage slowly fills a drum-drain until there is enough to be sensed with a liquid sensor. A trouble indication can then be given in the main control room.

Nozzles

Water mist nozzles must be protected against plugging by use of filters and screens. Annual checking and cleaning of screens or filters is important. Manufacturers have nozzles with different K factors, orifice diameters, and orifice patterns. A single system may utilize more than one type of nozzle. It is extremely important that the person who removes nozzles for inspection, or possibly replaces nozzles for whatever purpose, knows what characteristics each nozzle must have. Information on the type of nozzles and where they are installed is contained in the system documentation. As indicated earlier, the system documentation is the foundation for ongoing maintenance and inspection. The maintenance personnel must be familiar with the details of what types of nozzles were installed in the facility and should perform the following inspection:

1. Ensure that correct types of nozzles are installed.
2. Ensure that thermally activated nozzles with glass bulbs are the correct temperature rating for the location.
3. Remove several nozzles and inspect condition of screen and O-ring seals.
4. Ensure that structural members or stored materials will not obstruct the discharge pattern.
5. Ensure that nozzle orifices are not becoming plugged with external grime.
6. Ensure that nozzle dustcaps (if used) are in place and that they dislodge as required upon activation of the system.
7. Ensure that nozzles removed after flow testing are cleaned of grit, silt, sludge, or particles attached to the internal screens or filters.

Detection Equipment, Controls, Interlocks, and Ancillary Equipment

With the exception of water mist systems that are sprinkler equivalents and use thermally activated (automatic nozzles), water mist systems rely on a separate detection system to activate. Inspection and testing procedures therefore must include

TABLE 16-2 Inspection and Testing Recommendations for General System Components

Item (condition)	Activity	Frequency
System piping	Inspect, pipe, tube, hangers	Annually
Nozzles—position, obstructions	Inspect	Semiannually
Nozzles, interior—plugging, screens, O-ring seals	Inspect	Annually
Nozzles, exterior—dust covers dislodge on pressure	Test	Annually
Detectors	Inspect	As per *NFPA 72*
Control equipment (unsupervised) (functions, fuses, interfaces, primary power, remote alarm)	Test	Quarterly
Control equipment (supervised) (functions, fuses, interfaces, primary power, remote alarm)	Inspect	Semiannually
Control equipment, functional test	Test	Annually

the detection system. The type of detection is often dictated by the listing for the system. For example, turbine enclosures and machinery spaces listed by Factory Mutual are required to use thermal detectors. The listing also indicates that certain shutdown events must be synchronized with the release of the system. The ventilation system should be shut down, door holders released, dampers closed, and lubrication or fuel lines shut off. Obviously, the annual testing of a water mist system must confirm the proper operation of all events associated with release of the water mist system. See Table 16-2 for inspection and testing recommendations for general system components.

INSPECTION, TESTING, AND MAINTENANCE RECORDS

A written record of all inspection, testing, and maintenance activity on a system should be kept with the system documentation. Appendix of NFPA 25 provides examples of detailed forms for inspection and testing reports. The examples are based on standard sprinkler and standpipe systems and water tanks. As part of the water mist system manuals and documentation, the installer should provide a set of maintenance inspection report forms customized to the conditions and equipment installed. Items to be included in weekly, monthly, quarterly, semiannual, and annual inspections and tests should be identified on separate forms. The forms can be used as a checklist to ensure that all tasks are carried out.

SUMMARY

In the world of water-based fire protection and extinguishing systems, water mist systems are a relatively recent technology. Given that the first design and installation for these systems did not appear until 1996 and given the relatively small number of installations that have been completed thus far, it may be some time before the frequency for inspection, testing, and maintenance attributes for water mist systems is perfected. Water mist systems require the use of specifically designed nozzles that are now susceptible to blockage if the water supply is not adequately protected. Other systems utilize complete control circuits and loss controllers to initiate the sequence of system discharge. Like its related, but much different, sprinkler system counterpart, water mist systems require not only vigilance but also a skilled and knowledgeable individual to carry out some of the more complex procedures associated with the inspection of the system.

BIBLIOGRAPHY

NFPA Codes, Standards, and Recommended Practices

See the latest version of The NFPA Catalog for availability of current editions of the following documents.

NFPA 20, *Standard for the Installation of Stationary Pumps for Fire Protection*
NFPA 25, *Standard for the Inspection, Testing, and Maintenance of Water-Based Fire Protection Systems*
NFPA 70, *National Electrical Code®*
NFPA 72®, National Fire Alarm Code®
NFPA 750, *Standard on Water Mist Fire Protection Systems*

Special Agent Extinguishing Systems

CHAPTER 17

Gerard G. Back III

The most widely used special agent extinguishing systems are carbon dioxide, halogenated agents, and dry chemicals. Other clean agent extinguishing systems (i.e., halocarbons and inert gases) are discussed in Chapter 18, "Clean Agent Extinguishing Systems"; and wet chemical systems are discussed in Chapter 55, "Protection of Commercial Cooking Equipment." Because these systems dishazge a fixed agent quantity, it is important that all hazard areas be protected simultaneously.

Systems using these special agents have the following components: detection and control equipment, agent release devices, agent storage containers, agent distribution systems such as pipes and nozzles, and ancillary devices such as door closures and damper releases. These systems initially are designed for a defined hazard. However, changes in the hazard, such as the layout of the equipment being protected, changes to the enclosure surrounding the hazard, and changes in the type of fuel, could affect the system's ability to extinguish a fire.

GENERAL SYSTEM INFORMATION

Design Types

Special agent systems can be categorized into one of three basic types: total flooding systems, local application systems, and hand hose line systems.

Total Flooding Systems. Total flooding systems (Figure 17-1) discharge extinguishing agent into an enclosure to provide a uniform fire extinguishing concentration throughout the entire enclosure. Openings in the enclosure sometimes can be compensated for by providing automatic closure devices or by adding extinguishing agent. Note that although it is recognized that use of Halon 1301 and Halon 1211

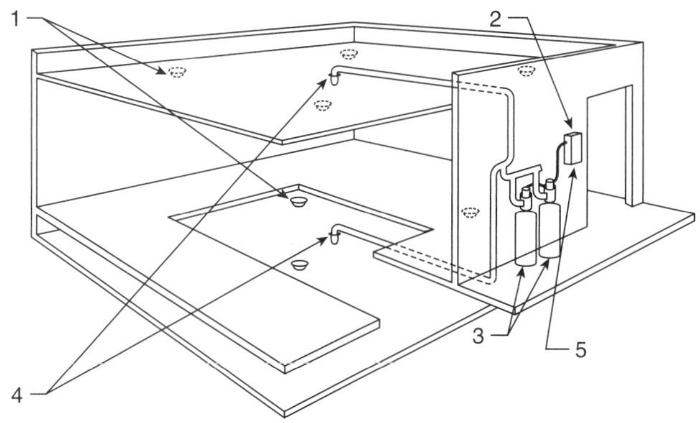

FIGURE 17-1 Total Flooding Halogenated Extinguishing Agent System Installed in Room with Raised Floor
Source: The Ansul Company.

1. Automatic fire detectors installed both in room proper and in underfloor area.
2. Control panel connected between fire detectors and cylinder release valves.
3. Storage containers for room proper and underfloor area.
4. Discharge nozzles installed both in room proper and in underfloor area.
5. Control panel might also sound alarms, close doors, and shut off power to the area.

Gerard Back, a senior engineer with Hughes Associates, Inc., has a B.S. in mechanical engineering and an M.S. in fire protection engineering, both from the University of Maryland. He has served as project manager/engineer on a wide range of R&D fire protection programs over the past 16 years. His current primary focus has been in the development and testing of total flooding halon replacement agents, both gaseous and other technologies.

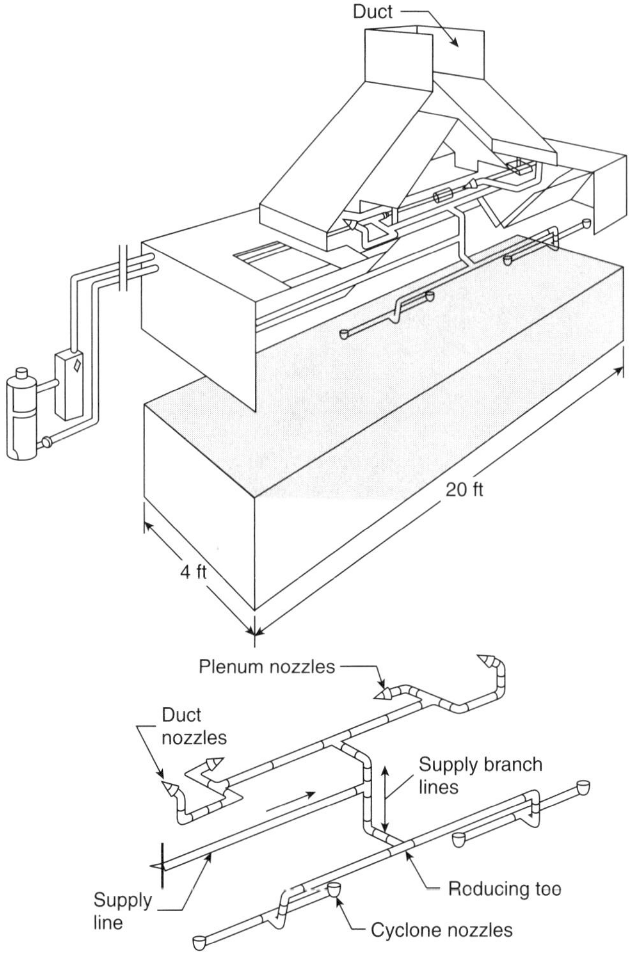

FIGURE 17-2 Typical, Single 30-Pound Cartridge-Operated Dry Chemical System with Fusible Links for Automatic Operation Engineered for Kitchen Range, Hood, Duct, and Fryer Fire Protection
Source: The Ansul Company.

systems has been curtailed for new installations, finding existing installations in service is still common.

Local Application Systems. Local application systems (Figure 17-2) discharge the agent directly onto the burning material or object being protected without relying on an enclosure to retain the agent. Dry chemical extinguishing systems are no longer utilized for protection of commercial cooking equipment. Replacement systems such as wet chemical or liquid agent systems are now more common. See Chapter 55 for a discussion of these systems. Local application systems are typically used to protect printing presses, dip and quench tanks, spray booths, oil-filled electric transformers, vapor vents, and so on.

Hand Hose Line Systems. Hand hose line systems (Figure 17-3) consist of a supply of extinguishing agent, such as carbon dioxide, Halon 1211, or dry chemical, and one or more hand hose lines that allow manual delivery of the agent to the fire. The hose lines are connected to the agent container either directly or by means of intermediate piping.

All three system types are available as either engineered or preengineered systems. An engineered system is one in which individual calculations and design are required to determine the agent flow rate, the size of the piping, nozzle pressures, and so on. A preengineered system, sometimes called a package system, is one in which minimum

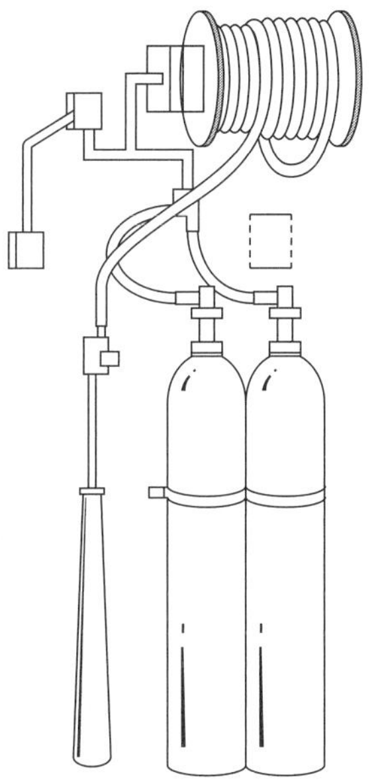

FIGURE 17-3 Carbon Dioxide Hand Hose Extinguishing System with Hose Mounted on a Reel

and maximum parameters have been predetermined and confirmed by an independent testing laboratory. Installation within the listed limits ensures adequate flow rate, pressure, and pattern coverage without individual calculation.

Inspection and Maintenance

These systems should be thoroughly inspected and tested for proper operation by competent personnel. A documented report with recommendations should be filed with the owner. The manufacturer, installing contractor, or other qualified organizations can provide regular service contracts.

The tables given in this chapter outline the main aspects of the inspection and maintenance procedure and should be used for guidance only. For more details on these procedures and on the frequency of the different verifications and tests, refer to the applicable NFPA standards (see Bibliography) and to the manufacturers' manuals. In addition to inspecting the system and the protected hazard, the personnel using and operating the system or any personnel who might be present during a fire should have adequate training.

CARBON DIOXIDE SYSTEMS

A number of properties make carbon dioxide a desirable fire extinguishing agent. It does not react with most substances, it does not conduct electricity, it provides its own discharge pressure, and it does not leave a residue.

Carbon dioxide is effective as an extinguishing agent primarily because it reduces the oxygen content to a point at which the atmosphere no longer supports combustion. Under suitable conditions, the available cooling effect also can be helpful, especially in local application.

Safety Considerations

Extinguishing concentrations of carbon dioxide can produce unconsciousness and death. Therefore, carbon dioxide total flooding systems should not be used in spaces that are normally occupied. To prevent accidental discharge, a "lock-out" is provided when persons not familiar with the systems and their operation are present in a protected space. Authorities responsible for continuity of fire protection must be notified of lock-out and subsequent restoration of the system. For total flooding hazards that are normally unoccupied but in which personnel might be present for maintenance or other purposes, there must be some means to warn the occupants of an impending discharge. This requires a time delay to ensure that they can be evacuated before the discharge occurs. It might be difficult to ensure evacuation if the space is large or if egress is impeded in any way by obstacles or complicated passageways.

Type of Storage

The carbon dioxide supply can be stored in high- or low-pressure containers. High-pressure containers (~850 psi [5860 kPa]), usually cylinders, are designed to store liquid carbon dioxide at ambient temperature. Because the temperature affects the pressure, it is important to store the cylinder in an area where temperatures will be within the listed limits. Low-pressure containers (~300 psi [2068 kPa]) are pressure vessels designed to maintain the temperature of carbon dioxide at about 0°F (−18°C) by means of insulation and refrigeration.

Inspection and Maintenance

Table 17-1 lists the main points of the inspection and maintenance procedure for carbon dioxide systems. Refer to NFPA 12, *Standard on Carbon Dioxide Extinguishing Systems,* and to the system manufacturer's manual for more complete information. Inspection and maintenance should be carried out by competent personnel only.

TABLE 17-1 Testing and Inspection Guidelines for Carbon Dioxide Systems

Verification or Test	Frequency
1. The system	
Overall physical appearance	Every 30 days
Any changes in the size or type of hazard	Every 6 mos
2. Supervised circuits	Every yr
3. Control panel (all functions)	Every yr
4. Power supply	Every yr
5. Emergency power	Every yr
6. Detectors (test, clean, and check wiring)	Every 6 mos
7. Verification of the time delay	Every yr
8. Alarms (operation, warning signs properly displayed)	Every yr
9. Selector valves (if applicable)	Every yr
10. Ancillary functions (closure of dampers and doors, shutdown of equipment, and so on)	Every yr
11. Manual releases (operation, accessibility)	Every yr
12. Piping (supports, blockage, integrity, and so on)	Every yr
13. Nozzles	Every yr
14. Containers (physical condition, weight)	Every 6 mos
15. Release devices	Every yr
16. Testing of hoses	Every 5 yrs
17. Hydrostatic testing of high-pressure containers	12 yrs; after a discharge if more than 5 yrs since last test.
18. Liquid level gauge of low-pressure containers	Every wk

For all of these verifications and tests, the actuating controls must be removed from the agent containers to avoid accidental discharge. Frequency is minimum recommended.

HALOGENATED AGENTS SYSTEMS

Halogenated extinguishing agents, or halons, have a number of unique fire protection qualities. In addition to their ability to extinguish flames, they leave no residue to clean up after a fire, and they do not cause thermal shock to delicate equipment. The two halons most widely used in North America are Halon 1301 [bromotrifluoromethane (CF_3Br)] and Halon 1211 [bromochlorodifluoromethane (CF_2ClBr)].

Fire protection halons are currently the focus of worldwide attention because they have been linked to the destruction of the Earth's stratospheric ozone layer. The global production of halons is being phased out by the Montreal Protocol, a far-reaching treaty signed by most countries of the world. Because future supplies of fire protection halons will be very limited, new installations of halon systems are unlikely. Meanwhile, many halon systems exist, and they must be inspected periodically and maintained.

Safety Considerations

Experience and testing have shown that personnel can be exposed to Halon 1301 and Halon 1211 vapors in low concentration for brief periods without serious risks. However, unnecessary exposure is not recommended. Exposure to high concentrations—10 percent for Halon 1301 and 4 percent for Halon 1211—can present a health hazard to personnel. Halon 1211 should not be used as a total flooding agent in enclosures that are normally occupied.

Type of Storage

Total flooding systems can have a modular design or a central storage design. In modular systems, a nozzle is connected to a halon container with little or no piping. Containers are located throughout the space to be protected so that the concentration of the agent will be uniform upon discharge. In central storage systems, halon containers are connected to a manifold, and the halon is delivered to discharge nozzles through a piping network.

Inspection and Maintenance

Table 17-2 gives the main items of the inspection and maintenance procedure for halogenated systems. Refer to NFPA 12A, *Standard on Halon 1301 Fire Extinguishing Systems* and to the system manufacturer's manual for more complete information. Full inspection and maintenance should be carried out by competent personnel only.

DRY CHEMICAL SYSTEMS

Dry chemical extinguishing agents are known as ordinary dry chemicals and multipurpose dry chemicals. The former are used to combat fires involving flammable liquids (Class B) and electrical equipment (Class C). The latter are effective on ordinary combustibles (Class A), on flammable liquids (Class B), and on electrical equipment (Class C).

Typical dry chemical agents include sodium bicarbonate, potassium bicarbonate, monoammonium phosphate, potassium chloride, or urea potassium bicarbonate as base materials. Dry chemical extinguishing agents should not be confused with dry powder agents, which were developed for use on combustible metals. Extinguishing agents used on combustible metal fires are discussed in Chapter 45, "Combustible Metals."

Extinguishing Properties

Ordinary dry chemical is used primarily to extinguish flammable liquids fires. However, its use might not result in permanent extinguishment if ignition sources, such

TABLE 17-2 Testing and Inspection Guidelines for Halogenated Agents Systems

Verification or test	Frequency
1. Detection and actuation system	Every 6 mos
Detectors checked and cleaned	
Supervision features checked	
Operation of actuating controls (removed from containers)	
Operation of manual operating controls	
2. Containers	
Visual examination	Every 6 mos
Verification of agent quantity and pressure	Every 6 mos
3. Piping and nozzles	
Visual verification for any evidence of corrosion or obstruction, and proper position and alignment of nozzles	Every 6 mos
Visual inspection of hoses	Every 6 mos
Hydrostatic testing of hoses	Every 5 yrs
4. Auxiliary equipment	Every 6 mos

For all of these verifications and tests, the actuating controls must be removed from the agent containers to avoid accidental discharge. Frequency is minimum recommended.

as hot metal surfaces, are still present. Dry chemicals might not extinguish deep-seated Class A fires and are not suitable for fires in materials that supply their own oxygen for combustion.

Dry chemicals should not be used in installations containing delicate electronic equipment because the insulating properties of the chemicals can render such equipment inoperative. Some dry chemicals are slightly corrosive, especially those with a monoammonium phosphate based, and should therefore be removed from all undamaged surfaces as soon as possible after extinguishment.

Safety Considerations

The discharge of dry chemicals can create hazards to personnel by reducing visibility and temporarily making breathing difficult.

Type of Storage

Dry chemicals are stored in pressure containers either at atmospheric pressure until the system is actuated or at the pressure of the internally stored expellant gas.

Inspection and Maintenance

Table 17-3 lists the main points of the inspection and maintenance procedure for dry chemical systems. Refer to NFPA 17, *Standard for Dry Chemical Extinguishing Systems,* and to the system manufacturer's manual for more complete information. Full inspection and maintenance should be carried out by competent personnel only.

SUMMARY

Carbon dioxide, halogenated agents, and dry chemicals are the three most widely used special agent extinguishing systems. Carbon dioxide extinguishing agents are effective because they reduce the oxygen content to a point at which the atmosphere no longer supports combustion. Halogenated extinguishing agents are unique in that they leave no residue after a fire and do not cause thermal shock to delicate

TABLE 17-3 Testing and Inspection Guidelines for Dry Chemical Systems

Verification or test	Frequency
1. Check location of the extinguishing system	Every mo
2. Manual actuators are not obstructed	Every mo
3. Tamper indicators and seals are intact	Every mo
4. Maintenance tag or certificate is in place	Every mo
5. No obvious physical damage	Every mo
6. Pressure gauge(s), if any, in operable range	Every mo
7. No modification of the hazard	Every 6 mo
8. Examine all detectors, containers, releasing devices, piping, hose assemblies, nozzles, alarms, and auxiliary equipment	Every 6 mo
9. Verify that agent distribution piping is not obstructed	Every 6 mos
10. Examine the dry chemical:	
Atmospheric pressure systems	Every 6 mo
Stored pressure systems	Every 6 yrs
11. Operational test of the system and all of its functions (excluding discharge)	Every 6 mos
12. Replacement of the fusible links	Every yr
13. Hydrostatic test	Every 12 yrs

Note: For all of these verifications and tests, the actuating controls must be removed from the agent containers to avoid accidental discharge. Frequency is minimum recommended.

equipment; they are, however, linked to the destruction of the ozone layer and are thus being phased out. Dry chemical agents are used primarily on flammable liquid fires. Personnel who inspect facilities where special agent extinguishing systems are used must become familiar with each system's design and components, safety considerations, types of storage, and required inspection and maintenance procedures. For details on these systems and maintenance procedures, inspectors should refer to the applicable NFPA standards (see Bibliography).

BIBLIOGRAPHY

NFPA Codes, Standards, and Recommended Practices

See the latest version of The NFPA Catalog for availability of current editions of the following documents.

NFPA 12, *Standard on Carbon Dioxide Extinguishing Systems*
NFPA 12A, *Standard on Halon 1301 Fire Extinguishing Systems*
NFPA 12B, *Standard on Halon 1211 Fire Extinguishing Systems*
NFPA 17, *Standard for Dry Chemical Extinguishing Systems*

Clean Agent Extinguishing Systems

CHAPTER 18

Eric W. Forssell and Scott A. Hill

Eric W. Forssell, a senior engineer with Hughes Associates, Inc., has an M.S. in chemical engineering from the University of Tennessee. He is a recognized expert in halon replacement technologies and has conducted many small and full-scale test series evaluating alternative agents.

Scott A. Hill earned a B.S. in Fire Protection Engineering from the University of Maryland and works as a fire protection engineer at Hughes Associates, Inc. He has conducted full-scale tests evaluating clean agent fire suppression systems and has helped develop engineering calculation software for their design.

The phaseout of halons has lead to the widespread development and use of non-halon-based clean agent extinguishing materials. This chapter examines the makeup and operation of these clean agents, discusses the safety considerations and storage requirements when dealing with them, and then looks at clean agent inspection and maintenance requirements.

Clean agents, which are defined as fire extinguishants that vaporize readily and leave no residue, include halocarbons and inert gases. Halocarbon agents are defined in NFPA 2001, *Standard on Clean Agent Fire Extinguishing Systems,* as extinguishing agents whose primary components include one or more organic compounds that contain one or more of the elements fluorine, chlorine, bromine, or iodine. Halocarbons include hydrofluorocarbons (HFCs), hydrochlorofluorocarbons (HCFCs), perfluorocarbons (PFCs and FCs), and fluoroiodocarbons (FICs). Inert gas agents are defined in NFPA 2001 as extinguishing agents that contain, as primary components, one or more of the gases helium, neon, argon, or nitrogen. Agents that consist of mixtures of these gases may also contain carbon dioxide as a secondary component.

OPERATIONS AND DISCHARGE OF CLEAN AGENT SYSTEMS

Clean agent systems are designed as total flooding systems to protect specific hazards. To extinguish a fire, the system must discharge a sufficient quantity of agent to develop and maintain the design concentration throughout the hazard area. The design concentration required varies depending on the hazard being protected. The quantity of agent needed to develop and maintain the design concentration is also dependent on the volume of the space, the ventilation configuration, and the integrity of the enclosure. Changes to the hazard or hazard area may adversely affect the ability of the clean agent system to extinguish a fire. The enclosure size and integrity and the hazard and use of the area must be monitored and evaluated carefully to ensure that the clean agent system will function properly. Because clean agent systems use a fixed quantity of agent, it is critical that the entire area that could be involved in a single fire be protected simultaneously.

Clean agent systems include detection and control equipment, release devices, storage containers, and distribution systems (Figure 18-1). They may also include ancillary equipment, such as automatic door closers and ventilation controls.

SAFETY CONSIDERATIONS

Safety should be a primary concern during installation, service, maintenance, testing, handling, and recharging of clean agent systems and clean agent containers. Because system equipment varies from manufacturer to manufacturer, the instructions and procedures provided in the equipment manufacturer's manual should be followed.

The improper handling of agent containers by unqualified personnel can be a cause of injury and property damage. Caution, therefore, should be exercised when clean agent containers are handled to minimize the potential for injury or property

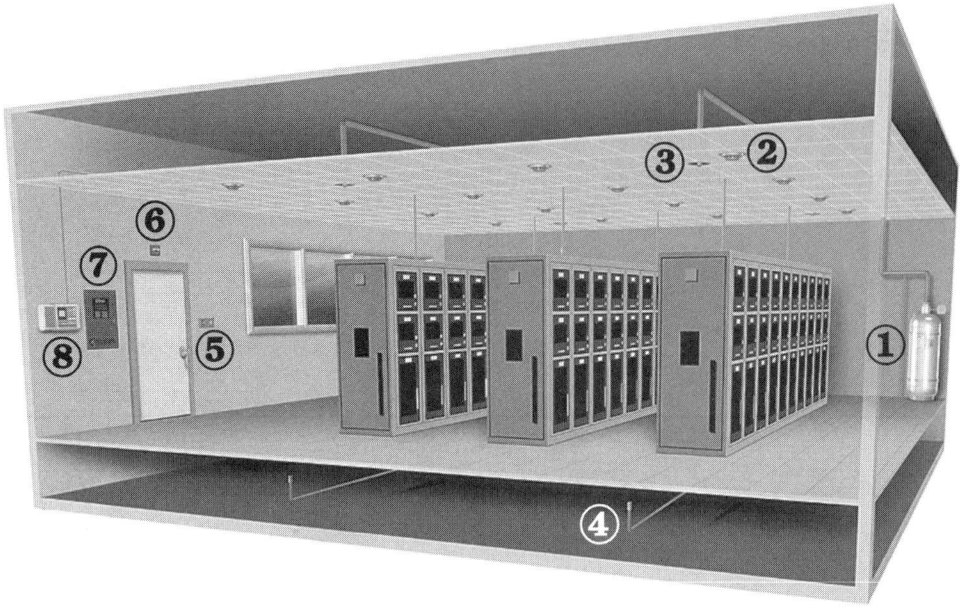

FIGURE 18-1 Typical Clean Agent System Installed in a Room with a Subfloor Area
Source: Courtesy of Fike Corporation

1. Storage container(s)
2. Automatic fire detectors
3. Discharge nozzles installed in room
4. Discharge nozzles installed in subfloor area
5. Manual release device
6. Audible and visual pre-discharge alarm device
7. Control panel
8. Air sampling detection unit (for cabinets)

damage. Careful handling may include the removal or the disabling of the actuator and the installation of a safety cap on the discharge outlet of the cylinder valve.

Any exposure to vaporizing halocarbon liquid should be avoided. As the liquid vaporizes, the temperature decreases rapidly and can cause frostbite burns to the skin. Because the liquid vaporizes rapidly in air, the hazard is limited to the area near the point of discharge. The rapid decrease in temperature may produce a light fog as the water vapor in the air condenses, resulting in a reduction of visibility.

In clean agent systems, the agent is discharged through the nozzles at a high velocity, which can create enough turbulence to move objects during discharge. The discharge of a clean agent system may also create a hazard either through exposure of personnel to the agent or through their exposure to products of thermal decomposition. Products of decomposition are present only if there is a fire.

Unnecessary exposure to the agent or to products of decomposition should be avoided, even though the design concentration for most agents is less than the no observed adverse effect level (NOAEL). For inert gas systems, design concentrations may result in oxygen concentrations of 12 percent or less in the protected enclosure. Provisions for the safety of all personnel working in these environments should be made.

Safety measures aimed at ensuring quick and safe evacuation from the area to avoid unnecessary exposure include, but are not limited to, providing predischarge alarms, maintaining adequate exit capacity, providing emergency lighting, and providing alarms within the hazard area that operate immediately upon detection of a fire. Additional safety measures include providing alarms at the entrances to the hazard area that sound until the atmosphere has been restored to normal conditions and posting waning signs at the entrances that describe the hazards of the installed system. Other safety measures may be necessary, based on the specific characteristics of the clean agent system and the protected hazard.

Clean agent systems should be inspected, tested, and maintained by competent personnel. In addition, personnel using and operating the system or working in the hazard area should be provided with adequate training. The training should include system operation and health and safety hazards associated with agent discharge.

TYPE OF STORAGE

Although halocarbon agents are typically superpressurized with nitrogen and stored in containers at 360 psi (2,482 kPa), agents may be stored at higher pressures or may not require a superpressurizing gas. The containers must be stored in areas where temperatures will be within the listed limits. Inert gases are typically stored in high-pressure containers at pressures in excess of 2,000 psi (13,790 kPa). The containers must be stored in areas where temperatures will be within the listed limits.

INSPECTION AND MAINTENANCE

Table 18-1 provides guidance on the inspection and maintenance requirements for clean agent systems. For detailed information regarding the inspection, testing, and maintenance requirements of clean agent systems, the inspector should consult NFPA 2001, *Standard on Clean Agent Fire Extinguishing Systems;* NFPA 12A, *Standard on Halon 1301 Fire Extinguishing Systems;* and the specifications of the hardware manufacturer.

As indicated on the table, the general condition of the system should be checked monthly to ensure that there has been no accidental damage or tampering. Every 6 months the quantity of clean agent and the container pressure should be checked. For halocarbon agents, the container must be refilled or replaced if the container shows a loss of agent quantity greater than 5 percent or a loss of pressure greater than 10 percent. For inert gas systems, the container should be refilled or replaced if the container pressure (adjusted for temperature) shows a loss greater than 5 percent.

TABLE 18-1 Inspection and Testing Guidelines for Clean Agent Systems

Verification or Test	Frequency
1. Thoroughly inspect general condition of the system	Monthly
2. Check agent quantity and container pressure	Semiannually
3. Inspect enclosure/hazard, check for changes to both	Annually
4. Test supervised circuits	Annually
5. Test control panel	Annually
6. Test power supply, including emergency power	Annually
7. Inspect, clean, and test detectors	Annually
8. Test time delay (if applicable)	Annually
9. Test auxiliary functions, such as alarms, ventilation controls, door closers	Annually
10. Test manual and automatic release devices	Annually
11. Inspect system components, including piping, nozzles, supports	Annually
12. Inspect system hose(s)	Annually
13. Test system hose(s)	Every 5 yrs
14. Complete external inspection of containers	Every 5 yrs
15. Perform hydrostatic testing of inert gas agent containers	After a discharge, if more than 5 yrs since last test

For all of these tests, the actuating controls must be removed from the container to prevent accidental discharge. Frequency is the minimum recommended.

Clean agent systems must be thoroughly inspected each year. This check includes an inspection of the enclosure for penetrations or other changes that adversely affect the ability of the system to develop and maintain the design concentration. It also includes an inspection of the distribution system, that is, the hoses, pipe, nozzles, and so on. Hoses must be tested or replaced, if the visual inspection indicates any deficiency.

Tests of detection and control equipment, release devices, and ancillary equipment should also be performed annually. These tests include the testing of the initiating circuits, control panel, power supply, automatic detectors, time delays, alarms, ventilation controls, door closers, and release devices (automatic and manual). If abort switches are provided, they should also be tested.

Every 5 years all system hose must be tested at 1½ times the maximum container pressure at 130°F (54.4°C). In addition, a complete external inspection of the containers, in accordance with Section 3 of CGA-6, *Standard for Visual Inspection of Steel Compressed Gas Cylinders,* is required every 5 years. This inspection includes checking the cylinder for corrosion, physical damage and distorction, or any other defect that may indicate the cylinder is unfit for service. The external inspection does not require containers to be emptied or stamped while under pressure.

And, finally, a container may not be refilled without it being retested, if it has been more than 5 years since the last test. For halocarbon agent containers, the retest may consist of a complete visual inspection in accordance with Section 3 of CGA-6, *Standard for Visual Inspection of Steel Compressed Gas Cylinders.* A hydrostatic test must be performed on inert gas agent containers before they are refilled, if it has been more than 5 years since the last hydrostatic test.

SUMMARY

With the phaseout of halons, nonhalon clean agents in the form of halocarbons and inert gases are enjoying widespread use. Clean agent systems are designed as total flooding systems to protect specific hazards. Unnecessary exposure to clean agents or the products of decomposition should be avoided. To ensure proper operation, clean agents systems should be inspected, tested, and maintained by well-trained, competent individuals. Inspectors can begin by consulting the table listing the key points of the inspection and maintenance procedure for clean agents. For more detailed information on specific inspection requirements, however, they should refer to NFPA 2001 and the system manufacturer's inspection and maintenance manual.

BIBLIOGRAPHY

CGA C-6, *Standard for Visual Inspection of Steel Compressed Gas Cylinders,* Compressed Gas Association, Arlington, VA, 1993.

NFPA Codes, Standards, and Recommended Practices

See the latest version of The NFPA Catalog for availability of current editisons of the following documents.

NFPA 12A, *Standard on Halon 1301 Fire Extinguishing Systems*
NFPA 2001, *Standard on Clean Agent Fire Extinguishing Systems*

Portable Fire Extinguishers

CHAPTER 19

Ralph J. Ouellette

CASE STUDY

Fire Destroys Lumber Warehouse

Employees unsuccessfully tried to contain a rapidly growing fire that destroyed a large combination lumber warehouse and retail hardware showroom in New Jersey.

The single-story, 31,800-ft^2 (2950-m^2) structure was constructed of unprotected wood framing and heavy timber. The warehouse, which measured 100 × 240 ft (30 × 73 m), was connected on its short side to a showroom and sales office that measured 60 × 130 ft (18 × 40 m). A hard-wired smoke detection system protected the showroom and office, but the warehouse had no such system. Nor did it have sprinklers.

The building was open for business when an employee noticed a slight haze at the warehouse ceiling. Thinking that it was exhaust from an operating fork lift, the man dismissed it and entered a room where windows were stored at the opposite end of the warehouse from the showroom. Here, he saw flames traveling up the wall. He immediately returned to the office, and someone called the fire department at 10:56 A.M.

Several employees then went to the window room with portable fire extinguishers, but the flames had spread to the ceiling and smoke was filling the warehouse. Realizing that the fire had spread beyond their control, they began to remove stock from the warehouse using handcarts.

The first fire fighter arrived in his own vehicle shortly after receiving a page. He ordered all the employees to evacuate when he saw that the fire had spread to a shed adjacent to the window room and to 30-lb (14-kg) propane cylinders used to power the fork lifts, which were exploding.

Several fire departments coordinated master streams, portable monitors, and hose lines to control the blaze, which spread throughout the warehouse and showroom. A water supply was established from a hydrant on the street side and from three drafting lines stretched to a nearby lake. Mutual-aid fire fighters also controlled several brush fires and a dumpster fire that were ignited by flying embers.

The building, valued at $1 million, and its contents, valued at $3 million, were a total loss. No one was injured in the blaze. The cause of the fire was undetermined.

Source: "Fire Watch," *NFPA Journal,* May/June, 1999.

Ralph J. Ouellette, a senior engineering technician with Hughes Associates, Inc., has, for the past 16 years, routinely designed, built, and set up test sites for small- and large-scale test fires for the evaluation of different nozzles, agents, and chemicals for fire suppression. Prior to joining Hughes Associates, Inc., he was a fire fighter for the Naval Research Laboratory, Chesapeake Beach, MD.

Portable fire extinguishers are installed in many occupancies to give the building occupants a means of fighting a fire manually. Not all occupancies are required to have portable fire extinguishers. Normally, a building code, NFPA *101*®, *Life Safety Code*®, or an insurance company requirement will have a provision that states,

"Portable fire extinguishers shall be installed in accordance with NFPA 10, *Standard for Portable Fire Extinguishers.*" Once it has been established that portable fire extinguishers are required, the inspector must ensure that they are properly selected, placed, and serviced.

In addition, the inspector should verify that a training program is present if occupants are expected to use portable extinguishers. It is recommended that this training program incorporate the "fight or flight" approach to using a fire extinguisher. A good reference for such a program is a video produced by NFPA entitled *Fire Extinguishers: Fight or Flight.* The intention of the fight-or-flight approach is to simply size up a fire and determine whether it is safe to fight with an extinguisher. Also included in this approach is guidance such as never fight a fire in a smoke-filled space, always fight the fire with your back to the escape route, and always call the fire department first before attempting to fight the fire with an extinguisher. All of this information, and more, is contained within this video.

SELECTING EXTINGUISHERS

The size and type of portable fire extinguisher is based on the total amount of Class A combustible materials, the total amount of Class B flammables, or, for some occupancies, a combination of both. For many areas, the extinguishing agent must also be compatible with energized electrical equipment. NFPA 10 provides the following criteria for determining the classification of hazards.

Light (Low) Hazard

Light hazard occupancies are locations in which the total amount of Class A combustible materials, including furnishings, decorations, and contents, is minor. These occupancies may include buildings or rooms occupied as offices, classrooms, churches, assembly halls, the guest rooms of hotels or motels, and so on. This classification anticipates that most of the contents of the occupancy are either noncombustible or that they have been arranged in such a manner that a fire is not likely to spread rapidly among them. Small amounts of Class B flammables used in duplicating machines, art departments, and the like are included, provided they are kept in closed containers and are stored safely.

Ordinary (Moderate) Hazard

Ordinary hazard occupancies are locations in which the total amount of Class A combustibles and Class B flammables are present in greater amounts than may be expected in light hazard occupancies. These occupancies may consist of dining areas, mercantile shops and their allied storage, light manufacturing facilities, research operations, auto showrooms, parking garages, the workshop or support service areas of light hazard occupancies, and warehouses containing Class I or Class II commodities, as defined by NFPA 230, *Standard for the Fire Protection of Storage.*

Extra (High) Hazard

Extra hazard occupancies are locations in which the total amount of Class A combustibles and Class B flammables present in storage, in production, or as finished products is over that expected in ordinary hazard occupancies. These occupancies may consist of woodworking shops; vehicle repair areas; aircraft and boat servicing facilities; cooking areas; product showrooms; convention center displays; and areas that house storage and manufacturing processes, such as painting, dipping, and coating. Also included is warehousing or in-process storage of commodities other than Class I and II commodities.

DISTRIBUTING EXTINGUISHERS

After the hazard classification of an occupancy has been determined, the portable extinguishers can be distributed. Extinguishers should be placed in locations that are readily available, provide easy access and are relatively free from temporary blockage, are near normal paths of travel, are near exits and entrances, and are free from the potential of physical damage.

Mounting Extinguishers

Most extinguishers are mounted on walls or columns by securely fastened hangers so that they are supported adequately, although some extinguishers are mounted in cabinets or wall recesses. In any case, the operating instructions must face outward, and the extinguisher should be placed so that it can be removed easily. Cabinets should be kept clean and dry.

In areas where extinguishers may become dislodged, brackets specifically designed to cope with this problem should be used. In areas such as warehouse aisles, where they are subject to physical damage, they should be protected from impact. In large open areas such as aircraft hangars, extinguishers can be mounted on movable pedestals or wheeled carts whose proper locations should be marked on the floor to maintain the pattern of distribution.

NFPA 10 specifies floor clearance and mounting heights, based on extinguisher weight. Extinguishers with a gross weight of no more than 40 lb (18.14 kg) should be installed so that the top of the extinguisher is not more than 5 ft (1.5 m) above the floor. Extinguishers with a gross weight greater than 40 lb (18.14 kg) (except wheeled types) should be installed so that the top of the extinguisher is not more than 3.5 ft (1.07 m) above the floor. In no case should the clearance between the bottom of the extinguisher and the floor be less than 4 in. (10.2 cm).

When extinguishers are mounted on industrial trucks, vehicles, boats, aircraft, trains, and so on, special mounting brackets, available from the manufacturer, should be used. It is important to install an extinguisher at a safe distance from a hazard so that it will not become involved in a fire.

Distribution for Class A Hazards

Table 19-1, which was taken from NFPA 10, provides the requirements for determining the minimum number and rating of extinguishers needed in any particular area to cope with Class A fires. Sometimes, extinguishers with ratings higher than those indicated in this table will be necessary, due to process hazards or building configuration. In no case, however, should the recommended maximum travel distance be exceeded.

TABLE 19-1 Fire Extinguisher Size and Placement for Class A Hazards

	Light (Low) Hazard Occupancy	Ordinary (Moderate) Hazard Occupancy	Extra (High) Hazard Occupancy
Minimum rated single extinguisher	2-A†	2-A†	4-A*
Maximum floor area per unit of A	3000 ft^2	1500 ft^2	1000 ft^2
Maximum floor area for extinguisher	11,250 ft^2**	11,250 ft^2**	11,250 ft^2**
Maximum travel distance to extinguisher	75 ft	75 ft	75 ft

For SI units: 1 ft = 0.305 m; 1 ft^2 = 0.0929 m^2

*Two 2½-gal (9.46-L) water-type extinguishers can be used to fulfill the requirements of one 4-A rated extinguisher.

**See E-3-3.

†Up to two water-type extinguishers, each with 1-A rating, can be used to fulfill the requirements of one 2-A rated extinguisher.

Source: NFPA 10, 1998, Table 3-2.1.

TABLE 19-2 Maximum Area to Be Protected per Extinguisher (Ft^2)*

Class A Rating Shown on Extinguisher	Light (Low) Hazard Occupancy	Ordinary (Moderate) Hazard Occupancy	Extra (High) Hazard Occupancy
1A	—	—	—
2A	6000	3000	—
3A	9000	4500	—
4A	11,250	6000	4000
6A	11,250	9000	6000
10A	11,250	11,250	10,000
20A	11,250	11,250	11,250
30A	11,250	11,250	11,250
40A	11,250	11,250	11,250

For SI unit: 1 ft^2 = 0.0929 m^2.

*11,250 ft^2 is considered a practical limit.

Source: NFPA 10, 1998, Table E-3.4.

The first step in calculating how many Class A extinguishers are needed is to determine whether an occupancy is a light, an ordinary, or an extra hazard occupancy. Next, the extinguisher rating should be matched with the occupancy hazard to determine the maximum area an extinguisher can protect. Table 19-1 also specifies the maximum travel distance, or actual walking distance, allowed; for Class A extinguishers, it is 75 ft (22.9 m). Thus, a 2.5-gal (9.5-L) stored-pressure water extinguisher rated 2-A will protect an area of 3000 square feet (279 m^2) in an ordinary hazard occupancy, but only 2000 square feet (186 m^2) in an extra hazard occupancy.

Table 19-2, which was taken from NFPA 10, was developed from Table 19-1 and summarizes what was intended by the second and third rows of Table 19-1. The following examples show how to place extinguishers in accordance with these tables.

Example 1 Determine the minimum number of extinguishers required for Class A fires according to maximum floor area per extinguisher listed in Table 19-1 given a building area of 67,500 ft^2 (6271 m^2). The dimensions of the outside walls of the building are 450 × 150 ft (137 × 465 m).

The maximum floor area per extinguisher for all occupancies is 11,250 ft^2 (1045 m^2), according to Table 19-1.

$$\frac{\text{Total floor area}}{\text{Maximum floor area per extinguisher}} = \frac{67{,}500\ \text{ft}^2}{11{,}250\ \text{ft}^2} = \frac{6271\ \text{m}^2}{1045\ \text{m}^2} = 6\ \text{Extinguishers}$$

A floor area of 11,250 ft^2 (1045 m^2) per extinguisher requires the extinguishers to have a Class A rating of at least 4-A for a light hazard occupancy, 10-A for an ordinary hazard occupancy, or 20-A for an extra hazard occupancy, according to Table 19-2. For this example, note that installing extinguishers with higher Class A ratings will not affect distribution or placement, due to the guidelines in Table 19-2 on the maximum area to be protected per extinguisher.

Placement of the extinguishers along the outside walls, in this example, would not be acceptable because the maximum travel distance would be exceeded (Figure 19-1). Relocation or additional extinguishers are needed.

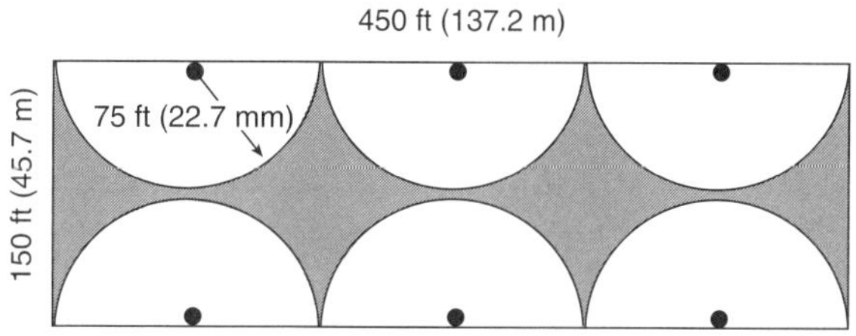

FIGURE 19-1 Diagrammatic Representation of Extinguishers Located Along Outside Walls of a 450 × 150-ft (137 × 46-m) Building. (The dots represent extinguishers. The shaded areas indicate "voids" that are farther than 75 ft [227 m] to the nearest extinguisher.)
Source: NFPA 10, 1998, Figure E-3.6.

Example 2 Using the building designated in Example 1, determine the minimum number of extinguishers required for Class A fires when a floor area per extinguisher of 6,000 ft^2 (557 m^2) is used.

$$\frac{\text{Total floor area}}{\text{Floor area per extinguisher}} = \frac{67{,}500\ \text{ft}^2}{6000\ \text{ft}^2} = \frac{6271\ \text{m}^2}{557\ \text{m}^2} = 12\ \text{Extinguishers}$$

A floor area of 6000 ft^2 (557 m^2) per extinguisher requires the extinguishers to have a Class A rating of at least 2-A for a light hazard occupancy, 4-A for an ordinary hazard occupancy, or 6-A for an extra hazard occupancy, according to Table 19-2.

Extinguishers could be mounted on exterior walls or on building columns or interior walls, as shown in Figure 19-2, and conform to both distribution and travel distance rules.

NFPA 10 also allows up to half the complement of extinguishers for Class A fires to be replaced by uniformly spaced small hose [1.5-in. (3.81 cm)] stations. However, the hose stations and the extinguishers should be located so that the hose stations do not replace more than one of every two extinguishers previously used.

Distribution for Class B Hazards

In areas in which flammable liquids are not expected to reach an appreciable depth, extinguishers should be provided according to Table 19-3, which was taken from NFPA 10. The basic maximum travel distance to Class B extinguishers is 50 ft (15.25 m) as opposed to 75 feet (22.9 m) for Class A extinguishers, because flammable liquids fires reach their maximum intensity almost immediately, and thus the extinguisher must be used earlier. With lower rated extinguishers, the travel distance drops to 30 ft (9.15 m).

Where flammable liquids are likely to reach an appreciable depth, a Class B-rated fire extinguisher must be provided on the basis of at least two numerical units of Class B extinguishing potential per square foot (0.0929 m^2) of flammable liquid surface of

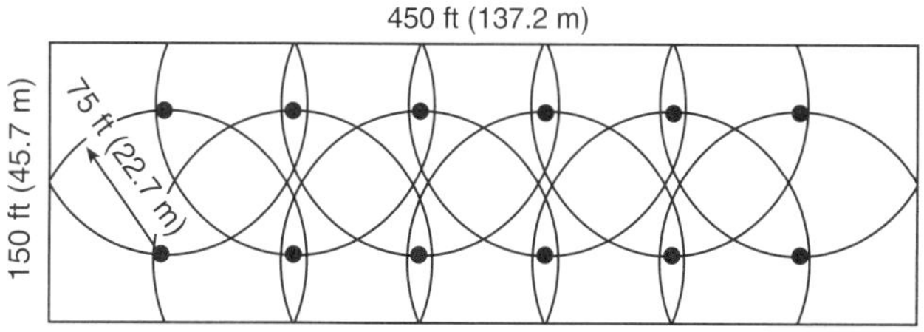

FIGURE 19-2 Configuration Representing 12 Fire Extinguishers Mounted on Building Columns or Interior Walls, in Which Requirements for Both Travel Distance and Fire Extinguisher Distribution Are Met
Source: NFPA 10, 1998, Figure E-3.8.

TABLE 19-3 Fire Extinguisher Size and Placement for Class B Hazards

Type of Hazard	Basic Minimum Extinguisher Rating	Maximum Travel Distance to Extinguishers (ft)	(m)
Light (low)	5-B	30	9.15
	10-B	50	15.25
Ordinary (moderate)	10-B	30	9.15
	20-B	50	15.25
Extra (high)	40-B	30	9.15
	80-B	50	15.25

The specified ratings do not imply that fires of the magnitudes indicated by these ratings will occur, but rather they are provided to give the operators more time and agent to handle difficult spill fires that could occur.

Source: NFPA 10, 1998, Table 3-3.1.

the largest tank hazard in the area. The travel distances specified by Table 19-3 also should be used to locate extinguishers to protect spot hazards. Sometimes, a single extinguisher can be installed to provide protection against several different hazards, provided that travel distances are not exceeded.

Extinguishers selected to protect cooking appliances in which combustible cooking media (vegetable or animal oils and fats) is present must be listed for Class K fires. Traditionally, sodium-bicarbonate- or potassium-bicarbonate-based dry-chemical-type fire extinguishers were recommended. However, these have been shown to be significantly less effective; therefore, only extinguishers listed specifically for Class K fires should be used. Extinguishers listed for Class K fires are intended to be used in conjunction with a fire protection system, which is to be activated prior to using the extinguisher. The maximum travel distance from the extinguisher to the hazard for Class K fires is 30 ft (9.15 m).

Distribution for Class C Hazards

Extinguishers with a Class C rating are installed where there is live electrical equipment. This sort of extinguisher contains a nonconducting agent, usually carbon dioxide, dry chemical, or halon.

Once the power to live electrical equipment has been cut off, the fire becomes a Class A or Class B fire, depending on the nature of the burning electrical equipment and the burning material in the vicinity.

Distribution for Class D Hazards

It is particularly important that the proper extinguishers be available for Class D fires. Because the properties of combustible metals differ, even a Class D extinguishing agent can be hazardous if it is used on the wrong metal. Agents should be chosen carefully according to the manufacturer's recommendations.

The amount of agent needed normally is figured according to the surface area of the metal plus the shape and form of the metal, which can contribute to the severity of the fire and cause the agent to "bake off." For example, fires in magnesium filings are more difficult to put out than fires in magnesium scrap, so more agent is needed for fires in magnesium filings. The maximum travel distance to all extinguishers for Class D fires is 75 ft (22.9 m).

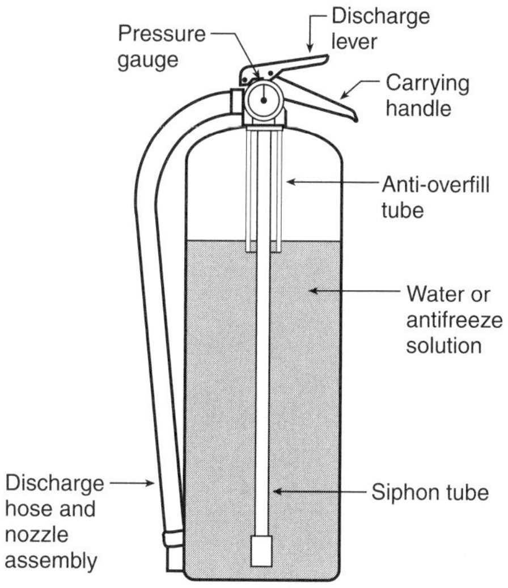

FIGURE 19-3 Stored-Pressure Water Extinguisher
Source: NFPA 10, 1998, Figure D-4.1.1.

TYPES OF FIRE EXTINGUISHERS

Listed fire extinguishers are classified into seven major types. These are stored pressure water extinguishers, dry chemical extinguishers, Halon 1211 extinguishers, carbon dioxide extinguishers, foam extinguishers, loaded stream extinguishers, and extinguishers and agents used on combustible metals.

Stored-pressure water extinguishers (Figure 19-3) use plain water as the agent, with stored pressure in the same chamber as the water. Typically, they have a 2.5-gal (9.46-L) capacity and a rating of 2-A.

Dry chemical extinguishers (Figures 19-4 and 19-5) are the stored-pressure type or the cartridge- or cylinder-operated type. Dry chemical extinguishing agents include ordinary dry chemical, which is sodium bicarbonate based; multipurpose dry chemical, which is monoammonium phosphate based; or dry chemicals with the other base ingredients of potassium bicarbonate, potassium chloride, and urea-based potassium bicarbonate.

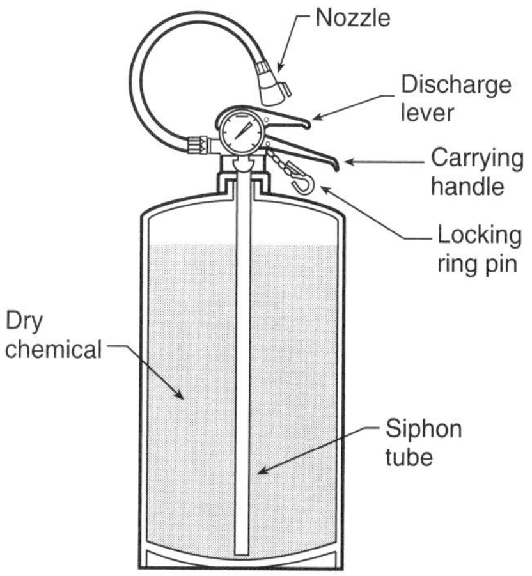

FIGURE 19-4 Stored-Pressure Dry Chemical Extinguisher
Source: NFPA 10, 1998, Figure D-4.5(a).

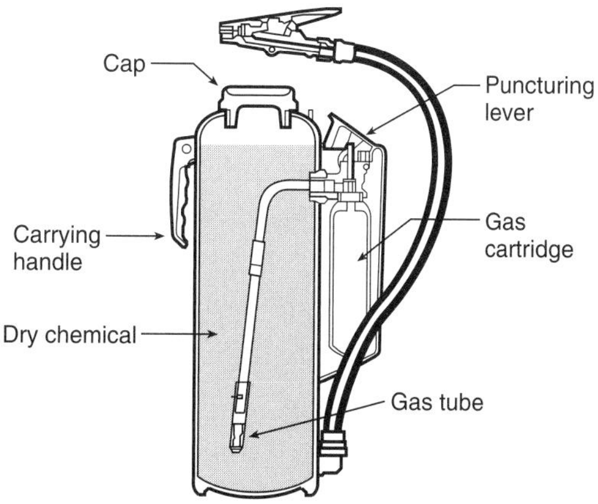

FIGURE 19-5 Cartridge-Operated Dry Chemical Extinguisher
Source: NFPA 10, 1998, Figure D-4.5(b).

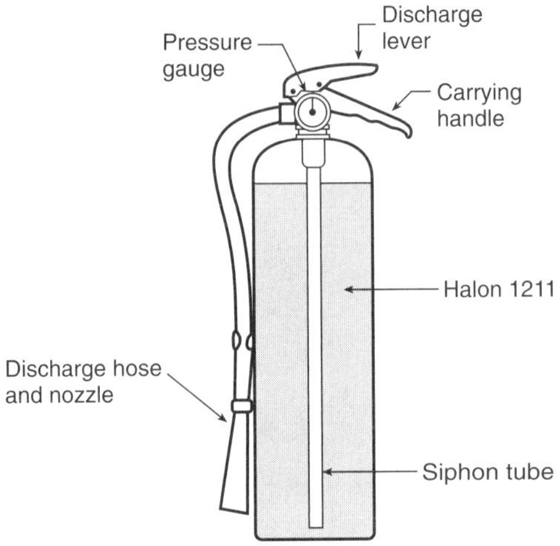

FIGURE 19-6 Halon 1211 and Halogenated Agent–Type Stored-Pressure Fire Extinguisher
Source: NFPA 10, 1998, Figure D-4.4.1.

Halon 1211 extinguishers (Figure 19-6) use Halon 1211 as the extinguishing agent and are the stored-pressure type. When using this type of extinguisher in confined spaces, precautions should be taken to avoid breathing the gases or vapors that are released. The agent is very "clean" (i.e., it leaves no residue), which is useful when it is used on electronic equipment. Ratings for Halon 1211 extinguishers are for Class B and C fires. However, larger units may carry Class A ratings.

Halon 1211 and other halogenated extinguishing agents have been linked to the destruction of the Earth's stratospheric ozone layer. The global production of halons is being gradually phased out by the Montreal Protocol; future supplies of halons are expected to be severely limited. Therefore, new installations of fire extinguishers should bear in mind the future supply and environmental considerations associated with Halon 1211 extinguishers. Existing installations of Halon 1211 extinguishers should continue to follow the suggested guidelines for inspection and maintenance.

Carbon dioxide (CO_2) extinguishers (Figure 19-7) use steel cylinders rated for 1800 psi (12.4 MPa) or higher. The carbon dioxide is stored as a liquid in the cylinder with a vapor space at the top. When the agent is discharged, it vaporizes quickly so the range is relatively short—3 to 8 ft (1 to 2.4 m). CO_2 is a nonconductor of electricity and can be effective on Class B and C fires. Typical ratings range from 5-B:C to 20-B:C.

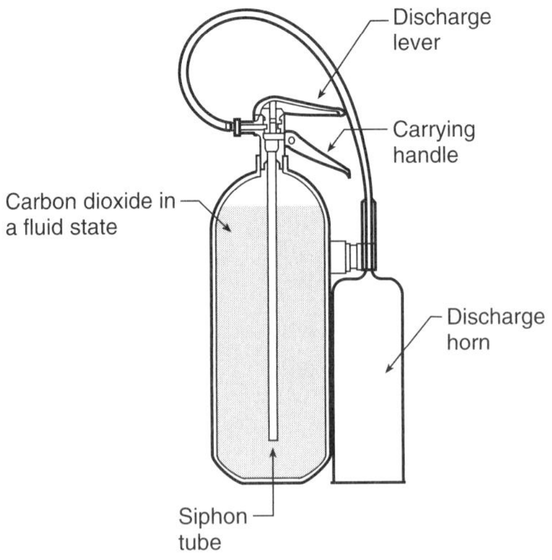

FIGURE 19-7 Carbon Dioxide Extinguisher
Source: NFPA 10, 1998, Figure D-4.3(a).

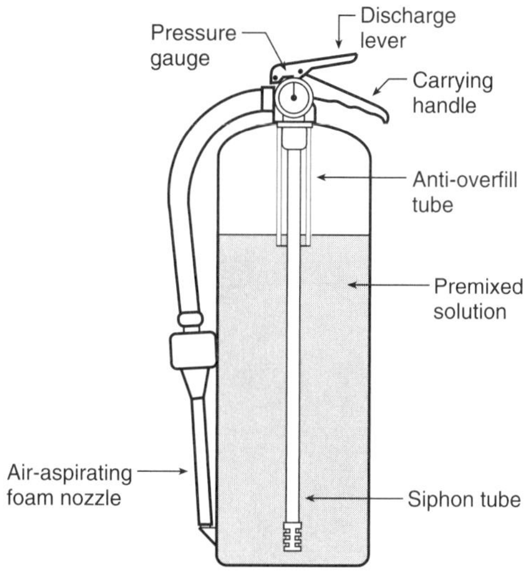

FIGURE 19-8 Stored-Pressure AFFF or FFFP Liquid Extinguisher
Source: NFPA 10, 1998, Figure D-4.2.1.

Foam fire extinguishers (Figure 19-8) are either the stored-pressure type or the cylinder-operated type and are intended for use on Class A and B fires. They contain an aqueous film-forming foam (AFFF) or film-forming fluoroprotein foam (FFFP) solution as the agent. The 2.5-gal (9.5-L) stored-pressure units are rated 3-A:20 to 40-B. The 33-gal (125-L) cylinder-operated units, rated 20-A:160B, are provided on wheels and are most commonly used in factories and warehouses.

Loaded stream extinguishers are stored-pressure water-based extinguishers that use antifreeze solutions for use in low temperatures. An additive consisting of alkali-metal salt solutions—the loaded stream—is added to the water. These extinguishers usually have a 2.5-gal (9.5-L) capacity and a rating of 2-A.

Agents for use on combustible metals can be applied from an extinguisher, a scoop, or a shovel. The metals on which these extinguishers and agents can be used are specified in the individual listings (see Chapter 45).

In addition to these major listings, there are listings for extinguishers containing dry chemical and halon mixtures, for miscellaneous extinguishers (such as those containing wetting agents), for special purpose extinguishers, and for pump tank water extinguishers. As of 1969, manufacture of inverting fire extinguishers was discontinued. These included soda-acid extinguishers and chemically generated foam units. Fire extinguishers of this type should not be hydrostatically tested. In fact, they should be destroyed. They can severely injure the operator because the sudden pressurization of a cylinder that occurs only when activated presents the potential for explosion.

Examples of the different classifications and ratings of portable extinguishers are shown in Table 19-4. The proper application of the extinguishers is illustrated in Figure 19-9.

INSPECTION, MAINTENANCE, AND HYDROSTATIC TESTING

Inspection and maintenance have very specific meanings within the context of portable fire extinguishers. According to NFPA 10, an "inspection" is a quick check intended to provide a reasonable assurance that an extinguisher is available, is fully charged, and is operable. This consists of determining if the extinguisher is in its designated place, that it has not been actuated or tampered with, and that there is no obvious physical damage or condition that would prevent its operation.

"Maintenance," on the other hand, is a thorough check of an extinguisher intended to give maximum assurance that it will operate effectively and safely. Maintenance includes a thorough examination and any necessary repair or replacement.

TABLE 19-4 Examples of Extinguisher Classifications and Ratings

Description	Rating
2.5 gal (9.5 L) water, stored pressure	2-A
20 lb (9.1 kg) carbon dioxide	10-B:C
5 lb (2.3 kg) dry chemical (ammonium phosphate)	2-A:10-B:C
10 lb (4.5 kg) dry chemical (sodium bicarbonate)	60-B:C
10 lb (4.5 kg) dry chemical (potassium bicarbonate)	80-B:C
125 lb (56.7 kg) dry chemical (ammonium phosphate)	40-A:240-B:C
33 gal (125 L) aqueous film-forming foam	20-A:160-B
5 lb (2.3 kg) Halon 1211	10-B:C
9 lb (4.1 kg) Halon 1211	1-A:10-B:C
1.5 lb (0.68 kg) Halon 1211/1301	1-B:C

It normally will reveal the need for hydrostatic testing. The manufacturer typically provides specific instructions regarding periodic examination and maintenance.

Maintenance must be performed annually by a servicing company or by a trained industrial safety or maintenance person. There are two lists of items in the appendix of NFPA 10 (Table A-4-4.2(a) and A-4-4.2(b)) that must be checked during maintenance.

FIGURE 19-9 Recommended Marking System
Source: NFPA 10, 1998, Figure B-2.1.

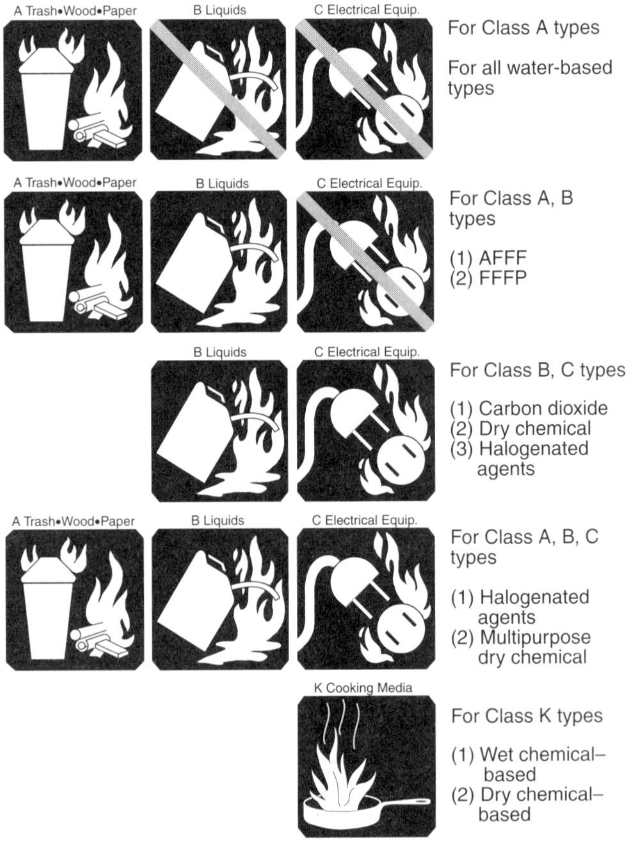

Note: Recommended colors, per PMS (Pantone Matching System) include the following:

BLUE — 299

RED — Warm Red

TABLE 19-5 Hydrostatic Test Interval for Extinguishers

Extinguisher Type	Test Interval (Years)
Stored-pressure water, loaded stream, and/or antifreeze	5
Wetting agent	5
AFFF (aqueous film-forming foam)	5
FFFP (film-forming fluoroprotein foam)	5
Dry chemical with stainless steel shells	5
Carbon dioxide	5
Wet chemical	5
Dry chemical, stored-pressure, with mild steel shells, brazed brass shells, or aluminum shells	12
Dry chemical, cartridge or cylinder operated, with mild steel shells	12
Halogenated agents	12
Dry powder, stored-pressure, cartridge or cylinder operated, with mild steel shells	12

Stored-pressure water extinguishers with fiberglass shells (pre1976) are prohibited from hydrostatic testing due to manufacturer's recall.

Source: NFPA 10, 1998, Table 5-2.

Cartridge-operated, cylinder-operated, and loaded stream extinguishers are the only types that are required to be examined internally on an annual basis. Extinguishers with a 5-year hydrostatic test interval are examined internally on the same 5-year basis. Those with a 12-year interval are examined internally on a 6-year basis.

Nonrechargeable extinguishers are neither examined internally nor hydrostatically tested. They are removed from service 12 years from the date of manufacture.

Maintenance tags or labels must be attached to fire extinguishers to indicate the month and year that they were last serviced. A separate label is required to record information on the 6-year teardown requirement.

"Hydrostatic testing" is performed by personnel who have been specifically trained. Untrained people should not attempt the procedure because serious safety hazards can easily develop.

The purpose of hydrostatically testing fire extinguishers is to protect against the unexpected failure of the cylinder. Table 19-5 provides the test intervals for fire extinguishers.

The cylinders of high-pressure extinguishers that pass the hydrostatic test must be stamped with the month, the year, and the DOT identification number. Low-pressure extinguishers are not stamped, but a self-destructive label listing the month, the year, and the identification of the person or company performing the test is affixed to the cylinder.

Inspection, maintenance, and hydrostatic testing must be carried out according to the minimum requirements established in NFPA 10 and in strict conformance with the manufactures' recommendations. The minimum frequency for inspections is at 30-day intervals.

SUMMARY

Portable fire extinguishers, which are found in many occupancies, give a building's occupants a chance to fight an incipient fire by hand. Once a facility has made the decision—either voluntarily, as required by the authority having jurisdiction, or by an insurance company—to install fire extinguishers, it must determine which type to install based on the classification of hazards.

The chapter describes in detail the seven types of extinguishers—stored-pressure water, dry chemical, Halon 1211, carbon dioxide, foam, loaded stream, and extinguishers used on combustible metals—and for what class of fire each is most effective. The required minimum number and rating of these extinguishers in any particular area is based on the type and amount of combustibles being protected or present in the protected area.

To ensure their effectiveness, portable fire extinguishers must be inspected monthly and must undergo a maintenance check by a servicing company or by a trained industrial safety or maintenance person annually. Doing so will help ensure their readiness as a first line of defense.

BIBLIOGRAPHY

Carson, W. G., and Klinker, R. L., *Fire Protection Systems: Inspection, Test and Maintenance Manual.* 3rd ed., NFPA, Quincy, MA, 2000.

Cote, A. E., ed., *Fire Protection Handbook,* 18th ed., NFPA, Quincy, MA, 1997.

"Rating and Fire Testing of Fire Extinguishers," *Fire Protection Equipment Directory,* issued annually, UL/ANSI 711, Underwriters Laboratories/American National Standards Institute.

NFPA Codes, Standards, and Recommended Practices

See the latest version of The NFPA Catalog for availability of current editions of the following documents.

NFPA 10, *Standard for Portable Fire Extinguishers*
NFPA 18, *Standard on Wetting Agents*
NFPA *101*®, *Life Safety Code*®
NFPA 230, *Standard for the Fire Protection of Storage*
NFPA 408, *Standard for Aircraft Hand Portable Fire Extinguishers*

Means of Egress

CHAPTER 20

Ron Coté

Chapter 7, "Means of Egress," of NFPA *101*®, *Life Safety Code*®, 2000 edition, is the most important fundamental chapter of the code. The occupancy chapters draw heavily on its concepts and detailed criteria. A stand-alone chapter on inspecting the means of egress might be a cornerstone around which the life safety-related portions of this manual could be structured. However, the occupancy chapters specify a comprehensive life safety package made up of a measured mix of egress, fire protection, building service, and interior finish requirements. Thus, no one should inspect only the means of egress. The occupancy chapters that follow in this manual correctly stress that the means of egress is only a part of the life safety package and that the package needs to be inspected as a whole. The detailed inspection checklist forms provided at the back of the book list the varied building features and systems that need to be evaluated in total. The following sections in this chapter highlight the facets of a life safety systems inspection that relate specifically to the means of egress.

INSPECTION TYPES

An inspection of the means of egress needs to be tailored to the purpose for which the inspection is being conducted. The inspection might be for any of the following purposes:

- Determining code compliance with the egress provisions in a jurisdiction that has adopted, for the first time, a given code—such as the NFPA *101, Life Safety Code*
- Signing off on the egress portion of a new construction or renovation project so that the certificate of occupancy can be issued
- Investigating the validity of a claim that the means of egress has been compromised by building operators
- Ensuring via periodic reexamination that code-compliant, egress-related features and systems are properly maintained
- Verifying that egress-related, code deficiencies that were noted during an earlier inspection have been corrected

Ron Coté is NFPA's principal life safety engineer. He has worked on six editions of the Life Safety Code® *and its handbook.*

ALL-INCLUSIVE INSPECTIONS

Inspections conducted for the purposes of items 1 (determining compliance on a newly adopted code) and 2 (signing off on a new construction or renovation project) are part of all-inclusive examinations and evaluations that involve plan review, calculations, and the application of detailed code criteria. Such inspections are beyond the scope of this manual. The preparation needed for all-inclusive inspections should include on-the-job training under the supervision of an experienced inspector, seminar attendance, and extensive studying of reference materials such as the *Life Safety Code® Handbook.*

REINSPECTIONS

Inspections conducted for the purposes of items 3, 4, and 5, which are all basically reinspections, involve the following:

1. Preparing oneself with a working knowledge of the code requirements against which the facility's means of egress features need to be compared
2. Assuming that the earlier comprehensive inspection, such as that conducted for purposes of issuing the certificate of occupancy, assured that the means of egress system was adequately sized (e.g., that its capacity does not need to be reevaluated provided that other features and uses have not changed)
3. Allowing intuition and gut reaction to serve as cues for when to stop and pay attention to a potential problem area
4. Identifying egress uses, practices, elements, and systems that might have changed since the facility was last inspected
5. Assimilating the effect that the changes have imposed on the overall level of life safety provided by the means of egress system prior to the changes
6. Incorporating the evaluation of the means of egress system into the overall inspection that will be complete only when it also includes consideration of the fire protection, building service, and interior finish requirements required by the appropriate occupancy chapters.

FAMILIARITY IN INSPECTIONS

Familiarity with a given facility, such as that gained by having conducted the recent periodic inspections of that building, will often make the inspection process easier. The inspector who is familiar with a facility experiences fewer distractions by factors unrelated to the inspection task. Observations can be more focused. An inspector who has previously visited the facility can more easily identify changes caused by reuse of the space, minor renovations, or laxness in housekeeping.

The inspector, however, needs to guard against the complacency associated with feeling too much at home within a facility. Also, a sole inspector at a given building might fall into the trap of focusing heavily on subjects of personal interest, while ignoring equally important subjects with which he or she has limited interest or expertise.

An effective inspection process might use two inspectors on an alternating basis for one facility. Each inspector would visit the site often enough to retain familiarity, yet not so often as to lose the inquisitive drive that often accompanies a new assignment. In effect, the two inspectors would see things through two different sets of eyes and would evaluate them using two different experience bases.

DEFINING MEANS OF EGRESS

The means of egress is everywhere, not just at the exit doors. It consists of the exit access, the exit, and the exit discharge. The exit access covers the vast majority of the floor area that serves the three-part means of egress. It is difficult to justify excluding any usable floor area from being within the egress system. Thus, a thorough inspection of the egress system necessitates viewing all building areas that occupants can be expected to occupy under both emergency and nonemergency uses.

Inspection Walk-Throughs

During each inspection, a complete walk-through will typically be needed. Even if the exit doors at the front of the building are correctly maintained in an unlocked condition, the inspector should not assume that the doors at the rear of the facility are similarly maintained. The doors at the rear might not be used on a day-to-day basis. They also create security problems because they are not watched as easily as are the doors at the main entrance. Thus, they are more prone to being locked.

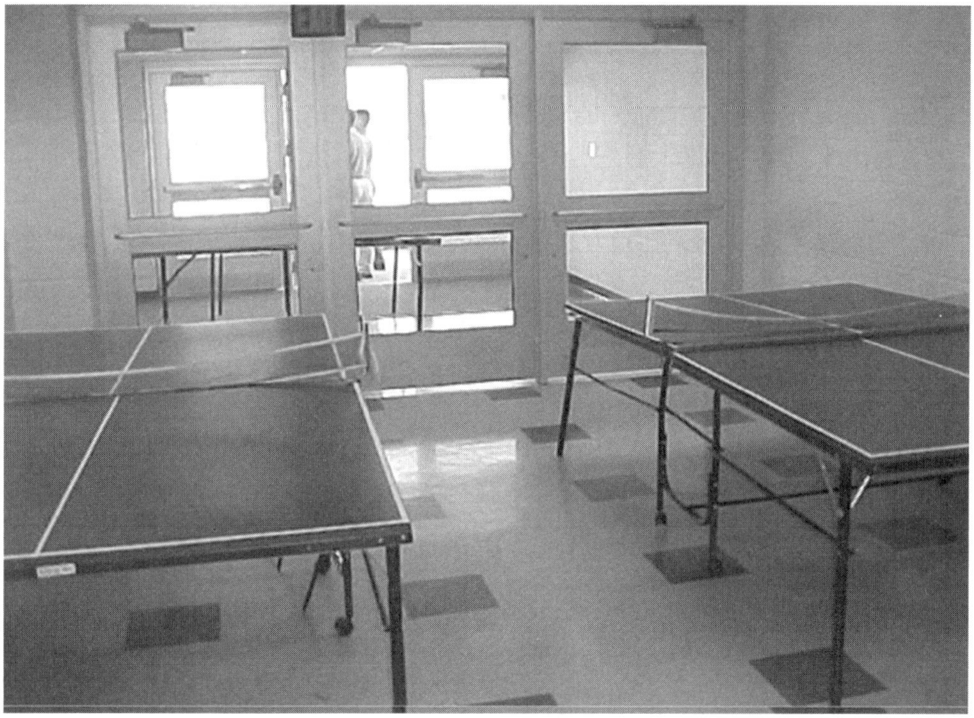

FIGURE 20-1 Tables Blocking Means of Egress
Source: Courtesy of Jon Nisja.

If the egress paths are maintained clear and unobstructed in one wing of the building, the inspector should not assume that the egress paths in another wing are similarly maintained (Figure 20-1). The orderliness noted in one area might be the result of one individual's or one department's efforts and might not reflect an enforced, facilitywide management policy.

An effective inspection of the egress system can only be made by physically traversing the myriad travel paths available to the building occupants. These paths also include those at the outside of the building that serve as exit discharge by connecting the exit to the public way. An inspection conducted in the summer in a locale known for its debilitating winter snowstorms cannot accurately evaluate whether the exterior exit discharge paths will be kept usable throughout the year, especially for those paths that are not used on a daily basis but are provided expressly for emergency egress.

Inspecting Out-of-Sight Features

Some of the egress paths, such as those within exit stair enclosures, might not be used on a day-to-day basis. The stair enclosure that is not used regularly typically loses its effectiveness as an egress path under two scenarios. The space is seen by occupants as unneeded, which leads to it becoming a catch-all for the storage of items used infrequently; or it becomes strewn with litter by occupants who make unauthorized use of the area, for example, for smoking breaks.

Various building features that serve as part of the required means of egress extend above suspended ceilings and cannot be inspected simply by walking through the occupied spaces of the building. A smoke barrier, if it is to do its job of preventing the migration of smoke from one smoke compartment to an adjacent compartment, must be continuous so as to run vertically from the floor slab to the underside of the floor or roof deck immediately above. To perform a proper inspection, the inspector will need to climb a ladder, remove a lay-in ceiling tile, and check above the ceiling for penetrations in the smoke barrier wall. The most common place to make penetrations in a smoke barrier is above the ceiling, where cables and conduit are typically hidden from the view of building occupants. The absence of penetrations in the portions of the smoke barrier that are positioned below the ceiling cannot be relied on as evidence of barrier continuity above the ceiling.

Adequate proof for considering a door to be a fire door usually rests with a label on the door that attests to the door's listing as a fire door by a recognized testing laboratory. If the inspector cannot see the listing label from the floor, he or she should not automatically assume that the door is not a fire door. Such labels might have been installed on the top edge of the door. Limited ceiling heights might prevent the inspector from standing on a ladder and looking down on the top edge of the door. The experienced inspector will, therefore, carry a small pocket mirror to ascertain the presence of, and information contained on, the fire door label attached to the top edge of the door.

MEANS OF EGRESS INSPECTION METHODS AND ASPECTS

Visual Observations

When inspecting the means of egress, visual observations alone are not adequate. If a building feature is meant to be operated, the inspector needs to operate it. For example, a seldom used egress door located in a wall at the building perimeter will often have seized up within its frame such that the average building occupant cannot exert the force necessary to free the door so that it can be opened.

Where a door has been equipped with delayed egress hardware, the inspector needs to test the specialized unlocking system operationally.

- Does pushing on the release device for 3 seconds initiate an irreversible process that permits the door to be opened after 15 seconds?
- Does an audible indicator at the door begin to sound at the end of the 3-second period so as to signify to the occupants that the system is functioning as intended?
- Does the door remain unlocked after it is has been opened and closed?

The inspector needs to confirm all these features by physically exercising the unlocking mechanism, not just by visually observing it.

Quantitative Versus Qualitative Evaluation

The inspector can often evaluate the opening force required for a given door against the maximum values specified by the code without having to measure the force with a scale. To the experienced inspector, the door opening force either feels right or it does not. However, where performance is suspected to be marginal, the inspector will need to use more quantitative methods. An inspection report citing that the door opening forces seem to be excessive will not be as convincing as one that establishes, for example, that the 35-lb force (155 N) that was measured during the inspection is in excess of the 30-lb force (133 N) permitted by the *Life Safety Code* for setting the door in motion once the latch has been retracted.

The inspector, when visiting a factory where the placement of new manufacturing equipment and mechanized conveyor lines has changed the egress path, might determine that the travel distance needs to be evaluated. In measuring the travel distance to the nearest exit, the inspector might pace off the distance using an exaggerated step that, through experience, has been honed to be an accurate indicator of 3 ft (1 m). In a warehouse, the inspector might make use of the fact that the columns supporting the roof structure are spaced on 40-ft (12-m) centers and might thus count the number of columns encountered along the egress path. In an office setting, the inspector might estimate the travel distance by counting the number of consistently sized, 2-ft (0.6-m) ceiling tiles positioned above the egress path. Each of these methods helps the inspector to avoid having to lay a tape measure at floor level along the actual path of travel.

Role-Playing

The inspector needs to role-play so as to ensure that the needs are met of all persons whom the code requirements are intended to serve. For example, in checking opening forces for a side-hinged and swinging door, the inspector should step into the smooth-soled shoes of a small person with low bodyweight and should not assume the role of a professional football linebacker with shoe cleats for added traction.

In role-playing, the inspector needs to guard against creating new requirements that are in excess of those that the inspector has been charged with using. In some cases, the minimum requirements as codified might not assure that every possible building occupant will be able to open an exit door. For example, persons with disabilities that force them to travel by wheelchair might not be able to exert the 30-lb force (133 N) permitted by the code for setting the door in motion once the latch has been retracted.

If the inspector works directly for the building operator and is given some latitude to evaluate the egress system against the needs of the neediest occupants, a recommendation to do more than that required by code might be in order. The inspector needs to keep in perspective the fact that codes, by their very nature, are compilations of minimum requirements.

Timing of Inspections

The timing of inspections depends on the use of the facility being inspected. For example, when evaluating whether a nightclub, rumored to be overcrowded, is operating with occupant loads in excess of the posted maximum, the inspector will need to visit the facility at night. An inspection during the inspector's normal daytime shift will not answer the question.

In another example, when determining whether the required minimum widths of egress aisles in a department store are being maintained, the inspector should visit the store in mid-December when the it is stocked to its fullest extent during the end-of-year holiday sales period.

In an atrium building that uses the balconies and walkways within the atrium as part of the required means of egress, a smoke-control system might be required by code. The smoke-control system becomes a feature that needs to be inspected as part of a means of egress inspection. The inspection and testing of a smoke-control system is complicated and time consuming. It is often done after normal business hours when the building has its lowest occupancy level. The presence of building occupants typically is not needed for a realistic smoke-control inspection. However, where the system is designed to rely on pressure differentials at barriers to control the migration of smoke, it might be necessary to wedge a door open to simulate the effect that an open door would have while it is being used, for example, by 150 occupants as part of the egress path.

Not all means of egress features can be evaluated by an instantaneous observation or measurement. Sometimes an elapsed time inspection is needed. The emergency lighting provisions contained in the *Life Safety Code* require that an annual test be conducted on battery-powered emergency lighting systems for not less than 1½ hours. The initial illumination level and a degraded level at the end of the 1½ hours must be measured. The inspector will need to be present, as a minimum, for the beginning and end of the test.

Rank-Ordering Code Violations

The product of a good inspection of the means of egress is a report from which informed actions can be taken. Code violations or deficiencies need to be rank-ordered in terms of importance. A locked egress door and an inoperative light bulb in an exit sign are not equal deficiencies; they have potential for very different effects on the safe egress of building occupants. Proposed solutions, their estimated costs, and analysis of the effect they will have on the overall life safety are additional features that will help to make the inspection report a useful tool for change.

SUMMARY

Means of egress inspections can be conducted for a variety of reasons. Because the egress system comprises almost all the occupiable space within a building, the inspector needs to walk through the entire building. Some egress features are easily observed; others are hidden and require closer examination.

BIBLIOGRAPHY

Coté, R., ed., *Life Safety Code® Handbook,* NFPA, Quincy, MA, 2000.

NFPA Codes, Standards, and Recommended Practices

See the latest version of The NFPA Catalog for availability of current editions of the following documents.

NFPA *101®*, *Life Safety Code®*

Interior Finish, Contents, and Furnishings

CHAPTER 21

Joseph M. Jardin

CASE STUDY

Man Dies in Egg Processing Plant Blaze

A fire of undetermined origin destroyed an egg processing building on the grounds of a California egg production facility, killing one farm worker and injuring another.

The single-story egg processing building, which measured 100 × 30 ft (30 × 9 m), was built of unprotected wood-frame construction covered with corrugated metal or plywood sheathing. The interior walls were covered with plywood, cardboard, or egg crates. The ceiling was flat except for where the ceiling opened to the roof under the roof's center ridge. The building had no fire detection or suppression equipment.

The facility was closed for the night when the fire department received several 911 calls at 12:38 A.M. to report the blaze. By the time fire fighters arrived, they found the processing building fully involved. They got the blaze under control in 24 minutes, preventing it from spreading to 10 nearby buildings. As they did so, they found the body of a 60-year-old farm worker who had died in the fire.

An hour and a half after they arrived at the scene, fire fighters were approached by a 36-year-old man whose face, neck, and hands were severely burned. As he was treated for his injuries, the man told fire fighters that he had been asleep in an illegal makeshift sleeping area in the building with the man who died when the fire broke out. The two men had apparently been living there for about a week. The fire quickly consumed the room and spread rapidly through the structure.

Workers load eggs at an egg processing plant that was destroyed in a blaze that killed one farm worker and injured another. (Photo courtesy of Lauriel Ward/*The Press Enterprise*)

Several factors contributed to this blaze. A substandard structure was being illegally and unsafely used as a residential property, the building lacked smoke detectors and sprinklers, and the building contained several code violations. Several other factors increased fire risk but could not be proven to have contributed to the fire. The building's wiring was exposed; it did not meet local code requirements; and it had reportedly been arced in the past. And the workers living there had consumed alcohol before the fire. In fact, the man who died had a blood alcohol level of 0.23 percent.

The building, valued at $50,000, and its contents, valued at $50,000, were a total loss. The property's owners were arrested on federal charges for employing undocumented aliens. They served prison terms, but the report was not clear about whether their sentences were for code violations or for hiring violations.

Source: "Fire Watch," *NFPA Journal,* November/December, 1998.

Joseph M. Jardin, a registered fire protection engineer, is a fire fighter with the New York City Fire Department, Rescue Co. No. 2. He is a former chair of the Safety to Life Residential Occupancies Committee and currently serves as a member of NFPA's Building Code Technical Correlating Committee.

Interior finishes have been a significant factor in rapid flame spread for many of the deadliest fires in recent decades. Inspectors, therefore, must be constantly aware of the different types of interior finishes that can be installed within a building.

Interior finishes traditionally are considered to consist of those materials or combinations of materials that form the exposed interior surfaces of walls and ceilings in a building. Interior floor finishes are the exposed floor surfaces of buildings and include floor coverings, such as carpets and floor tiles, that may be applied over or in lieu of a finished floor. Furnishings, which in some cases may be secured in place for functional reasons, should not be considered interior finish. Decorations and furnishings generally are not considered interior finishes but are handled separately. Furnishings are addressed later in this chapter.

Interior finish relates to a fire in four ways. It can affect the rate of a fire buildup to flashover conditions, contribute to fire extension by flame spread over its surface, add to the intensity of a fire by contributing additional fuel, and produce smoke and toxic gases that contribute to life hazard and property damage. Controlling the type of interior finish in a building based on occupancy type will ultimately increase the occupants' safety.

WALL AND CEILING FINISHES

Interior wall and ceiling finishes are the exposed interior surfaces of buildings, including, but not limited to, fixed or moveable walls and partitions, columns, and ceilings. Types of interior finish materials are numerous and include some commonly used materials such as plaster, gypsum wallboard, wood, plywood paneling, fiber ceiling tiles, plastics, and a variety of textile wall coverings.

For proper evaluation classifications, assemblies of interior wall and ceiling finishes should be tested under conditions that simulate their actual installation. If they are not tested under such conditions, the test findings may not be accurate. For example, the flame spread rating of an interior finish material tested without the adhesive that is actually used to apply it may be more favorable than it is when the adhesive is used.

The type and end use of a material determine the fire test it will undergo. A material that is used as a wall covering must have a flame spread rating. The same material used as a drape or a curtain must have a flame propagation resistance through vertical burn characteristics. And if the material is used on upholstered furniture, it must undergo an appropriate fire test for resistance to cigarette ignition.

Carpets and similar carpetlike materials cannot be arbitrarily applied to walls or ceilings. These materials may behave well from a fire development perspective when applied to floors, but when they are either oriented vertically or mounted on a ceiling, many will perform poorly. To be approved or permitted to be used as a textile wall or ceiling material, the material must be evaluated as a wall or ceiling covering rather than a floor covering. Fire test methods and approved applications for textile, expanded vinyl, and foamed plastic coverings are addressed later in this chapter.

Flame Spread and Smoke Developed Indices

Interior wall and ceiling finishes are generally classified in accordance with the results of actual fire tests, which record the flame spread and smoke development of the material. Flame spread ratings offer a general indication of the relative rate with which fire may spread across the surface of the material; smoke development is the degree of visual obscurity observed during this test.

The tests used for wall and ceiling finishes typically are measured in the Steiner Tunnel Test, also known as NFPA 255, *Standard Method of Test of Surface Burning*

Characteristics of Building Materials, and ASTM E-84, *Standard Test Method for Surface Burning Characteristics of Building Materials.* In this test a 24-ft long × 20-in. wide sample is mounted to the top of a 25-ft long "tunnel." A gas flame is ignited at one end and a regulated constant draft is maintained through the tunnel. Flame front progress is observed through side windows for a 10-minute period. Once the flame front progress has been charted, the area under a time–distance curve is used to determine the material's flame spread index. The flame spread index represents the relative rate at which flame spreads across the surface of this material as compared to its spread across the surface of inorganic reinforced cement board and red oak flooring, which exhibit flame spread indices of 0 and 100, respectively.

The flame spread and smoke developed rating is indexed into three classifications as follows:

Flame spread	Smoke developed index
Class "A": 0–25	0–450
Class "B": 26–75	0–450
Class "C": 76–200	0–450

The smoke developed index is based on obscuration, that is, on the ability to "see" illuminated exit signs. Smoke developed indices are derived from the flame spread evaluation of the Steiner Tunnel Test. The degree of smoke obscuration is measured by a photoelectric cell mounted opposite a light source in the tunnel vent pipe. Reduction in light due to passing smoke and particulate is read and recorded by the photoelectric cell, and this information is used to calculate the smoke developed index. The numerical value of 450 was chosen as the point at which the means of egress may become obscured. Some codes refer to "A," "B," and "C" as "I," "II," and "III," but the flame spread and smoke developed ratings are the same.

Occupancy classification dictates the permitted interior finish classifications. As indicated earlier, the classification of an interior finish is that of the basic material used either by itself or in combination with other materials. The material should be tested in exactly the same configuration in which it will be used in a facility.

Paint or wall coverings applied after the interior finish has been installed are not subject to interior finish requirements if they are no thicker than 1/28th of an inch. However, such materials would require a flame spread rating if they or their applications produced significant flame spread or smoke development in and of themselves. Multiple layers of wall coverings can contribute to a rapid fire growth and should be subject to the requirements for the interior finish of the type of occupancy in which they are used (Figure 21-1).

Textiles, Expanded Vinyl, and Cellular or Foamed Plastics

The application of textiles and expanded vinyl materials as wall and ceiling finishes is truly limited. If the materials are rated as Class A when tested in accordance with NFPA 255, they may be used as wall or ceiling finishes in the following circumstances:

- In areas protected by approved automatic sprinkler systems
- On partial height partitions no greater than three-fourths of the floor-to-ceiling height or less than 8 ft, whichever is less
- Where the finish item extends no more than 4 ft from the floor
- Where previously approved

FIGURE 21-1 DuPont Plaza Hotel Fire. Rapid fire growth was, in part, due to wall finish in room of origin.
Source: *Life Safety Code® Handbook,* NFPA, 2000, Exhibit 10.1.

An alternative test for this application is NFPA 265, *Standard Methods of Fire Tests for Evaluating Room Fire Growth Contribution of Textile Wall Coverings.* If the sample, when mounted as intended, passes this test, its use as a wall or ceiling finish is permitted.

The application of cellular or foamed plastic materials on interior walls and on ceilings should not be permitted unless fire tests can reasonably substantiate the combustibility characteristics of the materials for the particular uses under actual fire conditions. Cellular or foamed plastics may be used as trim if they do not exceed 10% of the wall or ceiling.

Trim or Incidental Finish

Trim or incidental finish applied to interior walls or ceilings may be of a Class C material, even when the interior wall or ceiling finish is required to be of a Class A or B material, as long as they are not applied to more than 10% of the aggregate wall area. The purpose of this provision is to permit the use of wood trim around doors and windows as a decoration or as functional molding, since this type of trim would be distributed uniformly throughout a room, rather than being concentrated in one area.

Fire-Retardant Coatings

Interior finish materials that do not have the appropriate interior flame spread rating can be modified by applying fire-retardant coatings. These coatings must be applied in accordance with the manufacturers' instructions and should possess the desired degree of permanency, remaining effective when in actual use.

Most fire-retardant paints and coatings require an application rate three to four times greater than that of ordinary paints. The treatment must be reapplied or renewed periodically because the retardant's overall effectiveness could be reduced by regular maintenance, washing, or cleaning. Fire-retardant treatments should comply with the requirements of NFPA 703, *Standard for Fire Retardant Impregnated Wood and Fire Retardant Coatings for Building Materials.*

INTERIOR FLOOR FINISHES

Interior floor finishes are the exposed floor surfaces of a building, including coverings, that are applied over the normal finished flooring or stairs, including risers. To obtain a rating for interior floor finishes, the Flooring Radiant Panel Test Method is used. This test method is also known as NFPA 253, *Standard Method of Test for Critical Radiant Flux of Floor Covering Systems Using a Radiant Heat Energy Source,* and ASTM E-648, *Standard Method of Test for Critical Radiant Flux of Floor Coverings Systems Using a Radiant Heat Energy Source.* The test measures burning that occurred on a sample. The results are then converted into a value, known as the "critical radiant flux," of Watts per square centimeter.

A Class I interior floor finish has a critical radiant flux of at least 0.45 W/cm^2, while a Class II interior floor finish has a minimum 0.22 W/cm^2. The greater the critical radiant flux value, the more resistant the floor finish is to ignition and flame propagation.

Since April 1971, the federal government has required that all carpets manufactured in the United States meet a flammability standard known as the Federal Flammability Standard, FF-1-70 Pill Test, or simply the Pill Test. In the Pill Test eight 9-in.-square sections of a carpet are secured in a test chamber. A methanamine tablet is ignited and placed in the center of the specimen. If the flame advances to any point within 1 in. of the edge of the sample, the specimen fails the test.

The appropriate floor finish rating is determined by the way in which the facility is used and the location of the finish within the facility—that is, in exits and corridors. Interior floor finish ratings only apply when floor finish presents an unusual hazard or where the floor finish requirements are specified based on occupancy type.

AUTOMATIC SPRINKLERS

The presence of an automatic sprinkler system in a facility provides a degree of safety that is incorporated into the interior finish requirements. The rating of a wall, ceiling, or floor finish can be reduced by one level if a sprinkler system is installed in the facility. Although floor finish items are permitted to drop to a nonrated material, wall and ceiling finish items cannot be reduced below a Class C-rated material.

FURNISHINGS AND DECORATIONS

Furnishings and decorations are not considered interior finishes when one is discussing flame spread characteristics. But these items can contribute fuel to a fire in an occupancy and are required under certain conditions to meet certain fire tests. Several occupancies restrict draperies, curtains, and other loosely hanging fabrics. Only those articles satisfying the applicable provisions of NFPA 701, *Standard Methods of Fire Tests for Flame Propagation of Textiles and Films,* are permitted.

Upholstered furniture and mattresses also contribute to the fire loading in a facility. Depending on the type of occupancy in which the furniture is placed, it may have to undergo a fire test. Many different types of tests are available to determine the fire characteristics of upholstered furniture. NFPA 260, *Standard Methods of Tests and Classification System for Cigarette Ignition Resistance of Components of Upholstered Furniture,* and NFPA 261, *Standard Method of Test for Determining Resistance of Mock-Up Upholstered Furniture Material Assemblies to Ignition by Smoldering Cigarettes,* are just two of these test standards. Depending on the way the occupants use the facility, upholstered furniture may be required to meet a certain threshold for rates of heat release. NFPA 266, *Standard Method of Test for Fire Characteristics of Upholstered Furniture Exposed to Flaming Ignition,* provides the necessary testing procedures to obtain these rates of heat release. NFPA 267, *Standard Method of Test for Fire Characteristics of Mattresses and Bedding Assemblies Exposed to Flaming Ignition Source,* provides the test criteria for determining heat release rates for mattress and bedding contents. As with interior finishes, upholstered furniture need not be tested if an automatic sprinkler system has been installed in the facility.

SUMMARY

The faster a fire develops, the greater the threat it presents to the occupants of a building. Interior finishes and furnishings exert a major influence on how fast a fire will develop and how intense it will become.

Interior wall and ceiling finishes are defined as the exposed interior surfaces of buildings—including walls, partitions, columns, and ceilings—and consist of a variety of commonly used building materials and wall coverings. Interior floor finishes, in turn, are the exposed floor surfaces of a building, including the coverings applied over the finished flooring or stairs.

To establish minimum safety standards, interior finishes are tested and classified according to flame spread and smoke developed ratings. The type of material and its end use determine the type of fire test it will undergo. In all cases the materials should be tested under conditions that simulate the actual installation; for example, if a textile is applied to a wall, it is considered an interior finish and must be tested with the adhesive that bonds it to the wall. Furnishings and decorations, such as curtains, draperies, and upholstered furniture, are not considered interior finishes, but because they contribute fuel to a fire, they must also, under certain conditions and depending on the occupancy, meet specific fire test requirements. When minimum criteria for interior finish requirements are established, the level of safety for building occupants is raised.

BIBLIOGRAPHY

Standard Specification for Polystyrene Molding and Extrusion Materials (PS). ASTM D-4549, ASTM, West Conshohocken, PA.

Standard Test Method for Combustible Properties of Treated Wood by the Fire-Tube Apparatus, ASTM E-69, American Society for Testing and Materials, West Conshohocken, PA.

Standard Test Method for Critical Radiant Flux of Floor-Covering Systems Using a Radiant Heat Energy Source, ASTM E-648, ASTM, West Conshohocken, PA.

Standard Test Method for Fire Retardancy of Paints (Cabinet Method), ASTM D-1360, American Society for Testing and Materials, West Conshohocken, PA.

Standard Test Method for Flame Height, Time of Burning, and Loss of Weight of Rigid Thermoset Cellular Plastics in a Vertical Position, ASTM D-3014, ASTM, West Conshohocken, PA.

Standard Test Method for Flammability of Finished Textile Floor Covering Materials, ASTM D-2859, American Society for Testing and Materials, West Conshohocken, PA.

Standard Test Method for Measuring the Minimum Oxygen Concentration to Support Candle-Like Combustion of Plastics (Oxygen Index), ASTM D-2863, ASTM, West Conshohocken, PA.

Standard Test Method for Rate of Burning and/or Extent and Time of Burning of Plastics in a Horizontal Position, ASTM D-635, ASTM, West Conshohocken, PA.

Standard Test Method for Surface Burning Characteristics of Building Materials, ASTM E-84, ASTM, West Conshohocken, PA.

Test for Surface Burning Characteristics of Building Materials, UL 723, Underwriters Laboratories, Inc., Northbrook, IL.

Test Method for Critical Radiant Flux of Floor-Covering Systems Using a Radiant Heat Energy Source, ASTM E-648, ASTM, West Conshohocken, PA.

Cote, A. E., ed., *Fire Protection Handbook,* 18th ed., NFPA, Quincy, MA, 1997.

Coté, R., ed., *Life Safety Code® Handbook,* NFPA, Quincy, MA, 2000.

"Carpets and Rugs—Notice of Standard," *Federal Register,* Vol. 35, No. 74 (April 16, 1970). Standard for the surface flammability of carpets and rugs (Pill Test).

NFPA Codes, Standards, and Recommended Practices

See the latest version of The NFPA Catalog for availability of current editions of the following documents.

NFPA 35, *Standard for the Manufacture of Organic Coatings*

NFPA *101*®, *Life Safety Code*®
NFPA 253, *Standard Method of Test for Critical Radiant Flux of Floor Covering Systems Using a Radiant Heat Energy Source*
NFPA 255, *Standard Method of Test of Surface Burning Characteristics of Building Materials*
NFPA 260, *Standard Methods of Tests and Classification System for Cigarette Ignition Resistance of Components of Upholstered Furniture*
NFPA 261, *Standard Method of Test for Determining Resistance of Mock-Up Upholstered Furniture Material Assemblies to Ignition by Smoldering Cigarettes*
NFPA 265, *Standard Methods of Fire Tests for Evaluating Room Fire Growth Contribution of Textile Wall Coverings*
NFPA 266, *Standard Method of Test for Fire Characteristics of Upholstered Furniture Exposed to Flaming Ignition*
NFPA 267, *Standard Method of Test for Fire Characteristics of Mattresses and Bedding Assemblies Exposed to Flaming Ignition Source*
NFPA 701, *Standard Methods of Fire Tests for Flame Propagation of Textiles and Films*
NFPA 703, *Standard for Fire Retardant Impregnated Wood and Fire Retardant Coatings for Building Materials*

SECTION 3 Occupancies

CHAPTER 22 Assembly Occupancies *Joseph Versteeg*

CHAPTER 23 Educational Occupancies *Joseph Versteeg*

CHAPTER 24 Day-Care Facilities *Joseph Versteeg*

CHAPTER 25 Health Care Facilities *Joseph M. Jardin*

CHAPTER 26 Ambulatory Health Care Facilities *Joseph M. Jardin*

CHAPTER 27 Detention and Correctional Occupancies *Joseph M. Jardin*

CHAPTER 28 Hotels *Joseph M. Jardin*

CHAPTER 29 Apartment Buildings *Joseph M. Jardin*

CHAPTER 30 Lodging or Rooming Houses *Joseph M. Jardin*

CHAPTER 31 Residential Board and Care Occupancies *Joseph M. Jardin*

CHAPTER 32 One- and Two-Family Dwellings *Joseph M. Jardin*

CHAPTER 33 Mercantile Occupancies *Joseph Versteeg*

CHAPTER 34 Business Occupancies *Joseph Versteeg*

CHAPTER 35 Industrial Occupancies *Joseph Versteeg*

CHAPTER 36 Storage Occupancies *Joseph Versteeg*

CHAPTER 37 Special Structures and High-Rise Buildings *Joseph Versteeg*

Assembly Occupancies

CHAPTER 22

Joseph Versteeg

CASE STUDY

Drapes Ignite in Theater

When a fire broke out in the showroom/theater of a large hotel and casino in Nevada, a sprinkler extinguished the blaze while employees safely evacuated the theater. Following the fire, an official noted that the safety evacuation plan worked exceptionally well.

The theater was part of a 17-story hotel and casino that was equipped with an automatic sprinkler system and a smoke exhaust system.

Staff watching a theatrical performance noticed flames spreading up drapes located approximately 12 ft (4 m) off the ground near the stage. Two of the employees tried to pull the burning drapes to the floor, as others activated the safety evacuation plan. Security immediately notified the fire department at 8:30 P.M, and two engines, a ladder company, and a chief officer responded. While still en route 30 seconds later, the commanding officer upgraded the response to a full assignment to include two more engines, another ladder company, and a rescue squad.

Fire fighters arrived within 5 minutes to find that the theater had been evacuated and the blaze extinguished by a sprinkler. Using the smoke exhaust system, they limited smoke damage to the theater and evaluated five employees for minor smoke inhalation.

Investigators determined that the fire began near a television monitor plugged into a wall-mounted electrical receptacle outlet. Neither the monitor, the cord, nor the plug showed signs of damage resulting from an electrical component fault or failure, however. Noting a similar monitor on the opposite wall of the stage, they found decorative drapes hanging over the monitor cord and electrical outlet. Like the fabric that had ignited, these drapes had a gold metal base, which later proved to conduct electricity. The investigators determined that the monitor's cord cap had not been completely inserted into the receptacle, exposing the blades, which energized the hanging fabric.

The hotel told the fire department that the drapes had been treated with a flame retardant. As a precaution, however, they were removed before the theater reopened.

Damage to the theater, valued at $500,000, was estimated at $10,000.

Source: "Fire Watch," *NFPA Journal*, May/June, 1999.

This chapter discusses items of special interest when an assembly occupancy is inspected. General inspection principles and items covered in other chapters apply, as well.

Joe Versteeg is a fire protection consultant based in Torrington, CT. He was with the Connecticut Department of Public Safety for over 20 years; most recently he was the commanding officer of the technical services section for the State Fire Marshal's Office. He also serves on several Life Safety Code® *committees.*

FIGURE 22-1 A Banquet Hall

Assembly occupancies are defined by NFPA *101*®, *Life Safety Code*® as buildings or portions of buildings in which 50 or more persons gather for such purposes as deliberation, worship, entertainment, dining, amusement, or awaiting transportation (Figure 22-1). The character of an assembly occupancy should never be assumed to remain constant, and neither should the occupant load.

Because the many legal ways in which as assembly occupancy can be used are so diverse, inspectors should thoroughly review the *Life Safety Code* before beginning an inspection to ensure that they understand the proper requirements for a particular assembly occupancy. Unlike most other occupancies, assembly occupancies encompass a wide range of uses, each of which necessitates different considerations.

It is important to note that occupancy of any room or space for assembly purposes by fewer than 50 persons in a building of another occupancy, such as, for example, a 30-person conference room in an office building, is considered incidental to the predominant occupancy. Although subject to the requirements governing the predominant occupancy, the occupant load of such small assembly uses is to be calculated based on the actual use of the space. For example, a small employee breakroom containing tables and chairs within a factory is to be calculated based on 15 ft^2/person (1.4 m^2/person).

OCCUPANCY CHARACTERISTICS

Changes of use, or the multiuse of assembly occupancies, could result in the application of provisions that normally might not be considered necessary. For example, a building used as a place of worship must meet certain, basic code requirements. Yet this same building also might be used for dining, dancing, or other purposes totally foreign to a place of worship, thus triggering the need to meet additional code requirements. Assembly occupancies in schools, such as multipurpose rooms, are rented or freely used for purposes other than education and often take on the character of exhibit halls. The use of available space in hotels, banquet rooms, shopping malls, and exhibit halls can also be very creative. When inspecting assembly occupancies, therefore, inspectors must be sure to ascertain all intended or possible uses.

INSPECTING THE PREMISES

As when they inspect any other type of facility, inspectors should be seen by the owner of the assembly occupancy as providing a service to them by conducting a fire and life safety inspection. Inspectors should always meet with the manager or owner of the establishment before beginning an inspection, and should encourage him or her to accompany them on the inspection.

Inspecting assembly occupancies is no different from inspecting other occupancies. First a general "once-over" inspection should be done to spot any immediate concerns and the floor area of the building should be determined through measurement. With this information, inspectors can figure out the occupant load of the facility. They should note any exterior violations, such as accumulations of trash or obstructed fire lanes. They should also point out to the manager or owner any misuse of extension cords, sloppy maintenance practices, or other areas of concern so that these can be corrected before the facility is reinspected.

When completing an inspection, inspectors should develop and file a sketch or drawing of the facility for future reference. This sketch should show the exterior and interior wall arrangements; the locations of all exit doors; the side yard, street, and property line clearances; and any other conditions of special hazard or consideration of special interest to assembly occupancies. The sketch should also identify any portable sliding or folding partitions used to divide rooms, as well as the occupant loads allowed for different room layouts, such as tables and chairs, theater-style seating, dancing, and so on. This sketch is essential for future use, not only as a reminder of existing conditions but, more importantly, as an easy reference to determine if any changes have occurred since the last inspection.

Inspectors should keep a permanent checklist to indicate the construction of interior and exterior walls, floor and roof coverings, the flame-spread ratings, the type of heating, the lighting and electrical systems and their conditions, and the available fire protection devices and systems. This checklist will be useful for future inspections.

Occupant Load

The intended use of the premises and the number of exits in excess of the minimum number required will influence the maximum allowable occupant load. If the use of the occupancy has changed since the last inspection, the maximum allowable occupant load probably has changed, too. If it is a multipurpose space, the inspector should review any changes that would affect the variations described in the original sketches and should change the sketches accordingly. The *Life Safety Code* contains requirements for calculating the occupant load. If the allowable load has changed in any way, the inspector should ask the owner to provide a new "maximum occupant load" sign and display it as required, provided the means of egress are still acceptable for the new occupant load.

Established occupant loads should be posted prominently to ensure that not only the owner, but also the manager, operator, and occupants, are aware of the limitations. Occupant loads for multipurpose rooms should be posted for each approved use, such as tables and chairs, theater seating, dancing, and so on. Posting load figures will also help the inspector determine whether the occupancy is overcrowded.

Means of Egress

Exiting is the most critical of all requirements for any assembly occupancy. While the **probability** of a fire in an assembly occupancy might be low, the **potential** for loss of life once a fire occurs is extremely high. A fire of any magnitude can easily result in a large number of injuries and deaths. Therefore, it is essential that assembly occupancies have enough egress capacity to accommodate the number of people likely to occupy the space and that they be properly located, easily accessible, and well maintained.

A life safety evaluation is an added requirement for assembly occupancies having an occupant load in excess of 6000 people. The life safety evaluation is a yearly assessment of the following conditions:

1. Nature of events and the participants and attendees
2. Access and egress movement, including crowd density problems
3. Medical emergencies
4. Fire hazards
5. Permanent and temporary structural systems
6. Severe weather conditions
7. Earthquakes
8. Civil or other disturbances
9. Hazardous material incidents within and near the facility
10. Relationships among facility management, event participants, emergency response agencies, and others having a role in the events accommodated in the facility

Inspectors must ensure that conditions altered since the last inspection have not compromised or blocked egress routes. If any alterations or renovations have been made since the last inspection, the inspector must be sure that travel distances to exits have not been increased beyond the maximum allowed. Where exit paths merge, the path of travel must be wide enough to accommodate the combined occupant load that can be expected to use the individual paths of travel before they merge.

All exit doors must open easily, with no more than 15 lb (6.8 kg) of force necessary on the panic bar to release the latch. Inspectors must ensure that exit doors are not chained or padlocked closed. Life safety requirements, particularly those relating to egress, must be maintained at *all* times.

Inspectors must not allow registration booths, head tables, projection screens, ticket booths, turnstiles, revolving doors, guide ropes, and so on to obstruct any means of egress. When loose chairs are provided, setting up and maintaining proper aisles is a particularly difficult problem. Normally, loose chairs must be ganged—that is, connected to each other—when the number of chairs exceeds 200.

The allowable configuration of aisles will vary depending on the type of seating provided—that is, banquet or conference-type tables, auditorium/theater seating arrangements, bleachers, grandstands, and so on. Spaces between and around such seating must be adequate to provide access to aisles.

In addition the egress routes, as well as the exit and directional exit signs, should be illuminated in both the normal and emergency modes.

Interior Finish

Another major issue is the flame spread rating of interior finish materials and the flammability of decorative materials, curtains, drapes, and similar finishings. Interior finish in stairways should always be Class A. In corridors and lobbies, it may be Class A or B. In the general assembly area itself, it may be Class A or B; however, in assembly occupancies of 300 or fewer persons, Class C is permitted. Only rated material is allowed. See Chapter 21, "Interior Finish, Contents, and Furnishings," for further information.

Combustible, decorative materials should be treated with a flame retardant. From a practical standpoint, flame-spread ratings are difficult to ascertain during a field inspection. If inspectors are unable to see any markings on the products, they should ask if the original construction, any subsequent installations, or manufacturers' test data are available. They should also check the inspection file for prior acceptance of existing materials.

It might be possible to obtain a sample of the decorative material from an unobtrusive location—along an inside seam, for example—and test it in a relatively wind-free location outside the building by placing the sample in a vertical position and setting a flame to the lower edge of the material. If charring does not occur beyond the flame and no flame or charring occurs after the flame has been removed, the product can be assumed to be reasonably safe. If charring, dripping, or flaming continues, however, the product is suspect and should be removed, replaced, or subjected to a standard fire test.

Building Services

The inherent sources of ignition in assembly occupancies also include air-conditioning, heating, and refrigeration units or systems; electrical wiring; and electrical appliances; as well as conditions that exist in commercial kitchens. Frying and deep-fat cooking constitute the greatest single danger. Because hood and duct fires are very common, the operating condition of the hood and vent extinguishing and exhaust systems should be inspected carefully. Hoods and vents should be examined to determine if there has been a buildup of flammable material. These areas must be surveyed and cleaned continually, sometimes daily. The inspector should make sure that the exhaust damper opens when the exhaust fan is operated.

The inspector should determine the type of heating system used in the facility and the type of fuel used in the heating system and should ask the following questions:

- Must the heating unit be separated from the rest of the building?
- Are the walls, ceiling, and floor of proper construction?
- Are all openings, including duct openings, properly protected?
- Are there any smoke detectors on the downstream side of filters in the air supply or return system?

If the heating system fuel is LP-Gas, the inspector should find out if the system has shutoff controls that activate automatically if the pilot light goes out. Is the system located where LP-Gas will pocket or become trapped in the building in the event of a gas leak? LP-Gas cylinders should never be stored or used inside except under very limited conditions. Where is the LP-Gas supply located? If supply tanks are used, the inspector should make sure they are properly installed, secured, protected, and safeguarded against tampering or accidental damage and that the cylinders are stamped and designed for use with LP-Gas.

If the fuel is a flammable or combustible liquid, the inspector should determine whether the door opening is diked. On gravity feed systems, the inspector should verify that there is an antisiphon device and should make sure there is a fusible shutoff device that will activate in the event of fire near the heating equipment.

When checking the electrical wiring and appliances, the inspector should determine whether any permanent installations have been made using temporary equipment and should ask the following questions:

- Are the electric circuits large enough to handle the expected load?
- Are the noncurrent-carrying metal parts of portable and fixed electrically operated equipment properly grounded?

- Have any electrical extension cords been approved for their intended use, and are they being used properly?

If the inspector has any doubts about these items, he or she should have the community's electrical inspector make the determination.

Smoking

Smoking is not always prohibited in assembly occupancies, with one exception: smoking is never allowed in theaters or assembly occupancies similar to theaters, such as facilities hosting stage shows and concerts. The prohibition of smoking in restaurants is becoming more popular, but this essentially is a health-related issue, not a fire-related issue.

SPECIAL SAFEGUARDS FOR UNIQUE OCCUPANCIES

Stages and Projection Rooms

Stages and enclosed platforms present unique hazards associated only with assembly occupancies, and they require special safeguards, such as protection of the proscenium wall, including the proscenium curtain; automatic sprinklers above and below the stage; and automatic venting. Motion picture projection rooms require special supply and exhaust air, egress, and port openings, all of which must be protected. They also require room enclosure and proper working space. Projection machines require individual exhaust capabilities, which vary with the type of equipment. Projection rooms in which cellulose nitrate film is used must comply with NFPA 40, *Standard for the Storage and Handling of Cellulose Nitrate Motion Picture Film.*

Exhibits and Trade Shows

Because promoters and exhibitors are often creative in what they want to do and the materials they want to use, exhibits and trade shows can be challenging to inspect (Figure 22-2). The inspector should review the products that will be displayed, as well as the exhibits, and closely review special provisions in the *Life Safety Code* for help with this difficult assignment.

FIGURE 22-2 Trade Show

When inspecting exhibits and trade shows, the inspector should make sure to have a plan that shows details of the area, the booth arrangement, the fire protection equipment, and so on. The exhibit booths' construction should be of noncombustible or limited combustible materials, and curtains, drapes, acoustical materials, decorations, and so on should be flame retardant. Multilevel booths and those over 300 ft^2 (27.9 m^2) should be sprinklered when in sprinklered occupancies. Access must be plainly visible, and the travel distance inside a booth to an exit access aisle should not exceed 50 ft (15 m).

Cartons and crates should be stored in a room separated from other portions of the building with construction that has a 1-hour fire resistance rating and sprinkler protection. Cooking devices should be limited in number and protected with sprinklers or some other form of extinguishing agent.

The inspector may also have to deal with motor vehicles displayed in this type of occupancy. The electrical system on vehicles should be disconnected to reduce ignition sources, and fuel tanks should be sealed. Fueling and defueling should not be allowed inside the structure.

Special Amusement Buildings

Special amusement buildings present yet another life safety problem because they generally entertain customers by confusing them. Nonetheless, the means of egress must be plainly visible and lighted during an emergency. Under certain conditions, smoke detection systems may be necessary. The *Life Safety Code* requires that every special amusement building be protected with automatic sprinklers. Moveable or portable special amusement buildings must also be protected, and the water supply must come from sources approved by the authority having jurisdiction.

The *Life Safety Code* contains specific criteria within the means of egress portion of the assembly occupancy chapters for calculating the minimum width of aisle accessways and aisles serving seating arranged in rows, as well as for seating at tables. In addition, specialized seating arrangements such as grandstands and folding and telescopic seating are contained within the special provisions portion of the assembly chapters.

The requirements governing grandstands, in addition to establishing minimum spacing dimensions to ensure safe egress, contain structural requirements for portable grandstands to safeguard against collapse; size limitations on wooden grandstands in the event of fire; and safeguards against fire for the area immediately beneath all types of grandstands.

OPEN FLAMES AND PYROTECHNICS

In addition to the sources of heat and open flame previously discussed, restaurant owners often use table candles to enhance the atmosphere. This practice should be discouraged. When it is permitted, however, the candles should be placed in stable containers or holders of noncombustible construction that are designed not to tip over easily. The inspector should test one of the typical candle holders to ensure that the flame does not come in contact with other combustible materials if it does tip over.

Table carts with open flames used as food warmers or for actual cooking are another potential source of ignition. In many cases, food on these tables is saturated with alcohol, which is then ignited. This activity is generally conducted very close to the restaurant patrons. There is no established means of protection against the obvious hazards of this practice, except prohibition.

Both the *Life Safety Code* and NFPA 58, *Liquefied Petroleum Gas Code*, limit the indoor use of portable butane-fueled appliances in restaurants and in attended commercial food catering operations where fueled by not more than two 10-oz (0.28-kg) LP-Gas capacity, nonrefillable butane containers that have a water capacity not exceeding 1.08 lb (0.4 kg) per container. Storage of cylinders is also limited to 24 containers, with an additional 24 permitted where protected by a 2-hour fire resistance–rated barrier. The practice of bringing large propane or butane containers indoors is common in restaurants and presents an extreme life safety hazard.

In places of worship, the congregation occasionally holds lighted candles and sometimes marches in procession with them. While limited use of candles by designated officials can be permitted for religious purposes, the general assembly should not be allowed to hold any open flame devices.

Pyrotechnics traditionally have been used on the stages and platforms of assembly occupancies particularly during magic acts or shows. With the advent of discos and rock concerts, however, there has been a dramatic increase in the use of pyrotechnic devices. This practice can create extremely hazardous conditions, depending on the type, volume, setting, and control exercised when they are used. See NFPA 1126, *Standard for the Use of Pyrotechnics before a Proximate Audience,* for more information on this subject. There is also the tendency to overcrowd such facilities beyond the occupant load allowed by permit. Limiting both of these conditions is essential to minimum life safety and this limitation must be enforced on a performance-by-performance basis.

FIRE PROTECTION SYSTEMS

On-site or built-in fire protection equipment includes portable fire extinguishers, interior standpipes and hose lines, automatic sprinkler systems, and fire alarm systems. New assembly occupancies with occupant loads in excess of 300 persons are required by the *Life Safety Code* to be fully sprinklered. Other methods of on-site or built-in fire protection may also be used, and each system or item should be reviewed to determine whether it is an approved method for the hazard protected. Inspectors should also review the general condition of fire protection systems and, where applicable, supervise performance tests.

When located in a building of mixed occupancy, the assembly occupancy might require separation from the remaining occupancies by fire-resistive assemblies of various ratings. Under these circumstances, openings in such assemblies must also be protected. In some instances, however, no separation is needed because the exposure hazard is low or nonexistent. In other instances, a 1-, 2-, 3-, or even a 4-hour fire resistance–rated separation will be necessary. The specifics of these requirements will depend on the applicable building code.

SUMMARY

This chapter has identified the criteria necessary for properly classifying assembly occupancies as well as having provided a discussion on aspects of occupant loads, means of egress, interior finish requirements, and fire protection systems. In addition, the chapter has examined areas requiring special attention, such as stages, exhibits and trade shows, special amusement buildings, the use of open flame devices and/or pyrotechnics.

BIBLIOGRAPHY

NFPA Codes, Standards, and Recommended Practices

See the latest version of The NFPA Catalog for availability of current editions of the following documents.

NFPA 40, *Standard for the Storage and Handling of Cellulose Nitrate Motion Picture Film*
NFPA 58, *Liquefied Petroleum Gas Code*
NFPA 96, *Standard for Ventilation, Control and Fire Protection of Commercial Cooking Operations*
NFPA *101*®, *Life Safety Code*®
NFPA 102, *Standard for Grandstands, Folding and Telescopic seating, Tents and Membrane Structures*
NFPA 1126, *Standard for the Use of Pyrotechnics before a Proximate Audience*

Educational Occupancies

CHAPTER 23

Joseph Versteeg

CASE STUDY

Fire in Concealed Space Damages Occupied School

The lack of fire detectors allowed a fire to spread in the attic and concealed spaces between the ceiling and roof of a single-story California elementary school, which measured 292 × 30 ft and contained eight classrooms. The school was constructed of unprotected wood framing with a stucco exterior. It had no fire detection system or sprinklers, and its manual fire alarm was out of service when the fire occurred.

A student walking past an unoccupied classroom discovered the fire, which was started by a short circuit or a circuit overload in fixed wiring in the attic. The fire department was called at 9:17 A.M. By the time fire fighters arrived, flames were coming from the building's roof, and two classrooms were fully involved. The fire burned down through the ceiling, igniting the heavy fuel load in the classroom below, which was being used as a library. Lack of fire stops in the attic allowed the blaze to spread horizontally and from the attic to another classroom and through the roof. The first classroom was unoccupied because the responsible teacher was out sick, so the fire grew undetected.

Teachers had reported multiple electrical malfunctions before the fire. The building, valued at $1 million dollars, sustained $365,000 in damage. Its contents, valued at $250,000, sustained damages of $75,000. There were no injuries.

Source: "Fire Watch," *NFPA Journal,* January/February, 1998.

Joe Versteeg is a fire protection consultant based in Torrington, CT. He was with the Connecticut Department of Public Safety for over 20 years; most recently he was the commanding officer of the technical services section for the State Fire Marshal's Office. He also serves on several Life Safety Code® *committees.*

NFPA *101®*, *Life Safety Code®*, defines educational occupancies as buildings used for gatherings of six or more persons, for 4 or more hours a day, or more than 12 hours a week, for the purpose of instruction through the twelfth grade. Educational occupancies include schools, academies, and kindergartens. Day-care facilities are not classified as educational occupancies; they must meet the requirements of Chapters 16 and 17 of the *Life Safety Code,* 2000 edition. Schools for levels beyond twelfth grade are not classified as educational occupancies. They must comply with the requirements for business, assembly, or other appropriate occupancies.

OCCUPANCY CHARACTERISTICS

The activities in educational occupancies can vary from education in a classroom with contents of a low fire hazard to work in a laboratory or shop area where the contents could pose a moderate or high fire hazard. Educational occupancies also typically contain assembly areas, such as auditoriums, cafeterias, and gymnasiums, where the fire hazard is low or moderate, the concentration of occupants is high, and persons using such assembly areas may not be familiar with the facility. Where such assembly areas have an occupant load of 50 or more persons, an assembly occupancy also exists within the same building (see Chapter 22, "Assembly Occupancies").

It is important for fire inspectors to fully understand the activities taking place and the number and ages of the occupants in the facility they are inspecting because this information will affect whether the facility is subject to the requirements for educational occupancies or those for day-care occupancies.

The inspector should review applicable codes and previous inspection reports before conducting the inspection. Existing facilities must meet the requirements of Chapter 15 of the *Life Safety Code*. New and renovated buildings and buildings whose occupancy classification has changed—such as from residential to educational or day care—must meet the requirements of Chapter 14 of the *Life Safety Code*. If only part of a building is renovated or the occupancy of that part changes, that portion and its means of egress must meet the requirements of Chapter 14 of the *Life Safety Code* and the rest of the building must meet the requirements of Chapter 15.

The inspector must cite all code deficiencies, such as exit enclosure doors that are propped open, even if they are corrected in the inspector's presence. If such violations continue to appear on future inspections, legal action or suspension of the license might be required to correct the problem.

The inspector should also review drawings of the facility before conducting the inspection. If none are available, the inspector should suggest that the owner get them from the architect for his or her own use in planning as well as for future inspections. In fact, this may be required by law, especially for public schools. Features such as the means of egress are sometimes more obvious on a plan than when they are viewed in the facility. The inspector should check that any conditions violating code have been permitted by the authority having jurisdiction under an "equivalency" agreement.

INSPECTING THE PREMISES

Occupant Load

Occupant load varies in educational occupancies. It is based on a minimum of one person for every 20 ft^2 (1.9 m^2) in classroom areas and one person for every 50 ft^2 (4.6 m^2) in shops, laboratories, and similar vocational rooms. In gymnasiums or cafeterias used for 50 or more persons, the occupant load is calculated using occupant load factors for assembly uses. The occupant load in such rooms not having fixed seating should be posted at the room's main entrance door. This posted maximum occupant load should be based on the available egress width for the rooms multiple uses and arrangement of furnishings.

Means of Egress

Corridors in schools are a major component of the means of egress, and it is common for them to be lined on either side by classrooms and education support rooms. The corridors should lead directly to exits or to other corridors that lead to exits. Usually corridors must be enclosed by either 1-hour or 20-minute fire resistance–rated walls. They must have self-closing latching doors. There are exceptions in the *Life Safety Code*, especially in sprinklered facilities. Transoms or other glazing that is part of a door assembly must be either wire glass or 20-minute fire protection–rated. Corridor wall windows that are not part of a door assembly must be 45-minute fire protection–rated fire window assemblies.

The most important action to be taken when fire occurs in a school is evacuation, thus it is imperative that the means of egress be maintained in a usable condition. Exit stairways must be unobstructed, and they cannot be used for any other purpose, especially storage. Combustible materials are often found stored under stairs or in stair enclosures.

Corridors, which should be at least 6 ft (1.8 m) wide, cannot be restricted. Often, when a special function is being held, tables are set up and coat racks on wheels are provided in the corridor. These items should not restrict the corridor and cannot be placed in such a way that they interfere with the means of egress. Athletic equipment stored along the walls of the gymnasium must not encroach on the means of egress.

Doors swinging into corridors must be arranged not to interfere with corridor travel. Corridor doors should not be wedged open; any wedges found should be removed and this code violation should be cited in the report.

Emergency lighting should be provided in interior stairs and corridors, assembly use spaces, portions of the building that are interior or windowless, and shops and laboratories.

The inspector should be sure banners, signs, and similar materials are flame retardant and do not obstruct or cover exit doors, signs, or other egress components. The inspector should note whether exit signs delineating the exits and paths to exits are illuminated and visible from any point where the exit or exit access path to reach the exit is not readily apparent to the inspector. Spotlight-type emergency lighting should be located so that it illuminates the exit signs; it should not shine in the eyes of those looking for exits.

Security gates or doors chained to secure an area used for events taking place after normal school hours from the remainder of the facility should not restrict the egress facilities needed for the in-use area.

Inspectors should make sure that exit doors designed to be kept closed, such as for enclosed stairways or through fire barrier walls, are self-closing or automatic closing by smoke detection. They should check that these doors close and latch freely. Inspectors should observe that all egress doors are accessible, unobstructed, and cannot be locked except as allowed by the appropriate code requirements. They should open each exit and exit access door to determine that it will open easily and leads to a public way or area of refuge. They should also check the exit discharges from outside the building to be sure that the exit discharges are not blocked and that the paths to public ways are clear.

Doors in required smoke barriers should not be wedged open. If they are, have the wedges removed and cite them in your report. If violations persist the inspector may require the installation of smoke detectors and magnetic hold-open devices. Smoke barriers should be continuous from outside wall to outside wall and floor slab to floor or roof deck above.

Preschool, kindergarten, and first-grade students must not be located above or below the floor of exit discharge. Second-grade students must not be more than one story above the floor of exit discharge.

Inspectors should review fire drill records to ensure that drills have been conducted regularly.

If there are any signs of remodeling or renovation, the inspector should check that all egress areas still meet requirements for egress capacity, travel distance, illumination, and marking. Travel distances to exits must still be within required limits and exits must not have been eliminated where they are required. If plans for renovations have to be reviewed in the jurisdiction, the inspector should require that they be submitted by a certain date and should enforce that deadline.

Windows

There must be a window or door for ventilation and rescue in each room occupied by students in buildings not protected by automatic sprinklers. This requirement does not apply to toilet rooms and offices. The window or door must be at least 20 in. (510 mm) wide and 24 in. (610 mm) high and have an area of at least 5.7 ft^2 (0.53 m^2).

Hazardous Areas

Fire experience has shown that hazardous areas in schools include janitor rooms, basements, boiler rooms, storerooms, and closets. Fires have also started in workshops, laboratories, classrooms, and auditoriums. In all educational occupancies, good housekeeping is basic to safety to life from fire.

Housekeeping in basements must be of the highest order. Storage should be confined to sprinklered areas or to separate storerooms that are properly segregated and equipped with fire doors. The inspector should check that the doors are kept closed.

Laboratories in which hazardous chemicals and flammable liquids and gases are used also deserve special attention. Their location and the amounts of materials stored in them should be noted, and this information should be passed on to the local fire department. Inspectors should make certain that flammable materials are stored in reasonable quantities in appropriate containers and cabinets. They should examine labels on containers of chemicals to determine whether any have exceeded their shelf life and whether any are unstable. In either case, the material should be removed from the laboratory. Inspectors should watch for indications of chemicals that should be kept segregated from others to prevent hazardous situations in the event of fire or accidental spills and should check that chemicals are discarded properly. Inspectors should also check that laboratories are equipped with suitable fire extinguishers and fire blankets. Laboratories must comply with the requirements of NFPA 45, *Standard on Fire Protection for Laboratories Using Chemicals.*

Educational and building maintenance workshops are also potential hazardous areas, so the inspector must pay particular attention to housekeeping in these areas. Oily waste should be kept in self-closing containers until it is removed from the building for disposal. The inspector should check sawdust removal equipment. Sawdust suspended in the air is an explosion hazard. Scrap material should be cleaned up after each class and safely stored until it is removed for disposal. The inspector should note whether equipment and machinery appear to be well maintained and in good condition.

If the occupancy has kitchen facilities, inspectors should examine them as they would a commercial or restaurant kitchen. They should check for grease accumulations on and around fryers, ranges, hoods, and filters and they should note whether the locations of cooking equipment and the construction of the hood and ducts meet code requirements. The fire protection equipment should be examined to determine that it is in good condition, charged, and within acceptable limits. The extinguishing agents in the portable fire extinguisher near the hood and in the fire suppression system must be compatible. The inspector should check that any deficiencies found in the fire protection system in previous inspections have since been corrected.

Interior Finish

Interior finish should be Class A in exits such as enclosed stairways; it can be either Class A or B elsewhere in the building. The inspector should ascertain that draperies, curtains, and similar furnishings and decorations have been treated with a flame retardant.

Teaching materials and students' artwork should not cover more than 20 percent of the wall area. The inspector should be especially concerned with seasonal decorations—such as those for Halloween or Thanksgiving—and the props used for theatrical presentations. Items such as cardboard, paper, or cloth houses are often used for Halloween parties, and combustible materials such as corn stalks, straw, or paper are sometimes used for decoration. Special effects to be used in theatrical productions must meet the requirements of all applicable codes and ordinances. Intumescent paints and surface coatings can be used to reduce the surface flame spread on interior finish. However, some of these coatings have a short life and must be reapplied frequently.

FIRE PROTECTION

In educational occupancies fire alarm systems are often disabled. Thus, the inspector should locate the fire alarm control box and should check whether

- The system is operational
- Any trouble lights are on
- The supervisory signal is silenced
- Backup batteries are in place and fully charged

The inspector should check the condition of each manual fire alarm box for signs of damage. The inspector should examine fire drill records to ensure that the building can be evacuated in a resonable time period. He or she should note whether there are any recurring problems when the building is being evacuated that might need to be corrected, whether the fire alarm signal is distinct from the signal to change classes, and when the system was last tested.

NONTRADITIONAL BUILDINGS

Flexible and Open-Plan Buildings

In addition to the features and conditions to be inspected in traditional school buildings, the inspector must also check additional items in flexible or open-plan buildings. Flexible and open-plan buildings are designed to have multiple teaching stations and may have movable corridor walls and partitions of full height.

The interior furnishings in open-plan schools may be arranged to designate the exits and paths of egress travel. The paths should be direct, not circuitous, and of sufficient width to accommodate the occupant load that will use the egress path. The inspector should determine that the layout has not been altered since the last inspection without the approval of the authority having jurisdiction.

Temporary Buildings

The use of modular or portable structures is increasing, which presents a unique problem. These structures are on school grounds, but they are detached and are sometimes located great distances from the main school buildings and on surfaces that would not support fire-fighting vehicles.

The school fire alarm system should be audible, visible, and capable of being activated from the modular structures. If the temporary buildings have two-way communication with the main school buildings and if they have a constantly attended receiving station from where an alarm can be sounded, they do not need manual fire alarm boxes unless required to have them by the authority having jurisdiction.

SUMMARY

This chapter has identified the criteria necessary for properly classifying educational occupancies as well as having provided a discussion on aspects of occupant loads, means of egress, windows for rescue and ventilation, the protection of hazardous areas, interior finish requirements, and fire protection systems. In addition, the chapter has examined nontraditional school arrangements such as flexible and open plan buildings and temporary buildings.

BIBLIOGRAPHY

NFPA Codes, Standards, and Recommended Practices

See the latest version of The NFPA Catalog for availability of current editions of the following documents.

NFPA 45, *Standard on Fire Protection for Laboratories using Chemicals*
NFPA *101*®, *Life Safety Code*®

Day-Care Facilities

CHAPTER 24

Joseph Versteeg

Day-care facilities are divided into three classes: day-care occupancies, group day-care homes, and family day-care homes. In all subclasses of day care, the clients receive care, maintenance, and supervision by persons other than relatives or legal guardians for less than 24 hours per day.

Day-care facilities are not subject to the requirements governing educational occupancies in that the primary purpose of a day-care facility is other than education. In addition, day-care facilities within places of religious worship are not required to comply with the day-care occupancy requirements if they provide day care only during the religious service. The concept behind this exemption is that the parents of the clients will be in the same building attending the religious service and can assist with client evacuation in case of emergency.

Day-care occupancies are those facilities having more than 12 clients. Because of the greater number of occupants, day-care occupancies are subject to requirements that limit their location within buildings based on construction type and the presence of automatic sprinklers. They are also subject to requirements governing means of egress, protection from hazards, and fire alarm systems. When day-care occupancies and day-care homes are located within buildings containing other occupancies, NFPA *101*®, *Life Safety Code*®, requires that they be separated from the other occupancy by 1-hour fire barriers.

Day-care homes are divided into two subclassifications. Group day-care homes provide care for 7 to 12 clients; family day-care homes provide care for 4 to 6 clients. Because of their smaller size, the location of day-care homes is not restricted based on construction. In addition, means of escape and smoke alarms are utilized in lieu of means of egress and smoke detection systems.

An important term used in the criteria for classification of day-care occupancies and homes is the word *client*. Earlier editions of the *Life Safety Code* classified day-care occupancies based on the presence of children; currently, *Life Safety Code*'s use of the term *client* extends such occupancies to adult day-care environments.

Joe Versteeg is a fire protection consultant based in Torrington, CT. He was with the Connecticut Department of Public Safety for over 20 years; most recently he was the commanding officer of the technical services section for the State Fire Marshal's Office. He also serves on several Life Safety Code® *committees.*

DAY-CARE OCCUPANCIES

The occupant load in a day-care occupancy is typically based on one person for each 35 ft^2 (3.3 m^2) of net floor area used for day-care purposes. However, the occupant load is permitted to be increased to a density greater than that calculated when such an increase is approved by the authority having jurisdiction based on a diagram demonstrating that all aisle, seating, and egress requirements continue to be satisfied.

Each client-occupied floor must have two remotely located exits. All egress doors are required to swing in the direction of egress travel when serving a room or area having an occupant load of 50 or more persons. When the day-care occupancy serves an occupant load of 100 or more persons, the method of releasing door latches and/or locks must be accomplished by panic or fire exit hardware. Special attention should be focused on the type of hardware installed on closet and bathroom doors to ensure that closet doors are openable from inside the closet and bathroom doors are openable by staff in the event of an emergency.

In buildings not protected by an approved supervised automatic sprinkler system, each room or space that is normally subject to occupancy by clients (other than bathrooms) must have a door or window opening directly to the outside. In new occupancies, the size of the window opening must be at least 20 in. (510 mm) wide and 24 in. (610 mm) high, and it must provide an opening of at least 5.7 ft^2 (0.53 m^2). Although a window of the same size is required in existing facilities, the *Life Safety Code* permits smaller-sized existing awning-type windows as well as alternative escape route options.

The inspector should determine the type of building construction and check that it meets the minimum construction requirements based on the height of the occupancy. When determining compliance, the inspector should remember that the *Life Safety Code* does not govern the construction type of the overall building, rather it establishes the permitted location of the occupancy within a building based on a particular type of construction. For example, in a nonsprinklered, two-story building of 1-hour or less construction, the first floor can be used for day care; however, day-care use would not be permitted on any levels above or below the level of exit discharge.

The required fire alarm system is to be initiated by manual means and by any detection and sprinkler systems required by other code provisions. In new facilities and in existing facilities caring for more than 100 clients, the fire alarm system should notify the fire department by the most practical method allowed. Smoke detectors must be connected to the building fire alarm system unless the facility is a single-room center. Existing facilities are subject to an additional exemption if all clients are capable of self-preservation and there are no sleeping facilities.

GROUP DAY-CARE AND FAMILY DAY-CARE HOMES

Because of their fewer number of clients, group day-care and family day-care homes typically occur in living units of apartment buildings or in private homes; they are, however, not restricted to residential buildings and can be found in all types of occupancies. In instances in which care is provided on a temporary basis, with the parent or guardian in close proximity, the room or area used for such a purpose is exempt from the day-care home requirements and is governed by the predominant occupancy classification.

Each floor of a day-care home should have two remote means of escape and the travel distance to the nearest exit must not exceed 150 ft (46 m). In addition, every room used for living, sleeping, or dining purposes is required to have a second means of escape that provides an unobstructed route to the outside, such as a door or qualifying window. The use of loft areas that are accessible only by ladders or folding stairs is prohibited. Often overlooked are the requirements governing the interior finish of walls and ceilings when the day-care home is located within a private residence. Interior finish in occupied spaces should be at least Class C.

Single-station smoke alarms are required on all levels of the home and in all rooms used for client sleeping in accordance with the household fire-warning requirements of *NFPA 72®*, *National Fire Alarm Code®*. Smoke alarms should be powered by the building's electrical system. Existing battery-powered alarms are permitted where proper testing, maintenance, and battery replacement can be documented to the authority having jurisdiction.

OPERATING PROCEDURES

In all instances the day-care facility is required to have in effect a written fire emergency response plan. The plan should be available to employees for review; and all employees, including temporary staff, should be instructed as to their duties and responsibilities in accordance with the plan. Monthly egress drills are also required.

At least once a month, a fire prevention inspection should be conducted by a trained member of the facility, and a copy of the latest inspection report should be posted in a conspicuous location. In addition, staff members should inspect all egress facilities daily to ensure they are free of obstructions or impediments and available for use in an emergency. There are also restrictions on the amount of artwork and teaching materials attached to walls and the presence of clothing in corridors.

SUMMARY

This chapter has identified the criteria necessary for properly classifying day-care facilities. It has also provided a discussion on aspects of occupant loads, means of egress, windows for rescue and ventilation, protection of hazardous areas, interior finish requirements, and fire protection systems. In addition to day-care occupancies, the chapter has examined the requirements governing group day-care homes as well as staff responsibilities.

BIBLIOGRAPHY

NFPA Codes, Standards, and Recommended Practices

See the latest version of The NFPA Catalog for availability of current editions of the following documents.

NFPA 72®, National Fire Alarm Code®
NFPA *101®, Life Safety Code®*

Health Care Facilities

CHAPTER 25

Joseph M. Jardin

CASE STUDY

Safety Systems Limit Hospital Fire Damage

A large Illinois hospital undergoing a sprinkler retrofit quickly discovered the benefits of upgrading its system when sprinklers confined a fire caused by careless smoking to the basement storage room in which it began.

The five-story hospital, which was of unprotected, noncombustible construction, was occupied and operating at the time of the fire. Smoke and heat detectors providing full coverage were monitored by a municipal fire alarm system. The system also covered the new wet-pipe sprinkler system, 80 percent of which had been installed at the time the fire occurred. When the sprinklers activated, the smoke doors operated, the building's air-handling units shut down, and the staff initiated its emergency response plan.

The fire department received an automatic alarm, followed by a call from hospital staff reporting a working fire, at 6:00 P.M. Arriving fire fighters found a single sprinkler operating in a basement storage room that had recently been stocked with palletized cardboard boxes of toilet paper. The sprinkler had confined the fire to its area of origin, so fire fighters extended a hose line from a nearby standpipe to the area of origin and extinguished it.

Fire damage was limited to the object of origin, while smoke damage extended to the room and adjacent hallway. Some smoke also migrated to the third floor through a loose fitting on a pneumatic tube delivery system. The hospital staff quickly moved patients out of the affected area.

The fire was started by a carelessly discarded cigarette, which ignited the wooden pallets, allowing flames to spread to the boxes of toilet paper. Officials estimated the loss at $5000. There were no injuries.

The fire department stated that, "as a result of all life safety systems operating as designed, damage was held at $5000, and there was no fire extension."

Officials further noted that, had it not been for the sprinklers, "heavy smoke and fire would have filled the entire...basement and would have caused significant disruption to their operation."

Source: "Fire Watch," *NFPA Journal,* January/February, 1998.

NFPA *101*®, *Life Safety Code*®, defines a health care occupancy as a building, or any portion thereof, used on a 24-hour basis to house or treat four or more people who cannot escape from a fire without assistance. The reasons these people may not be able to escape include physical or mental illness, age, and security measures that the occupants cannot directly control. The buildings or portions of buildings in question include hospitals or other medical institutions, nurseries, nursing homes, and limited care facilities.

Joseph M. Jardin, a registered fire protection engineer, is a fire fighter with the New York City Fire Department, Rescue Co. No. 2. He is a former chair of the Safety to Life Residential Occupancies Committee and currently serves as a member of NFPA's Building Code Technical Correlating Committee.

A limited care facility is a building or part of a building that is used on a 24-hour basis to house four or more persons who are incapable of self-preservation because of age or physical limitation due to accident, illness, or mental limitations, such as mental retardation or developmental disability, mental illness, or chemical dependency, but who are not receiving medical or nursing care.

The code also addresses the ambulatory health care center, which is a building or any part of a building that provides services or treatment for four or more patients that would temporarily render them incapable of self-preservation during an emergency without assistance from others. These facilities include hemodialysis units, freestanding emergency medical units, and outpatient surgical areas in which general anesthesia is used. Ambulatory health care facilities do not provide overnight sleeping accommodations; the other categories of health care facilities do. (See Chapter 26 for a discussion of ambulatory health care occupancies.)

DEFEND-IN-PLACE THEORY

Health care facilities pose exceptional problems when it comes to moving and evacuating people, especially along great distances and down stairways from the upper floors of multistory buildings, to a safe area outdoors. The basic features of health care fire protection, therefore, involve a limited amount of patient movement.

The residents of a health care facility can remain safe even when relatively close to a fire if the corridor walls have been constructed properly, if the appropriate smoke and fire barriers have been installed, if hazardous areas that are likely to sustain a well-developed fire are protected or enclosed, and if approved fire detection and suppression systems have been installed (Figure 25-1). In many cases, this defend-in-place theory is not only desirable, it is necessary, especially in hospital intensive care units, cardiac care units, and operating room suites, where moving a patient could result in major health complications or even death.

Installing automatic sprinkler protection throughout all new health care facilities also provides enhanced protection for patients and staff who may be intimately involved with fire ignition. Sprinkler system design includes use of listed quick response

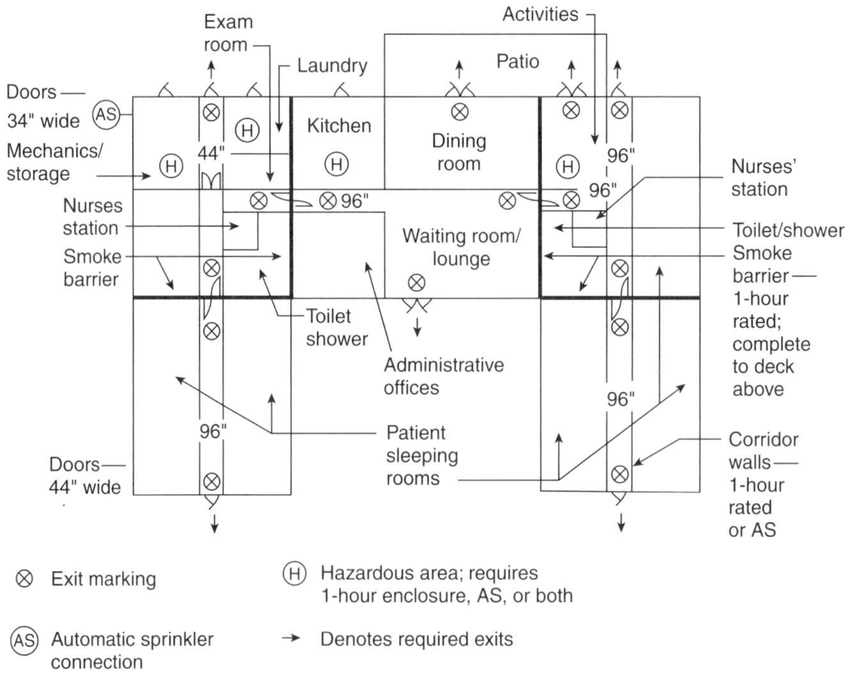

FIGURE 25-1 Typical Design of a Health Care Facility

and residential sprinklers installed throughout the smoke compartments containing patient sleeping rooms. The rapid activation of these specially designed sprinklers is intended to restrict the spread of fire and the associated products of combustion, thus reducing the need for extensive patient evacuation.

Ongoing and recently completed renovations should be reviewed for compliance with the *Life Safety Code*'s sprinkler provisions for renovated health care facilities. Each smoke compartment involved in a major renovation, alteration, or modernization requires sprinklers throughout. Listed quick response or residential sprinklers are required within those smoke compartments containing patient sleeping or treatment rooms.

PATIENT PROTECTION

The *Life Safety Code* emphasizes the protection and ultimate evacuation of persons from the immediate fire area to a safe area of refuge until the fire has been extinguished. The initial level of patient protection actually begins right in the patient's bedroom.

Draperies, curtains, furnishings, and decorations must be flame resistant. However, no restrictions currently apply to the actual clothing or bedding materials a patient may use.

If the fire is not contained within its immediate area of ignition, containment is attempted within the room. In new and sprinklered smoke compartments within existing buildings, corridors must be separated from all other areas by partitions that form a barrier to limit the transfer of smoke. In existing nonsprinklered smoke compartments, these partitions must have a fire-resistance rating of at least 30 minutes and must extend, through any concealed spaces, from the floor slab to the underside of the floor or roof slab above. Verifying the continuity of corridor walls to the floor deck above within the nonsprinklered smoke compartment will be very difficult and time consuming. Vision panels in these fire-rated walls are permitted only if they are made of approved, fixed fire window assemblies or of previously accepted wired glass construction, installed in steel or other approved metal frames, and limited in size to 1296 in^2 (8361.8 cm^2). Based on mandatory sprinkler provisions, corridor walls or associated vision panels in new buildings and sprinklered smoke compartments need not be fire rated.

Protection of Openings

Door openings in fire-rated corridor walls must be protected by solid bonded wood-core doors or approved assemblies that will resist the passage of fire for at least 20 minutes. These doors need not be fire-rated door assemblies, nor do they have to be equipped with self-closing devices. However, they must be equipped with approved positive latching hardware that will keep the door tightly closed. Vision panels for these doors can only be approved fixed fire window assemblies or previously accepted ¼-in. (0.64-cm) thick wired glass mounted in approved frames no larger than 1296 in^2. (8361.8 cm^2). As with corridor walls, corridor doors in fully sprinklered buildings need not be fire resistance rated, but they must be constructed so as to limit the transfer of smoke and be equipped with positive latches.

When other fire protection features are provided, fire ratings and, in some cases, the partitions themselves may be eliminated. Even in existing facilities, the installation of approved automatic sprinkler protection throughout the smoke compartment permits corridor partitions and the door openings within that compartment to be constructed of materials that resist only the passage of smoke; thus, they need not have a fire resistance rating. In addition, each smoke compartment may contain treatment rooms, spaces that are not used for patient sleeping rooms, hazardous areas, lounges, or waiting areas that may open directly to the exit corridor, provided the size of these areas is limited, they are supervised directly by the facility staff or by an electrically supervised automatic smoke detection system, and their furnishings are arranged so as not to obstruct access to the exits. However, devices that permit the passage of smoke, such as transfer grills or undercuts, can be installed only in doors to small areas, such as sink closets or bathrooms, that are unlikely to contain flammable or combustible materials. Small miscellaneous openings, including mail slots and pass-through windows, are also permitted in corridor vision panels or doors in smoke compartments that do not contain sleeping rooms, provided their size is limited and they are located in the lower portions of the wall.

During the fire safety inspection, the inspector should check the integrity of all the required fire-rated and smoke-resistant corridor wall and door assemblies and verify that all spaces open to the egress corridors are adequately arranged and protected.

Compartmentation

Because history has shown that smoke is the cause of most fire deaths, proper protection against smoke must be installed and properly maintained in a health care facility. All health care buildings must be subdivided into separate smoke compartments into which patients can be moved without having to leave the building or change floors. This requirement reduces the distance persons with limited mobility must be moved in order to be protected adequately.

In new facilities smoke barriers should be found on each floor used for inpatient sleeping or treatment as well as each floor with an occupant load of 50 or more (regardless of use). Existing facilities are required to subdivide every floor housing more than 30 sleeping patients.

In new buildings smoke barriers should have at least a 1-hour fire resistance rating; in existing buildings, they should have at least a 30-minute fire resistance rating. Smoke barriers must extend uninterrupted from outside wall to outside wall and from floor slab to floor or roof slab above, passing through all concealed spaces. A major problem in many health care facility inspections is verifying that unsealed penetrations have not smoke barriers. Smoke barriers must be positioned to provide at least 30 net ft^2 (2.79 m^2) per nursing home or hospital patient or 15 net ft^2 (1.39 m^2) per limited-care facility resident in a public access area. Each smoke compartment must be no larger than 22,500 ft^2 (2090 m^2), and the travel distance from any point in the building to a smoke barrier door must be no farther than 200 ft (61 m).

Openings in smoke barriers must be protected by substantial doors that are 1¾-in. (4.45-cm) solid bonded wood-core doors or are constructed to resist fire for at least 20 minutes. In new buildings these doors must be equipped with vision panels of approved fixed fire windows. Although these doors must be self-closing, they need not be a part of a rated fire door assembly and are not required to have positive latches. However, in new construction, appropriate rabbets, bevels, or astragals are required along the edges where the doors meet to prevent smoke from crossing the barrier. In most cases, these doors are held open by electromagnetic devices, which must be arranged to release when any component of the fire alarm system, including an approved smoke detector located near the doorway opening, activates. The inspector should check the construction, arrangement, and operation of all smoke-barrier doors and their associated release devices and automatic closers as part of each routine inspection.

Smoke dampers are required in nonfully ducted smoke barrier penetrations in both new and existing facilities, as well as in those ducted smoke barrier penetrations where sprinkler protection is not provided in each adjacent smoke compartment. The inspector should visually examine the smoke damper installations, and, if possible, check their operation. Specific requirements for these smoke barriers may be modified when an engineering smoke-control system is installed in accordance with NFPA 92A, *Recommended Practice for Smoke-Control Systems*.

In some cases subdividing a building into separate fire as well as smoke zones may be necessary. This is particularly true when buildings are of differing construction types, when existing portions of a building must be segregated from new additions, and when a building is more than one story high. These separations basically are the same as, and can be incorporated into, the smoke barrier, although the fire wall must be constructed to prevent the passage of fire as well as smoke. This requires the use of approved, labeled fire door and damper assemblies and is particularly important for vertical openings, such as stairways and service shafts; in some cases, they also are used to enclose hazardous areas. The inspector must examine the construction of horizontal and vertical opening protection to ensure that doors are equipped with self-closers and positive latching hardware and that doors are indeed closed, not wedged open.

The actual construction type of a building—that is, the combustibility and the fire resistance of the structure—plays a very important role in ensuring building integrity during a fire and allowing time to move and evacuate patients. The two major aspects of such construction involve the enclosure of hazardous areas and means of egress and the protection of building structural elements. Because actual construction specifications are developed when a building is designed, it the inspector's duty to ensure that design features are adhered to, the specified construction types are maintained, and alterations or modifications have not compromised any of the fire protection features, including construction type, fire-rated and smoke-resistant barriers and door assemblies.

A common maintenance issue has been the integrity of the structural fire protection. Often the concrete steel encasing or the sprayed-on fireproofing has been removed during renovations. The inspector should check for missing ceiling tiles because these are integral to the floor–ceiling and roof–ceiling assembly ratings. Sprinkler protection is also compromised when ceiling panels are not in place.

In each inspection, the inspector should check for the proper certification of all new interior finish materials, such as wall coverings and carpet, to ensure that they comply with the appropriate flame spread ratings and smoke developed values based on the area of installation and other fire protection features.

MEANS OF EGRESS

Although the objective of *Life Safety Code* requirements for health care facilities is basically to protect patients in place, the importance of required exits cannot be ignored. Each floor or fire section of the building must have at least two exits. Travel distances and egress capacities are contained in 18/19.2.6 of the code; increases are allowed for travel distance and capacities in fully sprinklered buildings.

Exits throughout the building should be accessible so that persons with impaired mobility can be moved in the event of a fire. Corridors should be clear and wide enough to relocate patients. New hospitals and nursing homes are required to provide corridors with 8 ft (2.4 m) of clear width; new psychiatric hospitals and limited care facilities need provide only 6-ft (1.8-m) wide corridors. Existing health care facilities are required to provide corridors with 4 ft (1.2 m) of clear width.

Corridor use should be restricted to pedestrian circulation. A pervasive problem is storage in the corridor. Generally, patients' charting stations, mobile medical carts, and wheeled waste and linen carts are acceptable within the corridor. However, the inspector should be wary of what appear to be permanently stored items such as spare beds and broken biomedical equipment.

The exit discharge must remain unobstructed. The use of exterior stairways and ramps and the operation of exterior doors must not be compromised by accumulations of snow, ice, or any type of exterior storage or parking.

Locks installed on exit doors must be of an approved type, and the staff must be able to open them quickly and easily for the rapid removal of occupants, either with the keys they carry at all times or by remote control. Thus the inspector must make sure that keys or some other means of unlocking secured egress doors are available to an appropriate number of on-duty staff to permit unimpeded relocation.

FIRE PROTECTION

Besides building construction features that serve to protect occupants from fire, all health care facilities must have a combination of systems to warn occupants, detect fires, and aid in fire control and extinguishment. Appropriate exit illumination, emergency lighting, and exit markings must be provided along all means of egress, and an approved manual fire alarm system must be installed. A special exception to the fire alarm requirements for health care facilities permits the installation of manual fire alarm boxes at nurses' control stations or other continuously attended staff locations, as long as such fire alarm boxes are visible, are continuously accessible, and meet all travel distance requirements (Figure 25-2). Although zoned and coded systems can be used, the operation of any fire alarm device must automatically provide a general alarm, perform all the control functions the device requires, and transmit an alarm automatically to the fire department by the most direct and reliable method approved by local regulations. The inspector should make an operational check of these systems or, at the very least, review their maintenance and test records.

Complete automatic sprinkler protection is required for all new facilities. Listed quick response or residential sprinklers should be installed throughout the smoke compartments containing patient sleeping rooms. Sprinkler protection is also required in smoke compartments in all existing facilities that have undergone renovation. Partial systems can be installed in place of the fire-rated enclosures of hazardous areas in existing facilities. However, areas housing high hazard contents in new and existing buildings must be separated by 1-hour construction and must be sprinklered. In existing unsprinklered facilities that are being renovated, altered, or modernized, complete sprinkler protection need only extend to the smoke compartments in which the work is being done.

Because so much depends on sprinkler operation, it is vital to install and maintain these systems correctly. Essentially, they must be installed in accordance with NFPA 13, *Standard for the Installation of Sprinkler Systems,* for light-hazard challenges.

FIGURE 25-2 Manual Fire Alarm Boxes
Source: *National Fire Alarm Code® Handbook*, NFPA, 1999, Exhibit 2.37; (a) SimplexGrinnell LP, Houston, TX, and Westminster, MA; (b) Edwards Systems Technology/EST, Cheshire, CT

(a)

(b)

Special installation specifications are permitted for systems with six sprinklers or less. All components, including the supervision of the main sprinkler control valves, must be interconnected electrically to the proper components of the fire alarm system.

To ensure an adequate means of first-aid fire fighting, portable fire extinguishers of an appropriate size and type should be provided at locations that are accessible to staff at all times. All building fire protection equipment, including fixed fire extinguishing systems for particular hazards such as kitchen hoods, cooking equipment, and specialized computer equipment must be marked to indicate that it has been inspected, tested, and maintained in accordance with fire code requirements and accepted engineering practices.

Health care fire safety depends on proper maintenance of all exits and fire protection equipment and on proper staff preparation. All utility, HVAC, and other service equipment should be installed and maintained in accordance with the applicable standards of the *National Fire Codes*®. Smoking regulations and evacuation plans must be adopted, implemented, and prominently posted throughout the facility. Fire exit drills must be conducted quarterly on each shift to familiarize facility personnel with the signals and the emergency actions required under varied conditions.

The inspector should review logbooks or databases that document fire protection systems inspection testing and maintenance and should ensure that the facility is keeping records concerning the required staff training and drill requirements. The inspector should also ensure the existence and proper location of the facility's written fire safety plan and he or she should review this plan.

Health care facilities may contain several other occupancies. For example, auditoriums, chapels, and cafeterias are considered assembly occupancies; laundries, boiler rooms, and maintenance shops are categorized as industrial occupancies. Other occupancies that are a section of, or are contiguous to, health care facilities may be classified as other occupancy types provided they are not meant to house, treat, or provide customary access to health care patients and are adequately separated from the health care occupancies by construction with a fire reistance rating of at least 2 hours. Consult the appropriate chapters in this book for help in inspecting these and the business occupancies—office and administration areas—of any health care facility. Because of the number and nature of the occupants, health care fire protection features must be maintained stringently to protect those who are unable to protect themselves.

SUMMARY

Patients within health care facilities are presumed to be incapable of self-preservation. Therefore, their safety depends on an appropriate combination of in-place fire and life safety features along with an acceptable staff response. In addition to verifying the adequacy of built-in features and systems, fire safety inspectors must carefully assess the facilities fire safety plan, paying particular attention to fire safety training and staff drilling. Accordingly, documentation of staff participation in drills along with inspection and maintenance logs must be reviewed.

BIBLIOGRAPHY

NFPA Codes, Standards, and Recommendations

See the latest version of The NFPA Catalog for availability of current editions of the following documents.

NFPA 13, *Standard for the Installation of Sprinkler Systems*

NFPA 25, *Standard for the Inspection, Testing, and Maintenance of Water-Based Fire Protection Systems*

NFPA 72®*, National Fire Alarm Code*®

NFPA 90A, *Standard for the Installation of Air Conditioning and Ventilating Systems*
NFPA 92A, *Recommended Practice for Smoke-Control Systems*
NFPA 96, *Standard for Ventilation Control and Fire Protection of Commercial Cooking Operations*
NFPA 99, *Standard for Health Care Facilities*
NFPA *101*®, *Life Safety Code*®

Ambulatory Health Care Facilities

CHAPTER 26

Joseph M. Jardin

Chapters 20 and 21 of the 2000 edition of NFPA *101*®, *Life Safety Code*®, contain the requirements for new and existing ambulatory health care occupancies, respectively. Previous editions of the *Life Safety Code* included the ambulatory health care facility requirements within the chapters dealing with health care occupancies.

Ambulatory health care facilities are defined as buildings or portions thereof used to provide services or treatment simultaneously to four or more patients on an outpatient basis, which renders them incapable of self-preservation. These outpatients require staff assistance as a result of the treatments provided and/or the anesthesia administered. Day surgical centers, outpatient surgical practices, group oral surgery centers, free-standing emergency medical centers, and hemodialysis centers are examples of ambulatory health care occupancies. Unlike health care occupancies, there is no intent on the part of the ambulatory health care center to provide the outpatients with overnight stay accommodations.

Where the intent is not to simultaneously render four or more patients incapable of self-preservation, such as with typical dental practices and physician's offices, an ambulatory health care occupancy classification is not appropriate. Conventional medical and dental practices are classified as business occupancies. Medical office buildings, which might consist of a collection of separate dental and medical offices, are treated as business occupancies rather than ambulatory health care occupancies. Even if the total number of "incapable" patients exceeds three among the collection of separate practices, as long as none of the individual practices exceeds three incapable patients, the entire building is classified as a business occupancy.

Similar to the concern for health care occupancies, patient movement and evacuation create problems for ambulatory health care facilities. The procedures involved, surgical and nonsurgical, combined with the administration of anesthesia places most of the responsibility for patient safety with the facility staff. Ambulatory health care occupancies are required to be compartmented using smoke barriers similar to those in health care occupancies. Also, similar to the health care occupancy approach, ambulatory health care facility staff members are required to participate in frequent fire drills.

To properly inspect ambulatory health care occupancies, the inspector must be familiar with business occupancy provisions. The *Life Safety Code* requires ambulatory health care facilities to comply with the chapters dealing with business occupancies as supplemented and modified by the ambulatory health care chapter provisions.

Joseph M. Jardin, a registered fire protection engineer, is a fire fighter with the New York City Fire Department, Rescue Co. No. 2. Mr. Jardin is a former chair of the Safety to Life Residential Occupancies Committee and currently serves as a member of NFPA's Building Code Technical Correlating Committee.

MEANS OF EGRESS

In general, the facility's means of egress should be arranged in accordance with the *Life Safety Code* provisions found in the business occupancy chapters. The inspector should ensure that all doors are operable in the egress direction and that furnishings or stored items are not blocking access to the means of egress doors. None of the required egress doors is permitted to be arbitrarily locked from the egress direction. Special locking arrangements (e.g., delayed egress and access-controlled egress doors), if provided, are permitted to be applied only to exterior doors. If delayed egress devices

FIGURE 26-1 Travel Distance in an Ambulatory Health Care Facility

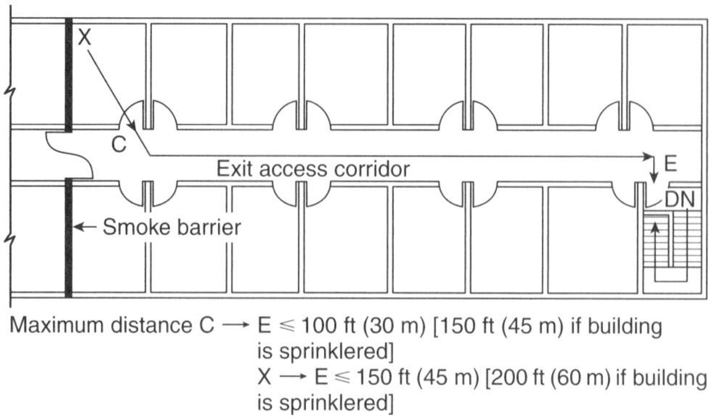

are present, the building should be either fully sprinklered or detectored. Automatic doors (normally held open) serving exit passageways, horizontal exits, or hazardous area enclosures are required to release upon manual fire alarm initiation in addition to the related smoke detector initiation.

Doors serving diagnostic and treatment areas should provide at least 32 in. (813 mm) of clear width. The inspector should evaluate corridor and passageway width to ensure there exists a minimum of 44 in. (1118 mm) of clear width. Each floor should be provided with at least two remotely situated exits. Rooms or suites exceeding 2500 ft^2 (232 m^2) are required to have at least two means of egress.

The inspector should evaluate the travel distance. In nonsprinklered facilities, the travel distance should not exceed 150 ft (45.7 m), with a maximum of 100 ft (30.5 m) from the room door to the exit. The travel distance within a sprinklered facility is permitted to be 200 ft (61 m) with a limit of 150 ft (45.7 m) from any room door to the exit (Figure 26-1).

A visual survey should be performed to ensure that all stairs, corridors, passageways, walkways, and ramps are normally illuminated and provided with emergency lighting. The inspector should perform a functional check of the emergency lighting system to verify operability and should confirm that exit signs are provided at the exits. Where there might be a degree of confusion, exit signs should also be placed within the means of egress in addition to those at the exits. The inspector should confirm that each sign is properly illuminated and that all the internally illuminated signs has operable bulbs.

PROTECTION FEATURES

Construction

In general, ambulatory health care centers are required to possess a minimum degree of fire resistance–rated construction. The permitted types of construction are based on the height of the building (number of stories) and whether the building is sprinkler protected. To determine the number of stories, inspectors should start counting at the primary level of discharge and end with the highest occupiable floor (Figure 26-2). Based on the number of stories, using the *Life Safety Code* (paragraph 20.1.6 for new buildings and 21.1.6 for existing buildings), inspectors should determine the permitted type(s) of construction. During the walk-through of the building, inspectors should verify that the structural elements (columns, beams, bearing walls, floors, and roof) are protected to the degree necessary to fall within a permitted building type.

Hazardous Areas

Hazardous areas, including but not limited to general storage, woodworking shops, paint shops, and boiler and furnace rooms, need to be addressed. These areas should

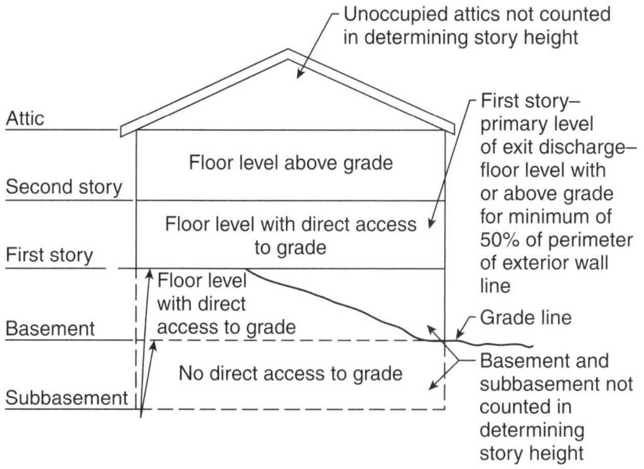

FIGURE 26-2 Determining Number of Stories in Ambulatory Health Care Centers

be separated either by 1-hour fire resistance–rated construction [with 45-minute fire protection-rated door(s)] or sprinkler protected. Even if the entire building is not sprinklered, the hazardous area is permitted to be protected by a limited area sprinkler system.

Laboratories

Laboratories considered to possess a severe hazard are required to be protected in accordance with NFPA 99, *Standard for Health Care Facilities.* According to NFPA 99, a severe hazard exists if the type and quantities of combustible materials is sufficient to breach a 1-hour fire resistance–rated fire barrier. This determination requires a somewhat subjective judgment. Inspectors should evaluate the types of chemicals used within the facility as well as the methods in which the chemicals are utilized, considering the flammability traits. They should also ensure that any flammable or combustible liquids are stored and used in accordance with NFPA 30, *Flammable and Combustible Liquids Code.*

Inspectors should determine whether a sufficient number of the appropriate type of portable fire extinguishers is located within the laboratory. In addition they should ensure that the fume hoods comply with the requirements of NFPA 99, *Standard for Health Care Facilities,* and NFPA 45, *Standard on Fire Protection for Laboratories Using Chemicals.*

Fire Alarm

Any building housing an ambulatory health care center is required to be equipped with a fire alarm. Inspectors should check to see if manual pull stations are properly located and are accessible, and they should verify the operation of the audible alarm appliances. The fire alarm systems in newer facilities are required also to provide for visible notification (for those occupants with hearing impairments) and to have in place an automatic mechanism to notify the fire department (either directly or through a central station).

Portable Fire Extinguishers

Inspectors should verify that portable fire extinguishers are properly distributed throughout the facility in accordance with NFPA 10, *Standard on Portable Fire Extinguishers.* Extinguisher locations should be readily identifiable. Inspectors should ensure that the

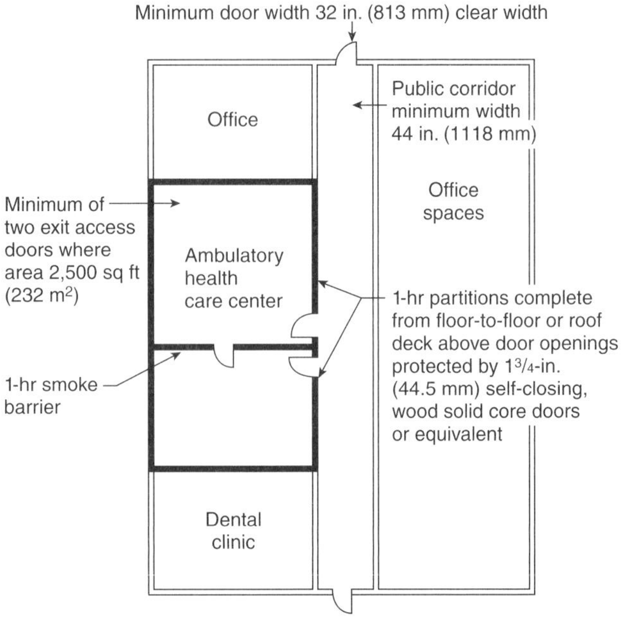

FIGURE 26-3 Smoke Barriers in Ambulatory Health Care Centers

proper extinguishing agent is being used, based on the predominant hazard, and they should verify that a monthly extinguisher inspection program is being conducted.

Smoke Barriers

Based on the potential need to protect ambulatory health care patients in place, smoke compartmentation is required. Ambulatory health care centers located in multi-tenanted buildings are required to be separated from the other tenants by 1-hour fire resistance–rated smoke barriers. The facilities themselves are required to be divided into at least two separate smoke compartments. Smoke-detectored facilities of less than 5,000 ft^2 (465 m^2) and sprinklered facilities of less than 10,000 ft^2 (929 m^2) are exempt from smoke compartmentation.

When inspecting smoke barriers, inspectors should ensure that the barriers are continuous from the floor to the underside of the floor or roof above. Ambulatory health care facility smoke barriers must also serve as 1-hour fire barriers. Smoke barrier doors must be self-closing and of at least 1¾-in. (44.5-mm) thick solid, bonded wood-core construction. Labeled 20-minute fire doors are also acceptable. These doors may be normally held open only if they are arranged to close automatically in response to a related smoke detector or the operation of a building detection or sprinkler system. The doors must close with no more than a ⅛-in. (3.2 mm) clearance between the door and the frame (Figure 26-3).

EMERGENCY RESPONSE ORGANIZATION AND TRAINING AND FACILITY MAINTENANCE ▶

Written Fire Safety Plan

Each ambulatory care facility is required to develop and maintain a written fire safety plan. A copy of the plan is required to be available at all times at either the telephone operator's position or the security office. Therefore, during the inspection, inspectors should verify that a copy of the plan is in place. They should review the plan to ensure that the plan provides for the following:

1. The use of alarms
2. Staff response to alarms

3. Fire confinement
4. Occupant evacuation of the immediate area
5. Occupant evacuation of individual smoke compartment
6. Total building evacuation
7. Fire extinguishment

Facility staff members must be instructed as to their responsibilities during a fire emergency, as outlined in the fire safety plan. Inspectors should review documentation of these staff instructions.

Fire Drills

The *Life Safety Code* requires that the facility conduct quarterly fire evacuation drills, one per shift. The inspector should review a record of these drills. Infirm or bedridden patients are not required to be moved during these drills. The purpose of the required drill is to cultivate appropriate staff reaction and response.

Smoking Regulations and Space Heaters

Inspectors should verify that the facility has a smoking policy in place and should ensure that smoking is prohibited in all areas where flammable liquids, combustible gases, or oxygen are used or stored. Signs should be posted in each such area. Only noncombustible ashtrays are permitted. Inspectors should ensure that no portable space heaters are being used in patient care areas.

SUMMARY

Ambulatory health care facilities are defined as buildings or portions thereof used to provide services or treatment simultaneously to four or more patients on an outpatient basis. As the result of the treatments provided, the patients are considered incapable of self-preservation and require staff assistance. As in health care occupancies, patient movement and evacuation, compartmentation using smoke barriers, and the need for frequent fire drills for facility staff members are all ambulatory health care facility issues. To properly inspect ambulatory health facilities, the inspector must be familiar with business occupancy provisions supplemented and modified by the ambulatory health care chapter provisions.

BIBLIOGRAPHY

NFPA Codes, Standards, and Recommended Practices

See the latest version of The NFPA Catalog for availability of current editions of the following documents.

NFPA 10, *Standard for Portable Fire Extinguishers*
NFPA 30, *Flammable and Combustible Liquids Code*
NFPA 45, *Standard on Fire Protection for Laboratories Using Chemicals*
NFPA 99, *Standard for Health Care Facilities*
NFPA *101*®, *Life Safety Code*®

Detention and Correctional Occupancies

CHAPTER 27

Joseph M. Jardin

CASE STUDY

Sprinkler Extinguishes Detention Center Blaze

A North Carolina detention center inmate who was protesting the center's no smoking policy started a blaze in his eight-person cell by piling up paper, blankets, and a mattress and lighting them with a match. Heat from the flames activated the cell's sidewall sprinkler, which extinguished the blaze.

The two-story, irregularly shaped, 39,426-ft^2 detention center was constructed of concrete and reinforced steel. It contained an automatic detection system; a full-coverage, wet-pipe sprinkler system; a smoke evacuation system; and portable fire extinguishers.

Two engines, a truck company, a squad, and a command vehicle responded to a water flow alarm at the center. An engine company and the squad investigated to find the blaze already extinguished. The truck company ventilated the building and conducted salvage operations, and the second engine supported the sprinkler system.

The property had a combined value of $9.5 million. Damage to the structure was estimated at $300 and to the contents at $150. No one was injured.

Source: "Fire Watch," *NFPA Journal,* January/February, 1997.

Detention and correctional occupancies are facilities in which occupants are confined or housed under some degree of restraint or security. These occupancies provide sleeping facilities for four or more residents who are prevented from taking action for self-preservation—that is, leaving the building—because of security measures that are not under their control. Therefore, detention and correctional facilities are required to be provided with 24-hour staffing. These occupancies include jails, detention centers, correctional institutions, reformatories, houses of correction, prerelease centers, work camps, training schools, and other residential-restrained care facilities.

Inspectors should recognize that detention and correctional occupancies present unique fire safety concerns by their very nature. Supervisory and operating personnel are substantially responsible for maintaining security within the facility in order to protect both the occupants and the general public. As a result, the personnel may be reluctant to initiate evacuation procedures or to take any other action that could compromise security, even in the case of a fire. It is, therefore, critical that detention and correctional facilities be designed, constructed, operated, and maintained so as to minimize the possibility of a fire.

Detention and correctional facilities may be a complex of buildings that serve a variety of purposes. The facility may include assembly occupancies, such as gymnasiums and auditoriums; business office areas; industrial shop areas; and storage occupancies. There also may be an infirmary or similar patient care areas that are classified as health

Joseph M. Jardin, a registered fire protection engineer, is a fire fighter with the New York City Fire Department, Rescue Co. No. 2. He is a former chair of the Safety Life to Residential Occupancies Committee and currently serves as a member of NFPA's Building Code Technical Correlating Committee.

care occupancies. Chapters 22 and 23 of the 2000 edition of NFPA *101*®, *Life Safety Code*®, primarily address the residential housing areas of the facility; other areas should meet the applicable *Life Safety Code* requirements for the appropriate occupancy classification. Where security measures require that the egress doors in those other occupancies be locked, however, a remote system for releasing the locks should be provided, or a sufficient number of attendants with keys must be available to promptly unlock the egress doors. Supervision in these areas must be continuous when the spaces are occupied. All keys required for unlocking doors within the means of egress need to be individually identifiable by both touch and sight.

Areas of a facility in which the egress doors are not locked are not classified as detention and correctional occupancies. An example of such an area is a halfway house or a prerelease area located in a larger detention complex. Facilities in which occupants are supervised but are not locked in are classified as residential occupancies and must comply with the appropriate residential occupancy requirements.

OCCUPANCY CLASSIFICATION

When conducting an inspection, the inspector should first determine the proper user category or use condition of the occupancy. The requirements for protecting the occupants increase as the degree of restraint increases. Because the primary difference between a detention and correctional occupancy and a residential occupancy is the presence of locks on the required means of egress that are not under the occupants' control, the inspector must understand the arrangement and management of the locking system. Five use conditions have been established that correspond to five degrees of restraint (Figure 27-1).

- *Use Condition I—Free Egress:* The occupants of such areas are permitted to move freely to the exterior. In other words, there are no locks that the occupants cannot control. Such an area is not considered a detention and correctional occupancy and is subject to the requirements of some other occupancy.
- *Use Condition II–Zoned Egress:* Free movement is allowed from sleeping areas from any occupied smoke compartment to other smoke compartments.
- *Use Condition III—Zoned Impeded Egress:* The occupants are allowed to move freely within any smoke compartment. Access to other smoke compartments are remotely unlocked.
- *Use Condition IV—Impeded Egress:* Occupants are locked in their rooms or cells, but the room doors can be remotely unlocked. The doors providing access to other smoke compartments also are remotely unlocked.
- *Use Condition V—Contained:* Occupants are locked in their rooms or cells by manually operated locks, and the staff must go physically to the room door to release them. The doors providing access to other smoke compartments are also manually locked.

REMOTE-CONTROLLED RELEASE

Remote-controlled release means that the locking mechanism can be released mechanically, electrically, pneumatically, or by some other means that is not located in the immediate area where the residents are housed. It does not mean that the door has to be opened, but only that the lock can be released. The remote location should provide sight and sound supervision of the resident living areas. Sight and sound supervision can be achieved by means of camera and communications systems. The inspector should be reasonably sure that the remote locking system will function as intended and that the staff can promptly gain access to the controls in an emergency.

An exception to the remote-controlled release requirement does permit the use of some manual locks in Use Condition IV facilities. No more than 10 manual

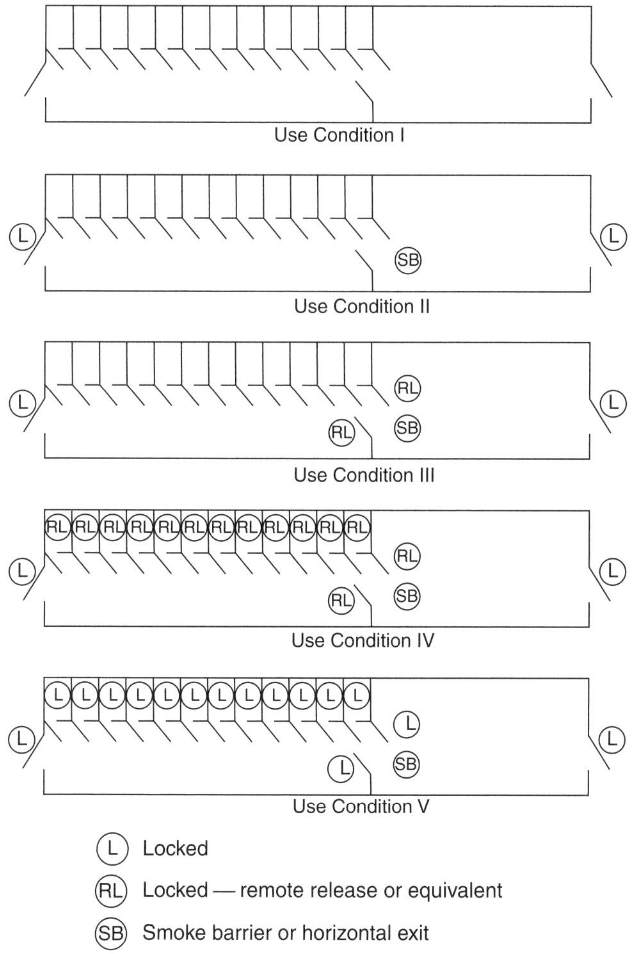

FIGURE 27-1. Detention and Correctional Use Conditions
Source: *Life Safety Code*®, NFPA, 2000, Figure A.23.1.4.1.

unlocking operations using no more than two separate keys are permitted to move the occupants promptly to some other smoke compartment. This exception allows small areas within a facility to have manual locks but still permits the entire facility to be classified as having remote control locks. The inspector should establish that keys are immediately available to staff and that the locks can be released promptly in an emergency.

The code requires that emergency power be provided for electrically operated doors and locks. The inspector should verify that the emergency power system is connected to all necessary devices and assess the system's reliability by reviewing its maintenance and testing. A manual mechanical redundant means also may be required to release the occupants should the emergency power system fail. Mechanically operated sliding doors or mechanically operated locks should have a manual mechanical means at each door to release and open the door.

To determine a facility's appropriate use condition, the inspector must fully understand its operation. For example, if the residents are free to move about the facility during the day when the inspection is being conducted but are locked down for 8 hours at night, the inspector must apply the use condition that reflects the more restricted movement. Each occupied area should be under continuous supervision, and the supervising personnel must be able to promptly release the occupants.

The inspector should also recognize that staffing levels may be much higher during the day when an inspection typically is made. Large numbers of administrative, supervisory, educational, and maintenance personnel probably will not be present at night; and security staffing levels may also be reduced at night. The administration should be required to demonstrate that there are enough operating personnel present on each shift to properly supervise all occupied areas and to perform effectively during a fire.

OCCUPANCY CHARACTERISTICS

Construction Type

The inspector should compare the construction type of the facility to that required by the *Life Safety Code,* based on facility height. Table 22.1.6.3 for new buildings and Table 23.1.6.3 for existing buildings provide the construction type parameters. In existing buildings, degree of sprinkler protection as well as use condition will influence the determination. Building construction type is classified in accordance with NFPA 220, *Standard on Types of Building Construction.*

Capacity

Generally, occupant load versus exit capacity is not a major consideration in detention and correctional occupancies. As noted in the *Life Safety Code,* the occupant load factor is 120 ft^2 (11.15 m^2) per person, reflecting a sparse population density. (This occupant load factor is not related to the minimum square footage requirements per resident in a room or cell specified by correctional standards.) The most appropriate way to determine the occupant load is to count the number of beds in a given housing area. Occupant load for the nonresident areas should be determined based on the use of the space (i.e., assembly use occupant load factors should be applied to gymnasiums and cafeterias). Once the occupant load is determined, ample capacity should be verified by the application of the appropriate capacity factor to the system of egress components [0.2 in. (0.51 cm) per person for level components and ramps and 0.3 in. (0.76 cm) per person for stairways].

Means of Egress

Detention and correctional occupancies allow various exceptions to standard egress requirements.

Horizontal Exits. Far greater attention is paid to horizontal exits in detention and correctional occupancies than in the other occupancy classifications. In fact, the code allows up to 100 percent of the required exits from a fire or smoke compartment to be horizontal exits. This provision reflects the philosophy that management generally will choose to move the residents within the facility rather than take them outside and possibly compromise security. When horizontal exits compose 100 percent of the required exits of a fire compartment, an exit other than a horizontal exit must be accessible in some other, though not necessarily adjacent, compartment, without requiring that the occupants return through the compartment where the fire originated. Figure 27-2 clarifies the intent of the code.

Sliding Doors. Sliding doors are permitted in detention and correctional facilities. The force required to slide the door to its fully open position must not exceed 50 lb (22.68 kg) with a perpendicular force of 50 lb (22.68 kg) against the door.

Exit Discharge. The exit may discharge into a fenced yard or court that is either outside or inside the facility. The outside area of refuge must provide at least 15 ft^2 (1.39 m^2) of space per person at least 50 ft (15.24 m) from the building.

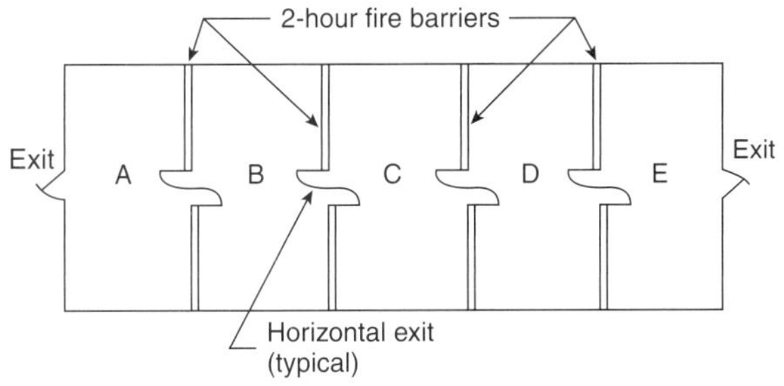

FIGURE 27-2 A Horizontal Exit System. One hundred percent of the required exits of fire Compartments B, C, and D are horizontal exits. The arrangement is in compliance as long as Compartments A and E have an approved exit other than a horizontal exit.

Day Rooms. Occupants may exit sleeping rooms through a day room or group activity space. In other words, one intervening room, such as a day room or activity space, is permitted between the sleeping room and the exit access corridor.

City and county jails are often housed in the same building as county or city offices or court facilities, or both. Where this is the case, the detention and correctional area must be separated from the other occupancies by construction with a minimum 2-hour fire-resistance rating. Horizontal exits are permitted into other contiguous occupancies, provided the other occupancies conform to the code requirements for detention and correctional facilities. If no high-hazard contents are present, however, the other occupancies may conform to the requirements of their appropriate occupancy classification.

For security purposes sally ports are often used in means of egress from detention areas. The two doors usually are interlocked to that only one of them can be opened at a time. This arrangement can obstruct the flow of persons out of the area and prevent a hose line from being stretched into the detention area. The *Life Safety Code* requires that an emergency override feature be provided so that both doors of a sally port can be opened at the same time in order to permit continuous and unobstructed passage in an emergency. The override feature should be tested to ensure that it operates properly.

As will frequently be the case for security purposes, all exits may discharge through the level of exit discharge. Where this occurs, the level of exit discharge must be subdivided so that no more than 50 percent of the exits discharge into a single fire compartment. In existing buildings, a smoke barrier may be used to subdivide the level of exit discharge if the level of discharge has approved automatic sprinkler protection.

Dead-end corridors are permitted up to a limit of 50 ft (15.24 m). In Use Condition V facilities (manual locks on all doors), however, this limit is reduced to 20 ft (6.1 m). No limits are specified for existing buildings, although dead-end corridors should be corrected whenever possible.

The use of exit signs in detention and correctional facilities is a contradiction because residents generally cannot use the exits until they are released. Therefore, exits signs may be omitted from resident sleeping areas. However, emergency lighting is required throughout the means of egress.

Hazardous Areas

Hazardous areas such as boiler rooms, kitchens, laundries, commissaries, storage rooms, trash rooms, and similar spaces require 1-hour rated enclosures with self-closing ¾-hour labeled fire door assemblies or automatic sprinkler protection, or both. Hazardous areas that are not considered incidental to the resident housing area must be separated by a 2-hour fire barrier and be provided with automatic sprinkler protection.

Protection of Openings

Vertical openings, such as stairways, ramps, elevators, chutes, and heating, ventilating, and air-conditioning (HVAC) shafts, must be enclosed with fire barriers, including fire-rated door assemblies. The required fire-resistance rating of the enclosure depends on a number of factors. For example, how high is the building? Is it a new or existing construction? Does the vertical opening serve as a required exit?

Various exceptions are permitted, some of which are unique to detention and correctional occupancies. Multilevel housing areas need not be enclosed if the distance between the lowest and highest floor level does not exceed 13 ft (3.96 m). With complete automatic sprinkler protection, unprotected openings are permitted in residential housing areas if the distance between the lowest and highest floor levels is no more than 23 ft (7.01 m). In neither instance is the actual number of levels restricted. The code is even more flexible for existing buildings, particularly where automatic sprinkler protection or an approved smoke control system is provided.

Interior Finish

In an effort to create a more habitable environment, facility designers may introduce finish materials that permit fires to develop rapidly. This is particularly true in correctional facilities that encourage self-help programs where the residents install paneling or other finish materials. Such areas require regular and thorough inspections to make sure that they comply with applicable code requirements for interior finish.

Interior finish also has proven to be a serious fire problem in padded cells. Some foam padding can develop extremely rapid and intense fires, and fire development is accentuated by the cell's small area and the insulating qualities of the foam. Although they are undesirable, padded cells are not prohibited. However, they are considered hazardous areas and require both a 1-hour rated enclosure with a self-closing ¾-hour fire door assembly and automatic sprinkler protection.

Contents

Because contents are the major fuel problem in detention and correctional facilities, the quantity and combustibility of furnishings and decorations must be strictly limited, particularly in the residential housing areas. The rates of fire development, smoke generation, and heat release are directly related to the quantity and nature of the interior furnishings. Since the time needed to release and move residents during a fire can be a major problem, it is essential to effectively control the fuel loading in order to slow fire development. Management policies must be reviewed to ensure that supervisory personnel understand the importance of this limitation and are practicing effective fuel control.

Requirements for the control of flammable contents can be found in the operating features section of Chapters 22 and 23 of the *Life Safety Code.* Combustible personal property allowed in sleeping rooms must be stored in closable metal lockers or fire-resistant containers. Furnishings such as mattresses and upholstered furniture must not be highly flammable. Specific flammability standards apply to furniture and mattresses in spaces that are not protected by automatic sprinklers. Resistance to cigarette ignition and the rate of heat release are of concern; hence, these articles are regulated. Curtains and draperies must be flame resistant, and wastebaskets must be of noncombustible or other approved materials. Combustible decorations are prohibited unless they are flame retardant. The facility administration also must control the use of heat-producing appliances, such as hot plates.

FIRE PROTECTION

Detention and correctional facilities must be equipped with an approved fire alarm system. Manual fire alarm stations are required in the normal path of travel near each exit, but the code does allow fire alarm activation stations to be placed within staff locations. Manual fire alarm stations also may be locked. In all instances, the inspector needs to verify that staff locations are occupied continuously and that keys are readily available to operating personnel so that the fire alarm system can be activated promptly if a fire is discovered. Evaluating the staffing levels and key distribution policies on all shifts is important.

Activation of the building fire alarm system must automatically, without delay, activate an audible alarm signal to alert the building occupants. With the exception of smoke detection initiation, presignal systems are not permitted. Typically, activation of the building fire alarm system must automatically transmit an alarm to notify the fire department by an approved means. However, the automatic fire department notification requirement is not mandated where a staff member is present at a constantly attended location who can promptly notify the fire department.

The residential housing areas of detention and correctional facilities must be equipped with an approved smoke detection system. However, the required extent of smoke detector coverage varies, depending on the facility's use condition and the presence or absence of automatic sprinkler protection. Automatic smoke detectors can be arranged to sound at a constantly attended location, and they need not sound the building fire alarm or transmit an alarm to the fire department. Smoke detectors also may be located in exhaust ducts from cells or behind grills to prevent them from being damaged or tampered with, as long as the design achieves the speed of detection prescribed by the code.

Sprinkler and Standpipe Protection

Automatic sprinkler protection may be required in detention and correctional facilities, depending on the building's construction type and height. Automatic sprinkler protection also may be provided to allow greater design flexibility, including the use of multilevel housing areas and atriums, and to reduce the required fire-resistance ratings of room and corridor separations. In addition, automatic sprinkler protection may be installed in existing nonconforming facilities to give them an acceptable level of safety. New Use Conditions II, III, IV, and V facilities must be fully sprinklered.

Class I standpipe systems are required in all detention and correctional facilities over two stories high. Class III standpipe systems, which are combined Class I and Class II systems, are required in facilities over two stories high that are not protected with automatic sprinklers. Portable fire extinguishers also are required. They may be located in staff areas or in locked cabinets if there is reasonable evidence that operating personnel will have prompt access to them when needed.

Verifying that fire protection equipment is operational and will function as designed when needed is essential. Documentation should be reviewed to verify that the equipment is being serviced, inspected, and tested by qualified personnel. The inspector should also witness operational tests of fire protection equipment during the course of the inspection and should check the fire alarm control panels and sprinkler control valves to verify that the systems have not been shut off.

Building Subdivision

One of the major protection features of the *Life Safety Code* is the requirement to subdivide the building into smoke compartments using approved smoke barriers. The smoke barriers provide areas of refuge into which the residents will be moved.

The inspector should determine the location of the smoke barriers to ensure that they do, in fact, subdivide the building as required. He or she can do this by reviewing a floor plan. Then the inspector should make a field check to ensure that the smoke barriers are of adequate construction and are continuous from slab to slab, that the doors are of adequate construction and are equipped with closers and the necessary hardware, and that HVAC ducts that penetrate the smoke barrier have dampers that will resist the passage of smoke. The inspector should also make sure that cross-corridor doors in smoke barriers have the required vision panels.

The *Life Safety Code* requires further subdivisions of the residential housing areas. These requirements, which can be found in the following table (Table 27-1) and also in Chapters 22 and 23 of the *Life Safety Code,* relate to the separation walls of the cell or room and the day space. If the facility has to be sprinklered because of its construction type, the sprinkler option can be used. If the code does not require a sprinkler system, management may choose whether to use sprinklers. This should be taken seriously because it can have a major impact on the construction and operation of the detention and correctional facility.

Each of the enclosing walls of the room or cell should be inspected for completeness and for the required fire resistance. Note that minimum 45-minute fire-rated glazing is now permitted in vision panels where the code previously mandated use of ¼-in. (0.64-cm) wired glass.

Building Services

The inspector should review the building services, including the HVAC systems. The *Life Safety Code* prohibits space heaters, and conducting inspections during cold weather

TABLE 27-1 Use Conditions and Required Fire Protection Features in Existing Detention and Correctional Occupancies

	Use Condition							
	II		III		IV		V	
Feature	**NS**	**AS**	**NS**	**AS**	**NS**	**AS**	**NS**	**AS**
Room to room separation	NR	NR	NR	NR	SR	NR	SR	SR
Room face to corridor separation	NR	NR	SR	NR	SR	NR	FR	SR
Room face to common space separation	NR	NR	NR/SR ≤50 ft >50 ft (≤15 m) (>15 m)	NR/SR ≤50 ft >50 ft (≤15 m) (>15 m)	STR	NR/SR ≤50 ft >50 ft (≤15 m) (>15 m)	SR	SR
Common space to corridor separation	SR	NR	SR	NR	SR	NR	FR	SR
Total openings in solid room face	120 in.^{2} (0.08 m^2)		120 in.^{2} (0.08 m^2)		120 in.^{2} (0.08 m^2)		120 in.^{2}* (0.08 m^2)	

AS, protected by automatic sprinklers; NS, not protected by automatic sprinklers; NR, no requirement; SR, smoke resistant; FR, fire rated (1 hour); FR(½), fire rated (½ hour);

*Closable from inside or 120 in.^{2} (0.08 m^2) with smoke control.

Source: NFPA *101*®, *Life Safty Code*®, 2002, Table 23.3.8.

will provide an opportunity to check for their use. In addition, management policies should be reviewed for any discussion of portable heaters.

All elevators, dumbwaiters, vertical conveyors, rubbish chutes, incinerators, and laundry chutes should be inspected for compliance with applicable codes. All chute doors should be self-closing, and the fire-resistive enclosures of vertical openings should be equipped with well-fitted, self-closing doors.

EMERGENCY PLANNING

Each detention and correctional occupancy must have a written fire emergency plan that is given to supervisory personnel and reviewed periodically with all operating personnel. This plan should establish specific procedures for discovering and isolating a fire, using alarm systems, responding to fire alarms, extinguishing fires, evacuating residents in immediate danger, notifying the fire department, and conducting zone evacuations. The plan should clearly establish authority and designate responsibilities, giving special consideration to procedures for night shifts, weekends, and holidays. The *Life Safety Code* requires that the facility review and coordinate the emergency plan with the fire department that is legally committed to serve the facility.

Realistic fire drills should be conducted periodically on each shift. These drills should include training on the use and location of the alarm systems, portable fire extinguishers, and related fire protection equipment. Fire emergency procedures should be reviewed with all new employees, and refresher training should be provided for all employees at least annually. Fire drills and employee training should be documented and available for review.

SUMMARY

Detention and correctional occupancies combine the concerns associated with sleeping occupants with the occupant's general lack of control relative to egress. This combination demands an astute facility review by the fire inspector. In order to assure that the defend-in-place concept is successful, particular attention must be focused on the compartmentation features as well as the staff's ability to protect the occupants.

BIBLIOGRAPHY

NFPA Codes, Standards, and Recommended Practices

See the latest version of The NFPA Catalog for availability of current editions of the following documents.

NFPA 90A, *Standard for the Installation of Air Conditioning and Ventilating Systems*
NFPA *101*®, *Life Safety Code*®
NFPA 220, *Standard on Types of Building Construction*

Hotels

CHAPTER 28

Joseph M. Jardin

CASE STUDY

Water Causes Electrical Short in Hotel

A toilet overflowed on the 18th floor of a Nevada high-rise hotel and casino, causing water to leak into a busway, short-circuiting transformer equipment on the seventh, eighth, and ninth floors, and igniting a fire on the eighth floor. Sprinklers quickly controlled the flames.

The hotel was built of fire-resistive construction. Smoke detectors and sprinklers provided full coverage.

Fire fighters responded to a 2:32 P.M. call to 911 and arrived to find the hotel's security staff evacuating the building. After hearing reports of smoke on the ninth floor and a confirmed fire on the eighth floor, they went up to investigate.

The eighth-floor fire began when the leaking water dripped into a concealed space that served as a main electrical channel, creating an electrical arc. The resulting flames involved the power transformer equipment and spread to the walls and ceiling, where a sprinkler activated and controlled the blaze. Crews on the seventh and ninth floors reported similar electrical shorts that caused smoke and water damage.

Fire, smoke, and water damaged the seventh through eighth floors for an estimated loss of $350,000. There were no injuries.

Source: "Fire Watch," *NFPA Journal*, March/April, 1999.

The term *hotel* is not specifically reserved for the modern fire-resistant high-rise building. Indeed, it may apply to any motel, inn, or club that provides sleeping accommodations for more than 16 people. As a result, hotels generally present a wide range of life safety and fire protection problems.

Inspecting hotels poses a special challenge because, in addition to providing sleeping accommodations for transient guests, hotels may house facilities such as meeting rooms, ballrooms, theatrical stages, kitchens, restaurants, storage rooms, maintenance shops, garages, offices, and retail shops, each of which may be classified as a different type of occupancy. For example, guest rooms are classified as residential occupancies; whereas ballrooms, theaters, and restaurants are assembly occupancies. Offices are business occupancies, parking garages and storage areas are storage occupancies, maintenance shops are industrial occupancies, and retail shops are mercantile occupancies.

If these different occupancies are separated and each has its own means of egress, they can be considered separate occupancies, according to NFPA *101*®, *Life Safety Code*®, 2000 edition. If they share the same means of egress, however, egress and fire protection requirements should comply with the most restrictive requirements of the occupancies involved.

Joseph M. Jardin, a registered fire protection engineer, is a fire fighter with the New York City Fire Department, Rescue Co. No. 2. He is a former chair of the Safety to Life Residential Occupancies Committee and currently serves as a member of NFPA's Building Code Technical Correlating Committee.

INSPECTION OBSERVATIONS

Before inspecting a hotel, the inspector should become familiar with the *Life Safety Code,* which contains requirements for means of egress and fire protection for new and existing hotels, as well as dormitories. The *Life Safety Code* also provides requirements for operating features, such as fire drills and employee training.

Means of Egress

Occupants must be able to evacuate quickly and safely from a hotel to reduce the potential for loss of life. The guest room corridors should have a minimum clear width of 44 in. (111.76 cm) and be free of obstructions. Exit signs should be readily visible and should clearly identify the path of travel to each exit. The *Life Safety Code* requires emergency lighting in any hotel with more than 25 rooms.

Occupants on each floor should have access to at least two separate exits. In newer buildings, these two exits are required to serve each floor. In general, the inspector should verify that at least two of the exits are considered remote in accordance with Chapter 7 of the *Life Safety Code* (i.e., a single fire will not block access to both exits). For recently constructed buildings, the inspector should verify that the exits are remote using the ½ diagonal distance convention (⅓ diagonal if sprinklered) and should ensure that corridor common paths of travel and dead ends are less than 35 ft (10.67 m) long [50 ft (15.24 m) in sprinklered buildings]. If guest rooms or suites exceed 2000 ft^2 (185.8 m^2), they are required to have two remotely situated exit access doors.

The inspector should confirm the integrity of the exit enclosures serving each floor. The construction of the enclosing walls should be intact and possess the appropriate fire-resistance rating (a 1-hour rating for all sprinklered buildings and nonsprinklered buildings of three or fewer stories and a 2-hour rating for all nonsprinklered buildings of three or more stories). The doors should function properly: They should swing open without the application of extreme force and should fully close and latch. The inspector should verify the presence of a fire door label indicating the required fire protection rating (e.g., 1 hour for 1-hour enclosures and 1½ hours for 2-hour enclosures). Stair landings as well as the area beneath the lowest flight should be free of storage or any other use other than egress.

The exit discharge area should be examined to ensure that the egress is free of impediments. The inspector should look for parked vehicles, trash containers, shrubbery, or fencing that might obstruct access to the public way.

All fire escapes or outside stairs, if provided, should be inspected from top to bottom. The fire escape must be attached securely to the outside of the building, and all handrails and guardrails should be secure. Counterweights on fire escapes should be free to swing when necessary to lower the stairs. The general condition of the structure and the quality of maintenance should be noted. There should be no accumulated trash beneath the outside stairways and fire escapes.

Means of Escape

The area within the guest room or suite is required to comply with the means of escape provisions of *Life Safety Code* Chapter 24, "One- and Two-Family Dwellings." Means of egress provisions apply to areas outside the guest room or suite. The means of escape provisions found in Chapter 32 of this book, "One- and Two-Family Dwellings," should be reviewed.

Interior

Top Floors. Interior inspection can start at the top or bottom of the hotel, but beginning at the roof will provide a bird's-eye view of the building. Restaurants and meeting rooms may be located on the top floors of hotels, particularly in high-rise hotels. These are assembly occupancies and should comply with the requirements of the *Life Safety Code* for new or existing assembly occupancies (see also Chapter 22 in this book). The occupant load of these occupancies should not exceed the available egress capacity. The inspector should keep in mind that the occupant load may have changed as the result of renovations.

Guest Room Floors. Generally, the guest room floors consist of guest rooms opening into a corridor. Protecting these corridors from smoke and fire is critical. Corridors should be properly fire resistance rated and complemented by doors with appropriate fire protection ratings. Most new hotels will be sprinklered; therefore the walls should posses a ½-hour rating. Nonsprinklered new hotels require 1-hour walls. Corridor walls in nonsprinklered existing buildings are required to have a ½-hour rating. If the existing building is sprinklered, then the walls must simply be smoke resisting. Corridor doors must be self-closing and latching. Generally, the doors should have a 20-minute fire protection rating; corridor doors in sprinklered existing buildings, however, must simply resist the passage of smoke.

If a fire occurs in a guest room and the guest evacuates the room, the guest room door must close to protect the corridor from heat and smoke at least for the time it takes the guests to evacuate. Thus, it is critical that the doors' self-closers work properly.

Transoms over the doors in corridor walls should be permanently fixed in the closed position. There should be no louvers in corridor walls, although the *Life Safety Code* does permit them if automatic sprinklers are installed or if smoke detectors in the corridors are arranged to shut down the fans that draw air into the corridor from the guest rooms. Transfer grilles must be located in the lower third of the corridor wall.

The inspector should examine a few guest rooms to ensure that their room doors are self-closing and that they close completely and latch. Each room should be equipped with a working smoke alarm, and fire safety information should be posted in each room.

The guest room smoke alarm is intended to set off an alarm in the room to alert the occupant. It is not intended to sound the alarm throughout the building. Because of the high incidence of nuisance alarms, these smoke alarms are not required to be connected to the building alarm system.

Rooms that contain trash chutes, laundry chutes, or trash collection areas should be enclosed in 1-hour fire-rated walls or be protected with automatic sprinklers. If there are sprinklers, the enclosing walls need not have a fire resistance rating, but they should be smoke resistant. Doors should be self-closing. Unless the building is sprinklered, spaces used to house vending or ice machines are required to be separated from the corridor.

The corridor interior finish should be Class A or B. Floor coverings should be Class I or II. New curtains, draperies, and similar furnishings should be flame retardant. These materials should be tested in accordance with NFPA 701, *Standard Methods of Fire Tests for Flame Propagation of Textiles and Films.*

Front Desk and Assembly Rooms. The front desk and its associated offices, restaurants, meeting rooms, and possibly ballroom are usually located on the ground floor. The occupant load of each of these assembly rooms should be posted conspicuously.

Meeting rooms and ballrooms generally have back-of-house service corridors, which the hotel staff uses to gain access to the rooms. They also may be used to store extra chairs, tables, and food carts. Some of these corridors are designed to be extra wide to accommodate these items; but the minimum required corridor width of 44 in. (111.76 cm) must still be maintained, even if they are not used in that way. Because this information on the required corridor width need not be posted, the inspector may have to determine the occupant load and available egress capacity.

The general housekeeping and maintenance of the building should be checked. Chances are the areas used by the public will be in fairly good condition, but the back-of-house service areas might not be.

Renovations or additions to the hotel may block exits, increase travel distance, decrease exit width, or disable exit signs or fire alarm equipment. The inspector should note any changes that have been made to the structure or its arrangements since the last inspection.

Fire-rated floor/ceiling assemblies are an important part of the building's fire resistance features. An unprotected opening in the ceiling may compromise the assembly's integrity. All the panels or tiles of the suspended ceilings that are part of a floor/ceiling or roof/ceiling assembly must be in place. If the details of construction are not noted on the previous inspection report, the inspector should check the building plans; the floor and roof design can usually be found on the architectural plans. The condition of electrical fixtures and wiring throughout the building should be checked to ensure that any flexible cords are used properly.

Service Areas. Storage rooms, building service equipment rooms, laundry rooms, maintenance shops, shipping and receiving areas, the employees' locker room, the employees' cafeteria, and a garage might be located in the basement. Some common hazards the inspector can expect to find in these occupancies include ordinary combustibles, such as paper and wood, which may be found in large quantities in storage rooms and in shipping and receiving areas, as well as paints and flammable solvents, which may be found in maintenance shops. Heat-producing equipment also presents hazards.

Good housekeeping practices are essential in service areas. Combustible trash should not be allowed to accumulate, and hazardous materials should be stored properly. Hand trucks and service carts should not be left in corridors and hallways because they can impede emergency evacuation.

Building service equipment rooms should not be used as storage areas. In building service equipment rooms and maintenance shops, the inspector should check machinery and appliances for frayed wiring, evidence of leaks, and signs of general deterioration. The clearances between heat-producing equipment and combustible materials should be checked to make sure they are adequate.

The inspector should inspect all the kitchens, including employee cafeteria kitchens. The inspection includes checking the hood exhaust filters and exhaust ducts for grease accumulations and examining the nozzles and valves on the range hood extinguishing system to make sure they are not blocked, are properly aligned over the cooking surfaces, and are not clogged with grease. The nozzles for the kitchen hood systems should be covered, as required by most manufacturers. The inspector should note the date the system was last inspected and serviced.

Stairwell doors are often locked from the inside for security reasons. The *Life Safety Code* permits this, as long as the doors unlock automatically upon fire alarm initiation, upon power loss, or when selective floors are arranged to permit reentry. The hardware arrangement of all stair doors should be examined.

FIRE PROTECTION

The *Life Safety Code* requires that hotels over three stories high with guest rooms that open into corridors have a fire alarm system. Buildings that are seven stories or higher also must have an annunciator to indicate the floor or area from which the alarm was transmitted. Manual fire alarm stations should be located adjacent to each exit, and a manual station should be located at the hotel desk or some other convenient location. Responsible hotel staff should continuously supervise this station. The fire alarm system must activate automatically, without delay, and sound the building's internal audible alarms. Presignal systems are permitted only with express permission of the authority having jurisdiction. The inspector should ensure there is a mechanism to provide fire department notification. A telephone accessible to the front desk clerk or building operator will suffice.

The fire alarm system should be tested to ensure that it operates properly and that it can be heard in the guest rooms when the guest room doors are closed. If conducting a test during the inspection is impractical, the system's maintenance and test records should be reviewed.

If the building is sprinklered, the sprinklers' general condition should be checked. This includes, but is not limited to, ensuring that floor-control valves are open and that sprinklers are neither painted nor obstructed. Painted sprinklers should be replaced. There should be a clear space of 18 in. (45.72 cm) below each sprinkler.

There should be portable fire extinguishers in hazardous areas of nonsprinklered buildings. During the inspection, the inspection tags on the extinguishers should be checked for the most recent inspection date. The inspector should ensure that they are properly mounted, fully charged, and undamaged. He or she should also look for dents in the containers or cylinders and cracks in hoses and nozzles.

Each hose station in any building equipped with a standpipe system should be checked. If the station consists of a valve and an outlet, the hose connection threads should be examined for damage and for signs of leakage. If the station is equipped with a hose for occupant use, the hose should be properly hung in the rack and the nozzle should be attached.

PLANNING AND TRAINING

The inspector should review the fire plan for the hotel, which should detail the duties of the staff in the event of a fire and spell out the procedures for immediately calling the fire department and for notifying and keeping the guests informed. The inspector should also review the fire drill records and ask employees randomly if they have actually been trained and have practiced the fire plan. See Sections 28.7 and 29.7 of the *Life Safety Code,* 2000 ed., for guidance on fire drills and plans.

SUMMARY

Sleeping occupants, who may very well be unaware of their surroundings, rely heavily on the in-place fire and life safety features. Experience has demonstrated that fatalities will result where there has been a failure in the basic life safety elements such as early warning and compartmentation. A majority of the traveling public pays little or no attention to their safety needs in the hotel environment. It is, therefore, incumbent upon the fire inspector to assure that adequate safety mechanisms are in place.

BIBLIOGRAPHY

NFPA Codes, Standards, and Recommended Practices

See the latest version of The NFPA Catalog for availability of current editions of the following documents.

NFPA 90A, *Standard for the Installation of Air Conditioning and Ventilating Systems*
NFPA 96, *Standard for Ventilation Control and Fire Protection of Commercial Cooking*
NFPA *101*®, *Life Safety Code*®
NFPA 701, *Standard Methods of Fire Tests for Flame Propagation of Textiles and Films*

Apartment Buildings

CHAPTER 29

Joseph M. Jardin

CASE STUDY

Smoking Fire Kills Elderly Woman

A 72-year-old woman who fell asleep while smoking died when the cigarette ignited a couch fire in her apartment.

The seven-story Washington apartment building, which was of fire-resistive construction, contained 125 units occupied by 140 residents, most of whom were elderly. A supervised smoke detection system and a sprinkler system of unreported coverage were connected to a fire department fire alarm system. However, the sprinkler system was not a factor in the fire.

The occupants and an inhouse manager were alerted to the fire when the fire detection system activated at 5:36 A.M. The manager ran from his apartment to the first floor to check the alarm panel, which indicated an alarm on the sixth floor. When he went to investigate, he found smoke on the sixth floor and relayed the information to the fire fighters who had responded to the automatic alarm.

The fire was confined to the living room of a sixth-floor apartment whose occupant was found dead on a couch near the area of origin. Investigators determined that the woman, who was intoxicated, had been smoking and had fallen asleep, dropping a cigarette between the cushions. The fire smoldered for some time before reaching ignition and triggering the hallway smoke detectors.

Damage to the building and its contents, valued at $7 million, was estimated at $25,000 in building losses and $25,000 in contents loss.

Source: "Fire Watch," *NFPA Journal,* May/June, 1999.

Apartment buildings are among the most difficult occupancies to inspect for many different reasons. The nature and character of the occupancy may vary considerably from building to building because of design, geographic location, and occupant age and social status. Another factor that must be considered is the edition of NFPA *101*®, *Life Safety Code*®, that was in force when the building was constructed. These factors, among many others, require the inspector to identify the particular issues and features of the fire protection system in a given building and conduct the inspection with knowledge of the various code requirements.

Joseph M. Jardin, a registered fire protection engineer, is a fire fighter with the New York City Fire Department, Rescue Co. No. 2. He is a former chair of the Safety to Life Residential Occupancies Committee and currently serves as a member of NFPA's Building Code Technical Correlating Committee.

CODE REQUIREMENTS

Chapters 30 and 31 of the *Life Safety Code,* 2000 edition, address the code requirements for new and existing apartment buildings. The inspector should become familiar with these requirements before conducting an inspection so that he or she can identify specific issues, such as adequacy of means of egress and amount and type of fire protection.

Most fire codes are maintenance codes and are only in force after a building has been completed. However, the *Life Safety Code* is written for both *new* and *existing* properties. Previous editions of the *Life Safety Code* listed four options for basic fire protection that could be applied to all new and existing apartment buildings as follows:

- Buildings without fire suppression or detection systems
- Buildings with complete automatic fire detection and notification systems
- Buildings with automatic sprinkler protection in selected areas
- Buildings protected throughout by an approved automatic sprinkler system

In addition to these options, the 1981 edition of the *Life Safety Code* had requirements for apartment buildings used as housing for the elderly. These requirements were adopted by HUD and used as a code for the construction of most of the federally subsidized housing for the elderly.

Editions of the *Life Safety Code* since 1991 no longer recognize those options in new construction. ***All new apartment buildings are required to be sprinklered.*** There are three alternatives to this rule and they are listed as an exception to 30.3.5.2 of the *Life Safety Code;* the inspector should be familiar with these provisions.

New Apartment Buildings

An *apartment building* is defined as any building containing three or more living units with independent cooking facilities. The building can be anything from a garden-type apartment to a high-rise building.

A sprinkler system meeting the requirements of NFPA 13R, *Standard for the Installation of Sprinkler Systems in Residential Occupancies Up to and Including Four Stories in Height,* is permitted in buildings up to four stories in height. The inspector should be aware of the exceptions for closets and bathrooms as well as of the differences between a system complying with the requirements of NFPA 13, *Standard for the Installation of Sprinkler Systems,* and a system complying with the requirements of NFPA 13R. The inspector should do the following:

1. Determine if the installation of the sprinkler system complies with code requirements
2. Witness a test of the operation of the sprinkler system or have the owner provide acceptable documentation of tests
3. Determine that all apartments have smoke alarms powered by the building electrical system
4. Be aware that in 30.2.1 of the *Life Safety Code* a distinction is made between a means of egress and a means of escape (see Section 24.2 of the *Life Safety Code* to become familiar with this provision)
5. Determine that hazardous areas are enclosed by a smoke-resistant enclosure even if they are sprinklered [laundries not greater than 100 ft^2 (9.29 m^2) located outside of living units are the only exception]
6. Determine that all new high-rise apartment buildings comply with Section 11.8, which requires automatic sprinklers, a voice alarm system, a standby power and light system, and a central control room

Although there are no special construction requirements for apartment buildings, knowledge of the building's construction can be of great value to the fire department. For example, fire fighters should be notified if there is truss construction in the roof or if the floor has lightweight floor trusses.

FIGURE 29-1 Fire in a High-Rise Apartment Building
Source: *Structural Fire Fighting*, NFPA, 2000, Figure 3.6.

Existing Apartment Buildings

All existing high-rise apartment buildings, with the exception of those buildings in which each dwelling unit is provided with exterior exit access, must be sprinklered or protected with an "engineered life safety system" approved by the authority having jurisdiction (Figure 29-1).

To make a thorough inspection, the inspector must see all official records, including records of previous inspections. Official records should indicate the option that was selected for protecting the occupants of the building when it was built or renovated. As subsequent inspections are made, a file of the reports will be provided for the new inspector. With this information the inspector will be in a better position to understand the level of protection that was originally provided and evaluate any changes that may have been made.

If there are no official records that identify a specific option or if the building was built before the code was established, the inspector should record sufficient information during the inspection to evaluate which option the building currently represents.

Table 29-1 summarizes the requirements for existing apartments based on the protection that is provided. The following are some general guidelines that may be helpful in inspecting apartment buildings with particular attention paid to what the *Life Safety Code* classifies as existing apartment buildings.

General Observations

INSPECTION OBSERVATIONS

The new inspection actually starts as the inspector approaches the building:

1. Is the distance to the nearest fire hydrant acceptable?
2. Is the hydrant on a public main or is it a private water system?
3. If it is a private hydrant, has it been tested to ensure the proper flow and pressure?
4. Was it installed according to NFPA 24, *Standard for the Installation of Private Fire Service Mains and Their Appurtenances?*
5. What is the water supply? (Many garden-type apartments have private water supplies.)
6. Are there obstructions to fire hydrants such as trees, shrubs, trash collectors, or new construction?
7. Is access to the building blocked by signs, marquees, or overhangs that could inhibit rescues?

TABLE 29-1 Alternative Requirements for Existing Apartment Buildings According to Protection Provided

	No Suppression or Detection System Option No. 1	Total Automatic Detection Option No. 2	Sprinkler Prot. in Select Areas Option No. 3	Auto. Ext. NFPA 13 (with Exceptions) Option No. 4
Exit access				
Travel distance	100 ft	150 ft	150 ft	200 ft
Smoke barrier req.	Req.	Req.	Req.	NR
Max. common path of travel (mod)	35 ft	35 ft	35 ft	50 ft
Max. dead end	50 ft	50 ft	50 ft	50 ft
Fire resistance				
Walls	½ hr	½ hr	½ hr	½ hr
Doors (fire protection rating)	20 min	20 min	SR	SR
Flame spread				
Walls and ceilings	A or B	A or B	A or B	A, B, or C
Floors	I or II	I or II	NR	NR
Exits—vertical				
Fire resistance walls				
1–3 stories	1 hr	1 hr	1 hr	1 hr
>3 stories	2 hr	2 hr	2 hr	1 hr
Smoke-proof enclosures				
=1 story <HR	NR	NR	NR	NR
HR	Req.	Req.	Req.	NR
Doors*				
1–3 stories	1 hr	1 hr	1 hr	1 hr
>3 stories	1½ hr	1½ hr	1½ hr	1 hr
Flame spread				
Walls and ceilings	A or B	A or B	A or B	A, B, or C
Floors	I or II	I or II	NR	NR
Habitable spaces				
Max. distance—door to corridor	75 ft	125 ft	75 ft	125 ft
Flame spread—walls and ceilings	A, B, or C	A, B, or C	A, B, or C	A, B, or C
Smoke alarms in unit	Req.	Req.	Req.	Req.
Door to corridor self-closing	Req.	Req.	Req.	Req.
Alarm system				
>3 stories or >11 units	Manual	Manual and auto	Manual and auto	Manual and auto
>2 stories or >50 units	Annunciator panel	Annunciator panel	Annunciator panel	Annunciator panel
Elevator				
ANSI Elevator Code	A17.3	A17.3	A17.3	A17.3

NR, no requirements; NA, not applicable; HR, high-rise buildings; Req., required; SR, smoke resisting.

*1¾-in. doors are permitted in non-high-rise Option 2, 3, and 4 buildings.

Source: NFPA *101*®, *Life Safety Code*®, 2000, Table A.31.1.

Older buildings could have old fire escapes in need of repair. They should be closely examined or the owner should be asked to provide documentation that a qualified person has approved them. Old fire escapes should not be confused with the safer, more substantial outside exit stairs that could have been added to remedy exit deficiencies where installing new interior stairs was not practical. The *Life Safety Code* gives requirements for acceptable construction of outside stairs. The inspector should make certain that materials posing a fire hazard are not stored under the outside stairs.

At the beginning of the inspection, the inspector should make sure that the use of the building has not changed or that another occupancy has not been added. If the building is no longer solely an apartment building but shares the premises with another type of occupancy, such as stores or offices, new safety hazards may have been introduced. The *Life Safety Code* requires that in mixed occupancies the most restrictive life safety requirements of the occupancies involved be applied if they are so intermingled that separate safeguards are impractical. Separate safeguards usually mean separation by fire-rated construction or automatic protection in the more hazardous occupancy. For example, if there is a drug store (mercantile occupancy) on the premises that was not there at the last inspection, either separation between the store and the apartments must be provided by means of construction having a fire resistance rating of at least 1 hour or automatic sprinkler protection must be provided for the store area.

Occupant Load

Although the occupant load is not required to be posted in a conspicuous place in apartment buildings, it is nevertheless an important consideration in the number of exits. The occupancy load for apartment buildings is one person for every 200 ft^2 (18.58 m^2) of gross floor area. Some apartments could have dormitory-type sleeping arrangements, such as several or many bunk beds in apartments occupied by students, which would appear to increase the occupant load well beyond the 200 ft^2 (18.58 m^2) per person criterion. If the occupancy is found to exceed this limit, then the adequacy of the exits must be assessed.

Exits

The inspector should check to make sure that all exit doors operate properly and that the means of egress are not blocked in any manner and are not hazardous in any other way. One of the greatest problems in existing apartment buildings is that through age or lack of maintenance, those life safety features originally constructed into the building have deteriorated to the point where they no longer function. In many apartment buildings it is quite common for tenants to put wedges under the doors that lead into stairwells to keep the doors open. Unapproved devices must not be allowed to hold the doors open. The inspector should examine the latching devices on all fire-rated doors and should check that self-closing devices function properly and completely close the door.

The accumulation of household goods and trash in corridors and stairwells is a common problem. Trash should be removed immediately so that egress out of the building can be safe and quick if a fire should occur. Many apartment buildings provide the tenants with storage areas; but even when such areas are not provided, some areas seem to evolve into storage spaces. These areas should be closely inspected. Almost anything, including outboard motors, flammable and combustible liquids, old tires, mattresses, and all kinds of furniture, will be found. These materials may have accumulated over many years and might cause the fuel load to exceed that which the building was designed to handle. Many times these spaces communicate with the means of egress or with vertical plumbing stacks, and there is almost always a lack of proper separation from the floor above. The code requires that these spaces be protected by 1-hour construction or by an automatic sprinkler system. In extreme cases both may be required.

Dwelling Units

Most of the time the inspector will not be able to enter the individual dwelling units. But he or she should inspect public areas, such as corridors, stairs, storage areas, utility areas, the building exterior, and exit doors. If central air conditioning is a part of the building, the inspector should check carefully for the accumulation of stored materials in the fan rooms. If renovations have been made, he or she should check to ensure that the distance to exits has not been violated.

The *Life Safety Code* requires that each unit in an apartment building be equipped with single station or multiple station smoke alarms that are continuously powered by the house electrical service; they may not be battery operated. This requirement is in addition to any sprinkler system or other detection system installed in the building. Installation of single station and multiple station smoke alarms must comply with *NFPA 72*®, *National Fire Alarm Code*®. *NFPA 72* requires that in addition to a smoke alarm located outside of each sleeping area, a smoke alarm must be installed on each story. If possible, the inspector should arrange with the building manager to look at the smoke alarms in a vacant apartment or take a random sample of one in every 10 apartments, which should be representative of smoke alarm installations throughout the building. The inspector should also ask to witness a test of the representative smoke alarm(s). This will at least show whether the manager knows how to conduct a test following the manufacturer's instructions. If the manager is not able to do this, the information is probably not being passed on to the tenants.

The inspector should also check individual apartments to be sure the entrance doors are self-closing. The *Life Safety Code* requires that doors between living units and corridors be self-closing. A fire occurring in an apartment could easily generate sufficient smoke, heat, and toxic gases to create untenable conditions in the corridor if air transfer grills were permitted. The *Life Safety Code*, therefore, does not permit transfer grills in the doors or walls that make up the corridor wall assembly.

Protection of Openings

In most instances vertical openings in apartment buildings are required to be enclosed or protected, but there are exceptions to this requirement. Where enclosure is required, it must be complete and continuous. Vertical openings are not just stairways; they include openings for rubbish chutes and building services as well as passages for pipes, conduits, ductwork, and elevator shafts, to name a few. The inspector should take some time to search for these openings. Doors to stairways and other vertical openings must be fire rated, self-closing, and positive latching.

The most common unprotected vertical openings are plumbing stacks and chases found in bathrooms and kitchens. Many times these shafts communicate from the boiler rooms, laundry rooms, and storage rooms up to the attic or roof. Chapter 8 of the *Life Safety Code* gives guidance on protection of all types of floor openings via fire-resistive shafts or fire stopping.

Elevator doors should be checked carefully to see if they would permit the spread of fire. Any openings found in elevator shafts should be repaired at once.

Where horizontal exits or smoke barriers are required, the cross-corridor doors should be checked for operation. The wall above the ceiling should be examined for unprotected penetrations. If there are suspended ceilings, the inspector should note whether all the ceiling tiles are in place. The ceiling could be part of a fire-rated floor/ceiling assembly, and missing tiles can badly compromise the effectiveness of the fire barrier.

If the building is of Type III, IV, or V construction, the existence of concealed combustible spaces is likely. Such spaces must be fire stopped in accordance with Chapter 8 of the *Life Safety Code*. In addition, in existing buildings, if Option 4 was used, these spaces will most likely require sprinkler protection. The inspector should check these spaces for fire stopping and sprinkler protection and should make sure they are not connected with unprotected vertical communications, such as pipe chases, stacks, wire runs, and so on.

Waste Chutes

In older apartment buildings, refuse chutes may be present and may have access openings in the corridors. The fire hazard is seriously increased and the fire integrity of the building is reduced with the continued use of these chutes. The refuse chute can become an avenue to transmit fire from floor to floor and from chute to floor. Nevertheless, these chute openings into corridors can be made relatively safe if the chute is properly enclosed and the service openings are properly maintained. All openings should be equipped with labeled, fire-rated hopper-type doors that are self-closing and positive latching. Gravity-type metal chutes are required to have automatic sprinkler protection in the terminal room, at the top, and at alternate floor levels. NFPA 82, *Standard on Incinerators, Waste, and Linen Handling Systems and Equipment,* gives requirements for the construction and maintenance of waste chutes.

The code requires service openings in waste chutes in newer buildings to be located in a room or compartment separated from other parts of the building by 1-hour fire-rated enclosures with a 45-minute fire door with closer and latch and sprinklered. The inspector should be sure he or she understands the method of handling household waste in the building and should inquire whether chute clogging has been a recurring problem. It could be that the chute, as originally designed, is too small to handle the type of trash that it must now accept, which could include compressed packages from kitchen compactors.

Interior Finish

Interior finish, or the exposed surface of walls and ceilings in exit enclosures, must be Class A in exits and Class A or B in corridors and lobbies. Class A, B, or C interior finish is allowed in living units and in lobbies and corridors that do not provide access to exits. The inspector should review Section 10.2 of the *Life Safety Code* before making decisions on interior finish. He or she should especially note multiple layers of vinyl wall covering and wallpaper and should require verification of the product if carpet is found on wall or ceiling surfaces.

Hazardous Areas

Particular attention should be paid to hazardous areas of the building, such as boiler and heater rooms, laundries, repair shops, refuse storage rooms, and general storage areas set aside for occupants' personal belongings, which are usually found in basements. If the hazardous area is not sprinklered, it must be separated from other parts of the building by walls and floor-ceiling construction with at least a 1-hour fire resistance rating and with openings protected by 45-minute rated fire doors with closers and latches. If the building is sprinklered, enclosures around the hazardous area can be of any reasonably smoke-resistant construction with or without a fire rating.

The general level of housekeeping in the service areas should be noted. The inspector should determine whether flammable liquids in the amounts allowed by the local fire prevention code are stored in approved containers in approved cabinets. If the building is equipped with a sprinkler system, the inspector should make sure that access to the main control valves is not blocked.

FIRE PROTECTION

Automatic Sprinkler Systems and Portable Fire Extinguishers

The best-protected apartments have complete automatic sprinkler systems. NFPA 13, 13R, and NFPA 25, *Standard for the Inspection, Testing, and Maintenance of Water-Based Fire Protection Systems,* should be used for guidance when sprinkler systems are inspected. In those areas to which the inspector has access, he or she should examine the sprinklers to make sure they have not been painted over or tampered with in ways that make their operation questionable. Painted sprinklers must be replaced. In storage areas there should be at least 18 in. (45.72 cm) of clearance between stored materials and sprinklers and no other obstruction to sprinklers. Sometimes partitions are put up without consideration of the effects on the sprinklers; partitions can create areas that are no longer reached by sprinkler discharge, thus seriously affecting the fire protection for the whole building.

The *Life Safety Code* requires portable fire extinguishers for hazardous areas in apartment buildings. Access to the extinguishers should not be blocked and the extinguishers should be clearly visible and properly hung. The inspector should note the date (usually written on an attached tag) when each extinguisher was last given a maintenance inspection. NFPA 10, *Standard for Portable Fire Extinguishers,* provides details on maintenance and placement of extinguishers.

Alarm Systems

In most cases the *Life Safety Code* requires that apartment buildings four or more stories in height or with 11 or more apartment units have a manual alarm system. Buildings that are more than three stories high and buildings that contain more than 50 dwelling units are additionally required to have annunciating panels. The requirements for the particular type of alarm system are established by the fire protection option selected for a particular apartment building. What is important to the inspector is that the alarm system be compatible with the present use of the building; thus the inspector should become familiar with requirements of the *Life Safety Code* for alarm systems in new and existing apartments.

Once the inspector has determined whether the system is adequate for the building, he or she should inquire about the testing and maintenance schedule that is followed for the system. (See the applicable NFPA standards for details.) If the inspector has concerns about the readiness of the system, he or she should arrange to witness the next scheduled test.

Visual examination of detector units from the floor could indicate some points of maintenance. Paint on the unit might indicate obstructed ports in the unit housing. Close examination also could reveal accumulations of dust or insects that might impede operation.

Lighting

The *Life Safety Code* requires that sufficient illumination be provided for the safe egress of occupants. Emergency lighting is required in apartment buildings with more than 12 living units or over three stories in height. Exit signs are to be provided in all apartment buildings, unless the living units have direct access to the outside. The inspector should become familiar with Chapter 7 of the code for details of all these requirements and should also be aware of 28.7 and 29.7 of the *Life Safety Code* regarding requirements for emergency instructions for apartment buildings.

SUMMARY

Early warning accomplished by the proper placement of operating smoke alarms is perhaps the primary factor in protecting occupants in any residential setting, including apartment buildings. The fire inspector should emphasize this fact to facility management as well as the apartment building residents.

BIBLIOGRAPHY

NFPA Codes, Standards, and Recommended Practices

See the latest version of The NFPA Catalog for availability of current editions of the following documents.

NFPA 10, *Standard for Portable Fire Extinguishers*
NFPA 13, *Standard for the Installation of Sprinkler Systems*
NFPA 13R, *Standard for the Installation of Sprinkler Systems in Residential Occupancies Up to and Including Four Stories in Height*
NFPA 24, *Standard for the Installation of Private Fire Service Mains and Their Appurtenances*
NFPA 25, *Standard for the Inspection, Testing, and Maintenance of Water-Based Fire Protection Systems*
NFPA 72®, National Fire Alarm Code®
NFPA 82, *Standard on Incinerators, Waste, and Linen Handling Systems and Equipment*
NFPA 90A, *Standard for the Installation of Air-Conditioning and Ventilating Systems*
NFPA *101®, Life Safety Code®*

Lodging or Rooming Houses

CHAPTER 30

Joseph M. Jardin

CASE STUDY

Drug User Ignites Deadly Fire

A drug user in Connecticut started a fire that killed a boarding house resident and injured five other people, including a police officer and two fire fighters, when the burning alcohol he used to smoke crack cocaine ignited a mattress. Fire spread to the hallway and upper floors through an open door and broken windows.

The three-and-a-half-story rooming house was of wood-frame construction with a brick exterior and measured 53 × 42 ft (16 × 13 m). There were hardwired smoke detectors in the hallways and battery-operated smoke alarms in each room. The building had no sprinklers.

A second-floor occupant was smoking crack cocaine when burning alcohol ignited his mattress, creating enough heat for the windows to fail, allowing the fire to spread to the upper floor. An open door also allowed the fire to spread into the hallway.

A passing police officer saw the fire and called in the alarm at 2:15 A.M. Operating smoke detectors and alarms alerted the building's residents.

Two occupants received smoke inhalation and leg injuries, and and a police officer suffered smoke inhalation. One fire fighter suffered neck burns and a second sustained neck and shoulder injuries. Estimated damage to the building, valued at $200,000, and to the contents, valued at $35,000, was $150,000 and $30,000, respectively.

Source: "Fire Watch" *NFPA Journal*, November/December, 1999.

Lodging or rooming houses, as defined in NFPA *101*®, *Life Safety Code*®, 2000 edition, include buildings with sleeping rooms that provide sleeping accommodations for a total of 16 or fewer persons on either a transient or permanent basis. Meals might or might not be available, but separate cooking facilities for individual occupants are not part of the arrangement. Examples of occupancies classified as lodging or rooming houses might include guest houses, bed and breakfasts, small inns or motels, and foster homes. It also can include small sleeping accommodations in other occupancies such as fire stations and coast guard stations. One- and two-family dwellings can accommodate up to three "outsiders" in rented rooms, which effectively narrows the scope of code requirements for lodging or rooming houses to only those facilities accommodating from 4 to 16 persons. The requirements for lodging or rooming houses apply to both new buildings and to existing or modified buildings.

Joseph M. Jardin, a registered fire protection engineer, is a fire fighter with the New York City Fire Department, Rescue Co. No. 2. He is a former chair of the Safety to Life Residential Occupancies Committee and currently serves as a member of NFPA's Building Code Technical Correlating Committee.

USE AND CODE REQUIREMENTS

Lodging or rooming houses are generally distinct occupancies, often located in a converted one- or two-family dwelling. Unfortunately, past experience has shown that these conversions may occur without proper review and approval, resulting in facilities that do not meet applicable code requirements. All too often, remodeling creates combustible voids and concealed spaces. This hazard is increased in older buildings if the remodeling process covers walls and ceilings that lack the proper firestopping features.

If a lodging or rooming house shares a building with another type of occupancy and the facilities are so intermingled that separate safeguards are impractical, the occupancies are considered to be mixed, and the most restrictive requirements of each of the occupancies involved are applicable throughout the entire facility. If the occupancies can be adequately separated, however, the requirements of each occupancy are to be separately applied.

As is true with all multiple-dwelling residential occupancies, lodging or rooming houses cannot be located above a nonresidential occupancy unless the dwelling occupancy and the exits from it are separated from the nonresidential occupancy by 1-hour fire-rated construction or the nonresidential occupancy is protected throughout by an approved automatic sprinkler system. Buildings with no more than two dwelling units above a nonresidential occupancy are permitted if the nonresidential occupancy is protected by automatic fire detection. No dwelling unit can have its sole means of egress through any nonresidential occupancy that is housed in the same building.

Several common problems are associated with lodging or rooming houses. One is the tendency to increase accommodations by altering the building's interior layout to create additional guest quarters. If the inspector suspects that this is the case, he or she should make sure that required exits have not been eliminated and that there are sleeping accommodations for no more than 16 persons. Another problem is a tendency to accept as residents individuals who may not be able to care adequately for themselves or to evacuate the building on their own in the event of a fire. Thus, occupants will have to be evaluated periodically to assess whether they are mentally and physically capable of self-preservation and have the appropriate escape capabilities. In either case, if the facility no longer meets the definition of a lodging or rooming house, it should be reclassified immediately to its applicable use. Refer to Chapters 3 and 6 of the *Life Safety Code* for additional information concerning occupancy classification.

INTERIOR FINISH

Interior finish has been proven to be a significant factor in fires involving lodging or rooming houses. This is particularly true in older buildings that were renovated or remodeled with extensive use of unrated paneling and combustible ceiling tile.

Interior finish in lodging and rooming houses is limited to Class A, B, or C in all areas. The *Life Safety Code* permits the use of fire-retardant paints to achieve required flame spread ratings. However, the inspector should study the code thoroughly to understand the limitations of this process and may also want to require proof of application and assurances that the manufacturer's guidelines have been followed strictly. Items that do not meet the required flame spread and smoke developed limitations pose an extremely serious problem.

If the inspector sees a combustible interior finish, he or she should ask to see all the information necessary to verify that the flame spread and smoke developed ratings of the materials used meet the requirements for acceptable interior finish. The classification—if there is one—of wall paneling will be stamped on the back of each sheet. In most instances the paneling classification is only valid if the paneling is attached to a noncombustible substrate, such as gypsum board or plaster.

Suspended ceiling tiles will not have such a stamp, but they may have a sticker indicating their classification. If they do not, the inspector will have to see the original packing paper for information pertaining to flame spread and smoke developed. Verifying that the ceiling tiles are acceptable might not be a problem in new construction, but identifying various interior finish materials in existing buildings may be virtually impossible. If this is the case, the inspector should insist that the material in question be tested by an approved testing laboratory or be protected with a fire-retardant coating.

PROTECTION OF EXITS

Before conducting an inspection, the inspector must understand that exit requirements for lodging or rooming houses and single-family dwellings are different from those of other occupancies. Most notably, exit requirements are referred to as means of escape rather than means of egress (see discussion in Chapter 32 of this book, "One- and Two-Family Dwellings"). Each sleeping room and living area is required to have access to both a primary and secondary means of escape. The primary means of escape for sleeping rooms above or below the primary discharge level are required to be protected interior stairs, exterior stairs, horizontal exits, or fire escape stairs for exiting facilities. Secondary means of escape are exempted for those rooms with doors leading directly to the outside at grade level or to an exterior stair. Two means of escape are required from every story of every new lodging or rooming house that exceeds 2000 ft^2 (185.8 m^2) in area or where the travel distance to the primary means of escape is more than 75 ft (22.86 m), unless the facility is protected throughout by an approved supervised automatic sprinkler system.

Vertical openings of lodging or rooming houses must be protected because occupants may be sleeping, thus unaware of a rapidly developing fire, and they may not be familiar with their means of escape due to their potentially transient nature. Generally, this requirement is met by the provision of 30-minute-rated enclosures of vertical openings along the primary exit route. An exception is permitted in buildings of three or fewer stories if they are equipped with an approved automatic sprinkler system. In this latter case, however, there still must be one primary means of escape from each sleeping area that does not pass through any open area in the lower levels.

The exit requirements for lodging and rooming houses are intended to provide for a higher level of safety than that found in single-family dwellings. Thus the inspection of the lodging or rooming house exiting scheme should verify that every sleeping room has access to a primary means of escape that provides a safe path of travel to the outside without traversing any corridor or space exposed to an unprotected vertical opening. Every sleeping room in a nonsprinklered facility should also have a secondary means of escape that meets the requirements of the *Life Safety Code.* The code exempts the requirement of a secondary escape if the building is equipped with an approved sprinkler system.

All sleeping rooms should be separated from the escape route corridors by smoke-resistant walls and doors. The doors must latch, and no louvers are permitted. The doors also must have self-closers or close automatically upon detection of smoke, unless the building has an automatic sprinkler system.

During the inspection, the inspector should make sure that the exits are not blocked or equipped with locks that do not meet code requirements. Delay release locks should be used only in facilities with a complete automatic sprinkler system or a complete fire detection system, provided not more than one device is found in any single egress path.

The inspector should check to be sure that the safe path of travel does not pass through an area or room that might be locked or blocked and that the entire path of travel, including exterior stairs and walks, is adequately illuminated. All doors and paths of travel through the means of escape should be at least 28 in. (71.12 cm) wide. Bathroom doors are permitted to be 24 in. (60.96 cm) wide.

Finally, the inspector should determine whether closet door latches can be opened easily from the inside and bathroom doors can be opened from the outside with an unlocking device.

BUILDING SERVICES

When inspecting a facility, the inspector should check four basic items relating to building services: housekeeping, electrical installations, heating appliances, and cooking operations.

Housekeeping

Housekeeping is a general barometer from which a great deal can be learned. Poor housekeeping generally reflects a lack of concern on the part of the owner or tenant. Storage areas in particular inevitably contain a large amount of combustible materials. The inspector should inspect these areas thoroughly for combustibles and for any unusual hazards, such as oil-based paints, thinners, and flammable liquids, which must be removed.

Electrical Installations

Improper or damaged electrical installations always present a fire hazard in buildings of any occupancy classification. Deficiencies to be looked for include overloaded circuits; open junction boxes; frayed wiring; overloaded extension cords and outlets; and improperly maintained appliances, fixtures, and heating appliances.

In many cases, the problems and corrective actions are clear. When in doubt about a condition involving an electrical installation, however, the inspector should request that it be inspected and approved by a local electrical inspector, a licensed electrician, or a reputable electrical inspection agency.

Heating Appliances

There is a wide range of heating appliances. With fixed permanent heater installations and portable heaters, the inspector should be concerned primarily with proper maintenance and clearance to combustibles. The inspector should also make sure that any extension cords used for electrical heaters are of the proper size, that only UL-listed heating devices are used and that gas unit heaters are installed properly. In addition, the inspector should verify that kerosene heaters are operated safely, taking into account their location with respect to means of egress, venting for combustion air, fuel storage, refueling, and use of the correct fuel. In general, unvented fuel-fired heaters are prohibited; gas space heaters complying with NFPA 54, *National Fuel Gas Code,* however, are permitted.

Cooking Operations

Cooking operations in lodging and rooming houses generally are limited in nature. However, control of individual hot plates or similar appliances is very important, considering the frequency of fires caused by these devices.

Larger cooking operations often use appliances that are not typically found in a person's home and that produce grease-laden vapors. In such cases the installation of these appliances must comply with the requirements of NFPA 96, *Standard on Ventilation Control and Fire Protection of Commercial Cooking Operations.*

Key among the items to inspect is the fixed extinguishing system protecting the cooking surfaces, plenum, and duct. Proof that the extinguishing system is inspected and serviced semiannually should be requested. The hood and duct should meet the minimum construction requirements, with liquid-tight continuous external weld seams and joints. A clearance of at least 18 in. must be maintained between all hoods and ducts and any combustibles. Deep-fat fryers should have secondary high-limit controls, hoods, and ducts. Grease-removal devices should be cleaned frequently to prevent grease from accumulating.

FIRE PROTECTION SYSTEMS

Detection and Alarm Systems

Detection and alarm systems in residential occupancies play an extremely important role in warning residents of a fire in time to evacuate the building safely. A fire alarm system complying with *NFPA 72®, National Fire Alarm Code®*, is required in lodging or rooming houses. However, it is up to the authority having jurisdiction to approve or reject the use of existing battery-powered smoke alarms, based on demonstrated testing, maintenance, and battery replacement programs. NFPA analysis of reported fires indicates that about one-third of all smoke alarms installed in homes are inoperative, usually because the batteries are dead or missing. It is extremely important that facility operators maintain a written log of periodic battery replacement and conscientiously conduct a testing program.

During the inspection, the inspector should locate and test the manual pull stations, which should be installed at the main exit or at the entrance to enclosed stairs on each floor. If the building is fully sprinklered, pull stations are not required to be distributed throughout the building. In this case the inspector should locate at least one manual means to initiate the fire alarm system. The inspector should make sure that the alarms are audible in all occupiable areas. Then the inspector should verify the presence of approved smoke alarms in each sleeping room and should make sure that they are mounted properly and are located in accordance with manufacturers' guidelines. Finally, the inspector should perform a functional test of the alarm/detection system and should document that such tests are conducted periodically.

Sprinkler Systems

Sprinkler systems are required in all new lodging and rooming houses in which a door from each sleeping room does not open directly to the outside at street or ground level or to an exterior stairway that leads to the ground. Existing lodging or rooming houses often do not have automatic sprinkler systems, even though current code alternatives and residential sprinkler systems make sprinklers a worthwhile and affordable investment.

If an automatic sprinkler system is to be used in the facility to compensate for otherwise deficient construction features, the proper NFPA 13, *Standard for the Installation of Sprinkler Systems,* classification of the sprinkler design is "light hazard," even though the *Life Safety Code* classification of the occupancy is "ordinary hazard." This classification makes the design and installation of automatic sprinklers an affordable consideration for many facilities and should be encouraged whenever possible.

If a sprinkler system is present, the inspector should verify that it is in service and that all control valves are open. Sprinkler systems that are required or that are used as an alternative method of protection, either for total or partial building coverage, must actuate the fire alarm system upon activation. This includes systems complying with the requirements of NFPA 13R, *Standard for the Installation of Sprinkler Systems in Residential Occupancies Up to and Including Four Stories in Height,* and NFPA 13D, *Standard for the Installation of Sprinkler Systems in One- and Two-Family Dwellings and Manufactured Homes.* The inspector should verify that all areas to be protected have adequate coverage, that there is adequate clearance between the sprinklers and any obstructions, and that the piping is properly protected against freezing.

SUMMARY

Sleeping occupants, who may very well be unaware of their surroundings, rely heavily on the in-place fire and life safety features. Experience has demonstrated that where there has been a failure in the basic life safety elements, such as early warning and compartmentation, fatalities will result. Particular attention must be focused on the availability and adequacy of escape routes. It is incumbent upon the fire inspector to assure that adequate safety mechanisms are in place.

BIBLIOGRAPHY

NFPA Codes, Standards, and Recommended Practices

See the latest version of The NFPA Catalog for availability of current editions of the following documents.

NFPA 13, *Standard for the Installation of Sprinkler Systems*

NFPA 13D, *Standard for the Installation of Sprinkler Systems in One- and Two-Family Dwellings and Manufactured Homes*

NFPA 13R, *Standard for the Installation of Sprinkler Systems in Residential Occupancies Up to and Including Four Stories in Height*

NFPA 54, *National Fuel Gas Code*

NFPA 72®, National Fire Alarm Code®

NFPA 96, *Standard on Ventilation Control and Fire Protection of Commercial Cooking Operations*

NFPA *101®, Life Safety Code®*

Residential Board and Care Occupancies

CHAPTER 31

Joseph M. Jardin

CASE STUDY

Unattended Candle Ignites Fire in Senior Housing

Smoke filled a Kentucky complex for older adults when an unattended candle ignited combustibles in a bedroom. The fire killed a 74-year-old occupant of the room of origin and filled the apartment with black smoke, causing extensive smoke damage. A smoke detection system detected the fire, which was too small to activate sprinklers.

The six-story building was constructed of protected steel and concrete. It had a monitored smoke detection system and a wet-pipe sprinkler system, as well as portable extinguishers in the hallways.

An automatic alarm activation at 2:48 P.M. prompted a private alarm company to dispatch the fire department. En route, dispatch updated the crews, confirming a fire in an occupied apartment. Arriving 4 minutes after the alarm, fire fighters found the alarms operating and noted light smoke. Directed by bystanders to a first-floor apartment, fire fighters saw heavy smoke when they opened the door and quickly closed it again.

Fire fighters set up two attack lines using two standpipe connections and advanced into the apartment to begin search and rescue opreations. They discovered a small fire in a bedroom and removed the occupant, who was on the bedroom floor. Fire fighters gave the victim to a medic unit for treatment. Meanwhile, the interior crews broke windows to ventilate and extinguish the remaining fire. Additional units, which were called as smoke filled the first-floor hallway and spread to the second floor, began to evacuate residents.

The fire was started by a candle left burning on a nightstand. The 74-years-old woman in the apartment of origin was handicapped and infirm, and the candle was often used to mask odors caused by a medical condition. She died of burns and smoke inhalation after being transported to the hospital.

According to fire fighters, the smoke alarms in the apartment were not sounding when they entered, and it appears that the slow-burning fire did not produce enough heat to activate the sprinklers.

Damage to the building was estimated at $15,000. Many of the occupants were treated for smoke inhalation but refused transport to the hospital.

Source: "Fire Watch," *NFPA Journal*, November/December, 1999.

Joseph M. Jardin, a registered fire protection engineer, is a fire fighter with the New York City Fire Department, Rescue Co. No. 2. He is a former chair of the Safety to Life Residential Occupancies Committee and currently serves as a member of NFPA's Building Code Technical Correlating Committee.

Residential board and care occupancies provide lodging, boarding, and personal care services for four or more residents who are unrelated by blood or marriage to the owner or operator of the facility. Residential board and care occupancies can also be called residential care or personal care homes, assisted living facilities, or group homes. These occupancies are distinguished from lodging houses because personal care services are provided to the residents. If nursing care is provided, the facility would be more appropriately considered a health care occupancy.

PERSONAL CARE VERSUS HEALTH CARE

The concept of personal care is critical in distinguishing a residential board and care occupancy from other occupancies. Personal care is intended to indicate that the owner or operator has responsibility for the safety and welfare of the residents. Personal care might include a daily awareness of the residents' whereabouts, the arrangement of appointments for residents, supervision of the residents' nutrition and medication, and the provision of transient medical care. For example, a residential board and care occupancy will often employ staff who will ensure that the residents eat properly and take their medication. The staff also might make doctor appointments for residents and notify a physician if a resident experiences some medical difficulty. However, unlike a health care occupancy, the staff will not actually administer or prescribe medication or treatment.

Examples of facilities that might be residential board and care occupancies include group homes for the physically or mentally disabled; rest homes for the aged; shelters for battered persons, unwed mothers, or runaways; and halfway houses for rehabilitated alcoholics, drug abusers, or prison parolees. In determining the appropriate occupancy classification, the inspector should be more concerned with the level of care provided, if any, than the name under which the facility operates. Additional guidance can be obtained from the license under which the facility operates, if such a license is required. The license might indicate whether the staff is permitted or expected to provide medical care, personal care, or no care. In NFPA *101*®, *Life Safety Code*®, 2000 edition, board and care occupancies are covered in Chapter 32 for new construction (including renovations and changes of occupancy) and in Chapter 33 for existing facilities.

FACILITY CLASSIFICATION

Residential board and care occupancies exist in a variety of configurations and sizes. Some facilities house a limited number of residents in a homelike arrangement. Many of these facilities are located in buildings that originally were single-family dwellings or small apartment buildings. Other facilities are located in larger facilities or multiple-building complexes, which appear and operate in a manner similar to an apartment building or hotel. Because certain protection features, such as smoke barriers, are not as functional in small buildings and because larger buildings usually are more difficult to evacuate, Chapters 32 and 33 of the *Life Safety Code* distinguish between small and large facilities. Small facilities are defined as board and care occupancies that provide sleeping accommodations for not more than 16 residents. This limit does not include staff members. Large facilities are defined as those providing sleeping accommodations for more than 16 residents.

EVACUATION CAPABILITY

When evaluating a residential board and care occupancy, the inspector must first evaluate the residents' capabilities to evacuate the building. For most occupancy classifications occupants are assumed to have similar capabilities from one facility to another. However, as the list of typical residential board and care occupancies indicates, the ability of the residents to evacuate the building can vary significantly. Prison parolees in a halfway house might have the ability to evacuate the building as quickly as the general population; however, residents in a group home for the mentally retarded might require assistance from staff or other residents.

The *Life Safety Code* defines three levels of evacuation capabilities for residential board and care occupancies as follows:

- *Prompt* evacuation capability is considered to be equivalent to the general population. If realistic fire drills are used to determine evacuation capability, the residents should be able to evacuate to a point of safety within 3 minutes.
- *Slow* evacuation capability indicates that the residents can move to a point of safety in a timely manner with some residents requiring assistance from the staff. If realistic fire drills are used, the residents should be capable of evacuating to a point of safety within 13 minutes.
- *Impractical* evacuation capability indicates that the residents cannot move reliably to a point of safety in a timely manner, even with assistance. As with the other levels, if fire drills are used, impractical evacuation capability would indicate that the residents cannot evacuate to a point of safety within 13 minutes.

If drills are used to determine evacuation capability, it is recommended that the facility conduct six drills per year on a bimonthly basis. At least two of these drills should be conducted during times when residents are sleeping. These drills should be carried out in conjunction with the authority having jurisdiction, and detailed drill records should be maintained.

Unless an acceptable methodology has been used to evaluate the evacuation capability of the residents, the inspector may assume they have a slow evacuation capability, provided they are all capable of traveling to a centralized dining room without staff assistance and the facility is staffed continuously. If the above two conditions are not met, the inspector should assume that the facility houses residents who have an impractical ability to evacuate unless the staff can demonstrate otherwise. One method that can be used to evaluate the residents is contained in Chapter 6 of NFPA 101A, *Guide on Alternative Approaches to Life Safety,* 2001 edition.

This discussion on evacuation capability refers to the capability of residents to move to a point of safety and not necessarily to evacuate to the exterior of the building. A *point of safety* is a location that meets one of the following criteria:

- A point of safety can be exterior to and away from the building. As such, if the means of egress system complies with the criteria for exit discharge to a public way, the exit discharge can be considered as the point of safety.
- If the building is protected with an approved automatic sprinkler system, a point of safety can be a code-complying exit enclosure or the other side of a smoke barrier that has a fire resistance rating of at least 30 minutes. The area also must have access to a means of escape or an exit, as permitted by the code, that does not require the residents to travel back through the smoke barrier to the area from which they were evacuated.
- If the building is of a construction type that has at least a 1-hour fire resistance rating [Type I, Type II (222), Type II (111), Type III (211), Type IV, or Type V (111)], a point of safety can be a code-complying exit enclosure or the other side of a smoke barrier that has a fire resistance rating of at least 30 minutes. The area also must have access to a means of escape or exit, as permitted by the code, that does not require the residents to travel back through the smoke barrier to the area from which they were evacuated.

OCCUPANCY CHARACTERISTICS

Residential board and care occupancies are similar in nature to other residential occupancies. The fire load within individual sleeping rooms is similar to that in sleeping rooms within dwellings. Many facilities provide some furnishings but permit the resident to bring personal furnishings such as chairs and tables into the rooms. The facilities also may include common living areas and lounges as well as a common dining area. Cooking usually is done at one central location, although residents might have access to a stove for personal use. Some residents might be responsible for their own laundry, although items such as bed linen might be laundered at a central location.

Facilities are usually staffed with personnel who supervise the residents and who might make appointments or plan activities for the residents. In some facilities, the staff live in a separate area or apartment within the facility. As such, the staff might not always be awake, alert, or directly supervising the residents.

Buildings that house occupancies other than residential board and care occupancies (mixed occupancies) must comply with the more restrictive provisions of the occupancies involved. As an alternative, the other occupancies can be separated from the residential board and care facility and its egress system by construction having a fire resistance rating of at least 2 hours. Also, Sections 32.4 and 33.4 of the *Life Safety Code* contain requirements that apply specifically to residential board and care facilities that are located within an apartment building.

Residential board and care facilities can be of virtually any type of construction. The vast majority of facilities, especially small ones, are located in single-family dwellings. There are no code-specified minimum construction requirements for small facilities. Construction types permitted by the *Life Safety Code* for large facilities depend on the size of the facility, the evacuation capability of the residents, the height of the facility, and the presence of automatic sprinkler protection.

MEANS OF EGRESS

Small Facilities

In small facilities the means of egress is more appropriately referred to as the means of escape because there are numerous deviations from the standard means of egress arrangements. (For a discussion of the terms *means of escape* and *means of egress,* the inspector should refer to Chapter 33 in the *Life Safety Code.*) Both the means of escape from sleeping rooms as well as from the facility in general must be evaluated. The means of escape requirements vary depending on the evacuation capability of the residents.

Each sleeping room and living area must have a primary means of escape that should lead to the point of safety without being exposed to any unprotected vertical openings. In facilities where residents are considered to have slow or impractical evacuation capabilities, the primary means of escape cannot be exposed to common living spaces such as living rooms or kitchens. By "exposed to," the code is referring to unprotected openings into such spaces. If the existing building is protected throughout with quick response or residential sprinklers, there is no restriction on the primary means of escape being exposed to common living spaces. (See vertical openings.)

When interior stairs are credited with serving as a primary means of escape, they should be protected or arranged to protect escaping occupants. A 30-minute stair separation is normally required; however, exceptions involving sprinkler protection, evacuation capability, and facility design may come into play.

Each sleeping room must also have access to a secondary means of escape. The most common arrangements involve either a window to the outside or a path of travel that is remote from the primary means of escape. The window must be operable from the inside without the use of tools and must provide a clear opening not less than 20 in. (50.8 cm) in width, 24 in. (60.96 cm) in height, and 5.7 ft^2 (0.53 m^2) in area. The bottom of the opening cannot be more than 44 in. (111.76 cm) above the floor. Other alternatives include passage through an adjacent space to which the residents have free access or the enclosure of the room with construction having a fire resistance rating of at least 30 minutes. Secondary means of escape are not required in rooms that have a door leading directly to the outside or in facilities that are protected with an automatic sprinkler system.

In addition to two means of escape from each room, the facility also should have two means of escape from each normally occupied story. Windows can only serve as one means of escape from the facility and only if the residents are considered to have prompt evacuation capabilities. In buildings that are protected with an automatic sprinkler system, a second means of escape is not required from each story as long as the entire facility still has two means of escape. This exception cannot be used in conjunction with the sprinkler exception for two means of escape from each sleeping room. Exit signs and emergency lighting are not required in small facilities.

Large Facilities

The means of egress from large facilities should be similar to the more traditional egress arrangement involving exit access doors to corridors that lead to standard exits. Dead-end corridors are limited to 50 ft (15.24 m). Common paths of travel are limited to 110 ft (33.53 m). If the building is sprinklered, the common path can be increased to 160 ft (48.77 m).

Travel within a room, suite, or living area to a corridor must be limited to 75 ft (22.86 m) unless the building is protected with an automatic sprinkler system, in which case the limit is 125 ft (38.1 m). The travel distance from the exit access door to an exit is limited to 100 ft (30.48 m). This travel distance can be increased to 200 ft (60.96 m) for exterior ways of exit access such as exterior balconies. Also, the *Life Safety Code* permits an increase in the travel distance to 200 ft (60.96 m) between room doors and exits if automatic sprinkler protection is provided.

Exit signs must be provided. Emergency lighting is required in facilities with more than 25 sleeping rooms. However, if each sleeping room has a direct exit to the outside of the building at ground level, no emergency lighting is required.

Egress Capacity

The occupant load factor for residential board and care occupancies is 200 ft^2 (18.58 m^2) per person. However, depending on the amount of common living and dining areas, the actual occupant load might be well in excess of that figure. In facilities in which the occupant load factor does not truly represent the anticipated occupancy, the maximum probable population of the room, space, or facility should be considered, and the maximum probable population must include residents, staff, and visitors. Hence, egress capacity should be evaluated accordingly. Corridors serving occupant loads of more than 50 are required to be at least 44 in. (111.76 cm) wide. Corridors serving 50 or fewer occupants are required to be only 36 in. (91.44 cm) wide.

PROTECTION OF VERTICAL OPENINGS

Small Facilities

In general, vertical openings must be enclosed with smoke partitions having a fire resistance rating of at least 30 minutes. Doors serving vertical openings should be capable of resisting fire for not less than 20 minutes. In facilities that are three stories or less in height, in which residents have prompt or slow capabilities to evacuate, unprotected vertical openings are permitted only if the facility is protected with an automatic sprinkler system. However, the primary means of escape must still be separated from all unprotected vertical openings. Therefore, if the opening is part of or is adjacent to the primary means of escape, a 30-minute fire-resistant cutoff must be provided. If the facility is protected throughout by quick response or residential sprinklers, this three-story opening can be completely unprotected as long as it is not used for the primary means of escape.

Large Facilities

In general, vertical openings are required to be enclosed with construction having a fire resistance rating of 1 or 2 hours, depending on the number of stories connected by the vertical shaft. In existing buildings 30-minute-rated enclosure walls are still permitted. In addition to the permitted use of atriums, other unprotected openings can be permitted depending on the number of stories connected, the openness of the area, and the provision of automatic sprinkler protection.

If exterior stairs serve the facility, the inspector must ensure that a fire that would block the interior stairs will not simultaneously block the exterior stairs. This might be accomplished by the protection of the openings exposing the exterior stairs or by the physical separation of the stairs from the building.

COMPARTMENTATION

Small Facilities

In general, corridor walls must have a fire-resistance rating of at least 30 minutes with corridor doors of at least 1¾-in. (4.45-cm) solid, bonded, wood-core construction. In facilities where residents can evacuate promptly or in facilities that are protected by automatic sprinkler systems, the corridor must be separated by smoke partitions. Corridor doors must be self-latching, with latches that will keep the door closed. Corridor doors must be self-closing or automatic closing unless the building is protected with an automatic sprinkler system.

Large Facilities

Sleeping rooms are required to be separated from corridors, kitchens, and living areas by smoke partitions. In general, these walls are required to have a 30-minute fire resistance rating protected by 20-minute-rated fire doors, which also must resist the passage of smoke. Existing 1¾-in. (4.45-cm) solid, bonded, wood-core doors are permitted in existing facilities. Also, if an existing facility is sprinklered, nonrated doors are acceptable. If the facility is a conversion, these walls and doors must simply resist the passage of smoke.

Corridor doors to sleeping rooms must be automatic closing upon activation of the smoke detection system because the operation of the facility could be such that the doors normally are open. In existing buildings if the doors have occupant-control locks such that the doors normally are closed for security or privacy purposes, the door can be self-closing. Corridor doors to other rooms must be self-closing or automatic closing. In existing buildings protected with an automatic sprinkler system, the corridor doors need not be self-closing or automatic closing unless required for another purpose, such as for an exit enclosure door or a door to a hazardous area.

Smoke barriers are required in existing, nonsprinklered large facilities with an aggregate corridor length of more than 150 ft (45.72 m) on the floor. Smoke barriers are also not required if each sleeping room is provided with exterior ways of exit access.

HAZARDOUS AREAS

Small Facilities

In small facilities, a hazardous area is considered to be one in which the fire threat is greater than that which is commonly found in one- and two-family dwellings and possesses the potential for a fully involved fire. Therefore, a typical kitchen arrangement would not necessarily constitute a hazardous area. However, rooms used for central storage of residents' belongings would most likely be considered a hazardous area. These areas must either be enclosed or protected by automatic sprinklers. If the hazardous area is on the same floor as and abuts or is in the primary means of escape, the enclosure must have a fire resistance rating of at least 1 hour; or if automatic sprinklers are provided in the enclosure, the enclosure must be smoke resistant. All other hazardous areas, such as basement storage areas, must be enclosed with construction having a fire resistance rating of 30 minutes unless automatic sprinklers are provided.

Large Facilities

Large facilities typically will contain more hazardous areas because of the increased centralization of services such as cooking, laundry, and storage. Boiler or heater rooms and repair areas also might pose a greater fire threat than those commonly found in one- and two-family dwellings. Such hazardous areas must be separated from all other parts of the building by construction having a fire resistance rating of at least 1 hour, or the area must be protected by automatic sprinklers.

DETECTION AND ALARM SYSTEMS

Small Facilities

New facilities should be equipped with a manual fire alarm system. In existing facilities a means should be provided by which the staff and residents can be alerted to a fire emergency. Although a standard fire alarm system can serve this function, the *Life Safety Code* permits other alternatives because of the size and arrangement of many small facilities. For example, if smoke alarms are interconnected, a manual fire alarm box can be provided on each floor, which can be activated to continuously sound the smoke alarms. Although not specified in the code, the manual fire alarm boxes should be located in a conveniently accessible location that is in the normal path of escape.

Within existing facilities a "system" consisting of alarms and manual switches that can serve to notify the occupants and staff might also be accepted. The size and arrangement of the facility should dictate what constitutes an acceptable alternate system. In accepting a system, the inspector must consider issues such as audibility, activation, power, secondary power, supervision, maintenance, and testing.

In order to minimize the impact of a fire on the means of escape, approved smoke alarms must be provided in all living rooms, day rooms, dens, and so on, and one alarm must be provided per floor, excluding crawl spaces and attics. The smoke alarms should be powered by the house electrical service. It must be verified that the source of power is not subject to control by a wall switch. If plug-in type smoke alarms are used, a restraining mechanism must be used to secure the plug. Additional guidance on smoke alarm installation can be found in *NFPA 72®*, *National Fire Alarm Code®*, which contains minimum requirements for the selection, installation, operation, and maintenance of the smoke alarms. The smoke alarms can be omitted in buildings protected by an automatic quick response or residential sprinklers system. In addition, single-station building-powered smoke alarms must be installed in all sleeping rooms unless the rooms are protected by quick response or residential sprinklers. However, sprinklers cannot be used as a substitute for both the sleeping room smoke alarms and the corridor and common space smoke alarms. Sleeping room smoke alarms do not have to be added if corridor and common space detectors are in place in existing buildings.

Large Facilities

In general, a fire alarm system must be installed in large facilities to permit the quick notification of staff, residents, and visitors of a fire emergency. The *Life Safety Code* contains some exceptions and variations that can be permitted in certain facilities. In new high-rise buildings a means also must be provided by which the occupants can be notified of the fire emergency by voice communication. The system should be arranged so that the proper evacuation or emergency instructions can be given to the buildings' occupants.

Corridors and common spaces must be provided with automatic smoke detectors that are connected to the fire alarm system. Detectors are not required in common spaces if the entire facility is protected with an automatic sprinkler system. In addition, single-station smoke alarms must be provided within sleeping rooms. These room smoke alarms must be powered by the building's electrical service unless the devices were installed previously, in which case battery-powered smoke alarms are permitted. However, battery-powered smoke alarms should be permitted only if the facility's testing, maintenance, and battery-replacement programs will ensure the reliability of the devices. In existing buildings that have an existing corridor smoke detection system, sleeping rooms are not required to have smoke alarms.

The inspector should ensure that a reliable means exists by which both staff and residents can notify the fire department. Public telephones are the most commonly used device to satisfy this criterion. Access to outside lines and the awareness of the correct telephone number should all be evaluated. The inspector should also evaluate the location of the fire alarm annunciator panel to ensure that its location is convenient and that it is accessible to fire department personnel when they arrive at the scene of a fire emergency.

FIRE SUPPRESSION EQUIPMENT

Using Chapters 32 and 33 of the *Life Safety Code,* the inspector needs to carefully ascertain whether automatic sprinklers are required for partial or complete building protection. For example, protection of hazardous areas can be accomplished with sprinklers serving the hazardous area only. Other code sections might apply to facilities protected throughout with an automatic sprinkler system. Finally, requirements such as allowable construction types depend on whether the entire building, which may contain other occupancies, is protected with an automatic sprinkler system.

Small Facilities

In new construction, which includes converted buildings, automatic sprinkler protection using quick response or residential sprinklers is required throughout new board and care facilities. In existing facilities, with the exception of facilities where residents' evacuation capabilities are considered impractical, automatic sprinkler protection is not required unless one or more of the exceptions related to sprinkler protection are contained within the facility. The code permits the use of NFPA 13D, *Standard for the Installation of Sprinkler Systems in One- and Two-Family Dwellings and Manufactured Homes;* NFPA 13, *Standard for the Installation of Sprinkler Systems;* or NFPA 13R, *Standard for the Installation of Sprinkler Systems in Residential Occupancies Up to and Including Four Stories in Height,* in such facilities. However, in facilities in which the residents have slow or impractical evacuation capabilities, the minimum water supply allowed for an NFPA 13D system is 30 minutes, and sprinklers must be present in all habitable areas and closets except small bathrooms [55 ft^2 (5.11 m^2)]. Activation of the sprinkler system must activate the building fire alarm system. Although portable fire extinguishers are not required, the inspector might recommend that extinguishers be provided in certain areas, if trained personnel are available to operate them.

Large Facilities

Automatic sprinkler protection using quick response or residential sprinklers is required throughout new board and care facilities, including converted buildings. In existing facilities automatic sprinkler protection is not required unless the building is a high rise or one or more of the exceptions related to sprinkler protection are contained within the facility. The inspector might recommend that consideration be given to sprinkler protection if code requirements such as door closers present a problem or are of concern to the facility owner or operator. If installed, the sprinkler system must comply with the requirements of NFPA 13 or NFPA 13R, except that sprinklers can be omitted from small closets [24 ft^2 (2.23 m^2)] and small bathrooms [55 ft^2 (5.11 m^2)]. Portable fire extinguishers must be provided near hazardous areas.

OPERATING FEATURES

Smoking should be restricted to areas where it is safe. The smoking policy, at a minimum, should discourage smoking in bed. In many facilities a smoking lounge is designated in order to minimize the potential for fires originating in sleeping rooms. Wherever smoking is permitted, proper noncombustible ashtrays or receptacles must be provided. Care must be exercised in the location of smoke detectors in such spaces.

Every facility should have a fire emergency plan that identifies the proper procedures to be followed by staff, residents, and visitors upon discovery or notification of a fire emergency. The plan should be evaluated at least six times a year through the conducting of fire exit drills. The drills must include the use of all designated means of escape or egress. If the means of escape involves windows, the residents need not actually climb out the windows. However, they should be required to go to the windows and open them. Staff personnel should participate in all fire exit drills. Experience with fires in board and care facilities has demonstrated that the lack of staff and resident training as well as the failure to familiarize the residents with egress routes other than the primary path have resulted in additional fatalities.

SUMMARY

As a group, the occupants of a residential board and care facility rely to a varying degree on staff assistance. This compounds the life safety concerns associated with sleeping occupancies. For these reasons, the fire inspector must verify the adequacy of in-place early warning systems as well as the availability of adequate escape routes. In this occupancy type, it is incumbent upon the inspector to actually observe fire drills in addition to reviewing the documentation of past drills.

BIBLIOGRAPHY

NFPA Codes, Standards, and Recommended Practices

See the latest version of The NFPA Catalog for availability of current editions of the following documents.

NFPA 13, *Standard for the Installation of Sprinkler Systems*
NFPA 13D, *Standard for the Installation of Sprinkler Systems in One- and Two-Family Dwellings and Manufactured Homes*
NFPA 13R, *Standard for the Installation of Sprinkler Systems in Residential Occupancies Up to and Including Four Stories in Height*
NFPA 72®, National Fire Alarm Code®
NFPA 90A, *Standard for the Installation of Air-Conditioning and Ventilating Systems*
NFPA *101®, Life Safety Code®*
NFPA 101A, *Guide on Alternative Approaches to Life Safety*

One- and Two-Family Dwellings

CHAPTER 32

Joseph M. Jardin

CASE STUDY

Electric Heater Fire Kills Four

Four occupants of a North Carolina house, including two children, died of smoke inhalation during a fire that began in the living room. Investigators believe a leaking kerosene fuel line may have soaked the carpet, contributing to the intensity of the blaze.

The single-family, wood-framed house was 35 ft (11 m) long and 24 ft (7 m) wide. Three bedrooms and a bathroom took up one end of the house, while a kitchen and living room with a rear-facing porch occupied the opposite end. At the time of the fire, the single-story house was occupied by a 7-year-old, an 8-year-old, their 69-year-old grandmother, and a 25-year-old woman. The house had no smoke alarms or sprinklers.

Fire fighters received a 911 call at 11:25 P.M. from the children's mother, who discovered the fire when she returned home. Fire fighters arrived two minutes later to find flames venting from the side and front living room windows.

The children's mother showed a fire fighter to a rear window, where she believed the victims were located. He entered the house through the window and immediately found the two children and their grandmother. With the help of several EMS personnel, he rescued the victims as additional fire fighters entered the house and brought the fire under control. The fourth victim was found in her bed in a front-facing bedroom.

The fire started in the living room near a couch where a portable electric space heater was operating. Investigators concluded that the heater had fallen against the upholstered couch, igniting the fabric, and that kerosene residue on the living room carpet had enhanced flame spread. The kerosene residue came from a leaking kerosene-fueled stove that had been turned off days before the fire. The grandmother's son had turned off the stove, intending to fix the line.

The house and its contents, which were valued at $50,000, and $10,000 respectively, were a total loss.

Source: "Fire Watch," *NFPA Journal*, January/February, 2001.

Four people died in this North Carolina house fire after a space heater fell against an upholstered sofa and ignited the fabric. The house and its contents were a total loss.
(Reprinted with permission of the *News & Record*, Greensboro, NC)

Joseph M. Jardin, a registered fire protection engineer, is a fire fighter with the New York City Fire Department, Rescue Co. No. 2. He is a former chair of the Safety to Life Residential Occupancies Committee and currently serves as a member of NFPA's Building Code Technical Correlating Committee.

NFPA *101*®, *Life Safety Code*®, 2000 edition, defines one- and two-family dwellings as buildings containing not more than two dwelling units in which each living unit is occupied by members of a single family, with no more than three outsiders, if any, accommodated in rented rooms. The *Life Safety Code* does not define the term *single family*. The definition of family is subject to federal, state, and local regulations. These regulations may not limit the definition of family to the traditional nuclear family model of a couple and their children.

The manner in which living units are separated from one another can determine their occupancy classification. For example, a row of six townhouses with complete vertical fire wall separations between each unit, including the attic spaces, and with independent means of escape from each unit can be considered as six individual units.

In most jurisdictions existing one- and two-family dwellings are not required to be inspected. New dwellings are usually inspected by the authority official having jurisdiction before they are occupied, but routine inspections are not performed after they have been occupied. However, homeowners frequently ask fire inspectors to conduct voluntary fire prevention inspections. Some jurisdictions mandate inspections, especially of smoke alarms, at the time a one- or two-family dwelling is sold or a bank may require an inspection when a home owner is refinancing.

OCCUPANCY CHARACTERISTICS

When a dwelling unit is located within a mixed-use occupancy, the most restrictive life safety requirements apply. The *Life Safety Code* lists specific requirements for mixed occupancies that include one- and two-family dwellings. Dwelling units may not have their sole means of egress through any nonresidential occupancy in the same building. Multiple-dwelling occupancies may not be located above nonresidential occupancies unless the dwelling occupancy and its exits are separated from the occupancy below by construction having a fire-resistance rating of at least 1 hour, or the occupancy below is protected by automatic sprinklers. In an existing mixed-occupancy arrangement, a building with not more than two dwelling units above another occupancy is permitted if the occupancy below is protected by an automatic fire detection system.

MEANS OF ESCAPE

A *means of escape* is a way out of a building that does not conform with the strict definition of means of egress but does provide a way out of residential dwelling units. Any dwelling unit with more than two rooms must have at least two means of escape from every bedroom and living area. Within the dwelling unit at least one means of escape must be a door or stairway with an unobstructed path of travel to the outside. Bedroom and living areas should be accessible by means other than ladders, folding stairs, or trapdoors. Stairs in the path of travel in the means of escape must be at least 36 in. (91 cm) wide. Winders and spiral stairs are permitted within a single living unit.

Most residential occupancies must have a second means of escape. It can be a door or stairway with an unobstructed path of travel to the outside or a window that can be opened from the inside without the use of tools. The window should provide a clear opening at least 20 in. (50.8 cm) wide, 24 in. (60.96 cm) high, and 5.7 ft^2 (0.53 m^2) in area. The bottom of the opening cannot be more than 44 in. (111.76 cm) above the floor.

However, a second means of escape is not required if the room has a door leading directly outside of the building to grade level or if the building is protected throughout by an approved automatic sprinkler system installed in accordance with NFPA 13D, *Standard for the Installation of Sprinkler Systems in One- and Two-Family Dwellings and Manufactured Homes.*

The means of escape from any room must not pass through another room or apartment that is not under the immediate control of the occupant or family of the first room, nor through a bathroom or other space subject to locking. For example, the second means of escape from a windowless living area could be through a bedroom and out the bedroom window as long as the bedroom door cannot be locked and the use of the bedroom is under the control of the occupant. Children must be able to open closet door latches from inside the closet, and locks on bathroom doors must be operable from the outside.

Doors in the path of travel of a means of escape must be at least 28 in. wide (71.12 cm), but bathroom doors can be 24 in. (60.96 cm) wide. Exterior exit doors must swing or slide open; they are not required to swing in the direction of exit travel. When the building is occupied, doors in any means of egress cannot be locked with double cylinder or dead-bolt locks that can only be unlocked with a key from the inside.

INTERIOR FINISH

Interior finish on walls and ceilings of occupied spaces must be Class A, B, or C. The flame spread and smoke developed characteristics of a material usually cannot be determined by visual inspection, so documentation certifying the characteristics of the interior finish must be provided by the installer or owner. Sometimes this information is printed on the backside of materials or on their packages.

FIRE PROTECTION

In new construction, approved, single station, building-powered smoke alarms must be installed in every sleeping room, outside of each separate sleeping area in the vicinity of the sleeping rooms, and on each level of the dwelling. Where two or more smoke alarms are required, they must be arranged such that the operation of any smoke alarm will cause all other smoke alarms in the dwelling to sound an alarm. Approved, building-powered smoke detection systems that provide for occupant notification are also acceptable in new construction. Refer to Chapter 13, "Fire Alarm Systems," of this book, as well as *NFPA 72®, National Fire Alarm Code®*, for additional information.

Battery-powered smoke alarms are permitted in existing one- and two-family dwellings. Smoke alarms are not required in every sleeping room of existing dwellings. However, single-station smoke alarms should be found in the hallways outside each sleeping room. If there is more than one story or the single story is rather large, additional smoke alarms will be needed.

VOLUNTARY INSPECTIONS

Fire inspections conducted in one- and two-family dwellings most frequently are voluntary fire prevention inspections done at the request of a homeowner to help evaluate fire safety. The following recommendations will help the homeowner maintain a fire-safe home.

Utilities

All gas, electric, and oil-fired utilities and appliances should be kept in good repair and serviced as needed. The homeowner should be instructed on the proper installation and use of electrical extension cords, portable heaters, fireplaces, and wood-burning stoves. Damaged light and appliance cords should be replaced. The inspector should emphasize that all major electrical work should be done by a licensed electrician.

Coal- and Wood-Burning Stoves

Coal- and wood-burning stoves warrant special attention because of their record as a cause of fires. Stoves should be inspected very carefully for the adequacy of the installation and the clearance of the stove and its chimney connector from combustibles.

The authority having jurisdiction might have specific requirements for the installation of stoves. If there are no specific local requirements, the inspector should ask the homeowner whether the stove was installed according to the manufacturer's instructions.

There should be a minimum clearance of 36 in. (91.44 cm) between the stove and combustible walls and ceilings. The stove should be positioned on a base of noncombustible material, such as metal or brick, extending at least 18 in. (45.72 cm) beyond the stove in all directions. The stovepipe, or connector, between the stove and chimney should be as short and straight as possible, and there should be at least 18 in. (45.72 cm) of clearance between the pipe and combustible surfaces.

Occupants should be reminded that creosote accumulations both in the connector and the chimney, which form very quickly in airtight chimneys that are not used properly, are a very dangerous fire hazard. They should be advised to have the chimney inspected frequently and cleaned when necessary to prevent chimney fires.

See NFPA 211, *Standard for Chimneys, Fireplaces, Vents, and Solid Fuel-Burning Appliances,* for requirements for proper clearances, and modifications that might be permitted, for heating appliances. The *Fire Protection Handbook* also has information on hazards and protection of coal- and wood-burning stoves.

Storage

The storage of flammable and combustible materials within the home should be controlled. The inspector should check the way in which paint, solvents, gasoline, and other materials are stored in the workshop areas. These materials must be stored away from ignition sources and preferably outside the home. Many homeowners do not understand that flammable liquids produce vapors that can be ignited by the furnace, hot water heater, or other devices.

There is no substitute for good housekeeping. Storage areas should be checked for accumulations of trash and large amounts of combustibles and these housekeeping problems should be brought to the attention of the occupant.

Detection Equipment

Smoke alarms should be installed in all one- and two-family dwellings. The inspector can recommend which type to install and where they should be placed. Ionization and photoelectric smoke alarms are comparable and either type can be used in residential occupancies. The smoke alarms should be located near the sleeping areas of the house, with at least one smoke alarm located on each floor, and they should be audible throughout the house. If the smoke alarms are plugged into a wall receptacle, the inspector should make sure the plug has a restraining means and the receptacle is not controlled by a wall switch that could shut off the power.

Because single-station battery-operated smoke alarms are so easy to install and are allowed in one- and two-family dwellings, they are the most common. The inspector should emphasize that the smoke alarms should be tested on a regular basis to ensure the batteries are working. The batteries should be replaced on a regular schedule, as well.

Fire Extinguishers

Homeowners will most likely have questions about type, size, location, and number of fire extinguishers. For this information, the inspector should refer to NFPA 10, *Standard for Portable Fire Extinguishers,* but he or she should assist the occupants by summarizing the different types, such as carbon dioxide, dry chemical, and multipurpose dry powder. The weight of the extinguisher should be a consideration in the decision of which size to buy. The inspector should advise the occupants to become familiar with the operation of the extinguishers and to practice using them. Occupants should also be referred to NFPA 10 and the manufacturer's recommendations for inspection, testing, and servicing of the extinguishers.

Residential Sprinklers

The use of residential sprinklers in one- and two-family dwellings is an increasing trend in some jurisdictions. All sprinklers should be inspected visually to ensure they are not painted over and that their discharge pattern is unobstructed. All water flow devices, alarms, pumps, water tanks, and other components of the system should be in proper operating condition. Valves should be inspected to ensure that they are open. The inspector should point out that, according to NFPA 13D, proper maintenance of a sprinkler system is the responsibility of the owner, who should understand how the system operates.

Fire Escape Plan

The inspector should help the homeowner establish a fire escape plan from each room of the house and should emphasize that once the plan is made, it should be practiced. The NFPA pamphlet "E.D.I.T.H. Exit Drills In the Home" is a good reference. The following principles of E.D.I.T.H. are simple and sound:

1. Have smoke alarms on each level of the house, and make sure they work.
2. Know two routes to the outside from all rooms, especially bedrooms.
3. Have everyone in the house memorize the fire department telephone number, and put the number on the telephones in the house.
4. Choose a place outdoors for everyone to meet for roll call.
5. Locate the closest telephone or emergency call box from which to report a fire in the home.
6. Never go back into a burning building.
7. Practice escape routines—testing closed doors for fire on the other side, crawling low under smoke, and getting out of bedroom windows.
8. Know what to do if occupants become trapped.

The inspector can help the family make an escape plan by reviewing potential escape routes from sleeping areas, pointing out alternatives that might be available, and demonstrating the proper techniques for testing doors for fire and for exiting through windows. The inspector's interest might help to convince a family that E.D.I.T.H. is serious business.

SUMMARY

A key element of protecting residents in the home is sound fire safety education. Fire inspectors should carry this message to homeowners and renters. Working smoke alarms, predetermined escape plans, and a fire-safe environment should be stressed.

BIBLIOGRAPHY

Cote, A. E., ed., *Fire Protection Handbook,* 18th ed., NFPA, Quincy, MA, 1997.

NFPA Codes, Standards, and Recommended Practices

See the latest version of The NFPA Catalog for availability of current editions of the following documents.

NFPA 10, *Standard for Portable Fire Extinguishers*
NFPA 13D, *Standard for the Installation of Sprinkler Systems in One- and Two-Family Dwellings and Manufactured Homes*
NFPA 72®, National Fire Alarm Code®
NFPA *101®*, *Life Safety Code®*
NFPA 211, *Standard for Chimneys, Fireplaces, Vents, and Solid Fuel-Burning Appliances*

Mercantile Occupancies

CHAPTER 33

Joseph Versteeg

CASE STUDY

Failure to Call Fire Department Allows Fire to Rekindle

Nearly 3 hours after employees thought they had extinguished an electrical fire, its smoldering remains reignited, and the fire spread to concealed spaces. The employees did not notify the fire department during the first incident, and by the time fire fighters arrived for the second incident, there was no hope of an interior attack.

The single-story building, which had concrete block walls and a combination wood-and-steel roof, covered a ground-floor area of 52,800 ft^2 (4,905 m^2). It had no smoke alarms or sprinklers. The Ohio business, which was operating, sold paint, wallpaper, and miscelleneous items.

A fire of undetermined electrical origin ignited a rear storage area and the roof at about 11:00 A.M. Employees detected the fire and thought they had extinguished it, so they never called the fire department. Almost 3 hours after the fire was first discovered, however, it rekindled and broke out through the roof. This time, employees did call fire fighters, but the fire was so advanced that no interior attack could be mounted.

The structure, valued at $1 million, and its contents, valued at $2 million, were a total loss. Four fire fighters suffered smoke inhalation.

Source: "Fire Watch," *NFPA Journal,* November/December, 1999.

Mercantile occupancies require more thorough inspections than other occupancies. They include shopping centers, department stores, drugstores, supermarkets, auction rooms, and any occupancy (or portion thereof) that is used for the display and sale of merchandise.

OCCUPANCY CHARACTERISTICS

The term *mercantile* encompasses many different types of materials and operations. Inspectors are just as apt to inspect a store dealing in glassware as one that sells a large number of paper products. Large department stores have a wide variety of products that react differently in a fire situation. In the past, the great majority of combustible material found within any mercantile occupancy has been Class A, which includes products made of wood, paper, or cloth. These days, though, there are more plastic items or plastic materials that are designed and manufactured to look like something else, such as plastic baby cribs that look, and even smell, like wood. Plastics in their various forms are introducing higher than normal fuel loads into sales and storage areas, and this fact should be considered when determining the overall fuel load because when plastics burn a more rapid fire growth can occur, resulting in production of heavy, thick, black, toxic smoke. Currently there is no restriction on the use of plastics in furniture or other industries, nor is there a restriction on the overall amount of the material permitted in a mercantile occupancy.

Joe Versteeg is a fire protection consultant based in Torrington, CT. He was with the Connecticut Department of Public Safety for over 20 years; most recently he was the commanding officer of the technical services section for the State Fire Marshal's Office. He also serves on several Life Safety Code® *committees.*

The separation and treatment of other occupancies found within or attached to a mercantile occupancy are other factors inspectors will have to be concerned with. In large shopping malls, a variety of different mercantile occupancies will often, in one way or another, connect to several assembly occupancies. In this case, inspectors must consider the different occupant load factors, for example, between a restaurant that may be found in the mall as compared to a department store. When inspecting the premises, inspectors should use the same walk-through process to thoroughly familiarize themselves with the building; after all, they may have to visit the building under fire conditions someday. With this approach, they will be able to make educated decisions as to the probable occupant load when a fire alarm does sound.

NFPA *101*®, *Life Safety Code*®, 2000 edition, separates mercantile occupancies into three categories. A Class A mercantile occupancy is any store having an aggregate *gross* sales area larger than 30,000 ft^2 (2787 m^2), or a store utilizing more than three floor levels for sales purposes. For example, a single-story store with 32,650-ft^2 (3033 m^2) of gross sales area is a Class A mercantile occupancy; a four-story sporting goods store with 20,000 ft^2 of (1858 m^2) gross sales area is also a Class A mercantile occupancy. Inspectors should also remember to measure the total square footage (gross) instead of only the floor area not covered with stock of some sort (net).

A Class B mercantile occupancy is any store with less than 30,000 ft^2 (2787 m^2), but greater than 3,000 ft^2 (279 m^2) of aggregate gross sales area or one that utilizes any balconies, mezzanines, or floors above or below the street floor for sales purposes. The exception to this is one of the examples used above for Class A mercantiles: a four-story, 20,000-ft^2 (1858-m^2) space that is Class A regardless of the size. A Class B mercantile then is the two-story, 25,000-ft^2 (2322 m^2) department store or the 4,000-ft^2 (372-m^2) drugstore.

Class C mercantile occupancies are all stores with 3000-ft^2 (279-m^2) or less of gross sales area that are *located on the street floor only.* If they are less than 3000-ft^2 (279-m^2) but above or below the street floor, they are Class B. Thus, an 850-ft^2 (79-m^2) tobacco store is a Class C occupancy if it is located on the street floor. If it is located below the street floor, it is a Class B occupancy. Being located above the street floor also would make it Class B, regardless of gross square footage.

In addition to the traditional store categories, the *Life Safety Code* also contains provisions for covered mall buildings and bulk merchandising retail buildings where the display of merchandise is on pallets, in piles, or on racks in excess of 12 ft (3.7 m) in height.

INSPECTING THE PREMISES

When beginning the inspection of any marcantile occupancy to determine compliance with the *Life Safety Code* and other pertinent *National Fire Codes*®, as well as the building codes and ordinances of the jurisdiction, the inspector should get an immediate and general idea of what level of maintenance is carried out by the store staff. If the area is somewhat cluttered, with questionable aisles and unswept floors, the chances are good that the entire store will look that way or worse. Poor house keeping is an indication of the general level of conscious fire safety behavior practiced in that particular store and of how much of a task the inspector faces.

Occupant Load

The occupant load for mercantile occupancies is covered in paragraphs 36./1.7 and 36.1.7 and 37.1.7 of the *Life Safety Code*: 30 ft^2 (2.8 m^2) of gross floor area of sales space per person on the street floor or sales floors below the street floor; 60 ft^2 (5.6 m^2) of gross floor area of sales space per person on upper floors used for sales; 100 ft^2 (9.3 m^2) per person on floors or portions of floors used only for offices; and 300 ft^2

($28\ m^2$) of gross floor area per person for those floors or portions thereof not open to the general public but used for storage, shipping, or receiving. The *Life Safety Code* has special provisions for malls. Although the occupant load is not specifically required to be posted in a conspicuous place within the business, it is a good idea to suggest doing so to the manager. That same occupant load should be recorded on the inspection sheet when prefire planning is conducted in the building so that the inspector will know what occupant load to expect in the event of a fire.

Means of Egress

Various types of means of egress are allowed from any mercantile occupancy, such as stairways, smokeproof towers, doors, ramps, and, in some cases, escalators. Still others require the approval of the authority having jurisdiction for very special applications, such as revolving doors or fire escape stairs. For the exact application of each type of exit, the inspector should refer to paragraphs 36.2.2 and 37.2.2 of the *Life Safety Code.*

Generally speaking, at least two exits must be provided and be accessible from every part of every floor and especially from floors below the street-level floor. The few exceptions to this requirement can be found in 36.2.4 and 37.2.4 of the *Life Safety Code*. Dead-end corridors should be avoided, and none are allowed to be more than 20 ft (6.1 m) long. However, they can be up to 50 ft (15 m) long if the building is completely sprinklered or if the occupancy is existing. Exits should be located as far apart as practical, but generally not closer to one another than one-half the longest diagonal distance of the space served. In a grocery, discount, or variety store where checkout stands and turnstiles are provided to restrict exiting, at least one-half of the required exits in both number and capacity must be provided in such a manner that they can be reached *without* having to go through the turnstiles and checkout stands. This requirement, the inspector will find, is commonly violated, although its purpose is clear. In a fire or other emergency, persons within the space must be able to exit quickly and easily, without any obstructions to that exit travel.

Generally speaking, all exit doors are required to swing in the direction of exit travel, particularly (1) when used in an exit enclosure (e.g., stairway), (2) when serving a high hazard area, or (3) when serving an occupant load of 50 or more persons.

Although special locking features are allowed in some configurations, the general rule is that all locking devices on exit doors must be operable without the use of a key or special knowledge, and the method of operation must be obvious even in darkness, with a single operation.

No occupants of any mercantile occupancy should have to travel more than 100 ft (30.5 m) [150 ft (45.7 m) in an existing mercantile occupancy] to find the exit nearest them. This distance can be increased to 200 ft (60 m) in those buildings protected throughout by an approved automatic sprinkler system. In some instances, exit access can pass through the storerooms of mercantile occupancies, but only if (1) at least one other means of egress is provided, (2) the storeroom is not subject to locking, (3) the main aisle through the storeroom is not less than 44 in. (1.1 m) wide and in the clear, and (4) the main path of travel through the storeroom is obvious, has fixed barriers, and is completely unobstructed.

Emergency lighting is required in all Class A and Class B mercantile occupancies. Class C stores, due to their small size and occupant load, are not required to have such installations. Emergency lighting should provide not less than 1 footcandle of light at the floor for 90 minutes, throughout the means of egress, in the event of failure of the normal lighting. For more specific details concerning emergency lighting, the inspector should refer to Section 7.9 in the *Life Safety Code.*

Protection of Openings

Vertical openings in all mercantile occupancies are required to be enclosed or protected in some manner, but the exceptions to this are numerous. The inspector should refer to paragraphs 36.3.1 and 37.3.1 of the *Life Safety Code* for each specific application.

Protection of Hazards

Any area of the space that creates a greater hazard than other areas of the occupancy is required to be separated from those other areas by construction having no less than a 1-hour fire resistance rating, or it must be protected by automatic sprinklers. Areas requiring this special protection include maintenance closets, fuel storage areas, maintenance shops, general storage areas, boiler or furnace rooms, and kitchens.

Any areas with contents considered to be highly hazardous, that is liable to burn with extreme rapidity or result in an explosion, are required to be both separated by construction with at least a 1-hour fire resistance rating, *and* must have complete automatic sprinkler protection. It is up to the authority having jurisdiction to determine what degree of hazard the contents represent and then to make a case for that decision.

Interior Finish

Interior finish for walls and ceilings is required to be either Class A or Class B, except that *existing* Class C interior finishes are allowed on walls only (not ceilings) and in existing Class C stores (see Chapter 21 for further discussion and flame-spread ratings). Inspectors should use reasonable discretion when determining what the existing finish is and, if it is noncompliant according to code, what reasonable methods should be required to correct the violation. There are no specific prohibitions concerning floor finishes. However, if the inspector finds a floor finish that is of an unusual hazard, 10.2.2.2 of the *Life Safety Code* gives the authority having jurisdiction power to regulate it.

FIRE PROTECTION

While walking through the complex, the inspector should determine which type of portable extinguishers are available, if they are fully serviced and operational, and if the number provided is sufficient for the space. The general rule is that at least one hand extinguisher of at least a 2A:10BC rating be available within a travel distance not exceeding 75 ft (23 m) and that at least one should be provided for every 3000 ft^2 (279 m^2) of floor space. The inspector should refer to the pertinent sections of NFPA 10, *Standard for Portable Fire Extinguishers*, for further information.

Alarm Systems

All Class A stores, covered malls, and bulk merchandising retail buildings are required to have a manual fire alarm system throughout the building. However, buildings protected throughout by an approved automatic fire detection and alarm initiation system or protected throughout by an approved automatic sprinkler system that provides alarm initiation are required to have only one manual fire alarm box.

An alarm system, if present, should have manual fire alarm boxes at each exit and should also activate the fire alarm system in the event one of the manual fire alarm boxes is activated. Whether required or not, the system should be maintained in an operational condition. If people see a manual fire alarm box, they naturally assume that it will work when needed.

Sprinkler Systems

Approved automatic sprinkler protection is required (1) in all mercantile buildings having a story over 15,000 ft^2 (1400 m^2) in area, (2) in all mercantile buildings exceeding 30,000 ft^2 (2800 m^2) in gross area, and (3) throughout all stories of the occupancy below the level of exit discharge having an area exceeding 2500 ft^2 (230 m^2) that are used for sales, storage, or handling of combustible goods or merchandise. The inspector should rely on paragraphs 36.3.5 and 37.3.5 of the *Life Safety Code* to ensure that those mercantile occupancies meeting the areas listed here are provided with full sprinkler protection. The requirements for new construction are more stringent. In addition, the *Life Safety Code* also requires covered mall buildings and bulk merchandising retail buildings to be protected by automatic sprinkler systems, regardless of size.

Covered Malls

Covered malls require special considerations when it comes to inspections and code compliance. For the most part, a covered mall and all the shops that open into it are required to be fully sprinklered. Sections 36.4.4 and 37.4.4 of the *Life Safety Code* provide numerous requirements for exit widths and distances to exits. A fire alarm system that is activated by the mall smoke detectors or automatic sprinkler system is required within a covered mall. Manual fire alarm boxes are not required, however. A smokes-control system must also be provided in covered malls.

If an area is required by code to have sprinkler protection and there is none, the inspector should require that it be installed. If sprinkler protection is in place, the inspector should check to see that no sprinklers are obstructed, that they are not painted, and that there are no sales stock decorations or signs hanging from the piping. The sprinkler discharge rather should not be obstructed by sales stock. All areas of the store should be protected by sprinkler protection, and the control valves and inspector's test pipes should be easily accessible by engine crews and test personnel.

The inspector should check the design densities under which the sprinkler system was installed. If the densities are not now sufficient for the products being protected, the system should be upgraded to ensure that the system will do what it is being counted on to do—namely, to control a fire. If the sprinkler system was designed as an ordinary hazard group 1 system but is now protecting large amounts of plastic material, or a much greater load of Class A material, or perhaps a storage area with flammable or combustible liquids, that system will not perform as expected, and it should be upgraded to meet the new demands. If the inspector is not sure whether the sprinkler system will perform as it was designed to, the fire department plans reviewer, fire marshal, fire chief, or building official should be consulted to ensure that the protection is still adequate.

If some other form of automatic extinguishing system is present in the facility (such as a dry chemical system inside a cooking hood), that system should be fully operational and should have been serviced within the last 6 months. All nozzles must be unobstructed, and cooking should be done only under the hood (see Chapter 55, "Protection of Commercial Cooking Equipment").

The store area must be neat, with aisles organized as required by the *Life Safety Code*. The storage or display of hazardous commodities, such as flammable liquids or gases, combustible liquids, pesticides, or oxidizers, should be in compliance with the appropriate NFPA code or standard.

SUMMARY

This chapter has identified the criteria necessary for properly classifying mercantile occupancies. It has also provided discussion on aspects of occupant loads, means of egress, vertical openings between floor levels, protection of hazardous areas interior finish requirements, and fire protection systems. In addition, the chapter has examined nontraditional store arrangements, such as covered mall buildings and bulk merchandising retail centers.

BIBLIOGRAPHY

NFPA Codes, Standards, and Recommended Practices

See the latest version of The NFPA Catalog for availability of current editions of the following documents.

NFPA 10, *Standard for Portable Fire Extinguishers*
NFPA 13, *Standard for the Installation of Sprinkler Systems*
NFPA 30, *Flammable and Combustible Liquids Code,*
NFPA 72®, National Fire Alarm Code®
NFPA *101®, Life Safety Code®*

Business Occupancies

CHAPTER 34

Joseph Versteeg

CASE STUDY

Prefire Inspection Aids in Fire Attack

Information obtained by fire officers during inspections and prefire planning may have saved the lives of a number of fire fighters when the roof of a 3½-story building in Massachusetts collapsed during a fire in the cockloft.

Built in 1895 as a two-story structure, the building was expanded by the addition of a third-floor assembly area during the 1920s. Used as a fraternal meeting hall, the structure was of unprotected ordinary construction. Its roof was supported by large, unprotected steel trusses sitting in notches in the exterior brickwalls. The building, which had been subdivided into stores on the first floor and offices on the second and third floors, was unsprinklered. It had a smoke detection system, but the system did not function at the time of the fire.

The chief of the local fire department was heading to work when a passerby stopped him and pointed out that the building was on fire. Seeing the gray smoke rolling from a third-floor window and the roof flashing, the chief radioed the station with the address and requested a full alarm assignment. By this time, the operator was also receiving 911 calls reporting the fire.

The first alarm was sounded at 8:33 A.M., sending two engines and a ladder truck to the scene. Establishing a water supply, the companies advanced a 1¾-in. handline to the third floor, while the ladder company positioned the aerial to the roof and opened skylights for ventilation. A second backup line was also advanced to the third floor, but both lines were repositioned when fire was found on the second floor.

Additional alarms were struck to support the interior operations and prevent the fire from spreading into the cockloft. The chief also called in another ladder company to help ventilate the upper floors, as the earlier mutual-aid companies established additional water supplies. Despite their efforts, fire fighters, operating multiple handlines and pulling ceilings and walls, reported that the fire had spread into the cockloft. At this time, the chief noticed that the smoke was getting darker and was starting to push from the building. When he began receiving interior benchmark reports of deteriorating conditions, he ordered all fire fighters to evacuate.

Within two minutes, the third-floor ceiling and roof partially collapsed. After a rapid intervention crew confirmed that all fire fighters on the scene had safely evacuated, the chief ordered additional alarms and prepared for a defensive attack using large–diameter deck guns, ladder pipes, and monitor nozzles.

The fire was brought under control about 2 hours after the initial alarm, although companies maintained a fire watch until late the next day. In all, 7 ladder companies and 14 engine companies from 16 communities were called in to help control the blaze, which did an estimated $1 million in

Joe Versteeg is a fire protection consultant based in Torrington, CT. He was with the Connecticut Department of Public Safety for over 20 years; most recently he was the commanding officer of the technical services section for the State Fire Marshal's Office. He also serves on several Life Safety Code® *committees.*

damage to the building and $400,000 to its contents. Exposures suffered only minor damage. There were no injuries.

An investigation by local, state, and federal officials determined that the fire began in the ceiling of a second-floor office in the rear of the building. The cause is undetermined, but flames spread through concealed spaces to voids in the walls and ceilings, then into the common cockloft.

The chief said that inspections and prefire planning made operations safer and more effective.

"Knowledge of roof hatches and skylights provided rapid ventilation to upper floors," he said, "as did knowledge of a large common cockloft with unprotected steel trusses as a collapse potential. Calling for several alarms and staging several ladder companies supplied with water in defensive positions provided a quick change in tactics leading to limited spread to exposures."

Source: "Fire Watch," *NFPA Journal,* January/February, 1998.

NFPA *101*®, *Life Safety Code*®, 2000 edition, defines business occupancies as those used for the transaction of business, the keeping of accounts and records, and similar purposes. They include general offices, doctors' offices, government offices, city halls, municipal office buildings, courthouses, outpatient medical clinics where patients are ambulatory, college and university classroom buildings with less than 50 occupants, and instructional laboratories. Business occupancies typicaly have large numbers of occupants during normal business hours and/very few occupants during nonworking hours.

OCCUPANCY CHARACTERISTICS

Business occupancies can be in buildings of any construction type permitted by local building codes; the *Life Safety Code* does not specify construction requirements for business occupancies. With the exception of parking structures and residential occupancies, the *Life Safety Code* does not require business occupancies to be separated from other occupancies. However, where business occupancies are mixed with other occupancies, local building codes might require occupancy separation, usually with at least 1-hour fire-resistive construction.

In mixed occupancies, the *Life Safety Code* requirements for both occupancies must be satisfied. In other words, the *Life Safety Code* requirements for both occupancies are applied simultaneously. Where there are conflicting requirements, the requirements affording the highest level of safety must be applied.

Business occupancies traditionally have been subdivided into many small office spaces. Although done for other reasons, these subdivisions compartmentalized an otherwise large floor area. The advent of open plan office space, however, has, for the most part, taken away these natural fire barriers. In an open-plan design, large floor areas are subdivided into cubicles using office furniture and partitions that do not extend from the floor to ceiling. Fire can spread more quickly from one work station to another because of the exposed combustibles. One advantage of the open plan arrangement, however, is that occupants are usually able to detect a fire quickly because they have an open view of the floor area.

Business occupancies are generally thought to have a light hazard fuel load. Although the fuel load in office space from wood furniture and trim has declined since the 1940s, it has increased due to the use of more paper in office operations and the use of more plastics and other synthetic materials in furnishings and equipment. Because of the increase in the use of synthetic material, mostly plastics, the concept of fuel load should be used carefully as a predictor of fire severity. Synthetic products often have high heat release rates, which causes fires of these materials to be more severe than those of an equal volume of wood.

In addition, business occupancies often have significant fuel loads that are not fixed or constant. They include delivered materials and furniture and trash that is removed from the buildings. These fuels are usually found in aisles and corridors, which increases the threat to life safety. A fire load analysis cannot be considered complete without an estimate of the transient fuel loads expected in the builiding.

Unenclosed vertical openings between floors have become an extremely popular design characteristic in both large and small office buildings. Floor openings, such as atriums, miniatriums, and convenience stairs, can be a problem or benefit, depending on their size and location in the building. Such openings present a fire safety problem because heat and smoke have the ability to readily spread from floor to floor. For this reason, the *Life Safety Code* contains requirements for the protection and permitted location of unprotected floor openings.

Although building occupancies typically have light fuel loads, large life loss fires can occur in them during normal working hours because occupant load can be up to one person for each 100 square feet (9.3 m^2) of gross floor area. The significant transient fuel loads also contribute to this potential problem.

Business occupancy floor plans change often, and these renovations can cause a properly designed means of egress to become compliance, out of obstructed, or blocked. Renovations can also be a source of ignition, increasing the chance of fire. One of the first tasks during an inspection is to determine if there will be or have been any recent changes or renovations to the building or changes to floor plans.

Means of Egress

INSPECTING THE PREMISES

The basic requirements of means of egress for other occupancies apply to business occupancies, as well. There should be two remote exits from every floor, and egress travel paths should be illuminated and identified by proper signage. The exits must be located in a way that will reasonably reduce the possibility of both exits being blocked by a single fire incident. In some instances, there can be only one exit but have specific limitation associated with it. Those circumstances are described in Chapters 38 and 39 of the *Life Safety Code*.

When two exits are required, they must be remotely located and they must have separate paths of travel. The access to the exits as well as their discharges must also be located so that a single fire would not block both exits. Two exits discharging through a common lobby do not meet this requirement. One of the most prevalent problems in business occupancies is excessive common paths of travel. See Chapter 20, "Means of Egress," for a discussion on common paths of travel.

If occupants must walk through corridors to reach exits, the walls separating the corridors from spaces must be 1-hour fire barriers in new occupancies. Exceptions to this requirement are permitted for some existing occupancies (depending on which code edition is in use), single-tenant spaces, and in spaces protected by automatic sprinklers. The *Life Safety Code* gives requirements for permissible corridor wall penetrations and required protection of openings.

The exit discharge is probably the most overlooked portion of the means of egress. All the requirements pertaining to an unobstructed path of travel and illumination apply to the exit discharge even though the exit discharge is generally outside the building leading to a public way. When emergency lighting is required, the exit discharge must also have some degree of emergency lighting. The exit discharge must be kept free of obstructions, and there must be a reliable method for preventing ice and snow from accumulating in areas subject to such weather conditions. Check the areas around the exit doors to determine if there is anything unusual that could cause problems in an emergency.

To be useful, exit doors must be accessible. Many times, in attempting to lay out an office for best space utilization, little consideration is given to maintaining clear access to exits. In open plan office spaces, the inspector should be aware of furniture arrangements that obstruct direct access to exit doors.

Exit doors must be capable of being opened from the occupied side at all times. In multiple tenant offices, doors are sometimes locked for security reasons with little thought given to life safety. Only locking devices capable of being opened by the person seeking egress are permitted. Special locking arrangements, including delayed egress and access-controlled doors, are permitted and can usually be arranged to solve security concerns.

The exit itself must be accessible and unobstructed. Locking of stair doors is a common problem that the inspector must look for. The inspector should walk the stairs from top to bottom to ensure the path of travel is unobstructed. Doors at the termination of exits must be obvious and openable. Stair doors should not be locked from the stair side so that people using them will not get trapped if there is a fire in the stairway. If stair doors are locked from the stair side, there must be some method of unlocking them, or some floors should be designated to remain unlocked.

The *Life Safety Code* permits two methods for reentry in business occupancies. The first is electric locking devices that automatically unlock (but not unlatch) doors, allowing reentry upon the activation of the building fire alarm system. The second method requires that at least two doors for reentry be provided in each stair, one of which must be at the top or the next-to-top floor. In stairs serving 5 stories or more, reentry doors must be provided so that there are no more than four intervening floors between unlocked doors.

Travel distance is also an important life safety requirement, but care must be taken to ensure that the travel distance requirement is met in a reasonable way. Some open floor plans can be arranged in such a way that the furniture creates a maze, which greatly hinders occupants who are trying to leave the area.

Exits must be marked in business occupancies, and exit signs should be placed to properly mark exits and access to exits that are otherwise not readily apparent. The *Life Safety Code* requires exit signs to be placed in corridors so that a person is not more than 100 ft (30 m) from an exit sign at any point in the corridor. In buildings with open floor plans, the exit signs must properly mark the path to the exit. In occupancies with floor plans using low height partitions, the signs might be visible from a greater distance, but the floor plan might prohibit direct travel to the exit. Thus it might be necessary to place some signs on the partitions.

Although internally illuminated exit signs are not required, all exit signs must be illuminated in some way. If emergency lighting is required, then the exit signs must also be illuminated by the emergency lighting source. Often, a placard-type sign lit by emergency lights will meet this requirement.

Protection of Openings

Openings in fire-rated walls should be protected with doors or other approved opening protective devices. The inspector should check the doors to ensure that they close and latch properly. Closing devices are required on fire doors, and they should function properly. A labeled door leaf, door frame, closer, and latch are required. The inspector should have any wedges used to hold doors open removed immediately.

Automatic hold-open-and-release devices should be used on fire doors that, for functional purposes, need to remain open daily.

Windows in fire barrier and doors must be of fire-rated glazing (usually wired) or an approved material in steel frames. There are also transparent glazing materials that can be used as a fire-resistive barrier. The inspector should review information available on such material carefully. In rated walls, windows of plain glass or wooden or aluminum frames should be replaced with proper assemblies.

Hazardous Areas

General storage rooms, boiler rooms, fuel storage rooms, janitor closets, and maintenance shops are considered hazardous areas and should be separated from the rest of the building by 1-hour fire resistant construction; openings should be protected by ¾-hour fire door assemblies. If hazardous areas in new and existing construction are protected by sprinklers, they need only to be enclosed by smoke-resistant walls, and the openings should be protected by self-closing, smoke-resistant doors.

Computer Rooms

Electronic data processing equipment has become both vital and commonplace in many businesses. This equipment can be highly sophisticated and extremely valuable; in fact, both the equipment and especially the data can be unique and may not be able to be replaced. NFPA 75, *Standard for the Protection of Electronic Computer/Data Processing Equipment,* contains more detailed requirements for fire protection for electronic computer/data processing equipment and computer areas.

Protection of Records

Evaluating the worth of records, no matter what kind they are, is a management responsibility in which the inspector's only responsibility is to provide information on possible exposure to loss. Therefore, the inspector should be prepared to tell management what different levels of protection are available for records of different value and volume.

More specific information and requirements for the protection of records is available in the *Fire Protection Handbook,* and in NFPA 232, *Standard for the Protection of Records.*

Waste Disposal

INSPECTING BUILDING SERVICES

Business occupancies can generate large amounts of waste paper. Usually this waste is removed from the general office areas at the end of the day. This collected waste can create a significant fuel load in corridors, freight elevator lobbies, or where it is held for disposal.

The principal concern is how waste material is handled after it is collected from the various offices and work stations. Smoking materials improperly discarded into trash collection containers have ignited other trash in the containers, and these fires have spread to other areas. Thus, special precautions to prevent this situation should be taken. Precautions include using specially designed container tops and discarding smoking materials into noncombustible cans and allowing them to cool before they are disposed of.

All waste should be removed from the building quickly. If it is stored within the building for short periods of time, the waste should be in proper containers or stored in specially designed rooms. The practice of piling up waste in plastic bags in elevator lobbies or corridors while awaiting its removal can have disastrous results. Once waste is removed from the building, it is commonly held in outside storage bins (dumpsters) for frequent removal from the premises. If that is the case, the inspector should observe the general condition of the outside storage area. It should be far enough away from the building so that it does not present a fire exposure to the building. Sturdy enclosures around trash storage areas can discourage vandalism and unauthorized dumping, both of which can lead to dumpster fires. With paper recycling more common in office buildings, centralized collection areas can also result in additional concentration of paper and hence more fuel than what may be commonly found in such areas.

Some buildings have waste chutes running from upper floors to a collection bin or room. These chutes should be protected by sprinklers. The walls, ceiling, and floor of the room must have a fire resistance rating equal to that of the chute. Chutes, which are floor openings, must have fire-rated hopper-type doors that are self-closing and positive latching. The inspector should check that the chute access doors work well, which may not be the case in older buildings. Newer buildings might have pneumatic waste-chute services, which require electrical interlocks between the inner and outer doors at the chute openings in order to properly function. NFPA 82, *Standard on Incinerators, Waste, and Linen Handling Systems and Equipment,* contains more specific requirements for trash chutes and rooms.

Cafeterias

Many large business occupancy buildings have their own cafeteria facilities and kitchens. The inspector should examine the kitchen equipment for evidence of grease accumulations in hoods and exhaust ducts; good duct installations will have cleaning and inspection openings.

Hoods, exhaust ducts, and grease removal devices must be protected by approved extinguishing systems, which usually consist of fixed-pipe carbon dioxide, dry chemical, or foam sprinkler or spray systems. If there are no special extinguishing systems, the cooking equipment might be served by a listed or labeled grease retractor, which may provide sufficient protection. The cooking surface of fat fryers, ranges, griddles, and broilers, which can be a source of ignition, also need to be protected by appropriate extinguishing systems.

The inspector should make sure that instructions for manually operating the fire extinguishing systems are posted conspicuously. Asking kitchen employees how the systems operate should give the inspector an indication of how familiar they are with the systems. The inspector should check the log to determine when the special systems were last serviced and inspected; they should be inspected and serviced every 6 months.

The inspector should make sure there are enough portable fire extinguishers suitable for Class B and Class C fires near the cooking equipment. If the fixed extinguisher uses a dry chemical extinguishing agent, the inspector should make sure that the portable fire extinguishers in the area are compatible with it. For example, if the fixed system uses a BC dry chemical, the portable extinguisher should also be BC, not ABC. The inspector should note when the portable extinguishers were last inspected and serviced.

See Chapter 55, "Protection of Commercial Cooking Equipment," and NFPA 96, *Standard for Ventilation Control and Fire Protection of Commercial Cooking Operations,* for a more complete discussion of and specific requirements for kitchen cooking equipment installations.

Shafts and Chases

The integrity of shafts and chases is as important in business occupancies as it is in any other occupancy. When new buildings are constructed, utility risers are usually enclosed in fire-rated shifts, or floor penetrations are sealed with appropriate fire-resistive materials. As buildings are used, new cable, conduit, or piping is often run between floors, and the new penetrations may not be properly sealed.

The inspector should check telephone and electric closets and mechanical shafts and risers to determine if there-are unprotected floor openings or penetrations through fire barrier walls. Such openings can be sealed with lightweight concrete or other suitable materials, such as silicon foam.

FIRE PROTECTION

Automatic Sprinklers

Automatic sprinklers are required in business occupancies that are higher than 75 ft (23 m) when measured from the lowest level of fire department access to the highest occupiable floor. The sprinkler system must be electrically supervised. In lieu of sprinklers, existing buildings can be equipped with an alternative equivalent system approved by the authority having jurisdiction.

The adequacy of sprinklers, or an equivalent system, depends on several basic conditions. The hazard severity must be analyzed to ensure that the protective system is sufficient to control a fire. The impact of the expected fire must be judged to determine its effect on the occupants and the rest of the building. The occupants' response, both first aid fire fighting and ability to escape, must be determined as part of the incident impact. The inspector can use this information to determine whether adequate protection is provided.

Sprinkler systems, although highly reliable, require regular testing and maintenance. The inspector should ensure that all sprinkler valves, including water supply, are open and locked or supervised in that position. Closed water supply valves are the most common cause of sprinkler system failures. The inspector should find out if there is a procedure for ensuring that sprinkler valves are closed only when properly authorized and turned on again following any maintenance or modification. Records of sprinkler system testing and maintenance should be checked. See Chapter 15, "Automatic Sprinkler and Other Water-Based Fire Protection Systems," for details on inspection of automatic sprinkler systems.

Because of legal liability issues, the jurisdiction should have a policy regarding testing systems: In general the building owner or representative should do all the testing. The inspector should witness any tests required to be performed as part of an inspection.

Alarm Systems

A manual fire alarm system and an emergency lighting system are required when a business occupancy is two or more stories above the level of exit discharge, when the occupant load above or below the level of exit discharge is 50 persons or more (300 or more in existing buildings) or when the occupant load of the building is 100 persons or more (1000 or more in existing buildings). In addition, at least one manual station is still required on premises that can activate the building fire alarm system. When the building is fully sprinklered or totally protected by an automatic fire alarm system, manual fire alarm boxes are not required if the sprinklers or automatic system cause the evacuation alarm to sound. As with automatic sprinkler systems and because of poor maintenance in many business occupancies, the inspector should just witness tests, with the building owner or representative doing the actual testing.

Portable Fire Extinguishers

Portable fire extinguishers are required in business occupancies. They should be used, placed, and maintained in accordance with NFPA 10, *Standard on Portable Fire Extinguishers,* and state and local laws. The inspector examine testing and maintenance records to ensure that the extinguishers are being properly tested and maintained.

Special Provisions

In addition to determining the degree of compliance with the requirements of Chapters 38 and 39 of the *Life Safety Code,* the inspector should be aware of the often-overlooked additional provisions pertaining to windowless and underground structures in Chapter 11 of the *Life Safety Code*. Because of the added concerns unique to these types of structures, requirements in addition to the those found within the business occupancy chapters of the *Life Safety Code* may be mandated for these structures concerning automatic sprinklers, emergency lighting, egress arrangements, and, in certain instances, smoke venting.

OPERATING FEATURES

Each occupancy chapter contains requirements on operating features. Specific requirements for business occupancies include having written emergency procedures and training all occupants in procedures. In addition, emergency egress and relocation drills must be held on a schedule acceptable to the local authorities. Employees of the building, usually the operating staff, must be trained in the proper use of portable fire extinguishers in the building.

SUMMARY

This chapter has identified the criteria necessary for properly classifying business occupancies. It has also provided discussion on aspects of occupant loads, means of egress, the protection of hazardous areas, interior finish requirements, protection of records, utilities, and fire protection systems.

BIBLIOGRAPHY

Cote, A. E., ed., *Fire Protection Handbook,* 18th ed., NFPA, Quincy, MA, 1997.

NFPA Codes, Standards, and Recommended Practices

See the latest version of The NFPA Catalog for availability of current editions of the following documents.

NFPA 72®, National Fire Alarm Code®
NFPA 75, *Standard for the Protection of Electronic Computer/Data Processing Equipment*
NFPA 82, *Standard on Incinerators Waste, and Linen Handling Systems and Equipment*
NFPA 86, *Standard for Ovens and Furnaces*
NFPA 96, *Standard for Ventilation Control and Fire Protection of Commercial Cooking Operations*
NFPA *101® Life Safety Code®*
NFPA 232, *Standard for the Protection of Records*

Industrial Occupancies

CHAPTER 35

Joseph Versteeg

CASE STUDY

Sprinklers Control Factory Blaze

Sprinklers controlled a fire in a large manufacturing plant in California that produced polyurethane foam.

The single-story, 372,100-ft^2 building was constructed of concrete tilt-up walls, a concrete floor, and a wooden laminated roof. A wet-pipe sprinkler system, which included in-rack sprinklers and a water flow alarm in the risers, protected the property.

The plant was in operation when a polyol pump's suction pressure dropped low enough to trigger an alarm and stop the pump motor. Plant personnel quickly overrode the alarm, but the three-second delay that occurred while the pump restarted allowed the polyol level to drop low enough to cause an exothermic chemical reaction. The product ignited, activating several sprinklers.

The central monitoring station notified the fire department of a water flow alarm in one of the risers, and fire fighters responded at 12:55 P.M. to find nine in-rack sprinklers controlling the blaze. They used on-site hoses to complete extinguishment.

The loss of the contents was estimated at $20,000. There was no structural damage, and no one was injured.

Source: "Fire Watch," *NFPA Journal,* May/June, 1998.

An industrial occupancy is any building, portion of a building, or group of buildings used for the manufacture, assembly, service, mixing, packaging, finishing, repair, treatment, or other processing of goods or commodities by a variety of operations or processes. Industrial occupancies include, but are not limited to, the following:

1. Chemical plants
2. Factories of all kinds
3. Food processing plants
4. Furniture manufacturers
5. Hangars (for servicing/maintenance)
6. Laboratories involving hazardous chemicals
7. Laundry and dry cleaning plants
8. Metalworking plants
9. Plastics manufacture and molding plants
10. Power plants
11. Refineries
12. Semiconductor manufacturing plants
13. Telephone exchanges
14. Woodworking plants

Joe Versteeg is a fire protection consultant based in Torrington, CT. He was with the Connecticut Department of Public Safety for over 20 years; most recently he was the commanding officer of the technical services section for the State Fire Marshal's Office. He also serves on several Life Safety Code® *committees.*

Each building or separated portion of an industrial building should be inspected in accordance with the requirements of its principal use, for example: warehouses as storage occupancies; offices as business occupancies; and auditoriums, cafeterias, and lunchrooms as assembly occupancies. Because of the complexity of industrial occupancies, the inspections can be time consuming.

CLASSIFICATION OF OCCUPANCY

Industrial occupancies are subclassified in NFPA *101*®, *Life Safety Code*®, 2000 edition, into three types of usage: general, special purpose, and high hazard.

- *General Industrial Occupancy:* This subclassification involves ordinary and low hazard manufacturing operations conducted in buildings of conventional design suitable for various types of manufacture. Also included are multistory buildings where floors are occupied by different tenants and, therefore, subject to possible use for types of manufacturing with a high density of employees.
- *Special Purpose Industrial Occupancy:* This subclassification includes low and ordinary hazard manufacturing operations in buildings that were designed for and suitable only for particular types of operations. This subclassification is characterized by a relatively low density of employees, with much of the area occupied by machinery or equipment.
- *High Hazard Industrial Occupancy:* Buildings in this subclassification include those having high hazard materials, processes, or contents. Incidental high hazard operations in low or ordinary hazard occupancies that are protected with automatic extinguishing systems or other protection (such as explosion suppression or venting) appropriate to a particular hazard are not considered high hazard occupancies overall.

Some of the common problems encountered in industrial occupancies include overcrowding, poor housekeeping, poor maintenance of electrical equipment and wiring, inadequate exit facilities, locked or blocked exits, misuse of flammable liquids and heat-producing appliances, and poor maintenance of fire protection systems and appliances. Code enforcement must be rigid and inspection thorough, with some emphasis on fire safety education and prefire planning.

INSPECTING THE PREMISES

Occupant Load

The occupant load in industrial occupancies is one person for each 100 ft^2 (9.3 m^2) of gross floor area, and exits that can provide this minimum capacity must be provided for them. In most plants, the space occupied by work benches, machinery, and equipment generally tends to keep the well within the ratio. "Sweat shops," however, tend to be the exception.

An increased occupant load is permitted if a floor plan showing that proper aisles and adequate exits are available to safely accommodate the increased occupant load is submitted to the authority having jurisdiction. In most instances, overcrowding will be fairly obvious to the inspector during the inspection. In a special purpose industrial occupancy, the occupant load is the maximum number of persons to occupy the area under any probable conditions.

Means of Egress

Requirements for exits in industrial occupancies are found in Chapter 40 of the *Life Safety Code.* Inspectors should be aware of the requirements because they are responsible for seeing that all portions of a means of egress are maintained in a safe and usable condition.

All exits must discharge to a clear and unobstructed path of travel to a public way. Where there is evidence of parked vehicles or other obstructions, signs or barriers should be erected to discourage the practice. Barriers should not obstruct the flow of persons exiting the building.

Exits must be clearly illuminated, identified, and accessible. The inspector should open every exit door to be sure that it is labeled when required, swings in the direction of egress travel when so required, and that self-closing or automatic-closing devices and mechanisms function properly. There can be no drop-offs on the other side of the door.

Where pilferage might be a problem, means other than locking are available to prevent unauthorized use of exits. The *Life Safety Code* permits the use of approved, listed, special locking arrangements on doors in industrial occupancies. All conditions set forth for their use must be followed. Delayed-egress locks and access-controlled egress device do not prevent the door from of opening; they merely delay opening and may require the sounding of an alarm. Special locking arrangements cannot be used in high hazard areas.

Where exit stairs are required to be enclosed, the enclosure and its protected openings must be of the proper fire resistance ratings. Handrails must be secure and stair treads and landings should be slip resistant. Stairways cannot be used for storage or any other purpose, and they must be illuminated.

Every worker must have access to not less than two remotely located exits. The path of travel must be clear, illuminated, unobstructed, and as direct as possible without exceeding maximum travel distances. Where the exit and path of travel are not clearly visible, signs must be provided to indicate the direction. A short common path of travel to two otherwise remote exits is permitted, except from an area of high hazard. Exit access must not pass through areas of high hazard. When evacuation must be delayed because of the need to safety shut down an operation, or for any other reason, the additional provisions governing ancillary facilities contained in Section 40.2.5.5 of the *Life Safety Code* should be met.

The inspector must ensure that all elements composing the means of egress remain in compliance with the requirements during periods of renovation and construction. This is especially important when partitions are erected to separate construction areas from work or production areas. Large loss fires have occurred in all types of occupancies during periods of construction, and industrial occupancies are particularly vulnerable due to their complexity and the work processes performed in them.

Emergency lighting is required in all facilities except those occupied only during daylight hours in which skylights or windows are arranged to provide, during those hours, the required level of illumination for all portions of the means of egress. The inspector should check the type of lighting used and review records of servicing and testing. If battery packs are used for an emergency power supply, there should be an indicator light to show full-charge condition and a test button to check its operability. The inspector must also be aware that the industrial occupancy chapter contains requirements for equipment walkways, platforms, ramps, and stairs that differ from the dimensional criteria of Chapter 7 of the *Life Safety Code.*

Protection of Openings

When inspecting industrial occupancies, the inspector must check on the integrity of fire barrier walls and fire-rated, floor/ceiling and roof/ceiling assemblies. With changing technology, changing operations and processes, and new tenants, industrial plants undergo revisions that create openings and holes through fire-rated assemblies. Pipes, electrical conduits, cable trays, and other penetrating items must be properly sealed and protected. Penetration seals must be made of approved or listed materials and be installed such that they maintain the fire rating of the wall or floor assembly in which they are installed.

Ductwork going through fire-rated assemblies must be equipped with fire dampers unless specifically exempted by code. Where dampers are prohibited, such as for exhaust systems for cooking appliances, such ductwork must not pass through rated assemblies or be properly enclosed.

Inspectors should check to see that fire doors are of the proper rating for the enclosure in which they are installed and that they are self-closing and positive catching. They should also check that automatic closure devices and mechanisms operate properly. Any obstructions that could interfere with the fire door closing completely must be removed (e.g., wooden wedges or the door being tied open). The inspector should examine the tracks of vertical sliding and roll-up doors for mechanical damage, especially when the openings are used by industrial trucks. Consideration should be given to installing guards to prevent stock from being piled up against the door or vehicles from striking it. The inspector should check that all doors get closed at the end of the business day and should inspect each door for evidence of excessive wear and tear, modifications to the door, or other defects that make its continued use suspect. Maintenance and testing of these doors should be done in accordance with NFPA 80.

The inspector should check all vertical openings, such as conveyors, elevators, stairs, dumbwaiters, and refuse chutes, for proper enclosure and to be sure that all openings are properly protected and of the proper fire rating. The inspector should also check that pipe chases and other vertical recesses are firestopped.

Fire shutters should be checked to ensure that they have proper automatic closing devices and that such mechanisms are operable. On building exteriors facing fire exposure hazards, doors and windows should be checked for rating and glazing. The glazing of all fire doors that are permitted to have glass should also be examined to ensure that the glazing is of the proper size and thickness and installed properly in acceptable frames.

Hazardous Materials

The inspector must determine the properties of all of the materials used in industrial plants and see that they are stored and handled safely. In order to do this, especially with chemicals, the inspector must have good reference sources. Because inspection is not an emergency activity, inspectors can record what is found, how it is stored and handled, and then do additional research. If the research indicates that special precautions are required, they will need to perform a follow-up inspection to ensure that such precautions are being taken. Caution should be used because the handling of the material—for example, the handling of flammable liquids that release vapors—can pose a significant potential for a hazard and should be corrected immediately.

Once a material has been identified, classified, and categorized by the degree of its physical and chemical properties, half the work has been done. The inspector must then determine if there are excessive amounts of materials for the fire area and for the provided level of protection. They must also determine the requirements for and the adequacy of the venting, if provided; whether electrical equipment has been classified properly; whether electrical wiring is in good condition and properly maintained; and whether there are ignition sources.

Production and process areas should contain only those amounts of hazardous materials that are necessary to the immediate process or operation. The maximum amount should be limited to the needs of one day or one shift, and then only when relatively small amounts are used. Inspectors must be sure to inspect the methods used to transfer hazardous materials from the shipping container or bulk storage area into the process or operation area. They should also look for possible ignition sources.

Many hazardous industrial processes have been fully evaluated, and standards have been established for their safe operation. The *National Fire Codes®* contain all of the standards for safe operation of the most common industrial processes and many that are not common. In addition, the NFPA *Industrial Fire Hazards Handbook* covers a broad range of fire hazards that are found in major industrial occupancies. Inspectors should use this material when evaluating hazardous industrial processes and also when conducting inspections.

In situations in which there is no established standard to follow, inspectors must use their judgment in identifying the process or operational hazards and determining whether they are being controlled properly. The protection afforded must be appropriate to the hazard or hazards, and ignition sources must be controlled. A relatively small hazardous process incidental to the main operations, such as a small paint spray booth, should not change the classification of the entire area to one of high hazard. Inspectors should look for the installation of a special extinguishing system, such as carbon dioxide or dry chemical, because very often it is required by a standard to provide protection of the process. They may also find that the installation of draft curtains or special venting arrangements are required.

In the chemical and allied industries, there are hundreds of different processes and thousands of variations that are well beyond the inspector's ability to evaluate. This is not to say that inspectors should skip inspecting these properties or only give them a superficial inspection. They should identify all of the chemicals used and their hazardous properties, and they should then place them in broad classifications, such as corrosive, flammable, combustible, unstable, or reactive, based on the degree of hazard. The inspector should ask to see a copy of the Material Safety Data Sheet (MSDS). An MSDS is provided by the manufacturer, compounder, or blender of the chemical and contains information about the chemical composition, physical and chemical properties, health and safety hazards, emergency response, and waste disposal of the material. Evaluating storage, transfer method, compatibility, and so on can give the inspector a good indication of whether safe practices are being followed.

Some general questions concerning the various processes can be asked:

- Is there an operator's manual?
- Is the operator trained?
- Does the manual cover the hazards of the materials, the safe and critical temperatures and pressures, the proper sequence for adding materials, the consequences for failure to follow a formula exactly?
- Does the process have fail-safe automatic controls?

Asking many additional questions can help the inspector make a limited evaluation.

To obtain meaningful information, inspectors must gain the confidence of management and show that they can be trusted with trade secrets and confidential information. Because there are trade secrets in every phase of industry, inspectors should not be insulted if they are asked to sign a pledge of confidentiality.

Storage

Outdoor Storage. The storage of materials outdoors is usually limited to those used in large quantities and those that are not susceptible to damage by weather. Storage practices should follow recommended safe practices.

All outdoor storage should be arranged so that it will not interfere with firefighting access to and around buildings and to the storage itself. If the stored materials are combustible, ignitible, or both, they should be far enough away from other buildings so that, if on fire, one will not be an exposure hazard to another. There should be sufficient fire hydrants and hose houses with fire lanes to make outdoor combustible storage accessible on all sides. Areas must be free of vegetation and other loose combustibles.

Indoor General Storage. Preferably, storage areas should be in separate buildings or in cutoff sections of buildings used for no other purpose.

If storage is incidental to the main use of the building, inspectors should still follow the general rules for storage occupancies (as noted in Chapter 36 of the book) as much as possible, but they must also make additional judgments as to safe practices. Hazardous materials in relatively small amounts should be stored with due regard for their hazardous properties: flammable liquids should be stored inside cut off storage rooms or cabinets; loose, highly combustible fibers should be stored in metal or metal-lined bins with automaticclosing covers; and pyroxylin plastics should be stored in vaults and tote boxes. Many of these materials have specific standards that address the proper storage applications.

The inspector must be certain that piles are stable and separated by adequate aisles, that clearance to sprinklers is maintained, and that materials being stored are compatible. Stock piled over 12 ft (3.7 m) in height and rack storage of material require special considerations. See NFPA 13, *Standard for the Installation of Sprinkler Systems,* for proper methods of storage and protection.

Idle pallets awaiting reuse, repair, or disposal can be a constant problem. They should never be stored in unsprinklered areas. When they are stored in sprinklered areas, the piles should cover a small area and be less than 8 ft (2.4 m) high. They should preferably be stored outdoors, well away from buildings and other storage areas.

Housekeeping and Maintenance. Poor housekeeping and maintenance practices can be the most frustrating problem the inspector will encounter and are probably the principal reasons for follow-up inspections. Improper housekeeping is not only a fire hazard, but it also indicates a lack of management commitment. Improper storage of materials and poor maintenance on pumps, piping, and exhaust systems can make floors slippery and atmospheres dusty and can interfere with the proper operation of fire protection equipment.

Industrial occupancies with good housekeeping and maintenance practices are relatively easy to inspect. As a general rule, where housekeeping and maintenance are a priority, most items of fire safety and protection will also be good, and less time will probably be needed to make a thorough inspection.

You should see that waste is removed properly and disposed of safely. Where waste has value as salvage, a safe collection area should be set apart and maintained in an orderly way. Chemical wastes must be disposed of in a manner that is safe for the environment and in accordance with state and federal regulations.

FIRE PROTECTION

When inspecting industrial plants, the inspector must be certain that existing fire protection systems and equipment are properly maintained and that portable fire extinguishers are properly located and are accessible. The locking pin should be in place and sealed, free from damage. There should be no foreign materials in hoses and nozzles that would interfere with their operation. Pressure gauges on extinguishers should indicate they are ready for use. The inspector should examine the tag for the last inspection and hydrostatic test dates. Extinguishers should be in cabinets or have covers when they are located in dusty or corrosive atmospheres. Their location should be clearly marked. Chapter 15, "Automatic Sprinkler and Other Water-Based Fire Protection Systems," addresses the requirements for the inspection, testing, and maintenance of water-based fire protection systems such as automatic sprinklers.

Water Supplies and Fire Pumps

Inspectors will have to rely on records and reports when inspecting water supplies because much of the piping and valves are buried. Industrial occupancies often have more than one source of water supply (e.g., tanks, ponds, and city connections), and each one will need to be examined. Inspectors should check aboveground portions for proper maintenance. They should make sure that all supply valves are open, gravity tanks work, fire hydrants are maintained, and so on. They should review records of water-flow tests, pump tests, valve-operating records for underground valves, and hydrant inspection reports. Signs of neglected maintenance will usually be obvious.

The inspector should check pump rooms to determine whether fire pumps and fire booster pumps are ready for operation if they are needed. The power should be on at the controller of electrically driven pumps, no trouble lights should be on, and the jockey pump should not run excessively or kick on too often. The fuel tanks of internal combustion drivers should be full or nearly so, and batteries should be fully charged with a trickle charger to keep them charged. The inspector should review records of pump testing and maintenance.

Sprinkler Systems

If sprinkler systems are to perform as they were intended to, periodic inspection and maintenance are essential. The inspector should inspect them visually and witness periodic tests. In general, hands-on testing should be the responsibility of the building owner or an authorized maintenance company.

Special Extinguishing Systems

Special extinguishing systems can consist of Halon 1301, clean agent, carbon dioxide, dry chemical, and foam systems. Inspectors must have a good idea of how the various systems operate. The inspection must be visual; the valves should not be tested or manipulated. Inspectors should check that the extinguishing agent used is suitable for the hazard(s) being protected and that a reserve supply is available, if required. They should check that actuating devices and alarms are operational, see that nozzles are clear and free of foreign matter, and determine that nozzle caps, where used, are free. Nozzles should be properly aimed and protected from damage.

When systems are of the total flooding type, all openings required to be closed on system actuation should be checked for proper operation. It is important that the hazard enclosure be properly sealed before system discharge. Piping, cable assemblies, valves, and manifolds should be checked for damage. Records of inspection, testing, and recharging should be examined to determine whether maintenance has been proper. Specific information on these systems can be found in the appropriate standard as code in the *National Fire Codes.*

Standpipe Hose Systems

In industrial occupancies, standpipes and hose stations are more often supplied from the sprinkler system than from a separate system. Therefore, inspectors should inspect these systems as they would sprinkler systems. They should check that all valves on the water supply are open and that the fire department connection is accessible. Threads should not be damaged and should be properly capped. Swivels should work freely, and threads should be compatible with those of the local fire department.

Inspectors should check hose cabinets or reels for proper installation, location, and accessibility. They should check the hose for signs of deterioration or need for reracking, and they should be sure that the attached nozzle works freely. Hoses and outlets should have the same threads as those of the local fire department, or there should be an adaptor in the cabinet or at the reel.

Fire Alarm Systems

The inspection of an alarm system should be visual. Testing should be the responsibility of thoroughly trained employees or an outside alarm service company. However, the inspector should review these test records as part of your inspection and should observe alarm-initiating devices for proper location, mechanical or electrical damage, painting, loading, or damage due to a corrosive atmosphere. Wiring should be in good condition and securely fastened. Control panels should be in a safe location and readily accessible. The "power on" light should be lit, and all trouble lights and signals should be off. Service and test records should be in the panel enclosure. When emergency power is required, batteries should be fully charged. Equipment should be free from dirt or grit that can find its way into delicate parts and contacts.

Manual fire alarm stations should also be inspected for signs of any problems. Wiring should be secure and in good condition, and there should be no tape, wire, string, or other encumbrance to the effective use of the system. The stations should be located along natural egress paths, near exits.

The inspector should check audible devices to see whether they have been tampered with, painted, or damaged and whether they can be heard above the ambient noise. The inspector should also check all records to determine that required servicing and testing has been done and should review records of all supervisory signal systems and alarm signal systems. Detection systems for actuating special extinguishing systems are usually serviced by an outside service company under contract.

SUMMARY

This chapter has identified the criteria necessary for properly classifying industrial occupancies. It has also provided discussion on aspects of occupant loads, means of egress, the protection of hazardous areas, interior finish requirements, and fire protection systems. In addition, the chapter has examined hazardous material storage and use.

BIBLIOGRAPHY

Cote, A. E., and Linville, J. L., eds., *Industrial Fire Hazards Handbook*, 3rd ed., NFPA, Quincy, MA, 1990.

NFPA Codes, Standards, and Recommended Practices

See the latest version of The NFPA Catalog for availability of current editions of the following documents.

NFPA 13, *Standard for the Installation of Sprinkler Systems*
NFPA 30, *Flammable and Combustible Liquids Code*
NFPA 30B, *Code for the Manufacture and Storage of Aerosol Products*
NFPA 32, *Standard for Drycleaning Plants*
NFPA 35, *Standard for the Manufacture of Organic Coatings*
NFPA 36, *Standard for Solvent Extraction Plants*

NFPA 61B, *Standard for the Prevention of Fires and Explosions in Grain Elevators and Facilities Handling Bulk Raw Agricultural Commodities*
NFPA 61C, *Standard for the Prevention of Fire and Dust Explosions in Feed Mills*
NFPA 61D, *Standard for the Prevention of Fire and Dust Explosions in the Milling of Agricultural Commodities for Human Consumption*
NFPA 65, *Standard for the Processing and Finishing of Aluminum*
NFPA 81, *Standard for Fur Storage, Fumigation and Cleaning*
NFPA 88B, *Standard for Repair Garages*
NFPA *101*®, *Life Safety Code*®
NFPA 120, *Standard for Coal Preparation Plants*
NFPA 480, *Standard for the Storage, Handling and Processing of Magnesium Solids and Powders*
NFPA 481, *Standard for the Production, Processing, Handling, and Storage of Titanium*
NFPA 482, *Standard for the Production, Processing, Handling, and Storage of Zirconium*
NFPA 505, *Fire Safety Standard for Powered Industrial Trucks Including Type Designations, Areas of Use, Maintenance, and Operation*
NFPA 513, *Standard for Motor Freight Terminals*
NFPA 651, *Standard for the Manufacture of Aluminum Powder*
NFPA 654, *Standard for the Prevention of Fire and Dust Explosions in the Chemical, Dye, Pharmaceutical, and Plastics Industries*
NFPA 664, *Standard for the Prevention of Fires and Explosions in Wood Processing and Woodworking Facilities*
NFPA 1124, *Code for the Manufacture, Transportation, and Storage of Fireworks*

Storage Occupancies

CHAPTER 36

Joseph Versteeg

CASE STUDY

Employees Control Blaze

Eight employees of a Michigan automotive parts warehouse suffered from smoke inhalation as they fought a fire ignited by sparks from arcing electrical equipment. According to fire officials, a lack of SCBA contributed to their injuries.

The 2,000,000-ft^2 (185,800-m^2), irregularly shaped warehouse was constructed of unprotected steel framing and concrete block walls. A steel-truss roof supported a metal deck covered by a rubber membrane. The one-story building had a wet-pipe sprinkler system and interior hose lines connected to a standpipe, but it had no detectors. Although the structure was being renovated, it was in full operation at the time of the blaze.

A company electrician was electrocuted when the metal conduit he was installing came in contact with an energized 480-V, 3000-A ceiling-mounted bussway. The resulting are sent off sparks that ignited paper, cardboard, and plastic wrapping material stored on the tops of two storage racks.

Several employees used interior hose lines to confine the fire to the racks, while another employee called 911 at 10:39 A.M. The employees were able to control the flames before they generated enough heat to activate the overhead sprinklers.

Damage to the building, valued at $7,024,500, and its contents, valued at $5,272,500, was estimated at $2000 and $110,000, respectively.

Source: "Fire Watch," *NFPA Journal,* November/December, 1998.

Storage occupancies are buildings or structures used to store or shelter goods, merchandise, products, vehicles, or animals. Examples are warehouses, freight terminals, parking garages, aircraft storage hangars, grain elevators, barns, and stables. These facilities may be separate and distinct facilities or part of a multiple-use occupancy. When storage is incidental to the main use of the structure, it should be classified as part of the main occupancy when determining life safety requirements.

Considerable judgment must be exercised when determining whether storage is incidental to the main use of the building. One consideration is the hazard classification of the contents stored in the area. If they are classified as high hazard, the room or space must be separated from the rest of the occupancy by fire-resistive construction that meets the requirements of the local code or, where no code has been adopted, NFPA 220, *Standard on Types of Building Construction.* In cases where the hazard is severe, both fire-resistive construction and automatic fire suppression might be required.

Joe Versteeg is a fire protection consultant based in Torrington, CT. He was with the Connecticut Department of Public Safety for over 20 years; most recently he was the commanding officer of the technical services section for the State Fire Marshal's Office. He also serves on several Life Safety Code® *committees.*

The inspector should be aware that storage occupancies or areas of storage occupancies that are used for packaging, labeling, sorting, special handling, or other operations that require an occupant load greater than that normally contemplated for storage must be classified as industrial occupancies when determining life safety requirements.

Parking garages, whether closed or open, aboveground or below, must also be classified as industrial occupancies if they contain an area in which repair operations are conducted. If the parking and repair sections are separated by 2-hour fire-rated construction, they can be treated separately. The inspector should be aware of the special requirements for underground and windowless structures, which are covered in Chapter 11 of NFPA *101*®, *Life Safety Code*®, 2000 edition.

OCCUPANCY CHARACTERISTICS

Storage occupancies can be classified as low, ordinary, or high hazard or a combination of these where mixed commodities are stored together. Where different degrees of hazard exist in the same structure and cannot be separated effectively, the requirements for the most hazardous classification govern. The authority having jurisdiction must use sound judgment when applying this principle of hazard classification. The *Life Safety Code* uses the ordinary hazard classification as the basis for general requirements. Most storage occupancies fall into this classification, although an increasing percentage are being classified as high hazard due to the rapid fire and smoke development that can be expected in some situations.

When looking at the overall fire hazard, inspectors should also consider building construction. Combustible building materials can affect the spread and development of fire, especially if there are combustible concealed spaces. Combustible insulation is a particular problem in certain storage facilities and represents a serious fire problem. Inspectors must determine specifically the type of any insulation present.

Modern developments in material handling have brought rapid changes to storage occupancies, including high-rack storage areas that can reach heights of 50 to 100 ft (15 to 30 m). Computer-controlled stacker cranes and robot-controlled material handlers are now being used to move materials. Regional distribution centers that cover several acres, which might contain two- or three-level mezzanines, are now being developed. Ministorage complexes that consist of rental spaces ranging from 40 to 400 ft^2 (3.71 to 37.2 m^2) in size are also being developed. These complexes, which consist of as many as 50 to 100 rental spaces in one building, often contain varying types and amounts of hazardous storage in one or more of the rental areas.

Storage occupancies can house raw materials, finished products, or goods in an intermediate stage of production, and these materials can be in bulk form, solid piles, palletized piles, or storage racks. Therefore, inspectors should remember that the storage arrangement can greatly affect fire behavior.

INSPECTING THE PREMISES

Contents

In determining life safety features and requirements in a storage occupancy, the inspector must first determine the hazard classification of the contents. Fire behavior will depend on the ease of ignition, rate of fire spread, and rate of heat release of the product itself. Products, however, are often complex items whose fuel content, arrangement, shape, and form affect their performance in a fire. A packaged product must be considered as a whole because that is the way it burns, so in classifying the contents, the inspector must examine the product, product container, and packaging material used.

Increasing amounts of plastics are now being used as part of the product and as part of the packaging. Bicycles have traditionally been all metal except for the tires, but now the frame and wheels of a bicycle may contain 50% or more plastic. Electrical and plumbing supplies have traditionally been metal, but now many of these supplies, including pipe, conduit, fittings, and junction boxes, are made of plastic. Washing machines typically have a limited amount of combustible parts in the machine assembly; however, today's typical packaging arrangement, the machine packed in a cardboard box surrounded with plastic foam, has made this commodity more hazardous even though the base commodity has not changed.

The inspector should become familiar with NFPA standards detailing requirements for the proper storage arrangement and level of protection for storage of specific items including flammable and combustible liquids, hanging garments, rolled paper, tires, and aerosol containers.

Occupant Load

A small number of people in relation to the total floor area will usually occupy a storage occupancy at any one time. Work patterns usually require employees to move throughout the structure using industrial trucks. In totally computerized warehouses, an even smaller number of occupants is present, which reduces the likelihood of early fire detection as well as available personnel to begin fire-fighting operations. Because of this, the *Life Safety Code* has no occupant load requirements for storage facilities. When establishing the occupant load for new and existing storage structures, the authority having jurisdiction will have to obtain in writing (from the building owner or occupant) the actual number of occupants expected in each occupied space, floor, or building. The authority having jurisdiction must then designate the number of occupants to be accommodated on every floor and in each room or space. Special attention needs to be given to parking garages, which, at given times, could be occupied by many people, such as at the end of a work day or when an entertainment event is over. it should be noted that due to the typical low density of a storage occupancy, egress capacity is rarely a problem if the minimum number and size of exits along with maximum travel distance limitations are met.

MEANS OF EGRESS

At least two separate means of egress, as remote from each other as possible, must be available from every floor in a storage structure. In smaller buildings a single exit is permitted as long as the common path of travel limitations are not exceeded. The inspector should inspect the *exit access* from within the building, the *exit* locations, and the arrangement of the *exit discharge* from the exit to a public way or street (see Chapter 20, "Means of Egress").

Periodically the storage arrangements in storage occupancies are modified to keep up with new technology and operations, and these modifications can significantly affect the components of the building's means of egress. Without proper planning, exits can become blocked by storage, travel distances significantly increased, dead-end corridors created, and even exit discharge adversely affected by building additions, altered security measures, or changes to property lines.

Exit Access and Travel Distances

All paths of travel from any part of the building must allow the occupants to travel safely, without obstructions, to the exits.

Travel distances to the exit locations must be as follows:

	Nonsprinklered	Sprinklered
A. Low hazard occupancies	no limit	no limit
Ordinary hazard occupancies	200 ft (60 m)	400 ft (122 m)
High hazard occupancies	75 ft (23 m)	100 ft* (30 m)
B. Aircraft hangars and grain or bulk storage elevators	Special requirements are noted in the *Life Safety Code*	
C. Parking garages (mechanical or exclusively parked by attendants)	*See distance listed in A above*	
Parking garages (customers park own cars)		
Below ground or closed structure	150 ft (45 m)	200 ft (60 m)
Open-air structure	300 ft (90 m)	400 ft* (120 m)

*Additional requirements are noted in the *Life Safety Code*

When repair operations are conducted within a parking garage, travel distances must meet the requirements for the industrial section, Chapter 40, of the *Life Safety Code.* If the repair operation area is separated by 2-hour fire-resistive construction, then the industrial requirements will apply to only the part that is used for repair operations.

Any rearrangement of the storage aisles or additions or changes to the rack systems made since the occupancy's last inspection can greatly affect travel distances to exits. When new mezzanine levels are added for additional storage space, the inspector must check that travel distances to exits are correct and that the appropriate number of sprinklers have been added to the new area.

When determining exit access, the inspector should look for areas where dead ends or common paths of travel are created by the storage arrangement. There is no limit to either in occupancies with a low hazard classification. In new or existing storage occupancies with an ordinary hazard classification, a dead end or common path of travel of up to 50 ft (15 m) is allowed and up to 100 ft (30 m) is allowed if the building is protected by an automatic sprinkler system. No dead-end conditions are allowed in areas that are classified as high hazard. In parking garages a dead end or common path of travel of up to 50 ft (15 m) is allowed.

Exits and Locations

At least two means of egress are required from all floors and areas of the building in storage occupancies classified as ordinary and high hazard, and they must be located so that a person can reach an exit location within the allowable travel distances.

In any enclosed parking garage in which there are gasoline pumps, there are several exit requirements. Travel away from the pumps in any direction should lead to an exit, and there must be no dead ends where people could be trapped by a fire originating at the pumps. The exits must lead to the outside of the building at the same level as the pumps or they must lead to stairs. Any story below the one housing the pumps must have exits directly to the outside by means of outside stairs or doors at ground level.

In aircraft storage and servicing areas, there must be exits at intervals of every 150 ft (45 m) on exterior walls of the hangar and every 100 ft (30 m) along interior fire walls when these walls serve as horizontal exits. The travel distance to reach the nearest exit from any point from a mezzanine floor located in an aircraft storage or servicing area must not exceed 75 ft (23 m). Such exits must lead directly to an enclosed stairwell discharging directly to the exterior, to a suitable cutoff area, or to outside stairs.

In grain or other bulk storage elevators, there should be two means of egress from all working levels of the head house. One must be stairs to the ground that are enclosed by a dust-resistant 1-hour fire-resistive enclosure. The second means of egress can be exterior stairs or a basket ladder-type fire escape that is accessible from all working levels of the head house and provides access either to ground level or to the top of an adjoining structure that provides a continuous path to another exterior stairway or basket ladder-type fire escape leading to the ground level. The underground spaces of an elevator must have at least two means of egress, one of which can be a ladder.

You should check all doors that serve as a required means of egress and are identified as exits for free and unobstructed operation to ensure that these doors are kept unlocked when the building is occupied. If locks requiring use of a key for operation from the inside of the building are used, make sure that a readily visible sign is posted next to the door on the egress side of the door stating, "This door to remain unlocked when the building is occupied." The locking device should be readily distinguishable if locked.

The inspector should make sure that exit doors located in a high hazard area swing in the direction of exit travel. In areas where flammable vapors or gases are present, or the possibility of an explosion exists, the inspector should make sure that exit doors are equipped with panic hardware.

In low and ordinary hazard areas that are protected throughout by an approved, supervised, automatic fire alarm or automatic sprinkler system, exit doors can be equipped with approved, listed, special locking devices that meet the requirements of Chapter 7 of the *Life Safety Code.*

The *Life Safety Code* permits horizontal sliding doors to be part of a means of egress. It also permits the use of a horizontal exit or smoke barrier. There are, however, special requirements in Chapter 7 of the *Life Safety Code.* It is quite common to find horizontal exits in storage occupancies due to the use of fire walls or barriers for compartmentation purposes. When the horizontal exit doorway is protected by a fire door on each side of the wall, one door must be swinging and the other can be an automatic-sliding fire door complying with specific requirements involving fusible links as detailed in Chapter 42 of the *Life Safety Code,* that must be kept open when the building is occupied.

In parking garages the opening for the passage of automobiles can serve as an exit from the street floor, provided that no door or shutter is installed in the opening.

In storage areas that contain low or ordinary hazard contents and have an occupant load of not more than 10 people, exit doors that are not sidehinged swinging are permitted.

Exit Discharge

The inspector must determine that there is a continuous path of travel from the building exit to a public way and that there is nothing in front of the exit door that would prevent it from working. The inspector must also make sure the path of travel from exits opening into an alley leads to the public way, is well marked, and is illuminated.

Identification of Exits

All required exits and paths of travel to an exit must be identified properly by signs that are readily visible from any direction of exit access. Where the exit or the way to reach it is not visible to the occupants, the path of travel should be marked so that no point in the route is more than 100 ft (30 m) from the nearest visible sign.

In large warehouses with high storage, exit identification can be a problem. Thus the inspector might want to suggest that exit signs be of sufficient size for visibility or that the travel paths to exits be painted on the floor.

The inspector should check that exit access routes are illuminated to allow the occupants to safely exit the building. If natural light is not available during the fire the building is occupied, the illumination must be continuous when the building is occupied. Emergency lighting is required in storage occupancies that are occupied at night or do not have exterior openings that would provide the required illumination during daylight hours. The inspector should check that emergency lighting operates when the normal lighting circuits for the affected area are turned off. When a generator is used to power emergency lighting, the generator should transfer power and should operate emergency lighting within 10 seconds. The inspector should check the records and, if possible, be present for a generator test to be sure it runs correctly. When checking battery-powered lighting units, the inspector should check for acid corrosion, the water level of wet-cell batteries, and that the unit is fully charged and operational.

Protection of Openings

Inspectors should check that fire doors operate correctly, that they close tightly, and that the self-closing devices work. They should also check the general condition of the doors for obvious damage. Nothing that would prevent counterbalance closing hardware from operating freely should be stored around the fire doors, and nothing should block the doors open. Inspectors should check that all door hardware and closing devices are lubricated and move freely. They should examine all fusible links associated with the closing hardware to see that they are positioned properly and have not been painted or wired together.

Inspectors must carefully check materials-handling conveyor systems that pass through fire walls. Is there any air-handling ductwork passing through fire walls? NFPA 90A, *Standard for the Installation of Air-Conditioning and Ventilating Systems,* requires any ductwork passing through a 2-hour fire wall to be protected by fire dampers. The inspector should check that all openings through the walls made for electrical cables or conduits are tightly sealed with a material that affords the same fire-resistance rating as the fire barrier. Where storage buildings are susceptible to exterior exposure problems, the inspector should check that fire shutters and wired-glass windows work and are properly placed. All wired glass that is missing or cracked should be replaced. The inspector should verify that installed roof vents are operating properly and that snow is not allowed to accumulate on the vent hatches during the winter.

GENERAL STORAGE PRACTICES

Indoor

In NFPA 230, *Standard for the Fire Protection of Storage,* commodity storage, which is defined as both pile and rack storage, has been grouped into four classes for ordinary commodities and three classes for plastics according to how easily automatic sprinklers will control a fire in them. The inspector should check this standard to determine the proper storage arrangement in buildings protected by automatic sprinkler systems.

Inspectors should make sure that materials that could be hazardous in combination are stored so they cannot come in contact with each other. They should verify that safe loads for floors are not exceeded. Floor loads for water-absorbent materials should be reduced to account for the added weight of water absorption during a fire. The inspector should check the clearance of stored material from sprinklers, heat ducts, unit heaters, duct furnances, flues, radiant space heaters, and lighting fixtures. The wall aisle space should be at least 24 in. (61 cm) in storage areas where materials that will expand with absorption of water are stored. Inspectors should verify that aisles are maintained to keep fire from spreading from one pile to another and to permit access for fire fighting, salvage, and removal or storage. They should also verify that all automatic sprinkler control valves, hose stations, and portable fire extinguishers are accessible and that there is free access to all fire protection equipment. All unused wood or plastic pallets should be kept outside or stored in stacks no higher than 6 ft (1.8 m). During the inspection, the inspector should check to see whether exterior access doors and windows are being blocked with storage that would affect fire-fighting operations and prevent access into the building.

Outdoor

The inspector should make sure that storage piles are not stacked too high and are in stable condition and that aisles are sufficiently maintained between individual piles, between piles and buildings, and between piles and boundary lines of the storage site. The inspector should also note whether the entire property is enclosed with a fence or some other means of keeping unauthorized persons from entering. There should be a gate to allow fire equipment to enter the area in the event of a fire. The storage yard should be free of unnecessary combustible materials, weeds, and grass, and any tarpaulins used to cover materials should be made of fire-retardant fabric.

HAZARDOUS MATERIALS

Many different materials with different hazards can be stored in a storage occupancy, or the entire occupancy can be used to store a specific hazardous material. Special requirements for the storage of hazardous materials are found in several *National Fire Codes*. The inspector should be able to recognize out-of-the-ordinary storage and refer to the appropriate standard to determine special storage arrangements and protection requirements. Examples of hazardous materials include rubber tires, plastic products, combustible fibers, paper and paper products, hanging garments, carpeting, pesticides, flammable liquids and gases, reactive chemicals, and flammable aerosol containers. Storage of aerosol containers should meet the requirements of NFPA 30B, *Code for the Manufacture and Storage of Aerosol Products*. All storage occupancies should be properly identified on the outside of the building using NFPA 704, *Standard System for the Identification of the Hazards of Materials for Emergency Response.*

Industrial Trucks

The inspector should determine that the industrial trucks being used are approved for use within the building for the hazard of the materials being stored. NFPA 505, *Fire Safety Standard for Powered Industrial Trucks Including Type Designations, Areas of Use, Conversions, Maintenance, and Operation,* designates the types of trucks that can be used in hazardous areas. A fire extinguisher that can be used on flammable liquid and electrical fires should be mounted on each truck. The inspector should check to see that the trucks are being maintained, that all refueling operations are conducted outside the building, and that fuel for the trucks is properly stored. The inspector should also examine the area where batteries are recharged for electrical trucks. Areas used for the repair of trucks should be separated from the storage area.

Hazardous Processes

The inspector should check the precautions management takes when a welding or cutting operation takes place in the storage area. In some cases, these operations should not be allowed at all until the hazardous materials are removed from the area. During welding operations, all combustible materials located below the operation should be removed or covered with a fire-retardant cover. Portable fire extinguishers and small hose lines should be laid out ready for operation. A fire watch should be present at all times during the operation and at least 30 minutes after the welding or cutting is completed.

If fuel pumps are located in a parking garage, the inspector should check that the dispensing unit and nozzle are approved and that no ignition sources are located within 20 ft (6 m) of the dispensing area. When the dispensing units are located below grade inside the building, the entire dispensing area must be protected with approved automatic sprinkler systems. The inspector should make sure there is mechanical ventilation for the dispensing area to remove flammable vapors and the mechanical ventilation system is electrically interlocked with the dispensing unit so that no dispensing can be conducted without the ventilation system being in operation.

HOUSEKEEPING

The inspector should look for debris and trash accumulated in out-of-the-way places and neglected corners. The level of fire safety is greatly improved when areas are kept clean and neat. All waste generated daily should be removed from the building and disposed of in a safe manner outside the building. The inspector should check for the accumulation of dust and lint on sprinklers, on fire door self-closing hardware, and around electrical motors and compressors. All containers used for the disposal of waste material must be made of noncombustible materials and have lids.

In grain storage buildings, the single most important fire prevention practice is effective daily removal of dust, which will collect everywhere. Housekeeping in this type of occupancy should be done consistently and carefully.

FIRE PROTECTION

Sprinkler Systems

A major factor in large fire losses in storage buildings has been the overtaxing of the automatic sprinkler system or associated water supply, which was or became improperly designed for the type of material stored and the storage arrangement. Fire losses have also occurred because the sprinkler system water supply has been shut off. Sprinkler plans should be checked to determine the hazard for which the system was designed to protect. During the inspection, the inspector should check whether the material being stored or the storage arrangement has changed in a way that would require redesign of the sprinkler system. Detailed inspection records should be kept indicating the type of material stored, pile arrangement, aisle width, storage methods, and height of storage materials. The inspector should pay special attention to buildings protected by older sprinkler systems designed for Class III or lower commodities that now store more hazardous materials or have more hazardous storage arrangements.

The inspector should verify that the sprinkler control valves are accessible, not blocked by storage, and in the open position. The inspector should also verify that the system has been maintained properly, is in working order, and that all alarms operate when tested, properly identifying the alarm/supervisory condition. The outside fire department sprinkler connection must not be blocked by storage. The inspector should check for any areas unprotected by sprinklers such as small office enclosures with ceilings, mezzanines, or blind combustible spaces.

Standpipe Hose Systems

All hose stations and standpipe connections must not be blocked by storage materials. There should be adequate hose stations so that all areas of the storage buildings can be reached by the hose stream. All hose stations should be identified properly. The inspector should check the condition of the hose and nozzles and should find out whether employees are expected to use this equipment and, if so, whether they are trained properly for using the standpipe hose system.

Fire Extinguishers

The inspector should determine that fire extinguishers are accessible, that they are the correct type for the hazard, and that their locations are identified. All extinguishers should be fully charged and inspected at least annually. Employees should be trained to use the extinguisher correctly.

Fire Pumps

The fire pump should be examined to determine whether it is being properly maintained and tested and whether it is being run weekly and tested by a competent contractor at least annually. The inspector should verify that all alarms operate when tested, and he or she should properly identify the alarm/supervisory condition. The inspector should determine whether the pump is set for automatic or manual operation and should check that all controls are working. The inspector should also determine that proper documentation of testing and maintenance is being maintained. Chapter 15, "Automatic Sprinkler and Other Water-Based Fire Protection Systems," addresses the requirements for the inspection, testing, and maintenance of water-based fire protection systems such as automatic sprinklers.

Alarm System

Because storage buildings are usually large open-floor areas that are occupied by only a few employees who are working in many different parts of the building, they should have a fire alarm system that, when operated, will sound an alarm at a continuously attended location so that some type of emergency action can be initiated. If the occupancy has a trained industrial fire brigade or an emergency prefire plan, there should be a means of notifying people in all areas of the building so that the brigade or action plan can be initiated.

The *Life Safety Code* requires a fire alarm system in storage buildings when they contain either ordinary or high hazard contents and have an aggregate floor area of more than 100,000 ft^2 (9300 m^2). This requirement would not apply if the occupancy is protected by an automatic extinguishing system.

A fire alarm system is also required in public parking garages except when the parking structure is classified as an open-air structure. When inspecting this type of alarm system, the inspector should make sure that the entire system and all functions are in operating order and that all initiating devices, such as manual fire alarm boxes, are identified and not blocked by storage. The inspector should also check that the notification alarm or signal is adequate to notify all employees that are part of an industrial fire brigade or to activate the established, emergency prefire plan.

SUMMARY

This chapter has identified the criteria necessary for properly classifying storage occupancies. It has also provided discussion on aspects of occupant loads, means of egress, the protection of vertical openings, and fire protection systems. In addition, the chapter has examined sale practices for indoor and outdoor general storage, hazardous materials and processes, aircraft hangers, and parking garages.

BIBLIOGRAPHY

NFPA Codes, Standards, and Recommended Practices

See the latest version of The NFPA Catalog for availability of current editions of the following documents.

NFPA 13, *Standard for the Installation of Sprinkler Systems*
NFPA 30, *Flammable and Combustible Liquids Code*
NFPA 30B, *Code for the Manufacture and Storage of Aerosol Products*
NFPA 42, *Code for the Storage of Pyroxylin Plastic*
NFPA 61, *Standard for the Prevention of Fire and Dust Explosions in Agricultural and Food Products Facilities*
NFPA 88A, *Standard for Parking Structures*
NFPA 90A, *Standard for the Installation of Air-Conditioning and Ventilating Systems*
NFPA *101*®, *Life Safety Code*®
NFPA 220, *Standard on Types of Building Construction*
NFPA 230, *Standard for the Fire Protection of Storage*
NFPA 480, *Standard for the Storage, Handling, and Processing of Magnesium Solids and Powders*
NFPA 481, *Standard for the Production, Processing, Handling, and Storage of Titanium*
NFPA 482, *Standard for the Production, Processing, Handling, and Storage of Zirconium*
NFPA 490, *Code for the Storage of Ammonium Nitrate*
NFPA 505, *Fire Safety Standard for Powered Industrial Trucks Including Type Designations, Areas of Use, Conversions, Maintenance, and Operation*
NFPA 513, *Standard for Motor Freight Terminals*
NFPA 704, *Standard System for the Identification of the Hazards of Materials for Emergency Response*
NFPA 1124, *Code for the Manufacture, Transportation, and Storage of Fireworks and Pyrotechnic Articles*

Special Structures and High-Rise Buildings

CHAPTER 37

Joseph Versteeg

CASE STUDY

Electrical Fire Occurs at Rockefeller Plaza

At approximately 4 A.M. an electrical fire occurred at Rockefeller Plaza in New York City. The Rockefeller Center is a complex of interconnected buildings, with the highest being 70 stories. The building in which the fire occurred was 11 stories high and was equipped with a fire alarm system composed of smoke detectors (ceiling and duct), pull stations, and flow switches. It was monitored by an on-site security staff. The ground floor and lower levels were sprinklered, but the upper stories were not. Fire standpipes were located in the stairwells and within the floors.

Several fires broke out in five remote locations, filling many areas of the building with smoke. The entire building was evacuated, live broadcasts from a major television network were interrupted, and traffic movement was impacted for several blocks around the building. Since the fire occurred early in the morning, relatively few people were in the building. Everyone was able to evacuate successfully.

This high-rise building fire presented a challenge for the fire department because of the varied locations of simultaneous fires and the confusing layout of the building. The fire department received a telephone report of the fire and dispatched a full first alarm assignment. Arriving fire fighters saw smoke coming from one of the upper floors of the building. The on-site building security staff, however, was not aware of the fire nor were there any alarms indicated on the fire alarm system. Fire fighters who entered the building were receiving reports from civilians of fire on floors 7 through 10. It was later determined that the area of origin was an electrical room on the fifth floor where service entered the building. Fire had become deeply entrenched in the electrical wiring system in five separate electrical rooms. Cabling within the rooms ignited, generating heavy quantities of smoke.

The fire department ultimately transmitted five alarms, bringing approximately 300 fire fighters to the scene. Despite the large amount of fire suppression resources that were committed, it took fire fighters approximately 4 hours to extinguish the fire. Five civilians and 12 fire fighters were injured. Significant contributing factors to the loss of property in this incident included the following:

1. Inadequate circuit protection
2. Lack of adequate space for electrical conductors to safely dissipate heat
3. Unprotected vertical and horizontal penetrations
4. Lack of sprinkler protection in the areas of the fires
5. Lack of smoke detection in the areas of the fires
6. Failure of the building fire alarm system to transmit the alarm

Joe Versteeg is a fire protection consultant based in Torrington, CT. He was with the Connecticut Department of Public Safety for over 20 years; most recently he was the commanding officer of the technical services section of the State Fire Marshal's Office. He also serves on several Life Safety Code® *committees.*

7. Confusing building layout
8. Multiple points of origin

Source: NFPA Fire Investigations Department, *Fire Investigation Summary, Rockefeller Center, New York City, October 10, 1996*, NFPA, Quincy, MA, 1996.

Chapter 11 of NFPA *101*®, *Life Safety Code*®, 2000 edition, governs open structures, piers, towers, windowless buildings, underground structures, vehicles, vessels, water-surrounded structures, membrane structures, tents, and high-rise buildings.

1. *Open structures:* Open structures are those in which operations, accompanied by their necessary equipment, are conducted in open air and not enclosed within building walls, such as those used for oil refining and chemical processing plants. Roofs or canopies that provide shelter without enclosing walls can be provided and would not be considered an enclosure.
2. *Piers:* Piers are structures projecting from the shore into a body of water.
3. *Towers:* Towers are independent structures or portions of buildings occupied for observation, signaling, or similar limited use and not open to general use.
4. *Windowless buildings:* Windowless buildings lack any means for direct access to the outside from the enclosing walls, or lack outside openings for ventilation or rescue through windows. (Exceptions for windowless buildings are discussed later in this chapter.)
5. *Underground structures:* Underground structures are structures or portions of a structure in which the story is below the level of exit discharge. (Exceptions for underground structures are discussed later in this chapter.)
6. *Vehicles:* Vehicles are any trailer, railroad car, streetcar, bus, or similar conveyance that is not mobile or is attached to a building or is permanently fixed to a foundation.
7. *Vessels:* Vessels are any ship, barge, or other vessel permanently fixed to a foundation or mooring or unable to get under way under its own power and occupied for purposes other than navigation.
8. *Water-surrounded structures:* Water-surrounded structures are structures completely surrounded by water.
9. *Membrane structures:* Membrane structures are buildings or portions of a building incorporating an air-inflated, air-supported tensioned-membrane, membrane roof or membrane-covered rigid frame to protect a habitable or usable space.
10. *Tents:* Tents are typically erected on a temporary basis and consist of a pliable covering supported by beams, columns, poles, arches, or by ropes and/or cables.
11. *High-rise buildings:* High-rise buildings are buildings having an occupiable story 75 ft (23 m) or more above the lowest level of fire department vehicle access.

The requirements unique to each of the different types of special structures identified within Chapter 11 of the *Life Safety Code* are to be applied in addition to the requirements mandated by the applicable occupancy chapter. One example is a bank records keeping facility located partly in the subbasement of a new office building and partly on the fourth floor of the same building. The Chapter 38 requirements applicable to new business occupancies must be met for both parts of the facility. Additionally, the portion in the subbasement must meet the provisions of Section 11.7 applicable to underground and windowless structures.

STRUCTURE CHARACTERISTICS

A typical occupancy in a special structure will fall under one of the occupancy classifications previously discussed. Thus, the means of egress and fire protection requirements for that occupancy apply, as do the additional requirements for the special structure. Examples of typical occupancies in special structures include the following:

- A restaurant located in a permanently fixed boat or ship. This is clearly an assembly occupancy (the occupancy classification), which is located in a ship (the special structure).
- A souvenir shop in an earth bunker or cave. This is a mercantile facility (the occupancy classification) in an earth enclosure (the special structure).
- A hotel supported on pilings along a waterfront that is connected to the mainland by access ramps. This is a hotel (the occupancy classification), which is housed on pilings (the special structure).

When inspecting a special structure, which can be difficult, the inspector must first determine which general occupancy classification it falls under. Then it will be easier to determine the structure's special features and the requisite applicable portions of Chapter 11 of the *Life Safety Code.* If the special structure has protection needs not fully addressed by the applicable occupancy chapter, the inspector should use the fundamental provisions given in Chapter 4 of the *Life Safety Code* to evaluate the constraints the structure's special features place on properly protecting the occupants.

SPECIAL STRUCTURES

This section explains the unusual conditions inspectors will see and the specific observations that they will have to make when inspecting special structures. They should use this information in conjunction with the information contained in chapters on the specific occupancy housed in the special structure.

Open Structures

Open structures are common in industrial operations. Therefore, most of the time the structure will come under the guidelines of those for industrial occupancies. The primary problem for the inspector will be determining whether the occupancy is a general, special purpose, or high hazard industrial occupancy. Typically, the high hazard industrial occupancy classification will be determined based on the hazard classification of the building's contents and processes.

Open-air structures are classified more commonly as special purpose industrial occupancies, which are defined in Chapter 40 of the *Life Safety Code.* Special purpose industrial occupancies have a relatively low density of employees, and much of an open-air structure is occupied by machinery or equipment. Typically, the open structure facilitates access to the equipment with platforms, gratings, stairs, and ladders. The structure might even have a roof to provide some protection from the elements (Figure 37-1).

It is difficult to determine when a structure is open and when it is enclosed. Many open structures will have some walls that are intended to shield the operations from environmental conditions or to segregate operations. The authority having jurisdiction must determine whether or not the open-air structure truly meets the definition for open structures or if the facility is, in fact, a building.

To determine whether the structure really is an open structure, the inspector must decide whether it would react as an enclosed building in the event of a fire. Walls acting in conjunction with a roof enclose the combustion process and products of combustion, thereby allowing the fire to spread both horizontally and vertically to unaffected areas within the building. However, an open structure allows the products

FIGURE 37-1 An Open Structure

of combustion to vent to the atmosphere instead of spreading to unaffected areas of the structure. Obviously, this is dependent on wind and climatic conditions, but the overall concept is valid. Could the walls cause the products of combustion during a fire to be directed or channeled to other unaffected portions of the structure, thus preventing them from venting to the atmosphere? If the answer to this question is "yes" or "probably," then the inspector might have to consider that portion of the structure as a building.

Once inspectors have classified the structure, they should conduct the inspection based on the occupancy classification(s) and the special provisions provided in Chapter 11, "Special Structures and High-Rise Buildings," of the *Life Safety Code.* There are no special provisions for open-air structures, but Chapter 11 contains several exceptions for this form of special structure. One of them permits open structures to have a single means of egress when they are occupied by no more than three persons and have a travel distance of not more than 200 ft (61 m).

A fire that occurs in an open structure usually does not pose as serious a threat to life as does a fire in a building, unless the structure contains high hazard operations. When you are inspecting an open structure that houses a high hazard operation, the inspector should refer to Chapter 40 of the *Life Safety Code* and carefully evaluate the life safety features of the open structure.

When high hazard operations are involved, the egress system usually becomes the biggest problem, especially on levels that are above grade. No simple generalizations can be made about this problem; therefore, each individual structure must be evaluated on its own merits and the level of protection being provided.

Piers

A pier, as defined by NFPA 307, *Standard for the Construction and Fire Protection of Marine Terminals, Piers, and Wharves,* is a structure, usually longer than it is wide, that projects from land into a body of water. It can be either open deck or have a superstructure. Contrary to its exact definition, piers can be constructed over land.

Occasionally, a designer will construct a pier of earth that pushes its way out from the main body of land into an area, which, for elevation reasons, might not be usable otherwise. A building that can be classified under one of the other categories covered in the *Life Safety Code* will be erected on this earthen pier. Although the pier is not made of traditional structural elements, it is still a pier. The inspector must first determine the occupancy requirements for the building and then assess the affects the unusual structure, the pier, will have on them. Many times a structure is located within what first appears to be an unusual circumstance, but it will still probably have the typical fire protection problems.

Chapter 3 of NFPA 307, 2000 edition, contains requirements for property conservation features of piers and wharves, and Chapter 30 of the *Life Safety Code* also discusses piers. An exception to the number of exits required is noted in Section 11.5.2 of Chapter 11: "Piers used exclusively to moor cargo vessels and to store materials shall be exempt from number of means of egress requirements where provided with proper mean of egress from structures thereon to the pier and a single means of access to the mainland, as appropriate with the pier's arrangement."

Aside from this exception, the appropriate occupancy chapter and any other applicable provisions of Chapter 11 would apply to piers.

Towers

A tower, as defined in Chapter 3 of the *Life Safety Code*, is "an enclosed independent structure or portion of a building with elevated levels for support of equipment or occupied for observation, control, operation, signaling, or similar limited use . . . " (Figure 37-2). An assembly occupancy or any other occupancy classification located on top of a tower must meet the requirements of the appropriate occupancy classification chapter of the *Life Safety Code*, not necessarily those of Chapter 11 of the *Life Safety Code*.

Chapter 11 of the *Life Safety Code* contains requirements for towers that are typically used for purposes such as forest fire observation, railroad signaling, industrial

FIGURE 37-2 A Tower

purposes, and aircraft control. Normally they are not occupied or are occupied by a limited number of persons who are capable of self-preservation. Usually there are no provisions for living or sleeping in such towers. When inspecting towers that meet the requirements of Chapter 11 of the *Life Safety Code,* the inspector must keep in mind the limited use and the nature and character of the specific occupancy involved. If the tower is used as an assembly occupancy or contains more than just a few occupants, it probably does not meet the requirements of Chapter 11 and should be reviewed wholly under the requirements of the appropriate occupancy chapter.

The *Life Safety Code* does not require towers to have any special features for means of egress, but it does provide several exceptions to general requirements for means of egress in Chapter 11, some of which are described briefly here.

The code permits ladders to be used as a means of egress when the tower is occupied by three or fewer people.

The capacity and width of the means of egress have to provide only for the expected number of persons occupying the tower. (See Chapter 11 of the *Life Safety Code* for all of the exceptions. Towers are permitted to have a single exit if the following conditions are met:

1. The tower is occupied by less than 25 persons.
2. The tower is not used for living or sleeping purposes.
3. The tower is of Type I, II, or IV construction.
4. The tower interior finish is Class A or B.
5. The tower has no combustible materials in, under, or in the immediate vicinity, except necessary furniture.
6. There are no high hazard occupancies in the tower or immediate vicinity.

Where the tower sits atop a building, additional criteria must be met to satisfy the provisions that permit a single exit from the tower portion of the building.

The average tower will meet these conditions, but some unusual towers, such as the Washington Monument and the Statue of Liberty, might not do so. When deciding whether the single exit provision applies, the inspector must keep in mind the actual configuration of the tower, the fuel load, the exposure of the tower, and the protection it has. Usually the use of the single exit provision can still be justified when an overall evaluation is completed.

During the inspection the inspector must determine if the tower is exposed to any combustible materials that might be under or in the immediate vicinity of the structure. The inspector or the authority having jurisdiction should not establish arbitrary requirements that would severely restrict the use of the tower; however, the inspector should ensure that the tower could not be exposed to an exterior fire severe enough to affect the means of egress system before the occupants could evacuate.

The inspector should also ensure that any high hazard occupancy located close to a tower will not threaten the integrity of the physical structure of the tower and the egress system. Included are not only hazards such as LP-Gas storage and explosive-prone occupancies, but also possible exposures to vehicles, such as flammable liquid transport carriers.

Windowless Buildings

The hazards of windowless buildings are very similar to those of underground structures. Therefore, the same concerns about underground structures would also relate to windowless structures. The significant difference from a code standpoint is that windowless structures are not required to have smoke-venting facilities. However, if the occupancy will have a high occupant load the inspector might suggest that a smoke-control and smoke-venting system be installed, if feasible, because it will greatly improve the level of life safety.

Typically, it is not difficult to have a good exit system in a windowless building. This type of structure lends itself to the use of exterior access panels for fire department and emergency use; however, access panels not meeting the requirements of the means of egress cannot be considered as exits.

The most common problem with windowless buildings is determining whether the building is actually considered windowless.

The definition of a windowless building has two parts, that is, it addresses two considerations. The first part of the definition states that a windowless building is a structure or a portion of a structure that lacks a means for direct access to the outside from the enclosing walls. This part of the definition involves determining the occupants' ability to exit the building. This is not typically a problem unless the designer intentionally deleted exits along the exterior of the building, which is commonly done for access control and security purposes. As important as security and access control are, they cannot be allowed to interfere with a minimum level of life safety in the building.

The second part of the definition addresses outside openings for ventilation or rescue through windows. The problem thus becomes determining how many windows a building can have before it is no longer considered a windowless structure.

The inspector should use judgment when determining whether the building is truly windowless. The inspector's decision should also take into consideration the nature and character of the occupancy and the actual occupancy classification. It is not as critical for the life safety requirements to be enforced in a windowless building in a storage occupancy as it is in a windowless building in an assembly occupancy.

Underground Structures

The *Life Safety Code* defines an underground structure as any structure or portions of a structure in which the story is below the level of exit discharge. At face value, any structure that has a floor level below the level at which the occupants exit is an underground structure. Since 1985, there are virtually no differences between an underground structure and a basement where, in both cases, the exit system requires the occupant to travel up from the occupied floor to exit.

The *Life Safety Code* provides additional guidance by stating that a structure is not considered an underground structure if escape openings of given minimum size are provided at given spacings along a required portion of the building's perimeter. See *Life Safety Code* Section 11.7 and Chapter 3's associated definition of "emergency access opening."

In summary, if there are sufficient openings around the outside that are clearly identifiable and usable as rescue and ventilation openings, then that floor level is not required to meet the requirements of an underground structure. During the inspection the inspector should keep in mind the following problems that arise in an underground structure: (1) safe egress, (2) accessibility for fire department and rescue operations, (3) ventilation and smoke control, and (4) control and suppression of fire. These issues will guide the inspector in identifying problems and where the appropriate code issues apply.

When inspecting an underground structure, the inspector should first look for compliance with all the requirements that apply to the same occupancy if it were housed in a structure erected on or above grade. The occupancy classifications, hazard levels of contents, occupant loads, exit systems, vertical openings, and other features of fire protection should be consistent with the appropriate occupancy chapter in the *Life Safety Code*.

Next the inspector should turn to Section 11.7 of the *Life Safety Code,* which gives additional special provisions for underground structures. Each exit from an underground structure with an occupant load over 100 persons and having a floor level more than 30 ft (9 m) below the level of exit discharge must have outside smoke-venting facilities or some other means of preventing the exits from becoming filled with smoke from any fire in the area served by the exits. Underground structures with an occupant load over 100 persons must have automatic smoke-venting facilities if areas in them have combustible contents, interior finish, or construction; however, there is an exception to this requirement for existing structures.

An underground structure with an occupant load over 50 persons (100 persons for existing) must be protected by an automatic sprinkler system installed in accordance with NFPA 13, *Standard for the Installation of Sprinkler Systems.*

Because there is virtually no natural light in an underground structure, this type of structure must have an emergency lighting system that meets the requirements of appropriate sections of the *Life Safety Code.* The inspector should verify that the emergency lighting system operates and provides a level of light consistent with the minimum specified by the code.

The inspector should also check the emergency power supply to ensure that all air-handling and other electrically operated emergency mechanical support equipment will function if there is a power failure in the event of a fire. The inspector should witness or ask to see results from tests of all smoke-venting systems in the means of egress to ensure they are working properly. The inspector should also ensure that all protection systems are being supervised.

In addition to meeting the specialized provisions of Section 11.7 detailed for underground structures, the underground structure must also comply with the provisions applicable to windowless structures.

Vehicles and Vessels

Occasionally inspectors will have to inspect an occupancy that is located in a vehicle or vessel. The occupancy classification most commonly encountered will be an assembly occupancy or a mercantile occupancy, but it could be another occupancy, depending on the type of vehicle or vessel involved. An aircraft, boat, or ship could have been taken out of service, placed on a foundation, and turned into a restaurant. One national restaurant chain places railroad cars on fixed railroad tracks that are placed on foundations. The cars are then arranged and altered to accommodate a dining establishment.

Following are some specific points to cover when an occupancy in a vehicle or vessel is inspected.

Permanent Foundation or Mooring. The vehicle or vessel that is no longer mobile and is permanently fixed to a foundation or mooring will determine how the actual occupancy is regulated under the codes. In the United States in the case of a floating restaurant that is aboard a traditionally functional ship, the occupancy typically would not be under the regulation of a fire department or building department inspection agency but would most likely be regulated by the U.S. Coast Guard.

Unusual Nature and Character. Although the vehicle or vessel might house a traditional assembly occupancy such as a restaurant, the unusual configuration of the structure will often cause the occupants to have to respond differently in the event of an emergency. In order to accommodate the occupancy in this vehicle or vessel and create a pleasant, attractive atmosphere, some renovations in direct contradiction to a more traditional design for the occupancy might be made. The changes, which make the structure marketable and attract customers, might compromise safety. Although the inspector should try to accommodate the unique features of the occupancy, these features must not compromise life safety.

Chapter 4 of the *Life Safety Code,* which contains the fundamental requirements of the code, will often be the only resource that the inspector has to judge a certain configuration or condition. This chapter will help him or her to interpret the intent of the occupancy and determine that the required level of life safety is provided for the occupants.

Means of Egress. Because the vehicle or vessel was often not originally designed for the occupancy that it now contains, it is unlikely that it will comply with the *Life Safety Code* requirements for the means of egress. There are also additional exceptions for structures fully surrounded by water, which are explained later in this chapter.

Typical and/or Special Hazard Areas. Because of the character of special structures, Section 11.6 of the *Life Safety Code* allows the authority having jurisdiction to require whatever is necessary to ensure that egress systems work properly. This could include requiring automatic water suppression systems, fire alarm systems, modified sprinkler systems, or compartmentation. This methodology will often offset the design problems generated by a special structure's unique character and provide a reasonable level of life safety.

The inspector should look for any unusual approaches to or applications of a protection system intended to provide an equivalent level of life safety. Whatever method is used must be clearly documented and properly installed and maintained to ensure it will be effective throughout the existence of the occupancy.

In addition to those items specified in Chapter 11 of the *Life Safety Code,* the inspector must also determine whether the hazards that existed in the vehicle or vessel before its present occupancy use have been mitigated or adequately protected. Items such as fuel tanks, hydraulic machinery, and electrical equipment not relevant to the structure's present use must be removed or adequately protected. Any materials, hazardous or not, stored in preexisting compartments must be identified and their potential contribution to the fuel load assessed. The inspector should check every area of the structure, looking carefully for any cosmetic feature that could conceal hazardous compartments or processes.

Because of the originally intended use of a vehicle or vessel, the designer of the present occupancy might not have considered how stable the structure would be in the event of a fire. Thus, the structure's supporting system might not perform as intended under fire conditions. Although a structure originally intended to be an assembly occupancy might be supported on a noncombustible foundation and fire-resistive supports, it would be unlikely to find a ship or vessel supported on a noncombustible foundation and structural supports.

Also, the structure's original foundations and supports may create concealed combustible spaces, and these spaces could contain emergency equipment and systems, which would adversely affect their integrity in a fire. When conducting the inspection, the inspector must determine that the structure's foundation meets code requirements, and he or she must assess its reaction in a fire and how this will affect the occupant's life safety.

Water-Surrounded Structures

Water-surrounded structures are completely surrounded by water, unlike piers where one side of the structure is attached to land. Typically, a water-surrounded structure can only be reached by a vehicle, but occasionally it might have a ramp that cannot qualify as a pier. Although Chapter 11 of the *Life Safety Code* contains no specific requirements for water-surrounded structures, it does have several exceptions that refer to U.S. Coast Guard regulations for water-surrounded structures.

Before inspecting a water-surrounded structure, the inspector should first determine whether the U.S. Coast guard has jurisdiction over it; if it does, the inspector should try to conduct a joint inspection with U.S. Coast Guard authorities. If this is not possible, then the inspector should try to obtain previous inspection reports from the U.S. Coast Guard.

The traditional occupancy requirements and features of fire protection remain the same for water-surrounded structures. The inspector should inspect the occupancy as if it were not surrounded by water and should then apply the special circumstances that are unique to the structure.

Membrane Structures

A membrane structure is defined in Chapter 3 of the *Life Safety Code* as "a building or portion of a building incorporating an air-inflated, air-supported, tensioned-membrane structure; a membrane roof; or a membrane-covered rigid frame to protect a habitable or usable space." Although the use of membrane structures on a temporary basis (less than 180 days) has been a common occurrence, it is becoming increasingly more common for these types of structures to be erected on a permanent basis (more than 180 days). They are often used for year-round sporting activities such as tennis or golf driving ranges, for music pavilions, or for rooftop dining.

As with any type of special structure, the use occurring within the structure must be properly classified and comply with the appropriate occupancy chapter. Once the occupancy has been determined to be compliant, the additional requirements regarding the structure need to be satisfied. The requirements governing permanent and temporary membrane structures are found within Sections 11.9 and 11.10, respectively, of the *Life Safety Code*.

In general, the *Life Safety Code* governs the flame resistance of the membrane material by requiring all exposed membrane materials to have a flame-spread index of Class A. In addition, membranes that compose the roof must have a roof covering classification as required by the local building code, when tested in accordance with NFPA 256, *Standard Method of Fire Tests of Roof Coverings*. The *Life Safety Code* also requires the design of tension membrane structures to be prepared by a licensed architect or engineer knowledgeable in tension-membrane construction, with consideration given to the various loads the structure will encounter.

Air-supported or air-inflated structures are subject to the same requirements that govern flame resistance as are tension-membrane structures. The former are also subject to requirements governing the reliability of the pressurization or inflation system. In all such structures, the *Life Safety Code* establishes limitations on the use of fuel-fired and electric heaters.

Tents

A tent is defined in Chapter 3 of the *Life Safety Code* as "a temporary structure, the covering of which is made of a pliable material that achieves its support by mechanical means such as beams, columns, poles, or arches, or by rope or cables, or both." Just as with any other structure, the occupancy occurring within the tent must be properly classified and determined to be compliant with the appropriate occupancy chapter.

FIGURE 37-3 A High-Rise Building

In addition to placing requirements on the flame resistance of the tent fabric and on the use of fuel-fired or electric heaters, the *Life Safety Code* also places prohibitions on the presence of dried vegetation and other combustibles within or near the tent. In the event that multiple tents are erected on a site or a tent is erected on a site containing a permanent structure, the *Life Safety Code* establishes minimum separation distances.

In addition to the requirements within the *Life Safety Code*, the inspector must be aware of the hazards associated with the particular use of a temporary tent. Large tents are often used for weekend craft fairs, where it is not uncommon to find artisans using open flame torches or individuals providing cooking demonstrations using propane-fueled appliances. Circus tents also use large tents and may feature acts incorporating open flames and pyrotechnics.

HIGH-RISE BUILDINGS

The additional protection features for a high-rise building, that is, a building having an occupiable story 75 ft (23 m) or more above the level of fire department vehicle access, are based on the fact that the height of the building poses unique challenges for both the fire department as well as the occupants (Figure 37-3). The height of the building typically exceeds the reach of most fire department aerial apparatus access, which causes a delay in the initial manual suppression activities. Also, occupants are often relocated within the building during an emergency as it may often be impractical for all occupants to egress in a timely manner.

SUMMARY

Occupancies in special structures are not easy to inspect, but it is imperative to inspect them in a logical manner using the appropriate occupancy chapter, along with Chapter 11 of the *Life Safety Code* and any other applicable NFPA standards. It is rare for only a single code to be applicable.

Even if the structure's building plans were processed through an appropriate plans review and approval process, it is very common for the structure and occupancy to change with time. Therefore the inspection results might differ from those of the original requirements or observations. Every inspection is worthy of the inspector's best professional expertise and judgment. The objective is to identify, interpret, and record all aspects relative to life safety, property conservation, and the preservation of the public welfare. A fire inspector is one of the few professionals whose product, when produced with experience, sound judgment, and good technical skills, can literally save lives.

BIBLIOGRAPHY

NFPA Codes, Standards, and Recommended Practices

See the latest version of The NFPA Catalog for availability of current editions of the following documents.

NFPA 13, *Standard for the Installation of Sprinkler Systems*
NFPA *101*®, *Life Safety Code*®
NFPA 256, *Standard Method of Fire Tests of Roof Coverings*
NFPA 307, *Standard for the Construction and Fire Protection of Marine Terminals, Piers, and Wharves*

SECTION 4 Process and Storage Hazards

CHAPTER 38 Waste-Handling and Processing Systems *Sharon S. Gilyeat*

CHAPTER 39 Radioactive Materials *Wayne D. Holmes*

CHAPTER 40 Materials-Handling Systems *L. Jeffrey Mattern*

CHAPTER 41 General Storage *Stephen R. Hoover*

CHAPTER 42 Storage and Handling of Flammable and Combustible Liquids *Orville M. "Bud" Slye, Jr.*

CHAPTER 43 Gas Hazards *Theodore C. Lemoff*

CHAPTER 44 Combustible Dusts *William J. Bradford*

CHAPTER 45 Combustible Metals *Carl H. Rivkin*

CHAPTER 46 Chemicals *John A. Davenport*

CHAPTER 47 Plastics and Rubber *Steve Younis*

CHAPTER 48 Explosives and Blasting Agents *Lon Santis*

CHAPTER 49 Fireworks and Pyrotechnics *Ken Kosanke*

CHAPTER 50 Heat-Utilization Equipment *Richard A. Gallagher*

CHAPTER 51 Spray Painting and Powder Coating *Don R. Scarbrough*

CHAPTER 52 Welding, Cutting, and Other Hot Work *August F. Manz and Amy Beasley Spencer*

CHAPTER 53 Hazards of Manufacturing Processes *L. Jeffrey Mattern*

CHAPTER 54 Aerosol Manufacturing and Storage *Michael Madden*

CHAPTER 55 Protection of Commercial Cooking Equipment *R. T. "Whitey" Leicht*

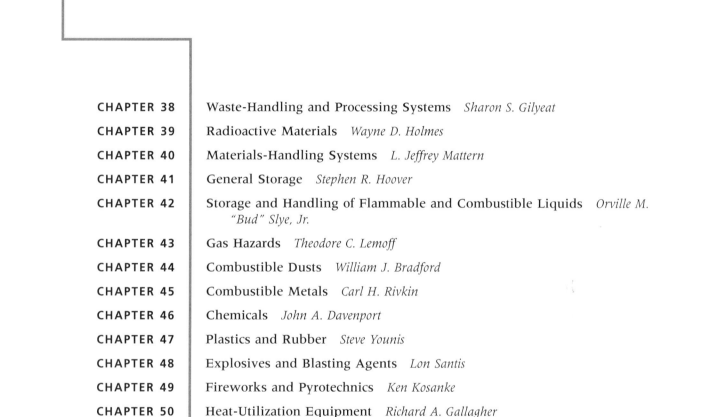

Waste-Handling and Processing Systems

CHAPTER 38

Sharon S. Gilyeat

CASE STUDY

Sprinkler Extinguishes Dumpster Fire in Public Works Building

A single sprinkler extinguished a dumpster fire in a Michigan public works building that started when waste chemicals placed in the dumpster reacted with each other or with other waste products and ignited.

The single-story, 36,000-ft^2 building had masonry walls, unprotected steel roof trusses, and a wood built-up roof. It housed employee facilities, a dog pound, and a place to store and repair heavy equipment. A central station alarm company monitored the building's wet-pipe sprinkler system.

The building was closed for the night when the alarm company called the fire department at 4:24 P.M. to report a water flow alarm. Fire fighters arrived three minutes later to find that an overhead sprinkler had extinguished the blaze.

Fire damage was limited to the area of origin, but smoke filled the building. The building, valued at approximately $2 million, was not damaged. Damage to its contents, valued at approximately $300,000, was estimated at $250.

Source: "Fire Watch," *NFPA Journal*, September/October, 1997.

Although the waste generated in the course of a normal day's activity is rarely a source of ignition, it can provide a significant source of fuel. Large accumulations of combustible waste represent a potentially serious fire hazard. Waste also may be contaminated with hazardous materials ranging from flammable solvents to highly toxic substances. Contaminated waste complicates the disposal process and further increases the possibility of unwanted fires.

Collecting and disposing of waste safely and efficiently dramatically reduces both the source of fuel and the fire hazard. Methods of handling and disposing of waste range from the simple to the sophisticated, from collecting it manually from waste baskets and disposing of it in small landfills to gathering it mechanically with pneumatic systems that service entire buildings and destroying it in huge, multistage incinerators.

Waste handling and processing are parts of an integrated waste management system. Waste handling is generally considered the portion of the system that is associated with the movement of generated waste. It does not include the storage, processing, and final disposal of the waste. Waste-handling equipment includes conveyors, chutes, carts, elevators, lifts, and other transportation vehicles. Wasteprocessing includes functions that alter the shape, size, uniformity, or consistency of the waste to facilitate handling, storage, and/or disposal. Waste-processing systems and equipment include compactors, shredders, crushers, pulpers, pulverizes, baggers, encapsulators, extruders, and dewatering devices.

*For approximately 6 years **Sharon S. Gilyeat**, P.E., worked as fire protection engineer and manager for the Bureau of Engraving and Printing and the U.S. Printing Office. Currently senior fire protection engineer at Koffel Associates, Inc., she has served on various NFPA technical committees and is chair of NFPA's Technical Committee on Incinerators and Waste Handling Systems.*

NFPA 82, *Standard on Incinerators and Waste and Linen Handling Systems and Equipment*, is the NFPA standard that contains design criteria for new incinerators and waste- and linen-handling systems. Existing incinerators and waste- and linen-handling systems are not required to be altered unless required by the provisions of another adopted code or standard. The owner must provide administrative, maintenance, and training programs that provide an equivalent level of fire protection and life safety if the existing systems do not meet NFPA 82 requirements.

In new construction NFPA 82 requires any room containing waste storage exceeding 1 yd^3 (0.765 m^3) must be enclosed in 2-hour rated construction. Doors to the room must be 1½-hour fire rated, self-closing, and self-latching. The storage room must be sprinkler protected in accordance with NFPA 13, *Standard for the Installation of Sprinkler Systems*.

WASTE-CHUTE AND HANDLING SYSTEMS

There are five basic types of waste-chute systems: general access gravity, limited access gravity, pneumatic, gravity-pneumatic, and multiloading pneumatic.

The general access gravity chute is an enclosed vertical passageway that leads to a building's storage or compacting room. The chute is accessible to all building occupants. The limited access gravity chute is similar, but its door is locked, and entry is limited to those who have a key.

Pneumatic waste-handling systems depend on airflow to move waste to a central collection point. This system is often used in hospitals and health care buildings to carry soiled linen, although it can be used for other types of waste. Pneumatic system risers may be vertically offset to fit the building or system design requirements. Where the riser is not totally enclosed in a shaft and penetrates a fire-rated floor, wall, or ceiling assembly, a fire damper is required at the penetration of the assembly. The dampers should be inspected to ensure that they are in good operating condition.

The gravity-pneumatic waste system uses a conventional gravity system to feed a collecting chamber, which in turn feeds a pneumatic waste-handling system.

The multiloading pneumatic system is similar to the pneumatic type except that entry into the system is through automatic or self-closing, positive latching inlet doors that are interlocked so only one can be open at any given time and they cannot be open when the material damper is open.

Waste Chutes

Waste chutes provide pathways along which flames and smoke may travel throughout a building. For this reason, they must be enclosed by a fire-rated barrier with rated fire doors and, depending on the type of chute, equipped with automatic sprinklers.

Waste chutes are often considered storage areas because they can become clogged with combustibles, thus presenting a fire hazard. The inspector should ask whether the chutes on the premises are frequently jammed with refuse. If they are, a thorough study of the volume and type of waste and the habits of chute users is warranted.

Gravity-type chutes are constructed of either masonry, refractory-lined steel, or stainless, galvanized, or aluminized steel. They must be enclosed by fire-rated construction to protect the building from a fire in the chute. Enclosure shafts penetrating three floors or less may be 1-hour fire resistance rated, otherwise a 2-hour fire resistance–rated shaft enclosure is required. The inspector should make sure the enclosing construction has not been damaged or penetrated in such a way that it is no longer a complete fire barrier.

Unlined metal chutes should also be protected by automatic sprinklers. Inspections should ensure sprinklers are located at the top and bottom of the chute and at alternate floors. These should be installed where they can be inspected and maintained yet remain out of the reach of vandals and falling objects.

Chutes are required to be vented to the exterior. Inspections should ensure that the chute extends at least 3 ft (0.92 m) above the roof of unprotected construction and be open to the atmosphere.

Chute Service Opening Rooms

For many years it was acceptable to have service openings to chutes direct from the corridor. This arrangement still exists and is generally accepted under existing conditions as long as the openings are kept secured. New construction requires a separate service opening room that is enclosed in construction equivalent to the rating of the chute enclosure. In no instance should the rating of service opening room be less than 1-hour fire resistance rating with ¾-hour-rated fire protection rated doors.

Chute Terminal Enclosures

The chute room or bin where the waste from chutes is collected must have a fire resistance rating at least equivalent to that required for the chute, and openings into the enclosure must be protected by fire doors suitable for the enclosure rating. Other codes and standards may prohibit the use of the chute discharge room for any other purpose; however, NFPA 82 does not limit the discharge room's use. The concentration of combustibles in such an area requires that it be equipped with automatic sprinklers.

Waste chutes should not discharge directly into incinerators because this increases the likelihood that they will serve as chimneys, spreading smoke and fire to other portions of the building.

INCINERATORS

Incineration significantly reduces the volume of wastes and, more importantly, the fire hazard associated with the accumulation of combustibles. Incinerators generally are of two types, either commercial–industrial or domestic.

Commercial–Industrial Incinerators

Commercial–institutional incinerators are mainly used for burning solid waste. Most commercial–industrial waste incineration facilities (Figure38-1) use either a rotary kiln, multiple-chamber, or a controlled-air unit, each of which operates on a different principle. Multiple-chamber units use high quantities of excess air, whereas controlled-air incinerators operate under starved-air conditions in the primary chamber (Figure 38-2). The fire hazards associated with each are essentially the same.

Incineration Fire Safety

By their very nature, incineration operations present significant fire hazards. Fire hazards result from the presence of volatile and combustible wastes, fuel-handling and combustion systems, high temperature flaming combustion, exhaust and ducting of high temperature gases, and handling of hot ashes. To minimize the fire hazard, the design and construction of incinerators must comply with NFPA 82. Inspections should ensure the provisions of NFPA 82 are met. Specifically, the following areas should be included in an inspection.

FIGURE 38-1 Typical Commercial-Industrial, Multiple-Chamber Incinerator

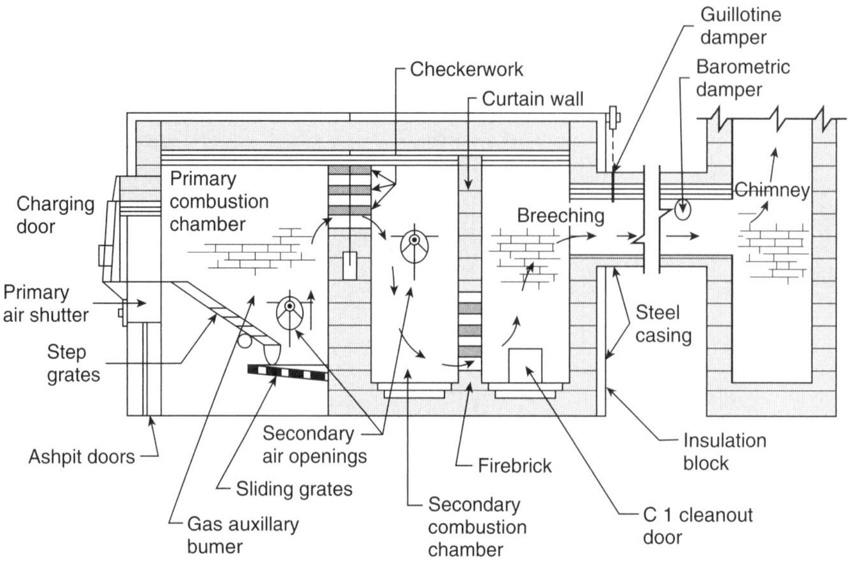

Incinerator and Waste-Handling Room. Ensure incinerators and related waste collection and handling equipment are enclosed in 2-hour fire-rated compartments equipped with self-closing 1½-hour fire doors. These compartments should be used for no other purpose.

Equipment Design and Construction. Ensure incinerators, breeching, stacks, and accessories that are subjected to high-temperature combustion reactions and gases are able to resist cracking, erosion, warping, and distortion. Outside surface temperatures to which an operator may be exposed should be no higher than 70°F above ambient. The incinerator also should have some form of explosion relief, which may be the chimney.

Layout and Arrangement. The facility design should ensure that waste and residue containers and the like do not block charging and clean-out operations or access to work areas and that waste material can be charged in a smooth, efficient manner. All parts of the incinerator should be accessible for cleaning, repair, and service, and clearances above the charging door and between the top and sides of the incinerator and

FIGURE 38-2 Cutaway of a Controlled-Air Incinerator

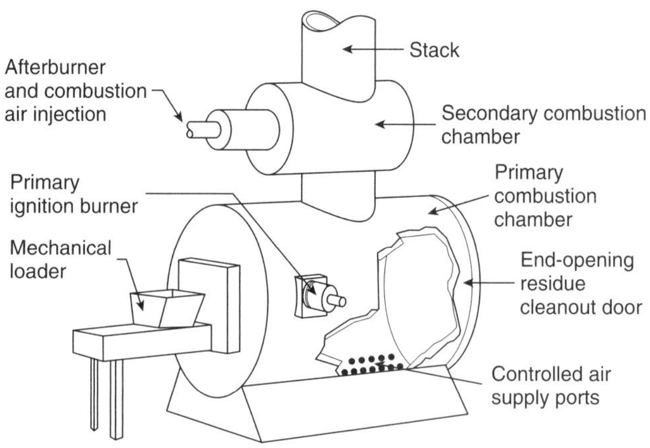

combustible materials should meet the requirements of applicable building codes and the specific requirements in NFPA 82.

- *Placement:* Ensure incinerator is placed on masonry or concrete foundation or noncombustible material with a fire rating of 3 hours or more. Incinerator combustion chambers must be elevated above concrete bearing surfaces via pedestals, cradles, skids, or other methods to provide for 4-inches (101.6-mm) clearance
- *Minimum Clearances from Incinerator to Combustible Materials:* Ensure clearances are as follows:
  - Sides and rear - 36 inches (914 mm)
  - Above - 48 inches (1220 mm)
  - Front - 8 feet (2.4 m)
- *Minimum Clearances from Incinerator to Walls or Ceiling:* 12 inches (305 mm), or 3 inches (76.2 mm) if construction is noncombustible

Charging Systems

"Charging" is the feeding or loading of waste materials into the incinerator. An improperly designed or operated charging system could permit flames and combustion products to escape from the incinerator and ignite waste materials nearby.

There are basically two types of incineration charging systems: manual and mechanical.

Manual Systems. As the name implies, waste materials in a manual system are loaded into the incinerator by hand. Manual charging generally is used in small-capacity units. Inspections should ensure:

- Charging doors are sized to accommodate largest load size
- Charging doors and frames are smooth and free of obstructions to prevent hang-up of waste materials
- Charging doors operate easily and have positive latching
- Charging door handles remain relatively cool
- Interlocks provided are between the charging door and the primary chamber burners to shut off burners when the door is opened

Mechanical Charging. Mechanical devices generally are used to load larger incinerators with small batches of waste at regular intervals. This protects the units against overcharging and provides for a continual and efficient combustion process.

One common mechanical charging system is the "hopper/ram assembly." Waste is loaded into a hopper, the hopper cover closes, a guillotine-type door opens, and a ram pushes the waste into the incinerator.

This particular mechanical system presents major fire and smoke problems against which you must guard. For example, waste can lodge under the fire door and prevent it from closing tightly. If this occurs, the lodged waste could ignite and spread fire back into the hopper, offering a path for fire spread. Thus, the fire door should be closed by power, not gravity, to seal the furnace opening as tightly as possible.

Another fire hazard associated with the hopper/ram assembly involves the ram itself. When the charging ram injects waste into the incinerator, its face is exposed to furnace heat. Eventually, this can cause the ram to heat up enough to ignite waste outside the incinerator. To avoid this problem, the charging ram should be cooled either by an internal water circulation system or a water spray system that quenches the ram face after every charging cycle.

The fire door on this type of mechanical charging system should be interlocked with the hopper cover, or outer door, to keep them from being opened at the same time while the incinerator is operating. There are a number of ways to protect against accidental ignition of waste materials in the hoppers. A fire detector with an audible alarm signal may be installed, or the system may be equipped with a high-temperature automatic sprinkler independent of the room sprinkler system. Another option is a manually activated water spray or quench valve or an emergency switch that would override the automatic charging cycle controls to inject the hopper contents immediately into the furnace.

Auxiliary Fuel Systems

Most incinerators have natural gas or distilled fuel oil burners to help burn a greater variety of waste, as well preheat the incinerator, ignite the waste, and destroy odors and smoke. Burners must be equipped with an electronic flame safeguard system that automatically shuts off the burner fuel supply if the burner fails to ignite, its flame is extinguished, or the furnace draft is insufficient.

WASTE COMPACTORS

Waste compactors use electromechanical-hydraulic means to reduce the volume of waste and to package it. Compactors may be either commercial–industrial or domestic.

Commercial–Industrial Compactors

There are four basic types of commercial–industrial compactor systems: the bulkhead compactor, the extruder, the carousel bag packer, and the container packer.

- *Bulkhead Compactor:* In this type of compactor, waste is squeezed in a chamber against a bulkhead. When a block of waste is ready for removal, a bag is installed and filled with the compacted material.
- *Extruder*: An extruder compacts waste by forcing it through a cylinder that has a restricted area and extruding it into a "slug," which is then broken off and bagged or placed in a container.
- *Carousel Bag Packer:* In this compactor, waste is squeezed into a bagged container.
- *Container Packers:* Container compactors compact waste directly into a bin, cart, or container. When it is full, the container can be either manually or mechanically removed from the compactor and taken out of the compaction area.

Chute Termination Bin

The waste chutes that serve compactors generally do not feed directly into the compactor but into a waste storage room, a chute connector, or an impact area. The potential fire hazard in the waste storage room can be minimized by installing sprinklers and a 2-hour fire-rated enclosure with 1½-hour doors at the openings.

Compactor Storage Rooms

Because compacted material is combustible, it will burn if ignited, even though its density will affect the rate of burning. For this reason, only minimal amounts of compacted materials should be stored in buildings. Compactors larger than 1 yd^3. (1.52 m^3) must be enclosed with 2-hour fire-rated wall, ceiling, and floor assemblies with 1½-hour fire doors.

OTHER WASTE-PROCESSING EQUIPMENT, INCLUDING SHREDDERS

Processing equipment changes the physical form or characteristics of waste. The major hazard associated with processing operations is the possibility of explosion, which may result from the ignition of combustible dust while it is operating. Ignition also may occur when ferrous materials, such as paper clips, are left in the material to be processed; this makes magnetic separators desirable.

Methods of protecting other processing operations from fire include dust collection, explosion suppression in shredder rooms, and explosion relief venting. Feed bins to the shredders and storage areas for the processed and unprocessed waste should be sprinkler protected, and all storage areas should be enclosed in 2-hour fire-rated rooms with 1½-hour fire doors.

SUMMARY

Waste-handling and processing are critical components of any facility management program. Processes and systems designed to reduce the amount or volume of waste, as well as to move the waste efficiently throughout the building, are important to the efficient operation of a facility. The unique fire hazards posed by the waste-handling and processing systems must be minimized by ensuring adequate fire protection and life safety through proper design and protection. NFPA 82 is a key resource for evaluating the protection of both new and existing waste-handling and processing systems.

BIBLIOGRAPHY

Cote, A. E., ed., *Fire Protection Handbook*, 18th ed., NFPA, Quincy, MA, 1997.

NFPA Codes, Standards, and Recommended Practices

See the latest version of The NFPA Catalog for availability of current editions of the following documents.

NFPA 13, *Standard for the Installation of Sprinkler Systems*
NFPA 68, *Guide for Venting of Deflagrations*
NFPA 80, *Standard for Fire Doors and Fire Windows*
NFPA 82, *Standard on Incinerators and Waste and Linen Handling Systems and Equipment*
NFPA 251, *Standard Methods of Tests of Fire Endurance of Building Construction and Materials*

Radioactive Materials

Wayne D. Holmes

CHAPTER 39

CASE STUDY

Two Major Plutonium Fires Occurred at Rocky Flats Plant

Two major plutonium fires occurred at the U.S. Department of Energy's Rocky Flats Plant (now the Rocky Flats Environmental Technology Site) near Denver, Colorado, the first in 1957 and the second in 1969. Buildings were modified and new safety procedures implemented as a direct result of these fires. The 1957 fire damaged Building 771, causing radiological contamination of much of the interior of the building. The fire spread from a glove box window on the fabrication line to the glove box exhaust filters, and the main filter plenum. The main fire was under control within 30 minutes of its discovery, but rekindled several times. Shortly after the fire was thought to be under control, flammable vapors collecting in the main exhaust duct exploded, spreading plutonium contamination throughout much of the building. Security officers discovered flames at around 10:10 P.M.; the fire was declared out by 2:00 A.M. the following day.

Prior to the 1957 fire, water was prohibited in the plutonium areas because of its moderating effect, potentially allowing a criticality event (spontaneous fission chain reaction) to occur. During the 1957 fire, water was used to extinguish burning combustible materials possibly contaminated with plutonium (i.e., Plexiglas and ducting materials in the exhaust plenum) without a criticality event or fatal consequences. As a result, standpipes and sprinkler systems were installed in plutonium-handling areas throughout the plant. Another result of this fire, which was propagated by combustible and flammable material, was that less flammable materials were investigated for use in glove box construction, specifically, a replacement for Plexiglas windows. Off-site release of plutonium into the atmosphere from the 1957 fire was estimated at approximately 1 g. No major injuries were reported as a result of the fire.

The second plutonium-caused fire occurred on May 11, 1969, in the Building 776/777 glove boxes. The first notice of the fire came at 2:29 P.M., when an alarm, triggered by a glove box overheat system, alerted firemen. No one was injured in the blaze, but some 33 employees were treated for contamination. The fire occurred from spontaneous ignition of a briquette of scrap plutonium alloy metal contained in a small metal can, probably without a lid. The 1969 fire was the first time that water was used directly on burning plutonium. (Note that in the 1957 fire, water was used to put out burning combustibles, not burning plutonium.) The fire resulted in $26.5 million in property loss. There was an estimated plutonium release from the building of 0.000012 g, all contained on the plant site. Decontamination of the area took approximately 2 years.

The fire changed the way that business was conducted at Rocky Flats and in the Atomic Energy Commission complex and possibly had international influences. Prior to the fire, there was little quality control. After the fire, the

Wayne D. Holmes, P.E., vice president of HSB Professional Loss Control, a fire protection engineering consulting firm, has an M.S. in fire protection engineering from WPI. He is a member of NFPA, SFPE, ASTM, and ASME; is chair of the NFPA Technical Committee on Nuclear Facilities; and serves on several other NFPA technical committees.

complex started applying multilayers of safety reviews and quality control, including the creation of an inert atmosphere in the glove boxes to prevent propagation of fires and the addition of water sprinklers and more fire walls. Because of their efforts, fire department personnel received a Group Presidential Citation for heroism in the 1969 fire for risking their own health and well-being to prevent a breach of the building, thus preventing plutonium contamination in the atmosphere.

Source: *http://www.rfets.gov*

Radioactive materials are substances that spontaneously decay, emitting energetic rays or particles in the process. Radioactive materials themselves present no unusual fire hazards, as their fire characteristics are no different from the fire characteristics of the nonradioactive form of the same element.

Various types of emitted radiation are capable of causing damage to living tissue. In particular, fire conditions involving such materials can cause the formation of vapors and smoke that contaminate the building of origin or neighboring buildings and outdoor areas with radioactive elements.

The nature of radioactive materials is such that their involvement in fires or explosions can impede the efficiency of fire-fighting personnel, thus causing increased potential for damage by radioactive contamination. The presence of radioactive materials can complicate a fire-fighting situation by presenting hazards unknown to the fire fighter and causing real or wrongly anticipated hazards to fire fighters that can inhibit fire-fighting operations. The dispersal of radioactive materials by fumes, smoke, water, or by the movement of personnel can cause a radiation contamination incident that can contribute significantly to the extent of damage, complicate cleanup and salvage operations, delay the restoration of normal operations, and affect personnel safety. The primary fire concern is to prevent the release or loss of control of these materials by fire or during fire extinguishment. This is especially important because radioactivity is not detectable by any of the human senses.

RADIOACTIVE MATERIALS

Most types of radioactive materials that are handled or stored in quantities sufficient to pose a hazard during an accident are restricted to government facilities or facilities licensed by the Nuclear Regulatory Commission (NRC). Radioactive materials used in consumer products such as ionization smoke detectors and tritium-activated emergency lights pose such a small risk that the end-user needs no special permits to use them. Ionization smoke detectors contain a small radioactive source that is safe in general use. However, ionization smoke detectors are considered hazardous waste, and unwanted detectors must be returned to the manufacturer for disposal rather than be discarded as normal waste. They may be damaged in a fire, but the normal products of combustion of most fires will far exceed the hazard added by the dispersal of radioactivity from these limited-use consumer products.

HAZARDS ANALYSIS REPORTS

NFPA standards recommend a safety analysis report, something major users and producers of radioactive materials, such as Department of Energy plants, power and research reactors licensed by the NRC, and fuel or waste handlers and processors, are required to have. This is a formal document that includes fire hazards analyses, failure analysis reviews, and descriptions of the most credible accidents, including fires, and the effects to be expected from each. These are openly available, and the inspector should review them as part of an initial or prefire planning inspection.

Facilities that use radioactive materials typically are required to demonstrate that the most probable accident scenarios will not adversely affect the public. This means that most facilities have automatic fire extinguishing systems, generally ordinary wet-pipe sprinkler systems.

Although nuclear facility managers sometimes express concern about water damage, the only real concern with water is that water is a neutron moderator. This means that a "criticality" accident may occur when some fissionable materials, such as enriched uranium or plutonium, are exposed to unlimited water.

A nuclear criticality incident should not be confused with a nuclear detonation. Radioactive materials in general use will not detonate. A nuclear criticality is generally a self-limiting occurrence generating large, short duration pulses of gamma radiation and neutrons. Criticality concerns are limited to highly regulated special nuclear materials, such as enriched uranium and plutonium, that will not be encountered by fire-fighting personnel in general residential, commercial, and industrial occupancies. However, the hazard can be calculated in advance and protection, such as automatic sprinklers, provided.

Today, the overwhelming majority of all nuclear facilities are protected by automatic sprinkler systems, many precisely because the dispersion of possibly contaminated water is desired to be controlled. It is far easier to collect and store water from a few sprinklers that open directly over a fire than it is to deal with a much greater volume of water from one or more hose streams directed by fire fighters through smoke and obstructions from some distance away.

PREFIRE PLANS

The inspector's review will form the primary input to the facility's prefire plan. To ensure safe and adequate emergency operations, the inspector should ascertain and discuss with knowledgeable people the types and quantities of materials present at the facility. The normal fire fighter turnout gear will provide appreciable protection against a significant amount of alpha and beta radiation, and the consistent use of self-contained breathing apparatus is the principle protection against internal contamination. The remaining threat from high-level gamma or neutron radiation is readily measurable with the proper instrumentation. Where such hazards exist, the inspector should review the radiation monitoring equipment available for both normal and emergency needs (Figure 39-1) and determine whether trained people will be available to assist the fire department during emergencies at any hour. The inspector must also ascertain the methods of containment available within the facility and the probable effectiveness of the containment during a fire.

Containment systems for radioactive materials include the usual fire doors and walls, ventilation enclosures and dampers, and sometimes special ventilation systems. In facilities from which contaminated air may be exhausted, the air commonly passes through a high-efficiency filter system. This may not have dampers in the ductwork in order to ensure that contamination will not back up into the plant. If this is the case, the filter system must be protected to ensure that fire will not damage the filters and allow contaminated air to escape. Filter systems typically consist of multiple banks of prefilters and mesh screens, some of which may be equipped with various combinations of automatic and manual water spray systems. The inspector should make sure that these systems adequately cover the filters and are in service. The supervision, inspection, and maintenance of these systems should be at least as thorough as they are for any other fire protection system in the plant.

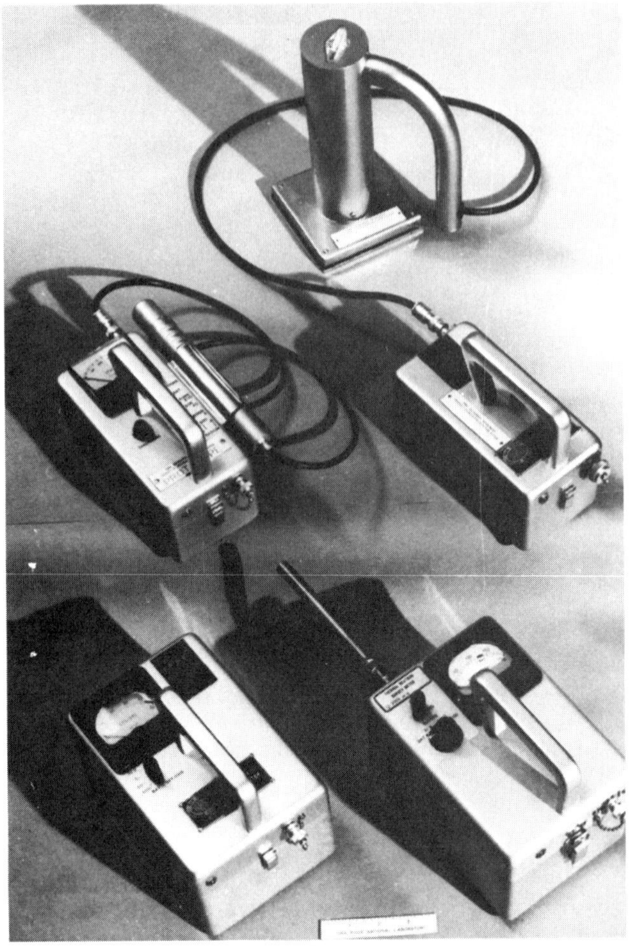

FIGURE 39-1 Portable Instruments Developed at Oak Ridge National Laboratory for Measuring Radioactive Emissions. The meters are for fast neutrons (bottom row at left), thermal neutrons (bottom row at right), beta-gamma (top row at left), and alpha particles (top row at right), with an alpha scintillation detector at the top.
Source: *Fire Protection Handbook,* NFPA, 1997, Figure 9-30C.

The inspector should also review the disposition of water that has come in contact with a possibly contaminated filter. Contaminated water should be drained to a holding tank, or the room should be diked or curbed so that the low-volume water discharge from a protection system will not spread throughout the premises.

Since radioactive materials in bulk storage generally are quite valuable, as well as hazardous, they usually are stored in vaults or safes for security as well as safety. Thus, the inspector must ascertain the integrity of all fire walls, doors, dampers, vaults, and ducts. The spread of contamination is not determined by the nature of the radiation but by the nature of the chemical form of the radioactive material and the way in which that chemical is affected by fire. Any opening through which smoke or hot gases can spread is an opening through which radioactive contamination can spread. While handheld meters make radiation detectable and the hazard more readily calculated, decontaminating a structure may require quite a bit of work; it may take much longer to remove radioactive contaminants than to remove smoke and soot.

The great importance of compartmentation and containment in nuclear facilities should not be underestimated. The 1969 Rocky Flats Plant fire resulted in a $26 million loss, of which $10 million alone was the cost of decontamination. This was the most costly contamination incident in the Atomic Energy Commission's (now part of the Department of Energy) history.

RADIOACTIVITY MACHINES

Machines, as well as radioactive materials, can pose radiation hazards. These machines may range from the small hospital X-ray machine to the multi-mile long particle accelerators used in nuclear research. All of them pose radiation hazards from the beam, but they all share a common characteristic: The radiation ceases when the beam is shut down.

These machines emit radiation only while in operation, and attempts to extinguish a fire in the immediate vicinity of the machine should be delayed until the machine power can be disconnected. Once shut off, radiation machines no longer create radiation. However, certain target materials become radioactive when exposed to atomic particles produced by these machines, and the target materials may continue to emit low-level radiation even after the machine is turned off. The inspector should ascertain that adequate interlocks and manual shutdowns are provided to shut the machine off in an emergency and that the shielding is adequate to protect emergency forces engaged in activities in any supporting structures.

In particle accelerators, the beam is directed at a target, which may produce other beam is be directed at other targets and eventually at a "beam stop" of high-density materials. These materials may have residual levels of radioactivity resulting from the creation of new radioactive elements. The inspector should determine the nature and extent of the target's radiation, as well as the protection afforded the target. In addition, the inspector should check out the experimental area surrounding a target. This area is the most frequently changing area of the facility and often the most cluttered. Thus it may be the most difficult area in which to fight a fire. It may also contain one-of-a-kind electrical experimental or monitoring equipment that presents additional ignition potentials.

SHIPPING AND STORAGE

Material Containers

Containers used to ship radioactive material must be strong and constructed well enough to withstand credible accidents. To this end, containers for high-level materials must be tested to determine that they can withstand impact, fire, and water submersion, in sequence, without leaking. Strangely enough, the history of shipping and shipping accidents makes the transportation accident of little concern to the fire service. Rather, the potential problem lies in the facility in which shipping containers are stored or through which they pass. Why? Because a container designed to survive the maximum fuel spill fire may be inadequate when open and exposed to the maximum fire in a general shipping depot or warehouse.

Transportation Facilities

When inspecting transportation facilities, the inspector should always ascertain whether they are likely to be used to store high-level radioactive shipping containers, such as cobalt gamma ray machines. This type of equipment includes a metallic Cobalt 60 source tightly encapsulated in steel and sealed sources in "beta gauge" thickness and measuring devices. There have been several instances of stainless steel encapsulated beta gauge sources surviving appreciable fire exposures without the release of the radioactive isotope contained in them. Unlike the electrical machine, these use a radioactive material as a power source. Thus, the unit is always highly radioactive and must be shielded. Since lead is a cheap material and is expensive to ship by virtue of its weight, it is often used to shield radioactive materials in storage. The presence of molten lead in the residue of a warehouse fire should be an immediate indicator of a possibly dangerous situation, and all overall operations should be suspended pending the completion of a thorough radiation survey. This also applies to any facility in which high-level radiation sources, protected by shielding, are used.

Hospitals

Hospitals are another occupancy in which the storage of radioactive materials—in this case radioisotopes—may present a hazard. Since radioisotopes are administered to patients in individual doses, the threat to a healthy fire fighter of exposure to radioactivity during a fire in a treatment area is minimal. Of greater concern are the areas in which these materials are stored and prepared. They may contain a considerable number of different isotopes and the quantity, as well as the combination of those with the highest activity and most volatility, may be very hazardous. Since patients cannot readily evacuate a hospital, separation and protection of storage areas is often more important for fire safety than it would be in an industrial or educational facility.

All other fire protection programs, procedures, and equipment applicable to a nonradiation facility are also applicable to the radiation facility. The inspector should have no illusions that some fire protection practices may not apply because of the special "radioactive" nature of the facility. In fact, the provision of standard fire protection practices, particularly automatic protection, is a necessity in facilities in which radiation is a significant additional hazard.

HAZARDOUS MATERIALS LABELS

All radioactive materials containers must be marked with appropriate warning signs. Samples of Department of Transportation shipping labels are shown in Figure 39-2. Containers used in facilities for interim storage should be labeled with the type of material they hold and the radiation levels measured at the surface of the container. These radiation levels can be readily translated into the maximum amount of time a person can remain adjacent to the container.

RADIATION PROTECTION

Considering the factors of time, distance, and shielding, it is generally unlikely that any reasonable fire-fighting time can result in an exposure level of concern to a fire fighter. Again, this can be readily calculated, and such considerations should be included in any prefire plan. In most facilities, individual laboratories or process rooms will also be marked to indicate the type and level of hazard within. Considering that radiation exposure levels are proportional to time, a fire fighter who may be exposed to radiation for 2 or 3 hours at the most should not be overly concerned about his or her health. Fire fighters and other personnel can take precautions to minimize exposure to radiation.

FIGURE 39-2 Examples of Department of Transportation Shipping Labels

Protection from External Radiation

In the case of external nuclear radiation, the dose and resulting injury to humans can be minimized in several ways:

- Only the smallest possible portion of the body should be exposed (e.g., the hands, rather than the entire body). Normal turnout gear can be very effective in stopping low energy radiation.
- The time spent in the hazardous area and, therefore, the time of exposure should be minimized.
- The intensity of the radiation received by a person can be minimized by maintaining the greatest possible distance from the radiation source or by placing a suitable (dense) materials between the radiation source and the person. Increasing the separation distance is very effective in reducing the exposure.

Protection from Internal Radiation

The possibility of radioactive materials entering the body can be reduced by wearing properly fitted protective face masks and clothing while in a hazardous area.

SUMMARY

Radioactive materials may emit radiation that may be damaging to living tissue. In the United States, highly hazardous radioactive materials are limited to government-owned and- operated facilities or facilities that are licensed by the federal government. For the most part, radioactive materials in commercial or consumer operations are relatively low hazard but do require some precautions.

The inspector should review the safety analysis report for any radioactive materials that may be present. It will contain information about the materials, their hazards, and their protection. Automatic sprinkler protection is an effective way of protecting radioactive material from fire.

Prefire planning of fire response, in conjunction with the safety and fire hazards analysis is important in protecting personnel and the environment. Prefire planning should include evaluation of required protection equipment for emergency response personnel, availability and use of radiation monitoring equipment, contamination control procedures, as well as general fire fighting procedures.

In addition to radioactive materials, there may be machines present that emit radiation when operating. The inspector should review the location, type, and operating procedures for radiation machines. It is important to note that these machines, unlike radioactive materials that constantly emit radiation, will cease to emit radiation when deenergized.

Containers for licensed radioactive materials, both for transportation and in situ storage, are constructed to withstand credible accidents, including fire, to prevent release of materials and radiation. In conjunction with the safety analysis report, inspectors should satisfy themselves that the potential fire exposure has not been increased beyond the limits postulated in the safety analysis report.

The inspector should be assured that there are adequate procedures to protect personnel, property, and the environment in the event that nuclear materials are exposed to fire. Emergency response personnel can limit their potential exposure to radiation by limiting the amount of time for which they may be exposed, by maintaining a safe distance from radiation sources, and by providing shielding between the source of radiation and themselves. Property and the environment can be protected by preventing fires from occurring; by protecting against fire development; and, in the event that a fire occurs, by confining any contamination that might be released.

With careful consideration of the types of radiation exposures, property safety analysis documentation, prefire planning for radiation protection and containment, in addition to normal fire protection considerations, the inspector can be reasonably assured of the safe function of facilities containing radiation hazards at the time of a fire emergency.

BIBLIOGRAPHY

Brannigan, F., *Living with Radiation: The Problems of the Nuclear Age for the Layman*, Energy Research & Development Administration, Division of Safety, Standards, & Compliance, Springfield, VA, 1976.

Cote, A. E., ed., *Fire Protection Handbook*, 18th ed., NFPA, Quincy, MA, 1997.

Cote, A. E., and Linville, J. L., eds., *Industrial Fire Hazards Handbook*, 3rd ed., NFPA, Quincy, MA, 1990. (See Chapters 7 and 40)

Purington, Robert G. and Patterson, H. Wade, *Handling Radiation Emergencies*, NFPA, Quincy, MA, 1977.

NFPA Codes, Standards, and Recommended Practices

See the latest version of The NFPA Catalog for availability of current editions of the following documents.

NFPA 801, *Standard for Fire Protection for Facilities Handling Radioactive Materials*

NFPA 804, *Standard for Fire Protection for Advanced Light Water Reactor Electric Generating Plants*

NFPA 805, *Performance-Based Standard for Fire Protection for Light Water Reactor Electric Generating Plants*

CHAPTER 40

Materials-Handling Systems

L. Jeffrey Mattern

Materials handling can be a complex process. It is important for the inspector to be aware of the type of commodities being brought into a facility and how they will be handled once in the facility. Additionally it is important for the inspector to understand how materials are transported from the receiving point to both the storage or area of use and to the shipping point. The type of material, its destination, and the transportation method can have an impact on the fire safety of a facility.

COMMODITY CLASSIFICATION

The materials, or as commonly referred to, the "commodity," used in the facility dictates the level of fire protection required to properly protect the building and its contents from major damage. When a new facility is constructed, the fire protection design is determined based on the commodity to be used or stored. Over the life of a facility, and a warehouse in particular, the commodity classification (see NFPA 13, *Standard for the Installation of Sprinkler Systems*) may change many times. These changes can result in the need for increased or possibly decreased fire protection. It is likely, however, that the evolution of commodity changes will increase, rather than decrease, the level of fire protection required. It is important for the fire inspector to properly examine what is currently being used or stored in the facility and then to determine the fire protection required based on the current standards.

Once the commodity classification is determined, it is important to remember that the fire protection should be designed based on the most hazardous commodity within the facility. In a warehouse, a commodity classification should not be "averaged." In a storage arrangement, even a small amount of a higher class commodity will burn as if the whole area is of that material.

STORAGE ARRANGEMENTS

Height of Storage

In both manufacturing and storage areas, an important condition to evaluate is the height of storage. As a warehouse facility matures, it often becomes overstocked. This condition can develop into potential fire protection problems. Frequently, the remedy sought by the warehouse is to increase the height of the storage in the facility. This is often accomplished by double stacking, or placing a second pallet load of stock on the very top tier, or top level, of storage. Double stacking may make the in situ fire protection design inadequate. The inspector will need to measure the storage height, combine that information with the commodity classification information to determine the level of fire protection required, and then compare the result with the original design criteria.

Aisle Storage

Another potential problem in a warehouse is the presence of aisle storage. All large-scale fire tests to determine the level of fire protection required have been conducted with aisles free and clear of storage. The presence of storage in aisles presents an

L. Jeffrey Mattern has worked for FM Global for over 30 years, where he serves as chief engineer/standards engineer of Mid-Atlantic Field Engineering and as worldwide coordinator of Arson and Fire Service Programs. He has been active on various NFPA committees and on the Standards Council since 1968, is a former member and past president of the Philadelphia/Delaware Valley chapter of SFPE, and is a current member of the SFPE Qualifications Board.

unbroken path for the fire to travel from rack to rack before automatic sprinklers can control the fire. The inspector should look for aisle storage and determine whether it is a routine method for handling an overstocked condition or a random occurrence. Neither reason is acceptable but the latter may be more manageable.

MATERIALS STAGING AND EQUIPMENT

Materials on Receiving Dock

As materials are received, they are typically staged on the receiving dock, awaiting movement into the racks or to the manufacturing area. Often, hazardous materials, which may include aerosols, flammable and combustible liquids, or even oxidizers, receive the same treatment. Seldom is a receiving dock designed for such hazardous materials. Hazardous materials, upon receipt, should be promptly identified and then expeditiously transported to the designated hazardous material storage area. The inspector should examine materials staged on the dock and should inquire about handling procedures for hazardous materials.

Materials on Shipping Dock

Similar conditions can exist on the shipping end of the process. Hazardous materials are often brought to the shipping dock to await loading on outbound trucks. Hazardous materials should not be brought to the shipping dock until they can be immediately loaded into the truck. As with the examination of the receiving dock, the inspector should examine materials staged on the dock and should inquire about handling procedures for these outgoing hazardous materials.

Materials-Handling Equipment

The inspector should be knowledgeable about the methods used to move materials from one place to another in manufacturing plants, warehouses, and other occupancies. Efficient handling of materials requires specialized equipment that can introduce serious fire and explosion hazards unless the equipment is properly selected, installed, protected, operated, and maintained. This equipment includes industrial trucks, mechanical stock conveyors, pneumatic conveyors, and cranes.

INDUSTRIAL TRUCKS

The most common industrial trucks used to handle materials are lift trucks of the fork or squeeze-clamp type. They are powered by compressed natural gas, diesel fuel, electric storage batteries, gasoline, or LP-Gas, which introduces an additional source of fuel into the facility and also provides an ignition source. This is especially true if easily ignited fibers, dusts, vapors, or gases are present in the area in which the truck is to be used. Unless these vehicles are properly maintained and used, they can be hazardous even if they are the appropriate type for the hazard.

Industrial Truck Markers

A system of easily recognized markers has been developed to help identify the different types of industrial trucks and the areas in which they can be used safely (Figure 40-1). The markers have black borders and lettering on a yellow background. The appropriate markers are affixed to both sides of a truck so that its type can be readily identified.

Markers of corresponding shape should be posted at the entrances to hazardous areas. When an industrial truck approaches a hazardous area, both the occupants of the facility and the driver will be able to see whether the truck can safely enter the area by comparing the markings affixed to the truck and those posted in the area.

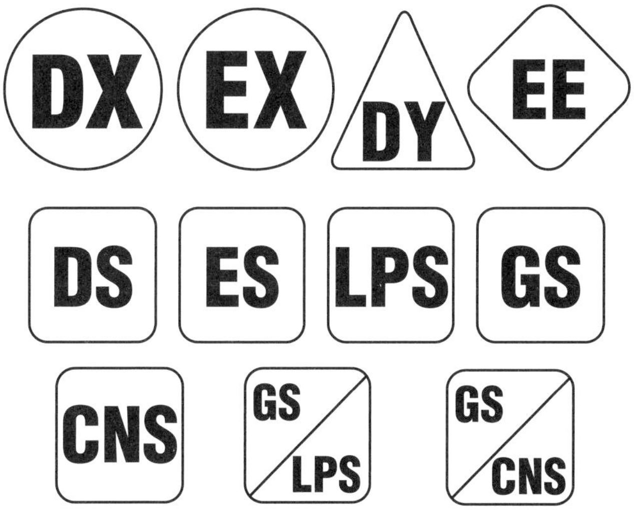

FIGURE 40-1 Markers Used to Identify Type of Industrial Truck
Source: NFPA 505, 1999, Figure 5-4.2.2.

NFPA 505, *Fire Safety Standard for Powered Industrial Trucks Including Type Designations, Areas of Use, Conversions, Maintenance, and Operation,* lists the different types of trucks and identifies the hazardous areas in which they may be used (see Table 40-1). Inspectors should be acquainted with the terms of NFPA 505 so they can provide guidance as to when the different types of trucks can and cannot be used. Many of the fires caused by industrial trucks have been the result of using the wrong trucks in the wrong locations.

Types of Trucks

Electric-Powered Trucks. Four types of electric-powered trucks are available, depending on the fire hazards present. These are the Type E, Type ES, Type EE, and Type EX. Type E trucks have the minimum necessary safeguards for use in ordinary hazard areas, Type ES have additional safeguards that prevent the ignition of fibers by controlling sparks emitted by the electrical system and limiting surface temperatures. In Type EE trucks, the electric motor and all other electrical equipment are completely enclosed so that they may be used in hazardous locations other than those that require Type EX trucks. Type EX trucks are explosion proof (Class 1, Group D) or dust tight (Class II, Group D) for use in areas in which explosive mixtures of flammable vapors or combustible dusts are likely during normal operations.

Gasoline-Powered Trucks. Two types of gasoline-fueled trucks are available, Type G and Type GS. Type G trucks have the minimum necessary safeguards for use in areas of ordinary fire hazard, and Type GS trucks have additional safeguards in the electrical, fuel, and exhaust systems for occupancies in which there are readily ignited combustible materials.

Diesel-Powered Trucks. Four types of diesel-fueled trucks are currently available. These are Type D, which are comparable in hazard to Type G gasoline-powered trucks; Type DS, which are comparable in hazard to Type GS gasoline-powered trucks; Type DY, which are equipped with additional safeguards to make them less hazardous than Type GS gasoline-powered trucks; and Type DX, which are diesel, powered units that can operate in specific atmospheres that contain flammable vapors, dusts and in some cases, fibers. Note that the surface and exhaust gas temperatures of Type DY trucks are limited, they have no electrical system, and they are equipped with other safeguards to minimize the fire hazard normally associated with internal combustion engines.

TABLE 40-1 Use of Powered Industrial Trucks

Locations	CNG-Powered		Diesel-Powered				Electric-Powered				Gasoline-Powered		LP-Gas-Powered		Dual Fuel–Powered			
	CN	CNS	D	DS	DY	DX	E	ES	EE	EX	G	GS	LP	LPS	G/CN	GS/CNS	G/LP	GS/LPS
Class I, Division 1																		
Group A	NA	NA	NA	NA	NA	NA	NA	NA	NA	NA	NA	NA	NA	NA	NA	NA	NA	NA
Group B	NA	NA	NA	NA	NA	NA	NA	NA	NA	NA	NA	NA	NA	NA	NA	NA	NA	NA
Group C	NA	NA	NA	NA	NA	NA	NA	NA	NA	NA	NA	NA	NA	NA	NA	NA	NA	NA
Group D	NA	NA	NA	NA	NA	A	NA	NA	NA	A	NA	NA	NA	NA	NA	NA	NA	NA
Class I, Division 2																		
Group A	NA	NA	NA	NA	K	K	NA	NA	K	K	NA	NA	NA	NA	NA	NA	NA	NA
Group B	NA	K	NA	K	K	K	NA	K	K	K	NA	K	NA	K	NA	K	NA	K
Group C	NA	K	NA	K	K	K	NA	K	K	K	NA	K	NA	K	NA	K	NA	K
Group D	NA	J	NA	J	A	A	NA	J	A	A	NA	J	NA	J	NA	J	NA	J
Class II, Division 1																		
Group E	NA	NA	NA	NA	NA	J	NA	NA	NA	J	NA	NA	NA	NA	NA	NA	NA	NA
Group F	NA	NA	NA	NA	NA	A[1]	NA	NA	NA	A[1]	NA	NA	NA	NA	NA	NA	NA	NA
Group G	NA	NA	NA	NA	NA	A	NA	NA	NA	A	NA	NA	NA	NA	NA	NA	NA	NA
Class II, Division 2																		
Group F[2]	NA	J	NA	J	A	A	NA	J	A	A	NA	J	NA	J	NA	J	NA	J
Group G	NA	J	NA	J	A	A	NA	J	A	A	NA	J	NA	J	NA	J	NA	J
Class III, Division 1	NA	J	NA	J	A	A	NA	J	A	A	NA	J	NA	J	NA	J	NA	J
Class III, Division 2	NA	A	NA	A	A	A	J	A	A	A	NA	A	NA	A	NA	A	NA	A
Ordinary (Unclassified)	A	A	A	A	A	A	A	A	A	A	A	A	A	A	A	A	A	A

[1]Where dust has resistivity of 10^5 ohm-cm, use J.

[2]Where dust has resistivity of 10^5 ohm-cm or greater.

A = Type truck authorized for location described

J = Type truck authorized for location described with approval of the authority having jurisdiction

K = Type truck authorized to be determined by the authority having jurisdiction

NA = Type truck not authorized in location described

Source: NFPA 505, 1999, Table 1-6.

LP-Gas-Powered Trucks. Two types of LP-Gas-fueled trucks are manufactured, Type LP and Type LPS. They are considered comparable in fire hazard to Type G and Type GS gasoline-powered trucks, respectively.

CNG-Powered Trucks. Compressed natural gas-powered trucks are still relatively uncommon and therefore are not covered here.

Dual-Fuel Trucks. There are two types of dual-fuel trucks that may be operated with either gasoline or LP-Gas. They are designated either Type G/LP, which are comparable in fire hazard to Types G and LP, or Type GS/LPS, which require the same safeguards against the hazards of exhaust, fuel, and electrical systems that Types GS and LPS do. Two other types of dual fuel-powered trucks are Type G/CN and Type GS/CNS, which can be operated with either gasoline or compressed natural gas.

Fire Hazards of Industrial Trucks

The major potential fire hazard associated with gasoline-, diesel-, LP-Gas-powered, and CNG-power trucks are fuel leaks ignited by the hot engine, the hot muffler, the ignition system, other electrical equipment, or other external ignition sources. This danger is lower for diesel trucks because diesel fuel has a higher flash point. Particular care must be exercised when storing and dispensing fuels, as with any flammable liquid or gas.

Comparatively few fires have occurred in battery-powered trucks. Nevertheless, electrical short circuits, hot resistors, and exploding batteries have been known to ignite fires. Frayed wiring and loose battery terminals also indicate potential trouble.

Careless and uninformed operation of industrial trucks contributes to property loss. The inspector should determine if there is an operator training program and that written operator rules are strictly enforced.

Also contributing to more serious fires are excessive material storage heights in the facility; careless handling of loads, especially containers of flammable liquids; and collision of the truck with sprinkler piping, fire doors, and other fire protection equipment. The provision of adequate, clear passageways for truck travel and clear warning of overhead obstructions and exposed piping will help to reduce accidents. The inspector should make particularly sure that sprinkler piping is well marked if there is any possibility that elevated loads will damage sprinklers or sprinkler piping.

Industrial Truck Maintenance

Proper maintenance of trucks is essential to good fire prevention. The inspector should check trucks carefully for signs of excessive wear and accumulations of lint, oil, and grease. A system of regularly scheduled maintenance based on engine-hour or motor-hour use can greatly reduce the danger of hazardous malfunctions. It is appropriate for the inspector to ask for and to review maintenance records as part of an inspection.

Trucks—particularly their fuel and ignition systems—should be maintained and repaired only in specially designated areas. Repairs should never be attempted within a hazardous area. Some basic points of good maintenance include filling water mufflers daily or as frequently as necessary to keep the water supply at 75 percent of the fill capacity, keeping muffler filters clean, and making sure that LP-Gas and CNG fuel containers are mounted securely to prevent them from jarring loose, slipping, or rotating. Any trucks seen giving off sparks or flames from the exhaust system or found to be operating at excessive temperatures should be removed from service immediately. The inspector should not permit them to return to service until the faults have been corrected. Each truck should be equipped with a portable extinguisher suitable for the fuel it uses.

Refueling and Recharging

Because almost half of the fires involving liquid-fueled trucks are caused by spillage during refueling, trucks should be refueled outdoors where there is minimal exposure to plant structures. The inspector should make sure the area is posted for "no smoking," that fuel-dispensing pumps are suitable for that use, and that scales for weighing the LP-Gas containers are accurately calibrated. LP containers should be stored outdoors, away from common fire exposures. The inspector should observe refueling operations.

Battery recharging operations also require special precautions. The corrosive chemical solutions (electrolytes) in the batteries present a chemical hazard to personnel. On charge, they give off hydrogen, which can accumulate to explosive concentrations if ventilation is inadequate. The inspector should note whether the recharging area includes some means for flushing and neutralizing spilled electrolytes and has a barrier for protecting charging apparatus. Most important, the inspector should make sure that the ventilation is adequate to carry hydrogen away from gassing batteries.

During the charging operation, vent caps should be kept in place to avoid electrolyte spray. The inspector should check to be sure that the vent caps are functioning and that battery covers remain open during charging to dissipate heat.

CONVEYING SYSTEMS

Mechanical conveyors and elevators are among the most commonly used equipment in materials handling. There are many types of mechanical conveyors, including the common belt conveyor. The belt conveyor presents two principal fire hazards: the material being carried, if it is combustible, and the combustible belt itself. The belt conveyor also may communicate fire from one building or area to another. Loss experience shows that automatic sprinkler or water spray protection may be warranted for important belt conveyors.

Fire Causes in Conveyors

Friction. One common cause of fire in conveyors is friction between the moving conveyor belt and a stuck roller or other object. Conveyors should be inspected regularly to detect belt slippage or defective rollers, and their moving parts should be lubricated on a regular basis. The use of fire-retardant belt material significantly reduces the belt's ignition potential. The inspector should determine if the belt is equipped with 20 percent slow-down interlock to shut down the conveyor if overloading or mechanical deficiency occurs.

Welding, Cutting, and Other Hot Work. Careless welding, cutting, or other hot work operations on or in the vicinity of a conveyor or its housing is another leading cause of fire. Hot slag can ignite combustibles on the belt, debris around it, or the belt itself. The inspector should determine if there is a hot permit system and that it is used (see Chapter 52, "Welding, Cutting, and Other Hot Work"). The permit system should be in accordance with NFPA 51B, *Standard for Fire Prevention During Welding, Cutting, and Other Hot Work).*

Dust. One factor that occurs frequently in fires involving mechanical conveyors is the dust that is produced when loose combustible materials are being conveyed. If dust generation is not prevented by the process arrangement, dust collection systems are needed.

Static Electricity. Static electricity is another hazard common to conveyors. Check to make sure all parts of the machines and conveyors are thoroughly bonded and grounded to minimize static discharges. Static electricity also can be controlled by the use of belts made of conductive material, the use of static collectors, or the application of conductive dressings to the belt surface (see NFPA 77, *Recommended Practice on Static Electricity*).

Other Causes. Discharging excessively hot materials from kilns, ovens, or furnaces onto a conveyor belt is a fire hazard in some plants, as is the spontaneous combustion of the materials being handled, such as coal. Electrical short circuits, smoking, and incendiarism are other potential causes of conveyor fires.

Other Mechanical Conveyors

Chain conveyors equipped with hooks and roller conveyors are commonly used in assembly lines, whereas screw conveyors, pan conveyors, and bucket conveyors are used for handling loose, hot, or molten materials. Belt conveyors are generally not suitable for handling materials over 150°F (66°C).

Bucket elevators, which are found most often in bulk processing plants, convey loads vertically and are susceptible to the same fire hazards as are other mechanical conveyors. These conveyors might need specialized fire protection if they handle combustible dust or materials.

Protection of Conveyor Openings

Protecting the openings in fire walls and floors through which conveyors pass is important in preventing the spread of fire. Many methods have been designed to protect openings for conveyors of different types, including water spray protection and fire doors with interlocks or counterweights to stop the flow of material. See Figures 40-2 and 40-3 for examples of opening protection for conveyors. It should be noted that the use of passive protection might be preferable to the use of suppression alone based on the type of material present in the area. (See the Bibliography for additional information on the protection of floor and wall openings.)

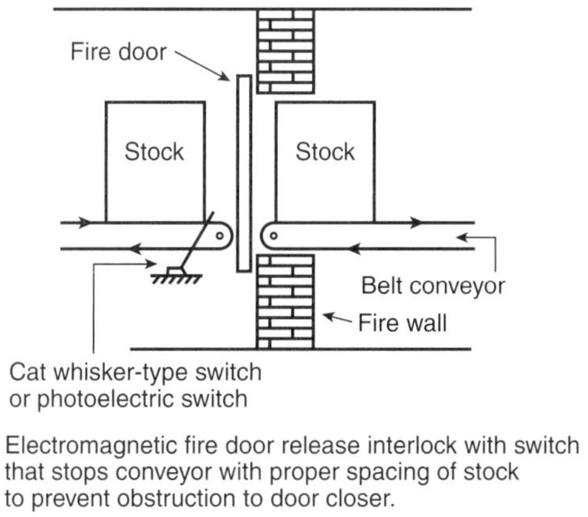

FIGURE 40-2 Protection of an Opening Where a Belt Conveyor Can Be Interrupted
Source: NFPA 80, 1999, Figure C-2.

FIGURE 40-3 Conveyor Carried Over a Fire Wall
Source: NFPA 80, 1999, Figure C-1.

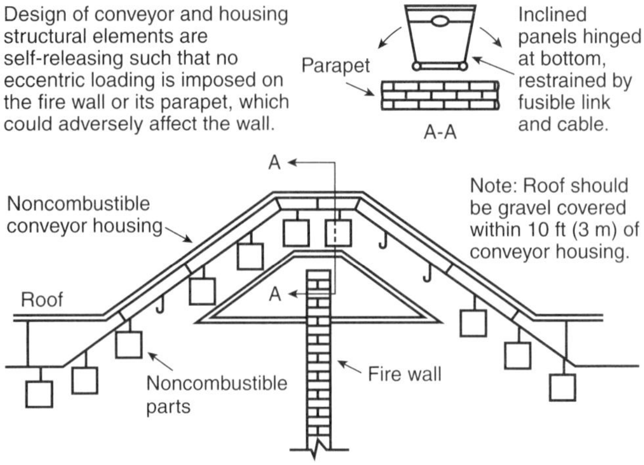

Pneumatic Conveyors

Pneumatic conveyor systems consist of enclosed tubing in which a material normally is transported by a stream of air with enough velocity to keep the conveyed material in motion. There are two principal types of pneumatic systems. The pressure type uses air at greater than atmospheric pressure; the suction type transports materials using air at less than atmospheric pressure.

In an air-conveying system, any dry collector containing a dust–air mixture in the explosive range must be considered a potential explosion hazard (see Chapter 40, "Combustible Dusts"). The inspector should look at the collector carefully to make sure it is in a safe location—preferably outdoors—to protect personnel. The collector should be constructed of conductive noncombustible materials and be completely bonded and grounded. Construction should be of nonferrous, nonsparking metal or nonmagnetic, nonsparking stainless steel.

Where conveying ducts are exposed to weather or moisture, the inspector should examine them carefully to make sure they are moisture tight. Moisture entering the system can react with dust, which generates heat and serves as a potential source of ignition. If the conveying gas-air mixture is relatively warm and the dust and collectors are relatively cold, the gas temperature could drop below the dew point, causing moisture to condense. If this is the case, it might be necessary to insulate the ducts and collectors or provide a heating system.

Explosion relief for ducts and collectors should always extend to the outside. They can be provided with antiflashback swing valves or rupture diaphragms. Fans and housings for fans that are used to move combustible solids or vapors should be constructed of conductive, nonferrous materials. It is important to note whether dust is drawn through the fan before it enters the final collector. The inspector should also note whether the fan bearings are equipped with suitable temperature-indicating devices wired with an alarm to alert occupants to overheating.

The inspector should determine by physical examination whether dust or other combustible materials are accumulating in the ducts. If so, it may be the result of inadequate carrying velocity within the ducts.

CRANES

Cranes are principally used to move heavy materials. Some cranes move along rails, including overhead traveling, gantry, tower, and bridge cranes. Overhead traveling cranes may be used either indoors or out; gantry, tower, and bridge cranes are used mainly outdoors. Outdoor cranes may be damaged by high winds.

Large cranes that move along rails generally have automatic or manual rail clamps or some means of anchorage, such as crane traps, wedges, and cables. The inspector should inquire about the method of anchorage.

Crane operators' cabs should be constructed of noncombustible materials. The inspector should check to make sure that they are free of oily waste, rubbish, and other combustibles. Some large cranes, as well as other high valued mobile equipment, have hydraulic oil systems and on-board fuel systems. A release of the pressurized oil or fuel can produce a severe fire within the power compartment. Often automatic fixed fire protection is provided within the power compartment and the cab. Each cab should be, at a minimum, equipped with portable fire extinguishers.

SUMMARY

When inspecting a facility's materials-handling system, the inspector must be aware of the types of material brought into the facility and how they are handled and transported within the facility from the receiving dock to the storage area or area of use and to the shipping dock. The materials must be examined to determine their classification and their fire protection requirement based on the current standards. Efficient and safe handling of materials requires specialized equipment. This equipment includes industrial trucks powered by electric storage batteries, gasoline, LP-Gas, diesel fuel, or compressed natural gas; various different types of conveying systems; and cranes. Because this specialized equipment can introduce fire and explosion hazards, the inspector must inspect it to ensure that that it is installed, protected, operated, and maintained properly.

BIBLIOGRAPHY

Cote, A. E., ed., *Fire Protection Handbook,* 18th ed., NFPA, Quincy, MA, 1997.

Cote, A. E., and Linville, J. L., eds., *Industrial Fire Hazards Handbook,* 3rd ed., NFPA, Quincy, MA, 1990.

Mitchell, D. W., et al., *Fire Hazard of Conveyor Belts,* Report of Investigations 7053, U.S. Bureau of Mines, Washington, DC, 1967.

NFPA Codes, Standards, and Recommended Practices

See the latest version of The NFPA Catalog for availability of current editions of the following documents.

NFPA 13, *Standard for the Installation of Sprinkler Systems*

NFPA 15, *Standard for Water Spray Fixed Systems for Fire Protection*

NFPA 51B, *Standard for Fire Prevention During Welding, Cutting, and Other Hot Work*

NFPA 68, *Guide for Venting of Deflagrations*

NFPA 69, *Standard on Explosion Prevention Systems*

NFPA 77, *Recommended Practice on Static Electricity*

NFPA 80, *Standard for Fire Doors and Fire Windows*

NFPA 498, *Standard for Safe Havens and Interchange Lots for Vehicles Transporting Explosives*

NFPA 505, *Fire Safety Standard for Powered Industrial Trucks Including Type Designations, Areas of Use, Conversions, Maintenance, and Operation*

NFPA 654, *Standard for the Prevention of Fire and Dust Explosions from the Manufacturing, Processing, and Handling of Combustible Particulate Solids*

General Storage

CHAPTER 41

Stephen R. Hoover

CASE STUDY

Products Stored Too Close to Lighting Ignites

Foam products stored in a large warehouse in Iowa ignited when they were placed too close to ceilling light fixtures. Improperly arranged storage also caused a number of additional sprinklers to activate before the fire was controlled.

The single-story warehouse measured 307 × 160 ft and was of unprotected, noncombustible construction. It was protected by two wet-pipe sprinkler systems that provided full coverage and a water-flow alarm. Two employees were in the warehouse at the time of the fire.

Overhead electrical lighting ignited foam rubber products stored near them at the top of a rack, causing several sprinklers to operate. The sprinkler activation, in turn, alerted the two employees, who called the fire department at 7:07 P.M. Arriving 4 minutes later, fire fighters found that the sprinklers had controlled the blaze, and they began using their hose lines to extinguish spot fires.

High storage was cited as the cause of the fire and as a contirbutor to fire spread and water damage, which was extensive, since the fire activated 361 sprinklers. The building, valued at $1 million, contained products valued at $500,00. The products were a total loss, and the building suffered $440,000 in damage. There were no injuries.

Source: "Fire Watch," *NFPA Journal,* January/February, 1998.

Stephen R. Hoover, P.E., has been in the fire protection industry for 40 years. During that time he has worked with the Industrial Risk Insurers, the HPR Department of Kemper Insurance, and the Loss Control Department of the Schirmer Engineering Corporation. He has been involved with various NFPA committees, including as chair of the Rubber Tires Committee.

Fire hazards found in storage occupancies are as varied as the products that are used in everyday life. The commodities being stored, the storage arrangements used, and the height of the storage all have an influence on the potential type of fire, on its intensity, and on the protection needed for the storage area. Collectively, these items, combined with the building construction, determine the type and level of protection necessary to protect the stored commodity. This chapter provides an overview of the information on the protection of general storage that is available from many sources.

COMMODITY CLASSIFICATION

The most crucial and often most difficult task is correctly assessing the fire potential of the commodity being stored. Many products are composed of a variety of materials, each having different burning characteristics. In addition, the performance of the commodity during a fire can be influenced significantly by the materials in which it is packed and the type of packaging being used. For example, a highly combustible product sealed inside a metal container might represent only a minimal fire hazard. Conversely, a metal product enclosed in a thick foamed polystyrene cocoon could pose a serious fire challenge. In most cases, the packaging material presents just as much of a threat to the structure as the actual stored product.

To help the warehouse manager and the fire inspector assess the fire hazards of stored materials, the sprinkler standard contains classification categories with attendant definitions and examples (see the Bibliography). These commodity classification systems must be studied carefully because the lower numerical or alphabetical designations indicate lesser hazards in some instances and signify greater hazards in others.

Most products found in a general-purpose warehouse can be categorized under the commodity classification system found in either NFPA 13, *Standard for the Installation of Sprinkler Systems,* or NFPA 230, *Standard for the Fire Protection of Storage.* This basic system establishes four categories—Class I, II, III, and IV—with Class I representing a minimal hazard and Class IV the greatest hazard. A commodity defined as Class I is noncombustible but presents a challenge to the fire protection system because of the packing container or the pallet on which it is stored. Since the product in the container or on the pallet is a noncombustible material, the installed fire protection measures are based on the fire hazard associated with the storage container or pallet. A Class III commodity, on the other hand, involves the storage of a combustible product in a combustible container.

The commodity classification, therefore, is governed by the types and amounts of materials (metal, wood, plastic, paper, etc.) that are part of the product and its packaging. However, in a storage facility the classification is also influenced by the makeup of the outer packaging or shipping container, the air space in the shipping container, the type of packing material used in the shipping container to prevent damage (e.g., newspaper and plastic "peanuts"), and the location of more hazardous materials in the container. For instance, in an example given in NFPA 13, a commodity that is an all metal product encased in expanded plastic packing that occupies 25 percent of the container behaves as an expanded plastic commodity in a fire.

For those products that are mostly or totally composed of plastics, a fifth description places plastics in three groups. Group A plastics represent the greatest level of fire hazard and Group C the lowest level. Further, Group B plastics are generally equated with the general Class IV category, whereas Group C plastics are generally equated with the Class III category. Since the protection measures for Group A plastics are so different, the use of a term such as "Class V commodities" has been avoided.

Since several NFPA standards have been combined to form NFPA 13 and NFPA 230, it seems desirable to mention commodities such as rolled paper, baled cotton, rubber tires, and forest products.

- Rolled paper classifications are provided as heavyweight, mediumweight, lightweight, and tissue paper. Heavyweight represents the minimum hazard and tissue paper the greatest hazard. Here, too, the wrapping or packaging (if any) can influence protection as, for instance, tissue paper can be protected as mediumweight paper when wrapped with one or two layers of heavyweight paper or with steel bands at the ends of the roll.
- Baled cotton does not have a classification breakdown.
- Forest products and scrap tires do not have classifications and are generally stored out-of-doors.
- Road-worthy rubber tires are stored in large warehouses or distribution centers but still do not have classifications, per se. Larger and heavier tires—for instance, off-road and earth-moving equipment tires—are harder to ignite. Otherwise the severity of a fire is influenced by the storage arrangement.

When determining any of these classifications, and hence the fire protection for the warehouse, the inspector must consider not only what is present on a given day but the potential for future changes in commodities or in the overall storage occupancy as well. The classic example in recent years has been the large-scale substitution of plastics for components that previously were made of metal, resulting in the need for major revisions or reinforcement in many warehouse sprinkler systems. This problem is further exacerbated by the tendency of some developers to build warehouses on speculation. No specific tenant or storage arrangement is considered at the time of construction; once the warehouse is completed, the owner hopes to find a tenant for the storage space.

STORAGE ARRANGEMENTS

For the purpose of fire protection, storage can be divided into four main categories: bulk, solid piling, palletized, and rack. The chief difference between the four major categories as far as their effect on fire behavior and fire control is concerned is the nature of the horizontal and vertical air spaces, or "flues," that the storage configurations create.

Bulk Storage

Powders, granules, pellets, chips, and flakes, all unpackaged, are the principal forms of materials that are stored in bulk. Silos, bins, tanks, and large piles on the floor are the usual storage methods.

Fires that start on the surface of large piles tend to burrow into the pile, particularly in the case of coarse particles, such as chips or pellets, making fire control with sprinklers very difficult. Having small hand-hose stations installed around the perimeter of the pile for early fire control is, therefore, advantageous.

Certain materials, such as wood chips or pulverized coal, can spontaneously combust in the interior of large piles or when trapped for long periods of time in the corners or under the interior baffles of tanks or silos. Fires of this nature can be extremely difficult to extinguish; removing the burning material from the pile or storage vessel might be the only way to put out the fire. Prefire planning should encompass this possibility and provide for the needed handling and transporting equipment, hose streams, and disposal area. Supplemental agents added to the water, such as low- and high-expansion foam concentrates, can sometimes help control and extinguish fires in these environments. Low-expansion foams can enhance the penetrability of the water into the material, while high-expansion foams can be used to control dust while the product is being unloaded.

Conveyor equipment, such as belt conveyors, air fluidizing through ducts, and bucket conveyors, agitates materials as they move to and from bulk storage. If this agitation produces combustible dust clouds, particularly in grain storage facilities, an explosion hazard may result. Automatic sprinklers frequently are needed in the housings around conveyor equipment. Select NFPA standards, such as NFPA 61, *Standard for the Prevention of Fires and Dust Explosions in Agricultural and Food Products Facilities*, provides special requirements for minimizing dust production.

Solid Piling Storage

Cartons, boxes, bales, bags, and so on that are in direct contact with each other to the full dimension of each pile make solid piles. These may be piled directly on the floor or the bottom layer may be placed on a pallet to prevent water damage. Air spaces, or flues, exist only where contact is imperfect or where a pile is close to, but not touching, another pile. Compared with palletized and rack storage, solid piling typically results in the slowest fire growth. Generally, the outer surfaces burn and it takes a long time before the interior of the pile becomes involved. However, if the outer surfaces possess rapid flame spread properties—as do bales of fibers such as cotton, for example—high solid piles can still be a severe hazard.

The height of the pile is governed by the weight of the commodity and by whether the commodity can be lifted without the need for forklifts, clamp trucks, or any other materials-handling equipment, which would create horizontal channels (air spaces) or vertical flues in the piles. As the pile increases in height, the weight of the uppermost portion of the pile cannot be allowed to crush the items at the bottom of the pile. The creation of horizontal channels or spaces and vertical flues changes the burning characteristics from those of a solid pile to those of palletized storage.

Palletized Storage

Palletized storage consists of unit loads mounted on pallets that can be stacked on top of each other. Each pallet is about 4 in. (101.8 mm) high and is usually made of wood, although some are made of metal, plastic, expanded plastic, and cardboard (Figure 41-1). The height of palletized storage is limited by the materials-handling equipment and the resistance to crushing at the lowest part of the pile. Pile heights are usually 30 ft (9.14 m) at the most. The increased hazard of palletized storage results from the horizontal air spaces formed by the pallets themselves within each layer of storage, as well as the vertical flues that exist between individual rows of palletized stacks.

Rack Storage

Rack storage consists of a structural framework supporting unit loads, generally on pallets (Figure 41-2). The height of storage racks is limited, potentially, only by the vertical reach of the materials-handling equipment, which, like the racks themselves, can be designed for great heights. In fully automated warehouses, racks are sometimes as high as 100 ft (30.48 m); some have even exceeded that height. In some installations, the steel storage racks are part of the building framing. These are referred to as rack-supported structures (Figure 41-3).

Even more than they do in palletized storage, the vertical and horizontal flue spaces that surround each unit load in rack storage promote rapid fire development and efficient combustion. As a result, rack storage normally requires greater sprinkler protection than equal heights of equivalent material stored in palletized or solid pile arrays. In some, but not all, cases, this protection may include the installation

FIGURE 41-1 Conventional Wood Pallet
Source: NFPA 230, 1999, Figure A-1-4.1(b).

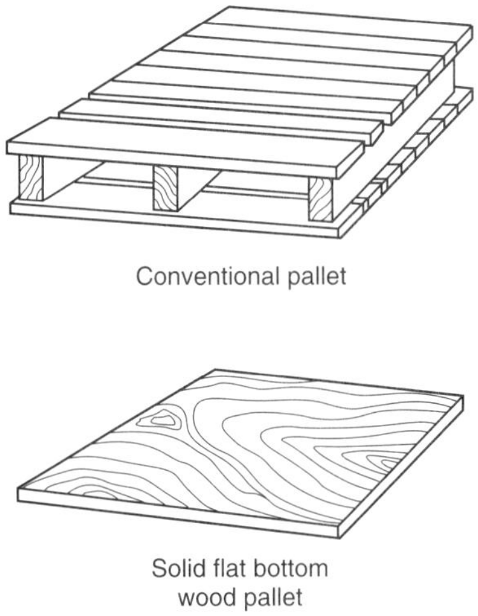

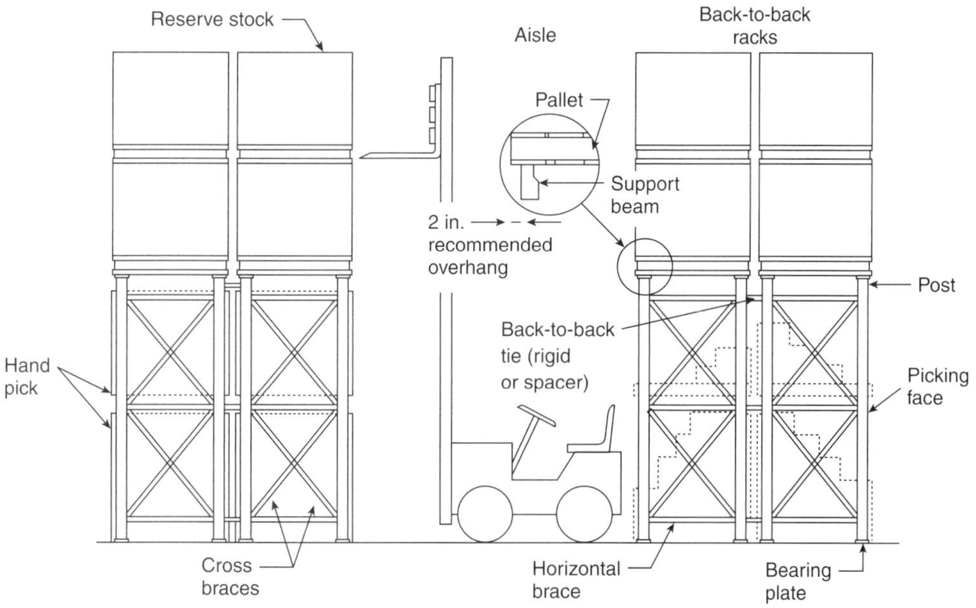

FIGURE 41-2 Common Arrangement of Double Row Racks with Palletized Storage Atop

of in-rack sprinklers. NFPA 13 uses decision tables to specify when in-rack sprinklers are necessary and what type of in-rack sprinklers, installed either in the longitudinal flue or at the rack face, are needed.

When rack storage is higher than 25 ft (7.62 m), special attention must be given to the problems that will exist in achieving fire salvage and overhaul. Although in-rack sprinkler protection can successfully control a fire that originates in, or spreads to, the upper reaches of high racks, full extinguishment cannot be expected.

Carousel Storage and Retrieval Systems

Carousel storage and retrieval systems are factory-built, motorized storage systems that revolve around a fixed base. Generally, the path of revolution has two long,

FIGURE 41-3 High Steel Storage Racks as Structural Support for Roofs and Walls of a Storage Facility

parallel sides connected by a round, short radius. They use fixed tracks with the motor mounted either on the top or the bottom track. Although most carousels travel horizontally, they may also travel vertically or both horizontally and vertically.

Products stored in the carousel are brought to a stationary picking station either manually or by computer command. Often the commodities are small parts used in assembly operations. There are no national standards available for the protection of this equipment.

FIRE CAUSES

Fires in storage areas are more likely to start with people than with any kind of equipment problems, although natural causes and exposure fires are important in this property class, too.

Leading Causes

Arson is the leading cause, and any lack of security, such as outside storage of combustibles, can tempt anyone from the juvenile firesetter to the profit-oriented arsonist. Warehouses with high values, desirable contents, and large concentrations of readily ignitable goods are natural targets, not only for arsonists, but also for thieves, some of whom may try to cover up their activities by setting fires when they leave. Among equipment-related causes, wiring, storage in close proximity to space heaters, and vehicles are worthy of attention.

In NFPA 13 and NFPA 230, approximately 90 percent of the protection criteria is centered on the installation of automatic sprinkler systems. Other preventative measures should also be in place. Smoking should be limited to well-defined, supervised, and safe locations and prohibited entirely in the storage areas. Cutting and welding operations must be handled only by means of a rigorous permit system that includes an authorized preinspection of the contemplated work area, removal of all combustibles to a safe distance, the use of flameproof tarpaulins, and the presence of an equipped firewatch during the work period and for at least 30 minutes after work ends. (See Chapter 52, "Welding, Cutting, and Other Hot Work" or NFPA 51B, *Standard for Fire Prevention During Welding, Cutting, and Other Hot Work,* for more information on hot work operations.) Forklift vehicles must be well maintained, and any combustion engine units must be refueled outside the storage areas.

Encapsulation

During encapsulation, a plastic sheet completely encloses the sides and top of a pallet load containing a combustible commodity, a combustible package, or a group of combustible commodities or packages. Because encapsulation decreases the possibility of prewetting, encapsulated commodities may allow greater fire spread. The plastic sheet protects the commodity from damage due to instability and exposure to the elements, but it also prevents cooling and prewetting of the commodity during a fire. The sprinkler water simply runs off the top and sides of the load, while at the same time allowing the fire to possibly spread under that same protective plastic.

FIRE PROTECTION FOR GOODS IN STORAGE

Automatic Sprinklers and Types of Storage

The installation of automatic sprinkler systems is the primary means for protecting the various commodities found in a warehouse. The type of storage will govern the specific parameters for design of the system. Because of the variety of challenges posed by the different types of commodities and storage arrangements, warehouse sprinkler systems must always be designed hydraulically in order to ensure the proper level of protection.

Types of Automatic Sprinklers

For conventional sprinkler systems using ½- or 17⁄32-in. (12.7- or 13.5-mm) orifice sprinklers, the design consists of supplying a specific controlling density (gallons per minute per square feet or liters per minute per square meters) over a prescribed area of application, referred to as the "remote area." This method is the "area-density" method. Methods for determining the appropriate density and area of application can be found in the applicable section of NFPA 13. The inspector should study these sections and NFPA 230 carefully to understand all the physical features that have an effect on fire development and the effectiveness of sprinkler protection. These features would include items such as the height of the storage, encapsulation of the stored product, clearance to sprinklers, the size of flue spaces, and the presence of obstructions.

***K* Factor.** K factor is a constant found in the formula $Q = K \sqrt{p}$ for flow through a specific orifice, where Q = flow in gpm (liters per minutes) and p = pressure in pounds per square inch (bars). For reference, the ½-in. (12.7-mm) sprinkler has a nominal *K* factor of 5.6 (7.6–8.4 SI) and the 17/32-in. (13.5-mm) sprinkler has a nominal *K* factor of 8 (10.7–11.8 SI). This difference is important because the *K* factor rather than the orifice size is becoming the method for identifying sprinklers.

Large-Drop Sprinkler. A major change in fire sprinkler design occurred when the large-drop sprinkler was developed. The large-drop sprinkler, which generates a higher percentage of large droplets that will penetrate a high velocity fire plume, made ceiling sprinkler protection more effective. This sprinkler has a 0.64-in. (15.9-mm) orifice and a nominal *K* factor of 11 (15.9–16.6 SI).

Extra Large Orifice (ELO) Sprinkler. The extra large orifice (ELO) sprinkler is another K-11 sprinkler, but it is not the large-drop type. The ELO sprinkler is available as a pendent or upright sprinkler and is used when the fire sprinkler system design calls for the area-density method of protection.

Early Suppression Fast Response Sprinkler. Another major development in fire sprinkler design was the early suppression fast response (ESFR) sprinkler. It is a K-14 (19.5–20.9 SI) sprinkler that delivers a high volume of water [100 gpm (4074.6 mm/min) at 50 psi (344.74 kPa) operating pressure] early in the development of the fire through its use of a fast-response operating element. Most of these sprinklers are pendant but there is at least one upright version.

Another ESFR sprinkler is the K-25 (34.9–38.7 SI). It uses a 12-sprinkler design and the pressure selection is based on the commodity class, storage arrangement, and building height. The advantage to using the K-25 sprinkler is that, because of its large orifice, it can achieve water flows equal to that of the K-14 ESFR but at lower operating pressures.

Neither the K-14 nor the K-25 sprinkler can be used in occupancies in which fires cannot be suppressed by water. At this time the K-25 sprinkler has only a small range of applications proved by large-scale testing.

Very Extra Large Orifice Sprinkler. Another K-14 sprinkler but not an ESFR one is the very extra large orifice (VELO) sprinkler. This standard spray sprinkler is a pendant sprinkler and is used in fire sprinkler systems that are designed for the area-density method.

Large-scale fire tests have demonstrated that this combination of heavy discharge and early response can protect up to 35 ft (10.69 mm) of racked or palletized storage of plastics, often with as few as 12 sprinklers operating. Recent testing has shown that these devices can protect higher combinations of stored product when used at higher operating pressures.

The advantage to using the K-11 large-drop sprinkler or the K-14 or K-25 sprinkler in their specific applications for rack storage is that in-rack sprinkler protection is not needed. The K-25 also may reduce the need for fire pumps. Both the K-14 and K-25 sprinklers are suppression mode rather than control mode sprinklers.

Another sprinkler that will provide adequate hydraulic design at lower pressures is the K-17 (23.1–25.4 SI) sprinkler. When this sprinkler is used, the design approach is the area-density method.

Hydrants, Hose Streams, and Extinguishers

Hydrants. Small storage facilities close to public hydrants present no special problem. In larger buildings, where the hydrants are not within 250 ft (76.2 m) of the building, however, hose lays from public hydrants to the far side of the building often can be impractical and ineffective. In these instances, private hydrants are needed. These hydrants should be located near points of entrance to the building to reduce the amount of hose needed.

Small Hose. Stations for 1½-in. (38.1-mm) hose should be located throughout the premises so that all areas can be reached. The hose can be supplied by adjacent wet-pipe sprinkler systems or by a separate piping system. These hose stations can be used for first-aid fire fighting or for overhaul operations by the fire department. Hose streams must be managed carefully so that they do not rob the sprinklers of water, which might cause the loss of fire control. This is done by requiring the designer to reserve a given quantity of water in the hydraulic calculations for use by the fire department for hose line application.

Portable Fire Extinguishers. Extinguishers are needed to fight Class A (wood, paper, fabrics), Class B (flammable liquid), Class C (electrical), and Class D (combustible metals) fires and to protect in-plant vehicles. Portable extinguishers are of limited value on storage fires because of their short duration and limited reach. Often, piles have to be pulled apart manually to extinguish a fire within them and hose lines are needed to control the fire during these mop-up operations.

STORAGE OF SPECIFIC MATERIALS

Some commodities deserve special mention because of the nature of the fire hazards involved.

Idle Pallets

Piles of wood or plastic pallets are a severe fire hazard. After they have been used for a short time, pallets can dry out, and their edges can fray and splinter. They can ignite easily, and, even when sprinklers are operating, the underside of the pallets can provide a shielded area where fire can grow and expand.

Generally, idle wood or plastic pallets should be stored out-of-doors or in a detached structure. When indoor storage of idle wooden pallets and nonexpanded polyethylene solid deck pallets is needed, then the piles are limited in height and in number to four stacks per pile, with each pile separated from the next by 8 ft (2.44 m) of clear space or 25 ft (7.62 m) of commodity. Within the limits given, standard spray sprinklers can be used for protection.

When indoor storage of idle plastic pallets is needed and cannot be cutoff from the other storage, then the piles are limited to 4 ft (1.22 m) in height and to two stacks in size. The piles must be separated from one another by 8 ft (2.44) of clear space or 25 ft (7.62 m) of commodity; and only a sprinkler system using high temperature sprinklers can be used.

However, if a 3-hour-rated cutoff room can be provided for the idle plastic pallet storage, then pile heights can be up to 12 ft (3.66 m). The room must be protected by sprinklers designed to a density of either 0.60 gpm/ft^2 (24.44 mm/min) over the entire room or 0.30 gpm/ft^2 (12.22 mm/min) over the entire room combined with a high expansion foam system.

Where large-drop sprinkler protection is provided, indoor storage of idle wooden pallets 20 ft (6.1 m) in height in a 30-ft- (9.14-m-) high building can be protected. Where ESFR sprinkler protection is provided, indoor storage of idle wooden and plastic pallets 25 ft (7.62 m) in height in a 30-ft- (9.14-m-) high building can be protected using K-14 sprinklers operating at 50 psi (344.74 kPa); storage of idle pallets 35 ft (10.67 m) in height in a 40-ft- (12.19-m) high building can be protected using K-14 sprinklers operating at 75 psi (517.11 kPa).

Rack storage of idle wooden and plastic pallets is not allowed by NFPA 13 unless ESFR sprinkler protection is provided. When ESFR protection is provided, then storage of idle pallets 25 ft (7.62 m) in height in a 30-ft- (9.14-m-) high building can be protected using K-14 sprinklers operating at 50 psi (344.74 kPa); storage of pallets 35 ft (10.67 m) in height in a 40-ft-(12.19-m) high building can be protected using K-14 sprinklers operating at 75 psi (517.11 kPa).

NFPA 13 allows idle indoor storage of nonwood (plastic) pallets that have demonstrated a fire hazard equal to or less than wood pallets and where they have been listed for this equivalency to be protected as a wood pallet. The Factory Mutual Research Corporation (FMRC) lists these pallets for idle pallet storage only. The Underwriters Laboratories (UL) lists them for both idle pallet storage and for use with a commodity. If these listed pallets were stored in racks, then the ESFR protection as noted earlier would have to be provided.

Rubber Tires

NFPA 13, *Standard for the Installation of Sprinkler Systems,* and NFPA 230, *Standard for the Fire Protection of Storage,* provide guidance for tire storage arrangement and the various methods of protection. Tires can be stored on-side or on-tread and in a laced-tire configuration directly on the floor or in fixed or portable racks. Tires in storage present a very challenging fire hazard. Though relatively hard to ignite, once started a tire fire generates a tremendous amount of heat and smoke and is extremely difficult to extinguish.

Fire tests used to establish design criteria were run without automatic smoke and heat venting. Some method of venting the building to remove smoke is needed to allow fire fighters to enter promptly and to allow salvage and overhaul to begin as soon as possible after the fire is brought under control.

Inspectors should remember the following specific items:

1. When tires are stored on-tread—whether on the floor or in a rack—the pile width should not exceed 50 ft (15.24 m). It is feasible to fight a fire in the wheel holes with hose streams from either aisle, and the effectiveness of the hose stream is about 25 ft (7.62 m) into the wheel hole.
2. Sometimes tires stored on-side on the floor are plastic wrapped for stability. Truck tires, both wrapped and unwrapped, were tested in large-scale fires at FMRC, and the wrapping had no detrimental effect on the sprinkler performance.

3. When tires are considered miscellaneous storage (a very specific definition found in Section 1.4.10 of NFPA 13) and are stored in accordance with NFPA 13, they can be protected by sprinkler systems designed to meet the requirements of ordinary and extra hazard occupancies.
4. Large-drop sprinkler protection is now available for virtually all rack storage arrangements, excluding laced-tire arrangements. Storage up to 25 ft (7.62 m) high in a 32-ft (9.75 m) building can be protected by 15 sprinklers operating at 75 psi (517.11 kPa) at the sprinkler.
5. ESFR sprinkler protection is now available for virtually all rack storage arrangements, excluding laced-tire arrangements. Storage up to 25 ft (7.62 m) high in a 30-ft (9.14-m) building can be protected by 12 sprinklers operating at 50 psi (344.74 kPa) at the sprinkler; storage up to 25 ft (7.62 m) high in a 35-ft (10.67 m) building can be protected by 12 sprinklers operating at 75 psi (517.11 kPa) at the sprinkler.
6. ESFR sprinkler protection is now available for laced-tire storage arrangements in open portable racks. Storage up to 25 ft (7.62 m) high in a 30-ft (10.67 m) building can be protected by 20 sprinklers operating at 75 psi (517.11 kPa) at the sprinkler. Note that the number of design sprinklers is 20, not the 12 normally used for ESFR systems.
7. K-25 sprinklers are also available for the protection of tires in a rack storage arrangement, excluding laced-tire arrangements. However, the K-25 sprinkler has been tested only with plastics as the commodity and has never been tested using rubber tires as the commodity.
8. Fire tests have shown that high-expansion foam in conjunction with sprinkler protection will extinguish tire fires.

Roll Paper

Roll paper can be stored on its side, on end, on pallets, or in racks. Rolls stored on end as separate columns are the most hazardous configuration and present the greatest challenge to sprinkler protection. NFPA 13, *Standard for the Installation of Sprinkler Systems,* and NFPA 230, *Standard for the Fire Protection of Storage,* provide guidance for storage arrangement and the various methods of protection. These standards classify the hazard of roll paper primarily according to the weight of the paper. The categories are heavyweight, mediumweight, and lightweight, with heavyweight paper representing the least hazard and lightweight, particularly tissue paper, the greatest. If mediumweight or lightweight papers are enclosed in a heavyweight paper wrapper on the sides and ends, however, the storage can be reduced one class—that is mediumweight paper can be protected as heavyweight; lightweight and tissue paper can be protected as mediumweight.

The storage array—closed, open, or standard—also makes a difference in the protection provided. A *closed array* is a vertical arrangement in which the distance between the columns is no more than 2 in. (50.8 mm) in one direction and 1 in. (25.4 mm) in the other. The *open array* is a vertical arrangement in which the distance between the columns is greater than the distances in either the closed or standard array. The *standard array* is a vertical arrangement in which the distance between the columns is 1 in. (25.4 mm) or less in one direction but over 2 in. (50.8 mm) in the other direction. Banding the ends of the rolls can also make a difference in the protection requirement.

The array is important to the protection provided as the closed array is the lesser hazard, requiring low densities and small remote areas. The open array is the greatest hazard and requires high densities and large areas.

Fire protection for roll paper is best provided by automatic wet-pipe sprinkler systems, although dry-pipe systems may be used for all weights of paper except tissue. Horizontal storage of heavyweight and mediumweight paper is protected as closed array vertical storage (the least hazardous). Otherwise all protection requirements provided by NFPA 13 are for vertical storage.

Extreme caution is necessary when fighting a fire in the vertical storage of roll tissue paper, something that should be noted in prefire planning procedures. As rolls of tissue sitting on the floor absorb water, the vertical stacks become unstable and are prone to collapse, scattering the heavy rolls in all directions.

Aerosols/Flammable Liquids

Flammable aerosols or liquids present special hazards, and their introduction into general purpose warehouses can spell disaster. The sprinkler protection probably will be inadequate for the hazard, and, as a result, any fire involving such storage will escalate so rapidly that it will be out of control before the fire brigade or fire department can mount an effective attack. The best protection measure is to store all flammable aerosols and liquids in separate cutoff areas and provide specialized protection to these areas.

Industry and insurance interests have established a classification system for aerosols that consists of three groups, Levels I, II, and III, with Level I representing the lowest hazard. NFPA 30B, *Code for the Manufacture and Storage of Aerosol Products,* governs the manufacturing, handling, and storage of aerosol products.

Flammable and combustible liquids are classified in NFPA 30, *Flammable and Combustible Liquids Code,* with Class IA representing the highest hazard and Class IIIB the lowest. New criteria were included in NFPA 30 in 1990 to govern the storage of some liquids in plastic containers and more changes have appeared in subsequent editions to clearly spell out the requirements for the display and storage of flammable and combustible liquids in mercantile occupancies and their associated warehouses, where the problem is most acute.

Refrigerated Storage

Temperatures in cold storage warehouses range from 40° to 65°F (4 to 18°C) for products, such as fruits, eggs, or nuts, that would be damaged by the lower temperatures of 0° to −35°F (−18° to −37°C) needed for frozen foods. In other facilities, the temperature may be kept as low as −60°F (−51°C).

Whether the building construction is combustible or not, its insulating materials, such as the widely used expanded plastics foamed polystyrene and polyurethane, generally are combustible. Exposed foamed plastics on the walls and ceilings of warehouses are an unacceptable fire hazard because of the potential they present for rapid fire spread over the surface and for heavy smoke generation. In general, these insulations should be covered with a cementitious plaster or one of several proprietary coatings listed specifically for this purpose. Some low-flame-spread foamed isocyanate, isocyanurate, and phenolic plastics have been tested and listed for ceiling and wall applications when protected only by an adhered foil or sheet metal covering.

Combustible materials found in cold storage warehouses include wood dunnage, wood pallets, wood boxes, fiberboard containers, wood baskets, waxed paper, heavy paper wrappings, and cloth wrappings. There generally are enough of these combustibles in warehouses to produce a fire that requires sprinkler protection. Sprinkler systems for these occupancies can be either preaction systems or dry-pipe systems. In some facilities, double interlock preaction systems are used. Since these systems require a number of events to occur before they discharge water, they further reduce the likelihood of inadvertent system operation.

Hanging Garments

There are no national standards available for the protection of hanging garments and this storage arrangement is generally not found in large warehouses or distribution centers. Mostly this configuration will be incidental to a retail occupancy, cleaners, and so on. Fire damage is often extreme even when the property is fully sprinkler protected.

Garments on hangers hung on pipe racks offer many surface areas available to burn and, therefore, the opportunity for a fast-developing, intense fire that could severely challenge sprinkler protection, particularly if the garments are stored more than two tiers high or if they are more than 10 ft (3.04 m) from the overhead sprinklers. Garments can be hung with or without plastic dust covers, although these covers are not particularly effective in reducing loss from soot, water, or odor damage.

Carpets

As with hanging garments, there are no national standards available for this particular commodity storage. The inspector, therefore, may need to acquire information from insurance companies or national associations that have developed internal standards for the protection of carpeting based on their own experience.

Rolls of carpeting, commonly 12 to 15 ft (3.65 to 4.57 m) long, are stored in deep shelving called racks, which are sometimes arranged back to back, so that the distances between aisles might total 30 ft (9.14 m). The rolls are stored individually in strong cardboard tubes on solid or slatted shelves or in racks arranged in cubicles. Each tier (shelf to shelf) is only about 2 to 3 ft (0.6 to 0.9 m) high, and there are as many as 10 tiers in a rack. Racks are frequently 100 or more ft (30.48 m) long.

This type of storage does not permit much water from ceiling sprinklers to penetrate the racks, so that in-rack sprinklers often are necessary. However, in-rack sprinkler protection at every tier usually is not economically feasible. As a result, fire may spread down the length of a rack in the unprotected tiers. This longitudinal fire spread is best curtailed by the use of vertical barriers at 24- to 30-ft (7.3- to 9.14-m) intervals. Another solution is to maintain well-defined, transverse flue spaces at the vertical rack supports. Because it is difficult for sprinkler discharge to reach fire in all tiers, manual fire fighting must be relied upon for final control and extinguishment. Small hose stations should be available at intervals sufficient to reach all portions of the storage arrays, and provisions should be made to vent smoke to aid in overhaul efforts.

Distilled Spirits

The distilling industry has different types of storage facilities.

Finished Goods, or Case Goods, Storage. In finished goods, or case goods, storage, which are protected in accordance with NFPA 30, *Flammable and Combustible Liquids Code,* spirits are stored in either glass or plastic bottles in cartons in either palletized or rack storage arrangements. Generally these products are 100 proof and lower (with some rare exceptions up to 150 proof).

Barrel Warehouses. In barrel warehouses spirits of 150 proof and greater are stored or matured in oak barrels for various periods of time to obtain the proper quality.

This warehousing varies from distiller to distiller and may include barrels stored on their sides in single-, double-, or multirow racks or on pallets, with four or six barrels per pallet. Barrel warehouses generally have few aisles and are usually filled to capacity and then locked for lengthy periods of time.

Fires in barrel warehouses are infrequent since operations are restricted to storage and loading and unloading operations. These warehouses must be kept clean and free from conditions that contribute to a fire. Areas of concern for fire include the following:

1. Lightning
2. Defective wiring
3. Burst barrels and the resulting flammable liquids spill
4. Class A fires resulting from poor housekeeping
5. Failure of lift trucks
6. Violation of smoking regulations
7. Uncontrolled cutting and welding
8. Exposure fires resulting from uncontrolled weed growth, trash accumulation, and so on

There are no national standards available for the barrel warehouse occupancy. The Distilled Spirits Council of the United States (DISCUS), however, publishes a guide, entitled *Safe Practices Guide for the Distilled Spirits Industry* (see the Bibliography).

Baled Cotton

Guidance is provided for the construction, storage arrangement, and protection of baled cotton warehouses in Appendix, NFPA 230, *Standard for the Fire Protection of Storage.* Baled cotton storage consists of natural seed fiber wrapped in burlap, woven polypropylene, or polyethylene and secured with steel, synthetic, or wire bands. "Naked" bales are bales that are unwrapped and secured only by wire or steel straps.

Baled cotton can be stored in racks, tiered, or untiered. In tiered storage bales are stored two or more high directly on the floor. The sprinkler protection provided in NFPA 13 is based on the storage arrangement, the height of storage, and whether a wet- or dry-pipe system is to be used. The basic design is based on nominal 15-ft (4.57 m) high storage for both tiered and rack storage. Storage that is restricted to 10-ft (3.04 m) nominal height by the ceiling or roof height is allowed to have reduced densities.

Special Occupancy Hazards

With the reorganization of NFPA 13, *Standard for the Installation of Sprinkler Systems,* this standard now includes the sprinkler requirements from some 40 other NFPA documents. Many of these occupancies do not fit into the general storage category, but where significant storage is encountered, guidance will have to be obtained from NFPA 13 and the following: NFPA 40, *Standard for the Storage and Handling of Cellulose Nitrate Motion Picture Film;* NFPA 42, *Code for the Storage of Pyroxylin Plastic;* NFPA 55, *Standard for the Storage, Use, and Handling of Compressed and Liquefied Gases in Portable Cylinders;* NFPA 430, *Code for the Storage of Liquid and Solid Oxidizers;* and NFPA 432, *Code for the Storage of Organic Peroxide Formulations.*

OUTDOOR STORAGE

Outdoor storage requires special protection. However, there are so many different elements to consider that no single set of rules can specify exactly what constitutes that protection. The best that can be done is to outline general principles and rely on the experience and judgment of those who apply them.

In general, outdoor storage sites should be level and firm underfoot with adequate clearances so that fire cannot spread to the site from other sources. Areas in which flooding and windstorms are problems should be avoided.

Some general principles also apply to site layout. Access to the yard and the piles in it must be made easy. Driveways should be of adequate width to permit fire apparatus to reach all portions of the yard. Aisles should be at least 10 ft (3.04 m) wide or as specified for the hazard, and main aisles or firebreaks can be used to subdivide the storage in unusually large storage areas or in moderate-sized yards with valuable commodities. The actual width of the aisles can be a matter of judgment, depending on the combustibility of the commodity, how it is stored, the height of piles, the distance from buildings, the wind conditions, the availability of fire-fighting forces, and so on. Particular emphasis should be placed on the control of potential ignition sources, such as refuse burners, overhead power lines, and acts of vandalism, as well as on the elimination of adverse factors, such as trash accumulations, weeds, and brush. Piles of materials that are stable under normal conditions can collapse during a fire and cause severe fire spread, particularly from flying brands. The fence surrounding the storage site should have an adequate number of gates. If public water flows are inadequate, private water storage facilities, pumps, or both might be needed.

Adequate public fire and police protection or the equivalent private fire brigade protection is a prime requirement for outside storage facilities. More recently, those agencies that enforce environmental regulation for the states have become involved in controlling some types of outdoor storage.

Monitor nozzles mounted on towers might be practical for adequate storage facilities, such as those in which lumber or wood chips are stored, and for facilities for which strong water supplies are available. Appropriate portable fire extinguishers placed at well-marked, strategic points throughout the storage area also are practical. If people are usually present, fully equipped hose houses can be provided, as long as personnel are trained to use hose lines.

Another important feature is the availability of some means of notifying the fire department. At the very least, a telephone should be available. Specific requirements can be found in NFPA 230's Appendix, *Protection of Outdoor Storage;* Appendix, *Protection of Baled Cotton History of Guidelines,* Part D-5; Appendix, *Guidelines for Storage of Forest Products;* and Appendix, *Guidelines for Outdoor Storage of Scrap Tires.*

STORAGE OF RECORDS

The explosion in information technology has increased the total volume of records that are generated and stored. In turn, the problems of record storage are intensified. Traditional storage methods such as insulated file cabinets, safes, and vaults are still viable; but the volume of records to be stored is often so large as to make traditional storage methods and the space needed for them impractical for the business owner. New methods of storage have been devised to solve the space problem resulting in off-premises storage that may often place not only the records at risk but the structures in which they are stored and other occupancies in that structure as well. Guidance can be found in NFPA 232, *Standard for the Protection of Records.*

Records Storage Options

Open-Shelf Storage. Open-shelf storage is often the records storage option chosen to make use of whatever space is available in the building. The records themselves are in file folders or cartons. This arrangement often results in a congested area with narrow aisles of 24 to 30 in. (610 to 762 mm) in width.

Plastic-Media Storage. Plastic media storage of records may constitute a greater hazard. Generally, acetate- and polyester-based tapes are considered at the same hazard level as paper. Polystyrene cases and reels, however, constitute a more severe hazard, are more difficult to protect, and may have to be limited in height and area or have special protection systems for them.

Mobile Compact Shelving. Mobile compact shelving is a storage method in which only one aisle at a time is exposed and can be accessed; all the rest of the shelves are pushed together. A fire in the exposed aisle would be controlled or extinguished by ceiling sprinklers. A fire elsewhere in this style of shelving would be a burrowing fire, sheltered from sprinkler discharge but slow growing. However, without external interference by the fire department, the burrowing fire could consume all records available to it.

Bulk Storage. Bulk storage usually includes very sizable records storage not contained in insulated cabinets, safes, or vaults. These facilities may range from small file rooms to large records centers occupying entire buildings. Virtually any storage method and container is used, and virtually anything (sheets of paper, film, tape, drawings, etc.) can be stored.

Risk Factors

Four risk factors need to be evaluated:

1. Can an exposure fire spread to the records storage area?
2. Can a fire start in the records storage area?
3. How much fuel does the records storage itself represent?
4. Are the records susceptible to indirect fire damage (heat, smoke, vapors, water, etc.)?

The first line of defense for virtually all records storage is sprinkler protection for the building. The sprinkler system combined with the storage method results in protection for the vital and important files. The storage methods, which include insulated records containers, fire-resistive safes, insulated file drawers, and insulated filing devices, vary in the degree of protection afforded from fire, heat, and physical impact. The hourly ratings for this equipment range from 30 minutes to 4 hours.

Vaults

The term *vault* refers to a completely fire-resistive enclosure up to 5000 ft^3 (142 m^3) in volume that is used exclusively to store vital records. Vaults usually contain a substantial fuel load. In many instances, in fact, the contents of a vault are more of a hazard than any external fire exposure. A fire in an unprotected vault can be disastrous unless it is discovered immediately and extinguished.

In the past, the only penetrations permitted in vaults were the door openings. This prohibited sprinklers, fire and smoke detection units wired to a master panel, and even fixed lighting systems. Vault standards were revised in 1986 to permit penetrations for sprinkler piping and conduits so vault contents can be better protected. Work stations and mechanical equipment, such as air-handling and cooling equipment, are not permitted in vaults, nor are ventilation penetrations. The ambient conditions within the vault must be controlled by means of indirect cooling and heating of the environment outside of the vault. In any event, water-type fire extinguishers or small hose, or both, should be available in an accessible location near the door of a vault. Sprinkler protection is required in oversized vaults up to 25,000 ft^3 (708 m^3) in volume.

File Rooms

File rooms are built as nearly like vaults as possible, but they are used for situations in which people work regularly with the records in the room. Thus, they usually have electric lights and steam or hot water heat. Any wall openings needed for air conditioning or ventilation must be equipped with fire dampers. Standard file rooms have a maximum ceiling height of 12 ft (3.65 m) and a maximum volume of 50,000 ft^3 (1416 m^3). They can be designed to a fire-resistance classification of 1, 2, 4, or 6 hours. File room doors can have vault door ratings or lesser ratings of 30 minutes or 1 hour. Noncombustible furniture and cabinets are allowed in the file room. Automatic sprinklers are desirable.

Archives and Records Centers

Bulk storage of paper records in separate buildings, in a major portion of a building, or in a room exceeding 50,000 ft^3 (1416 m^3) in volume requires special attention because of the large fire load it represents. The four basic factors that must be considered are exposure from nearby operations or buildings, the potential for ignition, the potential for fire spread, and the ability of the available fire control system to extinguish or control a fire with minimum damage to records.

Fire-resistive construction is essential to protect against exposure fires. Cleanliness, orderliness, and an absolute ban on smoking are the fundamentals of controlling the chances of ignition. The type of storage, whether steel cabinets or open-shelf system, governs the potential for fire spread. Open shelves present a wall of paper at the face of the shelves, up which fire can spread rapidly to involve other rows of shelves.

Automatic sprinklers, backed up by an early-warning detection system, provide good protection in view of the rapidity with which fire can spread in the large open areas customarily found for archival storage. Catwalks in aisles inhibit overhead sprinkler protection and might necessitate special provisions.

IDENTIFICATION OF MATERIALS: THE NFPA 704 SYSTEM

It is fairly easy to identify the fire hazards of materials commonly found in storage, such as wood, paper, fabrics, and LP-Gas cylinders. You know what to expect when one of these materials burns. However, there are literally thousands of combustible solids, flammable and combustible liquids, and liquefied and compressed gases for which the hazards are not so readily apparent. A system for identifying these hazards is needed so that the occupants can respond correctly in emergencies. One major approach is the NFPA 704 system.

NFPA 704, *Standard System for the Identification of the Hazards of Materials for Emergency Response,* provides usable and readily identifiable means of presenting information on the fire hazards of materials (Figure 41-4). The system covers only fixed installations, such as storage tanks, storage rooms, warehouses, and so on, as applied to industrial, commercial, and institutional facilities that manufacture, process, use, or store hazardous materials. It does not cover materials that are being transported. However, it can be used (and often is used) to mark individual containers once they reach their destination.

The system identifies the fire hazards of a material in three areas: health, flammability, and instability and indicates the relative severity of each hazard category with a numerical rating that ranges from 4, indicating severe risk, to 0, indicating minimal risk. It is important to understand, however, that the ratings are based on emergency situations, such as a fire or a spill. They provide not information on everyday hazards, which result from normal occupational exposure. This is especially true for the health hazard rating. Although many manufacturers include the NFPA 704 ratings on their material safety data sheets, the ratings are not meant to be used to evaluate the hazards presented by chronic exposure. The ratings are based on the inherent hazards of the materials, but they also take into account changes in behavior during a fire that could significantly exaggerate those hazards.

TABLE 41-1 Identification of the Fire Hazards of Materials

Signal	Identification of Health Hazard Color Code: BLUE Type of Possible Injury
4	Materials that, on very short exposure, can be lethal.
3	Materials that, on short exposure, could cause serious temporary or permanent injury.
2	Materials that, on intense or continued exposure, could cause temporary incapacitation or possible residual injury.
1	Materials that, on exposure, would cause significant irritation.
0	Materials that, on exposure under fire conditions, would offer no hazard beyond that of ordinary combustible material.
	Identification of Flammability Color Code: RED Susceptibility of Materials to Burning
4	Materials that will rapidly or completely vaporize at atmospheric pressure and normal ambient temperature, or that are readily dispersed in air and will burn readily.
3	Liquids and solids that can be ignited under almost all ambient temperature conditions.
2	Materials that must be moderately heated or exposed to relatively high ambient temperatures before ignition can occur.
1	Materials that must be preheated before ignition can occur.
0	Materials that will not burn.
	Identification of Instability Color Code: YELLOW Susceptibility to Release of Energy
4	Materials that, in themselves, are readily capable of detonation or of explosive decomposition or reaction at normal temperatures and pressures.
3	Materials that, in themselves, are capable of detonation or explosive decomposition or reaction but require a strong initiating source, or that must be heated under confinement before initiation; also materials that react explosively with water without requiring heat or confinement.
2	Materials that, in themselves, are normally unstable and readily undergo violent chemical change at elevated temperature and pressure; also materials that might react violently with water or that might form potentially explosive mixtures with water.
1	Materials that, in themselves, are normally stable but that can become unstable at elevated temperatures and pressures or that might react with water with some release of energy but not violently.
0	Materials that, in themselves, are normally stable, even under fire exposure conditions, and are not reactive with water.

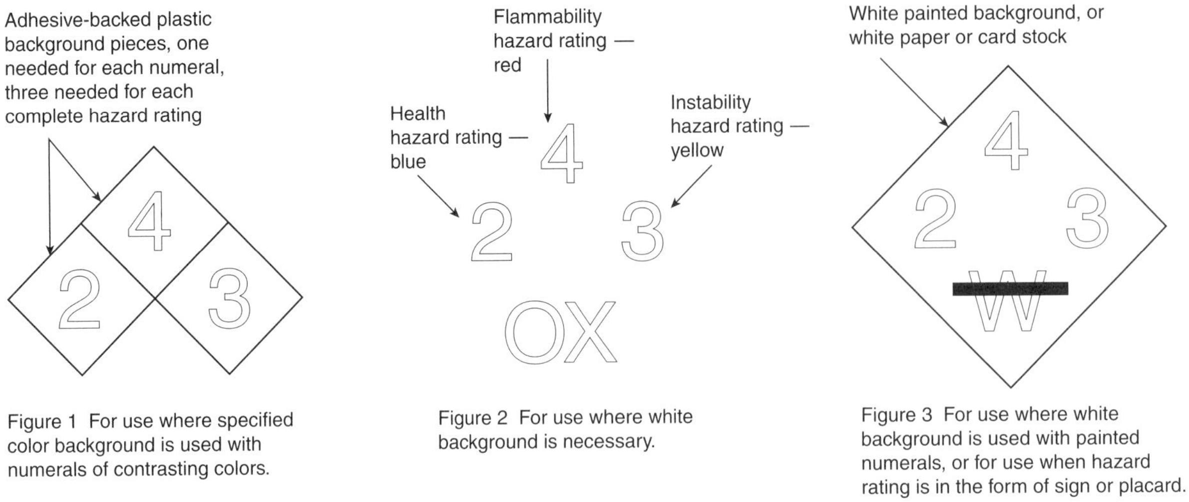
Adhesive-backed plastic background pieces, one needed for each numeral, three needed for each complete hazard rating
4
2
3
Figure 1 For use where specified color background is used with numerals of contrasting colors.
Flammability hazard rating — red
Health hazard rating — blue
Instability hazard rating — yellow
4
2
3
OX
Figure 2 For use where white background is necessary.
White painted background, or white paper or card stock
4
2
3
W
Figure 3 For use where white background is used with painted numerals, or for use when hazard rating is in the form of sign or placard.

FIGURE 41-4 Alternate Arrangements for Display of NFPA 704 Hazard Identification System
Source: NFPA 704, 1996, Figure 6-1.

The 704 diamond is illustrated in Figure 41-4. The health rating is always at the 9 o'clock position, the flammability rating is always at the 12 o'clock position, and the instability rating is always at the 3 o'clock position. Furthermore, each quadrant is identified by a colored background: blue for health hazard, red for flammability hazard, and yellow for instability hazard. Alternatively, the quadrant could be any convenient contrasting color, and the numerals themselves could be printed in the appropriate color described above for each hazard.

Special hazard identifiers go in the lower-most, or 6 o'clock, quadrant. No special color is assigned to this quadrant. The two recognized identifiers are W to indicate water reactivity and OX to indicate oxidizing ability.

In abbreviated form, the five degrees of risk for each of the three hazards are shown in Table 41-1. NFPA 704 should be consulted for more detailed definitions.

Recommended ratings for specific materials can be found in the *Fire Protection Guide to Hazardous Materials* (see Bibliography). Be advised that the ratings given for a chemical are accurate for only the "commercially pure" material, as shipped. The ratings might not be accurate for diluted mixtures. Furthermore, a mixture of two or more chemicals may have to be evaluated separately if it is to receive the proper NFPA 704 rating.

SUMMARY

General storage warehouses contain a large variety of products, ranging from noncombustible to flammable liquids and aerosols. These products and their storage arrangement influence the intensity of a fire in the facility and the protection needed to control or extinguish that fire.

There are many areas that are important to the inspector: the commodity classification; the storage arrangement for the commodity; the potential causes of a fire in a particular facility; the proper sprinkler system (given the large variety of sprinklers available, the inspector needs to know if the proper type has been used); the storage container; and, finally, whether the commodity is specific as to the protection available to it. Collectively, these factors, combined with quality construction, proper aisle spacing to reduce fire spread and to improve fire fighter access, controlled storage heights, and proper sprinkler system design for the hazards involved, increase the level of warehouse fire protection.

BIBLIOGRAPHY

Cote, A. E., ed., *Fire Protection Handbook,* 18th ed., NFPA, Quincy, MA, 1997.

Distilled Spirits Council of the United States, *Safe Practices Guide for the Distilled Spirits Industry,* 2nd ed., DISCUS, Washington, DC.

Emergency Operations in High-Rack Storage, Fire Protection Publications Department, Oklahoma State University, Stillwater, OK.

Schumann, T., "A New Protection Challenge in the Warehouse," *NFPA Update,* Issue 840, Vol. 1, No. 3 (December/January 2001), page 7.

Spencer, A. B., and Colonna, G. R., eds., *Fire Protection Guide to Hazardous Materials,* 13th ed., NFPA, Quincy, MA, 2002.

NFPA Codes, Standards, and Recommended Practices

See the latest version of The NFPA Catalog for availability of current editions of the following documents.

NFPA 10, *Portable Fire Extinguishers*

NFPA 13, *Standard for the Installation of Sprinkler Systems*

NFPA 30, *Flammable and Combustible Liquids*

NFPA 30B, *Code for the Manufacture and Storage of Aerosol Products*

NFPA 51B, *Standard for Fire Prevention During Welding, Cutting, and other Hot work*

NFPA 61, *Standard for the Prevention of Fires and Dust Explosiong in Agricultural and Food Products Facilites*

NFPA 230, *Standard for the Fire Protection of Storage*

NFPA 232, *Standard for the Protection of Records*

NFPA 395, *Standard for the Storage of Flammable and Combustible Liquids at Farms and Isolated Sites*

NFPA 704, *Standard System for the Identification of the Hazards of Materials for Emergency Response*

Storage and Handling of Flammable and Combustible Liquids

CHAPTER 42

Orville M. "Bud" Slye, Jr.

CASE STUDY

Electrical Short Ignites Transformer Fluid

Fire fighters, haz-mat teams, and power plant personnel worked for nearly 28 hours to extinguish a fire in a 138-kV step-down electrical transformer in the yard of a Texas utility company. There were several delays while the equipment was deenergized and the status of each component was tested and confirmed. Access to the interior of the transformer, which contained combustible components, was limited, and this further delayed extinguishment.

A 138-kV power transmission line fed the 138- to 69-kV step-down auto transformer, which was 41 ft long, 22 ft wide, and 17 ft high. The transformer was constructed of copper electrical conductors in transformer windings and incorporated paper and porcelain insulators. Hard maple was used as structural support for the transformer. The unit contained 24,525 gal of heat transfer liquid, which is classified as a Class III-B combustible liquid with a closed cup flashpoint of 295°F. The unit was located in a containment system designed to hold 1.5 times the volume of fluid in the transformer.

Plant personnel dialled 911 to report the fire at 3:56 P.M., and the dispatch supervisor sent a first-alarm haz-mat assignment consisting of an engine and a ladder company, two haz-mat engines, two haz-mat rescues, a haz-mat support unit, and six command officers. They arrived 4 minutes later and, noting heavy fire and smoke issuing from the transformer, established a large-diameter supply line and two 2-in. attack lines. The haz-mat teams simultaneously established hot, warm, and cold zones and a contamination corridor.

Although plant officials were able to shut off electrical power to the burning transformer within 10 minutes of ignition, fire operations were delayed nearly 2 hours as each circuit was inspected to confirm that the power had actually been shut down. As the heavily damaged transformer continued to burn, attack crews cooled nearby oil circuit breakers.

Once fire fighters confirmed that the equipment had been deenergized, crews applied water to cool it and aqueous film-forming foam (AFFF) to extinguish the fire, which was out in 27 minutes. Fire fighters continued to apply foam for nearly an hour, while crews created a secondary containment zone and monitored the air downwind of the fire.

A fire watch was assigned in case of reignition, and the other units were released.

A second haz-mat first-alarm was struck at 4:26 A.M. the next day, when the fire reignited. Although fire fighters had extinguished the blaze in the heat transfer fluid the first time around, wood and paper inside the transformer continued smoldering and eventually rekindled.

Orville M. "Bud" Slye, Jr., P.E., FSFPE, is president of Loss Control Associates, Inc., Langhorne, PA. A member of the NFPA 30 Correlating and Tank and Piping Committees, he has extensive experience in application of the code to flammable and combustible liquid facilities and equipment.

Working with plant crews, fire fighters devised a plan to drain the nearly 89,000 gal of heat transfer fluid that remained in the transformer and replace it with AFFF and water. The plan was successful, and the fire was finally extinguished at 8:00 P.M. the following evening. Fire fighters cooled the tank to a safe temperature, as crews monitored the vessel using infrared and ultraviolet equipment.

The transformer, valued at $2 million, was a total loss. The plant reimbursed the fire department $8896 for materials used to fight the fire and paid $35,000 in environmental remediation costs. There were no injuries.

Source: "Fire Watch," *NFPA Journal*, January/February, 1998.

The inspector must be familiar with the physical and fire hazard properties and risk associated with commonly used flammable and combustible liquids in order to determine of these liquids are stored and handled safely. The distinction between a flammable and a combustible liquid is somewhat arbitrary, and the terms have basically acquired the meanings of common usage. Strictly speaking, liquids either burn, in which case they are combustible, or they do not burn, in which case they are noncombustible. The word "flammable" connotes a greater fire risk than normal and serves to define a particular class of volatile combustible liquids, as is discussed later in this chapter. In this chapter, the word *liquid* means a liquid that burns, unless otherwise identified.

There are three important facts that apply to safety of flammable and combustible liquids. First, it is the vapor given off by the liquid that burns, not the liquid itself. It is important to visualize where vapors will travel once they are released. Second, the greater the tendency of a liquid to give off vapors—that is, the more volatile is the liquid—the greater the risk of fire. Although inspectors must pay particular attention to these liquids, they must also be alert for operations in which less volatile liquids are heated, thus increasing their propensity to generate vapors. And finally, the physical and fire hazard properties of liquids must be understood its order to be able to judge the risk of an individual operation.

PROPERTIES OF FLAMMABLE AND COMBUSTIBLE LIQUIDS

Physical Properties

Vapor Pressure. Vapor pressure is a measure of the pressure that a liquid exerts against the atmosphere above it. Just as the atmosphere exerts pressure on the surface of the liquid, the liquid pushes back. Vapor pressure normally is less than atmospheric pressure and is a measure of the liquid's tendency to *evaporate,* or move from the liquid to the gaseous state. This tendency is also referred to as *volatility,* which explains the use of the term *volatile* to describe liquids that evaporate very easily. The higher the vapor pressure, the greater the rate of evaporation and the lower the *boiling point.* Simply put, this means more vapors and increased fire risk. The inspector must ensure that reasonable measures, such as local exhaust ventilation, have been taken to control vapors and that sources of ignition are either removed from the area or controlled.

Boiling Point. The boiling point is the temperature at which the vapor pressure of a liquid equals atmospheric pressure. At this temperature, atmospheric pressure can no longer hold the liquid in the liquid state, and the liquid boils. A low boiling point is an indication of high vapor pressure and a high rate of evaporation.

Vapor Density. Vapor density, sometimes referred to as *vapor–air density,* is the ratio of the weight of a volume of pure vapor to the weight of an equal volume of dry air, both at the same temperature and pressure. Vapor density determines whether the vapors will rise or sink to the ground when released. A vapor density less than 1 means that the vapor will rise and dissipate very quickly. This reduces the risk of ignition somewhat, but the inspector should check for potential ignition sources that are in the likely path of the vapors. Such sources include ceiling-mounted or roof-level electrical equipment, light fixtures, and unit heaters.

A vapor density of 1 means the vapor is just as dense as air, and the same precaution applies.

A vapor density greater than 1 indicates that the pure vapor is denser than air and will sink from its point of release, tending to flow downward and to settle in low areas. Inside buildings, the inspector should pay close attention to below-grade areas, such as floor drains, sumps, trenches, and similar inadequately ventilated areas, including basements and crawl spaces. Outside, the inspector should note the grading of the surrounding terrain and try to determine where vapors will likely flow.

In the pure state, vapors from most flammable and combustible liquids are denser than air. However, pure vapor will only be given off at or above the boiling point of the liquid. For all other conditions, the vapor is mixed with some air, and the density is thereby proportionately changed, approaching that of air itself. Regardless, the inspector should understand that a mixture of vapor and air can be expected to travel at grade level, often for some distance, until it naturally disperses. Such mixtures can and have been known to ignite at some distance from their source, with flames spreading back to the source.

Specific Gravity. Specific gravity is the ratio of the density of one material to another, usually water. The density is expressed as weight per unit volume, for example, pounds per gallon or grams per liter. Most liquids—gasoline, for example—are not as dense as water and, if not miscible with water, will float on top. Water will float on the surface of liquids that are denser than water—Carbon disulfide, for example.

Water Solubility. Water solubility is a measure of the tendency of a liquid to dissolve in, or mix with water. It usually is expressed in terms of grams of liquid per 100 mL of water. Some liquids, such as acetone, certain alcohols, and amines, will mix completely with water, in all proportions. Other liquids, such as hydrocarbon fuels, will not mix with water at all.

Some flammable and combustible liquids are water-soluble, which is an important characteristic in determining fire-fighting tactics. Water-soluble liquids will destroy a blanket of ordinary fire-fighting foam; they require the use of special alcohol-resistant fire-fighting foams. If a burning water-soluble liquid can be contained, it is even possible to dilute it with water to a noncombustible mixture using available hose streams.

Viscosity. The viscosity of a liquid is a measure of its resistance to flow, or its "thickness." Obviously, a viscous liquid will flow slowly and will be easier to contain. However, the viscosity of a liquid depends on its temperature, and most liquids will "thin out" and flow more easily when heated. Certain plastic resins are an exception.

Temperature and Pressure Effects. Liquids are only slightly compressible, and they cannot expand indefinitely. Liquids will vaporize more rapidly as the temperature increases or as the pressure decreases.

NFPA 325, *Fire Hazard Properties of Flammable Liquids, Gases, and Volatile Solids,* contains information on most of the preceding properties for many flammable and combustible liquids.

Fire Hazard Properties

Flash Point. The flash point of a liquid is the minimum temperature at which it gives off vapor in sufficient concentration to form an ignitible mixture with air near the surface of the liquid. It is a direct measure of a liquid's volatility, or its tendency to vaporize. The lower the flash point, the greater the volatility and the greater the risk of fire. Flash point is determined using one of several different test procedures and apparatus that are specified in Subsection 1.7.4 of NFPA 30, *Flammable and Combustible Liquids Code,* 2000 edition.

Liquids with flash points at or below ambient temperatures are easy to ignite and burn quickly. Once the liquid is ignited, the spread of flame over the surface of the liquid will be rapid because the fire need not expend energy heating the liquid to generate more vapor. Again, gasoline is a familiar example of such liquids. Liquids with flash points above ambient temperature present less risk because they must be heated to generate enough vapor to become ignitible; they are more difficult to ignite and present less potential for the generation and spread of vapor. A common example of this type of liquid is home heating oil (Fuel Oil No. 2), which must be atomized to a fine mist before it is easily ignited.

In discussing flash point, reference is sometimes made to the "fire point" of a liquid. Fire point is the temperature at which ignition of vapors will result in continued burning. As the term "flash point" suggests, the vapors generated at that temperature will flash, but they will not necessarily continue to burn. While the difference between flash point and fire point has some significance in flash point tests, it is ignored in practice, and the flash point is used to classify the liquid and to characterize its hazard. Suffice it to say that the maxim "low flash, high hazard" applies.

Autoignition Temperature. Sometimes referred to as "spontaneous ignition temperature," "self-ignition temperature," or "autogenous ignition temperature," the autoignition temperature is the minimum temperature at which a liquid will self-ignite without an external source of ignition, such as a spark or pilot flame, under specified conditions and usually in air. In practice, autoignition results when an ignitible vapor–air mixture comes in contact with a hot surface or is introduced into a hot environment.

It is important to take autoignition temperature into account when selecting electrical equipment for areas in which ignitible vapor–air mixtures might be present. This is especially true of electrical equipment that heats with use, such as motors, transformers, and light fixtures. However, the inspector should be aware of any other equipment that might provide a hot surface, such as drying ovens, hot air ducts, and hot process piping.

Flammable Limits. The lower flammable limit is that concentration of combustible vapor in air *below* which propagation of a flame will not occur. The upper flammable limit is that concentration of combustible vapor in air *above* which propagation of flame will not occur. Between these limits, ignition is possible, and the concentrations between these limits is thus known as the "flammable range." Mixtures within the flammable range are said to be ignitible. A mixture whose concentration is below the lower flammable limit is said to be too lean to be ignited. Conversely, a mixture that is above the upper flammable limit is said to be too rich to be ignited. Bear in mind, however, that a too-rich mixture is not necessarily safe. Introduction of air could dilute the mixture into the flammable range.

The flammable limits are important in calculating the volume of clean air for ventilating spray booths, drying ovens, and other such pieces of equipment to prevent internal explosions. NFPA 86, *Standard for Ovens and Furnaces,* includes information for these calculations. Flammable limits also are important in determining the explosion hazards of confined spaces where vapors are present. NFPA 69, *Standard on Explosion Prevention Systems,* should be consulted for additional information. NFPA 325 contains information on most of the above properties for many flammable and combustible liquids.

CLASSIFYING AND IDENTIFYING FLAMMABLE AND COMBUSTIBLE LIQUIDS

Classification

The basic system for classifying liquids can be found in Section 1.7 of NFPA 30, *Flammable and Combustible Liquids Code.* For classification purposes, distinctions must be made between a gas, a liquid, and a solid. A gas is defined as a substance that has a vapor pressure of 40 psia (pounds per square inch, absolute) (276 kPa) or more at 100°F (37.8°C). Any substance whose vapor pressure is below this is considered a liquid. Liquids also have a specified fluidity and substances with less fluidity are treated as solids. The specified fluidity is that of 300 penetration asphalt, and the test procedure is described in ASTM D5, *Test for Penetration for Bituminous Materials.*

The broad categories of flammable and combustible liquids are defined as follows: flammable liquids have flash points below 100°F (37.8°C) and combustible liquids have flash points of 100°F (37.8°C) or more. In order to apply the fire protection requirements of NFPA 30 and other NFPA codes and standards, these two groups are further subdivided, as shown in Table 42-1.

The NFPA definitions of flammable liquid and combustible liquid are no longer the same as those of the U.S. Department of Transportation (DOT). DOT has adopted a new classification system that defines a flammable liquid as any liquid with a flash point that does not exceed 141°F (60.5°C). Under certain circumstances, most containers holding Class II liquids must be placarded to indicate they are storing flammable liquids (see Title 49 of the *Code of Federal Regulations,* Part 173).

The DOT definitions have very little effect on the proper application of fire protection codes and standards. Aside from its title and its definitions, NFPA 30 governs the storage and handling of liquids by their class designation. From a practical standpoint, the 1°F difference between the two systems at the boundary between Class II and Class IIIA is of little consequence.

DOT does not regulate Class IIIB liquids that do not meet any other definition for a hazardous material. Likewise, the U.S. Occupational Safety and Health Administration (see Title 29 of the *Code of Federal Regulations,* Part 1910) does not regulate Class IIIB liquids whose only hazard is combustibility. However, NFPA 30 *does* regulate Class IIIB liquids.

Identification

Identifying the class of a liquid can be difficult. The sense of smell is not reliable, and many liquids are sold under names that give no indication of their potential

TABLE 42-1 Classification of Flammable and Combustible Liquids

Classification	Flash Point °F (°C)	Boiling Point (°F)
A	Below 73°F (22.8°C)	Below 100°F (37.8°C)
B	Below 73°F (22.8°C)	100°F (37.8°C) and above
C	73° to 100°F (22.8° to 37.8°C)	100°F (37.8°C) and above
D	100° to 140°F (37.8° to 60°C)	100°F (37.8°C) and above
E	140° to 200°F (60° to 93°C)	100°F (37.8°C) and above
F	200°F (93°C) and above	100°F (37.8°C) and above

fire hazard. The label on the container or the placard on the transportation vehicle will be helpful, as will the UN or NA number assigned under DOT's hazardous materials transportation rules. In many cases, however, the inspector will have to rely on the Material Safety Data Sheet (MSDS) and any other information that must be available by federal and state right-to-know laws about all the hazardous materials on the premises. These sources include information about fire hazards, including flash point data, which can be used to determine the class of the liquid.

When in doubt, a flash point test will provide a positive determination. The inspector can make a cursory estimate of the hazard of a liquid by placing a sample of the liquid in a small metal, glass, or ceramic cup and trying to ignite the vapors. (Needless to say, this should be done in a safe location, such as in a laboratory hood or outdoors.) A flash of flame indicates that the liquid has a low flash point. However, the absence of a flame should not be accepted as proof that the liquid will not burn. There are liquids—some halogenated hydrocarbons, for example—that do not have flash points, yet will burn when subjected to a sufficiently intense ignition source. Although these liquids may not be easily ignited, they will add fuel to an established fire.

For consumer commodities, the container label that is required by the Federal Hazardous Substances Act (see Title 16, *Code of Federal Regulations,* Part 1500.43a) will provide useful information. The words "Danger. Extremely Flammable" mean that the liquid's flash point is 20°F (−6.7°C) or less. "Warning Flammable" means the flash point is in the 20° to 100°F (−6.7° to 37.8°C) range. "Caution. Combustible" means the flash point is in the 100° to 150°F (37.8° to 65.5°C) range. Absence of a label should not be considered as proof that the liquid will not burn.

STORING AND HANDLING FLAMMABLE AND COMBUSTIBLE LIQUIDS

The primary hazard involved in storing and handling flammable and combustible liquids is accidental release, followed by ignition of the spilled liquid. Accidental release usually results from container failure due to mishandling, corrosion, or puncture. Once a spill occurs, the likelihood of ignition increases significantly, and the resulting fire will present a greater challenge. If the spill takes place as a consequence of an existing fire, the severity of the incident will be magnified many times over.

Underground Tanks

Underground tanks are usually arranged with connections through the top of the tank shell and offer the most fire safe means of storing liquid. The tank is isolated from external fires; thus, it has only an extremely remote chance of internal ignition or physical damage. (Cases of internal explosions of underground storage tanks are almost unheard of and mostly occur during maintenance operations inside the tank.) However, underground storage tank systems are subject to leaks that can go undetected for long periods of time. Released liquid can flow into the basements of adjacent buildings or into other subterranean structures, resulting in very serious fire hazard conditions that must be dealt with immediately. The inspector should check NFPA 329, *Recommended Practice for Handling Releases of Flammable and Combustible Liquids,* for information on this subject.

Special care must be taken when installing underground storage tanks to prevent damage to the tank and its piping that might cause future leaks. Excavation during installation must not undermine adjacent structures, and the tank must be situated so that it is not subject to any static loads from these structrues. In areas subject to flooding or high groundwater levels, tanks must be secured to prevent them from being dislodged. Details of proper installation procedures are beyond the scope of this chapter. Chapter 2 of NFPA 30; the Petroleum Equipment Institute's

RP-100, *Recommended Practices for Installation of Underground Liquid Storage Systems;* and manufacturers' recommendations for specific tank types and designs contain guidence on underground tank installation.

Also, rules for underground tanks and piping systems are promulgated by the U.S. Environmental Protection Agency (EPA) in Title 40 of the *Code of Federal Regulations,* Part 280. Under these rules, tanks and piping systems must be either suitably protected to resist corrosion or constructed of corrosion-resistant materials. There also are very specific rules for leak monitoring systems and for periodic testing of the entire storage system for tightness. As a result of these rules, a modern underground storage tank installation is likely to include double-walled tanks and double-walled piping, a monitoring system to detect leaks from the primary containment, groundwater monitoring wells around the installation, and an impervious liner in the excavation.

Outside Aboveground Tanks

Outside aboveground storage tanks are suitable for storing almost any quantity of liquid, from several hundred gallons (liters) to a million barrels (1 barrel = 42 gal). Some of the various types of tanks used are shown in Figure 42-1. The smaller sizes, which will hold up to 50,000 gal (190 m^3) are typically factory built, while the

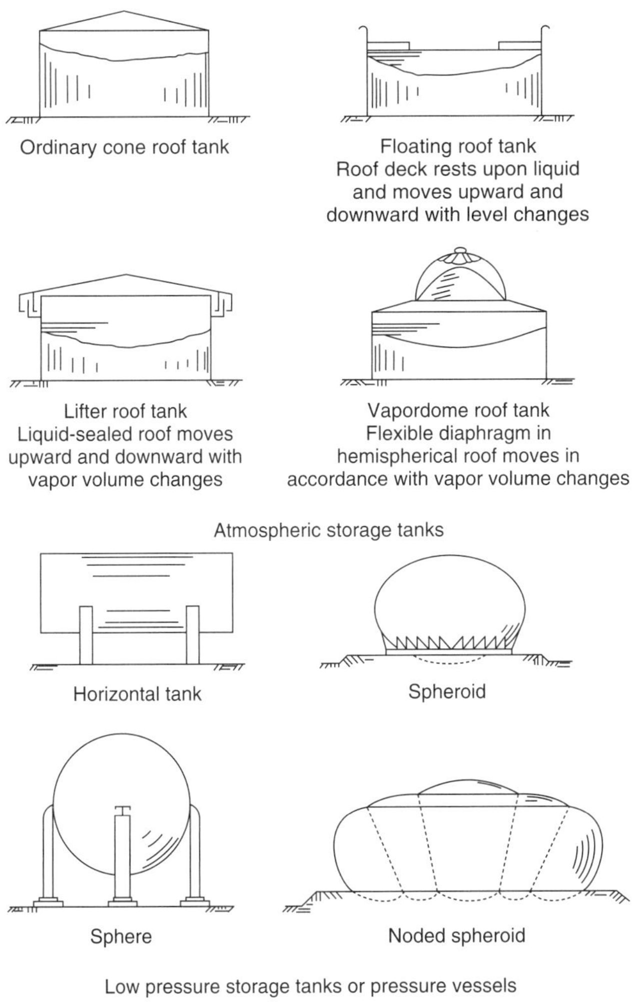

FIGURE 42-1 Storage Tanks for Flammable and Combustible Liquids

larger tanks are constructed on site. Chapter 2 of NFPA 30 applies in all cases. Essential features covered by NFPA 30 include proper design and construction; a well-engineered foundation; proper siting of the tank with respect to property lines, public roads, and nearby buildings; spill control, in the form of remote impounding, diking or use of secondary containment tanks; normal venting; emergency relief venting from fire exposure; overfill prevention; and substantial pipe connections with an adequate number of property placed valves to control flow in the event of fire or of breakage in the piping.

All of the above features should have been addressed at the design and construction stage, especially for the larger, site-built tanks. Consequently, close attention to the operating features and maintenance are a must for fire safety of the facility. Where spills are controlled by diking, the inspector should confirm that the dike can contain at least 100 percent of the contents of the largest tank within it, that the dikes are properly maintained, and that provisions have been made to drain accumulated rainwater from the dike. The inspector should reconfirm this any time another tank is installed within the dike. If the diked area has been roofed over, special considerations will be needed for fire safety (see the next section, "Tanks Inside Building"). Finally, fire protection systems on the site should be checked to determine proper operation.

The fire inspector is probably more involved with smaller aboveground tanks (Figure 42-2). The EPA's underground storage tank rules have been the impetus for abandoning small underground tanks in favor of aboveground tank systems that offer a number of desireable features, most directed at environmental protection. Some of these "new technology" aboveground tanks include one or more of the

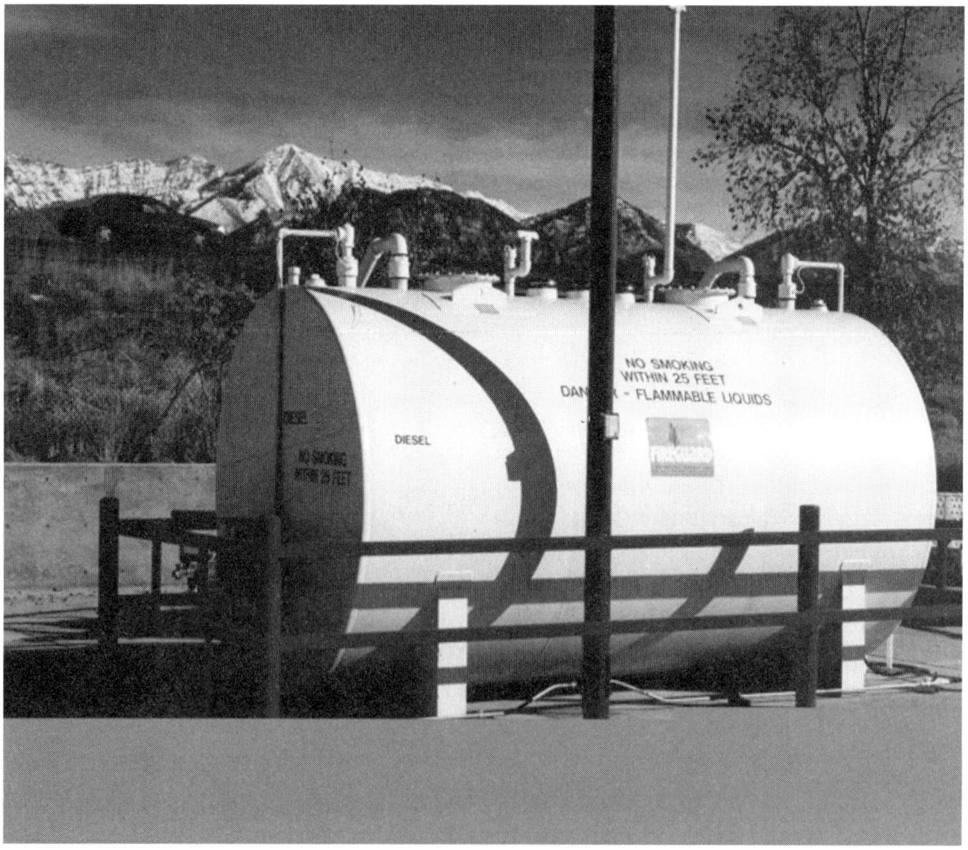

FIGURE 42-2 Outside Aboveground Storage Tank

following: integral spill pans, double-wall construction, and thermal insulation. These tanks typically are factory-built and seldom exceed 25,000 gal (95 m^3) capacity.

These tanks are installed in accordance with Chapter 2 of NFPA 30, as are the larger built-on-site tanks. The inspector might also wish to refer to the Petroleum Equipment Institute's RP200, *Recommended Practices for the Installation of Aboveground Storage Systems for Motor Vehicle Refueling*. This recommended practice includes detailed installation procedures specifically for these aboveground tanks.

NFPA 30A, *Code for Motor Fuel Dispensing Facilities and Repair Garages*, permits the use of such tanks at retail service stations. However, special requirements, in addition to those found in Chapter 2 of NFPA 30, apply. These special requirements distinguish between aboveground tanks in dikes, aboveground tanks installed in vaults, fire-resistant aboveground tanks, and protected aboveground tanks. The complete set of requirements can be found in Chapter 2 of NFPA 30A, 2000 edition.

NFPA 30 also permits some of the new technology tanks to be installed without diking or remote impounding, the two accepted means of providing spill control, provided the tanks do not exceed the 12,000-gal (45 m^3) capacity and meet certain other very specific requirements. Refer to the requirements for these tanks in Chapter 2 of NFPA 30.

Tanks Inside Buildings

Although not normally recommended, tanks often have to be installed inside buildings for process reasons or because the material stored must be kept under strict environmental control. Furthermore, many operators of small storage tank systems are choosing to build a roof over their tank storage areas or to enclose them entirely with a light-framed building. This is because many jurisdictions require that the rainwater that collects in the diked area be treated as a hazardous waste. Subsection 2.3.4 of NFPA 30 includes a very detailed set of requirements for tanks storing Class I, II, or IIIA liquids housed in buildings. The requirements are too detailed to be described here, and, as is the case with outside aboveground tanks, most are dealt with at the design and construction stage. The requirements include siting and construction of the building, ventilation, drainage, venting tanks, connections to tanks, electrical equipment, and fire prevention and control.

The inspector should ensure that there is enough access and egress to provide unobstructed movement of personnel and fire protection equipment and should verify that the necessary ventilation is fully operational and adequate. NFPA 30 requires that there be enough ventilation to keep the concentration of vapors below 25 percent of the lower flammable limit. Refer to Paragraph 2.3.4.4 of NFPA 30 for information on how this level of ventilation can be achieved. The inspector should also confirm that the drainage systems are adequate and operable. Tanks with weak roof-to-shell seams are not allowed inside buildings, and neither are floating roof tanks. Normal and emergency vents must terminate outside the building in all cases.

Storing fuel oil in inside tanks that hold not more than 660 gal (2500 L) with fill and vent connections outside, is an accepted practice under the requirements of NFPA 31, *Standard for the Installation of Oil-Burning Equipment*.

Portable Tanks and Intermediate Bulk Containers

NFPA 30 defines a portable tank as a closed vessel of 60 gal (227 L) to 660 gal (2500 L) capacity that is not intended for fixed installation. NFPA 30 now also recognizes the use of intermediate bulk containers (IBCs), which are similar to portable tanks, but have a maximum capacity of (793 gal) (3000 L). Both are regulated by DOT.

If suitably constructed, portable tanks and IBCs are considered more desirable than are smaller containers because they are equipped with pressure relief devices that will relieve internal pressure if the portable tank or IBC is exposed to fire, thus preventing catastrophic failure and complete release of the contents. Being specifically designed to be reused, they tend to be constructed more robustly than so some smaller containers. They are widely used in the automotive, chemical, and coatings industries. Not only are they a cost-effective way to transport, handle, and store liquids, but they can be directly connected to process equipment.

While portable tanks, as defined in NFPA 30 and as regulated by DOT are only constructed of metal, IBCs are constructed of metal; of thick-walled, rigid molded, self-supporting plastic; and of blow-molded plastic that is enclosed in of various types of metal overpacks, for structural support. As recently as the 1993 edition, NFPA 30 did *not permit plastic IBCs to* be stored in warehouses or inside storage areas. The 1996 edition allowed their use for Class II and Class III liquids only, but considered them to be unprotected, regardless of the presence of a fire protection system. The 2000 edition now includes fire protection system design criteria for plastic IBCs, based on full-scale fire testing sponsored by industry groups. Note, however, that they still are not allowed to be used for Class I liquids. See Table 42-2 for capacity limitations.

Portable tanks and IBCs that exceed 660 gal (2500 L) and 792 gal (3000 L), respectively, are considered by NFPA 30 to be fixed tanks and are governed by Chapter 2 of that code. Portable tanks are occasionally directly connected to process

TABLE 42-2 Maximum Allowable Size of Container, and Portable Tanks

	Flammable Liquids			Combustible Liquids	
Type	**Class IA**	**Class IB**	**Class IC**	**Class II**	**Class III**
Glass	1 pt	1 qt	1 gal	1 gal	5 gal
Metal (other than DOT drums) or approved plastic	1 gal	5 gal	5 gal	5 gal	5 gal
Safety cans	2 gal	5 gal	5 gal	5 gal	5 gal
Metal drum (DOT Specification)	60 gal	60 gal	60 gal	60 gal	60 gal
Approved metal portable tanks and IBCs	793 gal	793 gal	793 gal	793 gal	793 gal
Rigid plastic IBCs (UN 31H1 or 31H2) and composite IBCs (UN 31HZ1)	NP	NP	NP	793 gal	793 gal
Polyethylene DOT Specification 34, UN 1H1, or as authorized by DOT exemption	1 gal	5 gal*	5 gal*	60 gal	60 gal
Fiber drum NMFC or UFC Type 2A; Types 3A, 3B-H, or 3B-L; or Type 4A	NP	NP	NP	60 gal	60 gal

For SI units, 1 pt = 0.473 L; 1 qt = 0.95 L; 1 gal = 3.8 L.

NP—Not permitted.

*For Class IB and IC water-miscible liquids, the maximum allowable size of plastic container is 60 gal (227 L), if stored and protected in accordance with Table 4.8.2(g) of NFPA 30.

Source: Table 4-2.3, NFPA 30, *Flammable and Combustible Liquids Code,* NFPA, Quincy, MA, 2000.

equipment. In such cases, the installation also must meet the applicable requirements of Chapter 5 of NFPA 30.

Drums and Other Portable Containers

NFPA 30 recognizes the use of a number of different types of containers and specifies the maximum allowable sizes for the different liquid classes. Table 42-2 is taken from NFPA 30.

Drums and other portable containers are best stored outdoors away from buildings or in small detached storage buildings used only for such storage. Where drums and portable containers are stored inside buildings, they should be stored in a special storage room that is protected by an automatic extinguishing system. Drainage from drum storage facilities should be provided and arranged in a way that facilitates fire control.

Chapter 4 of NFPA 30 contains specific requirements for maximum quantities and maximum storage heights for container storage for a variety of occupancies and storage methods. Table 4.4.4.1 of NFPA 30 governs maximum storage height, maximum storage quantity per pile or rack section, and maximum total quantity for unprotected storage areas and storage areas that have substandard fire protection. Section 4.8 of NFPA 30 contains specific design criteria for sprinkler and foam-water sprinkler fire protection systems for numerous combinations of class of liquid, container type, storage arrangement and height, and maximum building ceiling height. These criteria are based on evaluation of more than 100 full-scale fire tests.

NFPA 30 now recognizes the use of moveable prefabricated hazardous materials storage buildings as an option for storing flammable and combustible liquids. If they meet the requirements for inside storage rooms, they can even be located within a building and shifted about on the factory floor as needs dictate. Use of these buildings outside is governed by a separate set of requirements in NFPA 30.

Safety Cans

Safety cans have a maximum capacity of 5 gal (19 L) and come equipped with a spring closing lid and spout cover so that the can will safely relieve internal pressure when subjected to fire exposure. The spring also ensures that the cover will snap closed if the container is dropped while being used. Figure 42-3 shows

FIGURE 42-3 Typical Safety Cans with Pouring Outlets with Tight-Fitting Caps or Valves Normally Closed by Springs

the types of containers that can be used for storing and dispensing small quantities of flammable liquids inside buildings. The safety can is not intended for use in settings where the periodic release of flammable vapors could create a hazardous atmosphere.

TRANSFERRING AND DISPENSING FLAMMABLE AND COMBUSTIBLE LIQUIDS

Flammable and combustible liquids can be transferred by gravity flow, by container-mounted pumps, or by a closed-pipe fixed pumping system. The latter is most often used for transferring large amounts of liquid and is the safest method of handling a large-quantity transferal.

Pumping Systems

Positive displacement pumps offer a tight shutoff and prevent the liquid from being siphoned when not in use. Centrifugal pumps do not provide a tight shutoff when taking suction under head, and siphoning of the liquid is possible. Valves should be provided to isolate pumps during maintenance or emergencies.

Gravity Systems

Gravity transfer is most often used to dispense liquids from containers. Dispensing valves should be of the self-closing type. Fire safety measures should include self-closing valves, and such valves should not be blocked open. To avoid vapor lock in the pumping systems, gravity flow is sometimes used where very volatile liquids are transferred from tanks. The inspector should make sure that isolation valves are provided in the dispensing lines from such tanks.

Compressed Gas Displacement Systems

Transfer of liquids by compressed gas displacement is allowed only under certain conditions. The vessels and containers between which transfer occurs and the transfer lines must be designed to withstand the anticipated operating pressure. Safety controls and pressure relief devices must be provided to prevent any part of the system from over pressuring. Compressed air can be used to transfer Class II and Class III liquids, but only if handling temperatures are not at or above the flash point of the liquid. Inert gas must be used for Class I liquids and for Class II or III liquids heated to their flash points. Compressed air cannot be used because it increases the possibility of a vapon–air explosion inside the container being pressurized.

Since compressed gas displacement keeps the system under constant pressure, a pipe failure or careless valve operation can cause a considerable amount of product to spill. Adequacy of the safety and operating procedures is important and should be verified.

Dispensing Systems

Dispensing systems generally involve the transfer of liquid from fixed piping systems, drums, or 5-gal (19 L) cans into smaller end-use containers. Because the release of some vapor is practically unavoidable, dispensing must take place in designated areas. This includes dispensing from safety cans through closed piping systems, hand pumps, or similar devices, which transfer by drawing liquid through an opening in the top of the tank or container, by means of gravity through a self-closing valve or faucet, through a hose equipped with self-closing valves, or by means of approved dispensing units, such as those used at service stations. Designated areas should be protected adequately and ventilated or properly segregated from adjacent hazards.

LOSS CONTROL GUIDELINES

The following basic loss control guidelines apply in principle to all operations in which flammable and combustible liquids are stored and handled.

Confinement of Liquids

The major objective of an effective loss control program is to confine liquids and vapors in their containers. A second objective is to minimize the effects of an accidental release. The following steps will help to achieve these goals:

1. Use equipment that is designed for flammable and combustible liquid storage. Such equipment should be vapor tight, should have the minimum number of openings necessary, and should be designed to relieve excess internal pressure to a safe location.
2. Equip open vessels and vessels with loose-fitting covers with overflow drains and emergency bottom dump drains that are piped to a safe location.
3. Handle small amounts of liquids in approved containers.
4. Provide adequate drainage systems to prevent the flow of liquids into adjacent work areas.

Ventilation

Ventilation is a loss prevention measure that can prevent the build-up of released vapors under normal operating conditions. However, ventilation cannot prevent ignitions where abnormal vapor releases occur.

The ventilation required for personnel health and safety greatly exceeds that required for fire safety. NFPA 30 requires that the ventilation necessary for fire safety be sufficient to ensure that the vapor concentration will not exceed 25 percent of the lower flammable limit. In addition to the traditional ventilation rate of 1 cfm/ft^2 (0.3 m^3/min/m^2) of floor area. NFPA 30 also allows design of the ventilation system based on calculation of the anticipated fugitive emission of vapors and on actual monitoring of the space for buildup of vapors.

The inspector should confirm that the ventilation system provides a sweeping action across the entire floor area and that the system exhausts to a safe location. The system should be interlocked so that the operation using flammable liquids will shut down if the ventilation is inadequate. Spot ventilation at the work site is also acceptable.

Control of Ignition Sources

All ignition sources should be controlled or eliminated in areas where flammable vapors could be present. Sources of ignition include open flames, heated surfaces, smoking, cutting and welding, frictional heat, static sparks, and radiant heat. Smoking, open flames, cutting and welding, and hot work should be controlled whether flammable vapors are present.

Specially classified electrical equipment may be needed in some areas (see Chapter 9, "Electrical Systems" of this book). Generally, electrical equipment of the explosion-proof type is used, but nonclassified equipment can be housed in a purged enclosure. Refer to NFPA 496, *Standard for Purged and Pressurized Enclosures for Electrical Equipment,* for further information. NFPA 497, *Recommended Practice for the*

Classification of Flammable Liquids, Gases or Vapors and of Hazardous (Classified) Locations for Electrical Installations in Chemical Process Areas, is a valuable resource for estimating the extent of the classified zone around vapor sources and for determining the proper type of electrical equipment to be provided.

Fire Protection

A wet-pipe automatic sprinkler system is a preferred basic fire control system in areas used to store and handle flammable and combustible liquids. Containerized storage may require special sprinkler installations, including in-rack, ceiling-level sprinklers, large-drop sprinklers, or ESFR sprinklers of special design. Storage tanks, vessels, and process equipment in chemical plants might require deluge water spray systems for cooling or fire control. Automatic or manually actuated foam extinguishing systems are also used in certain process and storage areas in which flammable and combustible liquids are handled, stored, and processed.

All tank foundations and supports should be of fire-resistive or protected steel construction. In small confined areas or inside special equipment or vessels, it might be desirable to provide special extinguishing systems to supplement the automatic sprinkler systems. Appropriate portable fire extinguishers are also necessary in the event of small liquid fires or fires in other combustibles.

Hydrants and small fire hose with adjustable stream nozzles should be provided in areas where flammable and combustible liquids are stored, handled, or used. Hose streams can be used to cool adjacent tanks and structures, to extinguish fires, and to clean up spills.

CODES AND STANDARDS FOR SPECIFIC OCCUPANCIES

In addition to NFPA 30, inspectors may want to consult the following NFPA codes and standards for requirements for specific occupancies.

1. NFPA 30A, *Code for Motor Fuel Dispensing Facilities and Repair Garages,* covers both retail automotive service stations and private fleet fuel dispensing systems. It also covers fuel dispensing at marine facilities. Although primarily directed at the fuel dispensing system itself and fire safety within the service station building, it also addresses fire safety requirements for "quick lube" facilities and the use of aboveground storage tanks. The requirements for siting of aboveground storage tanks differ markedly from those of NFPA 30.
2. NFPA 30B, *Code for the Manufacture and Storage of Aerosol Products.* The scope of this document is provided in its title. It also allows one to classify the degree of hazard of the aerosol product. Its requirements for storing aerosols is quite extensive, and it even provides fire protection requirements for mixed flammable liquid/flammable aerosol storage.
3. NFPA 31, *Standard for the Installation of Oil-Burning Equipment.* This standard covers the installation of liquid-fuel-burning equipment, such as oil burners and oil-fired water heaters. Chapter 7 of this standard covers the installation of the fuel oil tanks and allows some latitude, compared to NFPA 30. It addresses the typical home heating oil tank and the use of special enclosures for fuel oil storage tanks.
4. NFPA 32, *Standard for Drycleaning Plants.* This standard covers drycleaning operations, both at drycleaning plants and at self-serve establishments open to the public.
5. NFPA 33, *Standard for Spray Application Using Flammable or Combustible Materials.* This standard covers spray application of paints, coatings, and so on by means of compressed air, airless atomization, fluidized bed, and electrostatic methods.

It provides requirements for the equipment, the spray area, spray both construction, ventilation, liquid storage and handling, and special applications. It also provides specific requirements for limited finishing workstations, vehicle under coating and body lining, powder coating, and glass fiber- reinforced plastics.

6. NFPA 34, *Standard for Dipping and Coating Processes Using Flammable or Combustible Liquids.* This covers the location of dipping and coating processes, the construction of the equipment, ventilation requirements, liquid storage and handling, and operations and maintenance.
7. NFPA 35, *Standard for the Manufacture of Organic Coatings.* This standard is specific to the coatings manufacturing industry and also covers the manufacture of nitrocellulose-based coatings.
8. NFPA 36, *Standard for Solvent Extraction Plants.* This document is specific to the extraction of oil from oil-bearing seeds using hexane. It is very comprehensive and includes a description of the extraction process and the equipment used.
9. NFPA 37, *Standard for the Installation and Use of Stationary Combustion Engines and Gas Turbines.* This document covers the installation of engines and turbines and their fuel supplies for driving stationary equipment such as emergency generators and fire pumps. Its requirements for fuel storage differ from those of NFPA 30.
10. NFPA 45, *Standard on Fire Protection for Laboratories Using Chemicals.* This document covers the storage and handling of liquids in industrial and instructional laboratories.
11. NFPA 77, *Recommended Practice on Static Electricity.* This recommended practice provides guidance on proper bonding and grounding techniques to minimize ignition by static electric discharge.
12. NFPA 385, *Standard for Tank Vehicles for Flammable and Combustible Liquids.* This standard provides requirements for the construction of tank vehicles to be used in flammable and combustible liquids service. It also provides requirements for safe operation during loading and unloading.

SUMMARY

To determine whether flammable and combustible liquids are stored and handled safely, the inspector must be familiar with their physical and fire hazard properties and risks. Their physical properties include vapor pressure and density, boiling point, specific gravity, water solubility, viscosity, and temperature and pressure effects. Their fire hazard properties include flash point, autoignition temperature, and flammable limits.

The inspector must also be aware of the risks associated with the various containers in which flammable and combustible liquids can be stored and with how these liquids are handled. The storage containers include underground tanks, outside aboveground tanks, tanks inside buildings, portable tanks and intermediate bulk containers, drums and other portable containers, and safety cans. Flammable and combustible liquids can be transferred by gravity flow, by container-mounted pumps, or by a closed-pipe fixed pumping system.

In all operations in which flammable and combustible liquids are stored and handled, basic loss control guidelines apply. These are confinement of the liquids and vapors, ventilation, control of ignition sources, and fire protection in the form of automatic sprinkler system supplemented by special extinguishing systems when deemed appropriate.

BIBLIOGRAPHY

Benedetti, R. P., ed., *Flammable and Combustible Liquids Code Handbook,* 6th ed., NFPA, Quincy, MA, 1990.

Cote, A. E., ed., *Fire Protection Handbook,* 18th ed., NFPA, Quincy, MA, 1997.

Cote, A. E., and Linville, J. L., eds., *Industrial Fire Hazards Handbook,* 3rd ed., NFPA, Quincy, MA, 1990.
PEI/RP100-90, *Recommended Practices for Installation of Underground Liquid Storage Systems,* Petroleum Equipment Institute, Tulsa, OK, 1990.
PEI/RP200-92, *Recommended Practices for Installation of Aboveground Storage Systems for Motor Vehicle Refueling,* Petroleum Equipment Institute, Tulsa, OK, 1992.
Federal Regulations. These references are available from the U.S. Government Printing Office, Washington, DC.
Title 16, *Code of Federal Regulations,* Part 1500.43(a).
Title 29, *Code of Federal Regulations,* Part 1910.
Title 40, *Code of Federal Regulations,* Part 280.
Title 49, *Code of Federal Regulations,* Part 173.

NFPA Codes, Standards, and Recommended Practices

See the latest version of The NFPA Catalog for availability of current editions of the following documents.

NFPA 30, *Flammable and Combustible Liquids Code*
NFPA 30A, *Code for Motor Fuel Dispensing Facilities and Repair Garages*
NFPA 31, *Standard for the Installation of Oil-Burning Equipment*
NFPA 32, *Standard for Drycleaning Plants*
NFPA 33, *Standard for Spray Application Using Flammable or Combustible Materials*
NFPA 34, *Standard for Dipping and Coating Processes Using Flammable or Combustible Liquids*
NFPA 35, *Standard for the Manufacture of Organic Coatings*
NFPA 36, *Standard for Solvent Extraction Plants*
NFPA 37, *Standard for the Installation and Use of Stationary Combustion Engines and Gas Turbines*
NFPA 45, *Standard on Fire Protection for Laboratories Using Chemicals*
NFPA 69, *Standard on Explosion Prevention Systems*
NFPA 77, *Recommended Practice on Static Electricity*
NFPA 86, *Standard for Ovens and Furnaces*
NFPA 325, *Fire Hazard Properties of Flammable Liquids, Gases, and Volatile Solids*
NFPA 326, *Standard for the Safeguarding of Tanks and Containers for Entry, Cleaning, or Repair*
NFPA 329, *Recommended Practice for Handling Releases of Flammable and Combustible Liquids and Gases*
NFPA 385, *Standard for Tank Vehicles for Flammable and Combustible Liquids*
NFPA 496, *Standard for Purged and Pressurized Enclosures for Electrical Equipment*
NFPA 497, *Recommended Practice for the Classification of Flammable Liquids, Gases, or Vapors and of Hazardous (Classified) Locations for Electrical Installations in Chemical Process Areas*

Gas Hazards

CHAPTER 43

Theodore C. Lemoff

CASE STUDY

Fire Strikes Retail Propane Storage Facility

Two vehicles, including a liquid propane delivery vehicle, were involved in fire at a retail propane filling station in Georgia. The incident occurred when a hose fell off a rack and was jarred into the open position. The release of pressurized liquid propane caused the hose to whip around, hitting nearby metal components and striking a spark that ignited the leaking propane.

An employee who saw the fire start immediately called 911, and the fire department responded at 9:05 A.M. Employees tried to control the blaze with hand-held fire extinguishers, but they were ineffective.

Flames impinging on the large storage vessel caused the pressure relief valve to operate, venting more propane, which contributed to the fire. Several redundant shut-off valves were later found in the open position.

There were no reports on the dollar amount of damage, and no one was injured during the incident.

Source: "Fire Watch," *NFPA Journal,* November/December, 2000.

Of all three states of matter—gas, liquid, and solid—gas is the only one with no shape or volume of its own; it is the only one that expands to fill the container it is in. A gas always exerts pressure on its container. Gases are frequently compressed to facilitate shipment. Some gases, such as propane, butane, and ammonia, liquefy under moderate pressure and are stored in pressurized containers in both the liquid and gas states. Others will only liquefy at higher pressure. With both types of gases, any leak in the container will allow the material to escape until the pressure inside the container is reduced to the pressure of the atmosphere outside the container. Although the escaped gas can be treated, controlling the hazard by keeping the gas confined is easier than dealing with it if it is released. This chapter discusses the properties of gases, gas container safeguards and storage locations, and nonflammable medical gas systems.

Theodore C. Lemoff, P.E., is the principal gases engineer at NFPA, one of several positions he has held since 1985. He is the staff liaison to several NFPA technical committees, including Liquefied Petroleum Gas Code; Utility LP-Gas Plant Code; Production, Storage, and Handling of Liquefied Natural Gas; *and* National Fuel Gas Code.

PROPERTIES AND STORAGE OF GASES

Properties of Gases

The most dramatic evidence of the hazard that released flammable gases present is a fire or explosion resulting from the ignition of a gas–air mixture in a confined space. An accumulation of any gas, except oxygen and air, can cause asphyxiation by displacing the oxygen in the air in a room. Oxygen levels lower than 19.5 percent are considered hazardous by the U.S. Department of Labor's Occupational Safety and Health Administration (OSHA). When the concentration is reduced to 4 percent, death can occur in less than a minute. Enriched concentrations of oxygen in air also present health hazards; any concentration over 23.5 percent is considered hazardous by OSHA. The hazards include increased flammability of materials and physiological effects. Because the human senses cannot detect most gases, people will inhale a gas normally, as if it were air.

Gases can be hazardous in other ways as well. Some gases are chemically reactive either by themselves or on contact with other materials. Others are physiologically reactive and are toxic or can cause other health hazards.

Storage of Gases

Gases are distributed in pipelines and in containers. For economy of storage and transportation, gases are compressed or refrigerated to reduce their volume. There are three physical forms of gas storage: compressed gases, liquefied gases, and cryogenic liquids.

1. A *compressed gas (nonliquefied),* stored as gas under pressure, is defined by both the U.S. Department of Transportation (DOT) and NFPA 55, *Standard for Storage, Use, and Handling of Compressed and Liquefied Gases in Portable Cylinders,* as "a gas, other than in solution, that in a packaging under the charged pressure is entirely gaseous at a temperature of 68°F (20°C)."
2. A *liquefied gas,* stored as both liquid and gas under pressure (if the gas liquefies under moderate pressure), is defined as "a gas, other than in solution, that in a packaging under the charged pressure exists both as a liquid and a gas at a temperature of 68°F (20°C)."
3. A *cryogenic liquid,* a gas that is liquefied using a combination of pressure and low temperature and stored at low temperatures in insulated, pressurized containers, is defined as "a liquid having a boiling point lower than −150°F (−101°C) at 14.7 psia (an absolute pressure of 101 kPa)."

All gases are hazardous materials because of their pressure (compressed and liquefied gases) or temperature (cryogenic liquids). The volume of gas evolved from the leakage of a compressed, liquefied, or cryogenic gas is 200 to 850 times the stored volume.

All compressed and liquefied gases are stored at pressures of 100 psi (689 kPa) and higher. Cryogenic liquids are stored at their normal atmospheric boiling point in insulated containers that are also pressure vessels. Because heat leakage into the vessel is inevitable, these containers are designed to hold some pressure [25 to 100 psi (172 to 689 kPa), typically].

Despite the ability of these pressure vessels to contain pressure, the occasional operation of the pressure relief valve with some release of gas is normal. The most effective way to reduce gas hazards is to prevent containers from leaking by taking measures to maintain the integrity of the containers and by designing the containers to prevent failure.

GAS CONTAINER SAFEGUARDS

Gas containers are constructed to the same standards as those applied to pressure vessels. Portable pressure vessels include cylinders (Figure 43-1) and tanks that are part of cargo vehicles or railroad tank cars. They are constructed to U.S. Department of Transportation (DOT) or Transport Canada (TC) requirements in the United States and Canada, respectively, and to similar standards in other countries. Fixed pressure vessels (Figure 43-2) are constructed to the *Boiler and Pressure Vessel Code* of the American Society of Mechanical Engineers. Providing all the details of construction and inspection of these pressure vessels is beyond the scope of this manual.

FIGURE 43-1 DOT/TC Cylinders: Top: Typical Industrial Truck Motor (Engines) Fuel Cylinders; Bottom: Typical Stationary and Portable Cylinders for Residential and Commercial Uses
Source: *LP-Gas Code Handbook*, NFPA, 2001,Exhibit 2.1; Courtesy of Manchester Tank.

Pressure Relief Valves and Pressure Regulators

Regardless of the code used, pressure vessels must be constructed to contain the anticipated pressure of the gas stored. Because different gases are stored under different amounts of pressure, ensuring that the proper cylinders and tanks are being used is important. Some substitutions of gases and cylinders are acceptable; but if in doubt, a gas supplier can be contacted for more information.

Compressed Gases. The pressure in a gas container varies with both the temperature and the amount of gas contained. Compressed gases are filled to a predetermined pressure, depending on the construction of the container. As the gas is used, the pressure decreases proportionally. The gas also expands and contracts as the

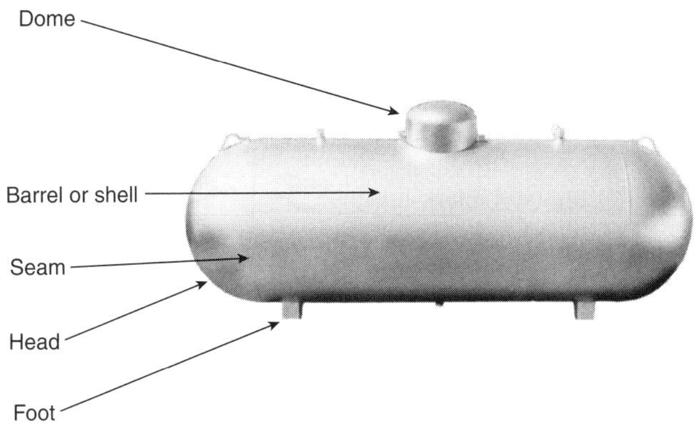

FIGURE 43-2 Fixed Pressure Vessel Used Commonly in Residential and Commercial Applications
Source: *LP-Gas Code Handbook*, NFPA, 2001,Exhibit 2.5; Courtesy of American Welding and Tank Company.

temperature increases and decreases. Safety devices are used to protect the container from overpressure failure. Compressed gas cylinders are usually equipped with a burst disk, which releases the contents of the container if the pressure exceeds a preset level. These cylinders are also equipped with fusible devices that release the entire contents of the cylinder if the temperature exceeds a preset value, as would be the case if the container were exposed to fire.

Liquefied Gases. Liquefied gas containers are not filled liquid full but have a vapor space to allow for expansion of the liquid. Liquefied gas containers are equipped with pressure relief valves, which open to reduce the pressure in a container if the preset pressure is exceeded. These valves close when the internal pressure returns to below the preset pressure. If the liquid expands to completely fill the container, the pressure will rise above the design pressure of the container, and the pressure relief valve will open to prevent container failure. This situation is not desirable because some of the gas will have been released. Depending on where the container is located, and what the properties of the gas are, this gas release may present problems. The pressure in a liquefied gas container is a function of the temperature of the liquid and vapor in the container. Information on the temperature-pressure properties of liquefied gases can be found in the Compressed Gas Association's *Handbook of Compressed Gases.*

Cryogenic Liquids. Cryogenic liquids are stored in insulated containers that are also designed to hold pressure. They are subject to the same potential expansion as liquefied gas cylinders, but because they are well insulated, fire exposure does not dominate the sizing of the pressure relief valve. Cryogenic liquids are stored at very low temperatures [below −130°F (−90°C)]. As a result, heat leakage into the cylinder is inevitable because all insulation will allow some heat transfer.

As the liquid warms, some vaporization occurs and the pressure in the cylinder increases. The pressure relief valve will open periodically to protect the cylinder. Because most of the cryogenic liquids are nonflammable (e.g., nitrogen and argon), this release is not a problem. Special care must be taken, however, with cryogenic oxygen cylinders to keep them separate from any flammable gases and to ensure that they are stored outdoors or in well-ventilated areas. Occasionally, liquid hydrogen cylinders will be stored. Such cylinders must be stored outdoors or indoors in accordance with NFPA 50A, *Standard for Gaseous Hydrogen Systems at Consumer Sites,* in buildings or rooms used only for hydrogen storage. The inspector should refer to NFPA 50A for storage requirement details.

Piping Systems. There is a possibility for accidental release of gases from the storage and piping system. These releases must be stopped if the released gas presents a controllable hazard. Release of unburned flammable gas in a building can lead to an explosion if the gas is ignited. Once ignited, the burning gas should not be extinguished unless the flow of flammable gas can be stopped. Outdoors, a vapor cloud ignition can cause extensive damage.

All compressed piped gas systems use one or more pressure regulators to maintain system pressure to accommodate gas users. These regulators often have pressure relief devices incorporated in them, or the pressure relief can be accomplished with a separate device. Both for flammable and nonflammable gases the regulator pressure relief opening or pressure relief device should be piped to the outdoors if the system is not already located outdoors. All gas containers have a manually operated valve at the gas outlet. An automatically operated valve may also be used. Where used, the inspector should check it for proper operation because it is often part of the safety system.

Boiler Liquid Expanding Vapor Explosion

Fire exposure can lead to container failure despite the proper operation of the pressure relief valve. The pressure relief valve is designed to open to release pressure above a safe level. In a fire, the container metal (usually steel) can be heated, especially if the flame is above the liquid level. As the metal is heated, its strength properties change. As a rule of thumb, metals lose their ability to hold pressure at about half their melting point in degrees Fahrenheit. Steel has a melting point of 2600°F (1427°C) or higher. Beginning at 1300°F (704°C), therefore, it weakens to the point at which it can no longer hold the pressure it was designed to hold. This temperature can easily be reached with direct flame impingement. Aluminum alloys melt at between 980°F (527°C) and 1200°F (649°C).

Cooling containers exposed to fire can prevent this failure mode, which is called a *boiling liquid expanding vapor explosion (BLEVE)*. Although no definitive studies have been done to confirm the fact, the BLEVE of a steel container within 10 minutes of flame impingement has not been observed to occur. Fire is not the only cause of BLEVE. BLEVE can occur if the container is no longer able to contain the pressure it was designed to. This can be caused by corrosion and mechanical reduction in container wall thickness. Inspectors should look for corrosion and gouging of container walls. If these are found, a pressure vessel engineer can be contacted to determine if a hazard exists.

Certification of Cylinders

Cylinders subject to DOT or TC standards are manufactured with a date stamped into the cylinder or its collar. Cylinders can be filled for 12 years from the date of manufacture. Cylinders can be filled and shipped up to that date. After the date is passed, the cylinder must be inspected and recertified in accordance with procedures stated in the DOT or TC standards. They can then be stamped with a date 5, 7, or 12 years from the date of recertification. The date is determined by the recertification method used. Note that the contents of cylinders with expired dates can be used with no limitation. The date applies only to the filling and shipping of cylinders. Inspector's should verify that out-of-date cylinders are not refilled.

Uses of Compressed Gases

The most common uses of compressed gases are as follows:

- As fuel in gas-burning appliances, industrial heating equipment, and motor vehicles, with air
- For cutting and welding, with oxygen
- For medical purposes, especially oxygen and nitrous oxide
- For shielding from oxygen in welding processes, using argon or other gases
- For entertainment purposes, in the form of flame effects in theme parks and concerts

All of these uses involve the release of gas from its container under controlled conditions. Specific equipment is used to control the flow and pressure of the gas to maintain safe conditions. Many smaller heating appliances have a temperature-sensing device in the flame to sense flame failure and to shut off the flow of gas if the flame is not present when it should be. Larger heating appliances use optical flame detectors that sense flame absence faster and can stop the gas flow more quickly.

Gas-Burning Appliances and Industrial Heating Equipment

Nearly all of the commercial gas-burning appliances and much industrial heating equipment are listed to the American National Standard Institute (ANSI) or Underwriters Laboratories (UL) standards. Special purpose and custom-made industrial gas appliances are often not listed and must be evaluated for safe design. Such an evaluation can be done by an independent testing laboratory, a gas engineer, or the gas utility company.

Oxygen-Fuel Gas Cutting and Welding

Oxygen-fuel gas cutting and welding systems have specially designed devices to control the gas flow and pressure and to prevent the addition of oxygen to a fuel cylinder or piping system and fuel to an oxygen cylinder or piping system. For more information on these devices, the inspector should refer to NFPA 51, *Standard for the Design and Installation of Oxygen-Fuel Gas Systems for Welding, Cutting, and Allied Processes.*

Entertainment

The use of fire as an entertainment effect has grown considerably since the most recent edition of this manual. NFPA has formed a committee on the subject, and a new standard, NFPA 160, *Standard for Flame Effects Before an Audience,* was issued in 1998. This standard provides minimum requirements for the storage of flammables, personnel safety, and flame safety in the entertainment field. The inspector should consult this document when evaluating and inspecting any entertainment flame effect.

GAS CONTAINER STORAGE LOCATION

Larger gas storage containers, which are usually constructed to the American Society of Mechanical Engineers (ASME) *Boiler and Pressure Vessel Code,* are normally located outdoors. Portable cylinders can be stored in many locations. NFPA 55, *Standard for the Storage, Use, and Handling of Compressed and Liquefied Gases in Portable Cylinders* provides requirements for cylinder storage. Storage of quantities of gas cylinders used in specific applications or for specific gases is covered in other NFPA documents, as shown in Table 43-1.

In addition, in the United States, storage of gases in industrial applications is covered by the regulations of the U.S. Department of Labor's Occupational Safety and Health Administration (OSHA). OSHA has adopted the NFPA codes and standards referenced in Table 43-1. Compliance with these codes and standards, therefore, provides OSHA compliance as well.

TABLE 43-1 NFPA Codes and Standards Related to Gas Storage

Gas	Code
Propane	NFPA 58, *Liquefied Petroleum Gas Code*
Hydrogen	NFPA 50A, *Standard for Gaseous Hydrogen Systems at Consumer Sites*
Oxygen and fuel gases used for cutting and welding	NFPA 51, *Standard for the Design and Installation of Oxygen-Fuel Gas Systems for Welding, Cutting, and Allied Processes*
Oxygen, other uses	NFPA 50, *Standard for Bulk Oxygen Systems at Consumer Sites*

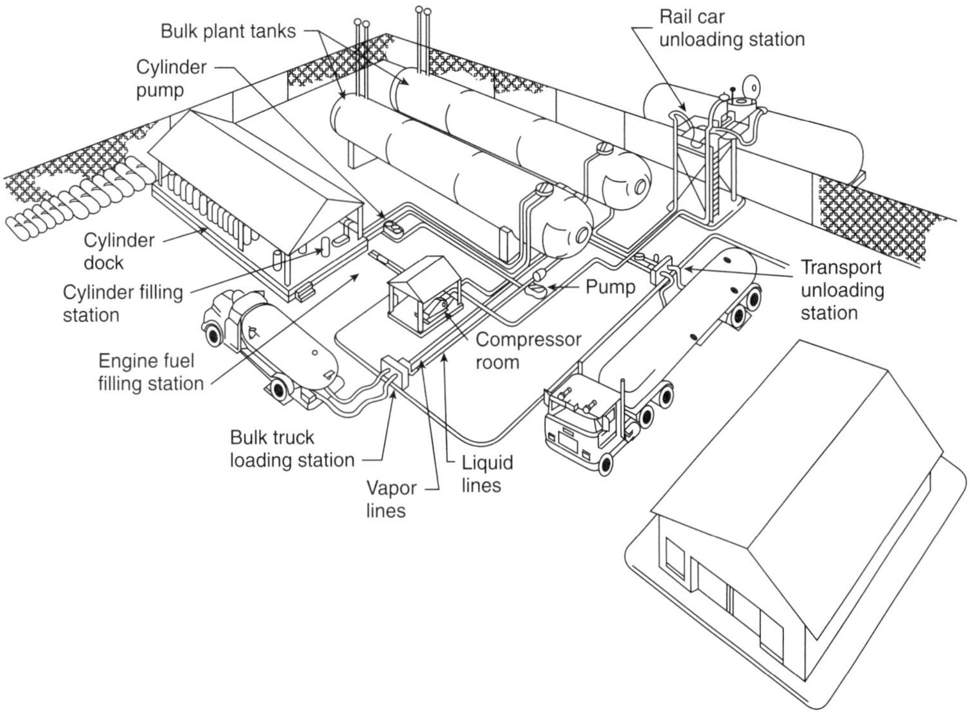

FIGURE 43-3 LP-Gas Bulk Plant

LP-Gas Bulk Plants

LP-Gas is transported by ships, barges, pipelines trucks and tank cars to bulk plants, which store and distribute LP-Gas to users. Bulk plants (Figure 43-3), of which there are a large number in the United States, are most likely to be located in suburban and rural areas. These plants must comply with NFPA 58, *Liquefied Petroleum Gas Code.*

Key Features of Bulk Plants

Tank Car Unloading. Tank cars on a rail spur, where rails have been installed, must be at least 50 ft (7.6 m) from the main line and must be enclosed with an industrial fence. The railroad tank car and the liquid loading or unloading piping must be equipped with valves to stop the flow of LP-Gas in the event of piping or hose failure.

Truck Unloading. The unloading point for transport trucks delivering LP-Gas should have a mechanical or physical guard to prevent damage to the piping and hose. Wheel chock blocks must be used. The truck and bulk plant piping system must be equipped to stop the flow of LP-Gas in the event of hose or piping failure. In addition, piping anchorage and an emergency shutoff valve are required to provide safety in the event a truck drives off with the delivery hose connected. These safety features are required in all bulk plants, irrespective of their installation date. Where needed, a shed to protect the driver or delivery attendant should be provided because the delivery must be monitored by a qualified person.

Bulk Plant Storage Tanks. Storage tanks must be installed on insulated steel or concrete foundations. A gauging device and pressure relief valves are required. Tank inlets and outlets must be equipped with internally mounted valves to automatically stop the flow of LP-Gas in the event of piping failure. NFPA 58 requires all storage tanks installed after July 1, 2003 to have remotely operated internal valves

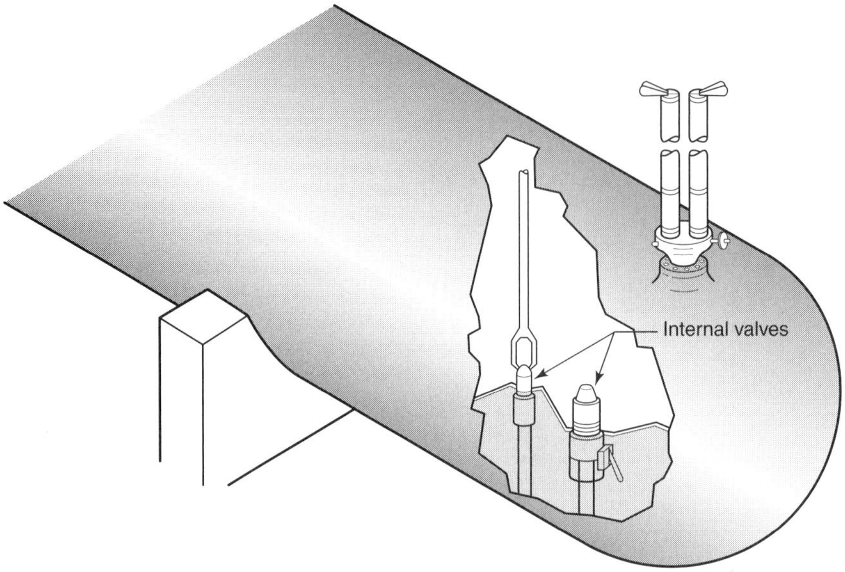

FIGURE 43-4 Internal Valves in Storage Tank
Source: Modified from *LP-Gas Code Handbook*, NFPA, 2001, Exhibit 3.34.

connected to the liquid space of the tank (Figure 43-4). This requirement will also be applicable to existing tanks in the year 2011.

Truck Loading. The loading area should be level and either paved or free of vegetation. Wheel chock blocks must be used.

Cylinder Filling. Cylinder filling can be done indoors or outdoors. When done indoors, the building or separate room used for filling must meet the requirements of NFPA 58 for buildings or structures housing LP-Gas distribution facilities. Included are limitation of sources of ignition, indirect heating, and minimum ventilation requirements.

Fire Safety Analysis. Many bulk plants have a fire safety analysis that was developed as part of the permitting process, as required by NFPA 58. NFPA 58 requires all bulk plants to have a fire safety analysis by February 2004.

Fire Protection. Where a fire protection system is installed at a bulk plant, it should be tested regularly for reliability. Fire extinguishers should be tested every 12 years. Water spray systems should be tested by flowing water at intervals required by NFPA 25. Tests should be witnessed.

Inspecting Bulk Plants

Inspectors who conduct inspections of bulk plants should be on the lookout for the following safety items:

1. Are there roads or is other access (e.g., shoreline access) provided for emergency vehicle access?
2. Is the gas stored at the plant odorized? The inspector should conduct a sniff test.
3. Are precautions taken against vehicle damage to tanks and piping?
4. If the plant is operated at night, is there adequate lighting for night operation?
5. Do storage containers have nameplates? Are they designed for 250 psi (1700 kPa), minimum?

6. Is the ground within 10 ft (3 m) of storage containers free of loose or piled combustible material and weeds and long dry grass? Note that live vegetation is permitted.
7. Does each storage container have a pressure relief valve? Is there a 7-ft (2-m) tailpipe for each pressure relief valve?
8. Are storage container supports concrete or insulated steel?
9. Are containers free of significant rust and pitting?
10. Does bulk truck and tank car loading and unloading piping have emergency shutoff valves with provision for local and remote operation, and does it have a fusible link? Does this piping have secure anchorage?
11. Do all storage container connections (except instruments, pressure relief valves, and plugged connections) have shutoff valves?
12. Does the plant have at least one fire extinguisher with at least 18 lb (8.2 kg) of agent and a B:C rating?

Medical Gases

NONFLAMMABLE MEDICAL GAS SYSTEMS

Nonflammable piped medical gas systems in a health care facility may include oxygen, nitrous oxide, nitrogen, helium, and carbon dioxide gases. Medical air and vacuum, generated by special compressor systems, and waste anesthesia gas disposal (WAGD) commonly use piped gas systems. Gas systems for specific occupancies are designated by "levels" relative to the nature of definitive patient care and life-sustaining dependency as defined within the occupancy chapters of NFPA 99, *Standard for Health Care Facilities.*

Because oxygen and nitrous oxide are nonflammable, their hazard as oxidizing agents is not always recognized. In large leaks both gases will lower the ignition temperature and accelerate the combustion of flammable and combustible liquids, gases, and solids.

Although containers storing cryogenic liquids have protective features, accidental spills and leakage can occur. Because liquid oxygen has a boiling point of −297°F (−183°C) and an 860:1 expansion ratio, potential hazards exist, including the following:

- The contact of skin tissue with the low temperature liquid
- The large volume of oxygen-enriched air that can be formed
- The ease of ignition of flammable and combustible materials in an oxygen-enriched atmosphere
- The explosive nature of spilled liquid oxygen when it is in contact with organic materials (including asphalt, wood, and carpet) and is subjected to shock, such as from a tool being dropped or someone walking on the liquid. This sensitivity exists until the spilled liquid has evaporated.

Medical Gases Safeguards

Valves and Alarm Systems. Valves and alarm systems are used with the piped systems to isolate areas during renovation and emergencies. Alarm systems are incorporated to identify operational problems, including valves that may have been accidentally closed and gas outages.

Ventilation and Construction. Compressed gas cylinder supplies are connected to the piping network through manifold systems. Storage areas are constructed to prevent a fire occurring outside the room from affecting the cylinders. Ventilation of the room is intended, in the event of leakage, to prevent oxidizing or inert gases from accumulating within the room or from being distributed to occupied spaces. Except for shipping crates or cylinder storage cartons and wooden rack construction for cylinder storage, combustible materials cannot be stored within the room. Piping system materials and installation methods are intended to provide system integrity against fire and the subsequent release of oxidizing gases into the fire environment.

Bulk Liquefied Oxygen Systems. Bulk liquefied oxygen (LOX) systems convert the LOX to gaseous oxygen for supplying health care facilities that use large quantities of oxygen for their piped systems. The storage part of LOX systems must meet the requirements of NFPA 50.

Small portable units containing LOX can be found in many health care facilities and private dwellings. These units enhance patient mobility by eliminating the need for people on respiratory therapy to carry a small compressed gas cylinder with them. The small LOX units are refilled daily, or more frequently, from a cryogenic reservoir provided by a medical gas supplier.

Hyperbaric Chambers. Hyperbaric chambers, which are pressurized vessels designed to contain 100 percent oxygen within the chamber, are used to treat decompression sickness (the bends), carbon monoxide poisoning, wounds, and other disorders. The increased availability of hyperbaric chambers can be attributed to the continued discovery of health benefits associated with concentrated levels of oxygen. Special requirements for construction and enclosures, electrical systems, fire protection, and furnishings are found in NFPA 99B, *Standard for Hypobaric Facilities.*

SUMMARY

Gases can be hazardous. Released flammable gases can cause fire or explosion; some gases are chemically reactive either by themselves or on contact with other materials; still other gases are physiologically reactive and are toxic or can cause other health hazards. Gases are distributed in pipelines and in containers. Containers and pressure vessels are constructed to contain the anticipated pressure of the gas stored. Different gases are stored under different amounts of pressure, therefore proper cylinders and tanks with the necessary pressure relief valves and pressure regulators must be used.

LP-Gas bulk plants store and distribute LP-Gas. Because of the hazard associated with these locations, plants must comply with the requirements of NFPA 58, *Liquefied Petroleum Gas Code,* regarding tank car unloading, truck unloading and loading, installation of storage tanks, cylinder filling, fire safety analysis, and fire protection.

Nonflammable medical gas systems also present a hazard. For example, oxygen and nitrous oxide are oxidizing agents and in large leaks they can lower ignition temperature and accelerate the combustion of flammable and combustible liquids, gases, and solids. Medical gases safeguards include valves and alarm systems, ventilation, and construction.

BIBLIOGRAPHY

Boiler and Pressure Vessel Code, Section VIII, American Society of Mechanical Engineers, 2001.

Code for Pressure Piping, ASME B 31, American Society of Mechanical Engineers, 345 E. 47th St., New York, NY.

Cote, A. E., ed., *Fire Protection Handbook,* 18th ed., NFPA, Quincy, MA, 1997.

Handbook of Compressed Gases, Compressed Gas Association, Chantilly, VA, 1999.

Cote, A. E., and Linville, J. L., eds., *Industrial Fire Hazards Handbook,* 3rd ed., NFPA, Quincy, MA, 1990.

Lemoff, T. C., ed., *LP-Gas Code Handbook,* 6th ed., NFPA, Quincy, MA, 2001.

NFPA Codes, Standards, and Recommended Practices

See the latest version of The NFPA Catalog for availability of current editions of the following documents.

NFPA 25, *Standard for the Inspection, Testing, and Maintenance of Water-Based Fire Protection Systems*
NFPA 50, *Standard for Bulk Oxygen Systems at Consumer Sites*
NFPA 50A, *Standard for Gaseous Hydrogen Systems at Consumer Sites*
NFPA 51, *Standard for the Design and Installation of Oxygen-Fuel Gas Systems for Welding, Cutting, and Allied Processes*
NFPA 55, *Standard for the Storage, Use, and Handling of Compressed and Liquefied Gases in Portable Cylinders*
NFPA 58, *Liquefied Petroleum Gas Code*
NFPA 99, *Standard for Health Care Facilities*
NFPA 99B, *Standard for Hypobaric Facilities*
NFPA 160, *Standard for Flame Effects Before an Audience*
NFPA 560, *Standard for the Storage, Handling, and Use of Ethylene Oxide for Sterilization and Fumigation*

Combustible Dusts

CHAPTER 44

William J. Bradford

CASE STUDY

Series of Explosions Kills Seven

The DeBruce Grain elevator in Wichita, Kansas, with a storage capacity of 20.7 million bushels of wheat, was the world's largest grain elevator. It consisted of 246 circular grain silos 30 ft in diameter and 120 ft in height, arranged in a linear array of 3 silos abreast. The 164 spaces between the silos, known as interstice silos, were also used for grain storage.

Midway between these silos was a headhouse, which housed four elevator legs and facilities to weigh and distribute grain into selected silos. The overall length of the silos plus headhouse was well over one-half mile. Across the top of the silos were two galleries. Elevated grain was carried horizontally in each gallery by belt from the headhouse out to a selected silo and dumped into that silo by means of a "tripper," which diverted the grain from the belt. Beneath the silos, four tunnels contained belts that carried grain discharged from the silos toward the elevator legs in the headhouse.

The initial DeBruce Grain elevator explosion—which set off a series of additional explosions of increasing severity—occurred when grain dust was ignited in a tunnel beneath the silos. The most probable ignition source was created when a concentrator roller bearing, which had seized due to no lubrication, caused the roller to lock into a static position as the conveyor belt continued to roll over it, raising the roller's temperature beyond that required to ignite layered grain dust inside the roller.

From the initial series of explosions in a tunnel, the blast and fire blew upward through the headhouse into the two galleries. Utilizing a crossover tunnel, the blast waves also rose vertically through silos and blew off many silo tops. Numerous silos as well as the headhouse were destroyed. Seven men were killed in the explosion.

Local fire and rescue responders were on-scene within 10 minutes after the explosion. Rescue efforts were, however, delayed because, in the absence of documented work assignments for all personnel and lack of worker knowledge of the facility's emergency action plan, the response team did not know the number of affected people and where in the facility they might be found. The height, length, and limited access of the structure presented rescuers with great challenges. The most severely inured had to be lifted from the top of 120-ft silos.

The explosion investigation team, commissioned by the U.S. Department of Labor's Occupational Safety and Health Administration (OSHA), concluded that of greater consequence than the seized roller bearing in causing the explosion was the facility's corporate decision to (1) allow massive amounts of fuel to continually be created and distributed throughout the elevator, awaiting any one of many possible sources of ignition; (2) forego repair and

William J. Bradford, P.E., is a loss prevention consultant. He is a registered professional engineer in Connecticut and Rhode Island, is a member of AIChE and SFPE, and serves on several NFPA technical committees.

restoration of long-failed grain dust control systems; and (3) abandon preventive maintenance of elevator equipment—particularly the grain conveyor and grain dust control systems.

Source: Based on "Report on Explosion of DeBruce Grain Elevator, Wichita, Kansas, 8 June 1998," commissioned by the U.S. Department of Labor, Occupational Safety and Health Administration (*http://www.osha.gov*).

Most finely divided combustible materials, including metals, are subject to rapid combustion when dispersed in air, usually resulting in destructive explosions. The ease with which these dust air mixtures ignite and their ability to propagate flame and generate damaging explosion pressures depend on a number of factors, such as particle size and shape, concentration, and moisture content. Although these factors cannot be addressed here in any detail, inspectors can find information on specific combustible dusts in the *Fire Protection Handbook* (see Bibliography).

For the purposes of this manual, a *dust particle* is defined as a piece of material 150 μ or less in diameter. Basically, this means any material, not necessarily of regular shape, that is fine enough to pass through a 100-mesh sieve. Particles larger than 100 mesh can be considered "grit" and might not even pose any unusual hazards. For example, sawdust is primarily a collection of variously sized chips of wood. Although some very finely divided particles are present, they do not contribute much to the explosion potential of the mass, and sawdust presents little more than a fire hazard. In contrast, the fine floury dust generated by a sanding operation can be readily ignited and can explode quite destructively. Some severe dust explosions have occurred in sanding operations in plywood and particle board plants.

The concept of a dust explosion is sometimes not well understood because an explosion can result from a number of different physical, mechanical and chemical processes. In terms of burning characteristics, the word *explosion* means either deflagration or detonation. A *deflagration* is a combustion reaction that travels slower than the speed of sound, a *detonation* is a combustion reaction that travels at or above the speed of sound. A dust explosion that is, in fact, a detonation is very unusual.

In a typical dust explosion, the flame front will travel 3.3 to 33 ft (1 to 10 m) per second. Although this is much faster than the flame speed in a typical fire, it is much slower than the pressure wave generated by the production and thermal expansion of the flammable gases. The pressure wave is responsible for most of the damage in a dust explosion and has two important characteristics: maximum pressure and maximum rate of pressure rise.

DUST EXPLOSIONS

Dust explosions can be very destructive, the most vivid examples being those that occur in large grain elevators, such as the one that occurred at the DeBruce Grain elevator in Wichita, Kansas, in 1998. In most cases, there actually is a series of explosions, and the first is frequently quite small. Although small in size and usually the result of a minor process malfunction, this first explosion is intense enough to dislodge any static dust on the walls, ledges, machinery, and other surfaces in the immediate vicinity. Because the pressure wave is moving much faster than the flame front, it knocks the dust loose and mixes it with air, creating a much larger dust cloud just in time to be ignited by the following flame front. This secondary explosion is much larger than the first and is usually very destructive. It can damage process equipment, easily destroy masonry walls, and trigger even larger subsequent explosions.

Rate of Pressure Rise

The rate of pressure rise is roughly the ratio of the peak pressure to the time interval during which the pressure is increased. It is the most important factor in determining the severity of a dust explosion. The size of explosion vents are largely determined by this rate of rise. For very rapid rates of rise, explosion vents might not be practical, and other protection devices might be necessary.

Maximum Explosion Pressure

The peak pressures most dust explosions reach under test conditions exceed 50 psi (345 kPa) although these peaks vary according to the particle size distribution, concentration, and other variables. When you consider that typical construction will withstand 1 psi (7 kPa) or less, it becomes evident that even the most inefficient dust explosion will do considerable damage. In fact, most dust explosions are far from optimal. Rarely is there uniform dispersion or particle-size distribution; the damage that typically results is done at well below the explosion pressures that could have developed.

Duration

Another factor in explosion severity is the duration of the explosion pressure. Consider the wall of the dust collector in which an explosion occurs. The wall "sees" a steady increase of pressure to some peak value over a finite time span, assuming it does not rupture. Then the pressure begins to subside. If the pressure is plotted against time, the area under the curve is the total impulse imparted to the wall by the explosion. It is the total impulse, rather than the peak pressure, that ultimately determines the damage. This partly explains why dust explosions tend to be more damaging than gas explosions, even though they build pressure more slowly and do not usually peak as high.

Confinement

The gaseous combustion products of a dust explosion expand at a rate as high as sonic velocity and, in so doing, exert significant pressures on the surrounding enclosure. Unless the enclosure is strong enough to withstand the peak pressure developed, it will fail. For this reason, process equipment and the buildings in which it is located must be protected by explosion vents or some type of explosion-prevention system (see NFPA 68, *Guide for Venting of Deflagrations,* and NFPA 69, *Standard on Explosion Prevention Systems*).

Inerting

Laboratory tests have shown that the reduction of the oxygen concentration or the introduction of an inert powder can reduce the maximum explosion potential. For most agricultural, plastic, and carbonaceous dusts, reducing the oxygen content below 10 percent by volume can prevent a dust explosion from occurring. Frequently, this is achieved by increasing the volume percentage of nitrogen or carbon dioxide in the enclosures or equipment handling the combustible dust.

For combustible metal dusts, such as aluminum or magnesium, however, this is not a satisfactory solution because these combustible metal dusts can react with these gases. In those cases, it would be necessary to use one of the inert noble gases, such as argon (see NFPA 69, *Standard on Explosion Prevention Systems*).

Introducing an inert powder to mitigate the effects of a dust explosion might not be practical because it takes relatively large amounts of the inert dust to be effective. This technique, referred to as "rock dusting," is limited mostly to coal mines, which generally require 65 percent of the total dust for a coal mine entry to be rock dust.

Evaluating the Hazard

When inspectors are confronted with a dust explosion hazard, they should find out as much as they can about the material itself and the process conditions including the size distribution of the dust particles. Next, they should compare the relative hazard of the dust.

The *Fire Protection Handbook* contains the explosion characteristics for common dusts. This information may also be available in some of the reports published by the U.S. Bureau of Mines, which are listed in the handbook and in the bibliography of this book.

If inspectors are confronted with a dust that has not been tested or are asked to evaluate one of the dusts described in the *Fire Protection Handbook,* they should have a sample tested, preferably in the spherical test vessels noted here. They will want to know the maximum explosion pressure, the maximum rate of pressure rise, the minimum explosion concentration, and the concentration yielding the highest value of maximum pressure. They might also want to determine the minimum ignition energy and the layer ignition temperature.

FIRE HAZARDS OF DUSTS

Combustible dusts are also a fire hazard because stationary deposits of dust provide an easy means for a fire to spread rapidly from its initial location. A flash fire of this nature could propagate so quickly that it might cause sprinklers to operate outside the sprinkler design area, limiting the system's ability to extinguish or control the fire.

Another characteristic of dust is its ability to act as thermal insulation. Thick deposits of dust on heat-producing equipment such as motors, shaft bearings, and similar components retard the flow of heat. The natural cooling effects of the equipment are therefore less efficient. Since the dust is organic, it will begin to degrade or carbonize, which tends to lower the ignition temperature of the dust. Eventually the dust will ignite or the equipment will fail because of high temperature, and a fire could result.

Particle Size

The smaller the particle size of the dust, the easier it is to ignite because the ratio of surface area to volume increases tremendously as particle size decreases. Because of their smaller mass, smaller particles are less able to absorb energy from an external source. Once the dust has been ignited, it radiates energy to nearby particles more quickly and efficiently. It is also true that a decrease in particle size will increase the rate of pressure rise during an explosion and decrease the minimum explosive concentration and the energy necessary for ignition. As a practical matter, however, a range of particle sizes will be present, and the behavior of a given dust sample will depend on particlesize distribution.

Concentration

A minimum concentration of dust in air below which propagating ignition will not occur is called the lower explosive limit (LEL). The LEL decreases with decreasing particle size. Plotting explosion pressures and rates of pressure rise against dust cloud concentration shows that both parameters are at a minimum value at the minimum explosive concentration, then rise to a peak value at a concentration near the so-called optimum concentration. However, it should be noted that the optimum concentration is generally considered to be the point at which the rate of rise is at a maximum. The maximum pressure does not always occur at the same point, varying very slightly. Once the optimum point is exceeded, both rate of pressure rise and explosion pressure decrease. Maximum explosive concentrations, described as the upper explosive limit, are very rarely determined and have no practical value.

The energy of the ignition source, the turbulence of the dust cloud, and the uniformity of dispersion all have some effect on the minimum explosive concentration, at least in laboratory tests. In the field, these influences are of secondary importance.

Moisture

The humidity of the air surrounding the dust particles has no effect on the course of a dust explosion. However, the moisture content of the dust particles can affect the potential for a dust explosion. A high moisture content tends to increase ignition temperature, ignition energy, and the minimum explosive concentration and tends to decrease the severity of the explosion. Again, these variations are of more interest in laboratory test work; they are of secondary importance in the field.

Other Factors

As would be expected, a decrease in the partial pressure of oxygen, either by vacuum or inert gas, will decrease the explosion hazard of a dust aerosol. In the production of metal powders, enough oxygen should be maintained in the system to permit controlled oxidation of the particle surface during the size reduction process. If this is not done, the metal particles will oxidize rapidly when exposed to air, and they may self-ignite.

The presence of a flammable gas in the dust–air aerosol greatly increases the hazard. Because they have a lower energy of ignition than the "pure" dust mixture, these so-called hybrid mixtures are easier to ignite, and they produce explosions that are much more violent than would normally be expected. The concentration of the flammable gas does not have to be at or above the lower flammable limit—considering the concentration of the flammable gas–air mixture alone, without the presence of the combustible dust—in order to achieve these effects. Such hybrid mixtures are encountered in fluid bed driers where a flammable solvent is being evaporated from a powder, for example, and must be provided with special safeguards, such as an inert gas atmosphere, which could reduce or even exclude the oxygen completely, or explosion protection systems.

Ignition Sources

Dust clouds and dust layers can be ignited by all the usual ignition sources, including open flames, electrical arcs, frictional and mechanical sparks, and hot surfaces. Ignition by electrical charge is usually ruled out. Dust clouds require 1 to 40 mJ of energy for ignition to occur, compared to the 0.20 to 1 mJ needed by flammable gases or vapors, making static discharge somewhat unlikely as an ignition source. The NFPA documents that deal with specific dust explosion hazards also address the control of ignition sources.

PREVENTION OF DUST EXPLOSIONS

Dust Control

Eliminating, or at least greatly reducing, the amount of airborne and static dust is the single most important means of preventing dust explosions. Dusts and materials that generate dust should be handled to the greatest extent possible in closed systems, either pneumatic or mechanical. In order to be effective, these systems must be dust tight. Inspectors should look for evidence of dust around the seams and joints of ducts or pipes and around access panels. If dust is evident, it may indicate a faulty gasket or mechanical damage to the conduit itself. They should make sure hatches on bins and tanks are closed securely.

Inspectors should keep in mind that these systems, whether they handle process material in bulk or collect fugitive dust, contain a dust/air mixture in the explosive range. They should make sure there are adequate provisions for keeping tramp metal out. One method to consider using is a magnetic separator.

Good housekeeping is imperative. Any surface, either horizontal or vertical, on which dust can accumulate should be kept clean. There are two rules of thumb for judging the adequacy of housekeeping: If the inspector cannot tell what color a surface is, it needs to be cleaned often. And if the dust deposits are thicker than a paper clip, cleaning is long overdue.

The frequency of cleaning will depend on conditions in the plant, such as the adequacy of dust collection systems and the tightness of the process equipment (Figure 44-1). Cleaning should be done by central vacuum systems. If portable systems are used, they should be suitable for Class II areas. Alternatively, soft brushes and dust pans made of conductive plastic or metal may be used.

Under no circumstances should dust be blown off with compressed air because doing this will just move the dust from one surface to another. Equipment that handles or produces dusts should be as tight as possible. Where equipment is loaded or unloaded by dumping, local dust collection pickups should be arranged. Dust control pickups should be initiated at drumming and bagging lines. Inspectors should

FIGURE 44-1 Typical Types of Dust-Collecting Equipment

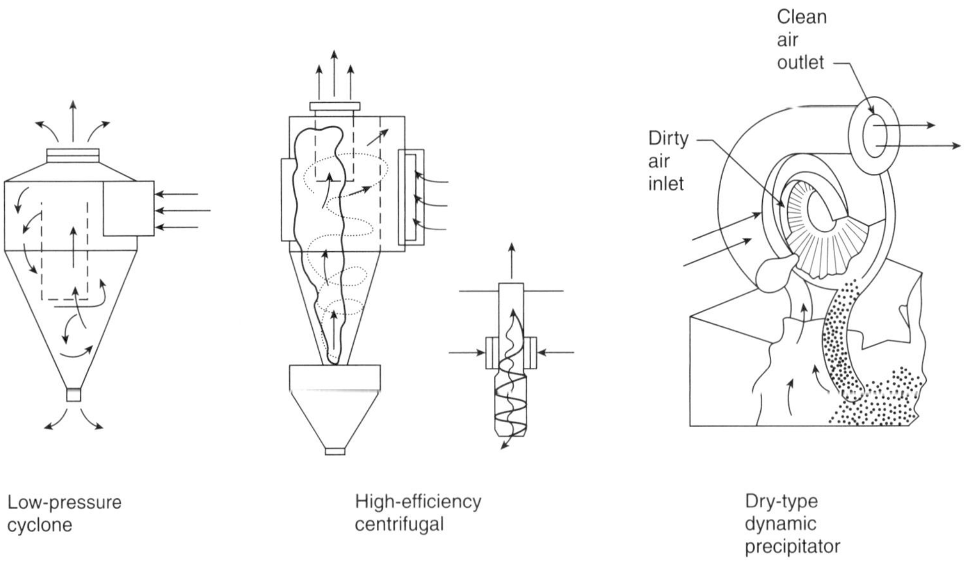

carefully review process schematics and the equipment itself to find those points at which dust aspirators might be needed. Such locations include hatch covers on bins, tanks, or vessels and frequently opened access panels.

Process Equipment

If a dust explosion is going to occur, chances are that it will begin in a piece of equipment. It is impossible to prevent dust clouds from forming around equipment unless the material is handled as a slurry or a damp, cakey solid. Equipment maintenance is very important in these situations. The inspector should look for similar signs of abuse on drag or en-masse conveyers. Conveyer belts should not show any sign of excessive wear, and idler rollers should be free spinning to avoid frictional heating and possible failure of the conveyer belt.

Certain pieces of equipment are especially susceptible to dust explosions. These include mills, pulverizers, dust collectors, cyclone separators, and the various types of dryers.

There are several ways in which this equipment can be protected. It may be designed to contain the expected explosion pressure. It also may be fitted with explosion vents or equipped with an explosion-suppression system. Or the equipment enclosure may be blanketed with inert gas. Again, a careful review of the process will help when choosing the most suitable option.

The inspector might need the help of qualified individuals to determine whether the equipment can withstand the expected overpressure. Explosion vents should never terminate inside the building. They should be located close to an outside wall and vented through a short, straight duct directly to the outside. Better yet, the vented equipment should be moved outside or to the roof of the building. Breather vents can terminate inside the building if they are fitted with filters. The inspector should check to make sure the filters are functioning properly.

Electrical Equipment

Electrical equipment in dust-prone areas that are well kept should be suitable for Class II, Division 2 locations, unless the dust happens to be electrically conductive. This would require that all electrical equipment be Division 1. The inspector should check to see that the conduit is properly sealed and that all electrical enclosures are tightly sealed. Process control equipment located within the process stream must be suitable for Class II, Division 2, or it must be intrinsically safe.

Ignition Sources

The most obvious ignition sources are smoking and welding operations. The inspector should make sure "no smoking" signs are posted in all operational areas and that special areas are set aside for safe smoking. A hot work permit system should be in place for welding, cutting, and other hot work operations (see Chapter 52, "Welding, Cutting, and Other Hot Work"). Ancillary equipment should be checked for overheating, for slipping or chafing drive belts, and for other mechanical defects that might cause sparks. If there is a possibility that tramp metal or other foreign objects may enter the equipment, magnetic separators or screens may be needed.

FIRE PROTECTION

Other chapters in this manual discuss what inspectors should look for when inspecting fire protection equipment—automatic sprinkler systems, hose connections, and portable fire extinguishers. In areas containing dust explosion hazards, the

FIGURE 44-2 Sprinkler Protection for a Bag-Type Dust Collector

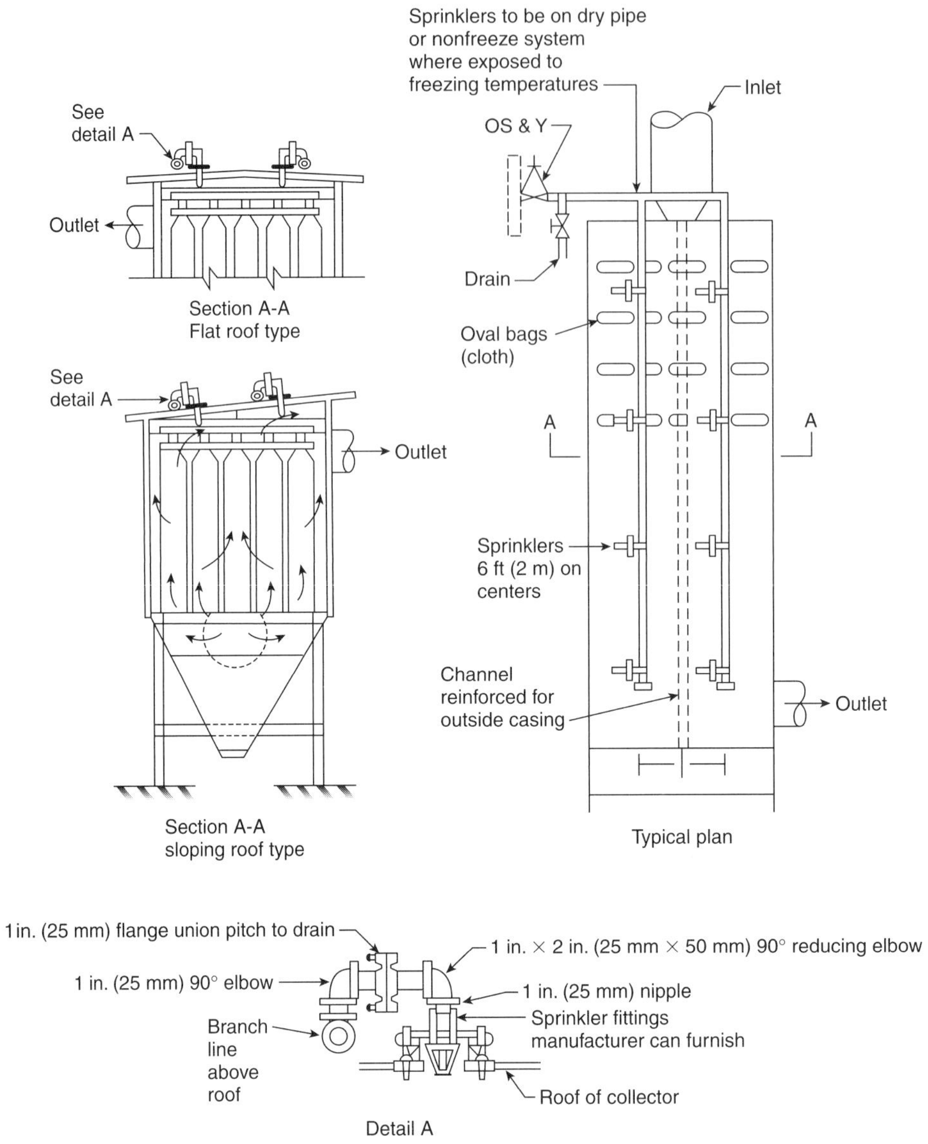

inspector must also check to be sure that all hose connections are equipped with fine spray nozzles and that special fog nozzles will be used where high voltage electrical equipment is located. (Figure 44-2 illustrates sprinkler protection for a bag-type dust collector.) The plant fire brigade should understand that a coarse or solid water stream could throw dust into suspension, thus causing a primary or secondary explosion.

Where metal dusts are involved, quantities of sand, talc, foundry flux, or other specifically approved, inert extinguishing agents should be available to smother small fires. The inspector should make sure that only the approved types of water or portable extinguishers are used.

SUMMARY

Dust explosions, the most vivid examples of which occur in large grain elevators such as the DeBruce Grain elevator explosion in 1998, can be very destructive. The pressure wave, generated by the production and thermal expansion of the flammable gases, is responsible for most of the damage in a dust explosion. Combustible dusts are also a fire hazard because stationary deposits of dust provide an easy means for a fire to spread rapidly from its initial location.

Eliminating or greatly reducing the amount of airborne and static dust is the single most important means of preventing dust explosions. Where possible, dust-tight closed systems should be used. Other dust reduction practices include good housekeeping and good maintenance of process equipment. Equipment especially susceptible to dust explosion can be protected, for example, by being fitted with explosion vents, equipped with an explosion-suppression system, or blanketed with inert gas. Ignition sources, such as smoking and welding, cutting, and other hot work operations, should be controlled.

Fire protection in facilities includes automatic sprinkler systems, hose connections, and portable fire extinguishers. In areas with dust explosion hazards, hose connections should be equipped with fine spray nozzles. In areas with metal dusts, sand, talc, foundry flux, or other approved inert extinguishing agent should be available.

BIBLIOGRAPHY

ASTM E1226-88, *Standard Test Method for Pressure & Rate of Pressure Rise for Combustible Dusts,* American Society for Testing & Materials, Philadelphia, PA. 1988.

Cashdollar, K. L., and Hertzberg, M., *Industrial Dust Explosions,* American Society for Testing and Materials, Philadelphia, PA, 1987.

Cote, A. E., ed., *Fire Protection Handbook,* 18th ed., NFPA, Quincy, MA, 1997.

Cross, J., and Farrer, D., *Dust Explosions,* Plenum Publishing Co., New York, NY, 1982.

Eckhoff, R. K., *Dust Explosions in the Process Industries,* Butterworth-Heinemann Ltd, Oxford, England, 1991.

NFPA Codes, Standards, and Recommended Practices

See the latest version of The NFPA Catalog for availability of current editions of the following documents.

NFPA 68, *Guide for Venting of Deflagrations*

NFPA 69, *Standard on Explosion Prevention Systems*

NFPA 499, *Recommended Practice for the Classification of Combustible Dusts and of Hazardous (Classified) Locations for Electrical Installations in Chemical Process Areas*

NFPA 654, *Standard for the Prevention of Fire and Dust Explosions from the Manufacturing, Processing, and Handling of Combustible Particulate Solids*

NFPA 664, *Standard for the Prevention of Fires and Explosions in Wood Processing and Woodworking Facilities*

Combustible Metals

CHAPTER 45

Carl H. Rivkin

Metals can burn. Some oxidize rapidly and reach flaming combustion; others oxidize so slowly that heat generated during oxidation dissipates before ignition occurs. Metals that can reach flaming combustion are considered combustible metals. The form of the metal is critical in determining its combustion characteristics. Metals in thin sections, as fine particles, or when molten can ignite easily and sustain combustion; in massive solid form, however, they are difficult to ignite. Certain metals, including the following, are of particular concern because of their combustion characteristics: magnesium, titanium, zirconium, hafnium, sodium, lithium, potassium, calcium, zinc, thorium, uranium, and plutonium.

Aluminum, iron, and steel normally are not combustible, although they can ignite and burn when finely divided. Clean, fine steel wool, for example, can be ignited. Dust clouds of most metals in air are explosive; and most metals are combustible in high oxygen concentrations. Particle size, shape, quantity, and alloy and ambient conditions are important factors in assessing the combustibility of metals. This chapter looks at the various combustible metals, their burning characteristics, their hazards, their storage, and their control.

GENERAL CHARACTERISTICS

Temperatures produced by burning metals are much higher than temperatures of flammable liquid fires (see Table 45-1). Some hot metals continue to burn in nitrogen, carbon dioxide, or steam atmospheres, whereas fires of other materials would be extinguished in those atmospheres.

Metals have different burning characteristics. Titanium produces little smoke; smoke from lithium is dense and profuse. Some water-moistened metal powders, such as zirconium, burn with near-explosive violence; yet the same powders, wet with oil, burn slowly. Sodium melts and flows; calcium does not. Some metals burn more readily after prolonged exposure to moist air and exposure to dry air can make them more difficult to ignite.

Combustible metals are used in an increasingly wide range of industries and applications, often as a substitute for steel because they are stronger and lighter. Titanium, for example, is used in submarine hulls. Aluminum, which is also lighter than steel, is another substitute. In addition, aluminum powders are used as pigments in paint manufacturing. There is an increasing demand in the electronic industry for a material that is light and strong. Such a metal may be used to fabricate parts for computers, cellular phones, and other personal electronic devices, where reduced weight is critical for use.

Aerospace, electronic component production, and pigment production are examples of industries that use combustible metals. The manufacture of components for these industries, however, may be contracted to small machine shops that may not have extensive experience handling combustible metals. With the increased use of combustible metals, inspectors must be aware of their special hazards and the industries that may use these metals.

Carl H. Rivkin, P.E., is senior chemical engineer on the NFPA engineering staff and the NFPA staff liaison to the Technical Committee on Combustible Metals and Metal Dusts. He has over 20 years of engineering experience that includes chemical process plant and regulatory experience with hazardous materials.

TABLE 45-1 Characteristics of Metals

	Solid	Dust	Melting Point (°C)	Boiling Point (°C)	Solid Metal Ignition (°C)
Aluminum	×		1220	4446	1031
Aluminum Dust		×			
Barium	×		1337	2084	347
Bronze Powder		×			
Calcium	×		1515	2624	1299
Calcium/Silicon Al		×			
Hafnium	×	×	4033	9750	
Iron	×	×	2795	5432	1706
Ferrochrome		×			
Ferromanganese		×			
Lithium	×		367	2437	356
Magnesium	×		1202	2030	1153
Magnesium Dust		×			
Manganese		×			
Molybdenum		×			
Niobium		×			
Plutonium	×		1155	6000	1112
Potassium	×		144	1400	156
Silicon		×			
Sodium	×		208	1616	239
Strontium	×		1425	2102	1328
Tantalum	×				
Tantalum Powder		×			
Thorium	×		3353	8132	932
Titanium	×		3141	5900	2899
Titanium Dust		×			
Uranium	×		2070	6900	6900
Zinc	×		786	1665	1652
Zinc Dust		×			
Zirconium	×		3326	6471	2552
Zirconium Powder		×			

Source: NFPA 484, 2002, Table A.1.1.3.

If the premises the inspector is about to inspect has processes involving metals other than iron and steel, he or she should, before the inspection, ascertain the

- Burning characteristics of the metals involved
- Quantities stored and handled
- Types of processes involved
- Arrangement made for the storage and handling of scrap
- Extinguishing agents used against combustible metal fires

Good sources of information on combustible metals are the *Fire Protection Handbook,* the *Industrial Fire Hazards Handbook,* and NFPA codes and standards (see Bibliography).

EXTINGUISHING COMBUSTIBLE METAL FIRES

Combustible metal (Class D) fires are difficult to extinguish. Common extinguishing agents do not work well on them; in many cases these agents violently increase combustion. Many different agents can be used to extinguish Class D fires, but one agent does not necessarily work on all metals. Some agents work with several metals, others with only one. Commercially available agents are known as dry powders. They should not be confused with dry chemical agents that are suitable for flammable liquid and live electrical equipment fires. Some powders, such as G-1 powder, Na-X powder, Lith-X powder, Met-L-X powder, go by their trade names, and others such as talcum powder, sand, graphite, sodium chloride, copper powder, and soda ash are known by their common names. Inspectors must understand the uses and limitations of the agent used. The references in the Bibliography will acquaint inspectors with the characteristics of different extinguishing agents.

Effectively controlling or putting out metal fires depends to a great degree on the method of application of the extinguishing agents and the training and experience of personnel who use them. In locations where combustible metals are present, inspectors should ask to see the supply of dry powder agents on hand, their location, and the tools available to spread them on burning metal. They should inquire about the training employees receive in extinguishing metal fires, because these fires involve techniques not commonly encountered in conventional fire fighting. Inspectors should emphasize that training is needed to get experience in techniques for specialized extinguishing agent application. Personnel responsible for controlling combustible metal fires should practice extinguishing fires of the metals used in their facility at an isolated outdoor location.

MAGNESIUM

The ignition temperature of pure magnesium in large pieces is close to its melting point of 1202°F (950°C), but magnesium ribbons and shavings can be ignited under certain conditions at about 950°F (510°C), and finely divided powder can be ignited at temperatures below 900°F (482°C). Magnesium is principally used in alloy form, and certain magnesium alloys can ignite at temperatures as low as 800°F (427°C). Thus, ignition temperatures can vary widely depending on the makeup of the alloys involved, as well as the size and shape of the metal. Finding out as much as possible about the alloys used and how the metal is processed will give the inspector a good idea of the degree of hazard associated with the metals at a given facility.

Process Hazards

Magnesium and its alloys are readily machinable; if the tools used are dull or deformed, frictional heat can ignite the chips and shavings created in machining operations. Machining magnesium is usually done dry, but any cutting fluids used must be of the mineral oil type. Water, water-soluble oils, and oils containing more than 0.2 percent fatty acids are hazardous and must not be used. The inspector should ensure that these cooling liquids are not used by mistake and that the machines and surrounding work area are clean. Magnesium fires have been known to occur in machine beds. Waste magnesium is best kept in clean, covered, dry, steel, or other noncombustible drums, and the drums should be removed from the building at regular intervals. The machine operator must have a fire extinguishing agent immediately available.

Magnesium grinding is possibly more hazardous than machining. Magnesium dust clouds, made up of minute fine particles that could result from unprotected grinding operations, can be explosive if there is an ignition source. The inspector should make sure that grinding equipment used on magnesium has physical barriers to separate the grinding operation from any ignition sources. An integral part of a good grinding installation is a liquid precipitation separator that converts the dust into a sludge (Figures 45-1 and 45-2). Fine particles from grinding generate hydrogen when submerged in water, but they cannot be ignited in this condition; however, fine particles resulting from grinding that are slightly wetted with water can generate enough heat to ignite spontaneously and burn violently because oxygen is extracted from the water with the release of hydrogen. Thus, grinding installations should have

FIGURE 45-1 Liquid Precipitation Separator for Fixed Dust-Producing Equipment
Source: NFPA 480, 1998, Figure 4-2.4.1(a).

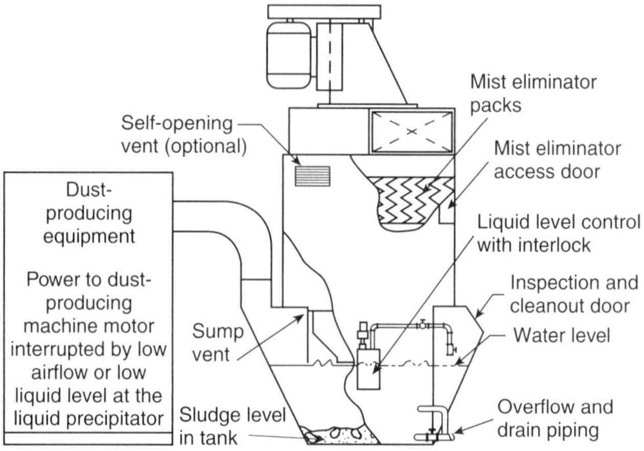

interlocks that permit the grinder to operate only if the exhaust blower and water spray are working to keep the magnesium fines fully wetted.

Storage

Storage buildings housing magnesium should preferably be noncombustible, and the magnesium should be segregated from combustible materials. If magnesium is stored in combustible buildings, the buildings must be protected throughout by automatic sprinklers. Dry fine particles should be stored in noncombustible containers in a fire-resistive storage building, although a room with explosion venting is also acceptable. However, wherever they are stored, the fine particles must remain dry to prevent the release of hydrogen. Scrap magnesium fine particles wetted with coolants should be stored outdoors in covered, vented, noncombustible containers because they can spontaneously heat and generate hydrogen through reaction with the coolant.

FIGURE 45-2 Liquid Precipitation Separator for Portable Dust-Producing Equipment
Source: NFPA 480, 1998, Figure 4-2.4.1(b).

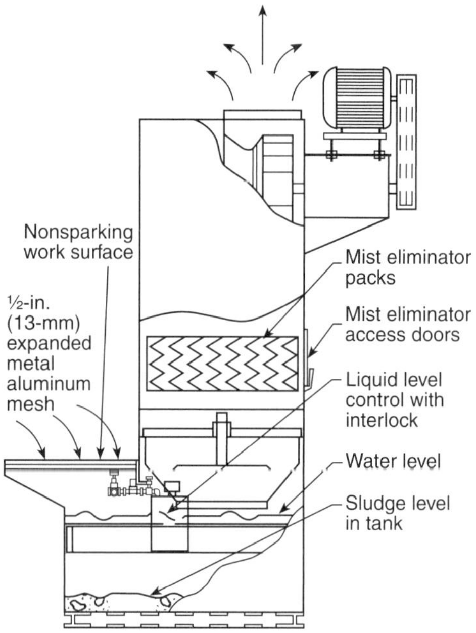

Fire Extinguishment

In areas where magnesium is machined, ground, or similarly processed, a supply of extinguishing powder approved for use on a magnesium fire must be kept within easy reach of all operators. The powder can be kept in closed containers with easily removable covers and a hand scoop or in approved portable extinguishers designed for use with these powders. The powders available include G-1, which is composed of screened, graphitized foundry coke to which an organic phosphate has been added; Metal Guard, which is the same in composition as G-1 powder; and Met-L-X, which has a sodium chloride base to which has been added ingredients such as tricalcium phosphate to improve flow characteristics, and metal stearates for water repellancy.

In general, water should not be used on magnesium chip fires because doing so can violently accelerate the fire. Automatic sprinklers, however, will extinguish the typical fire in a shop where quantities of chips are limited.

Castings, wrought products, and fabricated parts that are involved in a fire can be cooled and extinguished with coarse streams of water applied with a standard fire hose. Water spray-, water mist-, and water fog-type nozzles must not be used because they can accelerate a magnesium fire rather than cool it. Finally, care must be used when applying water to magnesium fires where quantities of molten metal are present; the steam formation and metal-water reactions can be explosive.

For magnesium fires in heat-treating furnaces, dry powder, foundry fluxes, or approved inert gas can be used to extinguish the fire. If dry powder is used, the burning metal should be removed from the furnace, if possible. Fluxes and gas can be applied successfully to the burning metal in the furnace. Water is not to be used to fight magnesium fires in furnaces. Because magnesium will burn in the presence of carbon dioxide and nitrogen and will react violently with halogenated extinguishing agents, these extinguishing materials should not be used.

TITANIUM

Titanium, with an ignition temperature ranging from 630°F (332°C) in dust-air clouds to 2900°F (1593°C) in solid castings, has hazard characteristics similar to magnesium. Castings and other massive pieces are not combustible under normal conditions; large pieces, however, can ignite spontaneously when they come in contact with liquid oxygen. This situation is unlikely because liquid oxygen is not used in most titanium production and fabrication operations. Small chips, fine turnings, and dust can ignite easily and burn with high heat release. Fine turnings and very thin chips have reportedly been ignited by a match. Titanium dust can also be ignited in atmospheres of carbon dioxide and nitrogen.

Process Hazards

Heat generated during machining, grinding, sawing, and drilling of titanium can be enough to ignite the turnings and chips formed in the operations. Consequently, water-based coolants should be used in ample quantity to remove heat, and cutting tools should be kept sharp. Fine particles should be removed from the work area regularly and stored in covered metal containers.

Grinding of titanium requires a dust-collecting system discharging into a liquid precipitation separator. Large quantities of cooling fluid should be used to keep the sparking down.

Descaling baths of mineral acids and molten alkali salts can cause violent reactions with titanium at abnormally high temperatures. Titanium sheets have also ignited when they have been removed for descaling baths.

Titanium melting furnaces present a special problem. Several severe explosions have occurred in titanium melting furnaces when water inadvertently entered the melting crucible during the melting operation. The use of sodium-potassium alloy (NaK) as a crucible coolant has been developed for both laboratory and commercial use. Although this reduces the danger of a furnace explosion, the handling of NaK presents its own hazards. Sodium reacts rapidly with water, producing hydrogen. Sodium also reacts with most halogens, halogenated hydrocarbons, and sulfuric acid.

Storage

Large pieces of titanium do not present storage hazards. Dry scrap fine particles and titanium sponge, however, should be stored in covered metal barrels that are kept well away from combustible materials. Moist scrap should be stored outdoors in covered metal barrels because the scrap could generate hydrogen, and scrap wet with oils could heat spontaneously.

Fire Extinguishment

Where operations generate titanium fine particles, dust, and turnings, an adequate supply of approved extinguishing powder should be within easy reach of all machine operators. The powder can be kept in closed containers with easily removable covers and a hand scoop, or in approved portable extinguishers. Incipient fires may be possible to extinguish. The safest procedure to follow with a fire involving small quantities of titanium powder is to ring the fire with the extinguishing powder and allow the fire to burn itself out. Care must be taken to prevent the formation of a titanium dust cloud when extinguishing powder is applied.

Tests conducted by Industrial Risk Insurers on titanium machinings in piles and in open drums showed water in a coarse spray was a safe and effective means of extinguishing fires in relatively small quantities of chips. Water-based (including foam), carbon dioxide, halon, and dry chemical extinguishers are not effective on titanium fires and should not be used.

ZIRCONIUM AND HAFNIUM

As is true of all metals, the combustibility of zirconium increases as the average particle size decreases. Other variables, such as moisture content, however, also affect how easily zirconium will ignite. In massive form, zirconium can withstand extremely high temperatures without igniting, but clouds of zirconium dust, in which the average particle size is 3 μ, have ignited at room temperature.

Massive pieces of zirconium do not ignite spontaneously under ordinary conditions, but they will ignite when an oxide-free surface is exposed to sufficiently high oxygen concentrations and pressure. Scrap chips, borings, and turnings can spontaneously heat and ignite if fine dust is present.

As with other combustible metals, the combustibility of hafnium is related to its size and shape. Large pieces are hard to ignite; turnings and chips ignite more easily. Hafnium is somewhat more reactive than titanium or zirconium of similar form. Unless inactivated, hafnium in sponge form can ignite spontaneously. When ignited, hafnium burns with very little flame, but it releases large quantities of heat.

Process Hazards

In general, processing recommendations for zirconium and hafnium are the same. Large flows of mineral oil- or water-based coolant are required to prevent dangerous heating during machining. Turnings should be collected frequently and stored under water in cans. When zirconium dust is generated, it should be captured in a liquid precipitation separator.

Zirconium powder should be handled under an inert liquid or in an inert atmosphere. If either zirconium or hafnium powder is handled in air, extreme care must be used because the small static charges generated can cause ignition.

Storage

Zirconium and hafnium castings do not require special storage precautions because massive pieces of the metal can withstand very high temperatures without igniting. Zirconium powder, however, is highly combustible, and it should be stored and shipped in 1-gal (2.2-L) containers with at least 25 percent water by volume.

Storerooms for zirconium powder should be of fire-resistive construction and equipped with explosion vents. Cans in storage should be physically separated from each other to limit fire spread. Cans should be checked for corrosion.

Fire Extinguishment

Fighting zirconium and hafnium fires requires the same approach as that recommended for titanium. Small quantities of the metals can be ringed with dry powder extinguishing agent and allowed to burn out. Fires in massive pieces of the metals can be fought with large quantities of water. Limited tests have indicated that water in spray form will have no adverse effect on burning zirconium turnings. Fires in enclosed spaces can be smothered with argon or helium.

ALKALI METALS: SODIUM, LITHIUM, POTASSIUM, AND SODIUM–POTASSIUM ALLOYS

The principal fire hazard with sodium is its rapid reaction with water. Hydrogen liberated in the reaction can be ignited by the heat of the reaction. Once sodium is ignited, it burns vigorously and forms dense clouds of caustic sodium-oxide fumes.

Lithium undergoes many of the same reactions as sodium; however, its reaction with water is not as vigorous, and not enough heat is generated to ignite the hydrogen given off in the reaction.

The fire hazards of potassium are very similar to those of sodium, but potassium is more reactive.

NaK is the term used for any of several sodium–potassium alloys. NaK alloys possess the same fire hazard properties as those of the component metals except that the reactions are more vigorous. All are liquids or melt near room temperature.

Process Hazards

A principal use of liquid sodium is as a heat-transfer medium. Where molten sodium is used in process equipment, steel pans should be located underneath to prevent contact and violent reactions of burning sodium with the moisture in concrete floors. There should be tray-type covers on the pans to catch the sodium and drain it into the pans through drilled holes. Any sodium flowing through the holes extinguishes itself in the pan. Information on sodium can be used as a guide in processing lithium, NaK, and potassium.

Storage

Because sodium reacts with water, it should be stored in drums and cases in a dry, fire-resistive room or building used exclusively for storing sodium. Because sprinklers cannot be used, combustible materials should not be stored in the same area as sodium. The inspector should check to see that no water or steam pipes are located in the storage area and that sufficient heat is maintained to prevent moisture condensation. Natural ventilation at a high spot in the room can vent any hydrogen that might be released by accidental contact of sodium with moisture. In general, storage recommendations for lithium, NaK, and potassium are the same as those for sodium.

Fire Extinguishment

Aqueous-based extinguishing agents should never be used on fires of sodium, potassium, or lithium because they would cause a violent reaction. The dry powders developed for metal fires and dry sand, dry sodium chloride, and dry soda ash are effective extinguishing agents. Sodium burning inside a piece of apparatus can usually be extinguished by the closing of all openings. Fire extinguishing recommendations for sodium fires also apply to lithium, NaK, and potassium fires. Copper powder is an extremely effective extinguishing agent for lithium fires because the copper will not react with the sodium and is an effective heat-transfer medium.

CALCIUM AND ZINC

The moisture in the air governs the flammability of calcium. If ignited in moist air, calcium burns without flowing. Finely divided calcium will ignite spontaneously in air.

Sheets, castings, or other massive forms of zinc do not present serious fire hazards because they are difficult to ignite. Once ignited, however, zinc shapes can burn vigorously, and zinc generates an appreciable amount of smoke when it burns.

NORMALLY NONCOMBUSTIBLE METALS

Aluminum

Aluminum, in ingots, fabricated parts, and other massive forms, has a sufficiently high ignition temperature so that its burning is not a factor in most fires. Aluminum only presents a special fire problem in powder or other finely divided forms. Under certain conditions, powdered or flaked aluminum can be explosive. Dry aluminum dust collection systems present a significant explosion hazard. The dust collector should be located outside of the building and exhaust air should not be recycled. Aluminum in contact with magnesium is much more combustible, possibly because they form an alloy at the interface.

Aluminum dust fires should be extinguished with approved dry powders. Nearly all vaporizing liquid fire extinguishing agents react violently with burning aluminum, usually intensifying the fire and sometimes exploding. Water hose streams should not be used because the impact of the water stream can lift enough dust into the air to produce a strong dust explosion. In addition, water reacting with aluminum can give off hydrogen.

Iron and Steel

Iron and steel are usually not considered to be combustible; however, steel in the form of fine steel wool or dust can be ignited in the presence of common ignition sources, such as a torch. Steel wool is more likely to ignite when it is saturated with a flammable solvent.

There have been reports of fires in piles of steel turnings and other fine scrap that presumably contained some oil or other material that promoted self-heating.

Spontaneous ignition of water-wetted borings and turnings in closed areas, such as ships' hulls, have been reported.

RADIOACTIVE METALS

Radioactive metals include those that occur naturally, such as uranium and thorium, and those that are produced artificially, such as plutonium and cobalt-60. It is important to remember, however, that radioactivity cannot be altered by fire, and radiation will continue wherever the radioactive metal spreads during a fire. Due to radioactive contamination, smoke from fires involving radioactive materials frequently causes more property damage than does the fire.

Uranium

Uranium is generally handled in such massive forms that it does not present a significant fire risk unless exposed to a severe and prolonged external fire. In finely divided form it ignites easily, and scrap from machining operations can ignite spontaneously. Dust from grinding has ignited, even under water, and fires have occurred spontaneously in drums of coarse scrap after prolonged exposure to moist air. Machine chips, if not adequately quenched with coolant, will self-ignite and burn in air from the heat of machining.

Thorium

Powdered thorium is usually compacted into small, solid pellets. In that form it can be stored safely or converted into alloys with other metals. Improperly compacted thorium pellets have slowly generated enough heat through absorption of oxygen and nitrogen from the air to raise the temperature of a steel container to red heat. Because of its low ignition temperature, powdered thorium should not be handled in air because the friction of the particles falling through air or against the edge of a glass container can ignite the powder electrostatically. Thorium powder is usually handled in a helium or argon atmosphere.

Plutonium

Plutonium can be ignited more easily than uranium. Normally it is handled by remote control under an inert gas or dry air atmosphere. In finely divided form, such as dusts and chips, plutonium is subject to spontaneous ignition in moist air.

Plutonium metal is never exposed intentionally to water, in part because of fire considerations. Plutonium, which ignites spontaneously, is normally allowed to burn under conditions limiting both fire and radiological contamination spread.

SUMMARY

Metals that can reach flaming combustion are considered combustible metals. Combustible metals are being increasingly used in industries and applications, often as a substitute for steel, because they are stronger and lighter. Because of their combustion characteristics, the following metals are of particular concern: magnesium, titanium, zirconium, hafnium, sodium, lithium, potassium, calcium, zinc, thorium, uranium, and plutonium.

Combustible metal (Class D) fires are difficult to extinguish. Common extinguishing agents do not work well with them and in many cases increase combustion. Controlling and extinguishing combustible metal fires depends on the method of application of the extinguishing agents and on the training and experience of personnel who use them. Described here are the physical characteristics of the various metals that determine their combustibility, their safe storage requirements, and the preferred method of fire extinguishment for each.

BIBLIOGRAPHY

Cote, A. E., ed., *Fire Protection Handbook,* 18th ed., NFPA, Quincy, MA, 1997.

Cote, A. E., and Linville, J. L., eds., *Industrial Fire Hazards Handbook,* 3rd ed., NFPA, Quincy, MA, 1990.

Purington, R. G., and Patterson, H. W., *Handling Radiation Emergencies,* NFPA, Quincy, MA, 1977.

NFPA Codes, Standards, and Recommended Practices

See the latest version of The NFPA Catalog for availability of current editions of the following documents.

NFPA 480, *Standard for the Storage, Handling, and Processing of Magnesium Solids and Powders*

NFPA 484, *Standard for Combustible Metals, Metal Powders, and Metal Dusts*

Chemicals

CHAPTER 46

John A. Davenport

CASE STUDY

Chemical Reaction Injures Five

Five people were injured when chlorine and antifreeze were inadvertently mixed, creating a chemical reaction that started a small fire in a workshop and warehouse in Texas.

The single-story structure, which measured 50 × 100 ft (15.2 × 30.5 m), was constructed of unprotected metal, with a wood floor and metal wall panels. A few dry chemical fire extinguishers were located in the building. There were no sprinklers or detectors.

At approximately 12:40 P.M., an employee was scooping chlorine from its container into a 1-gal (3.8-L) bucket that had previously held antifreeze when the chlorine reacted with residual antifreeze, causing a fire. Employees used two 2½-lb (1.1-kg) dry chemical fire extinguishers to knock down the fire, but each time, the reaction reignited the blaze. As the fire continued to grow, an employee called the fire department, 5 minutes after the initial reaction.

While waiting for the fire department, the workers used a 10-lb (4.5-kg) extinguisher and another 2½-lb (1.1-kg) extinguisher on the blaze, with limited success. They also retrieved a sewer jet machine containing 1000 gal (3285 L) of water and sprayed the container with it, limiting the fire somewhat.

Fire fighters wearing self-contained breathing apparatus entered the building, removed the container, and set up a positive-pressure fan to displace the fumes.

Five employees who fought the blaze were treated at a hospital for smoke inhalation and eye irritation and released. Damage was less than $100.

Source: "Fire Watch," *NFPA Journal*, September/October, 1998.

When chemicals are involved, safe and effective fire control measures require a knowledge of their hazardous properties. For the purpose of this discussion, chemicals are classified according to their combustibility, instability, reactivity with air or water, corrosivity, toxicity, oxidizing capability, and radioactivity.

Information on specific chemicals and chemical products can be found in materials safety data sheet (MSDS) documents. OSHA Regulations 29 CFR 1910.1200 requires many businesses that use or manufacture chemical products to have on file and available to employees MSDS documents that describe the safety precautions, including fire hazards, of the material. These are often available to fire inspectors if a request is made to the business owner.

John A. Davenport is a loss prevention consultant. He is a member of SFPE and a fellow of AIChE. He is also a member of several NFPA technical committees.

SOLID COMBUSTIBLE CHEMICALS

Several chemicals are solids at room temperature and are hazardous due to their combustibility. These solids can cause a fire or explosion through friction, absorption of moisture, or exposure to air or moderate heat. Some combustible solids and the conditions for their greatest hazard are shown in Table 46-1.

Aluminum and Magnesium

These metals, when in finely divided form—that is, when powdered or in the form of machine turnings—have the potential to be severe fire or dust explosion hazards. They are easily ignited, and the fire can only be controlled with special extinguishing agents.

Carbon Black

Carbon black is most hazardous immediately after it is manufactured. After thorough cooling and aging, it will not ignite spontaneously, although it can generate heat in the presence of oxidizable oils.

Nitroaniline

This combustible solid melts at 298°F (148°C) and has a flash point of 309°F (154°C). Explosive decomposition can occur during a fire.

Nitrochlorbenzenes

These isomers melt between 90°F (32°C) and 183°F (84°C) and have a flash point of 261°F (127°C). Their products of combustion are toxic.

TABLE 46-1 Typical Combustible Solids and Conditions for the Greatest Hazard

Danger from heating due to absorption of moisture	
Alkylaluminums	Phosphorous pentasulfide
Aluminum chloride	Potassium
Aluminum dust	Potassium peroxide
Calcium	Sodium
Calcium carbide	Sodium hydride
Calcium oxide	Sodium hydrosulfite
Lithium	Sodium peroxide
Lithium hydride	Tricholorisocyanuric acid
Magnesium (finely divided)	Zinc dust
Dangerous in air	
Diborane	Phosphorous (white)
Diethyl ether	Sodium hydride
Lithium	Zirconium dust
Danger from subjection to heat	
Antimony pentasulfide	Dinitrobenzene
Calcium	Phosphorus pentasulfide
Cellulose nitrate	Phosphorus sesquisulfide
Dinitroaniline	
All oxidizers when mixed with combustible or organic material.	

Sulfides

Most sulfides are easily ignited as they liberate hydrogen sulfide with water. Phosphorus pentasulfide can ignite spontaneously in the presence of moisture as it decomposes. Phosphorus sesquisulfide is highly flammable, with an ignition temperature of 212°F (100°C)

Sulfur

Molten sulfur has a flash point of 405°F (207°C) and an ignition temperature of 450°F (232°C). If impure it can liberate toxic and flammable hydrogen sulfide. Finely divided sulfur dust is an explosion hazard. Sulfur also forms explosive mixtures with powerful oxidizers such as chlorates and perchlorates.

Naphthalene

Naphthalene is combustible in both the solid and liquid forms, and its vapor and dust form explosive mixtures with air.

UNSTABLE CHEMICALS

Certain chemicals spontaneously polymerize or decompose with the liberation of heat when contaminated or even when pure. Such reactions can become violent.

Acetaldehyde

Acetaldehyde forms explosive peroxides when exposed to oxygen from the air. Aldehyde also can undergo an addition-type reaction with certain catalysts, and the reaction can be dangerous.

Ethylene Oxide

Ethylene oxide can polymerize or decompose violently when catalyzed by iron rust, aluminum, alkali metal hydroxides, and anhydrous chlorides of iron, tin, or aluminum. It also reacts vigorously with alcohols, organic and inorganic acids, and ammonia.

Hydrogen Cyanide

Hydrogen cyanide is both highly flammable and poisonous, and the liquid can polymerize explosively. The flash point is 0°F (−18°C), and the explosive range is from 5.6 to 40 percent.

Nitromethane

Nitromethane decomposes violently at 599°F (315°C) and 915 psig. There have been industrial accidents in which undiluted nitromethane was detonated by shock at room temperature.

Organic Peroxides

Organic peroxides are combustible and can decompose. Heat, shock, or even friction can cause many organic peroxides to decompose explosively. As a result, most organic peroxides are shipped, stored, and handled with inert diluents to reduce their decomposition potential. To prevent confinement under fire conditions, commercial organic formulations are often shipped in plastic bottles or paper bags. These materials should be stored in their original shipping containers until used.

Styrene

The polymerization reaction of styrene increases as temperature increases. Eventually, the reaction can become violent unless it is controlled.

WATER- AND AIR-REACTIVE CHEMICALS

The heat liberated during the reactions of certain chemicals with air or water can be great enough to ignite nearby combustibles. Materials that ignite spontencously on contact with air are termed *pyrophoric.*

Alkalies

Caustics or alkalies, although noncombustible, can react with water and generate enough heat to ignite combustibles.

Alkylaluminums

Alkylaminums are pyrophoric compounds and react violently with water. As a result, they generally are found in solutions with hydrocarbon solvents.

Anhydrides

Acid anhydrides are compounds of acids from which the water has been removed. They react with water, sometimes violently, to regenerate acids.

Carbide

Carbides of some metals react explosively on contact with water. Calcium carbide reacts with water to produce acetylene, and, without careful control, the heat of reaction will ignite the acetylene.

Charcoal

Under some conditions, charcoal reacts with air at a rate that will cause the charcoal to ignite.

Hydrides

Metal hydrides react with water to form hydrogen gas, which may be ignited by the heat of reaction.

Oxides

Oxides of some metals and nonmetals react with water to form alkalies and acids, respectively. Calcium oxide, or quicklime, reacts vigorously with water and liberates enough heat to cause some combustibles to ignite.

Phosphorus

There are two forms of phosphorus. White phosphorus is pyrophoric and extremely toxic. Red phosphorus, although less hazardous, is combustible and can be converted to the white form under fire conditions.

TABLE 46-2 Typical Corrosives

Antimony pentachloride	Hydrofluoric acid
Benzoyl chloride	Nitric acid
Bromine	Oleum
Chlorine	Perchloric acid (under 72%)
Chlorine trifluoride	Sodium hydroxide (caustic soda)
Chromic acid	Sulfur chloride
Fluorine	Sulfur trioxide
Hydrochloric acid	Sulfuric acid
Chlorosulfonic acid	

Sodium Hydrosulfite

On contact with moisture from the air, sodium hydrosulfite heats spontaneously and can ignite nearby combustibles.

CORROSIVE MATERIALS

Corrosive materials have a destructive effect on living tissues. Although some are oxidizers, they are classified separately to emphasize their injurious effect on contact or inhalation. Some corrosive materials are listed in Table 46-2.

Inorganic Acid

Concentrated aqueous solutions of the inorganic acids—sulfuric, nitric, and hydrochloric—are not combustible. The chief hazard is the danger of leakage and possible mixture with other chemicals or combustible materials nearby, which could create a serious fire or explosion hazard.

Halogens

The chemicals of the halogen family—fluorine, chlorine, bromine, and iodine—differ from each other, with their fire hazard potential decreasing in the order listed. Halogens are noncombustible and act as oxiders. They are also toxic chemicals.

TOXIC CHEMICALS

Most chemicals are not considered toxic unless taken orally or inhaled in relatively large quantities. However, a few can cause serious illness or death when they come in contact with the body or when they are swallowed or inhaled in small quantities. Gases and vapors tht are dangerous to life when mixed in small amounts in air are considered to be extremely dangerous poisons. Some of these chemicals are listed in Table 46-3.

OXIDIZING CHEMICALS

Most oxidizing chemicals are not combustible, but they can increase the ease of ignition of combustible materials and will increase the intensity of burning. Some oxidizing agents are susceptible to spontaneous decomposition. However, most are stable unless they are contaminated. If they are contaminated by organics, many of them can spontaneously ignite and may undergo explosive reactions. Some such chemicals, through by no means all, are listed in Table 46-4.

Nitrates

Inorganic nitrates are widely used in fertilizers, salt baths, and other industrial applications. When exposed to fire, they can melt and release oxygen, causing the

TABLE 46-3 Typical Toxic Chemicals and Conditions for the Greatest Hazard

Dangerous when small amounts are mixed in air	
Acrolein	Germane
Aluminum phosphide	Hydrogen cyanide
Boron trifluoride	Hydrogen sulfide
Choloropicrin	Methyl bromide
Cyanogen	Nickel carbonyl
Diborane	Phosgene
Fluorine	Phosphine
Dangerous toxins hazardous when taken orally or on contact	
Acrylonitrile	Endrin
Allyl alcohol	Ethyleneimine
Antimony pentafluoride	Methyl hydrazine
Arsenic compounds	Nicotine
Arsine	Nitrobenzene
Bromine pentafluoride	Nitrochlorobenzene
Chlorine	Parathion
Cyanides	Phenol (carbolic acid)
Diketene	Stibine
Dimethythydrazine	Tetraethyl lead
Dimethyl sulfate	Thionyl chloride

fire to intensify. Molten nitrates will react violently with organic materials. Solid steams of water used to fight nitrate fires may produce steam explosions. In addition, most nitrates are water soluble, and a solution of nitrates absorbed by a combustible material makes that material extremely easy to ignite once it has dried.

Nitric Acid

Although nitric acid is generally considered a corrosive, it will markedly increase the ease with which an organic material that comes in contact with it or its vapors ignites. The use of even dilute nitric acid solutions to clean organic residue from equipment can lead to an explosion.

Nitrites

Nitrites are more active than nitrates as oxidizing agents. Mixtures with combustible substances should not be subjected to the heat of a flame. Certain nitrites, notably ammonium nitrite, are explosive by themselves and must be avoided.

TABLE 46-4 Typical Oxidizing Materials

Bromates	Nitrates
Chlorates	Perchlorates
Chlorites	Peroxides
Hypochlorites	Permanganates
Hydrosulfites	Persulfates

Inorganic Peroxides

Sodium, potassium, and strontium peroxides react vigorously with water and release oxygen and large amounts of heat. If organic or other oxidizable material is present when such a reaction takes place, a fire or explosion is likely to occur.

Mixtures of barium peroxide and combustible or readily oxidizable materials are explosive. They are easily ignited by friction or a small amount of water.

Hydrogen peroxide is a strong oxidizing agent that can cause combustible materials with which it remains in contact to ignite, especially at concentrations above 35 percent. Contamination of a storage tank containing hydrogen peroxide can result in an explosion. At a concentration above about 92 percent, hydrogen peroxide can be exploded by shock.

Chlorates

When heated, chlorates liberate oxygen more readily than do nitrates. When mixed with combustible materials, they can ignite or explode spontaneously. Drums containing chlorates can explode when heated.

Chlorites

Sodium chlorite is a powerful oxidizing agent that forms explosive mixtures with combustible materials. When it comes in contact with strong acids, it releases explosive chlorine dioxide gas. Chlorites decompose explosively at lower temperatures than chlorates.

Dichromates

Among the dichromates, most of which are noncombustible, ammonium dichromate is the most hazardous. It may explode on contact with organic materials, and it decomposes at 338°F (170°C), releasing nitrogen gas. Closed containers rupture at the decomposition temperature. Other dichromates release oxygen when heated and react readily with oxidizable materials.

Hypochlorites

Calcium hypochlorite can cause combustible organic materials to ignite on contact. When heated, it gives off oxygen. When decomposing it liberates a large amount of heat. It is sold as a bleaching powder and, when concentrated, as a swimming pool disinfectant. Calcium hypochlorite swimming pool chemicals have been the cause of a number of retail store fires.

Chlorinated Isocyanurates

These chemicals, commonly used for pool sanitizers, include sodium dichloroisocyanurate, potassium dichloroisocyanurate, and trichloroisocyanuric acid (trichlor). They will react with organic materials such as oils and cause fires. Upon decomposition, they give off copious quantities of chlorine gas. On contact with water they give off small quantities of nitrogen trichloride, which will produce numerous small explosions resulting in popping noises. These materials have been involved in fires in retail stores.

Perchlorates

Perchlorates contain one more oxygen atom than chlorates do. They have similar properties but perchlorates are more stable than chlorates.

Ammonium perchlorate has great explosive sensitivity when contaminated with an oxidizable impurity. The pure material in finely divided form is classified as an explosive.

Permanganates

Mixtures of inorganic permanganates and combustible material can be ignited by friction, or they may ignite spontaneously in the presence of an inorganic acid.

Persulfates

Persulfates are strong oxidizing agents that can explode during a fire. An explosion can also follow an accidental mixture of a persulfate with a combustible material.

SAFETY PRECAUTIONS

Storage

Segregation. Good storage practice dictates that chemicals be segregated, that is, separated from other materials in storage, from processing and handling operations, and from incompatible materials. Segregation can take the form of isolating hazardous chemicals in a separate storage building or a separate room (by means of fire walls) or separating different classes of chemicals by intervening open space or a buffer of nonhazardous or inert material. The extent of segregation depends on the chemical's quantity, physical state, degree of incompatibility, and known behavior under fire conditions.

Hazard Identification. Each storage area should be marked with the name of the material and an easily understood means of identification of the material stored there. The hazard identification system described in NFPA 704, *Standard System for the Identification of the Hazards of Materials for Emergency Response,* should be used.

Handling and Transportation

Containers should be handled with care when they are shipped, transferred, and moved to and from storage. They should be checked to ensure that they are tightly sealed and are not damaged and for leaks and fugitive emissions. When containers are transported, they should be stable and in their normal upright condition. Containment is the key to safety when materials are moved in bulk handling systems. If local or spot exhaust ventilation is provided, the system should be checked to ensure it is operating properly. Safe transportation depends on knowing the properties of the chemicals as well as the conditions they will be exposed to during shipment.

Fire Protection

Automatic protection for storage should be employed wherever possible. The extinguishing agent used is determined by the chemical's reactivity, its toxicity, and its expected products of combustion.

SUMMARY

Safe and effective fire control measures for chemicals require knowledge of their hazardous properties. The groupings include solid combustible, unstable, water- and air-reactive, corrosive, toxic, and oxidizing chemicals. To safely store, handle, and transport chemicals, they should be segregated, clearly identified, and moved and transported carefully.

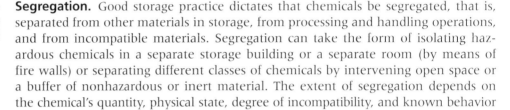

BIBLIOGRAPHY

Cote, A. E., ed., *Fire Protection Handbook,* 18th ed., NFPA, Quincy, MA, 1997.
Cote, A. E., and Linville, J. L., eds., *Industrial Fire Hazards Handbook.* 3rd ed., NFPA, Quincy, MA, 1990.
Spencer, A. B., and Colonna, G. R., eds., *Fire Protection Guide to Hazardous Materials,* 13th ed., NFPA, Quincy, MA, 2002.

NFPA Codes, Standards, and Recommended Practices

See the latest version of The NFPA Catalog for availability of current editions of the following documents.

NFPA 45, *Standard on Fire Protection for Laboratories Using Chemicals*
NFPA 430, *Code for the Storage of Liquid and Solid Oxidizers*
NFPA 432, *Code for the Storage of Organic Peroxide Formulations*
NFPA 490, *Code for the Storage of Ammonium Nitrate*
NFPA 495, *Explosive Materials Code*
NFPA 655, *Standard for Prevention of Sulfur Fires and Explosions*
NFPA 704, *Standard System for the Identification of the Hazards of Materials for Emergency Response*
NFPA 801, *Standard for Fire Protection for Facilities Handling Radioactive Materials*
NFPA 804, *Standard for Fire Protection for Advanced Light Water Reactor Electric Generating Plants*

Plastics and Rubber

Steve Younis

CHAPTER 47

CASE STUDY

Fire Damages Foam Manufacturing Plant

Polyurethane foam bungs self-ignited at a California foam manufacturing plant, starting a fire that damaged the structure before fire fighters and sprinklers could bring it under control.

The single-story, 415,440-ft^2 (38,596-m^2) structure, which was shared by two unrelated businesses, was of unprotected, ordinary construction with concrete tilt-up walls and a panelized wood truss roof. The fire occurred in a 141,680-ft^2 (13,163-m^2) section of the building that was protected by a smoke detection system and a wet-pipe sprinkler system, both monitored by a private alarm company.

The building was closed for the night and security guards were the only ones on the premises when the fire department received a water flow alarm at 8:40 P.M. The station dispatched a limited response, but when arriving companies saw smoke coming from the building, the response was upgraded to a full box assignment.

Fire fighters traced the alarm to the plant's curing room, where they found sprinklers controlling a fire involving approximately 10 polyurethane foam bungs. Using the building's hose lines, fire fighters had extinguished a few small smoldering fires when two more foam bungs 75 ft away started to burn. The fire crews stretched two 1¾-in. hose lines from their apparatus and then moved the burning bungs outside.

Once the fire appeared to be under control, fire fighters checked for signs of heating. After 45 minutes with no sign of fire, they shut down the sprinkler system to limit water damage to the building's other occupancy. Shortly afterward, however, several more bungs ignited, some of which were in hard-to-reach areas. Fire fighters turned the sprinkler system back on, but the fire began to make headway, forcing a defensive attack. Crews used fire-fighting foam to control the flames and protect exposure walls and activated the building's mechanical smoke removal system. In all, 42 sprinklers activated.

During the investigation, it was discovered that a newly shipped chemical used during the last manufacturing run of the day lacked a necessary "inhibitor package," which limits heat buildup in cured foam rubber to 430°F (221°C), just below its ignition temperature of 450°F (232°C). Without this inhibitor, the bungs self-ignited.

The blaze destroyed 75 foam bungs and damaged the 17 others in the room. Damage to the building, valued at $4.4 million, and its contents, valued at $2 million, was initially estimated at $50,000 and $420,000, respectively. However, the insurance company allowed for the possibility that the contents would prove to be a total loss. A security guard and three fire fighters suffered smoke inhalation injuries.

Source: "Fire Watch," *NFPA Journal,* November/December, 1998.

Steve Younis is a senior fire protection engineer for NFPA's Building Fire Protection and Life Safety Department, with fire protection experience in a wide variety of areas.

Plastics and rubber are all part of the family of materials called polymers, which are found throughout society in innumerable variations. Almost all are combustible to some extent, and some burn with extreme rapidity, producing large amounts of dense smoke. In most instances, it is impossible to determine the burning characteristics of a material simply by looking at it. Because the product appears in so many variations and its uses are so diverse, this chapter cannot present more than general guidelines that will tell the inspector what to look for and how to take the proper fire safety precautions. The inspector may have to refer to the handbooks, codes, and standards that are listed in the bibliography to obtain specific information on protection for a particular application.

GENERAL USES

Small amounts of plastics and rubber are seldom of concern. One foam cup hardly presents a significant fire threat. However, stacks of polystyrene foam stored in a warehouse could overpower a sprinkler system that has not been designed for the hazard. In general terms, plastics being stored and used require the same protection that similar quantities and arrangements of wood, cardboard, paper, fiberboard, natural fabrics, and other common cellulosic materials do. Their burning rates are often similar, but some materials and situations warrant extraordinary protective measures.

Foam plastic and foam rubber tend to burn faster and more intensely, generating more smoke and toxic gases, than other forms of the same polymer. Inhibitors, or flame retardants, can slow ignition or flaming, but these additives or other treatments are generally ineffectual in a fully developed fire. Ordinary sprinkler protection might be inadequate to protect lives or the building.

When used as wall and ceiling insulation in buildings, typical foam plastics, such as rigid polyurethane, polyisocyanurate, and polystyrene, must be completely covered on the inside of the building by a fire-resistant barrier, such as gypsum wallboard ½ in. (128 mm) or thicker. This is true even if the material has a low flame spread rating under standard test procedures. The inspector should consult NFPA *101*®, *Life Safety Code*®, and the local building code for details.

In accordance with the requirements of NFPA *101*, 2000 ed., Chapter 10, materials used for wall or ceiling finish should be tested and classified as being Class A, Class B, or Class C in accordance with NFPA 255, *Standard Method of Test of Surface Burning Characteristics of Building Materials* as follows:

- *Class A Interior Wall and Ceiling Finish.* Flame spread 0–25; smoke development 0-450. Includes any material classified at 25 or less on the flame spread test scale and 450 or less on the smoke test scale. Any element thereof, when so tested, should not continue to propagate fire.
- *Class B Interior Wall and Ceiling Finish.* Flame spread 26–75; smoke development 0-450. Includes any material classified at more than 25 but not more than 75 on the flame spread test scale and 450 or less on the smoke test scale.
- *Class C Interior Wall and Ceiling Finish.* Flame spread 76–200; smoke development 0-450. Includes any material classified at more than 75 but not more than 200 on the flame spread test scale and 450 or less on the smoke test scale.

For certain conditions or occupancies, the material may be required to meet the test protocol specified in NFPA 286, *Standard Methods of Fire Tests for Evaluating Contribution of Wall and Ceiling Interior Finish to Room Fire Growth:*

- Flames should not spread to the ceiling during the 40-kW exposure.
- During the 160-kW exposure, the following criteria should be met:

- Flame should not spread to the outer extremities of the sample on the 8×12 ft (2.4×3.7 m) wall.
- Flashover should not occur.
- For new installations, the total smoke released throughout the test should not exceed 35,315 ft^3 (1000 m^3).

Note that in accordance with NFPA *101,* cellular or foamed plastic should not be used as wall or ceiling finish unless they meet the specific requirements designated under the provision for these materials in Chapter 10 of NFPA *101.* The occupancy of the building or structure will dictate the wall and ceiling finish requirements. The inspector should consult the occupancy chapters within NFPA *101* for the requirements of the applicable occupancy.

Foam rubber and foam plastics in mattresses and furniture cushions can be of concern where people sleep, particularly in institutional occupancies. The cigarette ignition resistance of mattresses and mattress pads is controlled by Part 1632, *Code of Federal Regulations,* Title 16. Title 16 also contains standards for the flammability of clothing textiles, vinyl plastic film, children's sleepwear, and carpets and rugs. The requirements state that mattresses, unless located in rooms or spaces protected by an approved automatic sprinkler system, should have a char length not exceeding 2 in. (5.1 cm) when tested in accordance with Part 1632.

Also, in accordance with NFPA *101* where specifically required by the occupancy, unless the mattress is located in a room or space protected by an approved automatic sprinkler system, it should have limited rates of heat release when tested in accordance with NFPA 267, *Standard Method of Test for Fire Characteristics of Mattresses and Bedding Assemblies Exposed to Flaming Ignition Source,* or ASTM E1590, *Standard Method for Fire Testing of Real Scale Mattresses,* as follows:

- The peak rate of heat release for the mattress should not exceed 250 kW.
- The total energy released by the mattress during the first 5 minutes of the test should not exceed 40 MJ.

Where required by the applicable provisions of NFPA *101,* upholstered furniture and mattresses shall be resistant to a cigarette ignition (that is, smoldering) in accordance with the following:

- The components of the upholstered furniture, unless located in rooms or spaces protected by an approved automatic sprinkler system, shall meet the requirements for Class I when tested in accordance with NFPA 260, *Standard Methods of Tests and Classification System for Cigarette Ignition Resistance of Components of Upholstered Furniture.*
- Mocked-up composites of the upholstered furniture, unless located in rooms or spaces protected by an approved automatic sprinkler system, should have a char length not exceeding 1.5 in. (3.8 cm) when tested in accordance with NFPA 261, *Standard Method of Test for Determining Resistance of Mock-Up Upholstered Furniture Material Assemblies to Ignition by Smoldering Cigarettes.*

Also, in accordance with NFPA *101* where specifically required by the occupancy, upholstered furniture, unless the furniture is located in a room or space protected by an approved automatic sprinkler system, should have limited rates of heat release when tested in accordance with NFPA 266, *Standard Method of Test for Fire Characteristics of Upholstered Furniture Exposed to Flaming Ignition Source,* or with ASTM E1537, *Standard Method for Fire Testing of Real Scale Upholstered Furniture Items,* as follows:

- The peak rate of heat release for the single upholstered furniture item should not exceed 250 kW.

- The total energy released by the single upholstered furniture item during the first 5 minutes of the test should not exceed 40 MJ.

As an added requirement for specific individual occupancies within NFPA *101,* furnishings and contents made with foamed plastic materials that are unprotected from ignition should have a heat release rate not exceeding 100 kW when tested in accordance with UL 1975, *Standard for Fire Tests for Foamed Plastics Used for Decorative Purposes.*

Cellulose nitrate, also known as nitrocellulose or pyroxylin, is recognized as one of the most hazardous plastics. It begins to decompose dangerously at relatively low temperatures. Special rules have been written for its storage, fire protection, and handling. The inspector should consult NFPA 40, *Standard for the Storage and Handling of Cellulose Nitrate Film,* for film storage conditions, or NFPA 42, *Code for the Storage of Pyroxylin Plastic,* for other than cellulose nitrate film storage applications. Pyroxylin plastic is also largely used in the manufacture of lacquer products.

When inspectors have questions about the hazard of a material or product, they can rely on the standard fire test methods that are valid for the application. A number of tests are commonly used, either singly or in combination. However, one of them, the Steiner Tunnel Test, ASTM E84, which typically is used to evaluate the surface burning characteristics of interior finish materials, may not give a true indication of the actual hazard presented by many plastics. See the NFPA *Fire Protection Handbook* for more detailed information. The results of small-scale, "Bunsen burner" tests are often suspect when large quantities of plastic are being dealt with because the small flame does not simulate the much greater intensity of the fire exposure under real-life conditions. Inspectors should ask for proof that the materials have been tested and have passed the applicable fire test requirements.

Some general guidelines can be given for situations often encountered. Plastic glazing materials, such as Plexiglas®, Lucite®, and polycarbonates, should not be used in applications in which their combustibility is objectionable or for which a fire rating is required. This precludes the use of plastic glazing in firewalls, fire doors, fire-rated corridor partitions, and smoke barriers. To determine the requirements for fire-rated door and window assemblies that incorporate glazing materials, the inspector should consult NFPA 80, *Standard for Fire Doors and Fire Windows.*

Thin plastic laminates, such as Formica® and vinyl wall covering, can be considered to take on the burning characteristics of the substrate to which they are applied. For example, they would be low on gypsum wallboard or metal, moderate on wood particle board, and high on foam plastic. If in doubt, the inspector should ask for documentation of the flame spread rating that has been developed under such test methods as NFPA 255 and ASTM E84.

In air plenums and some other applications, the flammability of the plastic insulation on wiring that is not in a conduit is now often limited by code. Common industry materials used in wire and cable manufacture present a significant hazard in this application. This is especially significant where large quantities of cables are grouped or bundled together, as they are in cable trays and cable chases. These types of cable arrangements may extend both horizontally as well as vertically through a building. Self-propagating fires can occur under such conditions, with serious consequences in a significant percentage of occupancies ranging from power plants and telephone exchanges to everyday businesses and schools. With the ever-increasing demand for computer-based and informational technology, the installation of wire and cable has increased at an exponential rate.

Special protection, such as fire stops at floor and firewall openings, special coatings on the wiring, special flame-resistant insulation, and so on, might be necessary in these situations. Concerns have arisen over how the abandonment of such cable will effect fire spread and smoke development in these spaces. Whether requirements will dictate the removal of all accessible abandoned material from plenum spaces is currently being evaluated for incorporation into NFPA 90A, *Standard for the Installation of Air-Conditioning and Ventilating Systems,* and NFPA 90B, *Standard for the Installation of Warm Air Heating and Air-Conditioning Systems.* For specific details on plenum wire and cable material requirements, the inspector should refer to the provisions of NFPA 90A and NFPA 90B.

Because plastics and rubber are Class A materials, water-type portable fire extinguishers and hose lines are appropriate for fire fighting. Gaseous agent and multipurpose dry chemical portable fire extinguishers are suitable when used in accordance with their labeling. The inspector should consult NFPA 10, *Standard on Portable Fire Extinguishers,* for portable fire extinguisher requirements.

WAREHOUSING

In warehouses, the often intense fires that can develop rapidly in concentrations of plastics and rubber require the installation of specially engineered, high-density automatic sprinkler systems and, at times, other measures, such as smoke and heat vents. The inspector should verify that the fire protection plan takes into account storage of large quantities of items such as rubber tires, plastic packing materials, and plastic insulating materials.

Classification of Plastics and Rubber

NFPA 230, *Standard for the Fire Protection of Storage,* and NFPA 13, *Standard for the Installation of Sprinkler Systems,* are especially useful for evaluating the potential hazards and fire protection methods of different types of combustible materials stored in warehouses. Plastics and rubber are divided into three groups according to their potential for creating difficult-to-extinguish fires in stored products, as indicated in Table 47-1. Four commodity classifications are established, as follows:

- Class I encompasses essentially noncombustible products in paper or cardboard cartons or wrappings, with or without wooden pallets.
- Class II is composed of Class I products in wooden or multilayer cardboard containers, with or without pallets.
- Class III contains wood, paper, or natural cloth products or Group C plastics, with or without pallets. A limited amount of Group A or B plastics may also be included.
- Class IV is composed of any materials in Class I, II, or III, with an appreciable amount of Group A or B plastics in ordinary corrugated cartons or with Group A or B plastic packing, with or without pallets.

More detailed specifics on the proper classification for different commodities can be found in NFPA 230, *Standard for the Fire Protection of Storage,* and NFPA 13, *Standard for the Installation of Sprinklers.* Inspectors should also be aware of the requirements of NFPA 432, *Code for the Storage of Organic Peroxide Formulations.* These peroxides are often used as additives in plastics manufacturing and present special storage hazards. Additional guidance for the hazard presented by individual elastomeric materials is available in Chapter 41, "General Storage," of this manual, in the *Chemical Safety Data Sheets* issued by the manufacturers of plastics, and in Factory Mutual data sheets.

TABLE 47-1 Classification of Plastics and Rubber

Group A
Acrylonitrile-butadiene-styrene copolymer (ABS)
Acrylic (polymethyl methacrylate)
Acetal (polyformaldehyde)
Butyl rubber
Ethylene-propylene rubber (EPDM)
Fiberglass reinforced polyester (FRP)
Natural rubber (if expanded)
Nitrile rubber (acrylonitrile-butadiene rubber)
Polybutadiene
Polycarbonate
Polyester elastomer
Polyethylene
Polypropylene
Polystyrene
Polyurethane
Polyvinyl chloride (PVC, highly plasticized; for example, coated fabric, unsupported film)
Styrene acrylonitrile (SAN)
Styrene-butadiene rubber (SRB)
Thermoplastic polyester (PET)
Group B
Cellulosics (cellulose acetate, cellulose acetate butyrate, ethyl cellulose)
Chloroprene rubber
Fluoroplastics (ECTFE, ethylene-chlorotrifluoroethylene copolymer; ETFE, ethylene-tetrafluoroethylene copolymer; FEP, fluorinated ethylene propylene copolymer)
Natural rubber (not expanded)
Nylon (nylon 6, nylon 6/6)
Silicone rubber
Group C
Fluoroplastics (PCTFE, polychlorotrifluoroethylene; PTFE, polytetrafluoroethylene)
Melamine (melamine formaldehyde)
Phenolic
Polyvinyl chloride (PVC, rigid or lightly plasticized; for example, pipe, pipe fittings)
Polyvinylidene chloride (PVDC)
Polyvinyl fluoride (PVF)
Polyvinylidene fluoride (PVDF)
Urea (urea formaldehyde)

Note: These categories are based on unmodified plastic materials. The use of fire- or flame-retarding modifiers or the physical form of the material may change the classification.

Rubber Tire Storage

Chapter 6 of NFPA 230, 1999 edition, specifies the condition for storage of rubber tires for indoor facilities. The requirement applies to new facilities, but may be of value in assisting inspectors in evaluating existing storage facilities. Figures 47-1 through 47-5 depict most of the typical storage practices acceptable for tire storage, although others may be deemed acceptable through the evaluation of the fire safety provided.

Inspectors should be aware of building configurations when evaluating rubber tire storage. When evaluating steel construction in warehouses, note that for storage exceeding 15 through 20 ft (4.6 through 6 m) in height, columns should have 1-hour fire protection; for storage exceeding 20 ft (6 m) in height, columns should have 2-hour fire protection for the entire length of the column, including structural

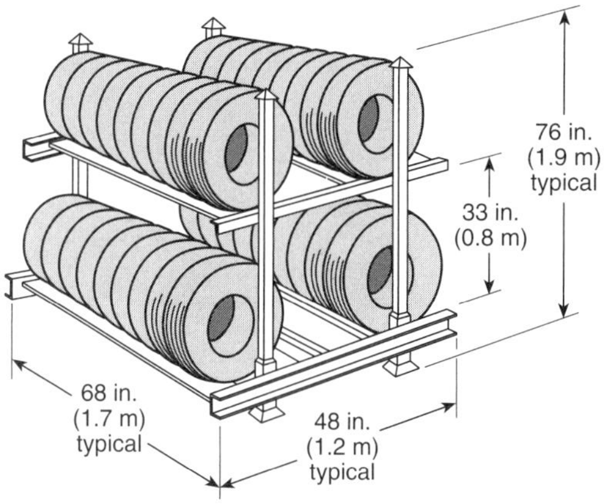

FIGURE 47-1 Open Portable Rack Tire Storage
Source: NFPA 230, 1999, Figure 6-2(c).

connections. Fire proofing is not required when a sprinkler system complying with NFPA 13 is installed that meets the requirements for this storage condition.

Inspectors should be aware that a 4-hour fire wall is required between tire warehouse and manufacturing areas (refer to NFPA 221, *Standard for Fire Walls and Fire Barrier Walls,* for design requirements). They should also be aware that piles of tires should not exceed 50 ft (15 m) wide and that the main aisle between tire piles should not be less than 8 ft (24 m) wide (consult Chapter 6 of NFPA 230 for further details).

Organic Peroxides

Current plastics manufacturing process involves the use of substances known as organic peroxides, which are additives used to enhance special properties within plastics. These peroxides also create a need for storage prior to the manufacturing

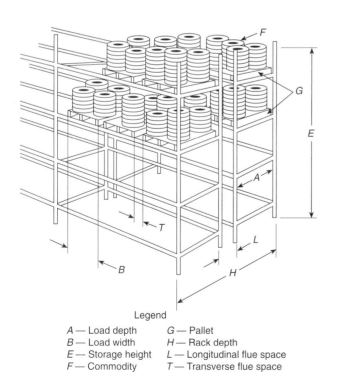

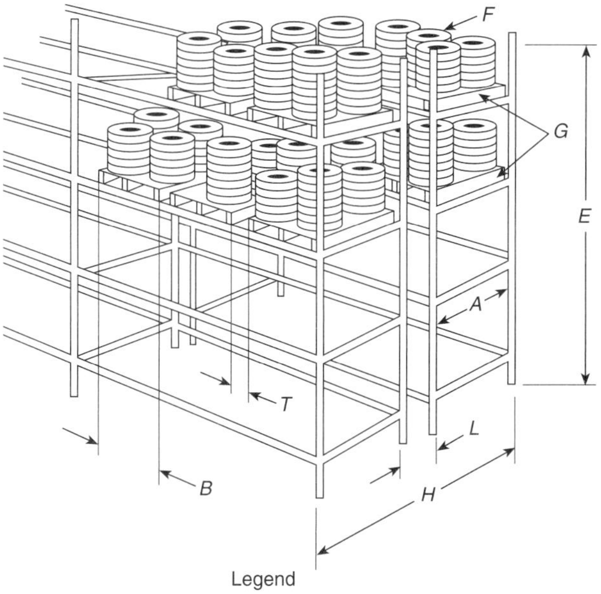

FIGURE 47-2 Double-Row Fixed Rack Tire Storage
Source: NFPA 230, 1999, Figure 6-2(d).

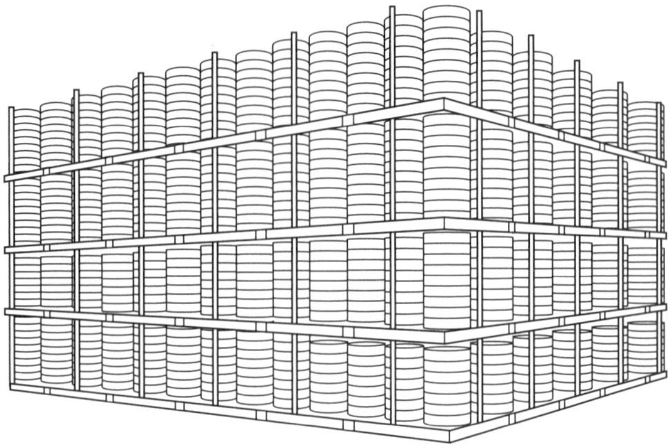

FIGURE 47-3 Palletized Portable Rack On-Side Tire Storage
Source: NFPA 230, 1999, Figure 6-2(e).

process. The inspector should be cognizant of the many variable requirements related to the storage of these products. A review of NFPA 432, which addresses the requirements for storing these chemicals, should be correlated against the conditions of the facility. Specific fire protection requirements are mandated depending on the conditions. There are also different parameters and requirements for segregated storage buildings.

MANUFACTURING AND FABRICATION

Operation Categories

In factories that produce plastics or plastic products, the operations are divided into three general categories—synthesizing, converting, and fabricating. Each of these has special fire problems the inspector must take into consideration.

Synthesizing. Synthesizing, or manufacturing, is the mixing of the basic plastic materials, or feedstock. Sometimes coloring agents or other substances are added.

Converting. Converting is the process of molding, extruding, or casting the plastic so that it will flow into a shape it will retain after cooling. Considerable heat is required for both of these processes.

Fabricating. Fabricating includes the process of bending, machining, cementing, decorating, and polishing plastic, sometimes using other materials that might be flammable or reactive. For example, the thermoplastic ABS is a combination of acrylonitrile, butadiene, and styrene. Two of these materials are low flash point flammable liquids, and the third is a gas with an ignition temperature of 804°F (429°C). A molding process might be used to combine a flammable resin with nonflammable glass fibers or with heat-resistant silicones.

FIGURE 47-4 On-Tread, On-Floor Tire Storage
Source: NFPA 230, 1999, Figure 6-2(f).

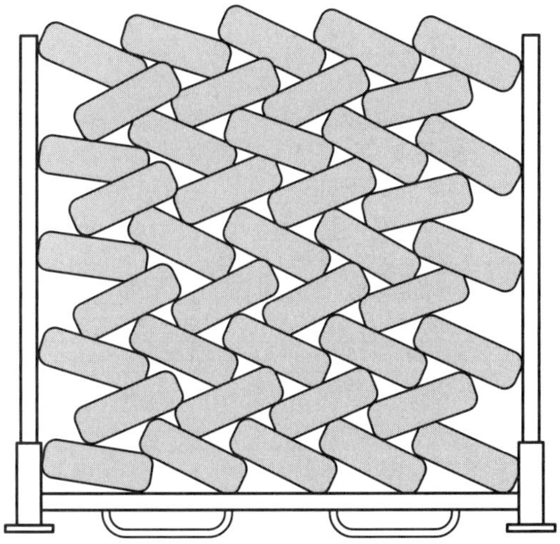

FIGURE 47-5 Laced Tire Storage
Source: NFPA 230, 1999, Figure 6-2(g).

Operation Hazards

Among the hazards to be considered for each of these operations are combustible dusts, flammable solvents, electrical faults, hydraulic fluids, and the storage and handling of large quantities of combustible raw materials and finished products. To aid in the evaluation of a situation, the inspector can consult the referenced standards and other documents published by NFPA, Underwriters Laboratories, Factory Mutual Research Corporation, and other reliable sources of technical information.

Suspensions of plastics and rubber dusts in air can form explosive mixtures, as is the case for organic materials in general. Special explosion prevention measures must be taken to prevent ignition or to minimize the dust concentrations in manufacturing and fabricating operations (see Chapter 44 in this book).

In the plastics manufacturing process, thermoplastic compounds are usually melted by heat, then forced into a mold or die for shaping. In their original state, these compounds can be in the form of pellets, granules, flakes, or powder, and each of these forms can produce dust. Thermosetting resins may be in the form of a liquid or a partially polymerized molding compound. Considerable heat is needed to mold either form.

The inspector should consider transfer operations of particle or powdered raw materials. These transfer operations involve the mechanical transfer of potentially combustible materials from hoppers to a mixing vessel and may present a significant risk.

Other manufacturing processes include blow molding, which produces hollow products, such as bottles, gas tanks, and carboys, and calendering, the process of converting thermoplastics into film or sheeting or applying a plastic coating to textiles or other materials. In casting, thermoplastics or thermosets are used to make products: The hot liquid is poured into a mold and then allowed to cool until it is solid. In coating, thermoplastic or thermosetting materials are applied to metal, wood, paper, glass, fabric, or ceramics.

In compounding, additives are mixed with resins by kneading mixers or screw extruders. Compression molding uses heat and pressure to squeeze or press material into a certain shape; in extrusion a screw thread or some other form of propulsion is used to shape a thermoplastic material into a continuous sheet, film, rod, cable, cord, or other product.

Foam plastics molding uses foam plastics in casting, calendering, coating, or rotational molding. High-pressure laminating uses heat and pressure to join materials. Injection molding involves the mixing of two or more high-pressure reactive streams

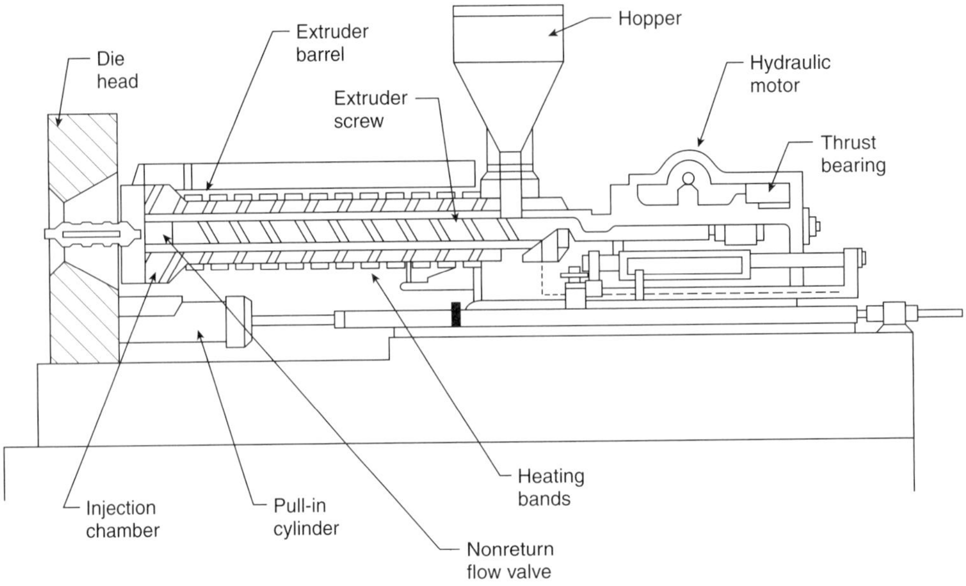

FIGURE 47-6 Reciprocating Screw Injection Molding Machine

in a mixing chamber and then the injecting of the mixture into a mold. Figure 47-6 illustrates a reciprocating screw injection molding machine in which plastic pellets are compacted, melted, and injected into a die, where the molten plastic is allowed to cool and harden.

Reinforced plastics processing combines resins with reinforcing materials. Rotational molding involves the movement of powdered plastic or molding granules in a moving, heated container. Transfer molding is the curing of thermosetting plastics in a mold under heat and pressure.

Molding and extrusion operations require temperatures of between 300°F (149°C) and 650°F (343°C), depending on the plastic being processed. Figure 47-7 illustrates a basic single-screw extruder in which plastic pellets are fed from a hopper, driven forward, and melted. The molten plastic is fed through the adapter into the die.

Because the upper temperatures are beyond the practical use of heat-transfer fluids, electrical-resistance heating is most commonly used. Sometimes, however, controllers do not operate correctly, and the temperatures become excessive. If plastic feedstock is allowed to remain in equipment, it might decompose under excessive or prolonged temperatures and release combustible gases. To reduce this hazard, the molding and extruding areas should be cleaned frequently.

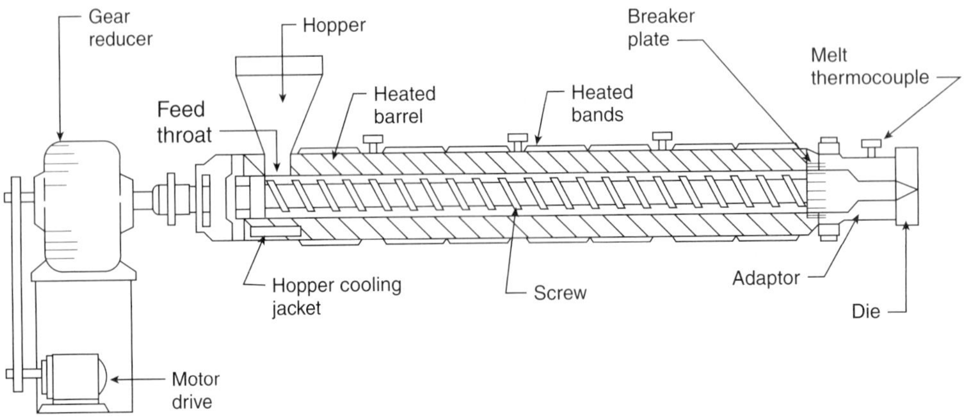

FIGURE 47-7 Basic Single-Screw Extruder

In all of these operations, the potential fire danger is influenced by the temperatures reached in these processes, the fire characteristics of the solids and liquids used, the type of portable and automatic fire extinguishing equipment, and the condition of manual and automatic fire alarms. Sources of heat include the equipment's operating temperatures, the electrical equipment and wiring, static sparks, friction, matches, and smoking materials.

Static sparks are common in plastic manufacturing because plastics are good electrical insulators. The movement of film across rolls or guides can generate sparks, as can transmission belts. The hazard is reduced if the equipment is correctly grounded and tinsel conductors are used on moving films or filaments.

Another potential ignition source is the temperature reached in hydraulic systems for clamping molds or providing pressure to the rams or screws that force plastic by compression, transfer, or injection molding. These temperatures might exceed the ignition temperature of some petroleum fluids.

FIRE PROTECTION

The burning of plastic is influenced considerably by its physical form. Molding pellets in bulk storage will burn differently from certain finished products, such as containers, polyvinyl envelopes, or insulated cables. Dust and some granules can flare rapidly on the surface of equipment, but a solid compound in a mold might be easy to extinguish.

Another potential danger may exist with the creation and emission of dust particles. Mechanical fabrication processes such as cutting, abrading, and sanding may produce significant concentrations of dust within the air. The emission of particulates or dust such as plastic creates a significant risk for a fast growth fire or for a potential deflagration within that area or facility. The inspector should be aware that plastic particulate emissions significantly increase the potential for fire spread and subsequently increase the potential for a major fire incident.

In most situations, water is the most appropriate extinguishing agent for fires involving plastics, and it should be available in large quantities. A plastics manufacturing plant should be equipped with automatic sprinklers, standpipe and hose systems, water-type portable extinguishers, and perhaps special automatic extinguishing systems for flammable liquids and electrical fires. Because of the many variables involved in plastics and their fire behavior, the arrangement of fixed extinguishing and explosion suppression systems and of portable extinguishing equipment should be designed specifically for each plant.

If large quantities of stored plastics are exposed to fire, fire fighters should direct hose streams onto them to cool them and prevent them from melting into the more flammable liquid state. At the same time, they must be careful not to agitate any dust that may be in the area, be it plastic dust, wood dust, or some other type of dust.

When a fire involves plastics, the potential for dense, noxious smoke must be considered, and the fire area should be evacuated. The fire officer in charge of responding fire fighters must be told immediately of the respiratory hazards and the need for protective breathing apparatus.

SUMMARY

Plastics and rubbers are part of the family of materials called polymers, almost all of which are combustible. Because the product appears in many variations, this chapter presented only general guidelines regarding the uses, warehousing, manufacture, and fabrication of plastics and rubbers, with special attention focused on rubber tire and organic peroxide storage. Although the burning of plastic is influenced by its physical form, water is, in most cases, the appropriate extinguishing agent for fires involving plastics. Automatic fire extinguishing and explosion suppression systems as well as portable extinguishing equipment should, however, be designed specifically for each facility depending on its conditions.

BIBLIOGRAPHY

Cote, A. E., ed., *Fire Protection Handbook,* 18th ed., NFPA, Quincy, MA, 1997.

Standard Method for Fire Testing of Real Scale Mattresses, ASTM E1590, American Society for Testing and Materials, West Conshocken, PA.

Standard Method for Fire Testing of Real Scale Upholstered Furniture Items, ASTM E1537, American Society for Testing and Materials, West Conshocken, PA.

Standard Test Method for Surface Burning Characteristics of Building Materials, ASTM E84, American Society for Testing and Materials, West Conshocken, PA.

NFPA Codes, Standards, and Recommended Practices

See the latest version of The NFPA Catalog for availability of current editions of the following documents.

NFPA 10, *Standard for Portable Fire Extinguishers*
NFPA 13, *Standard for the Installation of Sprinkler Systems*
NFPA 40, *Standard for the Storage and Handling of Cellulose Nitrate Film*
NFPA 42, *Code for the Storage of Pyroxylin Plastic*
NFPA 69, *Standard on Explosion Prevention Systems*
NFPA 70, *National Electrical Code®*
NFPA 80, *Standard for Fire Doors and Fire Windows*
NFPA 90A, *Standard for the Installation of Air-Conditioning and Ventilating Systems*
NFPA 90B, *Standard for the Installation of Warm Air Heating and Air-Conditioning Systems*
NFPA *101®*, *Life Safety Code®*
NFPA 221, *Standard for Fire Walls and Fire Bzrrier Walls*
NFPA 230, *Standard for the Fire Protection of Storage*
NFPA 255, *Standard Method of Test of Surface Burning Characteristics of Building Materials*
NFPA 260, *Standard Methods of Tests and Classification System for Cigarette Ignition Resistance of Components of Upholstered Furniture*
NFPA 261, *Standard Method of Test for Determining Resistance of Mock-Up Upholstered Furniture Material Assemblies to Ignition by Smoldering Cigarettes*
NFPA 266, *Standard Method of Test for Fire Characteristics of Upholstered Furniture Exposed to Flaming Ignition Source*
NFPA 267, *Standard Method of Test for Fire Characteristics of Mattresses and Bedding Assemblies Exposed to Flaming Ignition Source*
NFPA 286, *Standard Methods of Fire Tests for Evaluating Contribution of Wall and Ceiling Interior Finish to Room Fire Growth*
NFPA 432, *Code for the Storage of Organic Peroxide Formulations*
NFPA 654, *Standard for the Prevention of Fire and Dust Explosions from the Manufacturing, Processing, and Handling of Combustible Particulate Solids*
NFPA 704, *Standard System for the Identification of the Hazards of Materials for Emergency Response*

Explosives and Blasting Agents

CHAPTER 48

Lon Santis

CASE STUDY

Six Fire Fighter Fatalities in Construction Site Explosion

The Kansas City, Missouri, Fire Department lost six fire fighters and their vehicles—two entire pumper companies—in an explosion in 1988 that occurred while they were extinguishing a fire at a highway construction site. The fire involved a trailer/magazine containing blasting mixtures of ammonium nitrate and fuel oil, most containing aluminum pellets as well. In addition the fire involved two other vehicles and ultimately a second trailer/magazine that also exploded.

The fire fighters were not told specifically what was in the trailer/magazine but had been cautioned by the dispatcher about explosives on the site. Exactly what they suspected was in the trailers will probably never be known. The two captains and four fire fighters involved were highly experienced. Four of the six had attended National Fire Academy field courses on hazardous material identification. They also had DOT hazardous materials guidebooks in their vehicles.

However, because the trailers/magazines containing the blasting agents probably had no markings or placards indicating their contents, the crews may never have been sure of what was in them, especially since other prominently marked magazines were present and may have been thought to contain all the dangerous materials. The fire department was not aware of the presence of the trailers/magazines or their contents before the incident because of a lack of jurisdictional authority and because the city's fire prevention and protection code did not require the city engineer to notify the fire department that blasting permits had been issued. (This was immediately changed after the incident.) More importantly, the Kansas City Fire Department had no authority or responsibility to inspect the site because it was a state enclave.

Source: Federal Emergency Management Agency, U.S; Fire Administration National Fire Data Center. "Six Fire Fighter Fatalities in Construction Site Explosion, Kansas City, Missouri," *Technical Report Series,* November 29, 1988.

To maintain an effective level of fire protection in a plant that manufactures or transports explosive materials, inspectors must examine the entire plant, know the characteristics of the materials in it, and correct hazardous situations. Their task is to identify and correct all situations that are dangerous. When inspecting plants or motor terminals where explosives are present, inspectors should be concerned with general fire safety practices and with combinations of chemicals and liquids that have a wide range of sensitivity, potential flame, and explosive power. They should also know about the extent and efficiency of existing fire protection systems and the plans for controlling and responding to fires or explosions. Because sparks,

Lon Santis holds a B.S. and an M.S. in mining engineering from the University of Pittsburgh. He has worked for the U.S. Bureau of Mines (now NIOSH) in explosives safety research and is currently manager of technical services for the Institute of Makers of Explosives.

flames, impact, or the decomposition of explosive materials could cause catastrophic destruction, inspectors should stress to plant personnel the importance of good safety, good housekeeping, and good fire protection practices. Inspectors should also be familiar with the codes and standards on explosive materials.

DEFINITIONS OF TERMS

To understand potential problems, inspectors should be familiar with the terms and definitions that apply to explosive materials. The following are brief definitions of some important terms:

1. **Blasting Agent:** A material or mixture intended for blasting that meets the requirements of the U.S. Department of Transportation (DOT) "Hazardous Materials Regulations," as set forth in Title 49, *Code of Federal Regulations,* Parts 173.56, 173.57, and 173.58, Explosive 1.5D.
2. **Explosive:** Any chemical compound, mixture, or device, the primary or common purpose of which is to function by explosion (i.e., an extremely rapid release of gas and heat). The term includes, but is not limited to, dynamite, black powder, pellet powder, initiating explosives, detonators, safety fuses, squibs, detonating cords, igniter cords, and igniters.
3. **Composite Propellants:** A mixture consisting of an elastomeric-type fuel and an oxidizer. Composite propellants are used in gas generators and rocket motors.
4. **Detonating Cord:** A flexible cord containing a center core of high explosive used to initiate other explosives.
5. **Detonator:** Any device containing an initiating or primary explosive that is used for initiating detonation. A detonator is permitted to contain not more than 10 g of total explosive material per unit, excluding ignition or delay charges. The term includes, but is not limited to, electric detonators of the instantaneous and delay types, detonators for use with safety fuses, detonating cord delay connectors, and nonelectric detonators of the instantaneous and delay types that consist of a detonating cord, a shock tube, or any other replacement for electric leg wires.
6. **High Explosive Material:** Explosive material that is characterized by a very high rate of reaction, high pressure development, and the presence of a detonation wave.
7. **Low Explosive Materials:** Explosive material that is characterized by deflagration or a low rate of reaction and the development of low pressure.
8. **Oxidizing Material:** Any solid or liquid that may, by readily yielding oxygen, cause or enhance the combustion of other materials.
9. **Propellant:** An explosive that normally functions by deflagration and is used for propulsion purposes. It is classified by the U.S. Department of Transportation, "Hazardous Materials Regulations" as 1.1 (Class A) or 1.3 (Class B), depending on its susceptibility to detonation.
10. **Primer:** A unit, package, or cartridge of explosive material used to initiate other explosives or blasting agents; it contains a detonator or a detonating cord to which is attached a detonator designed to initiate the cord.
11. **Sensitivity:** A characteristic of an explosive material, classifying its ability to detonate upon receiving an external impulse such as impact shock, flame, or other influence that can cause explosive decomposition.
12. **Special Industrial Explosive Materials:** Shaped materials, sheet forms, and various other extrusions, pellets, and packages of high explosives used for high-energy-rate forming, expanding, and shaping in metal fabrication and for dismemberment and reduction of scrap metal.
13. **Water Gel:** Any explosive or blasting agent that contains a substantial portion of water.

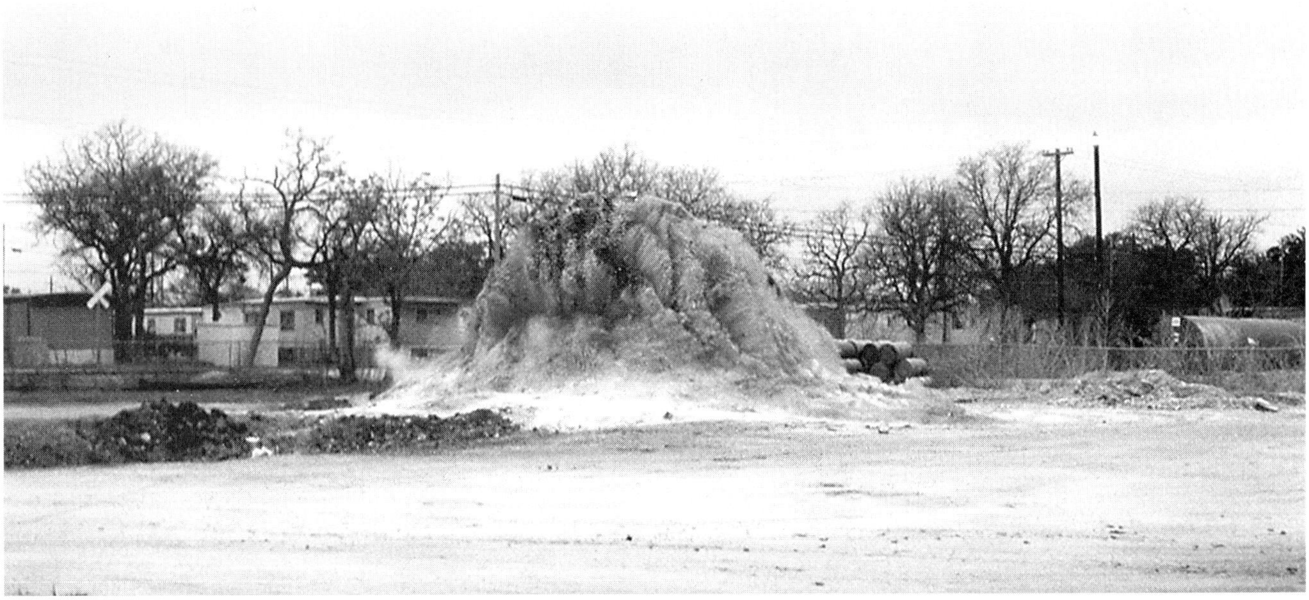

FIGURE 48-1 Dynamite-Triggered Explosion

In addition to these terms, the inspector should understand the classification of explosives as defined in the "Hazardous Materials Regulations" of the U.S. Department of Transportation. These classifications are as follows:

1. **Division 1.1:** Explosives that have a mass explosion hazard. A mass explosion is one that affects almost the entire load instantaneously. It includes, but is not limited to, detonating materials, because some substances, such as black powder, deflagrate violently. This division corresponds closely to the old Class A category. Examples include dynamite (Figure 48-1), desensitized nitroglycerin, lead azide, mercury fulminate, black powder, some detonators, and boosters.
2. **Division 1.2:** Explosives that have a projection hazard but not a mass explosion hazard. Explosives in this division correspond to the old Class A or Class B categories. Examples include certain types of ammunition and explosive components and devices.
3. **Division 1.3:** Explosives that have a fire hazard and either a minor blast hazard or a minor projection hazard or both, but not a mass explosion hazard. This division corresponds closely with the old Class B category. Examples include propellants such as smokeless powder.
4. **Division 1.4:** Explosives that present a minor explosion hazard. The explosive effects are largely confined to the package and no projection of fragments of appreciable size or range is to be expected. An external fire does not cause virtually instantaneous explosion of almost the entire contents of the package. This division corresponds closely with the old Class C category. It includes articles containing limited quantities of materials in Division 1.1 or 1.3 or both, such as small arms ammunition.
5. **Division 1.5:** Very insensitive explosives. This division comprises substances that have a mass explosion hazard but are so insensitive that there is very little probability of initiation or of transition from burning to detonation under normal conditions of transport. This division corresponds closely with the old blasting agent classification.

6. **Division 1.6:** Extremely insensitive articles that do not have a mass explosive hazard. This division comprises articles that contain only extremely insensitive detonating substances and that demonstrate a negligible probability of accidental initiation or propagation. This division has no clear-cut equivalent in the old U.S. classification system. It includes certain specialized ordnance items:
7. **Forbidden Explosives:** Explosives that are forbidden or not acceptable for transportation by common, contract, or private carriers; rail freight or rail express; highway; air; or water, in accordance with the regulations of DOT.

FIRE SAFETY IN MIXING PLANTS

Basic Principles of Fire Safety

Certain principles of fire safety are common to every area of manufacture, transportation, and storage of explosive materials. Each area and its equipment should be clean and free of deposits of the explosive materials. Smoking, matches, other smoking materials, open flames, spark-producing devices, and firearms should not be permitted in storage buildings or within 50 ft (15 m) of the building.

Buildings must be made of noncombustible materials or of sheet metal on wood studs. Floors of plants must be made of concrete or some other nonabsorbent material. There should be no drains or piping in the floor where molten materials can flow and be confined during a fire. Portable fire extinguishers and other appropriate equipment should be fully charged and readily available. Initiators and detonators should be kept separate from the explosive materials.

In all facilities, the inspector should ensure that the operator of the facility maintains an active training program in emergency procedures for all employees stationed at the facility. Written emergency instructions must be posted and readily accessible to all employees. Portable fire extinguishers should be placed in appropriate locations, and hoses can be connected to hydrants and standpipes.

Housekeeping

In addition to these principles, inspectors will have to consider the less obvious weaknesses or violations of fire safety, such as the relative cleanliness and order of the building interior and the workstations. The entire building should be cleaned thoroughly on a regular basis. Inspectors should check to see that aisles and other passageways are clear of obstructions. They should note whether there are accumulations of cloth, paper, or other combustibles and whether there are residues of flammable liquid or grease on equipment or on the floor. Inspectors should watch for accumulations of dust on walls, ledges, and equipment. The area around the plant must also be inspected. It should be cleared of brush, dried grass, leaves, and litter within at least 25 ft (7.6 m) of the building.

Inspectors should observe how unopened and emptied containers are stored and used, and they should look into emptied containers to check for residues of hazardous products. They should also check how containers are arranged and handled in the storage and shipping areas and should watch particularly for containers of oil or other flammable liquids and spills of these containers. Spilled materials and discarded containers must be removed and disposed of promptly. Empty ammonium nitrate bags must be disposed of daily in a safe manner.

Electrical Installations

Electrical installations must meet the requirements of NFPA 70, *National Electrical Code®*, for ordinary locations and must be designed to minimize corrosive damage. Inspectors should examine all electrical outlets and operating machinery for possible misuse or overloading and should inspect the main board for condition of fuses and circuit breakers. They should also check the operation of vehicles and lifting equipment.

Spark emission can be hazardous to materials in the plant; therefore, internal combustion engines used for generating electric power must be located outside the mixing building or must be shielded by a firewall. In addition, the inspector should check the heating equipment to verify that it does not produce flame or sparks inside the building.

Fire Protection

Inspectors should verify that portable fire extinguishers have been inspected and charged within the last year and are at their designated locations. Inspectors should also inspect the automatic extinguishing systems or explosion suppression equipment to ensure that they are in good condition and are ready to operate.

MANUFACTURING, TRANSPORTATION, AND STORAGE OF EXPLOSIVES

Inspectors should check conditions in areas used for manufacturing, transporting, and storing the explosive materials. Explosives within the confines of an explosive manufacturing plant should be separated according to the distances specified in Institute of Makers of Explosives (IME) SLP-3 Appendix. Explosives not in the process of being manufactured, transported, or used should be stored in appropriate magazines or in storage buildings that meet the requirements of NFPA 495, *Explosive Materials Code.*

Blasting Agents

Blasting agents are manufactured so that the final product is relatively insensitive; however, the materials from which they are made have their own hazards. A blasting agent consists of an oxidizer mixed with a fuel. Oxidizers readily yield oxygen to promote the combustion of organic matter or other fuel, so they must be processed and stored accordingly. Most oxidizers, including ammonium nitrate, are capable of detonating with about half the blast effect of explosives if they are heated under confinement that permits pressure buildup or if they are subject to a strong shock, such as from an explosion.

Fuel oil storage must be outside the mixing plant and located so that oil will drain away from the plant if the tank ruptures; the mixing building must be well ventilated. The inspector should check that emergency venting systems are operating correctly.

Mixing and packaging materials must be compatible with the composition of the blasting agent. The flash point of No. 2 fuel oil, which is 125°F (51.7°C), is the minimum flash pointpermissible for hydrocarbon liquid fuel for the agent mix.

Metal powders, such as aluminum, are sensitive to moisture and should be secured in covered containers. Solid fuels of small size, including metal powders, can create dust. Such dust should never be allowed to accumulate and can be removed either by vacuuming with appropriate nonsparking equipment or by washing with an appropriate solvent.

Ammonium Nitrate Storage

Facilities used for mixing, handling, and storing ammonium nitrate should have the same fire precautions as those used for other oxidizers. Because this compound is sensitive to contamination and heat, the inspector should watch for these two influences.

Buildings in which this compound is stored must not be taller than one story and should not have a basement, unless the basement is open on one side. There should be adequate ventilation or automatic emergency venting.

All flooring in storage and handling areas must be noncombustible or protected from impregnation by ammonium nitrate. There must be no open drains, traps, tunnels, pits, or pockets where the compound can accumulate in a fire.

Ammonium nitrate in storage should not exceed 130°F (54.4°C). Bags of ammonium nitrate should be stored at least 30 in. (76 cm) away from building walls and partitions. Storage piles should be no higher or wider than 20 ft (6.1 m) and should not exceed 50 ft (15 m) in length, unless the building is of noncombustible construction or protected by automatic sprinklers. Storage piles must be at least 3 ft (0.9 m) below the roof or beams overhead. Aisles must be at least 3 ft (0.9 m) wide, with at least one service or main aisle that is at least 4 ft (1.2 m) wide.

Bins for storing ammonium nitrate in bulk should be kept free of contaminating materials. Aluminum or wooden bins should be used because ammonium nitrate is corrosive and reactive in combination with iron, copper, lead, and zinc. The storage should be clearly identified by signs reading "AMMONIUM NITRATE" in letters that are at least 2 in. (5 cm) high.

This material must not be stored in piles higher than necessary. The pressure setting of the mass is affected by humidity and pellet quality as well as by temperature. Temperature variations between a diurnal low temperature and 90°F (32°C) as well as high atmospheric humidity can be hazardous to this product.

Ammonium nitrate can be affected by a wide range of contaminants, including flammable liquids, organic chemicals, and acids. These contaminants must be kept out of, or some distance from, the storage building. Ammonium nitrate should also be shielded from these contaminants. Areas where ammonium nitrate is stored, handled, and processed should be protected by automatic sprinkler systems, supplemented by portable fire extinguishers, standpipe systems, and fire hydrants.

MOTOR VEHICLE TERMINALS AND SAFE HAVENS

Explosives motor vehicle terminals include explosives interchange lots, explosives less-than-truckloads (LTL) lots, maintenance shops, driver rest facilities, or a combination of these facilities. Safe havens are secured areas specifically designated and approved in writing by local, state, or federal governmental authorities for the parking of vehicles containing Division 1.1, Division 1.2, or Division 1.3 materials (explosives). At each of these places, large quantities of sensitive explosive materials are brought near each other and can be vandalized or accidentally set on fire. Thus each of these places requires certain fire prevention measures.

Motor Vehicle Terminals

Interchange lots should be separated by at least 50 ft (15 m) from other facilities or from any fire hazard. A temporary holding facility conforming to the construction requirements for Type 1 or Type 2 magazines as described in NFPA 495, *Explosive Materials Code*, must be provided in the interchange lot. If detonators or other initiators are to be temporarily held at the same time as other explosives, then two temporary holding facilities must be required; one for detonators (initiators) and the second for the other explosives. These temporary facilities must be appropriately marked so that the interchange lot employees are aware of the location.

Safe Havens

Safe havens should not be located within 300 ft (91.5 m) of a bridge, tunnel, dwelling, building, or place where people work, congregate, or assemble. Weeds, underbrush, vegetation, or other combustible materials should be cleared for a distance of 25 ft (7.6 m) from the safe haven. The safe haven should be protected from unauthorized persons or trespassers through the use of warning signs, gates, and patrols. Spacing of not less than 5 ft (1.5 m) should be maintained between trailers parked side by side or back to back. Self-propelled explosives vehicles (i.e., anything with a motor, as opposed to a pushcart) should be at least 25 ft (7.6 m) from other explosives vehicles. Parking should be maintained so as not to require the moving of one vehicle in order to move another vehicle.

SUMMARY

Accidents with explosives can be prevented by careful inspection. A close relationship with the inspected entity is the first step to ensuring safety. An inspector should be familiar with all operational aspects of the inspected entity, especially any aspect relating to fire safety. Fire is responsible for most explosives facility accidents so controlling sources of fire and being prepared to deal with them is of utmost importance. Poor housekeeping is often not a hazard itself, but may be a sign of deeper safety problems and should cause an inspector concern. The inspector should always make sure plans are in place for effective emergency response.

BIBLIOGRAPHY

Cote, A. E., ed., *Fire Protection Handbook,* 18th ed., NFPA, Quincy, MA, 1997.

Safety Library Publication No. 3, *Suggested Code of Regulations for the Manufacture, Transportation, Storage, Sale, Possession and Use of Explosive Materials,* Institute of Makers of Explosives, Washington, DC, April 1996.

NFPA Codes, Standards, and Recommended Practices

See the latest version of The NFPA Catalog for availability of current editions of the following documents.

NFPA 70, *National Electrical Code®*

NFPA 490, *Code for the Storage of Ammonium Nitrate*

NFPA 495, *Explosive Materials Code*

NFPA 498, *Standard for Safe Havens and Interchange Lots for Vehicles Transporting Explosives*

Fireworks and Pyrotechnics

CHAPTER 49

Ken Kosanke

CASE STUDY

Fireworks Ignite Barge

Shortly after a public fireworks display began off a heavy steel barge anchored off the Massachusetts coast, a rocket exploded in the air prematurely, showering the barge and cardboard boxes containing reserve fireworks with sparks. Numerous fireworks detonated, burning all 10 people aboard the barge. Rescue personnel had to rescue the workers, including three who jumped into the ocean to escape.

The state fire marshal later cited several violations on board the barge. The deputy chief said two contributing factors were the improper storage of reserve fireworks and inadequate spacing between fireworks for separate displays.

Source: "Fire Watch," *NFPA Journal,* July/August, 1998.

Normal manufacturing and chemical safety issues are covered elsewhere in this manual; accordingly, this chapter addresses only special safety considerations for the manufacture and storage of pyrotechnic materials. As a group, facilities that manufacture or distribute fireworks or other civilian pyrotechnic products present a special and difficult challenge to inspectors because of the vast range of potential hazards associated with these products. During a fire, some pyrotechnic materials may burn with little violence, even in very large quantities; others may explode with a force rivaling high explosives, even in relatively small quantities. Some pyrotechnic materials are difficult to ignite, making accidental ignition extremely unlikely; others are so easily ignited that exceptional precautions must be exercised constantly during their preparation and use.

Obviously, the safety precautions needed may vary greatly from one type of facility to another and between different areas of the same facility. For these reasons, and because it is unusual for inspectors to be familiar with the characteristics of the vast number of different pyrotechnic materials, they should enlist the active cooperation and participation of the facility's safety manager as well as others with specialized knowledge in pyrotechnics.

The range of civilian pyrotechnic products is vast, and the differences in their manner of functioning is extreme. NFPA 1124, *Code for the Manufacture, Transportation, and Storage of Fireworks and Pyrotechnic Articles,* and NFPA 1125, *Code for the Manufacture of Model Rocket and High Power Rocket Motors,* address only fireworks, pyrotechnic articles, and model rocket motor manufacturing. Accordingly, in previous editions, this chapter specifically addressed only such manufacturing sites. Included was the suggestion that what constituted safe practice at fireworks, pyrotechnic article, and rocket motor manufacturing sites generally would also constitute safe

Ken Kosanke has a Ph.D. in chemistry and has been conducting research on pyrotechnics for 25 years. He serves on three NFPA technical committees dealing with pyrotechnics and explosives.

practice at other sites and for other civilian pyrotechnic materials and products. Although this advice is still good, work has recently begun to further define the requirements for retail and wholesale consumer fireworks sales locations. Thus inspectors will want to consult the new edition of NFPA 1124 (when it is available) for further guidance.

DEFINITIONS

Civilian Pyrotechnic Products

Civilian pyrotechnic products are commercial nonmilitary products deriving at least part of their energy from the burning or explosion of a pyrotechnic composition. Some examples are fireworks, model rocket motors, highway flares, signaling smoke devices, pest and predator control devices, pyrotechnic articles (theatrical special effects), and automobile airbags.

Consumer Fireworks

Consumer fireworks are intended for use by the general public where state and local laws permit. They contain limited quantities of pyrotechnic composition and include cone and cylindrical fountains, roman candles, small sky rockets, mines and shells, certain firecrackers and sparklers, and other devices (Figure 49-1). Consumer fireworks are classified as Explosives 1.4 G by the Department of Transportation (DOT). Unless specifically exempted, these devices must comply with construction, performance, labeling, and chemical composition regulations promulgated by the DOT and the Consumer Product Safety Commission (CPSC).

Display Fireworks

Display fireworks are larger and more powerful than consumer fireworks and are intended for use by only trained operators in outdoor displays for which the appropriate licenses or permits have been obtained. The DOT classifies display fireworks as Explosives 1.3 G.

Magazine

A magazine is a storage area for explosives regulated by the Bureau of Alcohol, Tobacco and Firearms (BATF). Most often this will be a specially constructed building situated some distance from normal work areas. However, magazines may also be smaller boxes or structures (portable magazines) and even specially constructed cabinets within work areas (indoor magazines).

FIGURE 49-1 Types of Consumer Fireworks

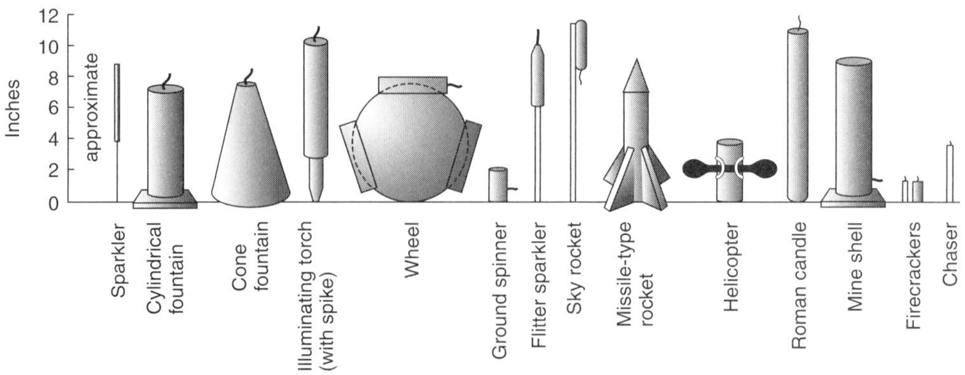

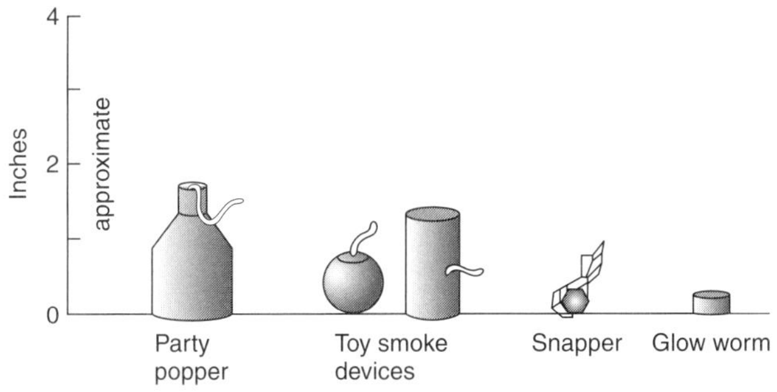

FIGURE 49-2 Types of Novelties and Trick Noisemakers

Model Rocket Motor

This solid propellant motor conforms to the standards for rocket motors as set forth in NFPA 1122, *Code for Model Rocketry*. Model rocket motors are classified as a Toy Propellant Device 1.4 G by DOT. By DOT exemption, however, most manufacturers may ship model rocket motors under the classification of flammable solid.

Novelties and Trick Noisemakers

Novelties and trick noisemakers are not considered fireworks for regulatory purposes. When test data are submitted to the DOT, the DOT may classify these devices as Explosive 1.4 S, or they may not regulate them as hazardous materials at all (Figure 49-2).

Process Building

A process building is any mixing building, any building in which pyrotechnic composition is pressed or otherwise prepared for finish or assembly, or any finishing or assembly building.

Pyrotechnic Article

A pyrotechnic article is a theatrical pyrotechnic special effect, intended for use by trained operators during stage performances for which the appropriate licenses or permits have been obtained. Pyrotechnic articles often produce effects similar to fireworks but with greater reliability, such that adequate safety of the viewing public can be accommodated using reduced separation distances. Like consumer fireworks, these items often contain rather limited quantities of pyrotechnic composition and are classified as Explosives 1.4 G by DOT. Unless specifically exempted, these items must comply with construction, performance, labeling, and chemical composition requirements of DOT and NFPA 1126, *Standard for the Use of Pyrotechnics before a Proximate Audience.*

Pyrotechnic Composition

A pyrotechnic composition is a chemical mixture containing fuel(s) and oxidizer(s) in the proper proportion that is designed to produce, when ignited, a visible or audible effect by a combustion or deflagration process. The effects include the production of light, color, smoke, motion, or noise.

Salute Powder

Salute powder is a pyrotechnic composition that makes a loud report when ignited and constitutes the sole pyrotechnic mixture in a salute. Flash powder is a type of salute powder.

PYROTECHNIC SAFETY

General Considerations

Obviously, inspectors examine a facility to evaluate its safety, and safety is said to be achieved when risk is reduced to an acceptable level. In essence, there are two components to risk: probability and consequence. Thus, risk can be managed either by minimizing the probability of an accident or by minimizing the consequences of that accident. In the manufacture, storage, and use of pyrotechnic materials, it is appropriate to look to minimizing both probability and consequence.

Generally, pyrotechnic accidents are the result of unintentional ignitions, and the consequence of an accident is directly related to the amount of material accidentally ignited and to the number of persons exposed to the accident. By introducing measures that reduce the chance of accidental ignitions and that keep the amount of pyrotechnic materials and the number of people in work areas to a minimum, a facility will remain relatively safe.

Concern for safety should be greatest during manufacturing operations such as mixing, pressing, and loading pyrotechnic compositions. This is the time when the probability of accidental ignition and the quantity of exposed pyrotechnic materials is the greatest and when propagation in the event of an ignition is likely to be most rapid. Accordingly, these operations deserve the most attention during inspection.

Once a pyrotechnic composition has been loaded into a container and is no longer exposed, the probability of an accidental ignition is greatly reduced, as is the rate of fire spread if there is an ignition. This is particularly true for products, such as consumer fireworks, pyrotechnic articles, and model rocket motors, that contain limited amounts of pyrotechnic composition. Tests on large quantities of consumer fireworks have shown that, while they may burn vigorously once ignited, they do not pose a mass explosion hazard. Thus, finishing operations such as labeling and packaging, during which little if any pyrotechnic composition is exposed, are lower hazard activities, as is the storage of completed consumer fireworks, pyrotechnic articles, and model rocket motors.

For larger pyrotechnic items, such as display fireworks, the probability of accidental ignition during finishing and storage is similarly reduced. However, larger pyrotechnic items have a greater individual explosion potential than consumer fireworks. Once started, a fire in such items will spread very rapidly, and an explosion is likely during a mass fire.

Inspection Guidelines—Manufacturing Sites

The U.S. Bureau of Alcohol, Tobacco and Firearms (BATF) regulates the commercial production of pyrotechnic composition, its use in manufacturing, and the storage and use of larger items, such as display fireworks. To help their inspectors prepare to examine fireworks facilities, the BATF has produced a videotape (see Bibliography). Any fire service inspector not familiar with pyrotechnic manufacturing would benefit from viewing this video before conducting an inspection.

While examining a facility, inspectors must follow the safety requirements established by that facility. For example, they may have to wear cotton clothing to minimize the generation of static electricity, limit the use of cameras and other battery-operated equipment and remove cigarette lighters and matches before entering the facility.

Cleanliness. Cleanliness and good housekeeping are critical in reducing the likelihood of an accidental ignition and the spread of a pyrotechnic accident, especially in process buildings and magazines. Significant quantities of pyrotechnic dust present a real danger; however, the nature of any dust that is present will often not be readily apparent. Nonetheless, any significant quantity of dust in a work area should be of concern and should generally be removed, whether it is a pyrotechnic dust or not. If it is not a pyrotechnic dust and does not represent an ignition hazard, it is still almost certain to represent an industrial hygiene threat if inhaled by personnel. Spills of pyrotechnic composition must be cleaned up immediately. Contaminated work clothing should not be worn outside the facility and should be laundered frequently.

Ignition Sources. Inspectors must consider all potential ignition sources in process buildings and magazines. In process buildings, electrical equipment must be suitable for use in hazardous locations in accordance with Article 502 of NFPA 70, *National Electrical Code®*, 2002 edition. There must be no open-flame heating and no high-temperature heating surfaces. Spark-producing tools and surfaces are of concern as a potential source of accidental ignition; however, in some cases, the use of small spark-producing tools, such a box-cutting knife, will be unavoidable. Areas used for testing components or completed devices by burning or firing should be sufficiently well separated or protected from manufacturing and storage areas such that the testing may proceed without risking a fire in those areas. Trash should not be allowed to accumulate anywhere in the facility, and brush and dry grass should be controlled.

Separation Distances. In large measure, it is the distance between work and storage areas and the public that determines whether and how fast a fire or explosion will spread from one work area to another or become a risk to public safety. The inspector should consult NFPA 1124 and NFPA 1125 for the recommended separation distance at fireworks, pyrotechnic articles, and rocket motor manufacturing and storage facilities. NFPA 1124 generally is applicable to facilities manufacturing other civilian pyrotechnic products, as well.

Quantity and Occupancy Limits. The severity of a pyrotechnic accident is directly related to the amount and explosive potential of the pyrotechnic composition involved and the number of people exposed. For this reason, the BATF has limited the amount of pyrotechnic composition and salute powder allowed in any process building to 500 lb (226.8 kg) and 10 lb (4.5 kg), respectively. However, the amount of pyrotechnic material present should always be the minimum practical amount; large quantities of raw materials and finished product should not be allowed to accumulate in any process building.

There should be no more people present in any process building or magazine than are needed to conduct the operation. A facility with a number of small process buildings, each with one or two persons working in them, is greatly preferred over a facility with large process buildings with many workers in each building.

Storage Requirements. At the end of the workday, all pyrotechnic materials must be stored properly. Display fireworks, high power rocket motors, and pyrotechnic compositions must be stored in magazines that comply with BATF regulations, while consumer fireworks and model rocket motors may be stored in a variety of structures. See NFPA 1124 and NFPA 1125 for guidance. Raw chemical materials also must be stored and handled safely.

Process Building Egress Requirements. A fire involving mass detonating bulk pyrotechnic composition or display fireworks is likely to spread so rapidly that sprinklers may not be effective. For this reason, the building should be constructed in such a way that personnel can leave their work areas easily and reach relative safety. Process buildings must be one story high with no basement. Buildings or rooms larger than 100 ft^2 (9.3 m^2) must have access to at least two exits, which should be located such that no point within the building or room is more than 25 ft (7.6 m) away.

Emergency Plan. There should be a review of the facility's emergency procedures and evacuation plan in case of fire or other pyrotechnic accident. The inspector should also verify that there has been compliance with any applicable so-called community-right-to-know requirements, such as for the storage of hazardous materials and the siting of magazines.

Inspection Guidelines—Consumer Fireworks Retail Sales Outlets

The information presented here is limited because the development of specific requirements is still at an early stage within the NFPA code development process.

Buildings. Buildings should have at least two exits remote from one another. The number of exits, the travel distance to exits, and the arrangement and configuration of consumer fireworks in the sales area should be such that easy egress can be made in the case of a fire emergency. Buildings should be located at least 100 ft (30.5 m) from liquid fuel dispensing facilities and should not have combustible weeds or trash near by. In buildings not used exclusively for the retail sale of consumer fireworks, only those types approved by the authority having jurisdiction are permitted. New buildings should be one story in height.

Quantities—New Buildings. Limits have been established for the quantity of consumer fireworks in new occupancies. In indoor sales areas, the quantity is limited to 50 lb (22.7 kg) net pyrotechnic composition [200 lb (90.7 kg) of fireworks, gross weight]. In buildings protected by an approved automatic sprinkler system, the quantity is limited to 100 lb (45.4 kg) net pyrotechnic composition [400 lb (181.4 kg) of fireworks, gross weight]. In open air mercantile operations as defined by NFPA *101*®, *Life Safety Code*®, the quantity is limited to 200 lb (90.7 kg) net pyrotechnic composition [800 lb (362.9 kg) of fireworks, gross weight]. Consumer fireworks in excess of these quantities is permitted providing that either public access to the fireworks is not permitted or the mercantile occupancy meets the requirements specified in NFPA *101* for when the contents are classed as high hazard.

Quantities—Existing Buildings. The only quantity limitations for consumer fireworks in existing buildings is that the total height of the retail display should not exceed 5 ft (1.5 m) and that the retail display area should not exceed 3000 ft^2 (279 m^2), unless it is protected throughout by an approved automatic sprinkler system.

SUMMARY

Obviously, in this inspection overview, it is not possible to present all of the relevant NFPA and various government agency requirements for inspections at fireworks and pyrotechnic sites. Similarly it is not possible to address the specific requirements for the full variety of possible inspection locations. Some of these sites will have such limited quantities of low hazard materials as to be mostly benign; other sites will pose major hazards because of the type and quantity of material. In most cases, using common sense and working cooperatively with the facility safety personnel may be sufficient to assure that a high level of facility and community safety has been achieved. However, when needed, consideration should be given to including a person on the inspection team with specific knowledge and experience with the type and level of hazards expected at the site. In addition, it is always appropriate to have other more complete and supporting codes and standards available at the time of the inspection (see Bibliography).

As a final thought, although not specifically mentioned in safety codes and standards, carelessness and especially complacency probably represent the greatest threat to safety at a fireworks or pyrotechnic site. Evidence of possible problems in these areas will often be exhibited in such simple and obvious things as a general lack of cleanliness, the presence of clutter in work areas, or a generally poor state of repair of buildings and equipment. These may seem like relatively minor safety deficiencies, but they may serve to indicate an inappropriate attitude regarding safety and the need to look more carefully for greater problems and deficiencies at that site.

BIBLIOGRAPHY

Bureau of Alcohol, Tobacco and Firearms, *Explosives Law and Regulations,* ATF Publication P5400.7, Washington, DC, 2000.

Bureau of Alcohol, Tobacco and Firearms, *Fireworks Manufacturing & Safety,* Washington, DC, 1992. (video)

Conkling, J. A., "American Fireworks Manufacturing: An Industry in Transition," *Fire Journal,* September 1986, pp. 41–47, 66–70.

Consumer Product Safety Commission, *Commerce and Trade,* Title 16, *Code of Federal Regulations,* Parts 1500 and 1507.

Department of Transportation, *Transportation,* Title 49, *Code of Federal Regulations,* Parts 100–177.

Poulton, T. J., and Kosanke, K. L., "Fireworks and Their Hazards," *Fire Engineering,* June 1995, pp. 49–64.

NFPA Codes, Standards, and Recommended Practices

See the latest version of The NFPA Catalog for availability of current editions of the following documents.

NFPA 70, *National Electrical Code®*

NFPA *101®*, *Life Safety Code®*

NFPA 430, *Code for the Storage of Liquid and Solid Oxidizers*

NFPA 495, *Explosive Materials Code*

NFPA 499, *Recommended Practice for the Classification of Combustible Dusts and of Hazardous (Classified) Locations for Electrical Installations in Chemical Process Areas*

NFPA 1122, *Code for Model Rocketry*

NFPA 1124, *Code for the Manufacture, Transportation, and Storage of Fireworks and Pyrotechnic Articles*

NFPA 1125, *Code for the Manufacture of Model Rocket and High Power Rocket Motors*

NFPA 1126, *Standard for the Use of Pyrotechnics before a Proximate Audience*

Heat-Utilization Equipment

CHAPTER 50

Richard A. Gallagher

CASE STUDY

Boiler Explodes at Automobile Plant

An explosion and fire occurred when a buildup of natural gas ignited in a giant boiler at the huge Ford Motor Company plant in Dearborn, Michigan, on February 1, 1999. The explosion initially killed 1 and injured more than 20 workers (the number of deaths rose subsequently). At the time of the explosion, workers were shutting down the boiler in readiness for an annual inspection. The force of the explosion split open the 60-ft–high furnace and blasted workers with heat as well as spraying them with superheated water that, experts estimate, likely topped 2000°F.

According to state officials, it was a "furnace explosion," with a buildup of natural gas in the firebox inside the boiler igniting. Hot slag was given as a possible heat source that could have detonated the natural gas buildup. The explosion and fire disrupted electricity, steam, mill water, and hot, high-pressure air service to a half dozen plants in the 1100-acre manufacturing complex. According to an expert, the incident could turn out to be one of the costliest single-site insurance losses in the United States.

Source: Information from various sources, including *www.cnn.com*, *www.msnbc.com*, *www.freep.com*, *www.detnews.com*, and *www.auto.com/industry*.

Many industrial processes employ heat-utilization equipment to provide energy to operate a process or heat to alter the physical characteristics of materials being processed. (For a discussion of building heating systems, see Chapter 10.) The fire and explosion hazards presented by heat-utilization equipment are related to their operation at elevated temperature in conjunction with their use of flammable and combustible fuels, flammable and combustible materials in process, and flammable special processing atmospheres. This chapter offers the inspector an understanding of the hazards associated with heat-utilization equipment and the features and controls that should be reviewed during the course of an inspection.

GENERAL

For the purpose of this inspection guide, heat-utilization equipment can be divided into two broad categories. *Boilers* are used for the generation of hot water or steam for process use. *Ovens* and *furnaces* are used for the commercial and industrial processing of materials above normal ambient temperature. NFPA uses the terms ovens and furnaces interchangeably.

Richard A. Gallagher has a degree in civil engineering from the University of Delaware, works for the Industrial Risk Insurers as a field services technical advisor, and has been a member of the NFPA Technical Committee on Ovens and Furnaces for over 10 years.

Boilers for Process Use

Many industrial facilities depend on boilers to generate hot water or steam to support manufacturing operations (Figure 50-1). For example, hot water may be used to heat a mixing vessel or steam can be used to drive a turbine that in turn drives

FIGURE 50-1 Installation of Combination Oil and Gas Firing in Large Electric Utility Boiler

an electric generator. The NFPA code on boilers addresses boilers fired at rates at or above 12,500,000 Btu/hr. For smaller boilers, the inspector will need to seek guidance from other codes or standards adopted by their jurisdiction.

As of the 2001 edition, NFPA has merged their various boiler standards into a single code, NFPA 85, *Boiler and Combustion Systems Hazards Code*. The purpose of this code is to establish minimum requirements for the design, installation, operation, and maintenance of boilers, their fuel-burning systems, and related systems to contribute to safe operation and the prevention of furnace explosions and implosions. Inspectors should consult this code for detailed guidance.

Ovens and Furnaces

Ovens and furnaces are used in a wide variety of industrial applications. Table 50-1 offers several examples.

In a batch oven (Figure 50-2), work is loaded, processed through a heating cycle, and then unloaded. In a continuous furnace (Figure 50-3), work enters one end of the furnace, is processed as it moves through the furnace, and then exits the furnace at the other end.

All ovens and furnaces share the hazards associated with their heating systems. Beyond this, however, ovens and furnaces are classified based on additional process hazards.

TABLE 50-1 Example of Industrial Applications

Curing lumber (lumber kiln)	Annealing glass bottles
Baking painted metal parts	Heating aluminum ingots
Curing rubber conveyor belts	Integral quench furnace
Drying cloth	Annealing steel, hydrogen atmosphere
Drying water washed metal parts	Treating metal parts in molten salts
Melting metals	Heating materials under vacuum

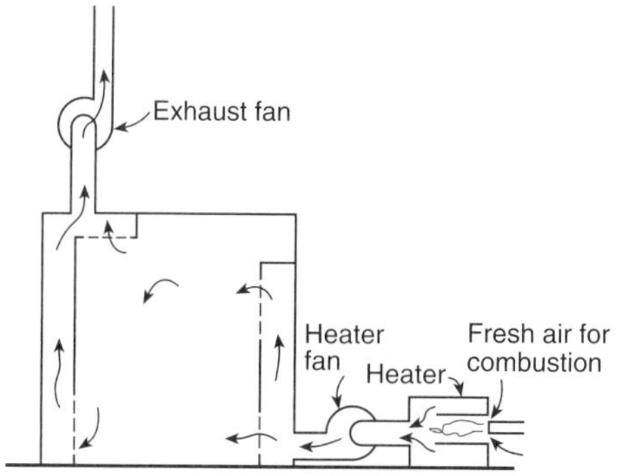

FIGURE 50-2 Direct-Fired Batch Oven with External, Nonrecirculating Heater
Source: NFPA 86, 1999, Figure A-2-1(a).

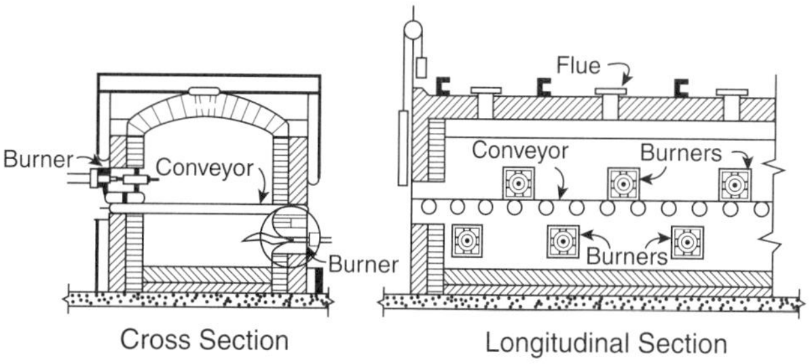

FIGURE 50-3 Direct-Fired Continuous Furnace with Multiple Internal Burners
Source: NFPA 86, 1985, Figure 1-4.19.

- **Class A:** Class A ovens or furnaces possess a fire or explosion hazard due to either the product they handle (e.g., cloth dryer) or the solvents vapors they generate during a drying or curing process (e.g., automobile paint bake oven).
- **Class B:** Class B ovens or furnaces noncombustible products (e.g., glass bottles) in air.
- **Class C:** Class C ovens or furnaces use a special atmosphere (e.g., nitrogen, hydrogen, and dissociated ammonia). The special atmosphere will exclude air to protect the work from oxidation. Also, the special atmosphere can alter or treat the surface of the work in process to change its physical characteristics (e.g., hardening of steel parts).
- **Class D:** Class D ovens or furnaces, referred to as vacuum furnaces, operate below atmospheric pressure at some point in their process cycle. Such furnaces may process any product or material. These furnaces can use special atmospheres introduced when the furnace is above or below atmospheric pressure.

NFPA develops three standards that address ovens and furnaces (Table 50-2). Because the heat processing of materials involve serious fire and explosion hazards, these standards have been developed to provide requirements on the location, arrangement, installation, control, and use of ovens and furnaces. Within the scope of these standards, an oven is any heated enclosure used for commercial and industrial processing of materials. This definition is extremely broad and can cover many processes. For detailed guidance, inspectors should consult the individual standards.

Boilers for Process Use

EQUIPMENT LOCATION

Boilers will typically be found in isolated or cutoff boiler rooms or detached powerhouses. They are not directly involved in the manufacturing process, making their separation from the rest of the plant possible. Because only the generated steam or

TABLE 50-2 NFPA Oven and Furnace Standards

NFPA 86	*Standard for Ovens and Furnaces*
NFPA 86C	*Standard for Industrial Furnaces Using a Special Processing Atmosphere*
NFPA 86D	*Standard for Industrial Furnaces Using Vacuum as an Atmosphere*

hot water is used in the process, the steam or hot water can be readily delivered by pipe to the point of use.

Ovens and Furnaces

Ovens and furnaces, in contrast, are directly associated with the processing of materials in production. Ovens and furnaces are typically located within a production area, separated only by some physical distance from other process equipment. When deciding on the location of an oven or furnace, planners consider factors such as the following:

- Where possible, locating the oven or furnace in a separate cutoff area
- Where a separate cutoff area is not feasible, providing adequate separation from paint spray booths, dipping or coating operations, external quench tanks, and combustible storage
- Where molten materials are involved (molten metal, molten glass, etc.), considering the effects of a release or spill of the molten material; providing containment systems to limit the spread of spills; and protecting combustible building construction, exposed steel building columns, or exposed steel furnace supports from thermal damage
- Where furnaces have integral quench tanks using combustible oils, providing spill containment and providing supplemental fixed fire protection (e.g., a carbon dioxide extinguishing system) in pits, at filter stations, immediately outside of the furnace vestibule door, and within any hood and associated exhaust ductwork over the furnace vestibule door

COMBUSTION CONTROLS

Burner Management Systems

While burner management systems for heat-utilization equipment will vary with regard to needed safety controls, they all share the common objective of initiating and maintaining a user-friendly fire. Safety controls are provided to manage the three sides of the fire triangle. Figure 50-4 shows a fire triangle and the typical controls associated with each side of the triangle. The flame detector (boiler term) or flame safeguard (oven and furnace term), which is the central safety control of the entire system, is shown in the center of the triangle.

Preignition Purge

In general the interior volume of any fuel fired heat-utilization equipment is purged with air before the ignition of burners is allowed. The objective is to remove accumulated fuel vapors or other flammable vapors that may have leaked into the equipment during the preceding idle period.

Single Burner Boilers. The preignition purge interval for a single burner boiler is based on the time needed to accomplish several air volume changes within the boiler. The number of air changes will depend on the type of boiler. For fire tube boilers (boilers in which the flame and products of combustion are contained within tubes surrounded by water being heated), four air changes must be provided. For water tube boilers (boilers in which the water is contained within tubes surrounded by the fire and products of combustion from the burner) eight air changes must be provided. The larger volume purge requirement for the water tube boiler is due to its internal configuration. The water tube boiler is essentially a large box that encloses the water-filled tubes. The larger volume purging is

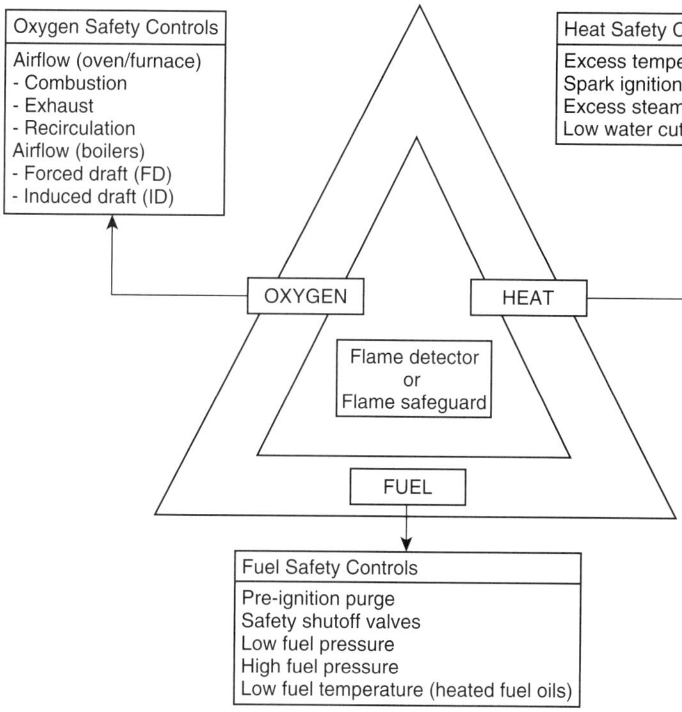

FIGURE 50-4 Heat, Fuel, and Oxygen Safety Controls Applied to Heat-Utilization Equipment

needed to flush out accumulated fuel that may have spread into the corners of the enclosure.

Multiple Burner Boilers. For multiple burner boilers, the required purge interval is 5 minutes or five volume changes, whichever is longer.

Ovens and Furnaces. For ovens and furnaces, four volume changes are needed to accomplish the purge.

Complete preignition purge intervals must occur before any attempt is made to ignite burners on heat-utilization equipment (boilers, ovens, or furnaces). Even when heat-utilization equipment experiences nuisance safety shutdowns, each restart must include another purge interval. The purge interval, especially repeated purge intervals, will encroach on valuable production time. If the encroachment becomes significant, operators or maintenance personnel may be tempted to reduce or eliminate the purge interval. Because tampering with the purge interval is a dangerous practice that can lead to explosions, inspectors should be on the lookout for signs of such tampering.

For heat-utilization equipment with a single burner, the purge interval may be regulated by the flame safeguard. With this arrangement, the purge interval is fixed and can only be altered by modifying the flame safeguard (e.g., replacing a plug-in purge time module). This arrangement greatly reduces the potential for purge interval tampering.

With more complex heat-utilization equipment—for example, multiple burner systems—a purge timer is typically provided that is physically external to the flame safeguard. Listed purge timers are adjustable devices (e.g., 0 to 30 minutes), and the adjustable feature introduces the opportunity for tampering. Finding purge timers adjusted to zero is not uncommon. Such a situation is a very serious and should be addressed promptly.

Fans

Fire tube boilers and some water tube boilers may have a single forced draft fan to provide combustion air. Larger water tube boilers, ovens, and furnaces often have multiple fans. Boilers fans are typically referred to as forced draft (FD) fans and induced draft (ID) fans. Oven and furnace fans are referred to as combustion air, exhaust air, and recirculation air fans. All fans needed for safe operation must be proven during the purge interval and during operation. The manufacturer of the heat utilization equipment should establish which fans are needed for purge and operation.

Proving Air Flow—Devices. Proving the operation of fans will be accomplished with devices such as the following:

- Pressure switch: A pressure switch will sense air pressure after the fan and will be actuated when the airflow from the fan is adequate.
- Suction switch: A suction switch will sense negative pressure before the fan and it will be activated when airflow into the fan is adequate.
- Differential pressure switch: A differential pressure switch will sense the pressure on each side of a fan or other feature such as an orifice plate. The differential pressure switch will be activated when the airflow through the fan or other feature is adequate.
- Airflow switch: An airflow switch has a paddle or sail that is inserted into the air stream. The switch is activated when the airflow past the paddle or sail is adequate.

Proving Air Flow—Device Setting. Many pressure switches, suction switches, and differential pressure switches are adjustable devices. Equipment records should clearly indicate the appropriate setting of each of these devices. If, during an inspection, they are found to have been adjusted to an extreme upscale or downscale setting, a question should be raised. The switch may never have been adjusted at the time of installation or the switch may simply have been adjusted to the extreme to resolve a nuisance trip problem.

Equipment Fuel Piping

The pipe, fittings, and devices that make up the fuel piping at a piece of heat-utilization equipment are referred to as the *fuel train*. Figures 50-5 and 50-6 illustrate typical arrangements for a single burner fuel gas train and a single burner fuel oil train, respectively.

For main and pilot fuel gas trains, double safety shutoff valves are always needed. For boilers covered in NFPA 85, the main burner and pilot fuel gas trains must also include a vent between the double safety shutoff valves. The vent line must be equipped with a normally open valve wired in series with the safety shutoff valves. When the safety shutoff valves are deenergized and closed, the vent valve will open. This will allow any fuel gas that leaks past the first safety shutoff valve to follow the vent line to atmosphere rather than attempt to force its way past the second safety shutoff valve into the idle unit.

The local authority having jurisdiction may not permit the use of normally open vents. In such a case, listed valve proving systems are permitted by NFPA 85. A valve proving system performs a pressure test on the safety shutoff valves during start-up to verify that the valves do not leak beyond permitted limits.

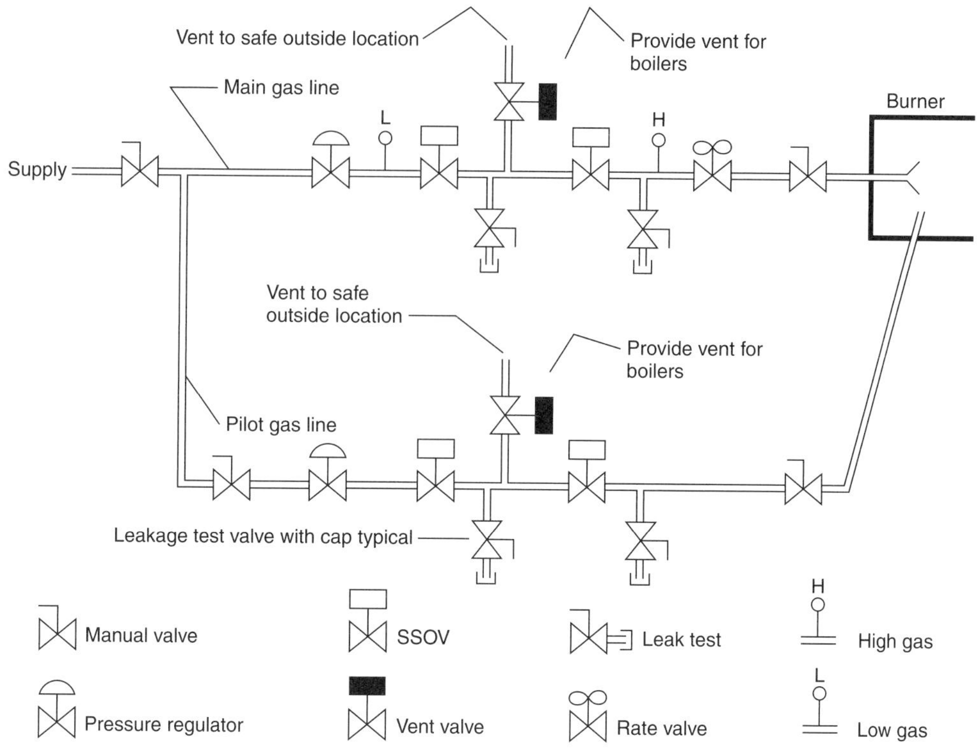

FIGURE 50-5 Typical Fuel Gas Train

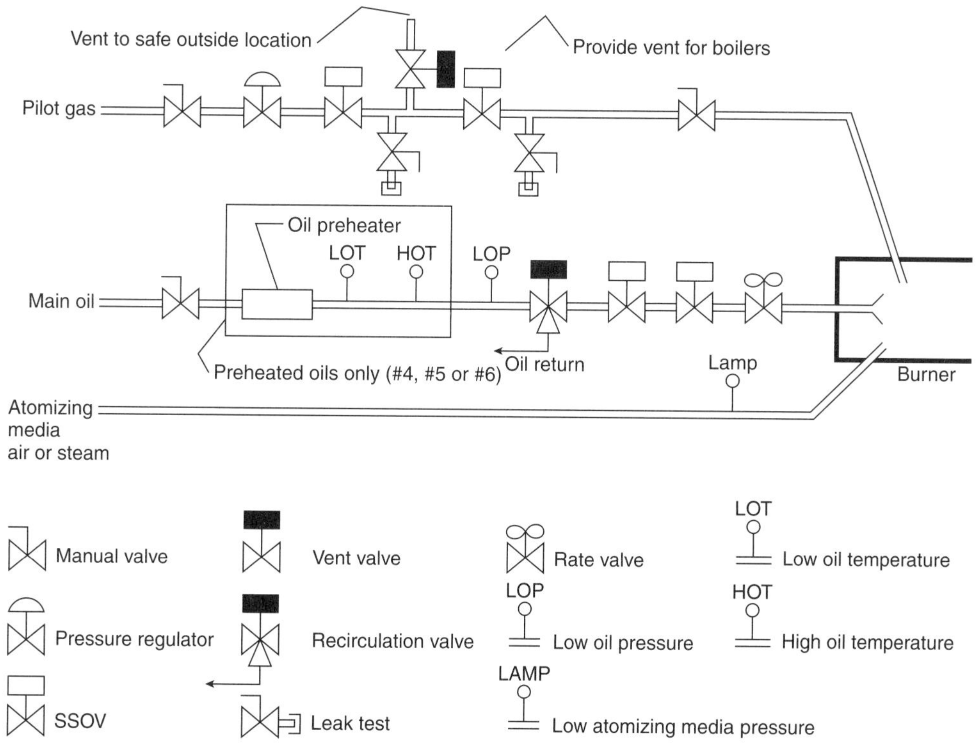

FIGURE 50-6 Typical Fuel Oil Train

In some instances, the inspector may encounter heat-utilization equipment fired on pulverized coal. Figure 50-7 illustrates one possible arrangement of a coal-fired system.

Fuel Gas Pressure Regulators and Fuel Gas Pressure Switches

Fuel gas pressure regulators and fuel gas pressure switches contain rubber diaphragms. Fuel-gas pressure acts on the bottom side of the diaphragm. A spring or microswitch, along with atmospheric pressure, acts on the top side. To maintain atmospheric pressure on the topside, a vent is provided in the device housing. This vent needs to be piped to a safe outdoor location so that fuel gas will not be released into a building should the diaphragm fail.

Flame Safeguards

Heat-utilization equipment that is equipped with just a single burner needs simple burner management. In particular, it needs only a single flame safeguard. It is not unusual, however, for equipment to have numerous burners. Numerous burners require numerous flame safeguards and increased safety control complexity.

Currently, four types of flame-sensing devices are listed for proving the presence of flame at a burner, as shown in Table 50-3.

FIGURE 50-7 Direct-Firing System for Pulverized Coal
Source: *Fire Protection Handbook*, NFPA, 1997, Figure 3-6l.

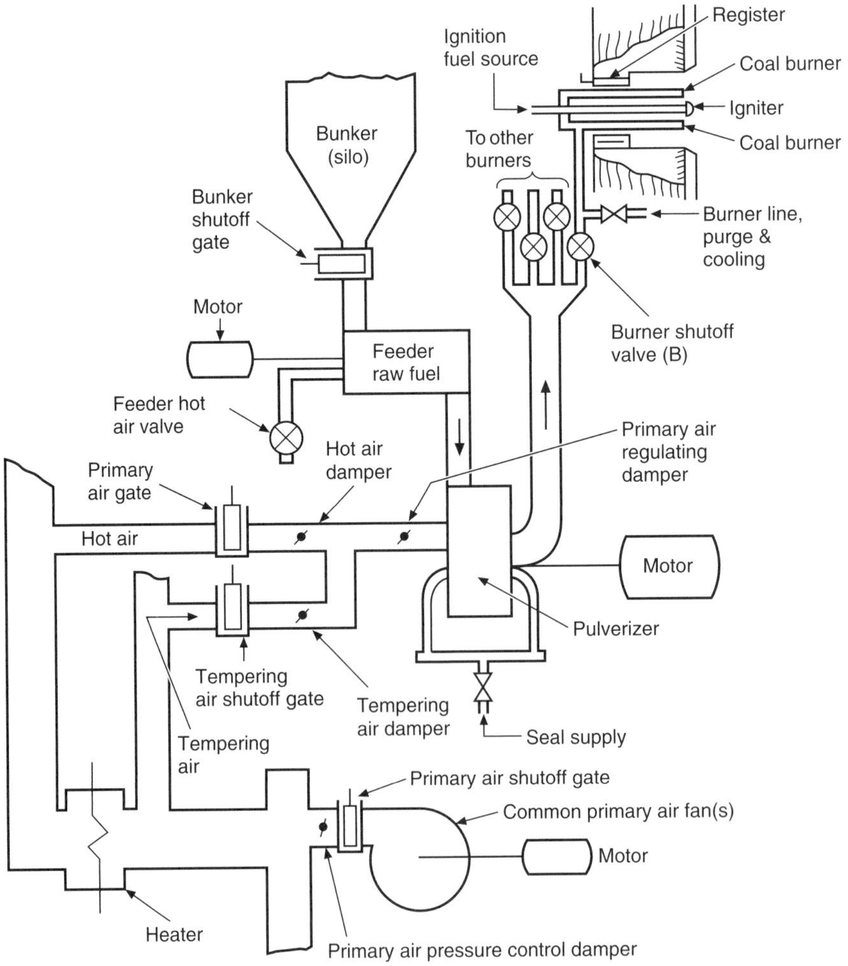

TABLE 50-3 Flame-Sensing Device

Type Scanner	Limitations	Problems
Ultraviolet (UV) Scanner	Fuel gas or oil	Can read spark as flame
Flame Rod	Fuel gas only	Oil will cause carbon buildup
Infrared (IR) Scanner	Fuel gas or oil	Can read refractory glow as flame
Photoelectric Cell	Oil only	Fuel gas is invisible

Although UV scanners are popular and very common flame sensing devices, they can fail. In some failure modes, the UV scanner may report the presence of flame whether flame is or is not present. This concern is addressed in two manners.

First, self-checking flame safeguards are available. These devices include a shutter assembly that will test the scanner by periodically closing the shutter while the scanner is in normal use. If the flame signal does not disappear when the shutter is closed, a safety shutdown occurs.

Second, most flame safeguards offer a "safe start check" diagnostic. During each burner start sequence, the safe start check diagnostic verifies that the UV scanner does not see flame. If the UV scanner is reporting flame, the start sequence is not allowed to proceed.

For boilers covered by NFPA 85, the use of a self checking UV scanner is required unless the boiler operates for periods less than 24 hours and a safe start check feature is provided.

For ovens and furnaces, it is a good practice to use a self-checking UV scanner where burner operate for more than 24 hours at a time; however, this feature is not required. In all cases, a safe start check feature is required.

Automatic Versus Supervised Manual

Many ovens, furnaces, and boilers (mostly smaller units with single burners) are arranged to cycle on and off automatically. The sequence is standardized. Off-the-shelf flame safeguards are available that will automatically provide the logic to step the unit through this standard start-up, operation, and shutdown sequence. During the operating phase, the flame safeguards can also allow the unit to modulate the burner firing rate as heat demands fluctuateand then shut down when the call for heat is satisfied.

Often larger boilers and most ovens and furnaces are arranged so that a trained operator will cycle the unit through a mandatory sequence of manual steps (supervised manual). This sequence is frequently too complex for off-the-shelf flame safeguards for the following reasons:

- There may be multiple burners requiring multiple flame safeguards.
- There may be more than one combustion air fan to start. One of the mandatory start-up steps completed by the trained operator may be the manual start of all fans needed for safe operation.
- There may be safety steps in the start-up sequence that are customized to a particular oven or furnace. These may include proving the position of furnace doors, proving the movement of furnace conveyor systems, and so on.

Although each step in the start-up sequence may be manually initiated, the supervised manual system logic require each step to be performed in the mandated sequence. The operator will not be able to deviate from this required sequence. The NFPA boiler code and the NFPA oven and furnace standards do not permit the use of manual operating systems where the required operating sequence is not enforced.

LISTING OF SAFETY DEVICES

The NFPA standards for heat-utilization equipment require that safety devices be listed. The official NFPA definition of *listed* is as follows:

> Equipment or materials included in a list published by an organization acceptable to the authority having jurisdiction and concerned with product evaluation, that maintains periodic inspection of production of listed equipment or materials and whose listing states either that the equipment or material meets appropriate standards or has been tested and found suitable for use in a specified manner.

Where a concern arises regarding the suitability of a safety device, an evaluation process is needed to resolve that concern. The following steps can be applied:

- **Step 1:** Is the device listed by a testing laboratory acceptable to the authority having jurisdiction?
- **Step 2:** Is the listed device being used for its intended purpose? Listed devices are specifically listed for an intended purpose. The listing organization will describe the intended purpose and in some cases limitations on its intended use in its listing book. This information should also be available from the device manufacturer. For example, if a safety shutoff valve is listed for use with only fuel oil, using that valve on a fuel gas line would be a violation of the listing requirement.
- **Step 3:** Is the device being used in accordance with the manufacturer's guidelines? Often the listing requirements will stipulate that the manufacture's guidelines be followed. These guidelines may describe how to install the device (e.g., a pressure switch must sit upright), limits on the conditions where the device is installed (e.g., maximum ambient temperature), and limits on the use of the device (e.g., maximum working pressure). These guidelines must be followed.

At times, a listed device will not be commercially available for the intended service. In these cases, the NFPA heat-utilization equipment standards allow the use of a nonlisted device. Its use, however, requires approval by the authority having jurisdiction. For example, there are no listed fuel gas safety shutoff valves over the size of 8 in. (203 mm). For valves 10 in. (254 mm) and larger, a nonlisted valve must be selected. The authority having jurisdiction may accept the nonlisted valve as long as it is from a manufacturer who produces smaller valves that are listed and as long as the selected nonlisted valve is built to the same standards as the smaller listed valves.

TRAINED OPERATOR

Where heat-utilization equipment is in use, a qualified operator is an important element in the overall safety system. Terms used in NFPA 86 to describe a qualified operator include *alert, competent, trained, knowledgeable, proficient,* and *effective*. The standard regards the operator be more than an attendant. When control systems fail, the trained operator must step in and take appropriate action to avoid an upset or explosion.

Written Operating Instructions

Operators must have access to written operating instructions at all times. Instruction must include normal start-up, normal shutdown, and emergency shutdown. Where necessary, cold start-up and warm start-up instructions are to be provided. A warm start-up may occur after a weekend where a furnace is not in use. Rather than shutting down and allowing the furnace to cool off, the furnace may be brought back to an idling temperature.

Operator Training

Operator error has been identified as a significant cause or contributing factor to upsets and explosions involving heat utilization equipment. These unwanted events usually result from operators taking actions that deviate from written operating instructions. Operators must be thoroughly instructed and trained on the written operating instructions. When off-normal conditions are detected, written emergency shutdown procedures should be implemented. Deviation from normal or emergency written operating instructions should not be permitted at the operator level.

As part of their training, operators must demonstrate their knowledge of equipment and its operation. Operators must receive ongoing training and testing to verify their maintenance of a high level of proficiency and effectiveness.

Operator Intervention

The cold start-up of heat-utilization equipment should occur under the supervision of a trained operator. When heat-utilization equipment experiences an automatic safety shutdown or a manual emergency shutdown by the operator, restart of the equipment is not permitted without maintenance personnel first identifying and correcting the cause of the off-normal shutdown.

INSPECTION, TESTING, AND MAINTENANCE

Heat-utilization equipment is equipped with numerous control and safety devices intended to maintain the equipment within safe operating boundaries or to cause safety shutdown. If these controls and safety devices are to be reliable, they must be periodically inspected and tested. Maintenance must then be provided as needed.

Establishing a Program

The NFPA standards for heat-utilization equipment place the responsibility for establishing an inspection, testing, and maintenance program with the equipment user. The program should identify the features to be inspected, tested, and maintained. Frequencies for each action are to be based on specific installation needs. Sample guidelines are presented in the annex of each of the NFPA standards listed in Table 50-2 as well as in NFPA 85.

Fuel Gas Safety Shutoff Valve Leak Testing Program

Leak testing of fuel gas safety shutoff valves is an essential program that is often overlooked by the user. A leak test program offers a control over one of the primary causes of explosions in heat-utilization equipment, that is, fuel leakage into idle equipment. The NFPA 86 series of standards requires an annual leak test of each fuel gas safety shutoff valve. NFPA 85 recommends a monthly leak test.

Written leak test procedures describing how the test is to be conducted should be provided. The typical leak test is the bubble test. In this test, one end of a flexible hose is connected to an outlet located between the safety shutoff valve to be tested and a downstream blocking valve. The other end of the flexible hose is submerged into a water bath for several minutes. The number of fuel gas bubbles per minute is measured. The written procedure will define a pass/fail leak standard in terms of bubbles per minute. No valve is perfect; therefore, a minimum acceptable leakage rate is established, for example, 10 bubbles per minute. Because the head

TABLE 50-4 Inspection, Testing, and Maintenance of a Safety Control

Inspection	Verify that the device is at the correct setting.
Testing	Verify that the device operates at the correct set point.
Maintenance	Recalibrate or replace the device if the displayed setting deviates from the actual setting beyond an acceptable limit.

of water above the end of the hose could create sufficient pressure to hold back any leaking gas, the test operator should avoid submerging the hose too deeply into the water bath during the test.

Safety Control Testing

A test procedure must be established for verifying the performance of each safety control. Off-normal conditions should be corrected promptly. Where the control has a design set point, the actions outlined in Table 50-4 should occur.

The inspection of safety devices should include an evaluation of their physical condition. Missing cover plates should be promptly replaced. Corroded devices or devices with obvious physical damage should be repaired or replaced. Tampering with devices must not be allowed. Evidence of tampering could include wire jumpers inserted across contacts, foreign material inserted between contacts, and devices adjusted to improper settings. When adjustable safety devices are found adjusted to extreme high or low settings, it is likely to be an indication that the device is not at its proper setting.

Class A Ovens

Solvent Issues. For Class A ovens that possess a solvent hazard—for example, ovens that drive off a flammable solvent from painted metal parts—there must be a safety design form that describes the solvent hazard the oven was designed to handle (Figure 50-8). The form should state the design solvents and the design solvent introduction rates. Inspections should verify that no unauthorized solvents are being used and that design solvent input rates are not being violated. Deviations should receive prompt attention.

Fire Protection. Class A ovens and their associated ductwork are required to be equipped with fixed fire protection systems such as automatic sprinklers or carbon dioxide in most cases. Where installed, fire protection systems must be inspected, tested, and maintained in accordance with the appropriate NFPA standard for the type of system involved.

HOUSEKEEPING

Combustible Materials

Combustibles materials such as stock and other storage should not be allowed close to heat-utilization equipment. Suitable clearance should be maintained at all times. The NFPA 86 series of standards offers a minimum required separation distance of 2½ ft (0.76 m).

Ovens

Explosion Relief Vents. Where ovens are equipped with explosion relief features, it is essential that these features be free and unobstructed so that they are operable

WARNING — Do not deviate from these nameplate conditions.

SOLVENTS USED ______________________________
For example, alcohol, naphtha, benzene, turpentine

SOLVENTS AND VOLATILES ENTERING OVEN ____________
Gal per batch or per hr

PURGING INTERVAL ______________________________
Minutes

OVEN TEMPERATURE, °F (°C) ______________________

EXHAUST BLOWER RATED FOR ____________ GALLONS (CUBIC METERS) OF SOLVENT PER HOUR OR BATCH AT MAXIMUM OPERATING TEMPERATURES OF ________ °F (°C)

MANUFACTURER'S SERIAL NUMBER __________________

MANUFACTURER'S NAME AND ADDRESS

____

____

____

SAFETY DESIGN FORM
FOR SOLVENT ATMOSPHERE OVENS

THIS OVEN IS DESIGNED FOR THE CONDITIONS AS INDICATED BELOW, AND IS APPROVED FOR SUCH USE ONLY

WARNING — Do Not Deviate From These Conditions

SOLVENTS USED ______________________________
For example, alcohol, naphtha, benzene, turpentine

SOLVENTS AND VOLATILES ENTERING OVEN ____________
Gal per batch or per hr

PURGING INTERVAL ______________________________
Minutes

OVEN TEMPERATURE, °F (°C) ______________________

EXHAUST BLOWER RATED FOR ____________ GALLONS (CUBIC METERS) OF SOLVENT PER HOUR OR BATCH AT MAXIMUM OPERATING TEMPERATURES OF ________ °F (°C)

MANUFACTURER'S SERIAL NUMBER __________________

MANUFACTURER'S NAME AND ADDRESS

____

____

____

Above information is for checking safe performance and is not a guarantee of this equipment in any form, implied or otherwise, between buyer and seller relative to its performance.

FIGURE 50-8 Recommended Safety Design Data Form
Source: NFPA 86, 1999, Figure A-1-7.3.

should an explosion occur. Work loads must not be place in front of doors that are designed to open in the event of an explosion. Beams, columns, piping, ductwork, stock, or other obstructions must not impair the functioning of explosion relief panels in oven walls or roof.

Ductwork. Oven ductwork exposed to combustible dusts or condensed flammable/combustible liquids requires periodic internal inspections and periodic cleaning. Suitable access to ductwork is required. The inspector should consult NFPA 91, *Standard for Exhaust Systems for Air Conveying of Vapors, Gases, Mists, and Noncombustible Particulate Solids,* for additional requirements. Frequencies for duct inspection and cleaning must be determined on a case-by-case basis.

OVENS AND AUTOMATIC SPRINKLERS

When an oven handles a combustible product, a fixed fire extinguishing system is needed within the oven to protect the unit. Chapter 10 of NFPA 86 addresses fire protection in general.

Where a closed sprinkler system is selected, the temperature rating of the selected sprinklers must be in accordance with the requirements of NFPA 13, *Standard for the Installation of Sprinkler Systems.* The selection will be based on the maximum operating temperature of the oven. Closed sprinklers are available for oven with operating temperatures up to 625°F (329°C).

If the oven operating temperature is above the boiling point of water (212°F, 100°C), a dry-pipe system will be needed. If a wet-pipe system were used, the water would be trying to boil within the pipes. This attempt to boil would be countered by the confining strength of the sprinkler system pipe and fittings. Excessive pressures would eventually lead to a failure at some point in the system.

Rubber-gasketed sprinkler fittings may not be appropriate. Such fittings must not be used at temperatures beyond their listing. Their maximum operating temperature is typically 150°F (66°C) unless specifically listed otherwise.

Some authorities having jurisdiction require dry-pipe systems to be made up of galvanized pipe. The use of galvanized pipe, however, should be avoided in ovens. Experience has shown that the galvanizing may flake off when exposed to oven temperatures. This flaking within the pipes could lead to obstructions or plugging.

FURNACES AND SPECIAL ATMOSPHERES

Class C furnaces and some Class D furnaces use a special processing atmospheres to protect the work in process from oxidation and to cause desired metallurgical changes to the surface of the work in process. Because products of combustion are usually detrimental to such furnaces, these furnaces are usually indirect fired—that is, the products of combustion do not come into contact with the work in process.

Protective atmospheres include those listed in Table 50-5 below. These protective atmospheres may be used alone or in different combinations. Concern develops where flammable atmospheres are used.

Special Atmosphere Sources

Stored Atmospheres. Hydrogen, nitrogen, and argon are atmosphere gases that will normally be supplied to the furnace from storage tanks. Hydrogen and nitrogen generators are also available for use at industrial facilities with large atmosphere gas demands. For requirements on the storage of hydrogen, inspectors should consult NFPA 50A, *Standard for Gaseous Hydrogen Systems at Consumer Sites,* or NFPA 50B, *Standard for Liquefied Hydrogen Systems at Consumer Sites.* (Also, see Chapter 43, "Gas Hazards," in this manual.)

Generated Atmospheres. Dissociated ammonia, exothermic gas, and endothermic gas are atmosphere gases that are produced in heated generators. An ammonia

TABLE 50-5 Protective Atmospheres

Stored Atmospheres	Generated Atmospheres	Liquids for Atmospheres
Hydrogen	Dissociated ammonia	Methanol
Nitrogen*	Exothermic gas	Anhydrous ammonia
Argon*	Endothermic gas	—

*Nonflammable gases.

dissociator heats ammonia (NH_3) to crack or dissociate its molecules into its constituent parts (25 percent nitrogen and 75 percent hydrogen). An exothermic gas generator will partially burn natural gas with air in a controlled ratio to form the desired atmosphere. An endothermic gas generator requires heat to complete the reaction of the gas and air generating the desired atmosphere. Requirements on the protection of these atmosphere generators can be found in NFPA 86C.

Liquids for Atmospheres. Methanol and anhydrous ammonia are liquids that will vaporize and dissociate to form the desired atmosphere when they are injected into a hot furnace.

Atmosphere Gas Introduction and Removal

Flammable special atmosphere gases are introduced to and removed from furnaces in either the burn-in/burn-out method or the purge-in/purge-out method.

Burn-In/Burn-Out Method. With the burn-in/burn-out method, the furnace must be above 1400°F (760°C) so that all flammables will be reliably ignited and consumed as they are introduced. A burning flame front will gradually move from the atmosphere inlet to the open furnace ends (in a continuous furnace) or to the effluent vent stack (in a batch furnace). Proven pilots will provide reliable means for maintaining combustion of the flammable atmosphere gas as it exits the furnace. A positive flow of atmosphere gas is needed to continually replenish the gas consumed at the exits or stacks.

Purge-In/Purge-Out Method. With the purge-in/purge-out method, nitrogen is used to purge the furnace before the flammable atmosphere is introduced. With the purge-in method, the nitrogen purge must provide five oven volume changes. Once this is accomplished, atmosphere sampling is conducted. When two consecutive manual samples confirm that there is less than 1 percent oxygen in the furnace atmosphere, the flammable atmosphere gas can be introduced. During furnace commissioning, the time needed to reach less than 1 percent oxygen can be measured. Thereafter, a purge timer can be used in conjunction with a nitrogen pressure switch to confirm the purge-in without manual sampling the atmosphere.

Regardless of which introduction and removal method is used, the operator must follow written instructions for normal atmosphere introduction, normal atmosphere removal, and emergency action. The emergency action typically involves the initiation of a nitrogen purge of the furnace.

Air Infiltration

When a special atmosphere furnace is operating under a flammable atmosphere, only the addition of air or oxygen is needed to cause an explosion. Air infiltration is controlled by keeping the furnace under positive pressure. The following are some conditions that can lead to air infiltration.

Furnace Safety Shutdown. When a safety shutdown occurs, the sudden interruption of heat will cause the furnace temperature to drop. This temperature drop will cause the furnace atmosphere to contract or shrink. As the special atmosphere flow may not able to maintain positive furnace pressure under this condition, an emergency nitrogen purge should be initiated when safety shutdown of the furnace heat source occurs.

Cold Work Load. The introduction of a cold work load into a furnace will cause the atmosphere to cool and shrink. This is normally accommodated by furnace design.

Liquids Injection. Where liquids are injected into a furnace to produce the desired atmosphere, any condition that causes the furnace temperature to fall below that needed to vaporize the liquid quickly causes a loss of internal furnace pressure. Minimum operating temperatures for liquid injected special atmospheres must be maintained at all times.

Visual Indicators

There are trends by furnace manufacturers (especially overseas manufacturers) to automate the introduction and removal of special furnace atmospheres. With automation, the maintenance of visual feedback indicators, such as flow meters, pressure gauges, and safety shutoff valve position indicators, is essential. During furnace upsets, the operator may need this visual information to verify that emergency actions are appropriate and effective.

SUMMARY

Heat-utilization equipment can be divided into two categories. Boilers, which generate hot water or steam to support manufacturing operations, are not directly involved in the manufacturing process and are, therefore, typically separated from the rest of the plant. In contrast, ovens and furnaces (the terms are used interchangeably), which process commercial and industrial materials above normal ambient temperatures, are directly involved in the process and are, necessarily, located within the production area.

Various devices and features help to initiate and maintain a friendly fire in heat-utilization equipment. These include preignition purges, fans for purges and operations, fuel piping with vents and safety shutoff valves, fuel gas pressure regulators and switches, and flame safeguards (including those with automatic capabilities). An important safety element is the thoroughly trained operator who has access to written operating instructions and knows how to perform cold start-ups and emergency shutdowns.

For safety and control devices to be reliable, they must be inspected, tested, and maintained periodically. Good housekeeping practices must also be instituted and enforced, and fire protection systems (i.e., the automatic sprinkler system) and safety practices (e.g., the correct use of protective atmospheres) must be examined.

This chapter has touched on the major relevant points regarding heat-utilization equipment; the inspector, however, is encouraged to remember that this discussion is general in nature. For comprehensive guidance, the inspector should consult the actual NFPA codes and standards.

BIBLIOGRAPHY

NFPA Codes, Standards, and Recommended Practices

See the latest version of The NFPA Catalog for availability of current editions of the following documents.

NFPA 13, *Standard for the Installation of Sprinkler Systems*
NFPA 50A, *Standard for Gaseous Hydrogen Systems at Consumer Sites*
NFPA 50B, *Standard for Liquefied Hydrogen Systems at Consumer Sites*
NFPA 85, *Boiler and Combustion Systems Hazards Code*
NFPA 86, *Standard for Ovens and Furnaces*
NFPA 86C, *Standard for Industrial Furnaces Using a Special Processing Atmosphere*
NFPA 86D, *Standard for Industrial Furnaces Using Vacuum as an Atmosphere*
NFPA 91, *Standard for Exhaust Systems for Air Conveying of Vapors, Gases, Mists, and Noncombustible Particulate Solids*

Spray Painting and Powder Coating

CHAPTER 51

Don R. Scarbrough

Fire loss was limited when a single sprinkler controlled a fire that began when an electric light fixture malfunctioned. The sprinkler activation also alerted fire fighters, since the building was closed for the night.

The single-story building in Florida was constructed of steel with metal walls and a metal roof deck. Covering 32,000 ft^2 (2973 m^2), it was protected by a wet-pipe sprinkler system monitored by a central station alarm company. The occupancy manufactured automobile and truck accessories.

Fire fighters responding to a water flow alarm at 6:32 A.M. initially saw no smoke or flames. After investigation, they saw smoke issuing from a rear corner and heard the water flow alarm sounding. The first-in officer then called for a full alarm assignment and entered the property using forcible entry tools to advance a 200-ft (61-m) hose line to the rear of the building. Once inside, fire fighters discovered that a single sprinkler had controlled the blaze. They quickly extinguished it and ventilated the building, searching for fire extension.

The blaze, which was traced to an undetermined malfunction in one of two fluorescent light fixtures in a paint spray booth, damaged a small area of the booth. The structure and contents, valued at $3 million, suffered losses estimated at $25,000. There were no injuries.

Source: "Fire Watch," *NFPA Journal*, May/June, 2001.

CASE STUDY

Sprinkler Controls Fire in Paint Spray Booth

Processes involving spray application of materials that are flammable or combustible—whether they are in the form of liquids or powders—involve the risk of fire. Since two sides of the fire triangle—air and fuel—are routinely present and already mixed, a source of ignition is the only thing needed to start a fire.

Although spray painting is the most familiar of these processes, similar equipment is used in industry for applying other materials including adhesives, lubricants, sealants, waxes, colorants, flavors, pharmaceuticals, and structural layers of products. Among these materials the fire hazards and means for controlling them are generally similar in principle and differ only in detail. NFPA 33, *Standard for Spray Application Using Flammable or Combustible Materials,* provides basic information and minimum standards for protecting these processes.

Don R. Scarbrough of Elyria, OH, is a consultant to the spray finishing industry and is a member of NFPA's Technical Committee on Finishing Processes and Technical Committee on Static Electricity. He is retired from Nordson Corp., a manufacturer of electrostatic coating systems.

HAZARDOUS CHARACTERISTICS OF SPRAY PAINTING PROCESSES

Solventborne Coatings

Flammable vapors are present at all stages of the spray painting process. They come from coating materials in storage, during preparation for the application process, from containers, and from the spray process itself.

Residues of the application process are readily ignitable, and resulting fires develop quickly and have a high rate of heat release. Residues of some process materials, particularly air dry enamels and varnishes, can spontaneously heat and ignite.

Containers for process liquids can leak or spill, and exposure of the containers to process fires can result in explosive rupture of containers, if they are not provided with adequate pressure relief.

Mixing some coating components with other components may cause a chemical reaction, which could lead to uncontrolled heating. Other components, such as organic peroxides, are inherently unstable, and they may decompose violently if they are contaminated or suddenly heated.

Cleanup solvents used to remove residues from spray apparatus and the spray booth are often flammable. Overspray residues and leftover materials from cleanup usually are classified as hazardous wastes.

Spray booth and exhaust ducts may be heated to incandescence in a residue fire and must be constructed to withstand severe fire conditions.

Waterborne Coatings

Some waterborne coatings are capable of releasing flammable vapors. These materials commonly exhibit a flash point, but generally they will not continue to burn after an initial flash.

Spray patterns and the liquid supplies used in the process are generally not ignitable in the condition in which they are used, but they may become ignitable if virtually all of the water content is evaporated.

Residues after these coatings have dried have burn characteristics comparable to those from solventborne coatings. Some residue accumulations may spontaneously heat and ignite. Materials used for cleaning residues from equipment and spray booths are sometimes flammable and should be handled with care.

Powder Coating

No flammable vapors are produced at any stage of powder coating—not in storage, in process, nor from residues.

Airborne dust within the spray pattern will burn vigorously, but residues deposited on surfaces are not readily ignitable. If the spray is interrupted immediately after ignition, the fire will extinguish itself without further effort. If powder feed to the spray guns is sustained, the flame will continue to heat surrounding materials, causing the collector filters and powder residing in the collector hopper to ignite.

Cyclone or bag-house type dust collectors and airborne dust in any confined enclosure can explode if ignited.

Excess or oversprayed powder is normally recycled to be reused in the process. After exhaust air passes through process filters, it is normally discharged back into the workplace.

Spray booths are commonly constructed of plastic materials to prevent overspray from adhering.

Differences Between Manual and Automatic Spray

Each coating process may be done with manual or automatic spray guns or with a combination of the two in the same operation. Manual and automatic spray devices operate in much the same way. Automatic operations, however, may involve numerous spray guns used within the same booth and may be unattended during operation. Unless there is a special flame detection apparatus, the spray guns may continue to operate after a fire has ignited, severely aggravating the results. Hand spray operators will notice a fire immediately and release gun triggers to interrupt the supply of sprayed fuel to the fire, leaving residue fire as the major remaining concern.

Electrostatics

In a production setting the coating process may be augmented with electrostatic application. In this arrangement, the spray device incorporates high voltage (between 50,000 and 120,000 V), which enhances deposition of the sprayed paint onto the work piece. In this high voltage environment, any electrically conductive object that is not grounded may discharge electrical sparks that could ignite a fire. Stringent grounding discipline is required.

SPRAY BOOTHS AND OPEN-FACE SPRAY AREAS

Spray Booths

Production spray painting of objects up to the size of automobiles is routinely done in a power-ventilated enclosure referred to as a spray booth. The objects may be brought to the booth manually or by a conveyor, and the booth can be any one of a wide variety of designs. Although the booth will typically enclose the object during painting, in some designs it will be adjacent to the object and will surround the object and the spray process with a controlled stream of air that will capture all vapors and overspray and carry them into a collector. No vapors or overspray should escape.

Spray booths range in size from larger than a railroad boxcar to smaller than a bread box. They must be constructed of materials that will withstand a nominal process fire without collapsing. Booths and exhaust ducts most commonly are constructed of steel of at least 18 MSG (1.3 mm) thick, but they may also be constructed of masonry or other materials. Aluminum or structural plastics are permitted for powder booths since residues are not readily ignitable and a fire is expected to last no longer than a second.

The principle components of a spray booth are the enclosure, the overspray collector, and the exhaust system. The interiors of all of these components are classified as Division 1 hazardous (classified) locations for purposes of applying NFPA 70, *National Electrical Code®*.

Open-Face Spray Booths

Spray enclosures most commonly are configured as boxes with one side open. The object to be painted is situated either within the box or at its open side during spraying and, if the booth operates properly, allover spray will be captured by the airflow and contained within the enclosure. Figure 51-1 shows delineation of the Division 2 hazardous (classified) area outside the openings of the spray booth.

Enclosed Spray Booth

Enclosures can also be designed as a tunnel, with openings at either end for entry and exit of a conveyor carrying objects, such as automobile bodies, through the process zone. Airflow in this type of booth is usually introduced through filters in the ceiling and exhausted through floor gratings into an exhaust plenum.

In larger models the booth will accommodate painters and spray equipment; in smaller models the painters and spray gun mounts will be outside spraying inward through openings of limited size (see Figure 51-2). See Figure 51-3 for delineation of the Division 2 hazardous (classified) area outside of openings.

To diminish the chance of fire spread, spray booths should be separated from other processes, from stored combustible materials, and from combustible structures by at least 3 ft (900 mm). If a side of the spray booth does not need to be accessed for maintenance purposes, it may be located closer to a masonry or fire-resistant wall.

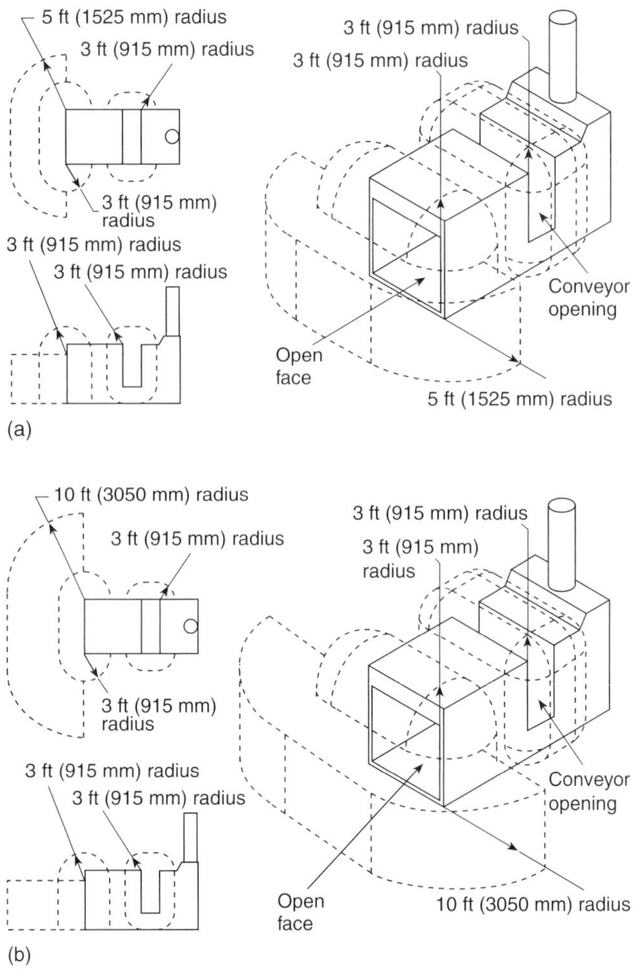

FIGURE 51-1 Class I or Class II, Division 2 Locations Adjacent to a Closed Top, Open-Faced or Open-Front Spray Booth. (a) Ventilation System Interlocked with Spray Equipment; (b) Ventilation System Not Interlocked with Spray Equipment.

Source: NFPA 33, 2000, Figures 4.3.2(a) and (b).

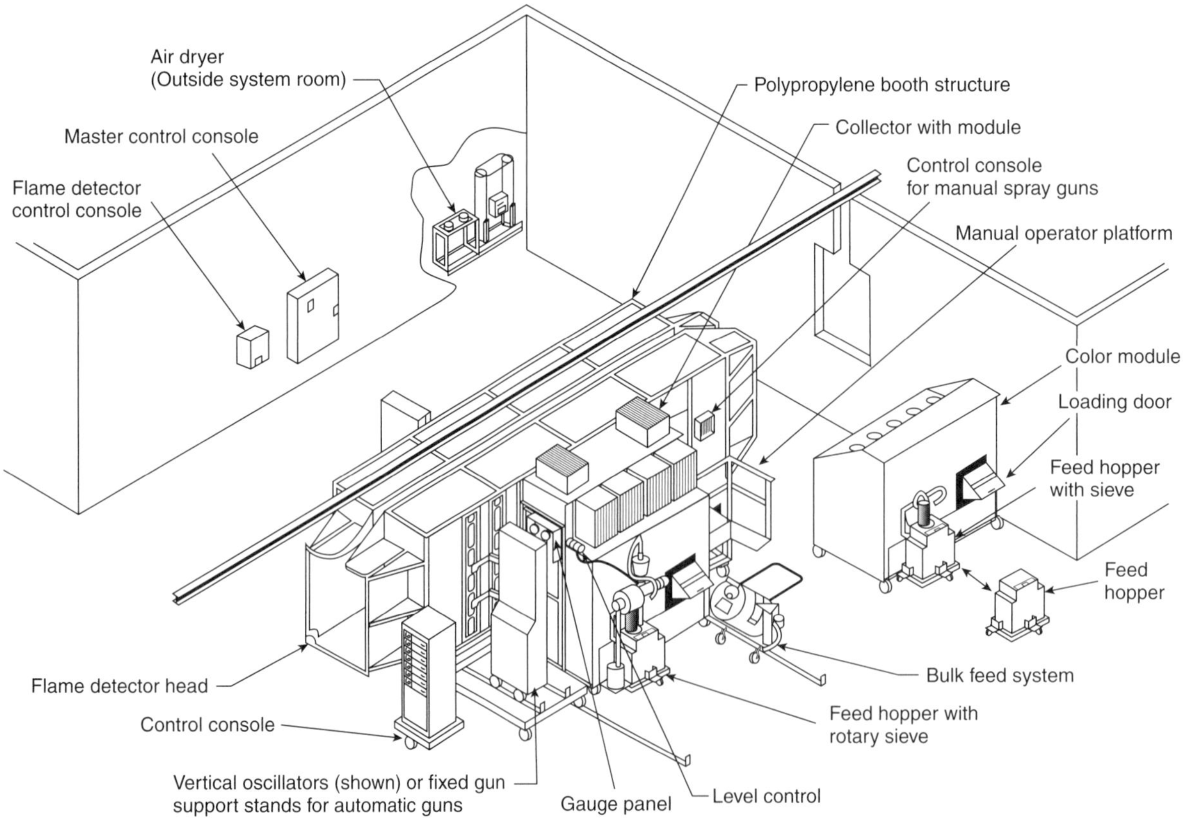

FIGURE 51-2 Integrated Powder Spray Booth/Recovery System

Source: Nordson.

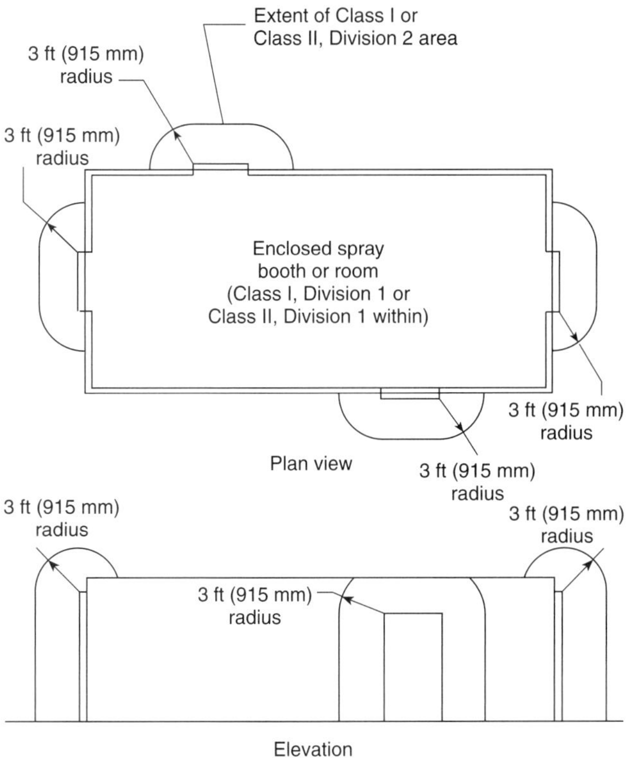

FIGURE 51-3 Class I or Class II, Division 2 Locations Adjacent to Openings in Enclosed spray Booth or Room.
Source: NFPA 33, 2000, Figure 4.3.4.

Processes that do not use a conveyor, such as automobile or furniture refinishing, can have booths that are totally enclosed during operation, with air intake through installed duct work or through a filter bank in a wall or a door. The Division 2 area outside the booth normally would extend only 3 ft (900 mm) from a door or opening.

LIMITED FINISHING WORKSTATION

A type of spray enclosure referred to as a "limited finishing workstation" is gaining acceptance for use in environments such as automotive repair shops. The total amount of material to be sprayed in any 8-hour period is limited by NFPA 33 to a maximum of 1 gal (4 L). See Figure 51-4 for a delineation of Division 2 hazardous (classified) area surrounding one of these units if spray apparatus is interlocked with ventilation, and Figure 51-5 if it is not interlocked.

The Overspray Collector

This component of the spray booth is intended to capture and remove particulate matter from the exhaust air stream, allowing vapors to pass and be discharged. They may use a dry collector or a water wash type for this purpose.

Maze-Type Dry Collectors

In its simplest form, this type of collector looks like a maze of steel panels or chains or one of folded paper through which the airstream is drawn, with the intention that particles will impact with the obstructions and be collected on them. This collector is not very efficient; it allows considerable amounts of contaminants to pass through the maze to subsequently be deposited in the exhaust duct, on the fan, and on the building roof around the stack discharge point.

FIGURE 51-4 Class I (or Class II), Division 2 Locations Adjacent to a Limited Finishing Workstation with Exhaust Ventilation Interlocked with Spray Equipment
Source: NFPA 33, 2000, Figure 12.3.5 (a).

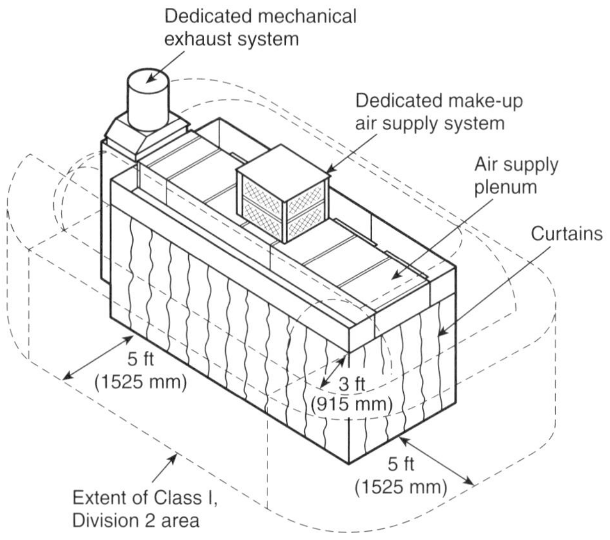

Dry-Filter Type

Dry filters made of shredded paper or of fiberglass mats are more efficient at trapping overspray. If cloth second-stage filters are used, virtually no solid contaminants are allowed to pass through to foul the stack.

In powder coating processes, the dry filters are engineered to stop all of the particulate material and not allow any to bypass into the exhaust system.

In liquid coating systems the residues that accumulate on dry collectors are usually readily ignitable and, especially on vertical filters, will burn violently. Fouled filters that have been removed from the spray booth but that are then stored in drums or boxes can spontaneously heat and ignite. Unless they are removed immediately from the building to a safe outside location, these filters must be immersed and stored in water until they are removed from the premises.

FIGURE 51-5 Class I (or Class II) Division 2 Locations Adjacent to a Limited Finishing Workstation with Exhaust Ventilation Not Interlocked with Spray Equipment
Source: NFPA 33, 2000, Figure 12.3.5 (b).

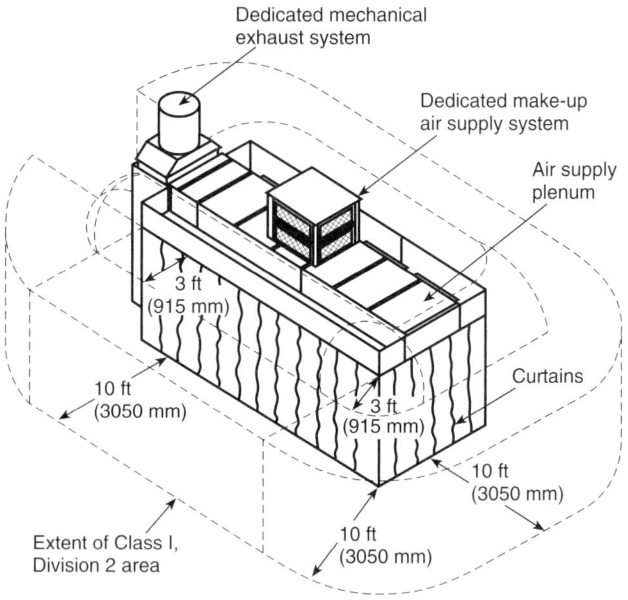

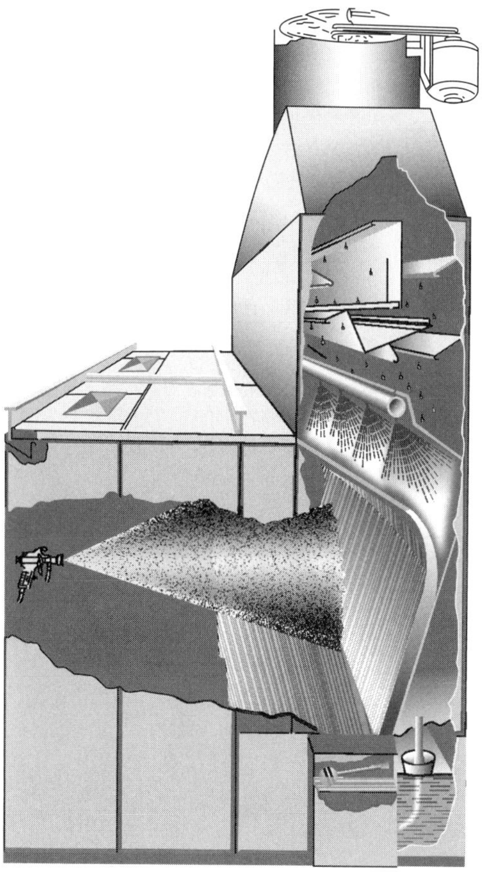

FIGURE 51-6 Water Wash Booth
Source: Team Blowtherm

Residues in a powder collector will not spontaneously ignite and they are not readily ignitable. Their hazardous characteristics are comparable to those of wheat flour or powdered sugar.

Water Wash Collectors

The water wash collector separates contaminants by passing the exhaust air stream through a series of sprays or waterfalls that scrub out particles of overspray and keep them submerged in a water tank. (Figure 51-6).

Special chemicals must be added to the water and maintained at the proper concentration to render the paint particles nonadhesive and thus prevent them from forming large sticky curds that can plug spray nozzles or adhere to overflow weirs and produce gaps in the waterfalls or scrubbers. Such gaps, if permitted to form, will permit overspray to pass through the collector and into the exhaust duct where it can accumulate as dry residue. As long as it is properly maintained, the water wash collector contributes substantially to fire safety by keeping residues submerged in water and unavailable as fuel for fire.

The Exhaust System

For liquid coating processes, this component of the spray booth comprises a plenum downstream of the collector, a fan, and ductwork leading to a discharge point outside the building. For powder coatings, the discharge is recirculated back to the work space within the building through a secondary set of filters.

The fan is preferably located at the discharge end of the duct and should be of nonsparking structure (AMCA Class C). Its rotating shaft should have heavy-duty bearings and the motor should not be exposed to the exhaust air stream unless listed for that service.

The exhaust system should have access doors or easily opened joints to facilitate inspection of the ductwork interior and cleanout of residues. The duct must be constructed of suitable materials and be supported to prevent collapse under fire conditions. (The duct can be heated to cherry red during a thermally drafted residue fire, substantially weakening the originally expected strength, and the load of water from internal sprinklers must be considered if horizontal duct runs are present.)

SPRAY ROOMS AND OPEN-FLOOR SPRAYING

Some objects to be painted, such as large, heavy machinery or structural steel fabricated components, are beyond the size that can be brought practically into a spray booth. They may be accommodated either in a spray room or under open-floor conditions.

Spray Room

A spray room differs in operation from a spray booth in that no attempt is made to provide exhaust air velocity that will carry overspray into a collector; powered exhaust is provided only to carry away vapors. Spraying may take place anywhere in the room and overspray is permitted to fall to the floor, from which it must be mechanically removed to prevent excess accumulation.

Spray rooms should be separated from other occupancies by structures with at least a 1-hour fire resistance rating. The entire interior of the spray room is classified as a Division I hazardous (classified) location, with Division 2 extending 3 ft (900 mm) outside of doors.

Exhaust systems associated with spray rooms are treated with the same fire control considerations as those associated with spray booths. Spray rooms should be separated from other occupancies of the building by construction with at least a 1-hour fire resistance rating.

Open-Floor Spraying

Another approach used for infrequent spraying of large objects involves spray painting on the open factory floor without an enclosure or a dedicated ventilation system. This type is referred to as open-floor spraying, and it requires the establishment of rather extensive surrounding areas classified as hazardous during operation. The Division 1 area is usually designated to be as extensive as the visible spray and the wet paint surfaces remaining on the object, and the Division 2 area extends 20 ft (6 m) further horizontally and 10 ft (3 m) vertically beyond the Division 1 area (see Figure 51-7).

Spray Apparatus

The devices that actually produce the paint spray are known as spray guns and exist in both handheld and automatic types (Figures 51-8 and 51-9). Paint is supplied to the spray device through a hose originating in a reservoir that may be a pressurized tank, or a "pressure pot," an unpressurized tank associated with a pump, or, for powder coating, a feed hopper with a pneumatic ejector type pump. Depending on the scale of the painting operation, this reservoir can be as small as 1 gal (4 L) or as large as 100 gal (400 L). Several reservoirs supplying different colors or types of paint might be associated with a single spray booth or room.

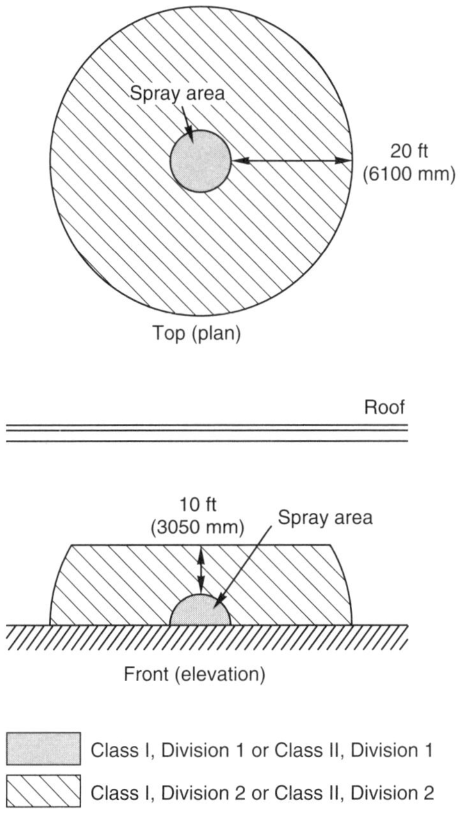

FIGURE 51-7 Class I or Class II, Division 2 Locations Adjacent to Unenclosed Spray Operation.
Source: NFPA 33, 2000, Figure 4.3.1.

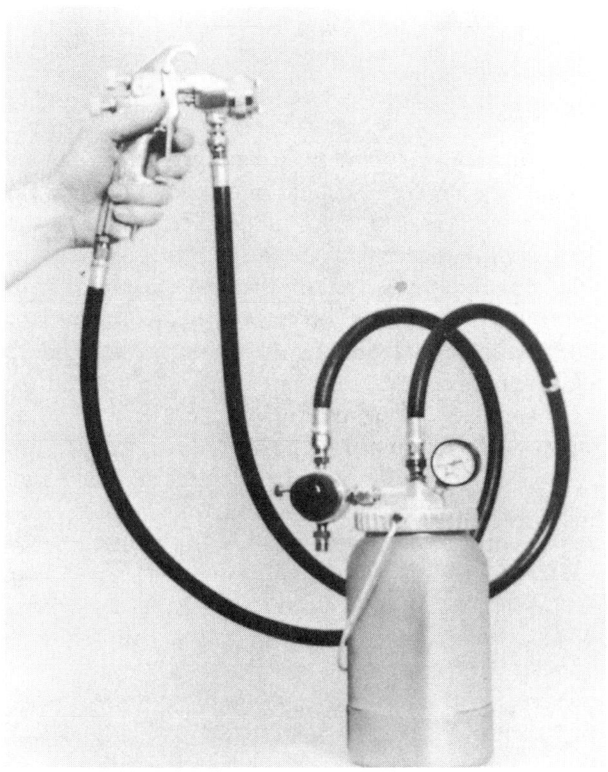

FIGURE 51-8 Hand Air Spray Gun with Fluid Supplied from Pressure Pot
Source: DeVilbiss.

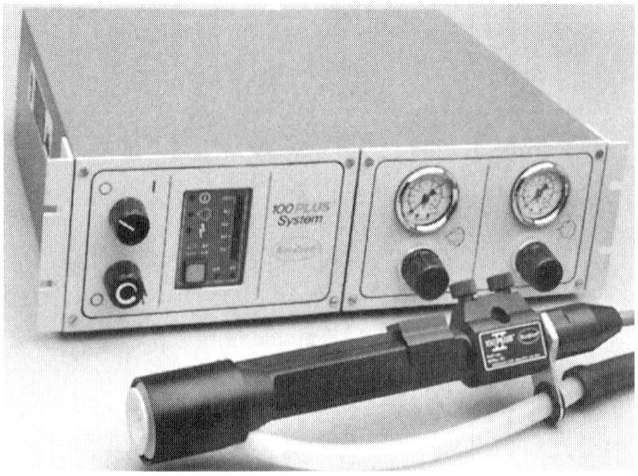

FIGURE 51-9 Automatic Powder Gun.
Source: Nordson

To lessen the effects of exposure during spray booth fire conditions, the supply reservoirs should be separated from the spray booth by several feet. Since hoses can be expected to burn off in a fire, pressurized paint supplies should be interlocked with a fire alarm to prevent the paint from being delivered into the spray booth during a fire.

In automatic operations at their simplest, the atomizers are mounted on "fixed gun stands" and are triggered on and off by a controller as the objects to be painted pass by on a conveyor. In increasing order of complexity, the guns are mounted in an array on a machine called a "reciprocator" to stroke back and forth, or up and down, like strokes of a paint brush as the work passes. The guns might be on constantly or might be triggered to spray only the work pieces. Some reciprocators function as two-axis or even three-axis machines. In the most complex arrangements, the guns are manipulated by a robot that can move the gun to any position within the area swept by its arm and point in any direction, just as if it were handheld. Triggering is controlled by a computer that also runs the movements of the robot.

INSPECTION CHECKLIST

Control of Ignition Sources

All recognized sources of ignition must be excluded from the area designated as hazardous, including open flames such as process heaters, pilot lights, torch welding and cutting operations, cigarette smoking, and space heaters; hot surfaces that might result from prior operations, heat treatment, welding and flame dry-off operations, shearing, drilling, and grinding operations, friction from binding conveyor components, from overheated bearings, or from hot surfaces of lamps; and sparks resulting from grinding or welding operations, flame cutting, or embers from poorly adjusted overhead fuel burning heaters.

Sparking from open electrical contacts or from faults must be avoided by ensuring that all electrical wiring and devices within the hazardous area are appropriately approved and listed.

Reactive finishing materials that could overheat if mixed inappropriately must be rigidly controlled in accordance with the process. For example, coating materials that heat after mixing as a result of a chemical reaction can remain at perfectly acceptable temperatures when spread as a thin film onto a product. The same materials could, however, heat sufficiently to burst into flames if mixed in a pail or bucket that has insufficient surface area to dissipate the heat.

Coating residues that can spontaneously heat, such as air-drying enamels and varnishes, must not be allowed to accumulate excessively in collector filters or in

FIGURE 51-10 Electrostatic Disk
Source: ITW Ransburg

wiping rags. Fouled filters that have been removed from spray booth collectors must be removed from the building immediately or immersed in water and kept immersed until removal to prevent spontaneous ignition.

Unstable materials such as organic peroxides and nitrocellulose materials must be meticulously managed to prevent contamination or excess temperatures that could cause spontaneous decomposition and ignition.

Sparking as a result of discharge of static electricity must be prevented by thoroughly grounding all electrically conductive objects in the process area. All spray equipment, flammable liquids handling equipment, spray booth components and associated apparatus, and the conveyors must have ground conductors connected to a common building ground.

Contact points on conveyor racks and rack-to-workpiece contacts must be kept clean and have a resistance to ground of less than 1 (MΩ) at all times.

The floor within the process area should be electrically conductive and shoes worn by all personnel entering the hazardous area should be of static-dissipative type. If the floor is covered by insulating residues, there must be alternative means for grounding personnel, such as a lanyard with a wrist strap.

Whether or not electrostatic apparatus is installed (see Figure 51-10) static electricity is a common source of ignition off lammable vapors in situations where overspray falls to the floor to form a sticky residue. In walking only three or four steps across the sticky residue, an individual can accumulate a high enough electrostatic voltage on his or her body to discharge an incendive (ignition-capable) spark when approaching a grounded object. In order to prevent this type of ignition, through prevention of static accumulation, it is critically important that a reliable means of grounding the human body be provided at all times.

CONTROL OF IGNITABLE ATMOSPHERES

Limitation of vapors and overspray to the smallest practicable area is accomplished through a combination of process enclosures and power ventilation systems. Ventilation systems should be checked frequently to ensure they are operating properly and that specified flow rates are being maintained.

Limitations of Fuel Quantity

The total quantity of fuel permitted in the process area must be restricted to the minimum needed for one shift or one day of operation.

Ready operating supplies should be kept in an area separate from the spray area to prevent their involvement in a spray booth fire.

Pressurized supplies of coating materials should have emergency shutoffs, both manual and automatic, interlocked with the fire alarm to prevent additional fuel from being poured into a spray booth fire when hoses are burned off.

Hose materials should have nominal resistance to fire. For example, polyethylene tubing will melt off immediately when exposed to fire, while rubber hoses will be more resistant, and Teflon® hoses even more resistant.

Overspray residues within the spray booths, the collector, and the exhaust ductwork must not be allowed to accumulate to a quantity that will produce a fire too big and too vigorous for the installed extinguishing system to suppress. Accumulation beyond approximately ⅛-in. (3 mm) thickness should draw careful consideration.

Use of a water wash overspray collector, which keeps collected residues submerged in water, is strongly favored over the dry filter alternative, which usually retains the residues in a configuration that will produce a severe fire when ignited.

Isolation of Process

The spray painting process should be separated from other manufacturing processes and occupancies, both to prevent ignition and to retard fire spread

Spray booths in which flammable materials are used and in which readily ignitable residues accumulate should be of fire-resistant construction. Spray rooms should have at least a 1-hour fire resistance rating separating them from other occupancies.

Spray booths should be of the appropriate type, size, and material for the process in which they are being used. Air velocity and volume must capture and contain overspray to the spray booth.

Exhaust systems that might accumulate readily ignitable residues must be structured and supported to prevent them from collapsing in a fire. They must have adequate clearance from combustible construction, such as at roof penetrations, and they should discharge free of the building to prevent vapors from reentering.

Explosion Prevention and Pressure Relief Venting

Since ignitable concentrations of vapors or combustible dust routinely exist in several parts of the process—in flammable liquid containers, within the spray process zone, within dust collectors—conditions that will produce an explosion must be prevented.

All flammable liquid reservoirs and closed containers of supply materials must have pressure release vents. To prevent explasive rupture from happening if exposed to a process fire, approved pressure/relief vent devices should be installed on all flammable liquid shipping containers over 5-gal (20 L) capacity.

Although ignitable concentrations are expected within the spray pattern, there must be sufficient exhaust airflow to prevent ignitable concentrations from being induced into the exhaust system. Flammable vapors should be diluted to less than 25 percent of the LFL (lower flammable limit) and ignitable dusts should be diluted to less than 50 percent of the MEC (minimum explosive concentration) in ducts leading to powder collectors that are connected to a separate collector, such as a cyclone or bag house, through ductwork. An interlock should prevent triggering of automatic spray guns unless the exhaust system is operating.

For powder coating systems, dust collectors that are integrated into the spray booths and are designed to eliminate the confinement necessary for formation of an explosion are preferable to more conventional dust collectors such as a cyclone or bag house, which is connected to ductwork and is capable of exploding.

Cyclones, bag houses, and long ductwork runs associated with powder coating facilities must have appropriate pressure release venting and ductwork to discharge relieved pressure and products of combustion to the exterior of the building. The maximum acceptable length of pressure relief ductwork is approximately 10 ft (3 m). All process and pressure relief ductwork associated with this type of equipment must be constructed to withstand the maximum pressure expected in an explosion. Ductwork must have welded longitudinal seams and bolted flange-type joints.

All automatic powder coating installations must have fast-acting flame detectors that will respond within a half second to shut down the process, sound an alarm, and close dampers in any associated ductwork between the booth and collector. Prompt shutoff of a supply of airborne combustible dust will extinguish a process fire before enough time has elapsed for embers to form that could be carried through the ductwork to ignite the collector. Dampers in the ductwork further contribute to interruption of the airflow which would be necessary to transport the embers.

Extinguishment

All spray booths are required to have approved automatic fire extinguishing systems installed. The systems should protect not only the spray process area but the overspray collector and the exhaust system as well. Although automatic sprinklers both inside the process enclosure and at ceiling level are recommended, other systems might be appropriate, with consideration given to the process, to the materials used, and to the surrounding circumstances.

Interlocks should shut down the process in the event of fire, paint hose rupture, conveyor wreck, ventilation failure, or similar event. They should be integrated with the fire alarm and the emergency stop control to shut down the spray process, interrupt all fuel supplies, and all energy inputs including electrical, compressed air, and hydraulic. The conveyor and all gun movers should be brought to a stop.

Automatic flame detectors with a response time of a half second or less should be installed in all automatic electrostatic systems for fluid or powder spray. Flames in spray patterns of both liquid and powder guns have very high heat release rates, which, if continued, will quickly heat residues and boost the rate of flame spread. Fast-acting flame detectors connected to the interlock almost instantly shut down the spray gun fires in the event of ignition, thus preventing residues from being heated and enhancing chances of prompt extinguishment.

In liquid spray systems where residues are expected to be readily ignitable, thus producing a persistent fire with large volumes of smoke, the air makeup and exhaust ventilation system should be kept in operation during fire conditions.

In powder coating operations where residues are *not* readily ignitable and no persistent fire is expected, the exhaust system should be shut down to prevent fanning of any flames that might then heat residues to ignition temperature.

In spray operations using flammable liquids and readily ignitable residues, smoke produced by a fire can completely fill the building or the room containing the spray operation in less than a minute, thereby threatening egress and making attack of the fire by fire fighters almost impossible. The installation of fire curtains and of smoke and heat vents over such operations is recommended to limit the opening of sprinklers to the area immediately above the fire and to aid in clearing smoke so that fire-fighting operations may be conducted.

A supply of appropriate hand extinguishers should be readily available to operating personnel in all spray-finishing operations.

Training

An ongoing training program should be in place for supervision and production employees. Topics to be addressed should include the finishing process, the material used, manufacturers' operating instructions for the apparatus, identification of hazards and means for controlling them, and emergency procedures. Records should be on hand with reference materials for the course content and memoranda of attendance records for each date that training has been conducted. There also should be a realistic program for refresher training.

Manufacturers instruction manuals for all process apparatus should be on hand.

Inspection of Apparatus and Procedures

Management-appointed inspectors should examine apparatus and should review procedures on a routine basis and after incidents, using a checklist to, ensure that all critical items are inspected.

Values determined during inspections—such as airflows, pressures, electrical current—should be compared with acceptable baseline values during nominal operation to aid in detecting misadjustments or deterioration of apparatus.

Reports of inspection results should be reviewed by management and retained in files for future comparison.

Maintenance

Any equipment damage or wear should be repaired promptly to the original manufacturer's specification. All repairs to equipment listed for hazardous area operation should be made using only replacement parts specified or supplied by the original manufacturer and in strict accordance with the manufacturer's instructions in order to maintain the validity of the safety listing and to assure continued operation in accordance with original specifications.

A permit system must be in place to control all hot work in spray operation areas and for all hot work performed on the spray booth, overspray collector, and exhaust system, including welding, torch cutting, drilling, and sawing operations. All residue should be removed before hot work begins on booth enclosure, collector, or ductwork.

Written instructions should be available describing in detail the procedure for removing readily ignitable residues from process equipment or floors. This procedure should specify clearly any chemical substances to be used, require the use of nonsparking scraping tools, require wetting of residues before scraping, and describe in detail the procedure for disposing of residues that have been removed.

A firewatch equipped with appropriate extinguishing equipment should be posted during all hot work and residue removal operations and for a reasonable time after those operations have been completed.

Storage and Handling of Materials and Residues

Written instructions should be on hand describing procedures for storing and handling as well as disposing of all hazardous process materials and residues.

Materials in excess of what is needed for the current day's production should be removed from the process are and kept in a storage area.

Residues should be placed in covered metal containers and removed immediately from the factory. Filters fouled with overspray residues should be immersed in water in metal containers and removed promptly from the factory.

Protective Clothing for Personnel

All persons entering the hazardous process area should be equipped with appropriate protective clothing. To protect from flash fire during process operations, maintenance, and clean-up, exposed personnel should be dressed in clothing providing maximum skin coverage and adequate resistance to flame. Textiles such as Nomex® IIIA or equivalent are most suitable for these tasks.

SUMMARY

The risk of fire can be kept to an acceptable level through a combination of efforts directed toward control of ignition sources, control of ignitable atmospheres (vapor or powder-air mixtures), limitation of available fuel quantities, isolation of the process, explosion prevention and pressure, relief venting, and provision of adequate means for extinguishment.

Simply stated, the person who adequately manages his or her process must do those things that will tend to prevent a fire from starting and those things that will limit the size of any fire that does occur to a scale that can be quickly extinguished with immediately available resources. If these things are done, the process can be operated with minimal risk to life and property.

BIBLIOGRAPHY

Instruction manuals from equipment suppliers

NFPA Codes, Standards, and Recommended Practices

See the latest version of The NFPA Catalog for availability of current editions of the following documents.

NFPA 33, *Standard for Spray Application Using Flammable or Combustible Materials*
NFPA 68, *Guide for Venting of Deflagrations*
NFPA 69, *Standard on Explosion Prevention Systems*
NFPA 70, *National Electrical Code®*
NFPA 77, *Recommended Practice on Static Electricity*
NFPA 91, *Standard for Exhaust Systems for Air Conveying of Vapors, Gases, Mists, and Noncombustible Particulate Solids*
NFPA 2112, *Standard on Flame-Resistant Garments for Protection of Industrial Personnel against Flash Fire*
NFPA 2113, *Standard on Selection, Care, Use, and Maintenance of Flame-Resistant Garments for Protection of Industrial Personnel against Flash Fire*

Welding, Cutting, and Other Hot Work

CHAPTER 52

August F. Manz and Amy Beasley Spencer

A plastics manufacturing plant in Texas was completely destroyed when cutting torch ignited cardboard, plastics, and other trash, and the fire spread rapidly to storage. A delay in fire department notification and a disabled sprinkler contributed to the huge loss.

The two-story plant had a steel frame, with a metal deck roof and masonry walls. It was 200 ft (61 m) long and 400 ft (122 m) wide. A wet-pipe system was inoperable, and its owners had been issued a notice to repair by fire officials. There were no smoke alarms, and the building was operating at the time of the fire.

Employees were using a cutting torch to remove a metal gate and overhead door assembly on a loading dock when the torch came into contact with the combustible trash. The resulting fire spread quickly while the employees tried to control it with handheld extinguishers before calling the fire department.

The department received a 911 call from the plant manager at 10:35 A.M. Arriving 2½ minutes later, the first company saw "a wall of fire" at one corner of the building.

Two fire fighters and two civilians were injured during the incident. The structure, valued at $1 million, and contents, valued at $1 million, were a total loss.

Source: "Fire Watch," *NFPA Journal*, May/June, 2000.

CASE STUDY

Delayed Alarm Leads to $2 Million Loss

From 1993 to 1997, there were 12,540 hot work fires on average per year, causing 23 civilian deaths, 414 civilian injuries, and $312.7 million in direct property damage.[1] These data should not surprise the experienced fire inspector, who has seen the hazard potential and the consequences of unsafe practices of these common operations. When compared to other types of hot work, cutting and welding were identified as the most dangerous types of hot work, causing on average 42 and 28 percent, respectively, of all nonresidential hot work fires.

Even though cutting and welding are the most dangerous types of hot work, an inspector would be remiss in not realizing that other types of hot work have the same hazards as cutting and welding and are controlled by the same precautions and regulatory requirements. Other types of hot work include heat treating, grinding, thawing pipe, using powder-driven fasteners, hot riveting, and welding and allied processes (open flame soldering, brazing, thermal spraying, oxygen cutting, plasma cutting, and arc cutting). Other similar applications, such as grinding, chipping, or abrasive blasting, that produce a spark, a flame, or heat should also be considered during an inspection.

[1] Special Data Information Package: *Torch Fires [2000]*. NFPA, June 2000.

August F. Manz, president of A. F. Manz Associates, Union, NJ, a consultant in welding and cutting safety and technology, is chair of ANSIZ49.1 technical committee and a past chair of NFPA's Technical Committee on Hot Work Operations. The author of several welding books and a contributor to safety chapters in more than two dozen other welding-related books, he holds an M.S. in electrical engineering from New Jersey Institute of Technology.

Amy Beasley Spencer, a senior chemical engineer with NFPA, is the NFPA staff liaison to NFPA 51B, the hot work fire prevention standard, and serves on the ANSI Z49.1 technical committee. She has a B.S. in chemical engineering from the University of New Hampshire and an M.S. in engineering management from Western New England College.

Hot work operations are hazardous for a number of reasons. First, they can use flames burning at temperatures of 4000° to 5000°F (2204° to 2760°C) or electric arcs that are estimated by some to be hotter than the surface of the sun. They frequently take place in areas where combustible materials are present, in cramped or elevated areas, or in areas cluttered with tools and equipment that may hinder movement during an emergency. Molten metal and sparks are created that can cause burns and other injuries. Many types of hot work emit radiation that can harm improperly protected eyes and skin.

Inspectors should understand how hot work operations can be performed to minimize the hazards and improve work place safety. They should also be aware of the regulations and standards that apply to these processes.

The National Fire Protection Association (NFPA) and the American Welding Society (AWS) publish a number of safety and health documents that are recognized by the industry as the major sources of information and that serve as the basis of many of the federal and state regulations. Two of the most important are ANSI Z49.1, *Safety in Welding, Cutting, and Allied Processes,* published by AWS, and NFPA 51B, *Standard for Fire Prevention During Welding, Cutting, and Other Hot Work,* published by NFPA. Earlier editions of these two documents are the source of regulations in OSHA 29 CFR 1910 Subpart Q, "Welding, Cutting, and Brazing."

PRECAUTIONS DURING HOT WORK

Inspectors should be particularly alert for circumstances that could lead to fires, explosions, or injury. Inspectors should be particularly vigilant in enforcing the following items because they are recognized to be major factors in hot work fires and injury.

Clothing and Other Personal Protective Equipment (PPE)

Eliminating or reducing a hazard by substituting a less hazardous procedure or chemical or by providing ventilation is always preferable. The situation is inherently safe when the hazard is not present. However, at times a hazard cannot be completely eliminated or satisfactorily reduced. In such cases personal protective equipment (PPE) must be used.

Operators need PPE to protect against many hazards, including, but not limited to, heat, UV radiation, infrared radiation, sparks, smoke, fumes, vapors, gases, dust, and electric shocks. Clothing should provide enough coverage to protect skin from radiation and spatter. During work, operators must wear gloves, long-sleeved shirts, aprons, coveralls, and at times leggings of leather or some other durable flame-resistant material. The outer garments should be free of oil and grease, which could be easily ignited. High-top safety shoes are recommended; sneaker-type or open-toed shoes are prohibited. Trousers should not be turned up or have outside cuffs; front pockets and rolled-up sleeves should be eliminated because they could trap sparks. Torn or frayed clothing can be ignited easily and should not be worn. During overhead hot work, operators need to wear fire-resistant head and shoulder covers and ear protection, such as earplugs, so the sparks or slag will not enter the ear canal.

Goggles or other eye protection must be worn during exposure to molten metal, sparks, and other flying materials. Face shields are considered a secondary form of eye protection and must be worn with either safety glasses or goggles. Filter lenses should be selected in accordance with ANSI/AWS F2.2, *Lens Shade Selector;* ANSI Z49.1 also provides a *Guide for Shade Numbers.* Contact lenses are acceptable to wear under eye protection; there has never been a reported case of lenses fusing to the eyes due to arc radiation, as is a common myth. All PPE should be kept clean and in good condition at all times and replaced as necessary.

At an industrial worksite, hazards other than hot work, such as radiation or noise, are often present. Any additional PPE donned for protection against these other hazards should also be appropriate—that is, flame resistant—for hot work. In 1997, a nuclear facility hot work operator donned PPE to protect not only against the hazards introduced by using a cutting torch, but also to protect against the hazards of radiation. He was wearing so much inappropriate PPE that it prevented him from knowing he was on fire as the hot work ignited his PPE, resulting in his death.[2]

Ventilation

The inspector should ensure that adequate ventilation is provided. The quantity and toxicity of hazardous fumes and gases cannot be easily classified because they depend on the process, the intensity of the heat source, the materials used, the base metal, as well as any coatings present on the base metal and other factors. The inspector should also check to be sure that the type of ventilation used does not pull the contaminants through the worker's breathing zone before removal. For example, local exhaust systems located overhead or behind the worker would, in many cases, pull the contaminants through the worker's breathing zone before removal. Operators should position themselves such that their breathing zone does not intersect with the fumes. When hazardous concentrations or unknown concentrations cannot be adequately removed, use of respirators is necessary. With any type of ventilation, inspectors should check where the contaminants are being transported. They should ask themselves, "Is it safe to discharge the contaminated air in that location?"

Designated Area Location of Hot Work

When possible, hot work should be confined to a specific area designed or approved for hot work, such as a maintenance shop or a detached outside location that is essentially free of flammables and combustibles and is suitably separated from adjacent areas. The intent of the designated area is to have a safe area for hot work to significantly reduce the hazards and to reduce the number of worksite inspections. However, work in some designated areas can be dangerous as well; designated areas are the most common areas for hot work fires, probably as a result of changing conditions, such as the temporary storage of combustibles. The inspector should carefully examine the designated area for combustibles or other hazards.

35-Foot (11-Meter) Rule (Nondesignated Area Location of Hot Work)

When hot work cannot be performed in a designated area, the nearby hazards, usually combustible materials in the immediate vicinity, should be transferred elsewhere. A good rule of thumb is that combustible materials should be moved at least 35 ft (11 m) in all directions from the work (Figure 52-1). Vertical fall distances must be much greater since sparks and slag have been known to fall hundreds of feet.

Inspectors should not be looking for only large combustible objects. Ensuring that paper, wood shavings, or textile fibers on the floor are swept up in the 35-ft (11-m) radius of the hot work is equally important. Accumulated dust or lint can also act as a fuse for fire, carrying it to other locations with combustibles. Combustible

[2] *Final Report: Type A Accident Investigation Report of the February 13, 1997 Welding/Cutting Fatality at the K-33 Building, K-25 Site, Oak Ridge, Tennessee* [1997]. Office of Oversight: Environment, Safety and Health, U.S. Department of Energy, April 1997.

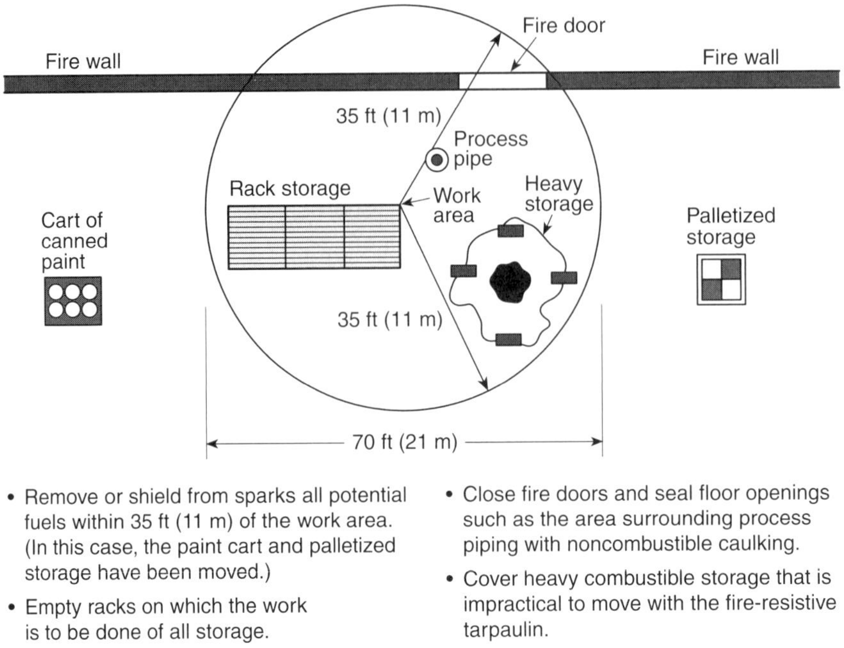
Fire wall
Fire door
Fire wall
35 ft (11 m)
Process pipe
Rack storage
Work area
Heavy storage
Cart of canned paint
Palletized storage
35 ft (11 m)
70 ft (21 m)
• Remove or shield from sparks all potential fuels within 35 ft (11 m) of the work area. (In this case, the paint cart and palletized storage have been moved.)
• Empty racks on which the work is to be done of all storage.
• Close fire doors and seal floor openings such as the area surrounding process piping with noncombustible caulking.
• Cover heavy combustible storage that is impractical to move with the fire-resistive tarpaulin.

FIGURE 52-1 Hot Work 35-Foot (11-m) Rule
Source: *Introduction to Employee Fire and Life Safety*, NFPA, 2001, Figure 10.9; Courtesy of Factory Mutual Insurance Company.

floors should be wet down, covered with damp sand, or be protected by noncombustible or fire-retardant shields. Where floors have been wet down, care should be taken to protect personnel operating arc welding or cutting equipment from possible electric shock and to protect all personnel from slipping hazards. Only when the combustible materials cannot be removed should the materials be protected with fire resistant tarps or shields.

Subtle Passages to Combustibles

Passages to combustibles are sometimes subtle. Any openings or cracks in the walls, floors, or ducts within 35 ft (11 m) of the site are potential travel passages for sparks, heat, and flames. Conveyor systems should be shut down or shielded during hot work. Because of heat transfer, hot work should be avoided on pipes or other metal structures that are in contact with combustible walls, partitions, ceilings, roofs, or other combustible materials, if the work is close enough to cause ignition by conduction. In addition to the areas beside the work, the areas above and below should be considered. For instance, heat from the hot work can travel up and down a pipe, igniting combustibles on another floor.

CONTROLS DURING HOT WORK

Hot Work Permit System

The hot work permit is the single most important item the inspector can use to determine if hot work safety procedures are being followed. The hot work permit is a document issued to authorize a specific job utilizing hot work. Using a hot work permit decision tree (Figure 52-2) will help determine if a permit is necessary. The purpose of a hot work permit is to evaluate whether all necessary safety precautions have been taken and whether controls for the hazards have been implemented. The permit is often set up in a site-specific checklist format as a reminder of site-specific hazards. Because the type of hot work being performed as well as the hot work environment varies, all permits will not look the same, have the same items, or contain the same information. In addition, a facility's hot work permit system will continue to evolve through use. Two sample permits are shown in Figures 52-3 and 52-4.

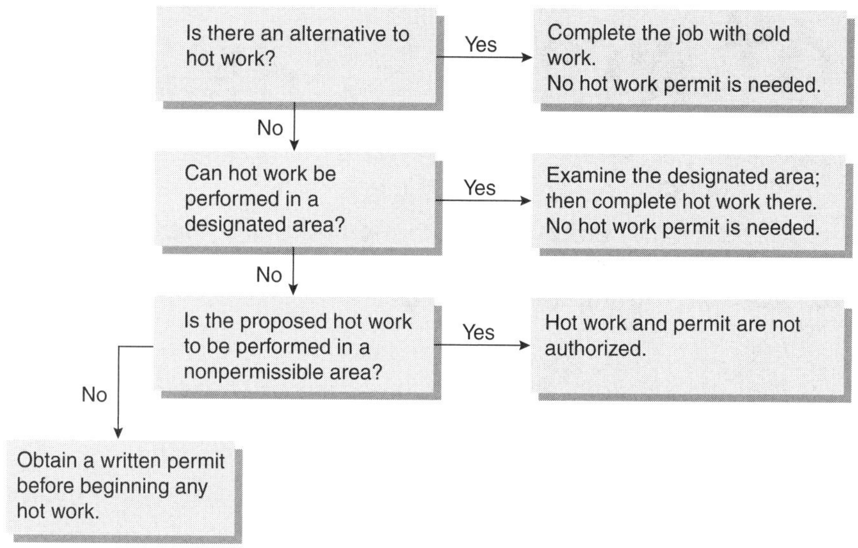

FIGURE 52-2 Hot Work Permit Decision Tree
Source: *Introduction to Employee Fire and Life Safety,* NFPA, 2001, Figure 10.8.

Side 1
HOT WORK PERMIT

Date ____________________

Building ____________________

Dept. ____________ Floor ____________

Work to be done ____________________

____

Special precautions ____________________

____

Is fire watch required? ____________________

The location where this work is to be done has been examined, necessary precautions taken, and permission is granted for this work. (See other side.)

Permit expires ____________________

Signed ____________________
Permit Authorizing Individual (PAI)

Time started ____________ Completed ____________

FINAL CHECK

Work area and all adjacent areas to which sparks and heat might have spread [including floors above and below and on opposite side of wall(s)] were inspected 30 minutes after the work was completed and were found firesafe.

Signed ____________________
Permit Authorizing Individual (PAI)

Side 2
ATTENTION

Before approving any hot work permit, the PAI shall inspect the work area and confirm that precautions have been taken to prevent fire in accordance with NFPA 51B.

PRECAUTIONS

- ❑ Sprinklers in service
- ❑ Hot work equipment in good repair

WITHIN 35 FT OF WORK

- ❑ Floors swept clean of combustibles
- ❑ Combustible floors wet down, covered with damp sand, metal, or other shields
- ❑ All wall and floor openings covered
- ❑ Covers suspended beneath work to collect sparks

WORK ON WALLS OR CEILINGS

- ❑ Construction noncombustible and without combustible covering
- ❑ Combustibles moved away from opposite side of wall

WORK ON ENCLOSED EQUIPMENT
(Tanks, containers, ducts, dust collectors, etc.)

- ❑ Equipment cleaned of all combustibles
- ❑ Containers purged of flammable vapors

FIRE WATCH

- ❑ To be provided during and 30 minutes after operation
- ❑ Supplied with a fully charged and operable fire extinguisher
- ❑ Trained in use of equipment and in sounding fire alarm

FINAL CHECK

- ❑ To be made 30 minutes after completion of any operation unless fire watch is provided

Signed ____________________
Permit Authorizing Individual (PAI)

FIGURE 52-3 Hot Work Permit Sample
Source: NFPA 51B, 1999, Figure A.3.3.1(a).

HOT WORK PERMIT

BEFORE INITIATING HOT WORK, CAN THIS JOB BE AVOIDED? IS THERE A SAFER WAY?

This Hot Work Permit is required for any temporary operation involving open flames or producing heat and/or sparks. This includes, but is not limited to: Brazing, Cutting, Grinding, Soldering, Torch Applied Roofing and Welding.

PART 1

INSTRUCTIONS

1. **Fire Safety Supervisor:**

   A. Verify precautions listed at right (or do not proceed with the work).

   B. Complete and retain Part 1.

   C. Issue Part 2 to person doing job.

HOT WORK BEING DONE BY ☐ EMPLOYEE ☐ CONTRACTOR..	
DATE	JOB NUMBER
LOCATION/BUILDING AND FLOOR	
NATURE OF JOB	
NAME OF PERSON DOING HOT WORK	

I verify the above location has been examined, the precautions checked on the Required Precautions Checklist have been taken to prevent fire, and permission is authorized for this work.

SIGNED (Firesafety Supervisor/Operations Supervisor)

PERMIT EXPIRES	DATE	TIME AM PM

NOTE: EMERGENCY NOTIFICATION ON BACK OF FORM. USE AS APPROPRIATE FOR YOUR FACILITY.

FM Global

2630 REV. 11/2000 ENGINEERING

REQUIRED PRECAUTIONS CHECKLIST

☐ Available sprinklers, hose streams and extinguishers are in service/operable.
☐ Hot Work equipment in good repair.

Requirements within 35 ft (11 m) of work

☐ Flammable liquids, dust, lint and oily deposits removed.
☐ Explosive atmosphere in area eliminated.
☐ Floors swept clean.
☐ Combustible floors wet down, covered with damp sand or fire-resistive sheets.
☐ Remove other combustibles where possible. Otherwise protect with fire-resistive tarpaulins or metal shields.
☐ All wall and floor openings covered.
☐ Fire-resistive tarpaulins suspended beneath work.
☐ Protect or shut down ducts and conveyors that might carry sparks to distant combustibles.

Work on walls, ceilings or roofs

☐ Construction is noncombustible and without combustible covering or insulation.
☐ Combustibles on other side of walls, ceilings or roofs are moved away.

Work on enclosed equipment

☐ Enclosed equipment cleaned of all combustibles.
☐ Containers purged of flammable liquids/vapors.
☐ Pressurized vessels, piping and equipment removed from service, isolated and vented.

Fire watch/Hot Work area monitoring

☐ Fire watch will be provided during and for 60 minutes after work, including any coffee or lunch breaks.
☐ Fire watch is supplied with suitable extinguishers, and where practical, a charged small hose.
☐ Fire watch is trained in use of equipment & in sounding alarm.
☐ Fire watch may be required in adjoining areas, above & below.
☐ Monitor Hot Work area for 4 hours after job is completed.

Other Precautions Taken:

☐ ....
....

FIGURE 52-4 Hot Work Permit Sample

Source: *Introduction to Employee Fire and Life Safety,* NFPA, 2001, Exhibit 10.2; © 2000 Factory Mutual Insurance Company. All rights reserved. Reprinted with permission.

Warning!

HOT WORK IN PROGRESS
WATCH FOR FIRE!

PART 2

INSTRUCTIONS

1. **Person doing Hot Work: Indicate time started and post permit at Hot Work location. After Hot Work, indicate time completed and leave permit posted for Fire Watch .**
2. **Fire Watch: Prior to leaving area, do final inspection, sign, leave permit posted and notify Firesafety Supervisor.**
3. **Monitor: After 4 hours, do final inspection, sign and return to Firesafety Supervisor.**

HOT WORK BEING DONE BY
☐ EMPLOYEE
☐ CONTRACTOR..

DATE	JOB NUMBER

LOCATION/BUILDING AND FLOOR

NATURE OF JOB

NAME OF PERSON DOING HOT WORK

I verify the above location has been examined, the precautions checked on the Required Precautions Checklist have been taken to prevent fire, and permission is authorized for this work.

SIGNED (Firesafety Supervisor/Operations Supervisor)

TIME STARTED ☐ AM ☐ PM	TIME FINISHED ☐ AM ☐ PM

PERMIT EXPIRES	DATE	TIME AM PM

FIRE WATCH SIGNOFF:
Work area & all adjacent areas to which sparks & heat might have spread were inspected during the watch period and were found fire safe.
Signed: ______________________________

FINAL CHECKUP:
Work area was monitored for 4 hours following Hot Work and found fire safe.
Signed: ______________________________

REQUIRED PRECAUTIONS CHECKLIST

☐ Available sprinklers, hose streams and extinguishers are in service/operable.
☐ Hot Work equipment in good repair.

Requirements within 35 ft (11 m) of work
☐ Flammable liquids, dust, lint and oily deposits removed.
☐ Explosive atmosphere in area eliminated.
☐ Floors swept clean.
☐ Combustible floors wet down, covered with damp sand or fire-resistive sheets.
☐ Remove other combustibles where possible. Otherwise protect with fire-resistive tarpaulins or metal shields.
☐ All wall and floor openings covered.
☐ Fire resistive tarpaulins suspended beneath work.
☐ Protect or shut down ducts and conveyors that might carry sparks to distant combustibles.

Work on walls, ceilings or roofs
☐ Construction is noncombustible and without combustible covering or insulation.
☐ Combustibles on other side of walls, ceilings or roofs are moved away.

Work on enclosed equipment
☐ Enclosed equipment cleaned of all combustibles.
☐ Containers purged of flammable liquids/vapors.
☐ Pressurized vessels, piping and equipment removed from service, isolated and vented.

Fire watch/Hot Work area monitoring
☐ Fire watch will be provided during and for 60 minutes after work, including any coffee or lunch breaks.
☐ Fire watch is supplied with suitable extinguishers, and where practical, a charged small hose.
☐ Fire watch is trained in use of equipment & in sounding alarm.
☐ Fire watch may be required in adjoining areas, above & below.
☐ Monitor Hot Work area for 4 hours after job is completed.

Other Precautions Taken:
☐ ....
....

FIGURE 52-4 *(Continued)*

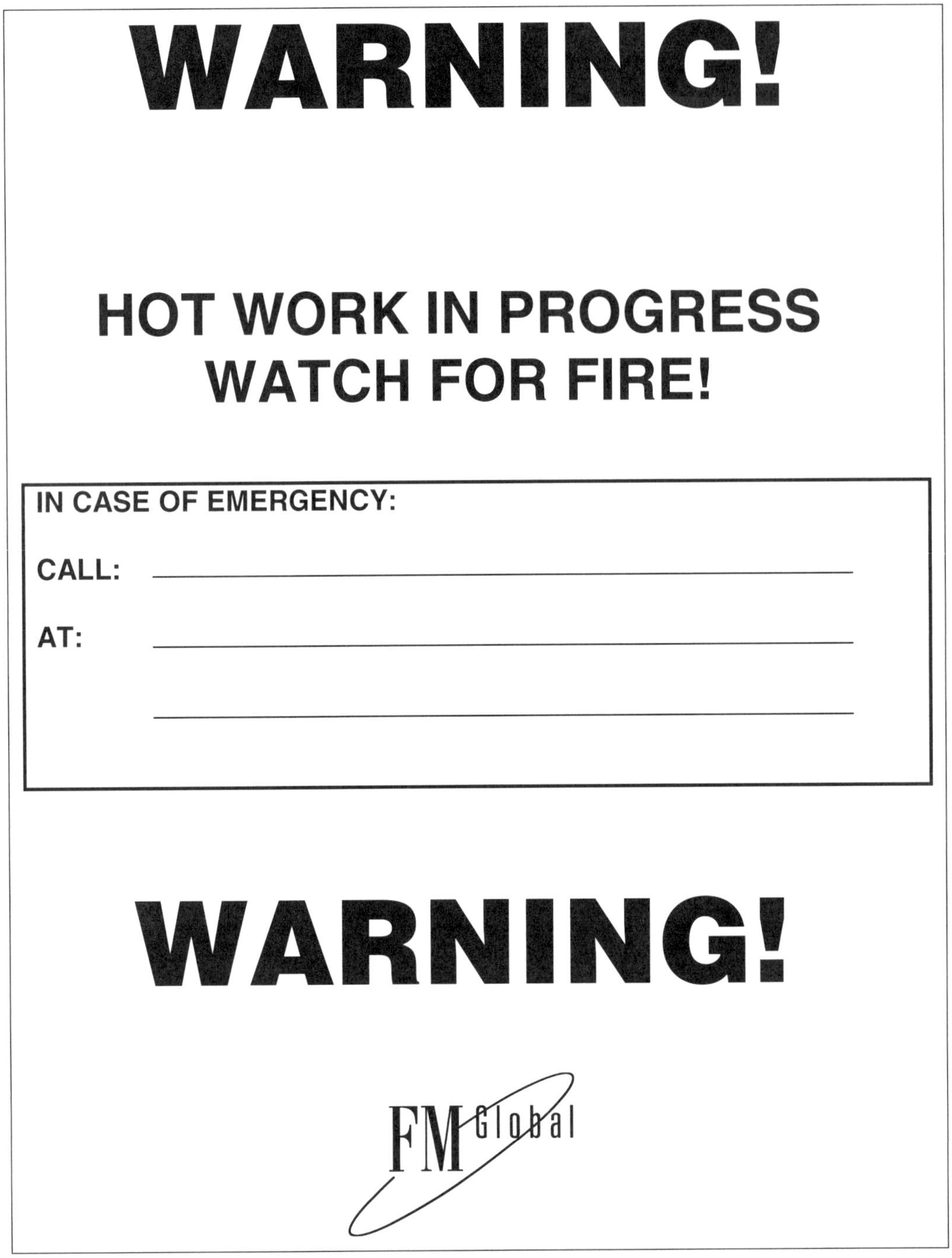
WARNING!

HOT WORK IN PROGRESS
WATCH FOR FIRE!

IN CASE OF EMERGENCY:

CALL: ______________________

AT: ______________________

____

WARNING!

FM Global

FIGURE 52-4 *(Continued)*

Although the issuer of the permit determines for how long the hot work permit is valid, in most cases the permit period will vary anywhere from at least a half hour to one full work shift, or eight hours. NFPA 51B recommends that the maximum period for a permit be 24 hours. Issuing the permit for longer periods of time can result in increasing the site fire hazards, because in a busy worksite, the conditions can change quickly. The permit is valid for only the conditions under which it was issued; it should be clearly posted. A permit should indicate if a fire watch is necessary. The inspector should review some of the previously issued permits to ensure it is fully filled out and contains the necessary information to ensure safety.

Fire Watch

To determine if a fire watch is needed, an inspector can use the fire watch decision tree (Figure 52-5). Although this decision tree allows hot work to be performed without a fire watch, many in the industry require a fire watch to be present during *all* hot work in nondesignated areas. This approach is a conservative one, however, and is not required by the regulations.

The job of the fire watch is to monitor the area and watch for fires from the hot work operation and to make sure others do not come into the area or move

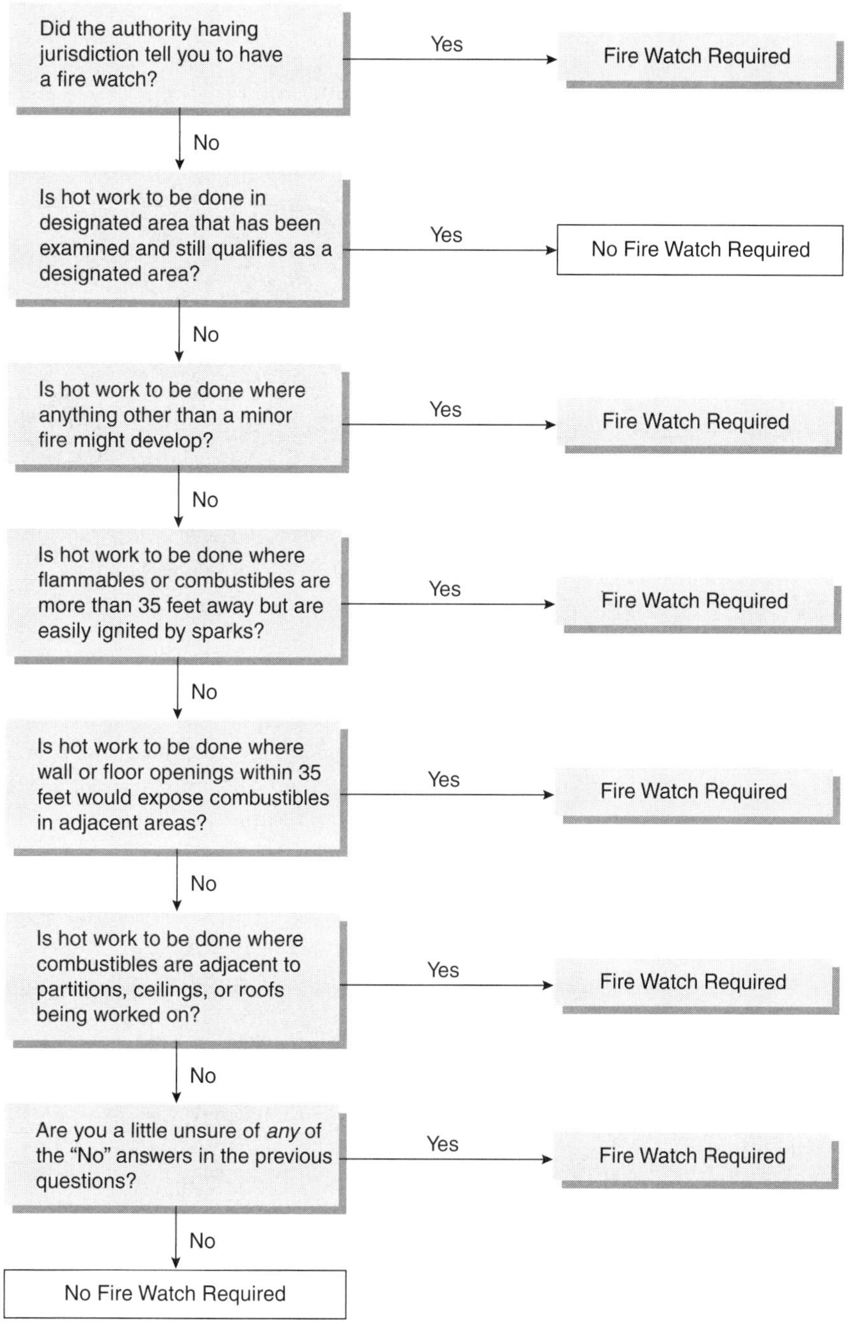

FIGURE 52-5 Fire Watch Decision Tree
Source: *Introduction to Employee and Life Safety,* NFPA, 2001, Figure 10.4.

combustible materials into the work zone during the hot work (Figure 52-6). The fire watch should have extinguishing equipment readily available in the area and be trained in its use. A good question for the inspector to ask the fire watch is whether the fire watch knows and understands that he or she is authorized to stop hot work if unsafe conditions develop. If the fire watch is not aware of this important responsibility, the fire watch has not been properly trained or the management does not have an understanding of the role of a fire watch. The fire watch should maintain a continuous watch during the course of the hot work and for at least 30 minutes after completion of the hot work. Any other assigned jobs should not interfere with the primary job of fire watch, which may include sweeping the immediate area or other minor jobs in the immediate vicinity.

More than one fire watch may be needed if the area to be watched is so great or the potential for fire within the area is so high that one individual alone cannot provide the desired protection. More than one fire watch is common when the work involves areas near elevators, mezzanines, atriums, oddly shaped workspaces, or where equipment is attached to walls or floors when the hidden area on the other side of the wall or floor may have combustibles nearby.

Sprinklers

If the heat of the hot work operation is likely to cause the automatic sprinkler system to operate, the sprinklers in the immediate vicinity can be temporarily covered with noncombustible material or damp cloths. The temporary protection must be sure to be removed when the operation is complete. Note, however, that hot work is prohibited in any area where the sprinkler system is impaired. Any place that was designed to require the protection of sprinklers is particularly vulnerable during hot work, a common source of fires. A few individual sprinklers covered on a temporary basis are not considered to be "impaired."

Hot Work Containers and Piping

The term *container* refers to jacketed vessels, pipes, tanks, covered parts, and other equivalent items, each of which should be considered to have contained flammable, toxic, explosive, or reactive material that the hot work could release or ignite. No container should be presumed to be clean and safe. Hot work should not be permitted on containers unless they have been inspected and certified as safe for the job. The assumption should not be made that the container will remain safe during the job; the heat of the work may release hazardous substances hidden in the crevices and corners of the container. For further information, the inspector should refer to ANSI/AWS F4.1, *Recommended Safe Practices for the Preparation for Welding and Cutting of Containers and Piping.*

EQUIPMENT

Regulators

Regulators are not interchangeable among the designated gas services at the site. They should be used only for the gas and pressure for which they are labeled. This will prevent them from failing and releasing potentially hazardous gases.

Oxygen gauges should be marked "Use No Oil." Oxygen-enriched fires ignite readily and burn with great violence, so it is imperative that damaged nuts and connections are replaced to avoid leaks and that all connections are checked with leak detection fluid before the system is used. Only qualified technicians should be allowed to repair regulators.

FIGURE 52-6 Fire Watch Monitoring a Hot Work Area
Source: *Introduction to Employee and Life Safety,* NFPA, 2001, Figure 10.5.

Torches and Other Hot Work Tools

The only torches that should be used are those that have been approved by a nationally recognized testing laboratory, meet nationally recognized standards, and have been tested and found to be safe. They are to be ignited only by friction lighters, stationary pilot flames, or some other suitable means of ignition. Matches are unsuitable because the torch gases can blow out their flames and create an explosion or health hazard by allowing too much gas to escape into the work area.

All tools must be shut off when they are not going to be used for a substantial period of time, such as during lunch or overnight. Remove the tools from the area when the work stops for the night, or, with a torch, make sure that the gas supply system has been turned off. Doing so will prevent gas, which can cause fires and explosions, from leaking from the torch and accumulating in the work area.

Hose

In the United States, fuel gas hoses are generally red and oxygen hoses are generally green. Black is used for inert gas and air hoses. These colors should not be mixed. Sometimes hoses are taped together for convenience and to prevent them from tangling. When this is the case, not more than 4 in. (10 cm) in each foot (30 cm) of hose should be covered by the tape. This will allow gases to escape when cracks and defects appear in the hose. The inspector should check the hoses for any damage that could cause leaks and should make sure that burned and damaged sections have been replaced or repaired. The inspector should make sure that all connections are tight and leak free. Gas connections should not be compatible with the breathing air connections.

Flash Arrestors and Protective Devices

Most flash arrestors and protective devices are optional. When an approved device, such as a reverse flow check valve or flash back arrestor, is used, it should be used and maintained in accordance with the manufacturer's instructions. These devices are not a substitute for proper system maintenance.

Cylinders

Cylinder contents must be labeled indelibly on the shoulder of the cylinder. Cylinders over 30 lb (0.45 kg) should have valve caps to prevent shearing the valves, resulting in the release of a gas. The valve protection cap is designed to take the blow in case the cylinder falls. Cylinders must be stored away from sources of heat, combustibles, and exits and at least 20 ft (6 m) from highly combustible materials. There are specific requirements in OSHA for capacities of various gases stored inside buildings, how specific gases should be stored, and which gases cannot be stored together. This chapter cannot cover all of the specifics, but it is important to know the requirements are there. NFPA 55, *Standard for the Storage, Use and Handling of Compressed and Liquefied Gases in Portable Cylinders,* provides detailed information. Note that almost all the cylinder requirements exempt oxygen and fuel gas cylinders "in use" that are connected with gauges and hoses for hot work. Cylinders should be chained or somehow restrained to keep them from falling. Gas cylinders are prohibited in confined spaces.

FIRE INCIDENT TRENDS: SCENARIOS FOR INSPECTOR VIGILANCE

If inspectors are aware of the trends for when and where hot work fires are most likely to occur, they can focus their efforts on these particular situations by querying the employees or by viewing the permits during inspections. Unless noted otherwise, the following data are taken from reports from 1993 to 1997 as reported in the NFPA Torch Report.

- *Perform inspections of cutting and welding operations:* When compared other types of hot work, the types of hot work most dangerous were identified as cutting and welding. Cutting torches and welding torches caused 42 and 28 percent of all nonresidential hot work fires, respectively; and 54 and 19 percent of nonresidential property damage, respectively.
- *Review the maintenance shop or other designated areas:* Surprisingly, the maintenance shop was the most common area for cutting and welding fires; and the roof was the most common area for other hot work fires. Concealed spaces, such as an attic or ceiling/roof assembly were the other most common areas.
- *Perform inspections when hot work is to be completed near structural members/framing and thermal/acoustical insulation:* The materials most frequently first ignited made up one-third of the data and included structural members/framing and thermal/acoustical insulation.
- *Perform inspections around break or shift change times:* The majority of hot work fires occur just before, during, or after lunch. There are also peaks in fires from 9:00 A.M. to 10:00 A.M., likely representing a morning break period; and again between 2:00 P.M. and 4:00 P.M., quitting time for many workers. As expected, the least likely time period for a fire due to hot work is between 2:00 A.M. and 5:00 A.M.
- *Pay special attention to completed hot work of outside contractors:* FM Global statistics show that from 1994 to 1999 the number of hot work fires in which outside contractors were involved was 66.5 percent, a 9 percent increase over the previous 5 years.

ADDITIONAL INFORMATIONAL RESOURCES

In addition to the standards and regulations noted earlier, there are numerous Internet sites for additional information. The OSHA regulations can be accessed through www.osha.gov. OSHA's website is easy to use; the homepage has a regulations and compliance link directly to the standard. Inspectors can download the regulation of interest into their word processing package, which allows them easy search capability and convenience. Information is also provided on the frequency and severity of code violations for a given section of the OSHA regulation.

AWS has a website at www.aws.org. In the welding industry news section, inspectors can find hot work health and safety fact sheets on topics such as specific process safety, electrical hazards, atmospheric hazards such as fumes, burn protection, ergonomics, confined spaces, and lockout/tagout.

SUMMARY

Inspectors should be especially vigilant when inspecting hot work procedures since hot work is frequently cited as the cause of many large loss fires. This connection between hot work and fires is not surprising: hot work tools are inherently dangerous because they are highly portable sources of ignition. If we take the time to examine the root cause of these hot work fires or near misses, we find that it often boils down to not following commonsense safe practices that have been in existence in regulations and standards, largely unchanged, for decades. Inspectors should become familiar with the most common hot work danger scenarios, as presented in this chapter. They should also take the time to carefully inspect the hot work procedures used on a daily basis. Inspectors should keep the dangers and the possible prevention strategies, such as hot work permits and fire watches, in mind. If they do so, they will likely spot and be able to help correct some glaring safety shortcomings that are often the result of the process having become routine for operators and safety personnel.

BIBLIOGRAPHY

Accident Prevention Manual for Industrial Operations: Engineering and Technology, 8th edition, National Safety Council, 1980.

ANSI/AWS F2.2, *Lens Shade Selector,* American Welding Society, Miami, FL.

ANSI/AWS F4.1, *Recommended Safe Practices for the Preparation for Welding and Cutting of Containers,* American Welding Society, Miami, FL.

ANSIZ49.1, *Guide for Shade Numbers,* American Welding Society, Miami, FL.

ANSIZ49.1, *Safety in Welding, Cutting, and Allied Processes,* American Welding Society, Miami, FL.

Cote, A. E., ed., *Fire Protection Handbook,* 18th ed., NFPA, Quincy, MA, 1997.

Cote, A. E., and Linville, J. L., eds., *Industrial Fire Hazards Handbook,* 3rd ed., NFPA, Quincy, MA, 1990.

NFPA Codes, Standards, and Recommended Practices

See the latest version of The NFPA Catalog for availability of current editions of the following documents.

NFPA 51, *Standard for the Design and Installation of Oxygen-Fuel Gas Systems for Welding, Cutting, and Allied Processes*

NFPA 51B, *Standard for Fire Prevention During Welding, Cutting, and Other Hot Work*

NFPA 55, *Standard for the Storage, Use and Handling of Compressed and Liquefied Gases in Portable Cylinders*

NFPA 306, *Standard for the Control of Gas Hazards on Vessels*

NFPA 326, *Standard for the Safeguarding of Tanks and Containers for Entry, Cleaning, or Repair*

Hazards of Manufacturing Processes

CHAPTER 53

L. Jeffrey Mattern

CASE STUDY

Sprinkler Extinguishes Blaze

A single sprinkler extinguished a fire in a vacuum form molding machine, greatly limiting losses at a multimillion-dollar bathtub manufacturing plant in Virginia.

The single-story building of unprotected metal construction had a smoke detector system and a dry-pipe sprinkler system that was monitored by a central station alarm company.

The plant was in full operation when the form machine malfunctioned and ignited an acrylic sheet being shaped into a tub. A sprinkler over the machine operated, triggering the water flow alarm. Smoke detectors also operated.

The alarm company and a worker at the plant called the fire department at 11:35 A.M. Fire fighters arrived to find that the sprinkler had extinguished the blaze. Damage to the plant, valued at $2.8 million, was estimated at $500. There were no injuries.

Source: "Fire Watch," *NFPA Journal,* July/August, 1997.

L. Jeffrey Mattern has worked at FM Global for over 30 years, where he serves as chief engineer/standards engineer Mid-Atlantic Field Engineering and as worldwide coordinator of Arson and Fire Service Programs. He was a volunteer fire fighter for 25 years and has been active on various NFPA committees and the Standards Council since 1968. He is also a former member and past president of the Philadelphia/Delaware Valley chapter of SFPE and a current member of the SFPE Qualifications Board.

Manufacturing processes are many and varied. Similar processes are used in a variety of industries. Though they may be used to accomplish different manufacturing objectives, they often operate in a similar manner, and their hazards are most often the same from industry to industry. In this chapter we discuss a number of different processes that can be found in common industries. However, these processes may also be found in industries not referenced. In general, the information provided here is applicable to that process regardless of the industry in which it is found. Processes involving flammable and combustible liquids, gases, and dusts present the most significant exposure to a fire or explosion incident.

HEAT TRANSFER FLUID SYSTEMS

Heat transfer fluid systems, sometimes referred to as hot oil systems, are used in a wide variety of manufacturing processes, such as textile, roofing materials, woodworking, and metalworking. These systems are used in lieu of steam heat systems because of the high-pressure requirements for a steam system and because these systems have the ability to maintain a constant temperature transfer media in these systems. The oil is usually heated to near its flash point and with some systems to above its flash point. A release of the oil can result in prompt ignition.

Hazard

The system consists of a boiler, which is used to maintain the oil at a constant temperature, a system of insulated pipes between the boiler and the point of use, return lines to the boiler, expansion tanks, and point-of-use couplings. The boiler is typically a gas- or oil-fired unit with an internal heat exchanger through which the oil circulates and is heated on its route to the point of use. The insulated piping between the boiler and the point of use is usually equipped with welded fittings. At the point of use there is often a coupling that allows connection to rotating rolls or other equipment. Other times, there is a stationary heat exchanger that gives off heat to the environment surrounding the process. To contain the heat, this environment is usually enclosed or partially enclosed, such as in the case of an oven. A bypass line arranged with automated valves is located ahead of the point-of-use area.

The hazard involving this type of equipment is multifold. It includes fires in the boiler as a result of an oil leak in the heat exchanger, leaks in oil piping between the boiler and the point of use, and leaks at the point of use.

Fire Protection

Protection for the boiler part of the process often includes the use of CO_2 or steam for discharge into the boiler firebox. This system is usually a manually operated system that is activated once the heating unit is shut off. Inspectors should review NFPA 12, *Standard on Carbon Dioxide Extinguishing Systems*, to familiarize themselves with the installation details for CO_2 systems.

A leaking heat exchanger within the boiler is detected by means of an alarm that senses an abnormal oil temperature rise at the exit point of the heat exchanger. Prompt deactivation of the heating system and introduction of the suppression agent will usually minimize the damage to the internal components of the boiler and, specifically, to the heat exchanger. Locating the boiler in a cutoff fire-rated room or a detached building provides for the best means to limit uncontrolled fire spread into the manufacturing area.

The second area of concern is the point-of-use area. This is the area where the vast majority of the oil leaks and subsequent fires occur. For this reason, heat detectors are provided at the point-of-use area. The purpose of the detectors is to sound an alarm at the time of a fire and to activate the bypass line ahead of the point-of-use area so that oil is diverted from the area of the leak.

There are only few instances of welded pipe failures and release of oil with a subsequent fire. There are, however, many occurrences in which oil has been released at point-of-use couplings, followed by prompt ignition and extensive damage to manufacturing equipment.

Automatic sprinkler protection is a particularly important feature for the boiler room and for the area surrounding the point of use. To reduce the frequency of leaks, a comprehensive equipment inspection and maintenance program is of critical importance. The inspection results should be submitted in writing to supervisory personnel for review and the correction of deficiencies.

The inspector should be giving special attention to the fire protection provided such as CO_2 protection in the heat exchanger but more importantly automatic sprinkler protection in all locations where oil is present. Additionally, it is important to note if the boiler is located in a cutoff, fire-rated room or a detached building.

METAL CLEANING AND PLATING OPERATIONS

Metal cleaning and plating operations are performed in a wide variety of industries. The process usually involves nonflammable and noncombustible liquids—such as acids, caustics, and detergent-laden water—and ordinary water for final rinsing. These liquids are most commonly heated either by steam or electric immersion heaters.

Process Hazard

None of these liquids is typically a fire hazard in itself. The fire hazard associated with metal cleaning is normally in the equipment used and not normally in the liquids used within the equipment. Because of the reactive nature of the acids and caustics, the liquids are contained within various types of plastic tanks. Many of the tanks are combustible. A few special plastics that do not spread a fire beyond the point of initial origin are now available and should preferably be used for such tanks.

Because of the fumes emitted during normal operations, extensive industrial exhaust systems are provided. These systems consist of hoods over the tanks or "lip" ventilation at the tank edge, connected to a system of ducts. The exhaust ducts from each tank are usually connected to a main duct that is, in turn, connected to a scrubber (Figure 53-1). The ducts and the scrubber can be constructed of a variety of plastic materials but are more commonly made of plastics such as polyvinyl chloride (PVC) or fiber-reinforced plastic (FRP). The inspector should consult NFPA 91, *Standard for Exhaust Systems for Air Conveying of Vapors, Gases, Mists, and Noncombustible Particulate Solids,* for requirements associated with this type of duct.

Fire Protection

Combustible plastic tanks are often ignited when electric immersion heaters malfunction or when the liquid level within the tank drops below the level of the heating element, resulting in the overheating and subsequent ignition of the tanks. Operators should make sure that electric immersion heaters are not left operating when the area is unattended. Providing low liquid level alarms and shutoff devices can greatly reduce the possibility of the overheating problem.

Automatic sprinkler protection within these ducts, unless listed as not requiring sprinkler protection, is an important consideration in protecting this process. Fires occurring in the tank or the ducts usually travel the full length of the exhaust system, consuming the tanks, ducts, and scrubber.

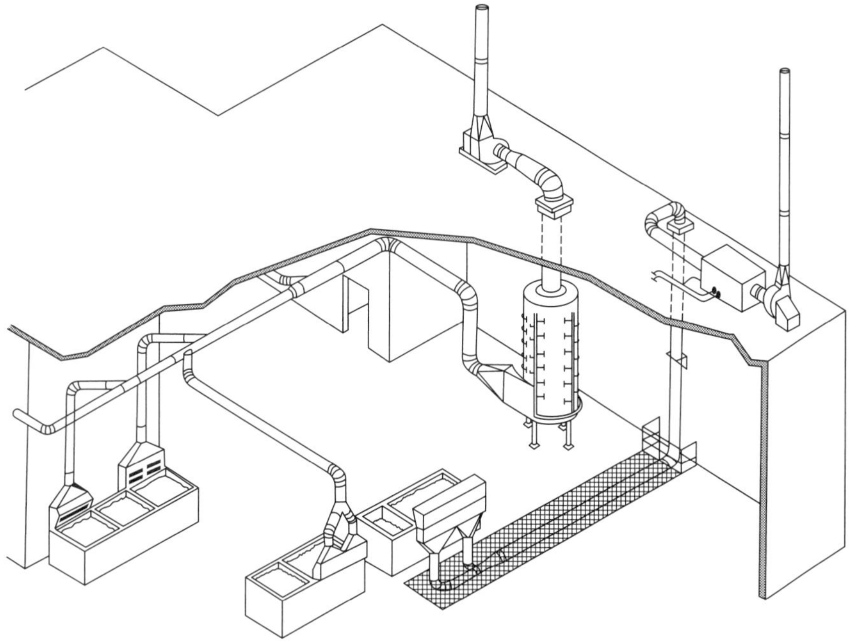

FIGURE 53-1 Exhaust System for One-Story Building Occupied by Various Types of Fume Hoods with Vertical-Type Fume Scrubber and Service Trench
Source: *Industrial Fire Hazards Handbook*, NFPA, 1990, Figure 51.7.

Certainly, another consideration is the control of spills of materials associated with this type of process. Under certain conditions, these materials can have an impact on the environment if they escape from the building. Containment arrangements surrounding tanks holding these materials or at points where the materials can exit the building should be investigated. Along these same lines, the discharge point of floor drains should be determined to limit the possibility of liquids escaping by this means.

OIL QUENCHING OPERATIONS

Oil quenching is conducted in conjunction with heat treating of metal parts. The purpose is to develop certain metallurgical qualities. Quenching can be conducted in tanks of a variety of sizes up to 1000 ft^2 (93 m^2) or even larger. The operation can consist of either a batch or a continuous operation. The oils typically have a flash point in excess of 300°F (149°C). The oil is normally maintained at a temperature in a range from ambient temperature to about 200°F (94°C). Some quenching operations include a heating system that may elevate oil temperatures to 400°F (204°C). When heated oil is present, the oil used typically has a flash point of some 500°F (260°C). Some systems also have temperature controlled by a cooling system.

Process Hazard

The type of oil, the size of the quench tank, the design of the tank, and the location of the tank in relation to the furnace determine the degree of hazard of the quenching operation. Another important consideration is the continual recirculation of the oil to prevent its overheating. The hazards associated with this operation focus on several possible areas:

- The hot metal part being lowered into the oil becomes hung up—partially submerged in the oil and partially above the oil surface—and is exposed to the atmosphere. The oil surrounding the hot metal part heats up to its flash point very rapidly and ignites.
- The hot metal part is submerged fully into the oil, which overheats due to the failure of the oil cooling system. The overheated oil reaches its flash point and ignites.
- Oil overflows the tank or splashes and comes in contact with the hot surface of the adjacent furnace and ignites.
- Water accumulates in the quench oil tank and when the hot metal part is loaded into the tank, the water is heated to its boiling temperature (212°F, 100°C) and converted to steam. The rapid expansion of water to steam at a potential expansion rate of 1:1760 causes the oil to be ejected from the tank. The oil is then often ignited on the hot surface of an adjacent furnace.

Fire Protection

The hazards of oil quenching can be mitigated by the following:

1. The provision of an emergency drain to a collection tank so that when a part becomes hung up, the oil can be quickly drained from the quench tank
2. The provision of an overflow drain to remove oil when it froths
3. The location of the quench oil tank a sufficient distance from hot surfaces to mitigate the possibility of oil ignition when oil splashes from the quench tank
4. The provision of a water detection device in the quench oil tank

5. The installation of automatic sprinkler protection over the tank and surrounding areas, which will provide a high level of protection
6. A carbon dioxide extinguishing system, which would provide additional protection

The inspector should consult NFPA 86, *Standard for Ovens and Furnaces,* regarding specific requirements. Also, see Chapter 50, "Heat-Utilization Equipment," in this manual for more information on furnace operations.

Certainly one of the more important aspects of reducing the hazard is operator training. Such training should include a good understanding of both normal operating procedures and emergency procedures. Additional important safety considerations include the existence of a prefire plan and the presence of either an incipient or structural fire brigade.

The inspector should investigate the type and flash point of the oil, the operating temperature of the oil if heated, the provision of emergency drains, provisions for containment of oil in the event of a spill or boilover, and provision of water detection devices within the quench tank. Of particular importance is the presence of a properly designed sprinkler system.

HYDRAULIC OIL OPERATED EQUIPMENT

Hydraulic oil operated equipment is used in virtually all types of manufacturing operations and is even present in many warehouses and distribution centers. In manufacturing operations it can be found in conjunction with extruders used for plastic blow molding and forming and for operating presses, lathes, and other equipment. In warehouses and distribution centers, hydraulic operated equipment can be found in conjunction with conveyor system elevators and with balers used for compressing and wrapping scrap cardboard and related materials.

Hydraulic oil systems can be internal to the piece of equipment being powered or they can be external and mounted on the floor adjacent to the equipment. Hydraulic oil systems can serve a single piece of equipment or they can service numerous pieces of equipment.

These systems typically use oils with flash points that range from mid 200°F (93°C) to high 200°F (93°C). The oil is contained in tanks ranging from 15 gal (0.06 m^3) to several hundred gallons (1.14 m^3 and greater) and in some cases is thousands of gallons (7.6 m^3 and greater). The oil is pressurized from as low as several hundred (21 bar and greater) to several thousand (200 bar and greater) pounds per square inch (psi). The oil can be transferred from the tank to the process equipment by hard piping; by rubber hose with an internal wire braiding; by rubber hose with a heavy external wire braid; or, as is often the case, by ordinary pressure-rated industrial rubber hose. Often, use of hard piping is not compatible with the flexibility needed to accommodate the movement of the process equipment. The priority of transfer is by hard piping, exterior wire braid hose, or, last, internal wire-braided hose.

Equipment Hazard

The rupture of a hose, the cracking of a fitting, or the separation of a coupling causes a release of fluid under high pressure. The release of the hydraulic oil fluid under high pressure can cause a large atomized oil cloud that can be readily ignited by many common ignition sources. Common ignition sources often present are hot surfaces of the process equipment, ceiling-mounted heating appliances, or other transient ignition sources. Once ignited, the pressurized fluid being discharged from the system will continue to burn, much like a blowtorch, until the hydraulic oil system is shut down or the oil is consumed.

Fire Protection

Low-level sensors in the oil tank can accomplish a relatively prompt shut down of the hydraulic oil system resulting from a discharge of fluid. However, with large oil systems and equally large oil tanks, it may take a considerable period of time until the oil reaches the level at which the sensor shuts off the system. Since more often than not, the atomized cloud will ignite, it is extremely important to shut down the oil supply to extinguish the flaming discharge. The shutdown can be best accomplished by the installation of heat detectors around the hydraulic oil operated equipment or by the addition of a cable located above the equipment between a microswitch at the oil pump motor controller and a fusible link mounted at a point above the equipment.

Another method of reducing the fire exposure of hydraulic oil systems is by the use of less hazardous hydraulic fluids. This type of fluid is difficult to ignite and once the ignition source is removed, the flaming is interrupted.

A comprehensive equipment inspection and maintenance program is critically important to reducing the frequency of leaks and equipment malfunctions. The inspection results should be submitted in writing to supervisory personnel for review and the correction of deficiencies.

The most important consideration for the inspector is the provision of automatic shutoff devices for the hydraulic oil pumps. The provision of such a device will promptly stop the flow of oil. Additionally, automatic sprinkler protection should be provided in all areas where flammable or combustible oils are present.

WOODWORKING

Hazard

Woodworking occupancies contain many manufacturing process hazards. These hazards are associated with the generation of wood dust from cutting and sanding operations, the use of flammable and combustible liquids in the finishing process, the use of industrial heating equipment and heat transfer fluid systems in the finish curing process, and sawdust and waste wood fired heating systems.

Wood dust is generated at virtually every step prior to the finishing process, with the dust becoming finer near the end of the prefinishing process. The dust must be recovered at each step. If dust is not adequately collected, fugitive dust fines will collect on high building members and on the top of equipment. These areas become difficult to access for cleaning purposes, and thus accumulations continue to develop. Even with good exhaust ventilation systems provided at points where dust is liberated, inspection of high-level building members; the tops of utility piping, ductwork, and conduits; and the tops of equipment is essential to detect accumulations of dust.

Such accumulations of dust can be shaken free—sometimes as the result of a primary explosion in equipment—and can be suspended in the air like an explosive cloud. This explosive dust cloud can be looking for an ignition source or can be ignited by the primary explosion in equipment that initially shook it free. Dust accumulations of considerably less than a half-inch can result in the potential for a flash fire or an explosion.

Fire Protection

The design of the dust recovery systems must be such that they have adequate capture velocity to lift and move the dust into the air stream and adequate carrying velocity to prevent the dust from falling out prior to reaching the cyclone. Ductwork should be inspected periodically to assure that dust is not accumulating in the duct. From the cyclone, the dust is dropped into a waste recovery bin. The dust is then either hauled away or fed into a wood waste burner.

Housekeeping on a regular frequency is important to help reduce the possibility of flash fire and explosions. The frequency of such cleaning is difficult to indicate specifically because of the great variability between the types of equipment used and the adequacy of the dust collection systems. Regular manual cleanup of dust residue on and around production equipment is important. This can be accomplished by using "soft" brooms to move the dust to floor level and then sweeping it into collection devices. Blowing down dust from upper levels is very undesirable because this action can create a dust cloud that can be ignited and result in a flash fire or an explosion. Dust accumulations on upper levels should be vacuum removed to eliminate the possibility of the formation of a dust cloud.

NFPA 664, *Standard for the Prevention of Fires and Explosions in Wood Processing and Woodworking Facilities,* and NFPA 654, *Standard for the Prevention of Fire and Dust Explosions from the Manufacturing, Processing, and Handling of Combustible Particulate Solids,* provide requirements for the protection of ductwork and other aspects of the occupancy. Details regarding spray finishing and thermal oil systems can be found elsewhere within this text.

The most important consideration for the inspector is to identify if there are dust accumulations at high levels of the building. These accumulations cannot be observed from floor level. Ladders or other specialized equipment must be used to inspect beams, girders, and equipment near roof level.

ELECTRIC DISCHARGE MACHINING

Process Hazard

Electric discharge machining, which is conducted on a stationary machine, is a common process in which the very precise cutting of metal parts is required. The machine has an open top coolant oil reservoir. The reservoir typically contains 30 gal (0.11 m^3) to a maximum of about 180 gal (0.68 m^3) of oil, with a flash point of approximately 280°F (138°C).

The metal part to be cut is mounted on a matrix at the base of the oil reservoir. The coolant oil is then introduced into the reservoir. Once the reservoir is full and the oil begins circulating from a clean oil tank through the reservoir and out to a dirty oil tank and filter, the electrode-cutting tool moves vertically from above the oil surface through the oil to the top of the part. The cutting is typically computer controlled. The electrode is activated and cuts through the metal part as programmed.

Fire Protection

The circulating oil is critical to cooling the part and the tool. Additionally, there are typically large volumes of oil used in the process because multiple machines are serviced by a single oil tank and filter system. It is extremely important that interlocks be provided for low as well as high oil levels.

The transfer of the oil from the tank and oil system to the machine reservoir is accomplished either by gravity flow or by pumping. In either case, the transfer should be through either steel pipe or exterior wire-braided hose. The oil tank and filter should be located in a contained or diked area. Fusible link safety shutoff valves are often provided on the feed line to the machine reservoir so that, in the event of a fire, the oil supply is shut off. Automatic sprinkler protection is an important consideration for areas having this type of equipment.

The inspector should identify that containment is provided for a potential spill, that low oil shutoff devices are provided on reservoirs, that proper hose or pipe transfer is provided, and that automatic sprinkler protection is provided.

AMMONIA REFRIGERATION SYSTEMS

Ammonia refrigeration systems are used in a variety of manufacturing processes in the food, chemical, and pharmaceutical industries. Additionally, they are used in refrigerated and freezer warehouses. Freon-type systems are also used for these same applications, but because large ammonia systems are more economical, the Freon-type systems are typically used for smaller refrigerated installations.

Hazard

Ammonia is a flammable and explosive refrigerant, with an explosive range of 16 to 22 percent. Although this explosive range is narrow and high, a catastrophic release can and has allowed ammonia to accumulate and enter this range, resulting in very large explosions (Figure 53-2). A favorable factor is that ammonia is lighter than air and can be vented to the atmosphere quite readily.

Fire Protection

Ammonia is typically circulated from a compressor room, often called the "engine room," to the point of production or warehouse refrigeration use. The compressors are usually internally oil lubricated with oil at low pressures (10 to 20 psi would be typical). There is also an oil separator that removes oil from the ammonia stream. Compressor rooms are typically located in a detached building or in a cutoff room on the exterior wall of a facility. Such a removed location is an important consideration because of the concern for the effects of an explosion within the compressor room. If the compressor room is not located in a detached building or cutoff room, an explosion can be all the more devastating because it can increase the amount of damage within the facility itself.

Because ammonia is explosive, consideration should be given to hazardous location electrical equipment. This equipment should be compatible with NFPA 70, *National Electrical Code*®. An alternative to hazardous location electrical equipment, as covered in ANSI/ASHRAE 15-1989, *Safety Code for Mechanical Refrigeration Systems,* allows for a roof-level emergency exhaust ventilation system, activated by an ammonia

FIGURE 53-2 Result of an Anhydrous Ammonia Explosion in Houston, TX, on December 11, 1983
Source: *Industrial Fire Hazards Handbook,* NFPA, 1990, Figure 50.5; Courtesy of Houston Fire Department.

detection system provided throughout the compressor room. This option is generally followed in industry. The rationale is that, as the result of the characteristics of ammonia (it is lighter than air and has a high and narrow explosion range), a rapid-activating, high-volume exhaust system will keep the ammonia concentration below its lower explosion level (LEL).

The two most important considerations for the inspector are the presence of emergency exhaust ventilation and the location of the "engine room" in a cutoff room on an exterior wall or located in a detached building.

PAINT FINISHING SYSTEMS

Paint finishing, which is one of the most common of manufacturing processes, can be found in a variety of industries, including woodworking and metalworking. The hazards associated with paint finishing systems focuses on the type of finishing material and how it is stored, prepared for use, transported, applied, and cured.

Finishing Materials

Finishing materials are usually in liquid form, although powder application finishing systems are found in increasing numbers in metal finishing industries. Both of these types of materials create unique hazards.

Process Hazard. Liquid finishing materials come in the form of flammable liquids, combustible liquids, and noncombustible liquids. Certainly, flammable liquids create the highest level of concern because these types of materials, once they are discharged from their container or transporting vessel and come into contact with air, create a flammable mixture that can be ignited immediately or shortly after the time of their release. All they need is an ignition source. The creation of static electric charges at the discharge point of the liquid can ignite some low flash point liquids.

Combustible liquids are liquids that must be preheated in order to release vapors that can be ignited. Typically, these liquids are not ignited immediately upon their release unless they are in the presence of a very strong ignition source, such as an exposure fire. More often than not, a combustible liquid fire results from the release of the liquid and the subsequent development of a pool of the liquid, with a delayed ignition occurring during the cleanup activity.

Noncombustible liquids are liquids that may have a flash point but lack a fire point. This is most often the case with water-based finishing liquids. To accurately evaluate the liquid, it is extremely important to know both the flash point and the fire point. The liquid's lack of a fire point does not by itself prevent the liquid from being a fire hazard. Processes using noncombustible liquids can have hazards associated with the liquid. Often, the dried-out residue of the liquid in the vicinity of the finishing activity can be combustible.

Finishing systems typically require large amounts of the liquid. When flammable and combustible liquids are used, these materials can be found in drums of less than 60-gal (0.23 m^3) capacity, in portable intermediate bulk containers (IBCs) up to 900-gal (3.4 m^3) capacity, and in stationary tanks often in the 10,000- (37.5 m^3) to 25,000-gal (95 m^3) capacity. The more of the material is present at the finishing process area, the greater is the hazard.

In smaller finishing processes, having drum storage right in the area where the finishing process is located is not uncommon. The old adage that it is acceptable to have a "shift supply" of drums on hand at the finish may not be applicable in many processes where six to eight different materials are required. The concern here is not only the presence of the flammable liquids in the finishing area but also the need to frequently transport drums to and from the finishing area. This transportation process increases the hazard because spills can occur anywhere on the way from the flammable liquid storage area to the finishing area. Sections of the facility between the drum storage area and the finishing area may not have automatic sprinkler protection designed for a flammable liquid type fire and may thus expose the entire facility to a catastrophic fire.

Fire Protection. Although IBCs greatly reduce the number of exposures resulting from transportation through a facility, they expose the facility to the dangers of large volume spills. Depending on the type of IBC (stainless steel, solid high-density plastic, thin plastic liner in wire basket containment, etc.), the propensity for a spill as a result of mechanical damage or an exposure fire changes dramatically. Additionally, mechanical damage to bottom drain piping or hoses supplying the liquid to the finishing process can result in a large uncontrolled spill. For this reason, a pump that is rated for the type of liquid involved and that is inserted in the top port of the IBC is a preferred means to transfer the liquid to the finishing process.

Regardless of whether drums or IBCs are used, potential spills must be contained. It is frequently argued that curbs are an obstruction to normal operations. To counter this argument, an *above finished floor* contained area with metal grating can be provided on which to place drums and IBCs so that any spill is contained. The containment size should be adequate to hold the largest possible spill as well as to allow for the discharge of sprinkler water.

The preferred means of transfer is to store the liquids in remote tanks and to pump them to the finishing area. This method allows liquids to be received by truck or rail and transferred to the tanks for storage. Fixed piping between the tanks and the finishing area then provides a vastly safer means to move the liquid. The design of the piping, its route through the plant, and the presence of fusible link safety shutoff valves at points between the tank storage and the finishing area are all important considerations.

Whether the bulk supplies of liquids are in drums, IBCs, or in tanks, they should be held in a cutoff or detached area until called upon to be transferred to the finishing area. A cutoff area should be of fire-resistant design suitable for the size and type of liquids to be stored within. Additionally, the area should have adequate containment, ventilation, ignition source control, and access limitation to those who are trained in the handling of flammable liquids.

The two most common methods of applying liquid finishing materials is by spray applying and by dipping and coating. Both of the methods introduce special hazards and require proper design in order to provide for safe operation. Spray applications are found in both woodworking and metalworking facilities; whereas dipping and coating is most often found in, though not limited to, metalworking facilities.

It is most important for the inspector to first determine the type of liquid that is being used. Once the inspector determines this, then he or she can assess the needs for automatic fire protection and containment for potential spills. The inspector should also be looking at the amount of liquid at the process area, whether it can be reduced, and whether it is properly contained. If large quantities of flammable and combustible liquids are needed at the plant, the inspector should ensure that storage areas are properly isolated from plant process areas.

Spray Application Finishing Systems

Spray finishing operations involve a variety of booths, including dry, water-wash, electrostatic, moving, downdraft, manually loaded, conveyor fed. Each type has it own unique operating features. The hazards associated with each type booth may vary somewhat, but the guidance provided focuses on the common hazards associated with spray finishing operations in general.

Process Hazard. Spray finishing places the liquid in the most ideal form for ignition—that is, as small droplets suspended in the air. All the ingredients for ignition are present, just waiting for an ignition source. Maintaining a very high degree of ignition source control is, therefore, extremely important.

Fire Protection. Finishing equipment, the part being finished (if the latter is a conductive material), and any conveyors should be grounded in accordance with the requirements of NFPA 33, *Standard for Spray Applications Using Flammable and Combustible Materials.* All electrical equipment in the finishing area should be designed, listed, and installed in accordance with NFPA 70, *National Electrical Code.*

Ventilation within the finishing area, whether it be a room or a single-spray booth, should be provided, with adequate ventilation to contain vapors as well as overspray to a defined area and to remove the accumulation of low-level vapors. Vapors should be maintained below 25 percent of the LEL, as defined in NFPA 33.

The best protection for spray finishing operations is automatic sprinkler protection. In accordance with NFPA 33, protection should be provided throughout the entire spray finishing areas well as in the booth.

The inspector should identify the presence of the proper type of electrical equipment in the areas where explosion vapors may be present. Additionally, adequate exhaust ventilation of vapors and automatic sprinkler protection are important considerations.

Dipping and Coating Finishing Systems

Process Hazard. Dipping and coating of materials can also provide significant exposure because of the presence of a typically large volume of the finishing liquid. Dipping and coating operations, which most often involve metal parts, increasingly use water-base liquids. Frequently these liquids do not display a fire point. If, however, the liquid is flammable or combustible, an appropriate degree of caution is required to reduce the possibility of ignition.

Fire Protection. If the finishing liquid is a flammable liquid, evolving vapors must be removed from the area through a means of exhaust ventilation. Removal is frequently accomplished by "lip" ventilation, located on the edge of the tank.

Containment is also an important consideration so that any tank overflow, whether due to overfilling or to sprinkler water discharge at the time of a fire, does not spread throughout the manufacturing area. NFPA 34, *Standard for Dipping and Coating Processes Using Flammable or Combustible Liquids,* should be consulted for specific details.

Because of the large volume of liquid used in this process, the liquid is typically pumped to the tank through a piping system rather than being transported in drums or IBCs. As discussed earlier, transporting drums and IBCs through the manufacturing area increases the possibility of a spill and a subsequent fire in an area where sprinkler protection is not designed for controlling a flammable liquid fire.

Sprinkler protection over the tank, over the surrounding area, and in any pits beneath the tanks, supplemented with a fixed protection systems for the tank proper, should be considered.

The inspector should be looking at the same basic flammable liquid controls as detailed earlier, namely proper identification of the characteristics of the liquid, containment of potential spills, and automatic sprinkler protection.

SUMMARY

Although manufacturing processes are all different, examining the fundamental characteristics of each is important. The most important step in analyzing a process that uses a liquid is to properly evaluate the liquid to clearly identify the hazards associated with its use. In a process in which a dust emitting material is used, the areas of the building must be examined for dust accumulations. If flammable gases or flammable liquid vapor are present, the ventilation must be evaluated. In all processes, fire protection includes an appropriate fire extinguishing system.

BIBLIOGRAPHY

ANSI/ASHRAE 15-1989, *Safety Code for Mechanical Refrigeration Systems*

Cote, A. E., and Linville, J. L., eds., *Industrial Fire Hazards Handbook,* 3rd ed., NFPA, Quincy, MA, 1990.

NFPA Codes, Standards, and Recommended Practices

See the latest version of The NFPA Catalog for availability of current editions of the following documents.

NFPA 12, *Standard on Carbon Dioxide Extinguishing Systems*

NFPA 33, *Standard for Spray Application Using Flammable and Combustible Materials*

NFPA 34, *Standard for Dipping and Coating Processes Using Flammable or Combustible Liquids*

NFPA 70, *National Electrical Code®*

NFPA 86, *Standard for Ovens and Furnaces*

NFPA 91, *Standard for Exhaust Systems for Air Conveying of Vapors, Gases, Mists, and Noncombustible Particulate Solids*

NFPA 654, *Standard for the Prevention of Fire and Dust Explosions from the Manufacturing, Processing, and Handling of Combustible Particulate Solids*

NFPA 664, *Standard for the Prevention of Fires and Explosions in Wood Processing and Woodworking Facilities*

Aerosol Manufacturing and Storage

CHAPTER 54

Michael Madden

CASE STUDY

Explosion Kills Employee

A flash fire and explosion quickly overcame a sprinkler system at a New Jersey recycling facility, destroying the building and killing an employee.

The single-story recycling center, which was of unprotected, noncombustible construction, measured 120 × 120 ft (37 × 37 m). A wet-pipe sprinkler system monitored by a private alarm company protected the building. There were no smoke detectors.

The plant had been in operation for half an hour when an explosion and flash fire occurred, quickly overcoming the fire protection system. Employees called 911 at 7:30 A.M. after the automatic alarm failed to transmit the signal to the alarm company.

Investigators determined that thousand of aerosol cans containing hair paint released their isobutane-propane propellant as they were crushed by the compactor. The propellant vapors were ignited when a propane-fired fork lift truck was started, and flames spread to nearby combustibles, causing a flash fire that activated more than 50 sprinklers.

The building, valued at $1.5 million, and its contents, valued at $500,000, were a total loss. A 36-year-old employee was killed and three other employees and a fire fighter were injured.

Source: "Fire Watch," *NFPA Journal,* January/February, 1999.

During the regular course of fire prevention inspections, a fire inspector will undoubtedly have to address fire protection issues regarding the packaging of aerosol products at aerosol manufacturing facilities and the storage of aerosol products in warehouses, retail stores, and business and industrial occupancies. The manufacturing and storage of aerosol products present some unique fire safety challenges, which require the application of various fire protection concepts.

AEROSOL DEFINED

An *aerosol* is a product that is dispensed from an aerosol container by a propellant. A large variety of consumer and industrial products are packaged as aerosol products. These products include personal care products, such as hair sprays and shaving creams; food products, such as whipped cream, snack spreads, and cooking oils; household and industrial cleaning products, such as disinfectants and deodorizers; automotive care products, such as lubricants and carburetor cleaners; paints and coatings; and pesticides.

The actual product dispensed from the aerosol container is called the *base product.* The base product itself can be a flammable or combustible liquid, which presents its own set of hazards in the manufacturing process. Aerosol propellants, which dispense the base product from the aerosol container, are liquefied or compressed

Michael Madden, P.E., is a fire protection engineer and principal in charge of the Los Angeles office of Gage-Babcock & Associates. He is also chair of the NFPA Technical Committee on Aerosol Products.

FIGURE 54-1 Cutaway View of Aerosol Container
Source: *Fire Protection Handbook,* NFPA, 1997, Figure 3-20A.

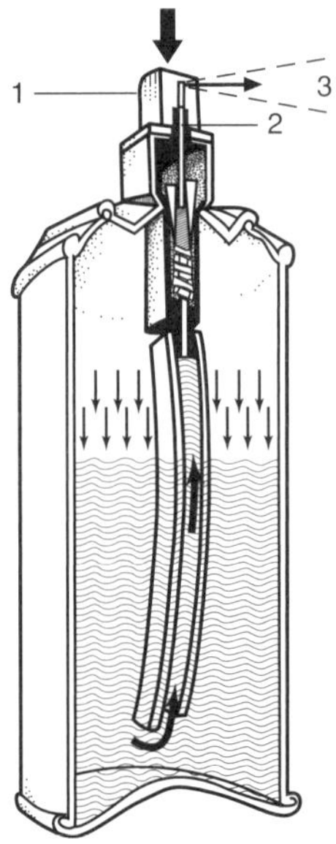

gases, with the most widely used propellants being hydrocarbon propellants, which are liquefied flammable gases.

The aerosol container is typically a high-strength metal container, with capacities ranging from a few ounces to 33.8 fl oz (1000 ml). Some small aerosol containers, up to approximately 4 fl oz (120 ml) are made of glass or plastic. The top and base of the aerosol container are typically domed or spherical, similar to the design of pressure vessels. Container working pressures range between 240 psi (1655 kPa) and 400 psi (2758 kPa). A typical aerosol container is shown in Figure 54-1. Note that when the plunger (1) is pressed, a hole in the valve (2) allows a pressurized mixture of product and propellant (3) to flow through the plunger's exit orifice.

Today, aerosol products are packaged by both large and small consumer product manufacturers, as well as by contract fillers. Contract fillers take the base product formulations supplied by the base product manufacturer and package the product in an aerosol container at the contract filler's facility. Aerosol filling operations are not limited to large manufacturing facilities in heavy industrial areas. Small contract filling facilities can be found in many communities.

AEROSOL PRODUCT FIRE PROTECTION STANDARDS

Requirements for the protection of aerosol manufacturing and storage facilities are found in NFPA 30B, *Code for the Manufacture and Storage of Aerosol Products.* The first edition of NFPA 30B was published in 1990. Prior to the publication of NFPA 30B, aerosol products were regulated under NFPA 30, *Flammable and Combustible Liquids Code,* as Class I-A flammable liquids. Protecting aerosol products as Class I-A flammable liquids resulted in what some considered excessive fire protection requirments. A series of fire test programs conducted in the 1980s by a group of aerosol manufacturers, in cooperation with sprinkler product manufacturers, identified that aerosol products present fire protection challenges different from those of flammable liquids and that the protection of aerosol products was dependent on the flammability of both the base product and the flammability of the aerosol propellant. This aerosol test program led to the development of an aerosol fire hazard classification system and the development of sprinkler design criteria to address the fire hazards presented by aerosol products.

CLASSIFICATION OF AEROSOL PRODUCTS

NFPA 30B classifies aerosol products into one of three categories, based on the combustibility of the product formulation. Sprinkler protection requirements, and aerosol product storage quantities and arrangement are based on this aerosol product classification system.

Aerosol products are classified based on a calculation of their chemical or theoretical heats of combustion, taking into account the base product and the aerosol propellant. The aerosol classification levels are as follows:

- *Level 1:* Those aerosol products with a total chemical heat of combustion equal to, or less than, 8600 Btu/lb (20 kJ/g)
- *Level 2:* Those aerosol products with total chemical heat of combustion greater than 8600 Btu/lb (20 kJ/g) and less than or equal to 13,000 Btu/lb (30 kJ/g)
- *Level 3:* Those aerosol products with total chemical heat of combustion greater than 13,000 Btu/lb (30 kJ/g).

Tables 54-1, 54-2, and 54-3 provide examples of the calculations utilized to determine the appropriate aerosol level classification. The Appendix to NFPA 30B contains a listing of common aerosol propellants and some of the more common base product constituents along with their chemical heats of combustion.

Although fire inspectors should have knowledge of the aerosol product classification system, they should not have to perform any classifications for the purpose of performing inspections of facilities. Information on aerosol product classification should be readily available from facility personnel.

NFPA 30B requires that the aerosol level classification be marked on the aerosol product cartons, so that the classification labeling is readily visible when the cartoned aerosol products are stored in a warehouse environment. The aerosol level classification system established in NFPA 30B is intended only for the application of fire protection requirements. The aerosol level classification is separate from any consumer product labeling requirements, which are intended as warnings for the end-use consumer. The terms "Flammable," "Highly Flammable," or "Extremely Flammable" labeled on the aerosol container itself is a warning for the consumer; it is not intended to classify the product for application of NFPA 30B requirements.

TABLE 54-1 Example of Level 1 Aerosol Product

Ingredient	Weight (%)	ΔH_c of Ingredient (kJ)	Weight % $\times \Delta H_c$ (kJ)
Isobutane	30	42.7	12.8
Water	69	0	0
Fragrance, etc.	1	43.7*	0.4
		Total =	13.2 kJ

Note: For U.S. customary units, 1 kJ = 0.95 Btu

*Since the fragrance constitutes a small proportion of the total, 43.7 kJ/g was used instead of actually determining or calculating the heat of combustion. In this example, the resulting classification of the aerosol product was not affected. However, with other products, this might not be the case and actual calculation of or testing for the heat of combustion might have to be done.

Source: NFPA 30B, 1998, Table A-1-7, Example 1.

TABLE 54-2 Example of Level 2 Aerosol Product

Ingredient	Weight (%)	ΔH_c of Ingredient (kJ)	Weight % $\times \Delta H_c$ (kJ)
Isobutane	20	42.7	8.5
Ethanol	60	25.5	15.3
Water	19	0	0
Fragrance, Surfactant, Corrosion Inhibitors, or other minor ingredients	1	43.7*	0.4
		Total =	24.2 kJ

Note: For U.S. customary units, 1 kJ = 0.95 Btu

*Since these minor ingredients constitute a small proportion of the total, 43.7 kJ/g was used instead of actually determining or calculating the heat of combustion. In this example, the resulting classification of the aerosol product was not affected. However, with other products, this might not be the case and actual calculation of or testing for the heat of combustion might have to be done.

Source: NFPA 30B, 1998, Table A-1-7, Example 2.

If aerosol product cartons are not marked with the appropriate aerosol classification in accordance with NFPA 30B, the inspector should treat the enclosed aerosol products as Level 3 aerosols.

AEROSOL MANUFACTURING FACILITIES

Aerosol product manufacturing facilities present several fire protection challenges. The challenges include the bulk storage of aerosol propellants, which typically include liquefied flammable gases; the blending and dispensing of base product formulations, which may be classified as flammable and combustible liquids themselves; and the filling and pressurizing of the aerosol container with the aerosol propellants. Aerosol manufacturing lines may be found in large and small consumer

TABLE 54-3 Example of Level 3 Aerosol Product

Ingredient	Weight (%)	ΔH_c of Ingredient (kJ)	Weight % $\times \Delta H_c$ (kJ)
Isobutane	25	42.7	10.7
Propane	10	43.7	4.4
Toluene	25	27.8	7.0
Acetone	15	27.9	4.2
Methyl Ethyl Ketone	15	30.7	4.6
Pigments (Titanium Dioxide), etc.	10	0	0
		Total =	30.9 kJ

Note: For U.S. customary units, 1 kJ = 0.95 Btu

Source: NFPA 30B, 1998, Table A-1-7, Example 3.

product manufacturing facilities, or they may be found in small contract filling operations throughout the community. Inspections of these facilities require a knowledge of flammable and combustible liquid storage and handling requirements (NFPA 30, *Flammable and Combustible Liquids Code*) and a knowledge of storage and use requirements for compressed gases (NFPA 54, *National Fuel Gas Code,* and NFPA 55, *Standard for the Storage, Use, and Handling of Compressed and Liquefied Gases in Portable Cylinders*) and liquefied flammable gases (NFPA 58, *Liquefied Petroleum Gas Code*).

Aerosol Propellant Tank Farms

Aerosol propellants are typically stored in bulk at aerosol product manufacturing facilities. The propellants are typically delivered to the facility by tank trucks or by railroad tank cars. Several different aerosol propellants may be stored at a manufacturing facility because of the different propellant requirements for different products. Aerosol propellants are selected based on the characteristics they provide for discharge of the base product. Most propellants used in the United States are hydrocarbon gases. Some special aerosol formulations may use noncombustible propellants, such as carbon dioxide.

The hydrocarbon propellants are typically blends of propane, butane, and pentane. Many of these propellant blends are identified by their vapor pressure at 70°F (21°C). For example, aerosol propellant A-31 refers to aerosol-grade isobutene, which has a vapor pressure of 31 psig (213.7 kPa). Hydrocarbon propellants are gaseous at ambient temperatures and pressures, and they condense to form liquids under moderate pressure or low temperatures. These propellants are typically stored as liquids in pressure vessels within a tank farm. Fire hazard properties of typical hydrocarbon propellants are indicated in Table 54-4.

TABLE 54-4 Fire Hazard Properties of Hydrocarbon Propellants

Chemical Name	CAS Number[1]	Chemical Heat of Combustion[2] ΔH_c, kJ/g
Acetone	67-64-1	27.7
Acrylic Resin	—	*
Alkyd Resin	—	*
Aluminum	7429-90-5	*
Asphalt	8052-42-4	22.7
Barium Sulfate	7727-43-7	0.0
Benzidine (Yellow)	92-87-5	*
Butane	106-97-8	43.3
2-Butoxyethanol	111-76-2	29.6
Butyl Benzyl Phthalate	85-68-7	31.5
Calcium Carbonate	1317-65-3	0.0
Carbon Black	1333-86-4	*
Carbon Dioxide	124-38-9	0.0
1-Chloro-1,1-Difluoroethane (HCFC 142b)	75-68-3	3.3
Chromium Hydroxide	1308-14-1	0.0
Corn Oil	8001-30-7	35.3
Diacetone Alcohol	123-42-2	35.1
1,1-Dichloro-1-Fluoroethane	1717-00-6	2.9
Diethylene Glycol Methyl Ether	112-34-5	33.0
1,1-Difluoroethane (HFC 152a)	75-37-6	6.3

(Continues)

TABLE 54-4 *Continued*

Chemical Name	CAS Number[1]	Chemical Heat of Combustion[2] ΔH_c,kJ/g
1,2-Dimethoxyethane	110-71-4	25.9
Dimethyl Ether	115-10-6	26.5
Dipropylene Glycol Methyl Ether	34590-94-8	32.2
Ethanol	64-17-15	24.7
Ethanol (95.6% Azeotrope)	64-17-15	23.6
2-Ethoxyethanol	110-80-5	25.9
2-Ethoxyethyl Acetate	111-15-9	30.9
Ethyl 3-Ethoxypropionate	763-69-9	32.0
Ethylbenzene	100-41-4	29.0
Ethylene Glycol	107-21-1	16.4
Ethylene Glycol Diacetate	111-55-7	32.0
Graphite	7782-42-5	*
Hexylene Glycol	107-41-5	28.5
Iron Oxide	1309-37-1	0.0
Isobutane, See 2-Methylpropane	—	
Isobutyl Alcohol	78-83-1	29.8
Isopropyl Acetate	108-21-4	25.5
Isopropyl Alcohol	67-63-0	27.4
Isopropyl Myristate	110-27-0	36.2
Isopropyl Palmitate	142-91-6	37.2
Kaolin Clay (Aluminum Silicate Hydroxide)	1332-58-7	0.0
Kerosene	8008-20-6	41.4
d-Limonene	5989-27-5	39.8
Liquids, Noncombustible/Nonflammable	—	0.0
Liquids, Noncontributory	—	*
Magnesium Silicate (Talc)	14807-96-6	0.0
Methanol	67-56-1	19.0
1-Methoxy-2-Propanol Acetate	108-65-6	30.9
Methyl Ethyl Ketone	78-93-3	30.6
Methyl Isopropyl Ketone	563-80-4	31.1
Methyl n-Amyl Ketone	110-43-0	35.0
Methylene Chloride	75-09-2	2.1
2-Methylpropane (Isobutane)	75-28-5	42.8
Mica (Mica Silicate)	12001-26-2	0.0
Mineral Oil	8012-95-1	31.5
Mineral Spirits (Petroleum Distillate)	64742-47-8	41.2
Mineral Spirits (Petroleum Distillate)	64742-88-7	41.2
N,N-Diethyl-m-Toluamide (Deet)	134-62-3	28.2
n-Butyl Acetate	123-86-4	27.6
n-Heptane	142-82-5	41.0
n-Hexane	110-54-3	41.1
n-Octyl Bicycloheptane Dicarboximide	113-48-4	30.0
Naphtha (High Flash)	8052-41-3	41.2
Naphtha (Petroleum Distillate)	8030-30-6	41.2
Naphtha, VM &P (Petroleum Distillate)	64742-95-6	41.2

TABLE 54-4 *Continued*

Chemical Name	CAS Number[1]	Chemical Heat of Combustion[2] ΔH_c,kJ/g
Naphtha, VM&P (Petroleum Distillate)	64742-48-9	41.2
Naphtha, VM&P (Petroleum Distillate)	64742-94-5	41.2
Nitrogen	7727-37-9	0.0
Paraffin (Wax)	8002-74-2	*
Pentane	109-66-0	41.9
Perchloroethylene (Tetrachloroethylene)	127-18-4	*
Petroleum Distillate	64741-65-7	41.2
Phthalocyanine Blue	147-14-8	*
Phthalocyanine Green	1328-53-6	*
Piperonyl Butoxide	51-03-6	32.0
Polyoxyethlene Sorbitan Oleate	9005-65-6	*
Polyoxyethylene (20) Sorbitan Monolaurate	9005-64-5	*
Propane	74-98-6	44.0
Propylene Glycol	57-55-6	20.5
sec-Butyl Alcohol	78-92-2	39.9
Silica (Crystalline)	—	0.0
Silica, Amorphous Hydrated	7631-86-9	0.0
Silicone Oil	63148-58-3	*
Silicone Oil	63148-62-9	*
Solids, Noncombustible/Nonflammable	—	0.0
Solids, Noncontributory	—	*
Sorbitan Monolaurate	1338-39-2	37.9
Sorbitan Monopalmitate	26266-57-9	37.9
Styrene Butadiene Rubber	25038-32-8	*
Tin Oxide (Stannic Oxide)	18252-10-5	0.0
Titanium Dioxide	13463-67-7	0.0
Toluene	108-88-3	28.4
Triacetin	102-76-1	35.4
1,1,1-Trichloroethane	71-55-6	*
Trichloroethylene	79-01-6	*
1,2,4-Trimethylbenzene (Pseudocumene)	95-63-6	27.5
Water	7732-18-5	0.0
Xylene	1330-20-7	27.4
Zinc Oxide	1314-13-2	0.0

*Materials that have either (1) a closed-cup flash point greater than 500°F (260°C), or (2) no fire point when tested in accordance with ASTM D 92, *Test Method for Flash and Fire Points by Cleveland Open Cup*, or (3) are combustible solids. Such materials contribute very little to the overall fire hazard of aerosol products in an actual fire, due to incomplete combustion or inconsistent burning behavior (i.e., the majority of the released material does not burn). Such materials are considered to be "noncontributory" to the overall determination of the product's level of classification. They can be ignored or they can be assigned a chemical heat of combustion (ΔH_c) of 0 kJ/g.

[1]Chemical Abstracts Service Registration Number.

[2]The theoretical heats of combustion and combustion efficiencies used to determine the chemical heats of combustion listed in this table are contained in the supporting documentation on file at NFPA.

Source: NFPA 30B, 1998, Table A-1-7(a).

The inspector should refer to the requirements of NFPA 58, *Liquefied Petroleum Gas Code,* which governs fire protection requirements for flammable propellant tank farms. NFPA 58 addresses the location of propellant tanks in relation to buildings, property lines, flammable and combustible liquids, and other hazards. NFPA 58 also addresses requirements for containers and associated equipment and piping, pressure relief valves, isolation valving, and requirements for transferring propellants from tank vehicles into the bulk storage tanks.

The inspector should review the layout of the tank farm with appropriate facility personnel to verify compliance with container location, spacing, and protection requirements and with requirements of tank vehicle unloading facilities contained in NFPA 58. The inspector should also examine the placement and operation of gas detection equipment and automatic shutoff valves provided at the tank farm.

One of the unique hazards presented by an aerosol propellant tank farm is that the propellants are not odorized, as is typically required for liquefied petroleum gases. The propellants are unodorized so that the propellant does not impart a disagreeable odor to the product. NFPA 58 allows liquefied petroleum gases to be unodorized when the odorization would be detrimental to the further processing or use of the gas. In the absence of an odorizing agent, combustible gas detection systems are often provided at propellant tank farms to aid in the detection of gas leakage.

The individual propellant storage tanks should be labeled with their actual contents, and tanks storing unodorized flammable gases should be labeled "UNODORIZED." While inspecting tank farms, inspectors should take note that all proper labeling procedures are followed.

The propellants are typically transferred from the tank farm to the aerosol filling area by fixed piping systems and propellant transfer pumps. Remotely actuated shutoff valves are provided at one or more locations between the tank farm and the aerosol charging or pump rooms to be able to automatically shut down propellant flow into the charging or pump rooms under emergency conditions. The inspector should always double check to make sure that the remote shutoff valves are provided and are in working condition.

Aerosol Filling Line

A typical aerosol packaging line is an automated line, incorporating conveyor systems and automated filling and charging systems. There are usually a number of personnel involved in the filling line operation to verify proper operation of the equipment, to remove damaged containers or product, and to shut down the line in case of a problem. The major areas of fire protection concern involve the base product filling operations when the base product is a flammable or combustible liquid; the gassing of the container, when the container is filled and pressurized by the propellant; and the hot water bath, which causes an increase in the pressure of the container, to test the containers ability to resist rupture due to its internal pressure.

By its very nature, aerosol filling lines utilizing flammable and combustible liquid base products and/or flammable propellants present significant fire safety challenges. During filling operations, some areas of the facility, such as the propellant charging room (also called the gas house), may be off limits to outside personnel. The aerosol line may need to be shut down in order to provide the inspector with access to such areas.

The dispensing of flammable and combustible liquid base products into aerosol containers is regulated by the requirements of NFPA 30. The inspector should review flammable and combustible liquid storage and handling operations to confirm that such operations are being conducted in accordance with the requirements of NFPA 30. Flammable product filling operations are typically located within the propellant charging room, where the filled container is charged with the propellant.

Propellant Charging and Pump Rooms

Propellant charging rooms present a significant fire and life safety risk because within the room aerosol containers are pressurized with a flammable gas at high pressures. The release of gas in the room under an upset condition may result in a flammable vapor–air mixture being formed. If the vapors are not dispersed by the ventilation system, ignition of the gas–air mixture can result in a deflagration.

Propellant pump rooms, which house the propellant charging pumps, present hazards similar to that of the charging room. Propellant charging pumps boost the liquid propellant to the pressures required by the propellant filler, usually between 300 psi (2068 kPa) and 1200 psi (8274 kPa).

Propellant charging rooms and pump rooms, therefore, are required to be separated from adjacent buildings and manufacturing areas and must be provided with damage-limiting construction and deflagration relief venting. NFPA 30B requires that flammable propellant charging and pump rooms be provided with either a wet-pipe or deluge automatic sprinkler system. NFPA 30B also requires that the charging room be provided with a deflagration suppression system, in accordance with NFPA 69, *Standard on Explosion Prevention Systems.*

All electrical equipment and wiring in flammable propellant charging and pump rooms are required to be suitable for Class I, Division 1 hazardous locations, in accordance with NFPA 70, *National Electrical Code®*. Gas detection systems are also required to be provided in the flammable propellant charging and pump rooms.

Flammable propellant charging and pump rooms are required to be provided with ventilation systems to keep the level of gas in the room below the LEL of the propellant. When gas is detected within the room at a level of 20 percent of the LEL, the gas detection system is required to sound an audible alarm, and the ventilation system switches to a higher "emergency" exhaust rate. If gas concentrations in the room reach 40 percent of the LEL, the system shuts down the propellant supply to the room, shuts down the aerosol filling line, sounds audible alarms, and maintains emergency ventilation rates. Operation of the fire suppression systems also initiates propellant supply and aerosol filling line shutdowns.

NFPA 30B contains detailed requirements for construction, fire protection, electrical safety equipment interlocks, emergency propellant line venting, ventilation system operation and controls, and gas detection system requirements. The inspector should request that the facility personnel review the operation and controls for the emergency ventilation systems and line shutdown and venting.

Aerosol Test Bath

Once the aerosol container is charged with propellant, and the container head and valve assembly is secured to the container, the aerosol product leaves the charging house and goes to the water test bath. The test bath water is heated with an automated heating system, provided with appropriate controls to prevent overheating of the bath. The aerosol containers are submerged in the test bath, and the container contents are allowed to increase in temperature, increasing the container's internal pressure. The test bath evaluates the container's ability to withstand pressure buildup. The test bath area is usually monitored by personnel who can locate, remove, and properly dispose of leaking containers.

From the test baths, the filed aerosol containers proceed to the packaging area where they are placed in cartons and assembled on pallets. The inspector must make sure full pallet loads of aerosol product are removed from the packaging area and transferred to the finished product warehouse since the sprinkler systems in the filling line area are usually not suitable for protection of large quantities of palletized aerosol products.

Aerosol Product Laboratories

Another area within aerosol product manufacturing facilities that presents a fire hazard is the aerosol product laboratory. The lab may be utilized for product formulation, and it may also be utilized for quality control testing of filled containers. NFPA 30B contains requirements applicable to the product testing and pilot filling operations found in these aerosol laboratories.

AEROSOL PRODUCT STORAGE

Fire History

The referenced documents in this chapter contain discussion on some significant fire losses involving the storage of aerosol products. When aerosol products are not adequately protected within a warehouse, a high challenge fire can develop within the aerosol product storage arrays. In some cases, containers rupturing due to exposure to fire have rocketed beyond the aerosol storage array, leading to ignition of other aerosol product storage or other combustible commodities. The fire history also indicates that aerosol products stored with, or adjacent to, flammable liquids also present a severe and challenging fire.

In the early 1980s, in response to several significant fire losses in which aerosols appear to have contributed to the fire loss, representatives from the aerosol industry, the insurance industry, and sprinkler manufacturers ran an extensive series of full-scale fire tests to develop appropriate sprinkler protection criteria for aerosol products. The sprinkler protection criteria contained in NFPA 30B is based on these full-scale aerosol product fire tests. Additional protection scenarios for aerosols have been added to NFPA 30B as additional full-scale fire tests have been run, and the test results have warranted changes to NFPA 30B.

Aerosol Product Cartons

An important consideration in protecting aerosol products in storage is whether the aerosols are contained within a cardboard carton. The current edition of NFPA 30B provides sprinkler protection criteria for only cartoned aerosols in storage, since full-scale fire tests have shown that the cardboard carton is instrumental in controlling fires in aerosol product storage arrays due to the pre-wetting of the carton by sprinkler water discharge. Although the carton itself is combustible, its ability to absorb water and to provide some shielding to the containers inside is important in controlling a fire in an aerosol storage array.

Future editions of NFPA 30B may contain sprinkler protection criteria for uncartoned or shrink-wrapped aerosol products as full-scale fire tests are being conducted to develop appropriate sprinkler protection criteria for these uncartoned and shrink-wrapped packaging methods. Preliminary testing results indicate that uncartoned and shrink-wrapped aerosols do present a more severe fire challenge than cartoned product, but adequately designed sprinkler protection and storage arrangement limitations can be utilized to control a fire in uncartoned aerosol product storage.

Storage Configurations

Aerosol products can be stored in general-purpose warehouses, in aerosol warehouses, and in flammable liquid storage areas and liquid warehouses, as defined in NFPA 30.

Aerosol product storage quantity limitations are based on the net weight of container contents, which includes the weight of the base product and the weight of the propellant. The net weight of the contents is required by other consumer product labeling requirements to be indicated on the container label. As a rule of thumb, a pallet load of aerosols is often approximated at a net weight of 1000 lb (4536 kg).

Requirements for the protection and arrangement of aerosol product storage arrays are based on the aerosol level classification. As indicated previously, the aerosol cartons are required to be marked with the aerosol level classification. The inspector should first determine the classifications of the aerosol products in storage to determine which protection requirements from NFPA 30B are applicable. If the cartons are not marked with the aerosol classification, the product should be treated as a Level 3 aerosol.

Level 1 aerosol products are considered equivalent to Class III storage commodity, as defined in NFPA 230, *Standard for the Fire Protection of Storage*. Therefore, Level 1 aerosols can be stored without limitation in warehouses, as long as the warehouse is protected in accordance with the appropriate protection requirements for Class III commodity storage. When assessing storage compliance, inspectors should be well aware of the requirements.

General Purpose Warehouses

In general-purpose warehouses, as defined in NFPA 230, Level 2 and Level 3 aerosol products can be stored in limited quantities, with the quantity allowed by NFPA 30B being dependent on the level of sprinkler protection provided and the level of segregation provided between the aerosol products and the remainder of the storage area. Very small quantities are allowed in unprotected warehouses. The allowable quantities increase as additional protection is provided.

The storage of Level 2 and Level 3 aerosol products within a segregated storage area in a general-purpose warehouse provides a defined aerosol product storage area and an additional level of protection against aerosols rocketing beyond their storage array under fire conditions. Segregation can be provided by chain-link fencing or by fire-resistive-rated partitions.

The total floor area permitted by NFPA 30B for aerosol storage in a segregated storage area in a general-purpose warehouse depends on the level of segregation provided. A 2-hour fire-resistive segregation allows a larger aerosol storage area than a 1-hour partition, and a 1-hour partition allows a larger storage area than a chain-link segregation.

Automatic-closing fire-rated door assemblies are required to be provided for protection of openings in fire-resistive-rated partitions; and automatic-closing chain-link fence closures or chain-link fence offsets need to be provided for chain-link fence enclosures to provide a complete segregation from other areas of the warehouse. In any case, sprinkler protection for the segregated storage area is required to be provided in accordance with the sprinkler protection criteria of NFPA 30B.

The storage of aerosol products in a general-purpose warehouse also requires that provisions be taken to limit the potential exposure of aerosol storage arrays to flammable liquid pool fires. This is accomplished by separating aerosols and flammable liquids by means of segregating partitions or by a 25-ft (7.62-m) separation distance.

Aerosol Warehouses

When the total quantity of aerosols exceeds the quantities allowed within general-purpose warehouses, the aerosol products are required to be stored within an aerosol warehouse. Aerosol warehouses are not limited as to the total quantity or total area devoted to the storage of aerosol products. Inspectors should be aware that aerosol warehouses are required to be in separate, detached buildings or separated from other occupancies through the use of free-standing, 4-hour fire walls with protected openings. Aerosol warehouses are required to be provided with sprinkler protection in accordance with the requirements of NFPA 30B.

Flammable Liquid Storage Rooms and Liquid Warehouses

As indicated previously, limited quantities of aerosols are permitted by NFPA 30B to be stored in flammable liquid storage rooms and liquid warehouses. Sprinkler protection is required to be based on the more restrictive of the aerosol sprinkler protection criteria or the applicable flammable and combustible liquids sprinkler protection criteria contained in NFPA 30.

Aerosol Sprinkler Protection Criteria

Sprinkler protection criteria for aerosol products recognize the high challenge fire potential posed by aerosol products in a storage occupancy. Level 1 aerosol products are considered equivalent to Class III storage commodities. As such, sprinkler protection for Class III commodity, as provided in NFPA 13, *Standard for the Installation of Sprinkler Systems,* is suitable for the protection of Level 1 aerosols. Sprinkler protection for Level 2 and Level 3 aerosol products is based on the aerosol storage configuration, the classification of the aerosols being protected, and the type of sprinkler system utilized. Specific sprinkler protection criteria for Level 2 and 3 aerosols are provided in NFPA 30B.

The inspector should refer to the sprinkler protection tables in NFPA 30B, which address both palletized and rack storage arrays. Separate protection criteria is provided for Level 2 and Level 3 aerosol products under various storage height and roof height conditions, as well as various sprinkler types. The aerosol sprinkler protection tables include criteria for spray sprinkler protection, early suppression fast response (ESFR) sprinkler protection, and large-drop sprinklers. For storage areas used for the storage of both Level 2 and Level 3 aerosols, the protection criteria for Level 3 aerosols needs to be utilized.

The protection tables for rack storage of aerosol products include criteria for in-rack sprinklers. Depending upon the storage height and aerosol classification, in-rack sprinklers at the face of the rack may be required in addition to in-rack sprinklers in the longitudinal flue. In addition, the use of in-rack sprinklers in conjunction with ESFR sprinkler protection at the ceiling is allowed by NFPA 30B, when certain conditions are met. As additional full-scale fire tests are run on aerosol product storage configurations and new sprinkler technologies, the number of available options for the protection of aerosol products in storage will increase.

Retail Display and Storage

NFPA 30B also provides requirements for protection of aerosol product storage and display in retail occupancies. Inspectors should be aware that in a typical retail setting, the total quantity of Level 2 and Level 3 aerosol products and the density of the aerosols in a display area are limited based on whether the store is sprinklered. They should also be cognizant of the fact that NFPA 30B requires that the aerosols on store shelving be removed from their combustible cartons in order to limit the exposure of the aerosols to large amounts of combustible materials. In the case of aerosols on retail display shelves, the carton provides no significant contribution to fire control.

The storage and display of aerosol products in bulk retail facilities also pose some interesting fire protection challenges. These stores usually incorporate some product displays on lower levels and palletized storage on racks above. NFPA 30B provides appropriate limits for such display and storage. Storage arrangement and protection for large quantities of aerosol products in these bulk retail facilities are required to comply with the same sprinkler system protection requirements as applied to warehouse facilities. Careful attention needs to be paid when aerosols are stored or displayed in the vicinity of flammable and combustible liquids, as might be found in the paint section of a home improvement bulk retail facility. Aerosols are required to be physically separated from flammable and combustible liquids by a physical separation of 25 ft (7.6 m), by a segregating wall, or by a noncombustible, liquid-tight barrier.

SUMMARY

Aerosol products are commonplace. They are found in our homes and businesses, and they incorporate a wide variety of consumer and industrial products. These products are manufactured at a many different manufacturing facilities, from the Fortune 500 consumer product manufacturer, to the small contract filler. Aerosol products are stored in warehousing and retail storage facilities throughout our communities. In the course of routine inspection duties, fire inspectors will likely come across aerosol storage in warehousing and retail facilities. An understanding of the aerosol protection requirements is important in assessing these facilities for compliance with code requirements. Effective fire protection and life safety precautions contained in the applicable NFPA standards, as well as widely accepted industry safety practices, can provide for the safe manufacture and storage of aerosol products in our communities.

BIBLIOGRAPHY

Cote, A. E., ed., *Fire Protection Handbook,* 18th ed., NFPA, Quincy, MA, 1997.

Cote, A. E., and Linville, J. L., eds., *Industrial Fire Hazards Handbook,* 3rd ed., NFPA, Quincy, MA, 1990.

NFPA Codes, Standards, and Recommended Practices

See the latest version of The NFPA Catalog for availability of current editions of the following documents.

NFPA 13, *Standard for the Installation of Sprinkler Systems*

NFPA 30, *Flammable and Combustible Liquids Code*

NFPA 30B, *Code for the Manufacture and Storage of Aerosol Products*

NFPA 54, *National Fuel Gas Code*

NFPA 55, *Standard for the Storage, Use, and Handling of Compressed and Liquefied Gases in Portable Cylinders*

NFPA 58, *Liquefied Petroleum Gas Code*

NFPA 69, *Standard on Explosion Prevention Systems.*

NFPA 70, *National Electrical Code®*

NFPA 230, *Standard for the Fire Protection of Storage*

Protection of Commercial Cooking Equipment

CHAPTER 55

R. T. "Whitey" Leicht

CASE STUDY

Fire in Concealed Spaces Damages Restaurant

A Michigan riverfront restaurant was badly damaged when a fire in a broiler exhaust hood spread to concealed spaces. The restaurant was operating at the time of the fire, but there were no injuries.

The two-story, unprotected, wood-frame structure had no fire or smoke detection system, but the exhaust hood system was protected by a dry chemical extinguishing system. There were no sprinklers.

Employees smelled smoke in the kitchen and went to investigate. Seeing flames above baffle filters over an operating broiler, they evacuated patrons and called the fire department. Two employees who stayed in the kitchen were initially able to control the flames using portable extinguishers, but smoke and fire persisted. By the time the extinguishing system began to operate, the fire had spread to roof vents.

When fire fighters arrived, heavy smoke was showing from the vents over the kitchen, and light smoke was coming from others. Interior conditions ranged from light smoke in the dining room to heavier smoke in the kitchen.

The chief directed the first engine company to the roof, which they ventilated, revealing that the fire had spread into the roof/ceiling void space. Despite 15 to 20 minutes of repeated efforts to located and successfully contain the fire, conditions continued to deteriorate, and the chief ordered the companies from the increasingly unstable roof. He then ordered the fire fighters to evacuate the building and set up a defensive attack. The fire was extinguished 8 hours later.

Investigators found that the blaze began in a blower motor in the broiler exhaust stack. Heat from the fire caused the motor supports to fail, dropping the blower into the stack and dislodging the connections. This allowed the fire to spread from the stack to the void and roof. A makeup air system in the kitchen failed to shut down when the dry chemical hood system activated, contributing to fire spread. Investigators believe that the extinguishing system was slow to activate because the blower was drawing heat and flames away from the fusible links in the hood.

Damage to the building, valued at $350,000, was estimated at $150,000. The contents, valued at $150,000, were a total loss.

Source: "Fire Watch," *NFPA Journal*, September/October, 1999.

R. T. "Whitey" Leicht is the senior fire protection specialist in the Delaware Fire Marshal's Office. He also serves on the executive board of the International Fire Marshals Association and is currently secretary of the Technical Committee for Ventilation, Control, and Fire Protection of Commercial Cooking Operations.

Prior to 1950 and the broad adoption or extensive use of any fire safety standard, there was a noticeably disproportionate number of fires involving commercial cooking equipment. The record also suggested that a high ratio of all restaurants would suffer a devastating fire. Although good fire protection practices since then have reduced the ratio, there are so many more cooking establishments today, that even the lower percentage still represents a very large number of fires.

Furthermore, as with any loss analysis, severity, not just frequency, has to be evaluated. Compared to 40 or 50 years ago, a fire in today's average cooking establishment will place a more demanding burden on fire-fighting forces and have a greater impact on a community's tax base, on the number of employees without jobs, and on higher cost to replace the building, furnishings, and equipment. The inspector's job is to contribute to the overall goal of preventing these unwanted incidents.

To be effective, inspectors can't just place their confidence in the initial installation. They have to ensure that all built-in systems are maintained. This includes equipment integrity, extinguishing system service, routine cleaning, implementation of management controls, and so on. However, in order for an establishment to be competitive, an endless number of non-fire safety issues demand the restaurant operator's attention. It is the inspector's responsibility to impress on the operator the fire hazards of the operation and the benefits of appropriate protection.

FIRE HAZARDS AND PROTECTION OF COOKING EQUIPMENT

Cooking grease, which is used in a liquid state, produces vapors at cooking temperatures. It is a dangerous fuel when ignited, producing heavy, acrid smoke, with potential temperatures of 2000°F (1093°C). Autoignition is possible. The intense heat generated from flames extending through the hood and filters and into the exhaust duct presents a serious threat to ignition of surrounding combustible materials and has the potential to spread fire and smoke throughout the establishment.

Fire protection consists of more than merely "squirting wet stuff at hot stuff"! Fire protection professionals acknowledge that fire protection, in its entirety, is a system. In commercial cooking operations, fire protection consists of four primary components: prevention, confinement, detection, and extinguishment.

Prevention

The two subparts of prevention are control of ignition sources and control of combustible materials. The human element plays a major role in the control of ignition sources. Restaurant management has to first commit to training employees in the proper operation of the cooking equipment to ensure that the equipment is not being operated at too high a temperature. Additionally, management has to commit to routine maintenance to ensure that defective equipment/motors do not become the culprit of an unwanted ignition source. Because accumulations of renegade grease deposits are a common fuel source for kitchen fires, the best means of controlling combustible materials is a conscientious hood-and-duct-cleaning program by competent individuals. The cleaning should extend to all areas where dangerous grease accumulations may be found, not just to the hood and duct areas (Figure 55-1).

Confinement

Confinement of fires in exhaust ducts is a topic that has been under a great deal of study. The original purpose of exhaust ducts was to vent excess heat and gases to the outside. This original intention did not, however, take into account that a fire might occur in the duct. As restaurants in buildings of combustible construction burnt down at a high rate in the 1940s and early 1950s, it became apparent that combustible concealed spaces needed to be protected from the duct. This can be accomplished by physical separation between the duct and combustible surfaces at modest distances or by enclosing the duct in a fire-rated chase.

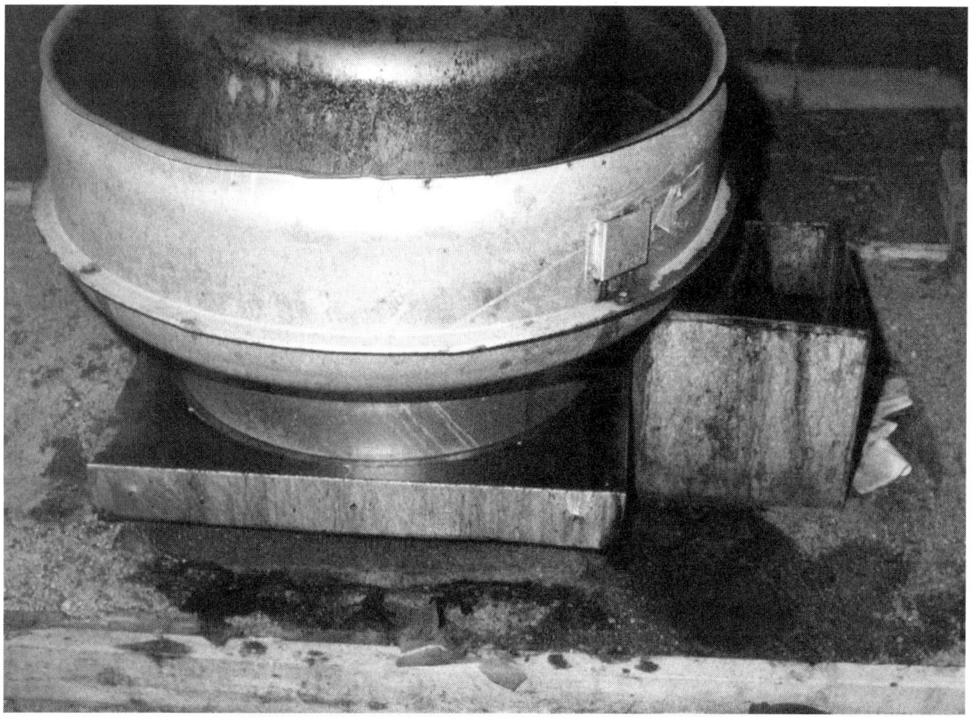

FIGURE 55-1 Kitchen Cooking Exhaust Hood Fan with Grease and Residue on Fan, Housing, and Roof Surface
Source: Courtesy of Jon Nisja.

Detection

Seeing, hearing, and smelling evidence of a fire are still the most reliable and effective means of detection. However, a means of detection is desired even when a human being is not present. Of great benefit in automatic detection are all the automatic appurtenances that accompany an automatic system. With automatic detection, employees and patrons can be alerted, fuel can be shut off, fire departments can be summoned, and extinguishing systems can be activated. Because many fire protection functions depend on it, ensuring that this equipment is not impaired is important.

Extinguishment

Extinguishment is either manual or automatic. It can be as simple as placing a cover over a container of flaming oil or of turning off the heat. As a fire progresses, the use of the appropriate portable fire extinguisher by a properly trained individual provides an excellent first-aid fire-fighting option. Over the years, automatic extinguishing systems—whether carbon dioxide, dry chemical, or wet chemical—have proved successful in fire control when they are properly maintained. In the event that an automatic extinguishing system dedicated to protect the ventilation equipment and cooking surfaces is not successful, the record indicates that automatic sprinkler systems provide the best fire protection to a building and its contents. Additionally, their water supply is considered to be inexhaustible. However, not all buildings are sprinkler protected.

THE INSPECTOR'S JOB

In order to *fully* evaluate a commercial cooking installation, the inspector can not wait until after the equipment is installed and operating because many of the areas of concern are established long before the installation takes place. Thus, the inspector's duty is not merely to verify that the operation is running smoothly and in a safe manner but to start with a review of the proposed installation. If the review is satisfactory, the inspector is next charged with the responsibility of assessing the installation for acceptance. Then, once the acceptance inspection is successful, the personnel have been properly trained, and the operation is running, the inspector conducts routine reinspections at some predetermined frequency.

The inspection of commercial cooking operations, therefore, is composed of three specific stages: review, acceptance, and reinspection. (This manual's Chapter 2, "Inspection Procedures," provides an overview of these elements.) During each of the three stages, the inspector examines the major components of the commercial cooking operation. Specifically, the inspection would include an examination of the following:

- Equipment and its arrangement
- Ventilation system, including the hood, the ducts, and the fan
- Fire protection system—whether automatic or manual
- General operation and maintenance

Of course some of the components are less important at one stage of the installation and more important at another stage. For instance, the general operation and maintenance is of little consequence during the review stage, and the layout of the ductwork is of little consequence during the annual and semiannual reinspection stage.

REVIEW AND FIRST INSPECTION CHECKLIST ITEMS

The following checklist items are normally part of the review stage for new installations. The same information, however, can also be applied during the first time a site is being inspected.

1. Except where an enclosure is required, are the hood, the grease removal devices, the exhaust fan, and the ductwork provided with a clearance of at least 18 in. (457 mm) to combustible material or at least 3 in. (76 mm) to limited-combustible material? If this distance is *not* provided, the inspector should check to see if an option for reduced clearance is being employed.
2. Is the hood composed of a minimum No. 18 MSG steel or minimum No. 20 MSG stainless steel? (See Figure 55-2 for different kinds of commercial cooking hoods.)
3. Is a fire-actuated damper, which is listed for such use, installed in the supply air plenum at each point where a supply air duct inlet or a supply air outlet penetrates the continuously welded shell of the hood assembly?
4. Are the proposed filters listed grease filters? The inspector should be aware that listed filters are not necessarily grease filters. Mesh type filters are typically used for HVAC or other applications, but are not acceptable for commercial cooking use.
5. Is the ductwork composed of a minimum No. 16 MSG steel or a minimum No. 18 MSG stainless steel?
6. Do the ducts pass through fire walls or fire barrier walls? The inspector should be aware that they are not permitted to do so. Ducts should lead as directly as practicable to the exterior of the building.
7. On horizontal ducts, is at least one of the openings 20 × 20 in. (508 × 508 mm) or greater?
8. Are the openings shown at the sides or top of the duct at the following locations:
   - At all changes of direction?
   - At 12-ft (3.7-m) intervals on horizontal ducts?
   - Within 3 ft (92 cm) of each side of the fan where the ductwork is connected to both sides of the fan?
9. Are the edges of the openings at least 1½ in. (38 mm) from the outside edges of the duct or seam?

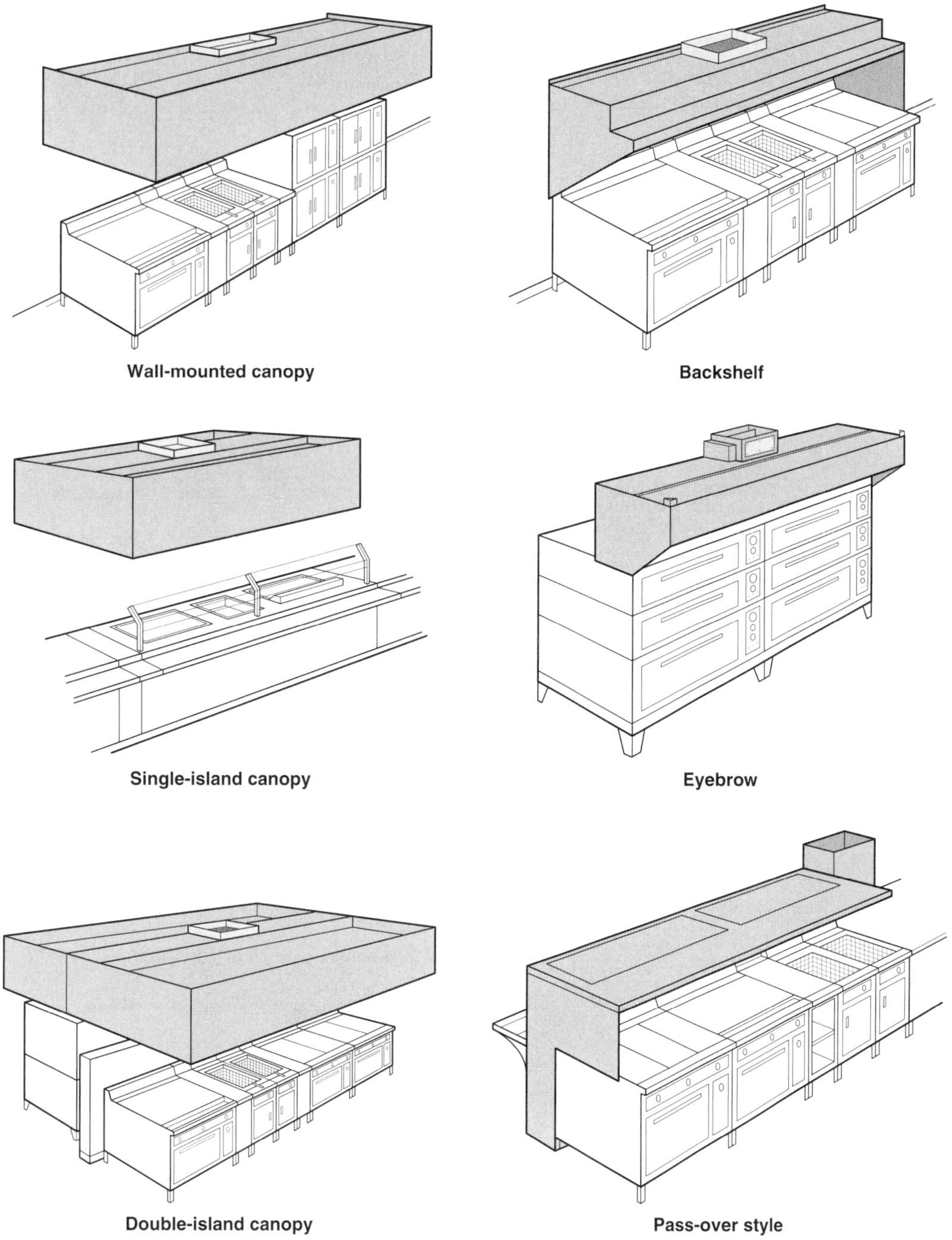

FIGURE 55-2 Commercial Cooking Hoods
Source: NFPA 96, 1998, Figure A-1-2.

10. In multistory buildings are interior duct installations enclosed in 1-hour fire-rated enclosure or in a 2-hour-rated enclosure if the building has four or more stories?

11. If the duct terminates at the roof, is the outlet a minimum of 10 ft (3.1 m) from the adjacent buildings, property lines, and air intakes. If this is not physically possible, is the outlet at least 3 ft (92 cm) above any air intake located within 10 ft (3.1 m) horizontally?

12. Is the exhaust flow directed up and away from the surface of the roof and a minimum of 40 in. (1 m) above the roof surface?

13. If the duct terminates through a wall, is the outlet a minimum of 10 ft (3.1 m) from the adjacent buildings, the property lines, the grade level, combustible

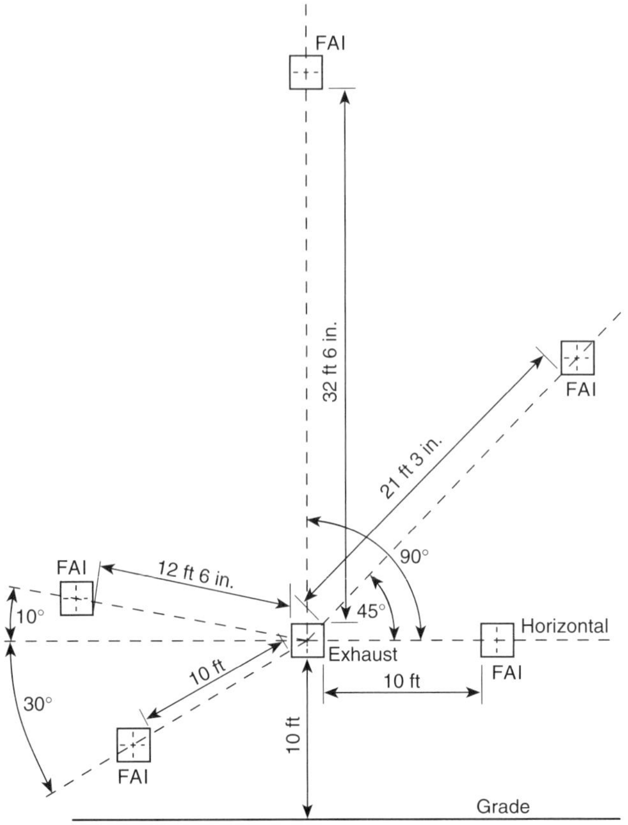

FIGURE 55-3 Exhaust Termination Distance from Fresh Air Intake or Operable Door or Window
Source: NFPA 96, 1998, Figure 4-8.3.1(a).

Fresh air intake (FAI) applies to any air intake, including an operable door or window.

Example:

FAI is same plane as exhaust or lower: 10 ft (min) between closest edges.

FAI above plane of exhaust: 10 ft + 0.25 ft per degree between closest edges

Note: For SI units, 1 ft = 0.305 m.

construction, electrical equipment or lines, and from the closest point of any air intake or operable door or window at or below the plane of the exhaust termination?

14. If any air intake or operable door or window is above the plane of the exhaust outlet, is it a minimum distance of 10 ft (3.1 m) plus 3 in. (76 mm) for each degree from horizontal (Figure 55-3)?

15. Verify that air velocity through any duct is not less then 1500 fpm (457.2 m/min). See Table 55-1.

TABLE 55-1 Air Velocity. To Determine the Maximum Duct Area That Will Still Allow a 1500 fpm (457.2 m/min) Velocity

Customary Units	SI Units
Divide the ventilation rate (Cubic Feet per Minute) by the required velocity (1500 Feet per Minute).	Divide the ventilation rate (meters per minute) by the required velocity (457.2 m/min)

To Determine the Minimum Fan Size Needed in Order to Achieve Minimum Velocity When the Duct Dimensions Are Given

Customary Units	SI Units
Step 1	
Multiply the width (inches) by the height (inches) of the cross section of the largest duct	Multiply the width (centimeters) by the height (centimeters) of the cross section of the largest duct
Step 2	
Divide the result by 144 in.^{2}/ft^2 to convert to square feet	Divide the result by 10,000 cm^2/m^2 to convert to square meters
Step 3	
Multiply the square foot area found in Step 2 by 1500 fpm	Multiply the square meter area found in Step 2 by 457.2 m/min

To Determine the Velocity to Which the Duct System Will Be Exposed When the Size of the Fan is Known

Customary Units	SI Units
Step 1	
Multiply the width (inches) by the height (inches) of the cross section of the largest duct	Multiply the width (centimeters) by the height (centimeters) of the cross section of the largest duct
Step 2	
Divide the result by 144 in^2/ft^2 to convert to square feet	Divide the result by 10,000 cm^2/m^2 to convert to square meters
Step 3	
Divide the fan size (Cubic Feet per Minute) by the square foot area found in Step 2.	Divide the fan size (meters per minute) by the square meter area found in Step 2
Step 4	
If the quotient is less than 1500 Feet per Minute the requirement is not being attained	If the quotient is less than 457.2 m/min, the requirement is not being attained

16. Is automatic fire extinguishing equipment provided for the grease removal devices, the hood exhaust plenums, the exhaust duct systems, and all the cooking equipment that produces grease-laden vapors?
17. If the water for a listed fixed baffle hood assembly is supplied from the domestic water supply, is the supply monitored?
18. Does the activation of the automatic fire extinguishing system shut off all sources of fuel or electric power to heat-producing equipment that requires protection by the system?
19. Does the activation of the automatic fire extinguishing system shut off all sources of fuel to gas appliances that do not require protection, but that are located under the same ventilating equipment?
20. Are the required shutoff devices provided with a manual reset feature?
21. Is a readily accessible means for the manual activation of the automatic fire extinguishing system located at least 42 in. (1067 mm) above the floor (but within reach) and located in an egress path (Figure 55-4)?
22. If the facility is (or will be) protected by a fire alarm system, will the activation of the automatic fire extinguishing system cause an alarm signal?
23. Are all deep fat fryers installed at least 16 in. (406 mm) from the surface flames of other cooking equipment? If not, is a noncombustible baffle, a

FIGURE 55-4 Three Standard Manual Activation Devices

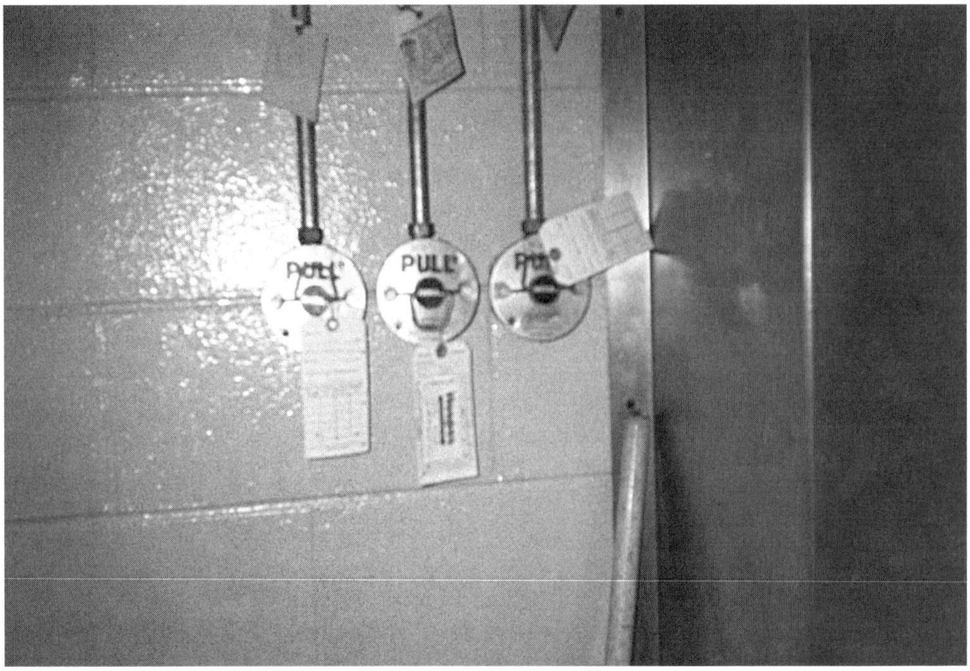

minimum of 8 in. (203 mm) in height, installed between the fryer and surface flames of the adjacent appliances?

IN-SERVICE INSPECTION CHECKLIST ITEMS

The following checklist items should be reviewed during any in-service inspections of the facility:

1. Are all interior surfaces of the exhaust system accessible for cleaning and inspection?
2. Does the clearance for the hood, grease removal devices, exhaust fans, and ductwork agree with the approved plans?
3. Are welds on seams and joints of the hood liquid tight, continuous, and on the exterior of the enclosure?
4. Is an access panel provided for any fire-actuated damper installed in the supply air plenum that would not otherwise be accessible?
5. Are only listed grease filters installed? Are they located at least the prescribed distance from the cooking surfaces? Are they installed at a 45-degree angle? Are they tight fitting and firmly held in place? Are they removable for cleaning?
6. Is the filter bank equipped with a drip tray beneath its lower edge pitched to drain into a maximum 1-gal (3.8 L) metal container?
7. Are the welds on the duct-to-hood collar and the seams and joints of the ducts liquid tight, continuous, and external to the duct?
8. Are the duct systems installed so as to avoid interconnection with any other building ventilation or exhaust system?
9. Are the duct systems installed so as to avoid forming dips or traps that might collect residues?
10. Is the exhaust flow from the outlet of a rooftop duct termination directed away from the roof surface and a minimum of 40 in. (1 m) above the roof surface?
11. Is the separation distance from the exhaust outlet to the adjacent buildings, the property lines, the grade level, combustible construction, electrical equipment

or lines, and the closest point of any air intake or operable door or window actually provided and does it agree with the approved plans?
12. Unless fan shutdown is required, does the hood exhaust fan continue to operate after the extinguishing system has been activated either by a listed component of the ventilation system or by the design of the extinguishing system?
13. Does a check confirm that no wiring system of any type is installed in ducts?
14. Does a check confirm that no motors, lights, or other electrical devices, except where specifically approved for such use, are installed in ducts or hoods?
15. Does a check confirm that the automatic fire extinguishing protection covers all grease removal devices, hood exhaust plenums, exhaust duct systems, and cooking equipment that produces grease-laden vapors?
16. Does a check confirm that water-wash fire protection provided in any fixed baffle hood is arranged to activate when the cooking equipment extinguishing system is activated?
17. If a single hazard area is protected by multiple automatic fire extinguishing systems, are the systems arranged for simultaneous operation upon actuation of any one of them?
18. Do all sources of fuel and/or electric power to cooking equipment requiring protection and to gas-fueled cooking equipment located under the same ventilating equipment as cooking equipment requiring protection shut off upon actuation of the automatic fire extinguishing system?
19. Does a check confirm that the fuel and/or power shut off devices require manual reset?
20. Is the means for the manual activation of the automatic fire extinguishing system located in an approved manner?
21. Where applicable, will activation of the automatic fire extinguishing system cause an alarm signal?
22. Are portable fire extinguishers of the appropriate type and proper size installed in the kitchen areas? The inspector should note that multipurpose dry chemical type extinguishers do not have any saponifying effect, whereas bicarbonate based compounds do. Saponification is desirable because when bicarbonates mix with "fatty" substances, it forms a soapy material that is noncombustible. However, with the recent advent of using vegetable oils for some dietary benefits in place of animal fat, saponification is less likely to occur, especially in fryers. Vegetable oils do not contain the amount of fatty acids needed to cause this transformation. Nevertheless, cooking appliances other than deep fat fryers can usually encounter substantial amounts of fatty acids since, in many cases, the products being cooked on these surfaces are meats. Regardless of whether the extinguishing agent is dry or wet chemical, it will be of some value as long as it is bicarbonate based.
23. Is a 16-in. (406 mm) space or an 8-in. (203-mm) baffle in place between deep fat fryers and adjacent cooking equipment?
24. Are all deep fat fryers equipped with a separate high temperature limit control in addition to the adjustable operating control? Is it set to shut off fuel or energy when the temperature reaches 475°F (246°C) at 1 in. (25 mm) below the fat surface? Although not required by NFPA 96, *Standard for Ventilation, Control, and Fire Protection of Commercial Cooking Operations* some fryers have a manual reset feature on the separate high-limit control.

PERIODIC INSPECTION STAGE

Once a baseline of fundamental information is established, periodic follow-up visits can be conducted to ensure that the various systems and features are in proper operating order. Most fires in commercial cooking equipment involve equipment whose installation was acceptable originally but has, over the years, become less compliant. The reason for this shift can usually be attributed to one or more of the following factors:

- The management becomes indifferent to new employee orientation.
- Fire protection equipment gets taken out of service inadvertently.
- Equipment and materials wear out or deteriorate.
- Preventative maintenance and service become lax.

The fire inspector's responsibilities include keeping on top of the situation through routine scheduled fire inspections. The following menu of items is not intended to be all inclusive. The inspector needs to evaluate the importance of the items and the risk they represent. The inspector then has to correlate his or her findings with available resources and the potential for improvement. An periodic inspection form can be a guideline for the inspector, provide a level of consistency, and serve as documentation for the file. The following items should be addressed during the inspection.

1. Is the exhaust system on while cooking equipment is operating? Cooking equipment should not be operating unless exhaust ventilation is being provided.
2. Is cooking equipment operating with filters missing from the filter-equipped exhaust hood? In filter-equipped exhaust hoods, filters should be in place when cooking equipment is operating.
3. Is cooking equipment operating while the fire extinguishing system is impaired? Cooking equipment should not be operating unless the fire extinguishing system is operative.
4. Is any physical damage noted to any material or product used for the purpose of reducing clearances? If there is, an evaluation has to be made to see if this affects the required protection. If it adversely affects the protection, further remedial action may be needed.
5. Are only listed grease filters being used and are they in place, tight fitting, and clean? The inspector should be aware that mesh filters are not permitted.
6. Are all grease removal devices, hood exhaust plenums, exhaust duct systems, and cooking equipment that produces grease-laden surfaces covered by the automatic fire extinguishing system? Note that grease removal devices, hood exhaust plenums, and exhaust ducts are permitted to be protected by a listed fixed baffle hood containing a constant or fire-actuated water-wash system.
7. Is a readily accessible means for manual activation of the automatic fire extinguishing system located at least 42 in. (1067 mm) above the floor but not so high that it might not be reached? Is it located in a normal and clear egress path?
8. Are portable fire extinguishers of the appropriate type and proper size installed in the kitchen areas?
9. Have the portable fire extinguishers been inspected by an approved person or company within the last 12 months and is there a form of documentation to this effect provided on the portable fire extinguisher unit?
10. Has an approved person or company inspected the fire extinguishing system within the last 6 months and is there a form of documentation to this effect provided in a conspicuous location?
11. Are hoods, grease removal devices, fans, ducts, and other appurtenances clean? Is an approved person or company doing the cleaning and is there a form of

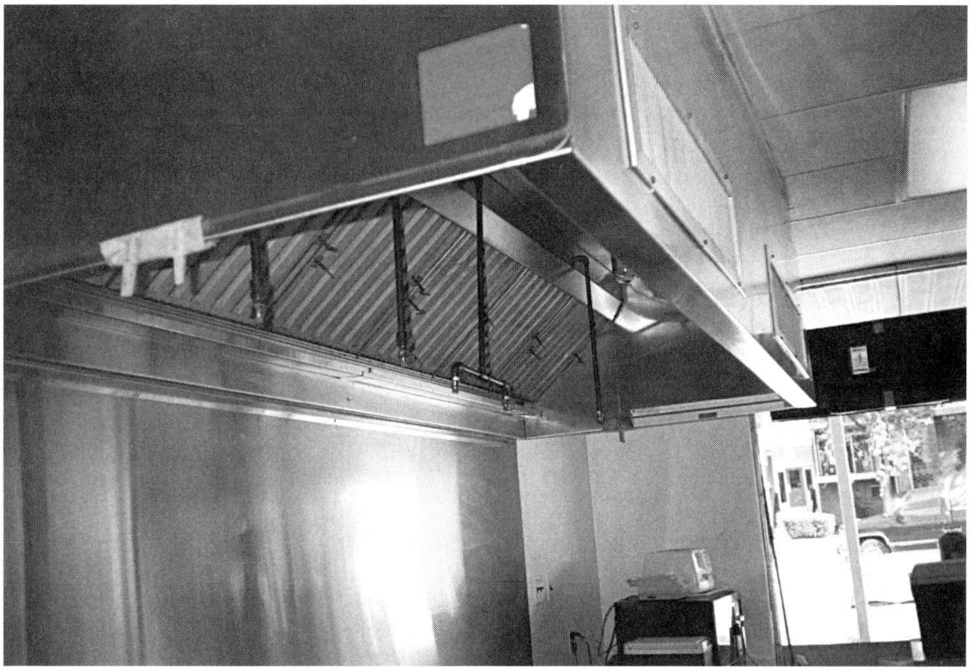

FIGURE 55-5 Clean, Wall-Mounted Canopy Hood, Displaying a Cleaning Certificate

documentation to this effect provided in a conspicuous location (Figure 55-5)? Although other information can be provided, the documentation should at least include the name of the servicing contractor and the date of the last cleaning.

12. Are nozzles of automatic extinguishing systems clean or otherwise protected from accumulation of grease buildup?

RECIRCULATING SYSTEMS

Recirculating systems are also known as ductless systems, ductless hoods, or ventless hoods. These units consist of cooking equipment and a hood with self-contained grease and odor control devices that are designed to clean the air sufficiently to allow the air to return to the surrounding interior environment. Naturally, these units do not have external exhaust. They must be listed for their particular design and use.

All recirculating systems must be listed and labeled to indicate the specific equipment allowed within its hood. In many cases, the hood is directly attached to the cooking equipment, thereby minimizing the chance of change within the equipment. The label should also indicate the ventilation rate, the type of filters used, and other system specifications.

In some units, interlocks prevent the operation of the system if filtration devices are not in place. All panels that access the interior grease-laden surfaces of the unit also have interlocks that prevent operation if the panels are not in place. In addition, special interlocks that monitor minimum airflow and minimum electrostatic particulate (ESP) filter performance will shut off the unit if the threshold limits are exceeded. The inspector can test the filter and panel sensor operation by removing one or more components to see if the unit will indeed shut down. The inspector can test the airflow interlock by restricting the hood filters with paper or cardboard to verify that the unit shuts off.

Recirculating systems have installation limitations based on surrounding combustible materials, return and supply air grills, lower ceilings, and other conditions that may present fire or health concerns. Most of these have been determined by the local authorities and will only be found in the final condition of approval for the system. These should be a part of the inspector's file and should be carefully reviewed at each inspection. The operator could, either intentionally or unintentionally, violate some of these conditions, with potentially serious consequences.

Every recirculating system should have a detailed operation and maintenance manual that not only covers all the standard material but also addresses special features and concerns. The inspector should review these features and concerns at each inspection. The recirculating units all have built-in fire extinguishing systems. Some have special nozzles that are not part of standard preengineered systems, and some use a sprinkler or nozzle with wet chemical. These nozzles are specially listed for the application and any differences that exist in the extinguishing system for recirculating cooking equipment are detailed in the manufacturer's system manual. The components and operations of the on-board fire extinguishing system are the same as those of conventional systems and should be reviewed in a like manner.

The operator should keep a log to record all maintenance. The log should be available for review by the authority having jurisdiction.

SOLID FUEL COOKING EQUIPMENT

Solid fuel cooking equipment presents unique challenges to adequate protection that is not encountered in conventional cooking equipment. First and foremost, the equipment probably creates grease-laden vapors that need to be vented. Second, it often generates smoke, carbon monoxide, creosote, and other by-products that typically result when wood is burned. Because the burning of wood or other solid fuels (charcoal, coal, etc.) can create an additional hazard, a more stringent level of safeguards needs to be provided. For instance, where solid fuel cooking is conducted, spark/ember arresters need to be provided. Additionally, terminating the exhaust duct through a wall is prohibited.

Unlike conventional appliances, exhaust ventilation over solid fuel-fired equipment can be either natural draft or power assisted. In many cases this exhaust ventilation is not permitted to be common with other exhaust systems. Also unlike conventional equipment, some solid fuel-burning equipment does not require automatic fire extinguishing protection, as long as the authority having jurisdiction approves. However, if automatic fire protection is required, it has to be in accordance with the appropriate standard, that is, NFPA 96.

All solid fuel appliances and solid fuel storage require a readily available source of water. The amount of water and the delivery mechanism depend on the size of the firebox.

Appliances with smaller fireboxes [less than 5 ft^3 (0.14 m^3)] and small piles of solid fuel require water-based fire extinguishers. For appliances with larger fireboxes [over 5 ft^3 (0.14 m^3)] and large piles of solid fuel, a water supply arrangement is needed to be able to produce 5 gpm (18.9 L/min) for an indefinite period of time. Such a stream is necessary so a continuous flow of water, with its heat absorbing effects, can be played on what could be a large flaming or smoldering wood mass.

The actual wording in the standard states that "a fixed water pipe system with a hose is to be provided in the immediate vicinity of the appliance." This has traditionally been achieved by a garden hose placed permanently onto a domestic water line and controlled by a normally closed globe valve. In lieu of a hose arrangement, larger solid fuel storage can be protected by an automatic sprinkler system.

The maintenance for this special equipment focuses on the potential of residue buildup, which can reduce the efficiency of the venting, and the potential of corrosion or physical wear, which could limit the flue's ability to contain combustion by-products. To operate solid fuel-fired equipment safely, minimal safety requirements should be adhered to. They include the following:

1. Installing solid fuel appliances on noncombustible flooring, with a 3-ft (92-cm) clearance from combustibles on each side and a 6-ft (1.8-m) clearance from any combustibles located above the appliance
2. Limiting the amount of solid fuel stored with the solid fuel appliance and solid fuel storage away from other materials, away from heat-producing equipment, and away from the path used to remove ashes
3. Igniting the fuel and stoking the fire safely
4. Removing ashes frequently and treating and discarding ashes safely
5. Ensuring that no device of any type is inserted in the flue pipe or chimney of a natural-draft solid fuel appliance
6. Ensuring that no solid fuel deep fat frying involves more than a quart of shortening and that no deep fat frying unit is located within 3 ft (92 cm) of a solid fuel appliance

SUMMARY

Commercial cooking is identified as a bona fide hazard. As such, commercial cooking establishments need to provide fire protection in order to eliminate, minimize, or at least control the associated hazards. Protecting this equipment should begin with the proposed installation, be followed up by appropriate systematic acceptance testing, and be routinely monitored through periodic inspections. All three of these procedures can be organized into manageable, methodical, and consistent processes. As with any hazard, some equipment, devices, or arrangements will be a departure from what is considered conventional. When inspectors encounter such special equipment—for example, solid fuel cooking equipment—they should verify that the minimal safety requirements are adhered to. Before and during an inspection, the inspector should become familiar with the relevant nationally recognized codes and standards, such as NFPA 96, *Standard for Ventilation, Control, and Fire Protection of Commercial Cooking Operations.*

BIBLIOGRAPHY

Cote, A. E., ed., *Fire Protection Handbook,* 18th ed., NFPA, Quincy, MA, 1997.

Leicht, R., *Commercial Cooking Protection—Understanding and Applying the NFPA Standard,* MidAtlantic Life Safety Conference, Laurel, MD, October 2000.

NFPA Codes, Standards, and Recommended Practices

See the latest version of The NFPA Catalog for availability of current editions of the following documents.

NFPA 96, *Standard for Ventilation, Control, and Fire Protection of Commercial Cooking Operations*

Inspection Forms

APPENDIX

Compiled by Wayne "Chip" G. Carson

An inspection is a visual check and generally does not involve testing or maintenance. The following selected forms, keyed to their respective chapters, are intended for use primarily as a memory jog. They are not meant to be a recitation of every code requirement. They are presented here, however, in reproducible format for actual use in the field. Both these inspection forms and the chapters they are linked to in the text should not be used for enforcement purposes. The inspector should verify all requirements in the relevant codes and standards.

Form A-1 Inspection Checklist to Accompany Chapter 2, "Inspection Procedures"
Form A-2 Inspection Checklist to Accompany Chapter 7, "Construction, Alteration, and Demolition Operations"
Form A-3 Inspection Checklist to Accompany Chapter 8, "Protection of Openings in Fire Subdivisions"
Form A-4 Inspection Checklist to Accompany Chapter 9, "Electrical Systems"
Form A-5 Inspection Checklist to Accompany Chapter 10, "Heating Systems"
Form A-6 Inspection Checklist to Accompany Chapter 11, "Air-Conditioning and Ventilating Systems"
Form A-7 Inspection Checklist to Accompany Chapter 12, "Smoke-Control Systems"
Form A-8 Inspection Checklist to Accompany Chapter 13, "Fire Alarm Systems"
Form A-9 Inspection Checklist to Accompany Chapter 14, "Water Supplies"
Form A-10 Inspection Checklist to Accompany Chapter 15, "Automatic Sprinkler and Other Water-Based Fire Protection Systems"
Form A-11 Inspection Checklist to Accompany Chapter 16, "Water Mist Systems"
Form A-12 Inspection Checklist to Accompany Chapter 17, "Special Agent Extinguishing Systems"
Form A-13 Inspection Checklist to Accompany Chapter 18, "Clean Agent Extinguishing Systems"
Form A-14 Inspection Checklist to Accompany Chapter 19, "Portable Fire Extinguishers"
Form A-15 Inspection Checklist to Accompany Chapter 21, "Interior Finish, Contents, and Furnishings"
Form A-16 Inspection Checklist to Accompany Chapter 22, "Assembly Occupancies"
Form A-17 Inspection Checklist to Accompany Chapter 23, "Educational Occupancies"
Form A-18 Inspection Checklist to Accompany Chapter 24, "Day-Care Facilities"
Form A-19 Inspection Checklist to Accompany Chapter 25, "Health Care Facilties"
Form A-20 Inspection Checklist to Accompany Chapter 26, "Ambulatory Health Care Facilities"
Form A-21 Inspection Checklist to Accompany Chapter 27, "Detention and Correctional Occupancies"
Form A-22 Inspection Checklist to Accompany Chapter 28, "Hotels"

Wayne "Chip" Carson is president of Carson Associates, Inc., a fire protection consulting firm in Warrenton, VA. He is secretary of the Technical Correlating Committee for NFPA 5000™, Building Construction and Safety Code™, *active on several NFPA committees, coauthor of* Fire Protection Systems: Inspection, Test & Maintenance Manual, *and contributor to the "In Compliance" column in* NFPA Journal.

Form A-23 Inspection Checklist to Accompany Chapter 29, "Apartment Buildings"
Form A-24 Inspection Checklist to Accompany Chapter 30, "Lodging or Rooming Houses"
Form A-25 Inspection Checklist to Accompany Chapter 31, "Residential Board and Care Occupancies"
Form A-26 Inspection Checklist to Accompany Chapter 32, "One- and Two-Family Dwellings"
Form A-27 Inspection Checklist to Accompany Chapter 33, "Mercantile Occupancies"
Form A-28 Inspection Checklist to Accompany Chapter 34, "Business Occupancies"
Form A-29 Inspection Checklist to Accompany Chapter 35, "Industrial Occupancies"
Form A-30 Inspection Checklist to Accompany Chapter 36, "Storage Occupancies"
Form A-31 Inspection Checklist to Accompany Chapter 37, "Special Structures and High-Rise Buildings"

Inspection Checklist

Inspection Procedures

PREINSPECTION CHECKLIST

Equipment: ______________________________

____

General

- ❑ Identification (photo ID)
- ❑ Business work hours

Clothing

- ❑ Coveralls
- ❑ Overshoes
- ❑ Boots

Personal Protective Equipment (PPE)

- ❑ Hard hat
- ❑ Safety shoes
- ❑ Safety glasses
- ❑ Gloves
- ❑ Ear protection
- ❑ Respiratory protection

Tools

- ❑ Flashlight
- ❑ Tape measure(s)
- ❑ Pad (graph paper) and pen or pencil
- ❑ Magnifying glass

Test gauges

- ❑ Combustible gas detector
- ❑ Pressure gauges
- ❑ Pitot tube or flow meter

Plans and Reports

- ❑ Previous reports
- ❑ Violation notices
- ❑ Previous surveys
- ❑ Applicable codes and standards

Notes: ______________________________

____

____

SITE INSPECTION

Property Name: ______________________________

Address: ______________________________

____

Occupancy Classification

- ❑ Assembly
- ❑ Educational
- ❑ Day care
- ❑ Health care
- ❑ Ambulatory health care
- ❑ Detention and correctional
- ❑ One- and two-family dwelling
- ❑ Lodging and rooming
- ❑ Hotel/Motel/Dormitory
- ❑ Apartment
- ❑ Residential board and care
- ❑ Mercantile
- ❑ Business
- ❑ Industrial
- ❑ Storage
- ❑ Mixed

Form A-1 Inspection Checklist to Accompany Chapter 2

Hazard of Contents

❑ Light (low) ❑ Ordinary (moderate) ❑ Extra (high)

❑ Mixed ❑ Special hazards

Exterior Survey

❑ Housekeeping and maintenance

Building construction type

❑ Type I (fire resistive) ❑ Type II (noncombustible) ❑ Type III (ordinary)

❑ Type IV (heavy timber) ❑ Type V (wood frame) ❑ Mixed

Construction problems

Building height ____________ feet ____________ stories

❑ Potential exposures ❑ Outdoor storage ❑ Hydrants

Fire department connection

❑ Vehicle access ❑ Is it obstructed? ❑ Is it identified?

❑ Drainage (flammable liquid and contaminated runoff)

❑ Fire lanes marked

Building Facilities

❑ HVAC systems ❑ Electrical systems

❑ Gas distribution systems ❑ Refuse handling systems

❑ Conveyor systems ❑ Elevators

Fire Detection and Alarm Systems

See Form A-8.

Fire Suppression Systems

See Form A-10.

Closing Interview

❑ Imminent fire safety hazards ❑ Maintenance issues

❑ Housekeeping issues ❑ Overall evaluation

Items to be researched:

❑ ____

❑ ____

❑ ____

Report

❑ Draft ❑ Review ❑ Final

Notes: ____

____

____

____

 (Page 2 of 2)

Form A-1 *Continued*

Inspection Checklist

Construction, Alterations, and Demolition Operations

Building: ______________________________

Address: ______________________________

Inspector: ______________________ **Date:** ______________

Date of Last Inspection: ______________ **Outstanding Violations:** ❑ Yes ❑ No

General

Is it a new construction?	❑ Yes	❑ No	
Is it a renovation?	❑ Yes	❑ No	
Is it a demolition?	❑ Yes	❑ No	
Is fire protection for construction shown on plans?	❑ Yes	❑ No	
Is it a high rise?	❑ Yes	❑ No	
Is it windowless?	❑ Yes	❑ No	
Is it underground?	❑ Yes	❑ No	

Site Preparation

Are roadways with all-weather surfaces suitable for fire apparatus?	❑ Yes	❑ No	
Are water supplies adequate?	❑ Yes	❑ No	
• Permanent	❑ Yes	❑ No	
• Temporary	❑ Yes	❑ No	

Process Hazards

Are welding facilities separated and is proper housekeeping maintained?	❑ Yes	❑ No	❑ N/A*
Is temporary heating equipment properly installed and maintained?	❑ Yes	❑ No	❑ N/A
Is flammable and combustible liquid storage per code?	❑ Yes	❑ No	❑ N/A
Are propane and other gases stored per code?	❑ Yes	❑ No	❑ N/A
Are there explosives?	❑ Yes	❑ No	
Has permit been obtained?	❑ Yes	❑ No	
Are they properly stored?	❑ Yes	❑ No	

Standpipes (Where Required)

Are they carried up with construction, floor by floor?	❑ Yes	❑ No	
• Are they permanent?	❑ Yes	❑ No	
• Are they temporary?	❑ Yes	❑ No	

**N/A (not applicable) means no such feature in the building.*

 (Page 1 of 3)

Form A-2 Inspection Checklist to Accompany Chapter 7

Renovations

Is egress maintained during renovation?	❑ Yes	❑ No	
Are sprinklers maintained during construction?	❑ Yes	❑ No	
Is firm alarm system maintained during construction?	❑ Yes	❑ No	

Underground Operations ❑ N/A

Are there written procedures for			
• Evacuation?	❑ Yes	❑ No	
• Inspections?	❑ Yes	❑ No	
• Emergency procedures?	❑ Yes	❑ No	
Is there a water supply whenever combustibles are present?	❑ Yes	❑ No	❑ N/A
Is extinguishment equipment for conveyors at			
• Head?	❑ Yes	❑ No	❑ N/A
• Tail?	❑ Yes	❑ No	❑ N/A
• Drive mechanism?	❑ Yes	❑ No	❑ N/A
• Take-up pulley?	❑ Yes	❑ No	❑ N/A
• <300 ft apart?	❑ Yes	❑ No	❑ N/A
Are Class I liquids improperly located underground or within 100 ft of portal?	❑ Yes	❑ No	❑ N/A
Are there check-in and check-out procedures?	❑ Yes	❑ No	❑ N/A
Are oil-filled transformers			
• Used?	❑ Yes	❑ No	
• Diked, vented, separated?	❑ Yes	❑ No	

Temporary Structures and Storage

Do they obstruct fire department access?	❑ Yes	❑ No	
Do they provide exposure to new construction?	❑ Yes	❑ No	
Is fabric in fabric structures flame resistant per NFPA 701?	❑ Yes	❑ No	❑ N/A

Housekeeping

Is trash removed regularly from buildings under construction?	❑ Yes	❑ No	
Are trash chutes			
• On outside of buildings?	❑ Yes	❑ No	❑ N/A
• Properly anchored?	❑ Yes	❑ No	❑ N/A
• Reasonably straight?	❑ Yes	❑ No	❑ N/A

 (Page 2 of 3)

Form A-2 *Continued*

Hot Work

Is open-flame and spark-producing equipment controlled?	❑ Yes	❑ No	
Is fire watch provided during and after hot work?	❑ Yes	❑ No	
Asphalt pots			
• Are they improperly located on roofs or under canopies?	❑ Yes	❑ No	
• Are they in proper working order?	❑ Yes	❑ No	
• Are fire extinguishers located within 25 ft?	❑ Yes	❑ No	
• Are mops properly stored after use?	❑ Yes	❑ No	

Portable Fire Extinguishers

Are they provided per code?	❑ Yes	❑ No	

Demolitions

Is gas turned off and are pipes capped as appropriate?	❑ Yes	❑ No	
Are standpipes demolished floor-by-floor?	❑ Yes	❑ No	❑ N/A
Are sprinkler systems maintained on floor-by-floor basis?	❑ Yes	❑ No	❑ N/A

Fire Walls

Are fire walls carried up with construction?	❑ Yes	❑ No	

Notes: ____

 (Page 3 of 3)

Form A-2 *Continued*

Inspection Checklist

Protection of Openings in Fire Barriers

Building: ______________________________

Address: ______________________________

Inspector: ______________________ **Date:** ______________

Date of Last Inspection: ______________ **Outstanding Violations:** ❑ **Yes** ❑ **No**

General

Were alterations/renovations made since last inspection?	❑ Yes	❑ No	

Stair Enclosures ❑ N/A

What is fire resistance of enclosure?	❑ 1 hr	❑ 2 hr	
Do doors have proper fire protection rating?	❑ ½ hr	❑ 1 hr	❑ 1½ hr
Are doors self-closing and latching?	❑ Yes	❑ No	❑ N/A*
Are penetrations limited to			
• Electrical conduit serving stair?	❑ Yes	❑ No	
• Ductwork necessary for independent stair pressurization?	❑ Yes	❑ No	
• Water or steam piping for stair heating/cooling?	❑ Yes	❑ No	
• Sprinkler and standpipe piping?	❑ Yes	❑ No	
Are penetrations properly protected?	❑ Yes	❑ No	
Are doors improperly wedged open?	❑ Yes	❑ No	

Utility Shafts ❑ N/A

What is fire resistance of enclosure?	❑ 1 hr	❑ 2 hr	
Do doors have proper fire protection rating?	❑ ½ hr	❑ 1 hr	❑ 1½ hr
Are doors self-closing and latching?	❑ Yes	❑ No	❑ N/A
Do ducts have dampers?	❑ Yes	❑ No	❑ N/A
Are doors improperly wedged open?	❑ Yes	❑ No	

Horizontal Exits ❑ N/A

Is fire resistance 2 hr?	❑ Yes	❑ No	
Are doors 1½-hr resistance rated?	❑ Yes	❑ No	
Are doors self-closing and latching?	❑ Yes	❑ No	
Are penetrations properly protected?	❑ Yes	❑ No	❑ N/A
Are doors improperly wedged open?	❑ Yes	❑ No	

**N/A (not applicable) means there's no such feature in the building.*

 (Page 1 of 3)

Form A-3 Inspection Checklist to Accompany Chapter 8

Fire Barrier ❑ N/A

Do fire barriers have proper fire resistance?	❑ ½ hr	❑ 1 hr	❑ 2 hr
Do doors have proper fire resistance?	❑ 20 min	❑ 1 hr	❑ 1½ hr
Are doors self-closing and latching?	❑ Yes	❑ No	
Are penetrations properly protected?	❑ Yes	❑ No	❑ N/A
Do ducts have dampers?	❑ Yes	❑ No	❑ N/A
Are doors improperly wedged open?	❑ Yes	❑ No	

Floor/Ceiling ❑ N/A

Are metal pipe penetrations properly protected?	❑ Yes	❑ No	❑ N/A
Are plastic pipe penetrations properly protected?	❑ Yes	❑ No	❑ N/A
Are wire and cable penetrations properly protected?	❑ Yes	❑ No	❑ N/A

Fire Windows ❑ N/A

Are fire windows installed in 1-hr or less fire-rated assembly?	❑ Yes	❑ No
Is glazing		
• Wired glass?	❑ Yes	❑ No
• Other labeled glazing?	❑ Yes	❑ No

Escalator Enclosures ❑ N/A

Are escalator enclosures properly protected:		
• Are enclosures fire resistive?	❑ Yes	❑ No
• Are sprinklers with draft curtains provided in fully sprinklered buildings?	❑ Yes	❑ No
• Are there deluge sprinklers with exhaust?	❑ Yes	❑ No
• Are there rolling shutters?	❑ Yes	❑ No
• Are doors improperly wedged open?	❑ Yes	❑ No

Elevator Shafts ❑ N/A

What is fire resistance of enclosure?	❑ 1 hr	❑ 2 hr	
Do doors have proper fire protection rating?	❑ ½ hr	❑ 1 hr	❑ 1½ hr

Mail or Laundry Chutes ❑ N/A

What is fire resistance of enclosure?	❑ ½ hr	❑ 1 hr	❑ 2 hr
Do doors have proper fire protection rating?	❑ ½ hr	❑ 1 hr	❑ 1½ hr
Are doors self-closing and latching?	❑ Yes	❑ No	❑ N/A

Corridor ❑ N/A

What is fire resistance of corridor?	❑ ½ hr	❑ 1 hr	❑ N/A
Are doors 20 min, self-closing, and latching?	❑ Yes	❑ No	
Are all penetrations properly sealed?	❑ Yes	❑ No	

Form A-3 *Continued*

Fire Walls ❑ N/A

Do fire walls have proper fire resistance?	❑ 2 hr	❑ 3 hr	❑ 4 hr
Do doors have proper fire protection rating?	❑ 1½ hr	❑ 3 hr	
Are doors self-closing and latching?	❑ Yes	❑ No	
Are penetrations properly protected?	❑ Yes	❑ No	❑ N/A
Do ducts have dampers?	❑ Yes	❑ No	❑ N/A
Are doors improperly wedged open?	❑ Yes	❑ No	

Smoke Barrier ❑ N/A

What is fire resistance of smoke barrier?	❑ ½ hr	❑ 1 hr	
What is the fire protection rating or thickness of doors?	❑ 20 min	❑ 1¾ in., solid	
Are doors self-closing and latching?	❑ Yes	❑ No	❑ N/A
Do ducts have dampers?	❑ Yes	❑ No	❑ N/A
Are doors improperly wedged open?	❑ Yes	❑ No	

Smoke Partition ❑ N/A

What is fire resistance of smoke partition?	❑ ½ hr	❑ Other _________
What is the fire protection rating or thickness of doors?	❑ 20 min.	❑ 1¾ in, solid
Are doors improperly wedged open?	❑ Yes	❑ No

Notes: ______________________________

 (Page 3 of 3)

Form A-3 *Continued*

Inspection Checklist

Electrical Systems

Building: ______________________________

Address: ______________________________

Inspector: ______________________ **Date:** ______________

Date of Last Inspection: ______________ **Outstanding Violations:** ❑ **Yes** ❑ **No**

General			
Is it a new construction?	❑ Yes	❑ No	
Is it a renovation?	❑ Yes	❑ No	
Is it an existing building?	❑ Yes	❑ No	
Conduits, Raceways, and Cables			
Are they badly deteriorated?	❑ Yes	❑ No	
Are they improperly supported?	❑ Yes	❑ No	
Are they terminated with proper fittings at boxes?	❑ Yes	❑ No	
Are they protected from mechanical damage where they pass through walls and floors?	❑ Yes	❑ No	❑ N/A*
Circuit Conductors			
Are terminals or surfaces of conduits discolored (indication of overloaded circuits)?	❑ Yes	❑ No	
Are conductors excessively hot?	❑ Yes	❑ No	
Circuit Breakers			
Are they discolored from overheating?	❑ Yes	❑ No	
Are they too hot to touch?	❑ Yes	❑ No	
Are GFCIs used for wet locations?	❑ Yes	❑ No	
Motors			
Are combustibles adequately separated?	❑ Yes	❑ No	
Are motor casing excessively hot?	❑ Yes	❑ No	
Dry Transformers			
Are they separated from combustible materials?	❑ Yes	❑ No	
Is area adequately ventilated?	❑ Yes	❑ No	
Are clearance requirements (marked on transformer) maintained?	❑ Yes	❑ No	

**N/A (not applicable) means there's no such feature in the building.*

(Page 1 of 3)

Form A-4 Inspection Checklist to Accompany Chapter 9

Static Electricity ❑ N/A

Are control measures adequately maintained:

• Humidification?	❑ Yes	❑ No	❑ N/A
• Bonding?	❑ Yes	❑ No	❑ N/A
• Grounding?	❑ Yes	❑ No	❑ N/A
• Ionization?	❑ Yes	❑ No	❑ N/A

Extension Cords

Used for temporary portable equipment only?	❑ Yes	❑ No	❑ N/A
Of proper size for use?	❑ Yes	❑ No	❑ N/A
Improperly used for mechanical support (except lamps)?	❑ Yes	❑ No	❑ N/A
Improperly exposed to mechanical damage, such as under carpets or where damaged by carts or foot traffic?	❑ Yes	❑ No	❑ N/A
Improperly attached to building surfaces?	❑ Yes	❑ No	❑ N/A
Improperly run through walls, doors, or windows?	❑ Yes	❑ No	❑ N/A
Improperly coiled or hanked?	❑ Yes	❑ No	❑ N/A

Grounding

Are metal cable armor, raceways, boxes, fittings, and electrical machinery properly grounded?	❑ Yes	❑ No	
Are ground clamps and connectors tight?	❑ Yes	❑ No	
Are connections corroded?	❑ Yes	❑ No	

Boxes and Cabinets

Complete enclosures with covers provided?	❑ Yes	❑ No	
Are switch and outlets cracked or broken?	❑ Yes	❑ No	
Are switch and outlet covers discolored from overheating?	❑ Yes	❑ No	
Are all knockouts in place or covered?	❑ Yes	❑ No	❑ N/A

Switchboards and Panelboards

Are covers provided?	❑ Yes	❑ No	❑ N/A
Is cage or barrier provided?	❑ Yes	❑ No	❑ N/A

Lamps and Light Fixtures

Is cord insulation cracked or missing?	❑ Yes	❑ No	
Are halogen lamps too near combustibles?	❑ Yes	❑ No	
Are any fixtures mounted directly to combustible ceilings?	❑ Yes	❑ No	
Are fixture globes or lenses discolored from overheating due to wrong lamp size?	❑ Yes	❑ No	

 (Page 2 of 3)

Form A-4 *Continued*

Oil-Filled Transformers

Are they installed in vault?	❑ Yes	❑ No	❑ N/A
Is room designed to contain spill?	❑ Yes	❑ No	
Are clearance requirements (marked on transformer) maintained?	❑ Yes	❑ No	

Lightning Protection ❑ N/A

Do conductors show excessive corrosion or damage?	❑ Yes	❑ No	❑ N/A
Are connections tight?	❑ Yes	❑ No	❑ N/A

Hazardous Locations ❑ N/A

Is Class I, Div. 1 equipment properly maintained?	❑ Yes	❑ No	❑ N/A
Is Class I, Div. 2 equipment properly maintained?	❑ Yes	❑ No	❑ N/A
Is Class II, Div. 1 equipment properly maintained?	❑ Yes	❑ No	❑ N/A
Is Class II, Div. 2 equipment properly maintained?	❑ Yes	❑ No	❑ N/A
Is Class III, Div. 1 equipment properly maintained?	❑ Yes	❑ No	❑ N/A
Is Class III, Div. 2 equipment properly maintained?	❑ Yes	❑ No	❑ N/A

Notes: ______________________________

(Page 3 of 3)

Form A-4 *Continued*

Inspection Checklist

Heating Systems

Building: ______________________________

Address: ______________________________

Inspector: ______________________ **Date:** ______________

Date of Last Inspection: ______________ **Outstanding Violations:** ❑ Yes ❑ No

General			
Is it a new construction?	❑ Yes	❑ No	
Is it a renovation?	❑ Yes	❑ No	
Is it an existing building?	❑ Yes	❑ No	
Fuel			
Oil	❑ Yes	❑ No	❑ N/A*
Gas	❑ Yes	❑ No	❑ N/A
LP-Gas	❑ Yes	❑ No	❑ N/A
Coal	❑ Yes	❑ No	❑ N/A
Wood	❑ Yes	❑ No	❑ N/A
Other: ______________			
Flue Pipes			
Any sign of corrosion?	❑ Yes	❑ No	
Is clearance from combustibles adequate?	❑ Yes	❑ No	
Furnaces			
Is clearance from combustibles adequate?	❑ Yes	❑ No	
Are gravity-type or forced warm-air type provided with limit switches to shut off fuel?	❑ Yes	❑ No	
Do oil- and gas-fired types have fuel shutoff just ahead of appliance?	❑ Yes	❑ No	❑ N/A
Portable Heaters			
Are they being used?	❑ Yes	❑ No	❑ N/A
Any damage to cords?	❑ Yes	❑ No	❑ N/A
Any unvented heaters used in			
• Educational occupancy?	❑ Yes	❑ No	❑ N/A
• Day-care facility?	❑ Yes	❑ No	❑ N/A
• Health care facility?	❑ Yes	❑ No	❑ N/A
• Residential occupancy, except one- and two-family?	❑ Yes	❑ No	❑ N/A
• Detention and correctional occupancy?	❑ Yes	❑ No	❑ N/A

**N/A (not applicable) means there's no such feature in the building.*

Form A-5 Inspection Checklist to Accompany Chapter 10

Steam and Hot Water Piping

Is hot water pipe clearance to combustibles adequate?	❑ Yes	❑ No	❑ N/A
Is steam pipe clearance to combustibles adequate?	❑ Yes	❑ No	❑ N/A

Burner Controls

Does operator's log indicate periodic testing of controls?	❑ Yes	❑ No	❑ N/A
Has pilot safety device been provided and tested?	❑ Yes	❑ No	❑ N/A

Boiler/Heater Rooms

Are they used for storage?	❑ Yes	❑ No	
Is housekeeping good?	❑ Yes	❑ No	
Are there any fuel leaks?	❑ Yes	❑ No	
Are rooms with LP-Gas adequately ventilated?	❑ Yes	❑ No	❑ N/A
Is adequate combustion air provided?	❑ Yes	❑ No	❑ N/A

Solid Fuel Appliances

Are clearances around appliance and chimney adequate?	❑ Yes	❑ No	❑ N/A

Masonry Chimneys

Is spark arrestor provided if solid fuels used?	❑ Yes	❑ No	❑ N/A
Is chimney clean?	❑ Yes	❑ No	
Is mortar loose or cracked?	❑ Yes	❑ No	
If solid fuel used, any oil- or gas-fired equipment connected to same flue?	❑ Yes	❑ No	
Are combustibles separated ≥2 in.?	❑ Yes	❑ No	❑ N/A

Factory-Built Chimney

Is it installed per manufacturer's instructions?	❑ Yes	❑ No	❑ N/A

Notes: ______________________________

Form A-5 *Continued*

Inspection Checklist

Air-Conditioning and Ventilating Systems

Building: ____________________

Address: ____________________

Inspector: ____________________ **Date:** ____________________

Date of Last Inspection: ____________________ **Outstanding Violations:** ❑ Yes ❑ No

General

Were building alterations/renovations made since last inspection?	❑ Yes	❑ No

Type of refrigerant used in system: ____________________

Air Intake

Is air intake protected with grille or screen?	❑ Yes	❑ No	❑ N/A*
Is intake duct reasonably clean?	❑ Yes	❑ No	❑ N/A

Conditioning Equipment

Is room required to be separated due to size of equipment, pressure vessels, etc.?	❑ Yes	❑ No	❑ N/A
Is room enclosure fire resistive?	❑ Yes	❑ No	❑ N/A
• 1 hr with ¾-hr doors?	❑ Yes	❑ No	❑ N/A
• 2 hr with1½-hr doors?	❑ Yes	❑ No	❑ N/A
Is room used as return air plenum?	❑ Yes	❑ No	
Is storage present (not permitted if room used as plenum)?	❑ Yes	❑ No	
Is room clean?	❑ Yes	❑ No	❑ N/A
If storage is in room, are sprinklers provided?	❑ Yes	❑ No	❑ N/A
Are fan belts enclosed?	❑ Yes	❑ No	❑ N/A
Are filters reasonably clean?	❑ Yes	❑ No	

Air Distribution Equipment

Do ducts passing through fire-rated barriers have dampers?	❑ Yes	❑ No	❑ N/A
Are dampers properly installed and maintained?	❑ Yes	❑ No	❑ N/A
Do ceilings that are part of fire-rated floor/ceiling assemblies have proper ceiling dampers?	❑ Yes	❑ No	❑ N/A
Is duct lining per code?	❑ Yes	❑ No	❑ N/A

**N/A (not applicable) means there's no such feature in the building.*

Notes: ____________________

____

 (Page 1 of 1)

Form A-6 Inspection Checklist to Accompany Chapter 11

Inspection Checklist

Smoke-Control Systems

Building: ______________________________

Address: ______________________________

Inspector: ______________________________ **Date:** ______________

Date of Last Inspection: ______________ **Outstanding Violations:** ❑ Yes ❑ No

General

Smoke-control system installed to protect: ______________________________

____

Were building alterations/renovations made since last inspection?	❑ Yes	❑ No	
Was any smoke-control system alteration made since last inspection?	❑ Yes	❑ No	

What is smoke-control system type?

❑ Purge

❑ Pressure sandwich

❑ Exhaust

❑ Other: ______________

Is original design description available?	❑ Yes	❑ No	❑ N/A*

If so, where: ______________________________

Is building fully protected with sprinklers?	❑ Yes	❑ No	

Controls

Are controls properly identified?	❑ Yes	❑ No	❑ N/A
Are operating instructions posted?	❑ Yes	❑ No	❑ N/A

Dedicated Systems

Is record of semiannual operational test provided?	❑ Yes	❑ No	❑ N/A

Nondedicated Systems

Is record of annual operational test provided?	❑ Yes	❑ No	❑ N/A

Periodic Tests and Maintenance

Are records of periodic tests and maintenance provided?	❑ Yes	❑ No	❑ N/A

**N/A (not applicable) means there's no such feature in the building.*

Notes: ______________________________

____

____

____

 (Page 1 of 1)

Form A-7 Inspection Checklist to Accompany Chapter 12

Inspection Checklist

Fire Alarm Systems

Building: ______________________________

Address: ______________________________

Inspector: ______________________ **Date:** ______________

Date of Last Inspection: ______________ **Outstanding Violations:** ❑ Yes ❑ No

General

Were building alterations/renovations made since last inspection?	❑ Yes	❑ No	
Was new alarm system added since last inspection?	❑ Yes	❑ No	
Were new detectors or alarms added since last inspection?	❑ Yes	❑ No	

Control Panel

Is green power light on?	❑ Yes	❑ No	
Are any trouble lights on?	❑ Yes	❑ No	
If yes, why? ______________			
Are supervisory lights on?	❑ Yes	❑ No	
If yes, why? ______________			
Does panel appear in good condition?	❑ Yes	❑ No	

Batteries

Are batteries in good condition without signs of corrosion?	❑ Yes	❑ No	

Fire Alarm Boxes (Manual Stations)

Are fire alarm boxes clear, unobstructed, and identified?	❑ Yes	❑ No	❑ N/A*

Fire Alarm Notification Appliances

Do number and location of fire alarm notification appliances appear adequate?	❑ Yes	❑ No	❑ N/A

Quarterly Tests Recorded

Test of fuses?	❑ Yes	❑ No	❑ N/A
Test of interfaced equipment?	❑ Yes	❑ No	❑ N/A
Test of panel lamps or LEDs?	❑ Yes	❑ No	❑ N/A
Test of supervisory signal devices (except tamper switches)?	❑ Yes	❑ No	❑ N/A
Test of off-premises transmission equipment?	❑ Yes	❑ No	❑ N/A

**N/A (not applicable) means there's no such feature in the building.*

(Page 1 of 2)

Form A-8 Inspection Checklist to Accompany Chapter 13

Semiannual Tests Recorded

Lead acid battery discharge test?	❑ Yes	❑ No	❑ N/A
Lead acid battery load test?	❑ Yes	❑ No	❑ N/A
Lead acid battery specific gravity test?	❑ Yes	❑ No	❑ N/A
Nickel-battery load voltage test?	❑ Yes	❑ No	❑ N/A
Radiant energy fire detectors (flame detectors)test?	❑ Yes	❑ No	❑ N/A
Valve tamper switches test?	❑ Yes	❑ No	❑ N/A
Waterflow devices test?	❑ Yes	❑ No	❑ N/A

Annual Tests Recorded

Test of panel functions?	❑ Yes	❑ No	❑ N/A
Test of transponders?	❑ Yes	❑ No	❑ N/A
Battery discharge test?	❑ Yes	❑ No	❑ N/A
Charger test?	❑ Yes	❑ No	❑ N/A
Control unit trouble signals test?	❑ Yes	❑ No	❑ N/A
Emergency voice communications equipment test?	❑ Yes	❑ No	❑ N/A
Remote annunciators test?	❑ Yes	❑ No	❑ N/A
Electromagnetic release devices test?	❑ Yes	❑ No	❑ N/A
Fixed extinguishing system switches test?	❑ Yes	❑ No	❑ N/A
Heat detectors test?	❑ Yes	❑ No	❑ N/A
Smoke detector sensitivity (See *NFPA 72,* 7-3.2.1) test?	❑ Yes	❑ No	❑ N/A
Alarm notification appliances test?	❑ Yes	❑ No	❑ N/A
Guard's tour equipment test?	❑ Yes	❑ No	❑ N/A

Notes: ______________________________

Form A-8 *Continued*

Inspection Checklist

Water Supplies

Building: ____________________

Address: ____________________

Inspector: ____________________ **Date:** ____________

Date of Last Inspection: ____________ **Outstanding Violations:** ❑ Yes ❑ No

General

Were building alterations/renovations made since last inspection?	❑ Yes	❑ No	
Any water supply system changes made since last inspection?	❑ Yes	❑ No	

Hose Houses

Are there hose houses?	❑ Yes	❑ No	
Have monthly inspections been recorded?	❑ Yes	❑ No	❑ N/A*

Dry Barrel Hydrants

Is dry barrel hydrant present?	❑ Yes	❑ No	
Is hydrant accessible?	❑ Yes	❑ No	❑ N/A
Are outlet caps in place?	❑ Yes	❑ No	❑ N/A
Are outlet threads damaged?	❑ Yes	❑ No	❑ N/A
Does hydrant show signs of leaking?	❑ Yes	❑ No	❑ N/A
Are there obvious cracks in hydrant barrel?	❑ Yes	❑ No	❑ N/A
Is operating nut in good condition?	❑ Yes	❑ No	❑ N/A

Wet Barrel Hydrants

Is wet barrel hydrant present?	❑ Yes	❑ No	
Is hydrant accessible?	❑ Yes	❑ No	❑ N/A
Are outlet caps in place?	❑ Yes	❑ No	❑ N/A
Are outlet threads damaged?	❑ Yes	❑ No	❑ N/A
Does hydrant show signs of leaking when opened?	❑ Yes	❑ No	❑ N/A
Are there obvious cracks in hydrant barrel?	❑ Yes	❑ No	❑ N/A
Is operating nut in good condition?	❑ Yes	❑ No	❑ N/A

Annual Tests Recorded

Test of hydrants opened and water flowed?	❑ Yes	❑ No

Annual Maintenance Recorded

Operating nut lubricated?	❑ Yes	❑ No
Outlet threads lubricated?	❑ Yes	❑ No

**N/A (not applicable) means there's no such feature in the building.*

Notes: ____________________

 (Page 1 of 1)

Form A-9 Inspection Checklist to Accompany Chapter 14

Inspection Checklist

Sprinkler Systems

Building: ______________________________

Address: ______________________________

Inspector: ______________________ **Date:** ______________

Date of Last Inspection: ______________ **Outstanding Violations:** ❑ Yes ❑ No

General

Date sprinklers installed: ______________

Were building alterations/renovations made since last inspection?	❑ Yes	❑ No	
Was new sprinkler system added since last inspection?	❑ Yes	❑ No	
Any sprinkler system alteration made since last inspection?	❑ Yes	❑ No	

What is system type?

- ❑ Wet
- ❑ Dry
- ❑ Preaction
- ❑ Deluge

Is building fully protected with sprinklers?	❑ Yes	❑ No	

If not, explain: ______________

Sprinkler Valves

Do sprinkler valves appear in good working order?	❑ Yes	❑ No	
Is dry pipe valve in heated enclosure?	❑ Yes	❑ No	❑ N/A*
Are spare sprinklers provided?	❑ Yes	❑ No	

Control Valves

Are control valves sealed?	❑ Yes	❑ No	❑ N/A
Are they locked?	❑ Yes	❑ No	❑ N/A
Do they have tamper switches?	❑ Yes	❑ No	❑ N/A

Fire Department Connections

Are fire department connections clear and unobstructed?	❑ Yes	❑ No	❑ N/A
Are protective caps in place?	❑ Yes	❑ No	❑ N/A
Are connections identified?	❑ Yes	❑ No	❑ N/A

Quarterly Inspections and Tests Recorded

Are quarterly inspections and tests recorded?	❑ Yes	❑ No	

**N/A (not applicable) means there's no such feature in the building.*

(Page 1 of 2)

Form A-10 Inspection Checklist to Accompany Chapter 15

Semiannual Inspections and Tests Recorded

Are semiannual inspections and tests recorded? ❑ Yes ❑ No

Annual Inspection and Tests Recorded

Are annual inspections and tests recorded? ❑ Yes ❑ No

3-Year Dry Pipe Full Flow Trip Tests Recorded

Are 3-year dry pipe full flow trip tests recorded? ❑ Yes ❑ No ❑ N/A

5-Year Alarm Valve Internal Inspection Tests Recorded

Are 5-year alarm valve internal inspection tests recorded? ❑ Yes ❑ No ❑ N/A

20 Years

Is sample of fast response sprinklers tested? ❑ Yes ❑ No ❑ N/A

50 Years

Is sample of standard response sprinklers tested? ❑ Yes ❑ No ❑ N/A

Notes: ______________________________

 (Page 2 of 2)

Form A-10 *Continued*

Inspection Checklist

Water Mist Systems

Building: ______________________________

Address: ______________________________

Inspector: ______________________________ **Date:** ______________

Date of Last Inspection: ______________ **Outstanding Violations:** ❑ Yes ❑ No

General

Date water mist system installed:______________

Were building alterations/renovations made since last inspection?	❑ Yes	❑ No	
Was a new water mist system added since last inspection?	❑ Yes	❑ No	
Was any water mist system alteration made since last inspection?	❑ Yes	❑ No	
Is building fully protected with water mist system?	❑ Yes	❑ No	

If not, explain:______________________

What is fire detection system type?

❑ Smoke ❑ Heat ❑ Flame ❑ Other

If "other," explain:______________________

Pressure Tanks

Are pressure tanks supervised?	❑ Yes	❑ No	❑ N/A*
Are they inspected weekly?	❑ Yes	❑ No	❑ N/A

Control Valves

Are control valves sealed?	❑ Yes	❑ No	❑ N/A
Are they locked?	❑ Yes	❑ No	❑ N/A
Do they have tamper switches?	❑ Yes	❑ No	❑ N/A

Fire Department Connections

Are fire department connections unobstructed?	❑ Yes	❑ No	❑ N/A
Are protective caps in place?	❑ Yes	❑ No	❑ N/A
Are connections identified?	❑ Yes	❑ No	❑ N/A

Quarterly Inspections and Tests Recorded

Are quarterly inspections and tests recorded? ❑ Yes ❑ No

Semiannual Inspections and Tests Recorded

Are semiannual inspections and tests recorded? ❑ Yes ❑ No

Annual Inspection and Tests Recorded

Are annual inspections and tests recorded? ❑ Yes ❑ No

**N/A (not applicable) means there's no such feature in the building.*

Notes: ______________________________

(Page 1 of 1)

Form A-11 Inspection Checklist to Accompany Chapter 16

Inspection Checklist

Special Agent Extinguishing Systems

Building: ______________________________

Address: ______________________________

Inspector: ______________________ **Date:** ____________

Date of Last Inspection: ____________ **Outstanding Violations:** ❑ **Yes** ❑ **No**

General

Type of special agent extinguishing system:

- ❑ Wet chemical
- ❑ Dry chemical
- ❑ Halon
- ❑ Carbon dioxide
- ❑ Other: ____________

What does the system protect? ______________________

Date system installed: ____________

Were building alterations/renovations made since last inspection?	❑ Yes	❑ No	
Was a new special agent extinguishing system added since last inspection?	❑ Yes	❑ No	
Was any system alteration made since last inspection?	❑ Yes	❑ No	
Is system connected to building fire alarm?	❑ Yes	❑ No	❑ N/A*

Automatic Shutdown

What is fuel source?

- ❑ Electricity
- ❑ Gas
- ❑ Other: ____________

Nozzles

Are caps in place?	❑ Yes	❑ No	❑ N/A
Are nozzles properly oriented to protect hazard?	❑ Yes	❑ No	❑ N/A
Are there signs of damage?	❑ Yes	❑ No	❑ N/A

Manual Releases

Are manual releases clear and unobstructed?	❑ Yes	❑ No	❑ N/A
Are they properly identified?	❑ Yes	❑ No	❑ N/A

System Pressure Gauges

Are system pressure gauges in proper operating range?	❑ Yes	❑ No	❑ N/A
Are they readily visible?	❑ Yes	❑ No	❑ N/A

**N/A (not applicable) means there's no such feature in the building.*

(Page 1 of 2)

Form A-12 Inspection Checklist to Accompany Chapter 17

Gaseous Agent Systems (Halon or CO_2)

Are room doors self-closing or automatic closing?	❑ Yes	❑ No	❑ N/A

Quarterly Inspections and Tests

Are quarterly inspections and tests recorded?	❑ Yes	❑ No

Semiannual Inspections and Tests

Are semiannual inspections and tests recorded?	❑ Yes	❑ No

Annual Inspections and Tests

Are annual inspections and tests recorded?	❑ Yes	❑ No

5-Year Tests

Are 5-year tests recorded?	❑ Yes	❑ No	❑ N/A
• Were CO_2 system hoses tested?	❑ Yes	❑ No	❑ N/A

12-Year Tests

Are 12-year tests recorded?	❑ Yes	❑ No	❑ N/A
• Were CO_2 system cylinders tested?	❑ Yes	❑ No	❑ N/A
• Were CO_2 systems discharge tested?	❑ Yes	❑ No	❑ N/A

Notes: ______________________________

Form A-12 *Continued*

Inspection Checklist

Clean Agent Extinguishing Systems

Building: ______________________________

Address: ______________________________

Inspector: ______________________ **Date:** ______________

Date of Last Inspection: ______________ **Outstanding Violations:** ❑ **Yes** ❑ **No**

General

Agent type: ______________________________

Clean agent extinguishing system protects: ______________________________

Date system installed: ______________________________

Were building alterations/renovations made since last inspection?	❑ Yes	❑ No	
Was new clean agent extinguishing system added since last inspection?	❑ Yes	❑ No	
Was any system alteration made since last inspection?	❑ Yes	❑ No	
Is system connected to building fire alarm?	❑ Yes	❑ No	❑ N/A*

Automatic Shutdown

What is fuel source?

❑ Electricity

❑ Gas

❑ Other: ______________________

Nozzles

Are caps in place?	❑ Yes	❑ No	❑ N/A
Are nozzles properly oriented to protect hazard?	❑ Yes	❑ No	❑ N/A
Are there signs of damage?	❑ Yes	❑ No	❑ N/A

Manual Releases

Are manual releases clear and unobstructed?	❑ Yes	❑ No	❑ N/A
Are they properly identified?	❑ Yes	❑ No	❑ N/A

System Pressure Gauges

Are system pressure gauges in proper operating range?	❑ Yes	❑ No	❑ N/A
Are they readily visible?	❑ Yes	❑ No	❑ N/A

Room Doors

Are room doors self-closing or automatic closing?	❑ Yes	❑ No	❑ N/A

Quarterly Inspections and Tests Recorded

Are quarterly inspections and tests recorded?	❑ Yes	❑ No

**N/A (not applicable) means there's no such feature in the building.*

(Page 1 of 2)

Form A-13 Inspection Checklist to Accompany Chapter 18

Semiannual Inspections and Tests Recorded

Are semiannual inspections and tests recorded?	❑ Yes	❑ No	

Annual Inspections and Tests Recorded

Are annual inspections and tests recorded?	❑ Yes	❑ No	

5-Year Tests

Are 5-year tests recorded?	❑ Yes	❑ No	❑ N/A
• Were system hoses tested?	❑ Yes	❑ No	❑ N/A

Notes: ______________________________

(Page 2 of 2)

Form A-13 *Continued*

Inspection Checklist

Portable Fire Extinguishers

Building: ______________________________

Address: ______________________________

Inspector: ______________________ **Date:** ______________

Date of Last Inspection: ______________ **Outstanding Violations:** ❑ Yes ❑ No

General

Are fire extinguishers provided?	❑ Yes	❑ No	

Mounting

Are extinguishers properly mounted?	❑ Yes	❑ No	
Are any extinguishers blocked?	❑ Yes	❑ No	
Does distribution appear adequate?	❑ Yes	❑ No	

Monthly Inspections

Are monthly inspections recorded?	❑ Yes	❑ No	❑ N/A*
• On tag on extinguisher?	❑ Yes	❑ No	❑ N/A
• On form?	❑ Yes	❑ No	❑ N/A

Annual Maintenance

Are records provided of annual maintenance?	❑ Yes	❑ No	

**N/A (not applicable) means there's no such feature in the building.*

Notes: ______________________________

(Page 1 of 1)

Form A-14 Inspection Checklist to Accompany Chapter 19

Inspection Checklist

Interior Finish, Contents, and Furnishings

Building: ______________________________

Address: ______________________________

Inspector: ______________________ **Date:** ______________

Date of Last Inspection: ______________ **Outstanding Violations:** ❑ Yes ❑ No

General

Is it a new construction?	❑ Yes	❑ No	
Is it a renovation?	❑ Yes	❑ No	
Is it an existing building?	❑ Yes	❑ No	

Interior Finish

Are wall and ceiling finishes per code?	❑ Yes	❑ No	❑ N/A*
• In exits Class	❑ A	❑ B	❑ C
• In corridors Class	❑ A	❑ B	❑ C
• In rooms Class	❑ A	❑ B	❑ C
Is floor finish per code?	❑ Yes	❑ No	❑ N/A
• Class	❑ I	❑ II	❑ N/A
Are fire-retardant coatings used and properly maintained (reapplied as required)?	❑ Yes	❑ No	❑ N/A

Textile and Expanded Vinyl Materials on Walls

Class A with sprinklers?	❑ Yes	❑ No	❑ N/A
Class A if used on partitions $\leq \frac{3}{4}$ floor-to-ceiling height?	❑ Yes	❑ No	❑ N/A
Class A if $\leq$4 ft high?	❑ Yes	❑ No	❑ N/A
Were they previously approved?	❑ Yes	❑ No	❑ N/A

Furnishings (Where Regulated)

Do they meet cigarette ignition resistant tests?	❑ Yes	❑ No	❑ N/A
Do they meet heat release rates?	❑ Yes	❑ No	❑ N/A

Decorations

Are any highly combustible?	❑ Yes	❑ No	❑ N/A
In educational occupancies are they <20% of wall area?	❑ Yes	❑ No	❑ N/A

**N/A (not applicable) means there's no such feature in the building.*

Notes: ______________________________

Form A-15 Inspection Checklist to Accompany Chapter 21

Inspection Checklist

Assembly Occupancies

Building: ______________________________

Address: ______________________________

Inspector: ______________________ **Date:** ______________

Date of Last Inspection: ______________ **Outstanding Violations:** ❑ Yes ❑ No

General

Were alterations/renovations made since last inspection?	❑ Yes	❑ No
Is building mixed occupancy?	❑ Yes	❑ No
What other occupancies? ______________		
Is building construction acceptable for height and occupancy?	❑ Yes	❑ No
Is it a high rise?	❑ Yes	❑ No
Is it windowless?	❑ Yes	❑ No
Is it underground?	❑ Yes	❑ No

Occupant Load and Exits

Is occupant load posted?	❑ Yes	❑ No
Are the exits per code?	❑ Yes	❑ No
Number of exits?	❑ 1 ❑ 2	❑ 3 ❑ 4 or more
Is egress capacity adequate?	❑ Yes	❑ No
What is fire rating of exit stair enclosure?	❑ 1 hr	❑ 2 hr
What is fire rating of exit stair doors?	❑ 1 hr	❑ $1\frac{1}{2}$ hr
• Are they self-closing?	❑ Yes	❑ No
• Latching?	❑ Yes	❑ No
Are exit enclosures free of storage?	❑ Yes	❑ No
Do 100% of exits discharge directly outside?	❑ Yes	❑ No
If not, do ≥50% discharge outside and is level of discharge sprinklered?	❑ Yes	❑ No
Is exit stair reentry per code?	❑ Yes	❑ No

Doors

Are doors blocked?	❑ Yes	❑ No
Are they locked?	❑ Yes	❑ No
Is ≤15-lb force required to release latch?	❑ Yes	❑ No
Do doors swing in direction of travel per code?	❑ Yes	❑ No
Is there panic hardware per code?	❑ Yes	❑ No

Form A-16 Inspection Checklist to Accompany Chapter 22

Egress Arrangement

Is egress clear and unobstructed?	❑ Yes	❑ No	
Are there any dead-end corridors?	❑ Yes	❑ No	
Is common path of travel within limits?	❑ Yes	❑ No	
Is travel through intervening rooms okay?	❑ Yes	❑ No	
Is egress blocked?	❑ Yes	❑ No	
Is aisle accessway width adequate?	❑ Yes	❑ No	
Is aisle width adequate?	❑ Yes	❑ No	

Travel Distance

Is travel distance per code?	❑ Yes	❑ No	

Emergency Lighting

Is emergency lighting per code?	❑ Yes	❑ No	
Is it tested monthly?	❑ Yes	❑ No	

Exit Marking

Is exit marking per code?	❑ Yes	❑ No	

Corridors

Is 1-hr rating required?	❑ Yes	❑ No	
Are corridor walls rated 1 hr with 20-min doors?	❑ Yes	❑ No	

Protection of Hazards

Are hazards protected by			
• Fire-rated enclosure?	❑ Yes	❑ No	
• Extinguishing system?	❑ Yes	❑ No	
• Self-closing door?	❑ Yes	❑ No	
Is kitchen cooking protected?	❑ Yes	❑ No	
Date kitchen hood and duct last cleaned: ____________			

Protection of Vertical Openings

Are vertical openings enclosed?	❑ Yes	❑ No	
Are elevators enclosed?	❑ Yes	❑ No	
Is atrium per code?	❑ Yes	❑ No	❑ N/A*
Are ≤3 levels open per code?	❑ Yes	❑ No	❑ N/A

Interior Finish

Are wall and ceiling finishes per code?	❑ Yes	❑ No	
• Are exits Class A?	❑ Yes	❑ No	
• Are corridors and lobbies Class A or B?	❑ Yes	❑ No	
• Is assembly >300 Class A or B?	❑ Yes	❑ No	
• Is assembly <300 Class A, B, or C?	❑ Yes	❑ No	
Are decorations per code?	❑ Yes	❑ No	❑ N/A
Are curtains/drapes per code?	❑ Yes	❑ No	❑ N/A

**N/A (not applicable) means there's no such feature in the building.*

Form A-16 *Continued*

Special Protection

Are chutes in good working order:			
• Trash chutes?	❑ Yes	❑ No	❑ N/A
• Laundry chutes?	❑ Yes	❑ No	❑ N/A
Is projection room properly protected?	❑ Yes	❑ No	❑ N/A
Is any equipment subject to rupture under or adjacent to exit stairs?	❑ Yes	❑ No	❑ N/A
Are stages per code?	❑ Yes	❑ No	❑ N/A
Are platforms per code?	❑ Yes	❑ No	❑ N/A
Are exhibits per code?	❑ Yes	❑ No	❑ N/A
Are special amusement buildings per code?	❑ Yes	❑ No	❑ N/A
Are open flames controlled per code?	❑ Yes	❑ No	❑ N/A
Are pyrotechnics controlled per code?	❑ Yes	❑ No	❑ N/A

Mezzanines

Is $\leq \frac{1}{3}$ of the mezzanine open area?	❑ Yes	❑ No
Is common path of travel on mezzanine per code?	❑ Yes	❑ No
If mezzanine is enclosed, is there second exit from mezzanine?	❑ Yes	❑ No

Operating Features

Are there crowd managers if >1000 occupants?	❑ Yes	❑ No
Are drills conducted?	❑ Yes	❑ No
Are employees instructed in fire extinguisher use?	❑ Yes	❑ No
Is announcement of exit locations made before each performance?	❑ Yes	❑ No
Is any clothing stored in corridors?	❑ Yes	❑ No

Detection and Alarm

Is it a manual alarm system?	❑ Yes	❑ No
Is there a fire detection system?	❑ Yes	❑ No
• Smoke detectors?	❑ Yes	❑ No
• Heat detectors?	❑ Yes	❑ No
Where: ________________________________		
Are there audible alarms?	❑ Yes	❑ No
Are there visual alarms?	❑ Yes	❑ No
Is there automatic fire department notification?	❑ Yes	❑ No

Form A-16 *Continued*

Extinguishment

Are there sprinklers throughout?	❑ Yes	❑ No
Partial sprinklers?	❑ Yes	❑ No
Where: ______________________________		
Is there a water flow alarm?	❑ Yes	❑ No
Are valves supervised?	❑ Yes	❑ No
❑ Electrical ❑ Locks ❑ Seal		
Other extinguishing systems:		
Type: ______________________________		
Where: ______________________________		
Standpipe?		
❑ Wet ❑ Dry ❑ None		
Fire pump?	❑ Yes	❑ No
Size: ________gpm @ ________ psi		
Date last tested: ______________________________		
Are fire extinguishers per code?	❑ Yes	❑ No

Building Utilities

Are utilities in good working order:		
Heat		
• Gas?	❑ Yes	❑ No
• Oil?	❑ Yes	❑ No
• Coal?	❑ Yes	❑ No
• Other?	❑ Yes	❑ No
Electrical installation?	❑ Yes	❑ No
Elevators		
• Elevator recall (Phase I)?	❑ Yes	❑ No
• Fire fighter control (Phase II)?	❑ Yes	❑ No
Emergency generator?	❑ Yes	❑ No
Size: ______________________________		
Date last tested: ______________________________		

Notes: ____

____

____

____

____

____

____

____

____

____

Form A-16 *Continued*

Inspection Checklist

Educational Occupancies

Building: ______________________

Address: ______________________

Inspector: ______________________ **Date:** ______________

Date of Last Inspection: ______________ **Outstanding Violations:** ❑ Yes ❑ No

General

Were alterations/renovations made since last inspection?	❑ Yes	❑ No
Is building mixed occupancy?	❑ Yes	❑ No
What other occupancies? ______________		
Is building construction acceptable for height and occupancy?	❑ Yes	❑ No
Is it a high rise?	❑ Yes	❑ No
Is it windowless?	❑ Yes	❑ No
Is it underground?	❑ Yes	❑ No

Occupant Load and Exits

Is ≤1st grade on ground floor?	❑ Yes	❑ No		
Is 2nd grade on ≤second floor?	❑ Yes	❑ No		
Are the exits per code?	❑ Yes	❑ No		
Number of exits?	❑ 1	❑ 2	❑ 3	❑ 4 or more
Is egress capacity adequate?	❑ Yes	❑ No		
What is fire rating of exit stair enclosure?	❑ 1 hr	❑ 2 hr		
What is fire rating of exit stair doors?	❑ 1 hr	❑ 1½ hr		
• Are they self-closing?	❑ Yes	❑ No		
• Latching?	❑ Yes	❑ No		
Are exit enclosures free of storage?	❑ Yes	❑ No		
Do 100% of exits discharge directly outside?	❑ Yes	❑ No		
If not, do ≥50% discharge outside and is level of discharge sprinklered?	❑ Yes	❑ No		
Is exit stair reentry per code?	❑ Yes	❑ No		

Doors

Are doors blocked?	❑ Yes	❑ No
Are they locked?	❑ Yes	❑ No
Is ≤15-lb force required to release latch?	❑ Yes	❑ No
Do doors swing in direction of travel per code?	❑ Yes	❑ No
Is there panic hardware per code?	❑ Yes	❑ No

Form A-17 Inspection Checklist to Accompany Chapter 23

Egress Arrangement

Is egress clear and unobstructed?	❑ Yes	❑ No	
Are dead-end corridors within limits?	❑ Yes	❑ No	
Is common path of travel within limits?	❑ Yes	❑ No	
Is travel through intervening rooms okay?	❑ Yes	❑ No	
Is egress blocked?	❑ Yes	❑ No	
Is aisle width adequate?	❑ Yes	❑ No	

Travel Distance

Is travel distance per code?	❑ Yes	❑ No	

Emergency Lighting

Is emergency lighting per code?	❑ Yes	❑ No	
Is it tested monthly?	❑ Yes	❑ No	

Exit Marking

Is exit marking per code?	❑ Yes	❑ No	

Corridors

Is 1-hr rating required?	❑ Yes	❑ No	
What is rating of corridor walls?	❑ ½ hr	❑ 1 hr	
Is rating of doors 20 min?	❑ Yes	❑ No	

Protection of Hazards

Are hazards protected by			
• Fire-rated enclosure?	❑ Yes	❑ No	
• Extinguishing system?	❑ Yes	❑ No	
• Self-closing door?	❑ Yes	❑ No	
Is kitchen cooking protected?	❑ Yes	❑ No	
Date kitchen hood and duct last cleaned: ____________			

Protection of Vertical Openings

Are vertical openings enclosed?	❑ Yes	❑ No	
Are elevators enclosed?	❑ Yes	❑ No	
Is atrium per code?	❑ Yes	❑ No	❑ N/A*
Are ≤3 levels open per code?	❑ Yes	❑ No	❑ N/A

Interior Finish

Is flame spread of wall and ceiling materials per code?	❑ Yes	❑ No	
Are decorations per code?	❑ Yes	❑ No	❑ N/A
Are curtains/drapes per code?	❑ Yes	❑ No	❑ N/A

**N/A (not applicable) means there's no such feature in the building.*

Form A-17 *Continued*

Special Protection

Are chutes in good working order:			
• Trash chutes?	❑ Yes	❑ No	❑ N/A
• Laundry chutes?	❑ Yes	❑ No	❑ N/A
Are stages per code?	❑ Yes	❑ No	❑ N/A
Are platforms per code?	❑ Yes	❑ No	❑ N/A
Are janitor closets sprinklered?	❑ Yes	❑ No	❑ N/A
Are rescue windows in each classroom per code?	❑ Yes	❑ No	
Are smoke barriers per code?	❑ Yes	❑ No	

Mezzanines

Is $\leq \frac{1}{3}$ of the mezzanine area open?	❑ Yes	❑ No
Is common path of travel on mezzanine per code?	❑ Yes	❑ No
If mezzanine is enclosed, is there second exit from mezzanine?	❑ Yes	❑ No

Operating Features

Is there a written emergency plan?	❑ Yes	❑ No
Are drills conducted?	❑ Yes	❑ No
Number of drills per school year: _______________		
Has evacuation relocation area been established?	❑ Yes	❑ No
Is any clothing stored in corridors?	❑ Yes	❑ No
Are artwork and teaching materials on walls ≤20% of wall area?	❑ Yes	❑ No
Is there daily inspection of exits?	❑ Yes	❑ No

Detection and Alarm

Is there a manual alarm system?	❑ Yes	❑ No
Is there a fire detection system?	❑ Yes	❑ No
• Smoke detectors?	❑ Yes	❑ No
• Heat detectors?	❑ Yes	❑ No
Where: ____________________________________		
Are there audible alarms?	❑ Yes	❑ No
Are there visual alarms?	❑ Yes	❑ No
Is there automatic fire department notification?	❑ Yes	❑ No

Form A-17 *Continued*

Extinguishment

Are there sprinklers throughout? ❑ Yes ❑ No

Partial sprinklers? ❑ Yes ❑ No

Where: ______________________________________

Is there a water flow alarm? ❑ Yes ❑ No

Are valves supervised? ❑ Yes ❑ No

❑ Electrical ❑ Locks ❑ Seal

Other extinguishing systems:

Type: ______________________________________

Where: ______________________________________

Standpipe?

❑ Wet ❑ Dry ❑ None

Fire pump? ❑ Yes ❑ No

Size: ________gpm @ ________ psi

Date last tested: ______________________________

Are fire extinguishers per code? ❑ Yes ❑ No

Building Utilities

Are utilities in good working order:

Heat

- Gas? ❑ Yes ❑ No
- Oil? ❑ Yes ❑ No
- Coal? ❑ Yes ❑ No
- Other? ❑ Yes ❑ No

Electrical installation? ❑ Yes ❑ No

Elevators

- Elevator recall (Phase I)? ❑ Yes ❑ No
- Fire fighter control (Phase II)? ❑ Yes ❑ No

Emergency generator? ❑ Yes ❑ No

Size: ______________________________________

Date last tested: ______________________________

Notes: ____

 (Page 4 of 4)

Form A-17 *Continued*

Inspection Checklist

Day-Care Facilities

Building: ______________________________

Address: ______________________________

Inspector: ______________________ **Date:** ______________

Date of Last Inspection: ______________ **Outstanding Violations:** ❑ Yes ❑ No

General

Were alterations/renovations made since last inspection?	❑ Yes	❑ No
Is building mixed occupancy?	❑ Yes	❑ No
What other occupancies? ______________		
Is building construction acceptable for height and occupancy?	❑ Yes	❑ No
Is it a high rise?	❑ Yes	❑ No
Is it windowless?	❑ Yes	❑ No
Is it underground?	❑ Yes	❑ No

What kind of day-care facility is it?

- ❑ Day care (≥13 clients)
- ❑ Group day care (7–12 clients)
- ❑ Family day care (4–6 clients)

Occupant Load and Exits

Is location of day care in building per code?	❑ Yes		❑ No	
Are the exits per code?	❑ Yes		❑ No	
Number of exits?	❑ 1	❑ 2	❑ 3	❑ 4 or more
Is egress capacity adequate?	❑ Yes		❑ No	
What is fire rating of exit stair enclosure?	❑ 1 hr		❑ 2 hr	
What is fire rating of exit stair doors?	❑ 1 hr		❑ 1½ hr	
• Are they self-closing?	❑ Yes		❑ No	
• Latching?	❑ Yes		❑ No	
Are exit enclosures free of storage?	❑ Yes		❑ No	
Do 100% of exits discharge directly outside?	❑ Yes		❑ No	
If not, do ≥50% discharge outside and is level of discharge sprinklered?	❑ Yes		❑ No	
Is exit stair reentry per code?	❑ Yes		❑ No	

Doors

Are doors blocked?	❑ Yes	❑ No
Are they locked?	❑ Yes	❑ No
Is ≤15-lb force required to release latch?	❑ Yes	❑ No
Do doors swing in direction of travel per code?	❑ Yes	❑ No
Is there panic hardware per code?	❑ Yes	❑ No

 (Page 1 of 4)

Form A-18 Inspection Checklist to Accompany Chapter 24

Egress Arrangement

Is egress clear and unobstructed?	❑ Yes	❑ No	
Are dead-end corridors within limits?	❑ Yes	❑ No	
Is common path of travel within limits?	❑ Yes	❑ No	
Is travel through intervening rooms okay?	❑ Yes	❑ No	
Is egress blocked?	❑ Yes	❑ No	
Is aisle width adequate?	❑ Yes	❑ No	

Travel Distance

Is travel distance per code?	❑ Yes	❑ No	

Emergency Lighting

Is emergency lighting per code?	❑ Yes	❑ No	
Is it tested monthly?	❑ Yes	❑ No	

Exit Marking

Is exit marking per code?	❑ Yes	❑ No	

Corridors

Is 1-hr rating required?	❑ Yes	❑ No	
What is rating of corridor walls?	❑ ½ hr	❑ 1 hr	
Is rating of doors 20 min?	❑ Yes	❑ No	

Protection of Hazards

Is kitchen cooking protected?	❑ Yes	❑ No	
Date kitchen hood and duct last cleaned: ____________			

Protection of Vertical Openings

Are vertical openings enclosed?	❑ Yes	❑ No	
Are elevators enclosed?	❑ Yes	❑ No	
Is atrium per code?	❑ Yes	❑ No	❑ N/A*
Are ≤3 levels open per code?	❑ Yes	❑ No	❑ N/A

Interior Finish

Is flame spread of wall and ceiling materials per code?	❑ Yes	❑ No	
Are decorations per code?	❑ Yes	❑ No	❑ N/A
Are curtains/drapes per code?	❑ Yes	❑ No	❑ N/A

Special Protection

Are chutes in good working order:			
• Trash chutes?	❑ Yes	❑ No	❑ N/A
• Laundry chutes?	❑ Yes	❑ No	❑ N/A
Are janitor closets sprinklered?	❑ Yes	❑ No	❑ N/A
Are rescue windows in each client-occupied room per code?	❑ Yes	❑ No	

**N/A (not applicable) means there's no such feature in the building.*

Form A-18 *Continued*

Mezzanines

Is ≤⅓ of the mezzanine area open?	❑ Yes	❑ No
Is common path of travel on mezzanine per code?	❑ Yes	❑ No
If mezzanine is enclosed, is there second exit from mezzanine?	❑ Yes	❑ No

Operating Features

Is there a written emergency plan?	❑ Yes	❑ No
Are drills conducted?	❑ Yes	❑ No
Number of drills per school year: ____________________		
Has evacuation relocation area been established?	❑ Yes	❑ No
Is any clothing stored in corridors?	❑ Yes	❑ No
Are artwork and teaching materials on walls ≤20% of wall area?	❑ Yes	❑ No
Is there daily inspection of exits?	❑ Yes	❑ No
Is there monthly fire inspection by trained staff?	❑ Yes	❑ No

Detection and Alarm

Is there a manual alarm system?	❑ Yes	❑ No
Is there a fire detection system?	❑ Yes	❑ No
• Smoke detectors?	❑ Yes	❑ No
• Heat detectors?	❑ Yes	❑ No
Where: ____		
Are there audible alarms?	❑ Yes	❑ No
Are there visual alarms?	❑ Yes	❑ No
Is there automatic fire department notification?	❑ Yes	❑ No

Extinguishment

Are there sprinklers throughout?	❑ Yes	❑ No
Partial sprinklers?	❑ Yes	❑ No
Where: ____		
Is there a water flow alarm?	❑ Yes	❑ No
Are valves supervised?	❑ Yes	❑ No
❑ Electrical ❑ Locks ❑ Seal		
Other extinguishing systems:		
Type: ____		
Where: ____		
Standpipe?		
❑ Wet ❑ Dry ❑ None		
Fire pump?	❑ Yes	❑ No
Size: ________gpm @ ________ psi		
Date last tested: ________________________________		
Are fire extinguishers per code?	❑ Yes	❑ No

Form-18 *Continued*

Building Utilities

Are utilities in good working order:

Heat

• Gas?	❑ Yes	❑ No
• Oil?	❑ Yes	❑ No
• Coal?	❑ Yes	❑ No
• Other?	❑ Yes	❑ No
Electrical installation?	❑ Yes	❑ No
Elevators		
• Elevator recall (Phase I)?	❑ Yes	❑ No
• Fire fighter control (Phase II)?	❑ Yes	❑ No
Emergency generator?	❑ Yes	❑ No

Size: ______________________________

Date last tested: ______________________________

Notes: ____

Form A-18 *Continued*

Inspection Checklist

Health Care Facilities

Building: ______________________________

Address: ______________________________

Inspector: ______________________ **Date:** ______________

Date of Last Inspection: ______________ **Outstanding Violations:** ❑ **Yes** ❑ **No**

General

Were alterations/renovations made since last inspection?	❑ Yes	❑ No
Is building mixed occupancy?	❑ Yes	❑ No
What other occupancies? ______________		
Is building construction acceptable for height and occupancy?	❑ Yes	❑ No
Is it a high rise?	❑ Yes	❑ No
Is it windowless?	❑ Yes	❑ No
Is it underground?	❑ Yes	❑ No

Occupant Load and Exits

Are the exits per code?	❑ Yes		❑ No	
Number of exits?	❑ 1	❑ 2	❑ 3	❑ 4 or more
Is egress capacity adequate?	❑ Yes		❑ No	
What is fire rating of exit stair enclosure?	❑ 1 hr		❑ 2 hr	
What is fire rating of exit stair doors?	❑ 1 hr		❑ $1\frac{1}{2}$ hr	
• Are they self-closing?	❑ Yes		❑ No	
• Latching?	❑ Yes		❑ No	
Are exit enclosures free of storage?	❑ Yes		❑ No	
Do 100% of exits discharge directly outside?	❑ Yes		❑ No	
If not, do ≥50% discharge outside and is level of discharge sprinklered?	❑ Yes		❑ No	
Is exit stair reentry per code?	❑ Yes		❑ No	

Doors

Are doors blocked?	❑ Yes	❑ No
Are they locked?	❑ Yes	❑ No
Is ≤15-lb force required to release latch?	❑ Yes	❑ No
Do doors swing in direction of travel per code?	❑ Yes	❑ No
Is panic hardware per code?	❑ Yes	❑ No

Form A-19 Inspection Checklist to Accompany Chapter 25

Egress Arrangement

Is egress clear and unobstructed?	❑ Yes	❑ No	
Are dead-end corridors within limits?	❑ Yes	❑ No	
Is common path of travel within limits?	❑ Yes	❑ No	
Is travel through intervening rooms okay?	❑ Yes	❑ No	
Is egress blocked?	❑ Yes	❑ No	

Travel Distance

Is travel distance per code?	❑ Yes	❑ No	

Emergency Lighting

Is emergency lighting per code?	❑ Yes	❑ No	
Is it tested monthly?	❑ Yes	❑ No	

Exit Marking

Is exit marking per code?	❑ Yes	❑ No	

Corridors

Is 1-hr rating required?	❑ Yes	❑ No	
Is rating 1-hr corridor walls with 20-min doors?	❑ Yes	❑ No	

Protection of Hazards

Are hazards protected by			
• Fire-rated enclosure?	❑ Yes	❑ No	
• Extinguishing system?	❑ Yes	❑ No	
• Self-closing door?	❑ Yes	❑ No	
Is kitchen cooking protected?	❑ Yes	❑ No	
Date kitchen hood and duct last cleaned: ____________			

Protection of Vertical Openings

Are vertical openings enclosed?	❑ Yes	❑ No	
Are elevators enclosed?	❑ Yes	❑ No	
Is atrium per code?	❑ Yes	❑ No	❑ N/A*
Are ≤3 levels open per code?	❑ Yes	❑ No	❑ N/A

Interior Finish

Is flame spread of wall and ceiling materials per code?	❑ Yes	❑ No	
Are curtains/drapes per code?	❑ Yes	❑ No	❑ N/A

**N/A (not applicable) means there's no such feature in the building.*

Form A-19 *Continued*

Special Protection

Are chutes in good working order:			
• Trash chutes?	❑ Yes	❑ No	❑ N/A
• Laundry chutes?	❑ Yes	❑ No	❑ N/A
Are laboratories protected per NFPA 99?	❑ Yes	❑ No	❑ N/A
Are anesthesia areas per NFPA 99?	❑ Yes	❑ No	❑ N/A
Are medical gases stored per NFPA 99?	❑ Yes	❑ No	❑ N/A
Are other occupancies separated by 2-hr fire-resistive construction?	❑ Yes	❑ No	❑ N/A
Are trash receptacles stored per code?	❑ Yes	❑ No	
Do patient rooms >1000 ft^2 have ≥2 means of egress?	❑ Yes	❑ No	
Do treatment rooms >5000 ft^2 have ≥2 means of egress?	❑ Yes	❑ No	
Are treatment suites ≤10,000 ft^2?	❑ Yes	❑ No	
Do patient room doors latch with			
• Positive latches?	❑ Yes	❑ No	
• Roller latches?	❑ Yes	❑ No	
Are smoke barriers provided?	❑ Yes	❑ No	
• Are doors 1¾ in. or 20 min.?	❑ Yes	❑ No	
• Are doors self- or automatic-closing?	❑ Yes	❑ No	
• Is gap between doors ≤⅛ in. or do they have astragals, bevel, or rabbit?	❑ Yes	❑ No	

Operating Features

Are drills conducted?	❑ Yes	❑ No
Frequency of drills: ______________________		
Are employees instructed in fire extinguisher use?	❑ Yes	❑ No
Is there a written emergency plan?	❑ Yes	❑ No

Detection and Alarm

Is there a manual alarm system?	❑ Yes	❑ No
Is there a fire detection system?	❑ Yes	❑ No
• Smoke detectors?	❑ Yes	❑ No
• Heat detectors?	❑ Yes	❑ No
Where: ______________________		
Are there audible alarms?	❑ Yes	❑ No
Are there visual alarms?	❑ Yes	❑ No
Is there automatic fire department notification?	❑ Yes	❑ No

(Page 3 of 4)

Form A-19 *Continued*

Extinguishment

Are there sprinklers throughout? ❑ Yes ❑ No

Partial sprinklers? ❑ Yes ❑ No

Where: ______________________________

Is there a water flow alarm? ❑ Yes ❑ No

Are valves supervised? ❑ Yes ❑ No

❑ Electrical ❑ Locks ❑ Seal

Other extinguishing systems:

Type: ______________________________

Where: ______________________________

Standpipe?

❑ Wet ❑ Dry ❑ None

Fire pump? ❑ Yes ❑ No

Size: ________gpm @ ________ psi

Date last tested: ______________________________

Are fire extinguishers per code? ❑ Yes ❑ No

Building Utilities

Are utilities in good working order:

Heat

- Gas? ❑ Yes ❑ No
- Oil? ❑ Yes ❑ No
- Coal? ❑ Yes ❑ No
- Other? ❑ Yes ❑ No

Electrical installation? ❑ Yes ❑ No

Elevators

- Elevator recall (Phase I)? ❑ Yes ❑ No
- Fire fighter control (Phase II)? ❑ Yes ❑ No

Emergency generator? ❑ Yes ❑ No

Size: ______________________________

Date last tested: ______________________________

Notes: ______________________________

Form A-19 *Continued*

Inspection Checklist
Ambulatory Health Care Facilities

Building: ____________________
Address: ____________________
Inspector: ____________________ **Date:** ____________________
Date of Last Inspection: ____________________ **Outstanding Violations:** ❑ Yes ❑ No

General

Were alterations/renovations made since last inspection?	❑ Yes	❑ No
Is building mixed occupancy?	❑ Yes	❑ No
What other occupancies? ____________________		
Is building construction acceptable for height and occupancy?	❑ Yes	❑ No
Is it a high rise?	❑ Yes	❑ No
Is it windowless?	❑ Yes	❑ No
Is it underground?	❑ Yes	❑ No

Occupant Load and Exits

Are the exits per code?	❑ Yes		❑ No	
Number of exits?	❑ 1	❑ 2	❑ 3	❑ 4 or more
Is egress capacity adequate?	❑ Yes		❑ No	
What is fire rating of exit stair enclosure?	❑ 1 hr		❑ 2 hr	
What is fire rating of exit stair doors?	❑ 1 hr		❑ $1\frac{1}{2}$ hr	
• Are they self-closing?	❑ Yes		❑ No	
• Latching?	❑ Yes		❑ No	
Are exit enclosures free of storage?	❑ Yes		❑ No	
Do 100% of exits discharge directly outside?	❑ Yes		❑ No	
If not, do ≥50% discharge outside and is level of discharge sprinklered?	❑ Yes		❑ No	
Is exit stair reentry per code?	❑ Yes		❑ No	

Doors

Are doors blocked?	❑ Yes	❑ No
Are they locked?	❑ Yes	❑ No
Is ≤15-lb force required to release latch?	❑ Yes	❑ No
Do doors swing in direction of travel per code?	❑ Yes	❑ No

Form A-20 Inspection Checklist to Accompany Chapter 26

Egress Arrangement

Is egress clear and unobstructed?	❑ Yes	❑ No	
Are dead-end corridors within limits?	❑ Yes	❑ No	
Is common path of travel within limits?	❑ Yes	❑ No	
Is travel through intervening rooms okay?	❑ Yes	❑ No	
Is egress blocked?	❑ Yes	❑ No	

Travel Distance

Is travel distance per code?	❑ Yes	❑ No	

Emergency Lighting

Is emergency lighting per code?	❑ Yes	❑ No	
Is it tested monthly?	❑ Yes	❑ No	

Exit Marking

Is exit marking per code?	❑ Yes	❑ No	

Corridors

Is 1-hr rating required?	❑ Yes	❑ No	
Is rating 1-hr corridor walls with 20-min doors?	❑ Yes	❑ No	

Protection of Hazards

Are hazards protected by			
• Fire-rated enclosure?	❑ Yes	❑ No	
• Extinguishing system?	❑ Yes	❑ No	
• Self-closing door?	❑ Yes	❑ No	
Is kitchen cooking protected?	❑ Yes	❑ No	
Date kitchen hood and duct last cleaned: ___________			

Protection of Vertical Openings

Are vertical openings enclosed?	❑ Yes	❑ No	
Are elevators enclosed?	❑ Yes	❑ No	
Is atrium per code?	❑ Yes	❑ No	❑ N/A*
Are ≤3 levels open per code?	❑ Yes	❑ No	❑ N/A

Interior Finish

Is flame spread of wall and ceiling materials per code?	❑ Yes	❑ No	
Are decorations per code?	❑ Yes	❑ No	❑ N/A
Are curtains/drapes per code?	❑ Yes	❑ No	❑ N/A

**N/A (not applicable) means there's no such feature in the building.*

Form A-20 *Continued*

Special Protection

Are chutes in good working order:

• Trash chutes?	❑ Yes	❑ No	❑ N/A
• Laundry chutes?	❑ Yes	❑ No	❑ N/A

Smoke barrier:

• Required if >5000 ft^2 and no smoke detection	❑ Yes	❑ No
• <5000 ft^2 with smoke detection	❑ Yes	❑ No
• 1-hr fire resistive	❑ Yes	❑ No
• Doors 1¾ in. self- or automatic-closing	❑ Yes	❑ No
Is ambulatory health care separated from other occupancies by 1 hr construction?	❑ Yes	❑ No

Operating Features

Is there a written emergency plan	❑ Yes	❑ No
Are drills conducted?	❑ Yes	❑ No

Frequency of drills: ______________________________

Have employees been instructed in fire extinguisher use?	❑ Yes	❑ No

Detection and Alarm

Is there a manual alarm system?	❑ Yes	❑ No
Is there a fire detection system?	❑ Yes	❑ No
• Smoke detectors?	❑ Yes	❑ No
• Heat detectors?	❑ Yes	❑ No

Where: ______________________________

Are there audible alarms?	❑ Yes	❑ No
Are there visual alarms?	❑ Yes	❑ No
Is there automatic fire department notification?	❑ Yes	❑ No

Extinguishment

Are there sprinklers throughout?	❑ Yes	❑ No
Partial sprinklers?	❑ Yes	❑ No

Where: ______________________________

Is there a water flow alarm?	❑ Yes	❑ No
Are valves supervised?	❑ Yes	❑ No

❑ Electrical ❑ Locks ❑ Seal

Other extinguishing systems:

Type: ______________________________

Where: ______________________________

Standpipe?

❑ Wet ❑ Dry ❑ None

Fire pump?	❑ Yes	❑ No

Size: ________gpm @ ________ psi

Date last tested: ______________________________

Are fire extinguishers per code?	❑ Yes	❑ No

Form A-20 *Continued*

Building Utilities

Are utilities in good working order:

Heat

• Gas?	❑ Yes	❑ No
• Oil?	❑ Yes	❑ No
• Coal?	❑ Yes	❑ No
• Other?	❑ Yes	❑ No
Electrical installation?	❑ Yes	❑ No
Elevators		
• Elevator recall (Phase I)?	❑ Yes	❑ No
• Fire fighter control (Phase II)?	❑ Yes	❑ No
Emergency generator?	❑ Yes	❑ No

Size: ______________________________

Date last tested: ______________________________

Notes: ______________________________

Form A-20 *Continued*

Inspection Checklist

Detention and Correctional Occupancies

Building: ____________________

Address: ____________________

Inspector: ____________________ **Date:** ____________________

Date of Last Inspection: ____________________ **Outstanding Violations:** ❑ Yes ❑ No

General

Are housing units separated from other occupancies by 2-hr construction?	❑ Yes	❑ No
Were alterations/renovations made since last inspection?	❑ Yes	❑ No
Is building mixed occupancy?	❑ Yes	❑ No
What other occupancies? ____________		
Is building construction acceptable for height and occupancy?	❑ Yes	❑ No
Is it a high rise?	❑ Yes	❑ No
Is it windowless?	❑ Yes	❑ No
Is it underground?	❑ Yes	❑ No

Occupant Load and Exits

Are the exits per code?	❑ Yes		❑ No	
Number of exits?	❑ 1	❑ 2	❑ 3	❑ 4 or more
Is egress capacity adequate?	❑ Yes		❑ No	
What is fire rating of exit stair enclosure?	❑ 1 hr		❑ 2 hr	
What is fire rating of exit stair doors?	❑ 1 hr		❑ 1½ hr	
• Are they self-closing?	❑ Yes		❑ No	
• Latching?	❑ Yes		❑ No	
Are exit enclosures free of storage?	❑ Yes		❑ No	
Do 100% of exits discharge directly outside?	❑ Yes		❑ No	
If not, do ≥50% discharge into 1 smoke compartment?	❑ Yes		❑ No	
Is exit stair reentry per code?	❑ Yes		❑ No	

Doors

Are doors blocked?	❑ Yes	❑ No
Are they locked?	❑ Yes	❑ No
Is ≤15-lb force required to release latch?	❑ Yes	❑ No
Do doors swing in direction of travel per code?	❑ Yes	❑ No
Is panic hardware per code?	❑ Yes	❑ No

Form A-21 Inspection Checklist to Accompany Chapter 27

Egress Arrangement

Is egress clear and unobstructed?	❑ Yes	❑ No	
Are dead-end corridors within limits?	❑ Yes	❑ No	
Is common path of travel within limits?	❑ Yes	❑ No	
Is travel through intervening rooms okay?	❑ Yes	❑ No	
Is egress blocked?	❑ Yes	❑ No	

Travel Distance

Is travel distance per code?	❑ Yes	❑ No	

Emergency Lighting

Is emergency lighting per code?	❑ Yes	❑ No	
Is it tested monthly?	❑ Yes	❑ No	

Exit Marking

Is exit marking per code?	❑ Yes	❑ No	

Corridors

1-hr rating required?	❑ Yes	❑ No	
What is rating for corridor walls?	❑ ½ hr	❑ 1 hr	
Are doors rated 20 min?	❑ Yes	❑ No	

Protection of Hazards

Are hazards protected by			
• Fire-rated enclosure?	❑ Yes	❑ No	
• Extinguishing system?	❑ Yes	❑ No	
• Self-closing door?	❑ Yes	❑ No	
Is kitchen cooking protected?	❑ Yes	❑ No	
Date kitchen hood and duct last cleaned: ___________			
Are padded cells protected?	❑ Yes	❑ No	❑ N/A*

Protection of Vertical Openings

Are vertical openings enclosed?	❑ Yes	❑ No	
Are elevators enclosed?	❑ Yes	❑ No	
Is atrium per code?	❑ Yes	❑ No	❑ N/A
Are ≤3 levels open per code?	❑ Yes	❑ No	❑ N/A

Interior Finish

Is flame spread of wall and ceiling materials per code?	❑ Yes	❑ No	
Are decorations per code?	❑ Yes	❑ No	❑ N/A
Are curtains/drapes per code?	❑ Yes	❑ No	❑ N/A

Special Protection

Are chutes in good working order:			
• Trash chutes?	❑ Yes	❑ No	❑ N/A
• Laundry chutes?	❑ Yes	❑ No	❑ N/A
Are smoke barriers per code?	❑ Yes	❑ No	❑ N/A
Is subdivision of housing unit per code?	❑ Yes	❑ No	

**N/A (not applicable) means there's no such feature in the building.*

Form A-21 *Continued*

Mezzanines

Is ≤⅓ of the mezzanine area open?	❑ Yes	❑ No
Is common path of travel on mezzanine per code?	❑ Yes	❑ No
If mezzanine is enclosed, is there second exit from mezzanine?	❑ Yes	❑ No

Operating Features

Is there a written emergency plan?	❑ Yes	❑ No
Are drills conducted?	❑ Yes	❑ No
Frequency of drills: ________________________		
Does staff have keys to release occupants?	❑ Yes	❑ No
Are keys identifiable by sight and touch?	❑ Yes	❑ No

Detection and Alarm

Is there a manual alarm system?	❑ Yes	❑ No
Is there a fire detection system?	❑ Yes	❑ No
• Smoke detectors?	❑ Yes	❑ No
• Heat detectors?	❑ Yes	❑ No
Where: ______________________________		
Are there audible alarms?	❑ Yes	❑ No
Are there visual alarms?	❑ Yes	❑ No
Is there automatic fire department notification?	❑ Yes	❑ No

Extinguishment

Are there sprinklers throughout?	❑ Yes	❑ No
Partial sprinklers?	❑ Yes	❑ No
Where: ______________________________		
Is there a water flow alarm?	❑ Yes	❑ No
Are valves supervised?	❑ Yes	❑ No
❑ Electrical ❑ Locks ❑ Seal		
Other extinguishing systems:		
Type: ______________________________		
Where: ______________________________		
Standpipe?		
❑ Wet ❑ Dry ❑ None		
Fire pump?	❑ Yes	❑ No
Size: ________gpm @ ________ psi		
Date last tested: ________________________		
Are fire extinguishers per code?	❑ Yes	❑ No

Form A-21 *Continued*

Building Utilities

Are utilities in good working order:

Heat

• Gas?	❑ Yes	❑ No
• Oil?	❑ Yes	❑ No
• Coal?	❑ Yes	❑ No
• Other?	❑ Yes	❑ No
Electrical installation?	❑ Yes	❑ No
Elevators		
• Elevator recall (Phase I)?	❑ Yes	❑ No
• Fire fighter control (Phase II)?	❑ Yes	❑ No
Emergency generator?	❑ Yes	❑ No

Size: ______________________________

Date last tested: ______________________________

Notes: ______________________________

Form A-21 *Continued*

Inspection Checklist

Hotels

Building: ______________________________

Address: ______________________________

Inspector: ______________________ **Date:** ______________

Date of Last Inspection: ______________ **Outstanding Violations:** ❑ Yes ❑ No

General

Were alterations/renovations made since last inspection?	❑ Yes		❑ No	
Is building mixed occupancy?	❑ Yes		❑ No	
What other occupancies? ______________				
Is building construction acceptable for height and occupancy?	❑ Yes		❑ No	
Is it a high rise?	❑ Yes		❑ No	
Is it windowless?	❑ Yes		❑ No	
Is it underground?	❑ Yes		❑ No	

Occupant Load and Exits

Are the exits per code?	❑ Yes		❑ No	
Number of exits?	❑ 1	❑ 2	❑ 3	❑ 4 or more
Is egress capacity adequate?	❑ Yes		❑ No	
What is fire rating of exit stair enclosure?	❑ 1 hr		❑ 2 hr	
What is fire rating of exit stair doors?	❑ 1 hr		❑ $1\frac{1}{2}$ hr	
• Are they self-closing?	❑ Yes		❑ No	
• Latching?	❑ Yes		❑ No	
Do rooms >2000 ft^2 have 2 egress doors?	❑ Yes		❑ No	
Are exit enclosures free of storage?	❑ Yes		❑ No	
Do 100% of exits discharge directly outside?	❑ Yes		❑ No	
If not, do ≥50% discharge outside and is level of discharge sprinklered?	❑ Yes		❑ No	
Is exit stair reentry per code?	❑ Yes		❑ No	

Doors

Are doors blocked?	❑ Yes	❑ No
Are they locked?	❑ Yes	❑ No
Is ≤15-lb force required to release latch?	❑ Yes	❑ No
Do doors swing in direction of travel per code?	❑ Yes	❑ No
Is panic hardware per code?	❑ Yes	❑ No

Form A-22 Inspection Checklist to Accompany Chapter 28

Egress Arrangement

Is egress clear and unobstructed?	❑ Yes	❑ No	
Are dead-end corridors within limits?	❑ Yes	❑ No	
Is common path of travel within limits?	❑ Yes	❑ No	
Is travel through intervening rooms okay?	❑ Yes	❑ No	
Is egress blocked?	❑ Yes	❑ No	

Travel Distance

Is travel distance per code?	❑ Yes	❑ No	

Emergency Lighting

Is emergency lighting per code?	❑ Yes	❑ No	
Is it tested monthly?	❑ Yes	❑ No	

Exit Marking

Is exit marking per code?	❑ Yes	❑ No	

Corridors

What is rating required?	❑ ½ hr	❑ 1 hr	
What is rating for corridor walls?	❑ ½ hr	❑ 1 hr	
Is door rating 20 min?	❑ Yes	❑ No	
Are room doors self-closing?	❑ Yes	❑ No	

Protection of Hazards

Are hazards protected by			
• Fire-rated enclosure?	❑ Yes	❑ No	
• Extinguishing system?	❑ Yes	❑ No	
• Self-closing door?	❑ Yes	❑ No	
Is kitchen cooking protected?	❑ Yes	❑ No	
Date kitchen hood and duct last cleaned: ___________			

Protection of Vertical Openings

Are vertical openings enclosed?	❑ Yes	❑ No	
Are elevators enclosed?	❑ Yes	❑ No	
Is atrium per code?	❑ Yes	❑ No	❑ N/A*
Are ≤3 levels open per code?	❑ Yes	❑ No	❑ N/A

Interior Finish

Is flame spread of wall and ceiling materials per code?	❑ Yes	❑ No	
Are decorations per code?	❑ Yes	❑ No	❑ N/A
Are curtains/drapes per code?	❑ Yes	❑ No	❑ N/A

**N/A (not applicable) means there's no such feature in the building.*

Form A-22 *Continued*

Special Protection

Are chutes in good working order:

- Trash chutes? ❑ Yes ❑ No ❑ N/A
- Laundry chutes? ❑ Yes ❑ No ❑ N/A

Are there rescue windows in each room? ❑ Yes ❑ No ❑ N/A

Are smoke barriers provided? ❑ Yes ❑ No

Mezzanines

Is $\leq \frac{1}{3}$ of the mezzanine area open? ❑ Yes ❑ No

Is common path of travel on mezzanine per code? ❑ Yes ❑ No

If mezzanine is enclosed, is there second exit from mezzanine? ❑ Yes ❑ No

---

Operating Features

Is fire plan posted in each guest room? ❑ Yes ❑ No

Are employees instructed in emergency duties? ❑ Yes ❑ No

Detection and Alarm

Is there a manual alarm system? ❑ Yes ❑ No

Is there a fire detection system? ❑ Yes ❑ No

- Smoke detectors? ❑ Yes ❑ No
- Heat detectors? ❑ Yes ❑ No

Where: ______________________________

Are there audible alarms? ❑ Yes ❑ No

Are there visual alarms? ❑ Yes ❑ No

Is there automatic fire department notification? ❑ Yes ❑ No

Are there smoke alarms in guest rooms? ❑ Yes ❑ No

Extinguishment

Are there sprinklers throughout? ❑ Yes ❑ No

Partial sprinklers? ❑ Yes ❑ No

Where: ______________________________

Is there a water flow alarm? ❑ Yes ❑ No

Are valves supervised? ❑ Yes ❑ No

❑ Electrical ❑ Locks ❑ Seal

Other extinguishing systems:

Type: ______________________________

Where: ______________________________

Standpipe?

❑ Wet ❑ Dry ❑ None

Fire pump? ❑ Yes ❑ No

Size: ________gpm @ ________ psi

Date last tested: ______________________________

Are fire extinguishers per code? ❑ Yes ❑ No

Form A-22 *Continued*

Building Utilities

Are utilities in good working order:

Heat

- Gas? ❑ Yes ❑ No
- Oil? ❑ Yes ❑ No
- Coal? ❑ Yes ❑ No
- Other? ❑ Yes ❑ No

Electrical installation? ❑ Yes ❑ No

Elevators

- Elevator recall (Phase I)? ❑ Yes ❑ No
- Fire fighter control (Phase II)? ❑ Yes ❑ No

Emergency generator? ❑ Yes ❑ No

Size: ____________________

Date last tested: ____________________

Notes: ____________________

 (Page 4 of 4)

Form A-22 *Continued*

Inspection Checklist

Apartment Buildings

Building: ______________________

Address: ______________________

Inspector: ______________________ **Date:** ______________________

Date of Last Inspection: ______________ **Outstanding Violations:** ❑ Yes ❑ No

General

Were alterations/renovations made since last inspection?	❑ Yes	❑ No
Is building mixed occupancy?	❑ Yes	❑ No
What other occupancies? ______________		
Is building construction acceptable for height and occupancy?	❑ Yes	❑ No
Is it a high rise?	❑ Yes	❑ No
Is it windowless?	❑ Yes	❑ No
Is it underground?	❑ Yes	❑ No

Occupant Load and Exits

Are the exits per code?	❑ Yes	❑ No		
Number of exits?	❑ 1	❑ 2	❑ 3	❑ 4 or more
Is egress capacity adequate?	❑ Yes	❑ No		
What is fire rating of exit stair enclosure?	❑ 1 hr	❑ 2 hr		
What is fire rating of exit stair doors?	❑ 1 hr	❑ $1\frac{1}{2}$ hr		
• Are they self-closing?	❑ Yes	❑ No		
• Latching?	❑ Yes	❑ No		
Are exit enclosures free of storage?	❑ Yes	❑ No		
Do 100% of exits discharge directly outside?	❑ Yes	❑ No		
If not, do ≥50% discharge outside and is level of discharge sprinklered?	❑ Yes	❑ No		
Is exit stair reentry per code?	❑ Yes	❑ No		

Doors

Are doors blocked?	❑ Yes	❑ No
Are they locked?	❑ Yes	❑ No
Is ≤15-lb force required to release latch?	❑ Yes	❑ No
Do doors swing in direction of travel per code?	❑ Yes	❑ No

Form A-23 Inspection Checklist to Accompany Chapter 29

Egress Arrangement

Is egress clear and unobstructed?	❑ Yes	❑ No	
Are dead-end corridors within limits?	❑ Yes	❑ No	
Is common path of travel within limits?	❑ Yes	❑ No	
Is travel through intervening rooms okay?	❑ Yes	❑ No	
Is egress blocked?	❑ Yes	❑ No	

Travel Distance

Is travel distance per code?	❑ Yes	❑ No	

Emergency Lighting

Is emergency lighting per code?	❑ Yes	❑ No	
Is it tested monthly?	❑ Yes	❑ No	

Exit Marking

Is exit marking per code?	❑ Yes	❑ No	

Corridors

What is rating required?	❑ 0 hr	❑ ½ hr	❑ 1 hr
What is rating for corridor walls?	❑ Resist smoke	❑ ½ hr	❑ 1 hr
What is rating for doors?	❑ Resist smoke	❑ 20 min	
Are room doors self-closing?	❑ Yes	❑ No	

Protection of Hazards

Are hazards protected by			
• Fire-rated enclosure?	❑ Yes	❑ No	
• Extinguishing system?	❑ Yes	❑ No	
• Self-closing door?	❑ Yes	❑ No	
Is kitchen cooking protected?	❑ Yes	❑ No	
Date kitchen hood and duct last cleaned: ___________			

Protection of Vertical Openings

Are vertical openings enclosed?	❑ Yes	❑ No	
Are elevators enclosed?	❑ Yes	❑ No	
Is atrium per code?	❑ Yes	❑ No	❑ N/A*
Are ≤3 levels open per code?	❑ Yes	❑ No	❑ N/A

Interior Finish

Is flame spread of wall and ceiling materials per code?	❑ Yes	❑ No	

❑ Class A
❑ Class B
❑ Class C

**N/A (not applicable) means there's no such feature in the building.*

Form A-23 *Continued*

Special Protection

Are chutes in good working order:

- Trash chutes? ❑ Yes ❑ No ❑ N/A
- Laundry chutes? ❑ Yes ❑ No ❑ N/A

Mezzanines

Is $\leq \frac{1}{3}$ of the mezzanine area open? ❑ Yes ❑ No ❑ N/A

Is common path of travel on mezzanine per code? ❑ Yes ❑ No ❑ N/A

If mezzanine is enclosed, is there second exit from mezzanine? ❑ Yes ❑ No ❑ N/A

Operating Features

Are emergency instructions provided annually to residents? ❑ Yes ❑ No

Detection and Alarm

Is there a manual alarm system? ❑ Yes ❑ No

Is there a fire detection system? ❑ Yes ❑ No

- Smoke detectors? ❑ Yes ❑ No
- Heat detectors? ❑ Yes ❑ No

Where: ________________________

Are there audible alarms? ❑ Yes ❑ No

Are there visual alarms? ❑ Yes ❑ No

Is there automatic fire department notification? ❑ Yes ❑ No

Are there smoke alarms in apartments? ❑ Yes ❑ No

Extinguishment

Are there sprinklers throughout? ❑ Yes ❑ No

Partial sprinklers? ❑ Yes ❑ No

Where: ________________________

Is there a water flow alarm? ❑ Yes ❑ No

Are valves supervised? ❑ Yes ❑ No

❑ Electrical ❑ Locks ❑ Seal

Other extinguishing systems:

Type: ________________________

Where: ________________________

Standpipe?

❑ Wet ❑ Dry ❑ None

Fire pump? ❑ Yes ❑ No

Size: ________gpm @ ________ psi

Date last tested: ________________________

Are fire extinguishers per code? ❑ Yes ❑ No

Form A-23 *Continued*

Building Utilities

Are utilities in good working order:

Heat

• Gas?	❑ Yes	❑ No
• Oil?	❑ Yes	❑ No
• Coal?	❑ Yes	❑ No
• Other?	❑ Yes	❑ No
Electrical installation?	❑ Yes	❑ No
Elevators		
• Elevator recall (Phase I)?	❑ Yes	❑ No
• Fire fighter control (Phase II)?	❑ Yes	❑ No
Emergency generator?	❑ Yes	❑ No

Size: ______________________________

Date last tested: ______________________________

Notes: ______________________________

Form A-23 *Continued*

Inspection Checklist

Lodging or Rooming Houses

Building: ______________________________

Address: ______________________________

Inspector: ______________________ **Date:** ______________

Date of Last Inspection: ______________ **Outstanding Violations:** ❑ **Yes** ❑ **No**

General

Were alterations/renovations made since last inspection?	❑ Yes	❑ No	
Is building mixed occupancy?	❑ Yes	❑ No	
What other occupancies? ______________			
Is building construction acceptable for height and occupancy?	❑ Yes	❑ No	
Is it a high rise?	❑ Yes	❑ No	
Is it windowless?	❑ Yes	❑ No	
Is it underground?	❑ Yes	❑ No	

Occupant Load and Exits

Number of sleeping beds: ______________			
Does every story >2000 ft^2 or travel to primary means of escape >75 ft have 2 primary means of escape?	❑ Yes	❑ No	❑ N/A*
Is there secondary means of escape from each sleeping room and living area?	❑ Yes	❑ No	
Is stair enclosure per code?	❑ Yes	❑ No	
Are stair doors per code?	❑ Yes	❑ No	

Doors

Are doors blocked?	❑ Yes	❑ No
Are they locked?	❑ Yes	❑ No
Is ≤15-lb force required to release latch?	❑ Yes	❑ No
Do doors swing in direction of egress per code?	❑ Yes	❑ No

Egress Arrangement

Is egress clear and unobstructed?	❑ Yes	❑ No
Is travel through intervening rooms okay?	❑ Yes	❑ No
Is egress blocked?	❑ Yes	❑ No

Travel Distance ❑ N/A

Emergency Lighting ❑ N/A

Exit Marking ❑ N/A

**N/A (not applicable) means there's no such feature in the building.*

 (Page 1 of 3)

Form A-24 Inspection Checklist to Accompany Chapter 30

Corridors

Are corridor walls smoke resistant?	❑ Yes	❑ No	
Are doors self- or automatic-closing?	❑ Yes	❑ No	

Protection of Hazards ❑ N/A

Protection of Vertical Openings

Is any primary escape route exposed to vertical opening?	❑ Yes	❑ No	
Are vertical openings enclosed in 20-min fire resistance?	❑ Yes	❑ No	

Interior Finish

Is flame spread of wall and ceiling materials per code?	❑ Yes	❑ No	
❑ Class A			
❑ Class B			
❑ Class C			

Special Protection

Are chutes in good working order:			
• Trash chutes?	❑ Yes	❑ No	❑ N/A
• Laundry chutes?	❑ Yes	❑ No	❑ N/A
Are fireplaces per code?	❑ Yes	❑ No	❑ N/A
If windows are used for secondary means of escape			
• Are they >5.7 ft^2?	❑ Yes	❑ No	❑ N/A
• Are they ≥minimum height and width?	❑ Yes	❑ No	
• Are they operable?	❑ Yes	❑ No	❑ N/A
Are doors to closets and bathrooms operable from inside and outside?	❑ Yes	❑ No	

Mezzanines

Is ≤⅓ of the mezzanine area open?	❑ Yes	❑ No	❑ N/A
Is common path of travel on mezzanine per code?	❑ Yes	❑ No	❑ N/A
If mezzanine is enclosed, is there second exit from mezzanine?	❑ Yes	❑ No	❑ N/A

Operating Features ❑ N/A

Detection and Alarm

Is there a manual alarm system?	❑ Yes	❑ No	
Is there a fire detection system?	❑ Yes	❑ No	
• Smoke detectors?	❑ Yes	❑ No	
• Heat detectors?	❑ Yes	❑ No	
Where: ______________________________			
Are there audible alarms?	❑ Yes	❑ No	
Are there visual alarms?	❑ Yes	❑ No	
Is there automatic fire department notification?	❑ Yes	❑ No	
Are there smoke alarms in sleeping rooms?	❑ Yes	❑ No	
Are smoke alarms audible in each sleeping/living room?	❑ Yes	❑ No	

Form A-24 *Continued*

Extinguishment

Are there sprinklers throughout? ❑ Yes ❑ No

Partial sprinklers? ❑ Yes ❑ No

Where: ________________________________

Is there a water flow alarm? ❑ Yes ❑ No

Are valves supervised? ❑ Yes ❑ No

❑ Electrical ❑ Locks ❑ Seal

Other extinguishing systems:

Type: ________________________________

Where: ________________________________

Standpipe?

❑ Wet ❑ Dry ❑ None

Fire pump? ❑ Yes ❑ No

Size: ________gpm @ ________ psi

Date last tested: ________________________

Are fire extinguishers per code? ❑ Yes ❑ No

Building Utilities

Are utilities in good working order:

Heat

- Gas? ❑ Yes ❑ No
- Oil? ❑ Yes ❑ No
- Coal? ❑ Yes ❑ No
- Other? ❑ Yes ❑ No

Electrical installation? ❑ Yes ❑ No

Elevators

- Elevator recall (Phase I)? ❑ Yes ❑ No
- Fire fighter control (Phase II)? ❑ Yes ❑ No

Emergency generator? ❑ Yes ❑ No

Size: ________________________________

Date last tested: ________________________

Notes: ____

 (Page 3 of 3)

Form A-24 *Continued*

Inspection Checklist

Board and Care Occupancies

Building: ______________________________

Address: ______________________________

Inspector: ______________________ **Date:** ______________

Date of Last Inspection: ____________ **Outstanding Violations:** ❑ **Yes** ❑ **No**

General

Any alterations/renovations since last inspection?	❑ Yes	❑ No		
Is building mixed occupancy?	❑ Yes	❑ No		
What other occupancies? ____________				
❑ Prompt ❑ Slow ❑ Impractical				
❑ Small ❑ Large ❑ Apartment building				
Is building construction acceptable for height and occupancy?	❑ Yes	❑ No		
Is it a high rise?	❑ Yes	❑ No		
Is it windowless?	❑ Yes	❑ No		
Is it underground?	❑ Yes	❑ No		

Occupant Load and Exits

Is number of means of escape per code?	❑ Yes	❑ No		❑ N/A*
Is number of exits per code?	❑ Yes	❑ No		❑ N/A # ______
Is egress capacity adequate?	❑ Yes	❑ No		
What is fire rating of egress stair enclosure?	❑ ½ hr	❑ 1 hr		❑ 2 hr
What is fire rating of egress stair doors?	❑ 20 min	❑ ¾ hr	❑ 1 hr	❑ 1½ hr
• Are they self-closing?	❑ Yes	❑ No		
• Latching?	❑ Yes	❑ No		
Are exit enclosures free of storage?	❑ Yes	❑ No		
Do 100% of exits discharge directly outside?	❑ Yes	❑ No		
If not, do ≥50% discharge outside and is level of discharge sprinklered?	❑ Yes	❑ No		
Is exit stair reentry per code?	❑ Yes	❑ No		

Doors

Are doors blocked?	❑ Yes	❑ No
Are they locked?	❑ Yes	❑ No
Is ≤15-lb force required to release latch?	❑ Yes	❑ No
Do doors swing in direction of travel per code?	❑ Yes	❑ No

**N/A (not applicable) means there's no such feature in the building.*

(Page 1 of 4)

Form A-25 Inspection Checklist to Accompany Chapter 31

Egress Arrangement

Is egress clear and unobstructed?	❑ Yes	❑ No	
Are dead-end corridors within limits?	❑ Yes	❑ No	
Is common path of travel within limits?	❑ Yes	❑ No	
Is travel through intervening rooms okay?	❑ Yes	❑ No	
Is egress blocked?	❑ Yes	❑ No	

Travel Distance

Is travel distance per code?	❑ Yes	❑ No	

Emergency Lighting

Is emergency lighting per code?	❑ Yes	❑ No	
Is it tested monthly?	❑ Yes	❑ No	

Exit Marking

Is exit marking per code?	❑ Yes	❑ No	

Corridors

What is rating required?	❑ 0 hr	❑ ½ hr	❑ 1 hr
What is rating for corridor walls?	❑ Resist smoke	❑ ½ hr	❑ 1 hr
What is rating for doors?	❑ 20 min	❑ 1¾ in. thick	
Are room doors self- or automatic-closing?	❑ Yes	❑ No	

Protection of Hazards

Are hazards protected by

• Fire-rated enclosure?	❑ Yes	❑ No	
• Extinguishing system?	❑ Yes	❑ No	
• Self-closing door?	❑ Yes	❑ No	
Is kitchen cooking protected?	❑ Yes	❑ No	

Date kitchen hood and duct last cleaned: ____________

Protection of Vertical Openings

Are vertical openings enclosed?	❑ Yes	❑ No	
What is rating?	❑ ½ hr	❑ 1 hr	❑ 2 hr
Are elevators enclosed?	❑ Yes	❑ No	

Interior Finish

Is flame spread of wall and ceiling materials per code?	❑ Yes	❑ No	

- ❑ Class A
- ❑ Class B
- ❑ Class C

Are curtains, draperies, and similar loosely hanging materials per code?	❑ Yes	❑ No	❑ N/A
Is newly introduced furniture cigarette ignition resistant and does it have limited heat release per code?	❑ Yes	❑ No	❑ N/A with AS

 (Page 2 of 4)

Form A-25 *Continued*

Special Protection

Are chutes in good working order:			
• Trash chutes?	❑ Yes	❑ No	❑ N/A
• Laundry chutes?	❑ Yes	❑ No	❑ N/A
Are fireplaces per code?	❑ Yes	❑ No	❑ N/A
If windows used for secondary means of escape, are they			
• >5.7 ft^2?	❑ Yes	❑ No	❑ N/A
• Operable?	❑ Yes	❑ No	❑ N/A
Are doors to closets and bathrooms operable from inside and outside?	❑ Yes	❑ No	

Mezzanines

Is ≤⅓ of the mezzanine area open?	❑ Yes	❑ No	❑ N/A
Is common path of travel on mezzanine per code?	❑ Yes	❑ No	❑ N/A
If mezzanine is enclosed, is there second exit from mezzanine?	❑ Yes	❑ No	❑ N/A

Operating Features

Is there a written emergency plan?	❑ Yes	❑ No
Are staff and residents trained in emergency procedures?	❑ Yes	❑ No
Are fire drills conducted?	❑ Yes	❑ No
Frequency of drills: ______________________________		

Detection and Alarm

Is there a manual alarm system?	❑ Yes	❑ No
Is there a fire detection system?	❑ Yes	❑ No
• Smoke detectors?	❑ Yes	❑ No
• Heat detectors?	❑ Yes	❑ No
Where: ______________________________		
Are there audible alarms?	❑ Yes	❑ No
Are there visual alarms?	❑ Yes	❑ No
Is there automatic fire department notification?	❑ Yes	❑ No
Are there smoke alarms in sleeping rooms?	❑ Yes	❑ No

Form A-25 *Continued*

Extinguishment

Are there sprinklers throughout?	❑ Yes	❑ No
Partial sprinklers?	❑ Yes	❑ No
Where: ______________________		
Is there a water flow alarm?	❑ Yes	❑ No
Are valves supervised?	❑ Yes	❑ No
❑ Electrical ❑ Locks ❑ Seal		
Other extinguishing systems:		
Type: ______________________		
Where: ______________________		
Standpipe?		
❑ Wet ❑ Dry ❑ None		
Fire pump?	❑ Yes	❑ No
Size: ________gpm @ ________ psi		
Date last tested: ______________________		
Are fire extinguishers per code?	❑ Yes	❑ No

Building Utilities

Are utilities in good working order:		
Heat		
• Gas?	❑ Yes	❑ No
• Oil?	❑ Yes	❑ No
• Coal?	❑ Yes	❑ No
• Other?	❑ Yes	❑ No
Electrical installation?	❑ Yes	❑ No
Elevators		
• Elevator recall (Phase I)?	❑ Yes	❑ No
• Fire fighter control (Phase II)?	❑ Yes	❑ No
Emergency generator?	❑ Yes	❑ No
Size: ______________________		
Date last tested: ______________________		

Notes: ______________________

(Page 4 of 4)

Form A-25 *Continued*

Inspection Checklist

One- and Two-Family Dwellings

Building: ____________________

Address: ____________________

Inspector: ____________________ **Date:** ____________________

Date of Last Inspection: ____________________ **Outstanding Violations:** ❑ **Yes** ❑ **No**

General

Were alterations/renovations made since last inspection?	❑ Yes	❑ No
Is building mixed occupancy?	❑ Yes	❑ No
What other occupancies? ____________________		
Is it a high rise?	❑ Yes	❑ No
Is it windowless?	❑ Yes	❑ No
Is it underground?	❑ Yes	❑ No

Occupant Load and Exits

Is there secondary means of escape from each sleeping room and living area?	❑ Yes	❑ No
OR		
Are sprinklers provided?	❑ Yes	❑ No
Is egress capacity adequate?	❑ Yes	❑ No
What is rating for exit stair enclosure?	❑ 1 hr	❑ 2 hr
What is rating for exit stair doors?	❑ 1 hr	❑ $1\frac{1}{2}$ hr
• Are they self-closing?	❑ Yes	❑ No
• Latching?	❑ Yes	❑ No
Are exit enclosures free of storage?	❑ Yes	❑ No
Do 100% of exits discharge directly outside?	❑ Yes	❑ No
If not, do ≥50% discharge outside and is level of discharge sprinklered?	❑ Yes	❑ No
Is exit stair reentry per code?	❑ Yes	❑ No

Doors

Are doors blocked?	❑ Yes	❑ No
Are they locked?	❑ Yes	❑ No
Is ≤15-lb force required to release latch?	❑ Yes	❑ No

Egress Arrangement

Is egress clear and unobstructed?	❑ Yes	❑ No
Is travel through intervening rooms okay?	❑ Yes	❑ No
Is egress blocked?	❑ Yes	❑ No

Form A-26 Inspection Checklist to Accompany Chapter 32

Travel Distance ❑ N/A*

Emergency Lighting ❑ N/A

Exit Marking ❑ N/A

Corridors ❑ N/A

Protection of Hazards ❑ N/A

Protection of Vertical Openings ❑ N/A

Interior Finish

Is flame spread of wall and ceiling materials per code?	❑ Yes	❑ No	

Special Protection

Are fireplaces per code?	❑ Yes	❑ No	❑ N/A
If windows used for secondary means of escape			
• Are they >5.7 ft^2?	❑ Yes	❑ No	❑ N/A
• ≥Minimum height and width?	❑ Yes	❑ No	
• Operable?	❑ Yes	❑ No	❑ N/A
Are doors to closets and bathrooms operable from inside and outside?	❑ Yes	❑ No	

Mezzanines ❑ N/A

Operating Features

Do occupants have escape plan?	❑ Yes	❑ No
Do occupants have emergency numbers near phone?	❑ Yes	❑ No
Do occupants have a meeting place?	❑ Yes	❑ No

Detection and Alarm

Is there a manual alarm system?	❑ Yes	❑ No
Is there a fire detection system?	❑ Yes	❑ No
• Smoke detectors?	❑ Yes	❑ No
• Heat detectors?	❑ Yes	❑ No
Where: ______________________		
Are there audible alarms?	❑ Yes	❑ No
Are there visual alarms?	❑ Yes	❑ No
Is there automatic fire department notification?	❑ Yes	❑ No
Smoke alarms:		
• In each sleeping room?	❑ Yes	❑ No
• In each living area?	❑ Yes	❑ No
❑ Battery		
❑ House powered		

**N/A (not applicable) means there's no such feature in the building.*

 (Page 2 of 3)

Form A-26 *Continued*

Extinguishment

Are there sprinklers throughout? ❑ Yes ❑ No

Partial sprinklers? ❑ Yes ❑ No

Where: ____________________

Is there a water flow alarm? ❑ Yes ❑ No

Are valves supervised? ❑ Yes ❑ No

❑ Electrical ❑ Locks ❑ Seal

Other extinguishing systems:

Type: ____________________

Where: ____________________

Building Utilities

Are utilities in good working order:

Heat

- Gas? ❑ Yes ❑ No
- Oil? ❑ Yes ❑ No
- Coal? ❑ Yes ❑ No
- Other? ❑ Yes ❑ No

Electrical installation? ❑ Yes ❑ No

Notes: ____________________

Form A-26 *Continued*

Inspection Checklist

Mercantile Occupancies

Building: ______________________________

Address: ______________________________

Inspector: ______________________ **Date:** ______________

Date of Last Inspection: ______________ **Outstanding Violations:** ❑ Yes ❑ No

General

Occupany subclassification:

❑ Class A ❑ Class B ❑ Class C

❑ Covered mall ❑ Anchor store ❑ Bulk merchandising retail

Were alterations/renovations made since last inspection?	❑ Yes	❑ No	
Is building mixed occupancy?	❑ Yes	❑ No	
What other occupancies? ______________			
Is building construction acceptable for height and occupancy?	❑ Yes	❑ No	
Is it a high rise?	❑ Yes	❑ No	
Is it windowless?	❑ Yes	❑ No	
Is it underground?	❑ Yes	❑ No	

Occupant Load and Exits

Are the exits per code?	❑ Yes	❑ No	
Number of exits?	❑ 1 ❑ 2	❑ 3 ❑ 4 or more	
Is ≤50% of egress through checkout stands?	❑ Yes	❑ No	❑ N/A*
Is egress capacity adequate?	❑ Yes	❑ No	
What is fire rating of egress stair enclosure?	❑ 1 hr	❑ 2 hr	
What is fire rating of egress stair doors?	❑ 1 hr	❑ 1½ hr	
• Are they self-closing?	❑ Yes	❑ No	
• Latching?	❑ Yes	❑ No	
Are exit enclosures free of storage?	❑ Yes	❑ No	
Do 100% of exits discharge directly outside?	❑ Yes	❑ No	
If not, do ≥50% discharge outside and is level of discharge sprinklered?	❑ Yes	❑ No	
Is exit stair reentry per code?	❑ Yes	❑ No	

**N/A (not applicable) means there's no such feature in the building.*

Form A-27 Inspection Checklist to Accompany Chapter 33

Doors

Are doors blocked?	❑ Yes	❑ No	
Are they locked?	❑ Yes	❑ No	
Is ≤15-lb force required to release latch?	❑ Yes	❑ No	
Do doors swing in direction of travel per code?	❑ Yes	❑ No	

Egress Arrangement

Is egress clear and unobstructed?	❑ Yes	❑ No	
Are dead-end corridors within limits?	❑ Yes	❑ No	
Is travel through intervening rooms okay?	❑ Yes	❑ No	
Is egress blocked?	❑ Yes	❑ No	

Travel Distance

Is travel distance per code?	❑ Yes	❑ No	

Emergency Lighting

Is emergency lighting per code?	❑ Yes	❑ No	
Is it tested monthly?	❑ Yes	❑ No	

Exit Marking

Is exit marking per code?	❑ Yes	❑ No	

Corridors

Is corridor rating required?	❑ 0 hr	❑ ½ hr	❑ 1 hr
What is rating for corridor walls?	❑ Resist smoke	❑ ½ hr	❑ 1 hr
What is rating for corridor doors?	❑ Resist smoke	❑ 20 min	
Are doors self-closing?	❑ Yes	❑ No	

Protection of Hazards

Are hazards protected by			
• Fire-rated enclosure?	❑ Yes	❑ No	
• Extinguishing system?	❑ Yes	❑ No	
• Self-closing door?	❑ Yes	❑ No	
Is kitchen cooking protected?	❑ Yes	❑ No	
Date kitchen hood and duct last cleaned: ____________			

Protection of Vertical Openings

Are vertical openings enclosed?	❑ Yes	❑ No	
Are elevators enclosed?	❑ Yes	❑ No	
Is atrium per code?	❑ Yes	❑ No	❑ N/A
Are ≤3 levels open per code?	❑ Yes	❑ No	❑ N/A

Interior Finish

Is flame spread of wall and ceiling materials per code?	❑ Yes	❑ No	
❑ Class A			
❑ Class B			
❑ Class C			

Form A-27 *Continued*

Special Protection

Are chutes in good working order:

• Trash chutes?	❑ Yes	❑ No	❑ N/A
• Laundry chutes?	❑ Yes	❑ No	❑ N/A
Are parking structures separated per code?	❑ Yes	❑ No	❑ N/A

Mezzanines

Is ≤⅓ of the mezzanine area open?	❑ Yes	❑ No	❑ N/A
Is common path of travel on mezzanine per code?	❑ Yes	❑ No	❑ N/A
If mezzanine is enclosed, is there second exit from mezzanine?	❑ Yes	❑ No	❑ N/A

Operating Features

Are employees trained in egress procedure?	❑ Yes	❑ No
Are employees trained in fire extinguisher use?	❑ Yes	❑ No

Detection and Alarm

Is it a manual alarm system?	❑ Yes	❑ No
Is there a fire detection system?	❑ Yes	❑ No
• Smoke detectors?	❑ Yes	❑ No
• Heat detectors?	❑ Yes	❑ No

Where: ______________________________

Are there audible alarms?	❑ Yes	❑ No
Are there visual alarms?	❑ Yes	❑ No
Is there automatic fire department notification?	❑ Yes	❑ No

Extinguishment

Are there sprinklers throughout?	❑ Yes	❑ No
Partial sprinklers?	❑ Yes	❑ No

Where: ______________________________

Is there a water flow alarm?	❑ Yes	❑ No
Are valves supervised?	❑ Yes	❑ No

❑ Electrical ❑ Locks ❑ Seal

Other extinguishing systems:

Type: ______________________________

Where: ______________________________

Standpipe:

❑ Wet ❑ Dry ❑ None

Fire pump? ❑ Yes ❑ No

Size: ________gpm @ ________ psi

Date last tested: ______________________________

Are fire extinguishers per code? ❑ Yes ❑ No

 (Page 3 of 4)

Form A-27 *Continued*

Building Utilities

Are utilities in good working order:

Heat

- Gas? ❑ Yes ❑ No
- Oil? ❑ Yes ❑ No
- Coal? ❑ Yes ❑ No
- Other? ❑ Yes ❑ No

Electrical installation? ❑ Yes ❑ No

Elevators

- Elevator recall (Phase I)? ❑ Yes ❑ No
- Fire fighter control (Phase II)? ❑ Yes ❑ No

Emergency generator? ❑ Yes ❑ No

Size: ______________________________

Date last tested: ______________________________

Is smoke removal system per code? ❑ Yes ❑ No ❑ N/A

Notes: ______________________________

Form A-27 *Continued*

Inspection Checklist

Business Occupancies

Building: ____________________

Address: ____________________

Inspector: ____________________ **Date:** ____________________

Date of Last Inspection: ____________________ **Outstanding Violations:** ❑ Yes ❑ No

General

Were alterations/renovations made since last inspection?	❑ Yes	❑ No
Is building mixed occupancy?	❑ Yes	❑ No
What other occupancies? ____________________		
Is building construction acceptable for height and occupancy?	❑ Yes	❑ No
Is it a high rise?	❑ Yes	❑ No
Is it windowless?	❑ Yes	❑ No
Is it underground?	❑ Yes	❑ No

Occupant Load and Exits

Are the exits per code?	❑ Yes	❑ No		
Number of exits?	❑ 1	❑ 2	❑ 3	❑ 4 or more
Is egress capacity adequate?	❑ Yes	❑ No		
What is fire rating of exit stair enclosure?	❑ 1 hr	❑ 2 hr		
What is fire rating of exit stair doors?	❑ 1 hr	❑ $1\frac{1}{2}$ hr		
• Are they self-closing?	❑ Yes	❑ No		
• Latching?	❑ Yes	❑ No		
Are exit enclosures free of storage?	❑ Yes	❑ No		
Do 100% of exits discharge directly outside?	❑ Yes	❑ No		
If not, do ≥50% discharge outside and is level of discharge sprinklered?	❑ Yes	❑ No		
Is exit stair reentry per code?	❑ Yes	❑ No		

Doors

Are doors blocked?	❑ Yes	❑ No
Are they locked?	❑ Yes	❑ No
Is ≤15-lb force required to release latch?	❑ Yes	❑ No
Do doors swing in direction of travel per code?	❑ Yes	❑ No

Egress Arrangement

Is egress clear and unobstructed?	❑ Yes	❑ No
Are dead-end corridors within limits?	❑ Yes	❑ No
Is common path of travel within limits?	❑ Yes	❑ No
Is travel through intervening rooms okay?	❑ Yes	❑ No
Is egress blocked?	❑ Yes	❑ No

Form A-28 Inspection Checklist to Accompany Chapter 34

Travel Distance

Is travel distance per code?	❑ Yes	❑ No	

Emergency Lighting

Is emergency lighting per code?	❑ Yes	❑ No	
Is it tested monthly?	❑ Yes	❑ No	❑ N/A*

Exit Marking

Is exit marking per code?	❑ Yes	❑ No	

Corridors

Is corridor rating required?	❑ 0 hr	❑ ½ hr	❑ 1 hr
What is rating for corridor walls?	❑ 1 hr	❑ Ohter ________	
Is rating for corridor doors 20 min?	❑ Yes	❑ No	
Are corridor doors self-closing?	❑ Yes	❑ No	

Protection of Hazards

Are hazards protected by			
• Fire-rated enclosure?	❑ Yes	❑ No	
• Extinguishing system?	❑ Yes	❑ No	
• Self-closing door?	❑ Yes	❑ No	
Is kitchen cooking protected?	❑ Yes	❑ No	
Date kitchen hood and duct last cleaned: __________			

Protection of Vertical Openings

Are vertical openings enclosed?	❑ Yes	❑ No	
Are elevators enclosed?	❑ Yes	❑ No	
Is atrium per code?	❑ Yes	❑ No	❑ N/A
Are ≤3 levels open per code?	❑ Yes	❑ No	❑ N/A

Interior Finish

Is flame spread of wall and ceiling materials per code?	❑ Yes	❑ No	
❑ Class A			
❑ Class B			
❑ Class C			

Special Protection

Are chutes in good working order:			
• Trash chutes?	❑ Yes	❑ No	❑ N/A
• Laundry chutes?	❑ Yes	❑ No	❑ N/A
Are parking structures separated per code?	❑ Yes	❑ No	❑ N/A

**N/A (not applicable) means there's no such feature in the building.*

Form A-28 *Continued*

Mezzanines

Is ≤⅓ of the mezzanine area open?	❑ Yes	❑ No
Is common path of travel on mezzanine per code?	❑ Yes	❑ No
If mezzanine is enclosed, is there second exit from mezzanine?	❑ Yes	❑ No

Operating Features

Is there a written emergency plan?	❑ Yes	❑ No
Are designated employees (if any) instructed in extinguisher use?	❑ Yes	❑ No
Are fire drills conducted?	❑ Yes	❑ No

Detection and Alarm

Is there a manual alarm system?	❑ Yes	❑ No
Is there a fire detection system?	❑ Yes	❑ No
• Smoke detectors?	❑ Yes	❑ No
• Heat detectors?	❑ Yes	❑ No
Where: ______________________		
Are there audible alarms?	❑ Yes	❑ No
Are there visual alarms?	❑ Yes	❑ No
Is there automatic fire department notification?	❑ Yes	❑ No

Extinguishment

Are there sprinklers throughout?	❑ Yes	❑ No
Partial sprinklers?	❑ Yes	❑ No
Where: ______________________		
Is there a water flow alarm?	❑ Yes	❑ No
Are valves supervised?	❑ Yes	❑ No
❑ Electrical ❑ Locks ❑ Seal		
Other extinguishing systems:		
Type: ______________________		
Where: ______________________		
Standpipe:		
❑ Wet ❑ Dry ❑ None		
Fire pump?	❑ Yes	❑ No
Size: ________gpm @ ________ psi		
Date last tested: ______________________		
Are fire extinguishers per code?	❑ Yes	❑ No

Form A-28 *Continued*

Building Utilities

Are utilities in good working order:

Heat

- Gas? ❑ Yes ❑ No
- Oil? ❑ Yes ❑ No
- Coal? ❑ Yes ❑ No
- Other? ❑ Yes ❑ No

Electrical installation? ❑ Yes ❑ No

Elevators

- Elevator recall (Phase I)? ❑ Yes ❑ No
- Fire fighter control (Phase II)? ❑ Yes ❑ No

Emergency generator? ❑ Yes ❑ No

Size: ______________________________

Date last tested: ______________________

Notes: ______________________________

 (Page 4 of 4)

Form A-28 *Continued*

Inspection Checklist

Industrial Occupancies

Building: ____________________

Address: ____________________

Inspector: ____________________ **Date:** ____________________

Date of Last Inspection: ____________________ **Outstanding Violations:** ❑ Yes ❑ No

General

Were alterations/renovations made since last inspection?	❑ Yes	❑ No
Is building mixed occupancy?	❑ Yes	❑ No
What other occupancies? ____________________		
Is building construction acceptable for height and occupancy?	❑ Yes	❑ No
Is it a high rise?	❑ Yes	❑ No
Is it windowless?	❑ Yes	❑ No
Is it underground?	❑ Yes	❑ No

Occupant Load and Exits

Are the exits per code?	❑ Yes	❑ No		
Number of exits?	❑ 1	❑ 2	❑ 3	❑ 4 or more
Is egress capacity adequate?	❑ Yes	❑ No		
What is fire rating of exit stair enclosure?	❑ 1 hr	❑ 2 hr		
What is fire rating of exit stair doors?	❑ 1 hr	❑ 1½ hr		
• Are they self-closing?	❑ Yes	❑ No		
• Latching?	❑ Yes	❑ No		
Are exit enclosures free of storage?	❑ Yes	❑ No		
Do 100% of exits discharge directly outside?	❑ Yes	❑ No		
If not, do >50% discharge outside and is level of discharge sprinklered?	❑ Yes	❑ No		
Is exit stair reentry per code?	❑ Yes	❑ No		

Doors

Are doors blocked?	❑ Yes	❑ No
Are they locked?	❑ Yes	❑ No
Is ≤15-lb force required to release latch?	❑ Yes	❑ No
Do doors swing in direction of travel per code?	❑ Yes	❑ No

Form A-29 Inspection Checklist to Accompany Chapter 35

Egress Arrangement

Is egress clear and unobstructed?	❑ Yes	❑ No	
Are dead-end corridors within limits?	❑ Yes	❑ No	
Is common path of travel within limits?	❑ Yes	❑ No	
Is travel through intervening rooms okay?	❑ Yes	❑ No	
Is egress blocked?	❑ Yes	❑ No	

Travel Distance

Is travel distance per code?	❑ Yes	❑ No	

Emergency Lighting

Is emergency lighting per code?	❑ Yes	❑ No	
Is it tested monthly?	❑ Yes	❑ No	❑ N/A*

Exit Marking

Is exit marking per code?	❑ Yes	❑ No	

Corridors

Is corridor rating required?	❑ 0 hr	❑ ½ hr	❑ 1 hr
What is rating for corridor walls?	❑ 1 hr	❑ Other _________	
Is rating for corridor doors 20 min?	❑ Yes	❑ No	
Are corridor doors self-closing?	❑ Yes	❑ No	

Protection of Hazards

Are hazards protected by			
• Fire-rated enclosure?	❑ Yes	❑ No	
• Extinguishing system?	❑ Yes	❑ No	
• Self-closing door?	❑ Yes	❑ No	

Protection of Vertical Openings

Are vertical openings enclosed?	❑ Yes	❑ No	
Are elevators enclosed?	❑ Yes	❑ No	
Is atrium per code?	❑ Yes	❑ No	❑ N/A
Are <3 levels open per code?	❑ Yes	❑ No	❑ N/A

Interior Finish

Is flame spread of wall and ceiling materials per code?	❑ Yes	❑ No	
❑ Class A			
❑ Class B			
❑ Class C			

Special Protection

Are chutes in good working order:			
• Trash chutes?	❑ Yes	❑ No	❑ N/A
• Laundry chutes?	❑ Yes	❑ No	❑ N/A

**N/A (not applicable) means there's no such feature in the building.*

FORM A-29 *Continued*

Mezzanines

Is ≤⅓ of the mezzanine area open?	❑ Yes	❑ No	❑ N/A
Is common path of travel on mezzanine per code?	❑ Yes	❑ No	❑ N/A
If mezzanine is enclosed, is there second exit from mezzanine?	❑ Yes	❑ No	❑ N/A

Operating Features ❑ N/A

Detection and Alarm

Is it a manual alarm system?	❑ Yes	❑ No
Is it a fire detection system?	❑ Yes	❑ No
• Smoke detectors?	❑ Yes	❑ No
• Heat detectors?	❑ Yes	❑ No
Where: ______________________		
Are there audible alarms?	❑ Yes	❑ No
Are there visual alarms?	❑ Yes	❑ No
Is there automatic fire department notification?	❑ Yes	❑ No

Extinguishment

Are there sprinklers throughout?	❑ Yes	❑ No
Partial sprinklers?	❑ Yes	❑ No
Where: ______________________		
Is there a water flow alarm?	❑ Yes	❑ No
Are valves supervised?	❑ Yes	❑ No
❑ Electrical ❑ Locks ❑ Seal		
Other extinguishing systems:		
Type: ______________________		
Where: ______________________		
Standpipe:		
❑ Wet ❑ Dry ❑ None		
Fire pump?	❑ Yes	❑ No
Size: ________gpm @ ________ psi		
Date last tested: ______________________		
Are fire extinguishers per code?	❑ Yes	❑ No

FORM A-29 *Continued*

Building Utilities

Are utilities in good working order:

Heat

• Gas?	❑ Yes	❑ No
• Oil?	❑ Yes	❑ No
• Coal?	❑ Yes	❑ No
• Other?	❑ Yes	❑ No
Electrical installation?	❑ Yes	❑ No
Elevators		
• Elevator recall (Phase I)?	❑ Yes	❑ No
• Fire fighter control (Phase II)?	❑ Yes	❑ No
Emergency generator?	❑ Yes	❑ No

Size: ______________________________

Date last tested: ______________________________

Notes: ______________________________

FORM A-29 *Continued*

Inspection Checklist

Storage Occupancies

Building: ______________________________

Address: ______________________________

Inspector: ______________________ **Date:** ______________________

Date of Last Inspection: ______________ **Outstanding Violations:** ❑ **Yes** ❑ **No**

General

Were alterations/renovations made since last inspection?	❑ Yes	❑ No
Is building mixed occupancy?	❑ Yes	❑ No
What other occupancies? ______________		
Is building construction acceptable for height and occupancy?	❑ Yes	❑ No
Is it a high rise?	❑ Yes	❑ No
Is it windowless?	❑ Yes	❑ No
Is it underground?	❑ Yes	❑ No

Occupant Load and Exits

Are the exits per code?	❑ Yes		❑ No	
Number of exits?	❑ 1	❑ 2	❑ 3	❑ 4 or more
Is egress capacity adequate?	❑ Yes		❑ No	
What is fire rating of exit stair enclosure?	❑ 1 hr		❑ 2 hr	
What is fire rating of exit stair doors?	❑ 1 hr		❑ $1\frac{1}{2}$ hr	
• Are they self-closing?	❑ Yes		❑ No	
• Latching?	❑ Yes		❑ No	
Are exit enclosures free of storage?	❑ Yes		❑ No	
Do 100% of exits discharge directly outside?	❑ Yes		❑ No	
If not, do >50% discharge outside and is level of discharge sprinklered?	❑ Yes		❑ No	
Is exit stair reentry per code?	❑ Yes		❑ No	

Doors

Are doors blocked?	❑ Yes	❑ No
Are they locked?	❑ Yes	❑ No
Is ≤15-lb force required to release latch?	❑ Yes	❑ No
Do doors swing in direction of travel per code?	❑ Yes	❑ No

FORM A-30 Inspection Checklist to Accompany Chapter 36

Egress Arrangement

Is egress clear and unobstructed?	❑ Yes	❑ No	
Are dead-end corridors within limit?	❑ Yes	❑ No	
Is common path of travel within limits?	❑ Yes	❑ No	
Is travel through intervening rooms okay?	❑ Yes	❑ No	
Is egress blocked?	❑ Yes	❑ No	

Travel Distance

Is travel distance per code?	❑ Yes	❑ No	

Emergency Lighting

Is emergency lighting per code?	❑ Yes	❑ No	
Is it tested monthly?	❑ Yes	❑ No	

Exit Marking

Is exit marking per code?	❑ Yes	❑ No	

Corridors

Is corridor rating required?	❑ Yes	❑ No	
What is rating for corridor walls?	❑ ½ hr	❑ 1 hr	
Is rating for corridor doors 20 min?	❑ Yes	❑ No	
Are corridor doors self-closing?	❑ Yes	❑ No	

Protection of Hazards

Are hazards protected by

• Fire-rated enclosure?	❑ Yes	❑ No	
• Extinguishing system?	❑ Yes	❑ No	
• Self-closing door?	❑ Yes	❑ No	
Is kitchen protected?	❑ Yes	❑ No	

Date kitchen hood and duct last cleaned: ___________

Protection of Vertical Openings

Are vertical openings enclosed?	❑ Yes	❑ No	
Are elevators enclosed?	❑ Yes	❑ No	
Is atrium per code?	❑ Yes	❑ No	❑ N/A*
Are <3 levels open per code?	❑ Yes	❑ No	❑ N/A

Interior Finish

Is flame spread of wall and ceiling materials per code?	❑ Yes	❑ No	

Special Protection

Are trash chutes in good working order?	❑ Yes	❑ No	❑ N/A

**N/A (not applicable) means there's no such feature in the building.*

FORM A-30 *Continued*

Mezzanines

Is ≤⅓ of the mezzanine area open?	❑ Yes	❑ No
Is common path of travel on mezzanine per code?	❑ Yes	❑ No
If mezzanine is enclosed, is there second exit from mezzanine?	❑ Yes	❑ No

Operating Features ❑ N/A

Detection and Alarm

Is it a manual alarm system?	❑ Yes	❑ No
Is there a fire detection system?	❑ Yes	❑ No
• Smoke detectors?	❑ Yes	❑ No
• Heat detectors?	❑ Yes	❑ No
Where: ______________________		
Are there audible alarms?	❑ Yes	❑ No
Are there visual alarms?	❑ Yes	❑ No
Is there automatic fire department notification?	❑ Yes	❑ No

Extinguishment

Are there sprinklers throughout?	❑ Yes	❑ No
Partial sprinklers?	❑ Yes	❑ No
Where: ______________________		
Is there a water flow alarm?	❑ Yes	❑ No
Are valves supervised?	❑ Yes	❑ No
❑ Electrical ❑ Locks ❑ Seal		
Other extinguishing systems:		
Type: ______________________		
Where: ______________________		
Standpipe:		
❑ Wet ❑ Dry ❑ None		
Fire pump?	❑ Yes	❑ No
Size: ________gpm @ ________ psi		
Date last tested: ______________________		
Are fire extinguishers per code?	❑ Yes	❑ No

FORM A-30 *Continued*

Building Utilities

Are utilities in good working order:

Heat		
• Gas?	❑ Yes	❑ No
• Oil?	❑ Yes	❑ No
• Coal?	❑ Yes	❑ No
• Other?	❑ Yes	❑ No
Electrical installation?	❑ Yes	❑ No
Elevators		
• Elevator recall (Phase I)?	❑ Yes	❑ No
• Fire fighter control (Phase II)?	❑ Yes	❑ No
Emergency generator?	❑ Yes	❑ No

Size: ______________________

Date last tested: ______________________

Notes: ______________________

(Page 4 of 4)

FORM A-30 *Continued*

Inspection Checklist

Special Structures and High-Rise Buildings

Building: ______________________________

Address: ______________________________

Inspector: ____________________ **Date:** ____________________

Date of Last Inspection: ____________________ **Outstanding Violations:** ❑ **Yes** ❑ **No**

General

Were alterations/renovations made since last inspection?	❑ Yes	❑ No
Is building mixed occupancy?	❑ Yes	❑ No
What other occupancies? ____________________		
Is building construction acceptable for height and occupancy?	❑ Yes	❑ No
Is it a high rise?	❑ Yes	❑ No
Is it windowless?	❑ Yes	❑ No
Is it underground?	❑ Yes	❑ No

Occupant Load and Exits

Are the exits per code?	❑ Yes	❑ No
Number of exits?	❑ 1 ❑ 2	❑ 3 ❑ 4 or more
Is egress capacity adequate?	❑ Yes	❑ No
What is fire rating of exit stair enclosure?	❑ 1 hr	❑ 2 hr
What is fire rating of exit stair doors?	❑ 1 hr	❑ 1½ hr
• Are they self-closing?	❑ Yes	❑ No
• Latching?	❑ Yes	❑ No
Are exit enclosures free of storage?	❑ Yes	❑ No
Do 100% of exits discharge directly outside?	❑ Yes	❑ No
If not, do ≥50% discharge outside and is level of discharge sprinklered?	❑ Yes	❑ No
Is exit stair reentry per code?	❑ Yes	❑ No

Doors

Are doors blocked?	❑ Yes	❑ No
Are they locked?	❑ Yes	❑ No
Is ≤15-lb force required to release latch?	❑ Yes	❑ No
Do doors swing in direction of travel per code?	❑ Yes	❑ No

FORM A-31 Inspection Checklist to Accompany Chapter 37

Egress Arrangement

Is egress clear and unobstructed?	❑ Yes	❑ No	
Are dead-end corridors within limits?	❑ Yes	❑ No	
Is common path of travel within limits?	❑ Yes	❑ No	
Is travel through intervening rooms okay?	❑ Yes	❑ No	
Is egress blocked?	❑ Yes	❑ No	

Travel Distance

Is travel distance per code?	❑ Yes	❑ No	

Emergency Lighting

Is emergency lighting per code?	❑ Yes	❑ No	
Is it tested monthly?	❑ Yes	❑ No	

Exit Marking

Is exit marking per code?	❑ Yes	❑ No	

Corridors

Is corridor rating required?	❑ Yes	❑ No	
What is rating for corridor walls?	❑ ½ hr	❑ 1 hr	
Is rating for corridor doors 20 min?	❑ Yes	❑ No	
Are corridor doors self-closing?	❑ Yes	❑ No	

Protection of Hazards

Are hazards protected by			
• Fire-rated enclosure?	❑ Yes	❑ No	
• Extinguishing system?	❑ Yes	❑ No	
• Self-closing door?	❑ Yes	❑ No	
Is kitchen protected?	❑ Yes	❑ No	
Date kitchen hood and duct last cleaned: _____________			

Protection of Vertical Openings

Are vertical openings enclosed?	❑ Yes	❑ No	
Are elevators enclosed?	❑ Yes	❑ No	
Is atrium per code?	❑ Yes	❑ No	❑ N/A*
Are ≤3 levels open per code?	❑ Yes	❑ No	❑ N/A

Interior Finish

Is flame spread of wall and ceiling materials per code?	❑ Yes	❑ No	

**N/A (not applicable) means there's no such feature in the building.*

FORM A-31 *Continued*

Special Protection

Are chutes in good working order:

- Trash chutes? ❑ Yes ❑ No ❑ N/A
- Laundry chutes? ❑ Yes ❑ No ❑ N/A

Are smoke barriers per code? ❑ Yes ❑ No

Mezzanines

Is ≤⅓ of the mezzanine area open? ❑ Yes ❑ No

Is common path of travel on mezzanine per code? ❑ Yes ❑ No

If mezzanine is enclosed, is there second exit from mezzanine? ❑ Yes ❑ No

Operating Features

Is there a written emergency plan? ❑ Yes ❑ No

Are fire drills conducted? ❑ Yes ❑ No

Number of drills per year: ____________________

Detection and Alarm

Is there a manual alarm system? ❑ Yes ❑ No

Is there a fire detection system? ❑ Yes ❑ No

- Smoke detectors? ❑ Yes ❑ No
- Heat detectors? ❑ Yes ❑ No

Where: ______________________________

Are there audible alarms? ❑ Yes ❑ No

Are there visual alarms? ❑ Yes ❑ No

Is there automatic fire department notification? ❑ Yes ❑ No

Extinguishment

Are there sprinklers throughout? ❑ Yes ❑ No

Partial sprinklers? ❑ Yes ❑ No

Where: ______________________________

Is there a water flow alarm? ❑ Yes ❑ No

Are valves supervised? ❑ Yes ❑ No

❑ Electrical ❑ Locks ❑ Seal

Other extinguishing systems:

Type: ______________________________

Where: ______________________________

Standpipe?

❑ Wet ❑ Dry ❑ None

Fire pump? ❑ Yes ❑ No

Size: ________gpm @ ________ psi

Date last tested: ______________________________

Are fire extinguishers per code? ❑ Yes ❑ No

FORM A-31 *Continued*

Building Utilities

Are utilities in good working order:

Heat

• Gas?	❑ Yes	❑ No
• Oil?	❑ Yes	❑ No
• Coal?	❑ Yes	❑ No
• Other?	❑ Yes	❑ No
Electrical installation?	❑ Yes	❑ No
Elevators		
• Elevator recall (Phase I)?	❑ Yes	❑ No
• Fire fighter control (Phase II)?	❑ Yes	❑ No
Emergency generator?	❑ Yes	❑ No

Size: ______________________________

Date last tested: ______________________________

Notes: ______________________________

FORM A-31 *Continued*

Index

A
Abrasive blasting, 533
Acetaldehyde, 465
Acetone, 417
Aerosol containers, 351
Aerosol propellant tank farms, 563–566
Aerosols
base product of, 559
classification of, 561–562
containers for, 560
definition of, 559–560
filling lines for, 566–567
fire protection standards for, 560
hazards of, 559
leak hazards of, 567–568
manufacturing facilities for, 562–568
manufacturing of, 428
product cartons for, 568
product laboratories for, 568
propellant charging and pump rooms for, 567
propellant tank farms for, 562–566
retail display of, 570–571
storage of, 405, 568–571
configurations for, 569
fire history of, 568
sprinkler systems for, 570
warehouses for, 570
test baths for, 567–568
AFCI (arc-fault circuit-interrupters), 85
AFFF (aqueous film-forming foam), 221
in electrical fires, 415–416
Air distribution equipment
inspection checklist for, 602
Air intake system, 105
Air terminals, 91
Air-cleaning equipment, 107–108
Air-conditioning systems
combustible dust in, 110
components of, 105–108
inspection checklist for, 602
installation of, 106
maintenance for, 110
smoke control in, 109
ventilation systems for, 109–110
Airflow method, of smoke-control systems, 114, 118–119
Airflow switches, 506
Air-inflated structures, 364
Air-sampling smoke detectors, 129
Air-supported structures, 364
Alarm systems
for ambulatory health care occupancies, 269
inspection checklist for, 634
annunciation for, 127
for apartment buildings, 296
inspection checklist for, 646
approvals and documentation for, 126
for assembly occupancies
inspection checklist for, 618
audible evacuation code for, 124
building safety functions of, 138
for business occupancies, 333–334
inspection checklist for, 664
categories of signals, 125
combination alarm systems and, 127
control unit panels for, 127
control units for, 124
coverage of, 134–135
cross zoning, 134–135
for day-care facilities, 256
inspection checklist for, 626
design requirements of, 125–126
for detention and correctional occupancies, 279
inspection checklist for, 638
for educational occupancies
inspection checklist for, 622
fire detectors in, 138
general system requirements
equipment listing, 126–127
for health care occupancies
inspection checklist for, 630
for hotels, 286–287
inspection checklist for, 642
for industrial occupancies, 342
inspection checklist for, 668
initiating devices for, 124, 127–134
inspection checklist for, 604–605
inspection of, 13
for lodging or rooming houses, 303
inspection checklist for, 649
for mercantile occupancies, 324–325
inspection checklist for, 660
notification appliances for, 124
offsite notification for, 138–139
for one- and two-family dwellings
inspection checklist for, 656
for parking garages, 353
power sources for, 127
for residential board and care occupancies, 311–312
inspection checklist for, 653
for residential occupancies, 139–140
for special structures and high rise buildings
inspection checklist for, 676
for storage occupancies, 354
inspection checklist for, 672
supervising stations of, 125
suppression system actuation of, 138
system components for, 124
test standards for, 118
wiring for, 140–142
Alarm-initiating devices, 121
Algae growth, in water storage tanks, 194
Alkali metals
characteristics of, 459
fire extinguishment of, 460
process hazards of, 459
storage of, 459–460
Alkali-metal salt solutions, 221
Alkalis, 466
Alkylaluminums, 466
Allied processes, 533
Alpha scintillation detectors, 380
Alterations
construction of, 63–64
inspection checklist for, 591–593
Aluminum
dust, 446
fire hazards of, 460, 464
Aluminum alloys, melting point of, 435
Ambulatory health care occupancies, 260
definition of, 267
fire alarms for, 269
fire drills for, 271
inspection checklist for, 632–635
laboratories in, 269
portable fire extinguishers for, 269–270
protection features of, 268–270
smoke barriers for, 270
smoking regulations for, 271
space heaters for, 271
written fire safety plans for, 270–271
American National Standard Institute (ANSI)
and gas-burning appliances and industrial heating equipment, 436
American Society of Mechanical Engineers (ASME)
Boiler and Pressure Vessel Code, 436
American Welding Society (AWS), 534

Amines, water solubility of, 417
Ammonia, anhydrous
in furnaces, 514–515
Ammonia dissociators, 514–515
Ammonia refrigeration systems, 554–555
fire protection for, 554–555
hazards of, 554
Ammonium dichromate, 469
Ammonium nitrate, 485, 489–490
Ammonium perchlorate, 470
Anhydrides, 466
Anhydrous ammonia, 554
Annunciation, for fire alarm systems, 127
Annunciators
for hotels, 286
ANSI. *See* American National Standards Institute (ANSI)
Ansul Company, The, 199, 200
Antifreeze
in fire protection systems, 144
hazards of, 463
in portable fire extinguishers, 221
sprinkler systems *vs.* wet-pipe sprinkler systems, 162–163
Apartment buildings, 295–296
code requirements for, 289–290
construction of, 52
definition of, 289
dwelling units in, 294
existing, 291
alternative requirements for, 292
exits for, 293–294
fire protection systems for, 296–297
hazards in, 67
inspection checklist for, 644–647
inspection observations for, 291–196
interior finishes for, 295
lighting for, 297
mercantile occupancies in, 293
new, 290
occupant load for, 293
protection of openings for, 204
smoking in, 289
waste chutes for, 295
Appliances, gas-burning, 436
Arc cutting, 533
Arc-fault circuit-interrupters (AFCI), 80, 85
Archives, for records storage, 410
Arcing, electrical fires from, 79, 80
Area separation walls. *See* Fire barriers
Argon, in furnaces, 514
Arson, in storage occupancies, 400
ASHRAE, 108
Askarel, 86
Asphalt pots
inspection checklist for, 593
Asphyxiation, 431
Assembly occupancies
building services for, 245
characteristics of, 242
definition of, 241–242
egress, means of, 243–244
electrical wiring and appliances in, 245–246
fire protection systems for, 248
heating systems for, 245
inspection checklist for, 616–619
inspection of, 243
interior finishes for, 244–245
LP-Gas in, 245
occupant load for, 243
smoking in, 246
theaters, 241
ASTM E84 (Steiner Tunnel Test), 233, 476
Atomic Energy Commission, 380
Atriums, 229
Attire, for fire inspector, 7
Audible notification appliances, 135–136
Audible private mode signals, 136
Audible public mode signals, 135–136
Authority having jurisdiction (AHJ), 126
Autoignition temperature, of flammable and combustible liquids, 418
Automatic closing fire doors, 73, 74
inspection of, 75
Automatic deluge water curtain
for protecting escalators, 71
Automatic dry-pipe system, 172. *See also* Standpipe and hose systems
Automatic fire detectors, 127–128
Automatic Sprinkler Systems Handbook, 71
Automatic sprinkler systems. *See* Sprinkler systems
Automatic wet systems, 172. *See also* Standpipe and hose systems
Average ambient sound level, 135

B

Backflow prevention devices, 145
Backflow protection, for public water supplies, 143–144
Bacteriological growth, in water storage tanks, 194
Baled cotton, 396
storage of, 407
Ballrooms. *See* Assembly occupancies
Basic single-screw extruder, 482
BATF (Bureau of Alcohol, Tobacco and Firearms), 496
Battery charging operations, 390
Beams, 40
fire-resistive ratings for, 50
heat detectors and, 133
smoke detectors and, 130–131
Bearing walls, 39, 40, 42
fire-resistive ratings for, 50
Below-grade areas, vapor hazards in, 417
Belt conveyors, 65
Beta gauge, 381
Blasting agents, 486, 489
Blasting caps, 485
BLEVE (boiler liquid expanding vapor explosion), 435
Blow molding, 481
Board and care facilities. *See* Residential board and care facilities
Boiler and Pressure Vessel Code, 432, 436
Boiler rooms
inspection checklist for, 601
Boiler-furnaces. See Boilers
Boilers
definition of, 501–502
hazards of, 501
in heat transfer fluid systems, 547
for large buildings, 95
location of, 503–504
multiple burner, 505
single burner, 504–505
types of, 96–100
Boiling liquid expanding vapor explosion (BLEVE), 435
Boiling point, of flammable and combustible liquids, 416
Boiling rooms
for detention and correction occupancies, 277
Bonding
of lightning down conductors, 91–92
vs. grounding, 89–90
Boxes, in electrical systems
inspection checklist for, 598
Break lines, in pump driven water mist systems, 190–191
Bromination, of stored water in mist systems, 194
Bromochlorodifluoromethane (halon 1211), 199–200
Bromotrifluoromethane (halon 1301), 199–200
Bucket elevators, 391
Building automation systems (BAS), 118
Building care and maintenance, 24–25
Building Construction for the Fire Service, 39, 46
Building construction. *See* Construction
Building facilities, 11
inspection checklist for, 590
Building separation walls. *See* Fire barriers
Building services
for business occupancies, 331–333
fire inspector's knowledge of, 4
for storage occupancies, 352
Bulk liquefied oxygen (LOX) systems, 440
Bulk plants
cylinder filling of LP-gas at, 438
features of, 437–438

fire protection systems for, 438
fire safety analysis of, 438
inspections of, 438–439
for LP-gas, 437
and portable fire extinguishers, 439
pressure relief valves in, 439
sniff tests at, 438
storage tanks for LP-gas at, 437–438
tank car unloading of LP-gas at, 437
truck loading, 438
truck unloading of LP-gas at, 437
Bulk storage, 397
elevators, 349
of records, 409
Bureau of Alcohol, Tobacco and Firearms (BATF), 496
Burner controls, 96–100
inspection checklist for, 601
Burner management systems, 504
Burner safeties, 101
Burst disks, 433
Bus bars, 81
Business occupancies
alarm systems for, 333–334
automatic sprinklers for, 333
cafeterias in, 332
characteristics of, 328–329
computer rooms in, 331
definition of, 328
hazardous areas in, 331
inspection checklist for, 662–665
means of egress for, 329–330
operating features of, 334
portable fire extinguishers for, 334
prefire planning for, 327–328
protection of openings for, 330–331
protection of records in, 331
waste disposal for, 331–332
windowless and underground structures in, 334
Butane-fueled appliances, in assembly occupancies, 247

C

Cabinets, for electrical systems
inspection checklist for, 598
Cables
common faults in, 80
for electrical systems
inspection checklist for, 597
Cafeterias, for business occupancies, 332
Calcium
burning characteristics of, 453
hazards of, 460
Calcium hypochlorite, 469
Calcium oxide, 466
Calendering, 481
Candles
in assembly occupancies, 247–248
in residential board and care occupancies, 305
Carbide, 466
Carbon black, 464
Carbon dioxide
in nonflammable medical gas systems, 439
portable fire extinguishers, for Class C hazards, 218
Carbon dioxide extinguishing systems, 200, 201
for industrial occupancies, 339, 341
inspection and maintenance of, 202
inspection and testing frequency of, 202
safety considerations of, 201–202
types of storage for, 202
Carbon dioxide portable fire extinguishers, 220
Carbon disulfide, 417
Carbon monoxide, 12
from solid fuel cooking equipment, 584
Carbon monoxide poisoning, 440
Carousel storage systems, 399–400
Carpeting, 235
for use on walls and ceilings, 232
Carpets, storage of, 406
Cartridge fuses, 84–85
Cartridge-operated dry chemical extinguishers, 219
Cat whisker-type switches, 391
Cathodic protection, 152. *See also* Grounding
Ceiling, 232–234
Ceiling finishes, 474
Ceiling height, smoke detector location and, 129–130
Ceiling tiles, suspended, 300–301
Ceilings, temperature classification of, 133
Cellulose nitrate, 476
Centrifugal fire pump systems, 148, 189
Chafing strips, 76
Chain conveyors, 391
Charcoal
fire hazards of, 466
for solid fuel cooking equipment, 584
Charging systems, for incinerators
manual systems, 373
mechanical charging, 373–374
Chases, for business occupancies, 333
Chemical waste disposal, for industrial occupancies, 340
Chemicals. *See also specific chemicals*
corrosive, 467
fire hazards of, 463
fire protection for, 470
handling and transportation of, 470
hazard identification of, 470
oxidizing, 467–470
safety precautions for, 470
segregation of, 470
solid combustible, 464–465
storage of, 470
toxic, 467, 468
unstable, 465–466
water- and air-reactive, 466–467
Chimneys, 102–104
connectors for, 101–102
inspection checklist for, 601
for one- and two-family dwellings, 318
Chlorates, 27
fire hazards of, 469
Chlorinated isocyanurates, 469
Chlorination, of stored water in mist systems, 194
Chlorine, 463
Chlorites, 469
Chutes. *See also* Waste chutes
for demolition, 64
fire protection for, 70
service opening rooms for, 371
terminal enclosures for, 371
Circuit breakers, 80–81, 81, 85
inspection checklist for, 597
Circuit conductors
common faults in, 81
inspection checklist for, 597
Circuits, Class A, in fire alarm system wiring, 140–142
Class K fires, portable fire extinguishers for, 218
Clean agent systems
fire alarm systems and, 134
for industrial occupancies, 341
inspection and maintenance of, 209–210
inspection and testing frequency, 210
inspection checklist for, 612–613
installation of, 208
operations and discharge of, 207
refilling storage tanks for, 210
safety considerations of, 208–209
thermal decomposition in, 209
training in, 210
types of storage for, 209
Clean agents, 207
Cleaning solvents, storage of, 24
Closed array storage, 404
Closing devices, for fire doors, 73, 74
Closing interviews
inspection checklist for, 590
Clothing
inspection checklist for, 589
CO_2 protection
for heat transfer fluid systems, 548
Coal
storage of, 97, 397
Coal-burning stoves, 97, 100, 317–318
Coatings
corrosion-resistant, for sprinkler systems, 160–161
Cobalt 60, 381

Code of Federal Regulations
and classification of flammable and combustible liquids, 419
and underground tanks, 421
Columns, 50
Combination alarm systems, 127
Combustible dust
in air-conditioning systems, 110
Combustible fibers
in industrial occupancies, 340
Combustible liquids
classification of, 419
compressed gas displacement systems for, 426
confinement of, 427
control of ignition sources of, 427–428
definition of, 416
dipping and coating processes and, 428
dispensing systems for, 426–427
fire hazard properties of
autoignition temperature, 418
flammable limits, 418–419
flash point, 418
fire protection, 428
foam extinguishing systems and, 169
gravity systems for, 426
identification of, 419–420
leaks of, 420
loss control guidelines for, 427–428
outside aboveground tanks for waste storage and, 27
in paint finishing systems, 555
physical properties of
boiling point, 416
specific gravity, 417
temperature and pressure effects, 417
vapor density, 417
vapor pressure, 416
viscosity, 417
water solubility, 417
pumping systems for, 426
spray applications of, 428
storage of, 405
storing and handling of
drums and other portable containers, 425
hazards of, 420
intermediate bulk containers, 423–425
outside aboveground tanks, 421–423
portable tanks, 423–425
safety cans, 425–426
tanks inside buildings, 423
underground tanks, 420–421
tank vehicles for, 428
in underground operations, 65
ventilation of, 427
vs. flammable liquids, 416
Combustible materials
around heating systems, 101
electrical risks and, 88–89
and heat-utilization equipment, 512
removal during renovations and alterations, 64
Combustible metals
characteristics of, 453–454
explanation of, 453
extinguishing fires of, 454–455
Combustion air, 101–102
Commercial cooking equipment
cleaning certificates for, 583
confinement of fires of, 574
detection of fires of, 575
exhaust ducts and equipment for, 24–25
exhaust fans, 574–575
exhaust termination distance for, 578
extinguishment of fires of, 575
fans for, 579
hazards of, 573–575
in-service inspections, 580
inspection of, 575–583
manual activation devices, 580
periodic inspection stage for, 581–583
prevention practices for, 574
solid fuel, 584–585
Commercial-industrial incinerators, 371, 372
Commissaries, for detention and correction occupancies, 277
Commodity classification, for materials-handling systems, 385
Commodity storage, 350
Compartmentalization
in health care occupancies, 262–263
in residential board and care occupancies, 310
in underground operations, 65
Composite propellants, 486
Compressed gas, 432, 433–434
in pump driven water mist systems, 192
securing of, 25
uses of, 435
in water mist systems, 186
Compressed gas displacement systems, for flammable and combustible liquids, 426
Compressed gas-driven water mist systems, 185–188
Compression molding, 481
Computer rooms, in business occupancies, 331
Concealed detectors, 134
Conduits
common faults in, 80
inspection checklist for, 597
Construction
of ambulatory health care occupancies, 268
of apartment buildings, 290
classification of, 10–11, 49–50
of detention and correctional occupancies, 276
fire-resistive ratings of, 50
fuel storage during, 62
of health care occupancies, 263
inspection checklist for, 591–593
of medical gas systems, 439–440
mixed type, 56
problems, inspection checklist for, 590
site preparation, 60
structural elements in, 39–46
theft and vandalism during, 61
types
exterior protected, combustible (III), 52–53
fire-resistive (I), 51
heavy timber (IV), 53–55
high hazard, 49
limited combustible, 49
low hazard, 49
noncombustible, 49
noncombustible (II), 51–52
ordinary hazard, 49
steel-frames, 40
wood frame (V), 55–56
Containment systems, for radioactive materials, 379–380
Control Units for Fire-Protective Signaling Systems, 118
Control valves, for water supplies, 145–146
Controlled-air incinerators, 371, 372
Conveyor systems, 390–392
dust hazards of, 449
fire causes in
dust, 391
friction, 390
other, 391
static electricity, 391
welding, cutting, and other hot work, 390
hazards of, 397
protection of openings for, 391
for spray booths, 519–520
Cooking grease, 574
Cooking hoods, 577
Cooking operations
dry chemical systems for, 200
in lodging or rooming houses, 302–303
portable fire extinguishers for, 218
Correction orders, 15
Correctional facilities. *See* Detention and correctional occupancies
Corridors
for ambulatory health care occupancies
inspection checklist for, 633
for apartment buildings
inspection checklist for, 645

for assembly occupancies
inspection checklist for, 617
for business occupancies
inspection checklist for, 663
for day-care facilities
inspection checklist for, 625
for detention and correctional occupancies, 277
inspection checklist for, 637
for educational occupancies, 250
inspection checklist for, 621
for health care occupancies, 263
inspection checklist for, 629
for hotels, 284, 285–286
inspection checklist for, 641
for industrial occupancies
inspection checklist for, 667
inspection checklist for, 595
for lodging or rooming houses
inspection checklist for, 649
for mercantile occupancies, 323
inspection checklist for, 659
for residential board and care occupancies, 309
inspection checklist for, 652
for special structures and high rise buildings
inspection checklist for, 675
for storage occupancies
inspection checklist for, 671
Corrosion, in pump driven water mist systems, 182
Corrosion inhibitors, in fire protection systems, 144
Corrosion-resistant coatings, in sprinkler systems, 160–161
Corrosive materials, 467
Cotton, 396
Covered malls, 325. *See also* Mercantile occupancies
Cranes, 392–393
Creosote
accumulation of, 318
from solid fuel cooking equipment, 584
Critical radiant flux, 235
Cryogenic gases, 432, 434
Cryogenic liquids, 439
Cutting operations. *See also* Hot work
fire incident trends in, 544
hazards in, 59, 62, 533
near conveying systems, 390
overview of, 533–534
oxygen and, 435
oxygen-fuel gas in, 436
Cyclone separators, 449

D

DACR (digital alarm communicator receiver), 138
DACT (digital alarm communicator transmitter), 138–139
Daily inspections, 16
Dampers
for commercial cooking equipment, 576
for health care occupancies, 262–263
for pressurized smoke control systems, 117
Day rooms, in detention and correctional occupancies, 277
Day-care facilities, 255–256
definition of, 255
inspection checklist for, 624–627
means of egress for, 255–256
occupant load for, 255
operating procedures for, 256–257
and unvented portable kerosene heaters, 96
Dead loads, 39
DeBruce Grain explosion, 443–444
Decompression sickness, 440
Decorations
hazards of, 235
inspection checklist for, 615
Dedicated branch circuits, 127
Deep fat fryers, 580, 581
Defend-in-place theory, 260–261
Deflagration, 444
Deluge sprinkler systems, 164
testing of, 166–167
Demolition operations, 64
inspection checklist for, 591–593
Department of Energy, 378, 380
Department of Transportation (DOT)
and classification of flammable and combustible liquids, 419
gas container requirements of, 432
Hazardous Materials Regulations and, 486
and radioactive materials, 382
Descaling baths, hazards of, 457
Detection equipment
for one- and two-family dwellings, 318
for water mist systems, 197
Detection systems, 303. *See also* Smoke detectors
for lodging or rooming houses
Detention and correctional occupancies
building services for, 280–281
building subdivisions in, 279–280
capacity of, 276
characteristics of, 276–278
classification of, 274
construction type of, 276
contents of, 278
definition of, 273–274
fire protection for, 279
fire safety concerns for, 273–274
hazardous areas of, 277
inspection checklist for, 636–639
interior finishes for, 278
means of egress for, 276–277
occupant load for, 276
protection of openings of, 278
remote-controlled release in, 274–276
smoke control for, 117
and unvented portable kerosene heaters, 96
use conditions in, 275, 280
Detonating cord, 486
Detonation, 444
Detonator, 486
Dichromate, 469
Differential pressure switches, for proving airflow, 506
Digital alarm communicator receiver (DACR), 138
Digital alarm communicator systems, for offsite notification, 138–139
Digital alarm communicator transmitter (DACT), 138–139
Dikes, for controlling hazardous spills, 422
Dipping and coating finishing systems
fire protection for, 557
of flammable or combustible liquids, 428
process hazards of, 557
Direct-fired batch ovens, 503
Direct-fired continuous furnaces, 503
Discharge
calculation of, 154
coefficient of, 154, 155
Discharge testing
of water mist systems, 195–196
DISCUS (Distilled Spirits Council of the United States), 407
Dispensing systems, for flammable and combustible liquids, 426–427
Distilled spirits
storage of, 406–407
Distilled Spirits Council of the United States (DISCUS), 407
Distribution equipment, for air-conditioning systems, 108–110
Door-opening force, 116
Doors
for ambulatory health care occupancies, 268
inspection checklist for, 632
for apartment buildings, 293
inspection checklist for, 644
for assembly occupancies
inspection checklist for, 616
for business occupancies
inspection checklist for, 662
clearance under, 77
for day-care facilities
inspection checklist for, 624
for detention and correctional occupancies
inspection checklist for, 636
for educational occupancies, 251
inspection checklist for, 620

fire protection ratings for, 71
for health care occupancies, 261
inspection checklist for, 628
for hotels, 284
inspection checklist for, 640
for industrial occupancies
inspection checklist for, 666
inspection of, in means of egress, 228
for lodging or rooming houses
inspection checklist for, 648
in means of egress, 226–227
for mercantile occupancies, 323
inspection checklist for, 659
for one- and two-family dwellings, 317
inspection checklist for, 655
for residential board and care occupancies
inspection checklist for, 651
for special structures and high rise buildings
inspection checklist for, 674
for storage occupancies, 349
inspection checklist for, 670
DOT. *See* Department of Transportation (DOT)
DOT/TC cylinders, 433
Down conductors, 91–92
Driveways, fire access for, 408
Drums, for storing flammable and combustible liquids, 425
Dry barrel hydrants
inspection checklist for, 606
Dry chemical portable fire extinguishers
for Class C hazards, 218
stored-pressure, 219
Dry chemical systems, 200
for business occupancies, 332
extinguishing properties of, 203–204
hazards of, 204
for industrial occupancies, 339, 341
inspection and maintenance of, 204
inspection and testing frequency of, 204
limitations of, 204
overview of, 203
testing intervals of, 161
types of storage for, 204
Dry powders, for extinguishing metal fires, 454–455
Dry rot, in fire doors, 76
Dry transformers
inspection checklist for, 597
Drycleaning plants, 428
Dry-filter collectors, 522–523
Drying ovens, autoignition hazards of, 418
Dry-pipe sprinkler systems
for refrigerated storage, 405–406
testing of, 166
vs. wet-pipe sprinkler systems, 164
Dry-type transformers, 85–86
Duct furnaces
for storage occupancies, 351
Duct systems
for commercial cooking operations, 24–25
fire protection for, 78
Ducts
for commercial cooking equipment, 576–578
for heat distribution, 100
for ovens, 513
for storage occupancies, 351
Ductwork
for industrial occupancies, 338
for storage occupancies, 350
Dumpsters
for business occupancies, 332
hazards in, 369
placement of, 27
Dust
collecting equipment, 448
concentration of, 447
control of, 448–449
in conveying systems, 391
and conveyor systems, 449
definition of, 444
and electrical equipment, 449
fire hazards of, 446–448
and flammable gases, 447
ignition sources for, 449
lower explosive limit (LEL) of, 447
moisture and, 447
particle size of, 446
plastic and rubber, explosive hazards of, 481
plastics manufacturing, fire hazards of, 483
in pneumatic conveying systems, 392
and process equipment, 449
process equipment for, 449
pyrotechnic, 497
storage hazards of, 397
in woodworking operations, 552–553
zirconium, 458
Dust, areas with
accumulation in, 88
combustible, fire hazards of, 25
high hazard examples of, 11
Dust collectors
bag-type, 450
dust hazards of, 449
and powder coatings, 518
Dust explosions, 443–444
confinement of, 445
duration of, 445
evaluating hazards of, 445–446
explanation of, 444–445
ignition sources for, 447–448
inerting of, 445–446
maximum explosion pressure of, 445
prevention of, 448–449
rate of pressure rise in, 445
Dust-collecting equipment, 25
Dwelling units
in apartment buildings, 294
Dynamite, 487

E

Early suppression fast response (ESFR) automatic sprinkler systems, 401
and pallet storage, 403
and rubber tire storage, 404
Earthquake areas
fire pump inspections in, 150
water storage tanks in, 152
Edison base plug fuses, 84
E.D.I.T.H. (Exit Drills in the Home), 319
Educational occupancies
characteristics of, 249–250
construction of, 52
definition of, 249
fire protection in, 253
flexible and open-plan buildings in, 253
hazardous areas in, 252
hazards in, 249
hazards of contents classification in, 12
inspection checklist for, 620–623
interior finishes for, 252–253
kitchen facilities in, 252
laboratories in, 252
means of egress for, 250–251
occupant load for, 250
and unvented portable kerosene heaters, 96
windows for, 251
EGFPD (Equipment Ground Fault Protective Devices), 85
Egress, means of
all-inclusive inspections, 225
for ambulatory health care occupancies, 267–268
inspection checklist for, 633
for apartment buildings
inspection checklist for, 645
for assembly occupancies, 243–244
inspection checklist for, 617
for business occupancies, 329–330
inspection checklist for, 662
for day-care facilities, 255–256
inspection checklist for, 625
defining, 226–227
for detention and correctional occupancies, 276–277
inspection checklist for, 637
doors in, 226–227
inspection of, 228
for educational occupancies, 250–251
inspection checklist for, 621
familiarity in inspections of, 226
fire doors in, 228

for group and family day-care homes, 256
for health care occupancies, 263–264
inspection checklist for, 629
for hotels, 284
inspection checklist for, 641
for industrial occupancies, 336–337
inspection checklist for, 667
inspecting out-of-sight features of, 227–228
inspection types of, 225
inspection walk-throughs, 226–227
for mercantile occupancies, 323
inspection checklist for, 659
observations in, 228
for one- and two-family dwellings
inspection checklist for, 655
qualitative vs. quantitative evaluation of, 228–229
rank-ordering code violations, 229
reinspections, 226
for residential board and care occupancies, 308–309
inspection checklist for, 652
role-playing in, 229
for special structures and high rise buildings
inspection checklist for, 675
for storage occupancies, 347–350
inspection checklist for, 671
timing of inspections of, 229
for vehicles and vessels, 362–363
weather and, 227
Egress paths
snow and, 28
Elastomeric coatings, for roofs, 45
Elastomers, fire resistant, 69
Electric discharge machining
fire protection for, 553
process hazards of, 553
Electrical boxes, common faults in, 81
Electrical cabinets, common faults in, 81
Electrical equipment
and dust hazards, 449
grounding for, 83
for underground operations, 65
water spray sprinkler systems and, 165
Electrical fires, 355–356
from arcing, 79, 80
and dry chemical systems, 203
Electrical hazardous areas
Class I
Division 1, 88
Division 2, 88
Zone 0, 89
Zone 1, 89
Zone 2, 89
Class II
Division 1, 88
Division 2, 88
Class III
Division 1, 89
Division 2, 89
Electrical heaters, portable, 96
Electrical inspectors, 80
Electrical installations
for lodging or rooming houses, 302
Electrical neutralizers, 90
Electrical systems, 80–92
inspection checklist for, 597–599
Electrical wiring and appliances
for assembly occupancies, 245–246
Electrolytes, hazards of, 390
Electromagnetic fire door release interlocks, 391
Electronic data processing equipment, 331
Electrostatic disks, 527
Electrostatic particulate (ESP) filters, 583–584
Electrostatic precipitation, 108
Elevator shafts
inspection checklist for, 595
Elevators
bulk storage, 349
smoke control for, 116–117
ELO (extra large orifice) automatic sprinkler systems, 401
Emergency lighting
for ambulatory health care occupancies
inspection checklist for, 633
for apartment buildings
inspection checklist for, 645
for assembly occupancies
inspection checklist for, 617
for business occupancies
inspection checklist for, 663
for day-care facilities
inspection checklist for, 625
for detention and correctional occupancies, 277
inspection checklist for, 637
for educational occupancies
inspection checklist for, 621
for health care occupancies
inspection checklist for, 629
for hotels
inspection checklist for, 641
for industrial occupancies, 337
inspection checklist for, 667
for mercantile occupancies, 323
inspection checklist for, 659
for residential board and care occupancies
inspection checklist for, 652
for special structures and high rise buildings
inspection checklist for, 675
for storage occupancies, 350
inspection checklist for, 671
for underground structures, 362
Emergency planning
for ambulatory health care occupancies, 270–271
for business occupancies, 334
for detention and correctional occupancies, 281
for one- and two-family dwellings, 319
for pyrotechnics, 498
for residential board and care occupancies, 313
Emergency power
for detention and correctional occupancies, 275
Emergency voice/alarm communications
notification equipment, 135–136
Encapsulation, in storage, 400
Endothermic gas, 514–515
Environmental loads, 39
EPA (Environmental Protection Agency), and underground tanks, 421
Equipment, 372. *See also specific equipment*
of fire inspector, 7
Equipment Ground Fault Protective Devices (EGFPD), 85
Escalator enclosures
inspection checklist for, 595
Escalators, fire protection for, 70–71
Escape, means of. *See also* Egress, means of
for hotels, 284
in lodging or rooming houses, 301
in one- and two-family dwellings, 315–316
ESFR sprinkler systems. *See* Early suppression fast response (ESFR) automatic sprinkler systems
Ethylene oxide, 465
Evacuation procedures, notification appliances in, 135
Evaporation, of flammable and combustible liquids, 416
Event matrix, 121
sample of, 120
Exhaust fans
inspection of, 118
Exhaust method, of smoke-control systems, 114, 118
Exhaust systems
for commercial cooking operations, 24–25
for protecting escalators, 71
Exhaust termination distance, for commercial cooking equipment, 578
Exhibits, in assembly occupancies, 246–247
Exit discharges. *See* Exits
Exit doors
for assembly occupancies, 244
for business occupancies, 330–331

Exit Drills in the Home (E.D.I.T.H.), 319
Exit signs
  for business occupancies, 330
  for detention and correctional facilities, 277
  for hotels, 284
  for large residential board and care occupancies, 309
  for storage occupancies, 349–350
Exits
  for ambulatory health care occupancies
    inspection checklist for, 632
  for apartment buildings, 293–294
    inspection checklist for, 644
  for assembly occupancies
    inspection checklist for, 616
  for business occupancies, 329–330
    inspection checklist for, 662
  for day-care facilities
    inspection checklist for, 624
  for detention and correctional occupancies, 276
    inspection checklist for, 636
  for educational occupancies
    inspection checklist for, 620
  for health care occupancies
    inspection checklist for, 628
  horizontal
    inspection checklist for, 594
  for hotels
    inspection checklist for, 640
  for industrial occupancies, 337
    inspection checklist for, 666
  inspection of, 10–11
  for lodging or rooming houses
    inspection checklist for, 648
  for mercantile occupancies
    inspection checklist for, 658
  for one- and two-family dwellings
    inspection checklist for, 655
  for residential board and care occupancies
    inspection checklist for, 651
  for special structures and high rise buildings
    inspection checklist for, 674
  for storage occupancies, 347–348, 348–349, 349
    inspection checklist for, 670
Exothermic gas, 514–515
Expanded vinyl materials
  inspection checklist for, 615
Explosion relief vents
  for ovens, 512–513
  for waste-processing equipment, including shredders, 375
Explosion suppression
for waste-processing equipment, including shredders, 375
Explosive Materials Code, 64
Explosives
  classification of, 487–488
  forbidden, 488
  manufacturing of, 489–490
  mixing plants, 488–489
    electrical installations for, 488–489
    fire protection for, 489
    housekeeping for, 488
  motor vehicle terminals for, 490
  overview of, 485–486
  safe havens from, 490–491
  storage of, 489–490
  transportation of, 489–490
Extension cords
  common faults in, 79, 81
  inspection checklist for, 598
Exterior inspection, 10
  of water supplies, 143–144
Extinguishing equipment, for commercial cooking equipment, 575
Extra hazard occupancies, 159
Extra large orifice (ELO) automatic sprinkler systems, 401

F

Fabrics. *See* Textiles
Factory Mutual Research Corporation (FMRC)
  hazards of plastics manufacturing and, 481
  and pallet storage, 403
Factory-built chimneys, 102, 103
Factory-Made Air Ducts and Connectors, 108
Family day-care homes, 255, 256
Family dwellings
  characteristics of, 315
  coal- and wood-burning stoves for, 317–318
  definition of, 316
  detection equipment for, 318
  fire extinguishers for, 318–319
  fire protection systems for, 317
  hazards in, 315
  inspection checklist for, 655–657
  interior finishes for, 317
  means of escape in, 315–316
  sprinkler systems for, 319
  storage in, 318
  utilities for, 317
  voluntary inspections of, 317–319
Fans, 107
  for commercial cooking equipment, 579
Farms
  storage of flammable and combustible liquids in, 428
Federal Hazardous Substances Act, 420
FF-1-70 Pill Test, 235
FFFP (film-forming fluoroprotein foam), 221
Fiber reinforced plastic (FRP), 549
FIC (fluoroiodocarbons), 207
File rooms, 410
Film-forming fluoroprotein foam (FFFP), 221
Filters
  air, 107–108
  for pump driven water mist systems, 191–192
  for radioactive materials, 379–380
Fire alarm control boxes, for educational occupancies, 253
Fire alarm signals, 125
Fire alarm speaker appliances, 135
Fire alarm systems. *See* Alarm systems
Fire barrier walls, 67
  for air distribution systems, 108
  construction of, 63
  fire resistance ratings of, 43
  for industrial occupancies, 337
  vs. fire walls, 68
Fire barriers
  inspection checklist for, 595
  protection of openings in
    inspection checklist for, 594–596
Fire dampers, 108–109
Fire departments
  connections for
    inspection checklist for, 590
    inspection of, 146
Fire detection systems, 13
  around escalators, 71
  for boiler rooms, 96
  for fire alarm systems, 138
  for fire suppression systems, 138
  maintaining files for, 35
  for residential board and care occupancies, 311–312
Fire doors, 70
  assemblies, fire protection for, 71–76
  construction of, 73
  for general purpose warehouses, 569
  identifying of, 228
  for industrial occupancies, 338
  ratings of, 72
  for storage occupancies, 350
  vision panels in, 76
  for waste-processing equipment, including shredders, 375
Fire drills
  for ambulatory health care occupancies, 271
  for business occupancies, 334
  for detention and correctional occupancies, 281
  for hotels, 287
  for one- and two-family dwellings, 319
  for residential board and care occupancies, 307, 313
Fire escapes
  for apartment buildings, 293
  for hotels, 284
Fire extinguishers, portable
  access to, 20
  for ambulatory health care occupancies, 269–270
  for apartment buildings, 296

for bulk plants, 439
for business occupancies, 332, 334
Class A
minimum number and rating for, 216–217
placement on outside walls of, 216–217
Class B
maximum travel distances to, 217–218
for commercial cooking equipment, 581
during construction, 63
for construction, alterations, and demolition operations
inspection checklist for, 593
for detention and correctional occupancies, 279
distribution of
for Class A hazards, 215–217
for Class B hazards, 217–218
for Class C hazards, 218
for Class D hazards, 218
for dust areas, 449–450
for explosives mixing plants, 489–489
halon, 221
for health care occupancies, 265
for hotels, 287
hydrostatic testing of, 221–223
for industrial occupancies, 340–341
for industrial trucks, 351
inspection checklist for, 614
inverted, discontinuation of, 221
for mercantile occupancies, 324
for one- and two-family dwellings, 318–319
for outdoor storage areas, 408
overview of, 213–214
for plastics and rubber, 477
for plastics manufacturing, 483
for residential board and care occupancies, 312, 313
for storage occupancies, 352, 354
types of, 219–221
extra (high) hazard, 214
light (low) hazard, 214
ordinary (moderate) hazard, 214
Fire extinguishing systems
for ambulatory health care occupancies
inspection checklist for, 634
for apartment buildings
inspection checklist for, 646
for business occupancies
inspection checklist for, 664
for cooking operations
in lodging or rooming houses, 302–303
for day-care facilities
inspection checklist for, 626
for detention and correctional occupancies
inspection checklist for, 638
for educational occupancies
inspection checklist for, 622
for health care occupancies
inspection checklist for, 631
for hotels
inspection checklist for, 642
for industrial occupancies
inspection checklist for, 668
for lodging or rooming houses
inspection checklist for, 650
for mercantile occupancies
inspection checklist for, 660
for one- and two-family dwellings
inspection checklist for, 657
for residential board and care occupancies
inspection checklist for, 654
for special structures and high rise buildings
inspection checklist for, 676
for storage occupancies
inspection checklist for, 672
Fire fighter service
Phase II, 10
Fire hoses
inspection requirements for, 172–174
testing requirements for, 174
Fire Inspection and Code Enforcement, 33
Fire inspector
attire of, 7
authority of, 3–4, 14
characteristics of, 3
communication skills of, 3
equipment for inspection procedures of, 7
knowledge of, 4–5
physical condition of, 3
Fire lanes, 10
snow and, 28
Fire partitions. *See* Fire barrier walls
Fire point, 418
Fire protection
for detention and correctional occupancies, 279
for educational occupancies, 253
for health care occupancies, 263–265
Fire protection equipment
snow and, 28
Fire Protection Guide to Hazardous Materials, 412
Fire Protection Handbook, 39, 46, 101
and combustible metals, 454
and dust explosions, 446
and plastics, 476
Fire protection ratings
for doors and associated support construction, 71
Fire protection systems. *See also under specific occupancies*
for apartment buildings, 296–297
for assembly occupancies, 248
for business occupancies, 333–334
for chemicals, 470
for dust areas, 449–450
for explosives mixing plants, 489
fire inspector's knowledge of, 4–5
for flammable and combustible liquids, 428
for goods in storage, 400–402
for hotels, 286–287
for industrial occupancies, 340–342
for mercantile occupancies, 324–325
for one- and two-family dwellings, 317
for plastics manufacturing, 483
for storage occupancies, 353–354
for underground operations, 65
Fire pump assemblies, 148
Fire pumps
inspection of, 148–150
for storage occupancies, 354
types of, 148
Fire ratings
Class A, 215–217
Class B, 217–218
Class C, 218
Class D, 218, 454–455
Class K
portable fire extinguishers for, 218
Fire Resistance Design Manual, 44
Fire Resistance Directory, 41, 44
Fire resistance rating
of fire barrier walls, 43
of fire walls, 43
of floor/ceiling assemblies, 44
Fire retardant coatings
on interior finishes, 234
Fire retardant paints
for lodging or rooming houses, 300
Fire safety
for explosives mixing plants, 488
for incinerators, 371–372
Fire safety plans, written, for ambulatory health care occupancies, 270–271
Fire shutters, 76
for industrial occupancies, 338
for storage occupancies, 350
Fire suppression systems
for boiler rooms, 96
fire detectors in, 138
inspection of, 13
maintaining files for, 35
for residential board and care occupancies, 312–313
Fire walls, 42, 43, 67. *See also* Fire barrier walls
composition of, 68
for construction, alterations, and demolition operations
inspection checklist for, 593
construction of, 63
for conveyor systems, 392
for demolition, 64

inspection checklist for, 596
for storage occupancies, 350
vs. fire barrier walls, 68
Fire watchers
during renovations and alterations, 64
for storage occupancies, 352
for welding and cutting processes, 62
Fire windows, 76–77
inspection checklist for, 595
Firebreaks
for outdoor storage areas, 408
Fire-protective coating, 41–42
Fire-resistance ratings
of constructions, 50
Fire-resistive construction, 49–50
ratings, history of, 49–50
requirements for, 50
Fire-resistive walls, 10
Fire-retardant coverings
Class A, 45
Firewood
storage of, 97
Fireworks. *See also* Pyrotechnics
consumer, 494
definitions of, 494–495
display, 494
hazards of, 493
magazine, 494
overview of, 493–494
retail sales outlets for, 498
Fixed fire protection, for demolition, 64
Fixed-pipe carbon dioxide systems, for business occupancies, 332
Fixed-temperature heat detectors, 132
Flame detectors, 134
Flame effects, 436
Flame retardants, for decorative materials, 245
Flame sensing devices, 508–509
Flame spread and smoke developed indices
for wall and ceiling finishes, 232–233
Flame spread ratings
for wall and ceiling finishes, 232–233
Flammability hazard identification, 411
Flammable and Combustible Liquids Code, 62, 418
Flammable gases. *See also* Gases
and dust, 447
water spray sprinkler systems and, 165
Flammable limits, of flammable and combustible liquids, 418–419
Flammable liquids
Class B portable fire extinguishers and, 217–218
classification of, 419
cleaning solvents, 24
compressed gas displacement systems for, 426
confinement of, 427
containers for
bonding of, 90
control of ignition sources of, 427–428
definition of, 416
dipping and coating processes of, 428
dispensing systems for, 426–427
disposal of, 22
distilled spirits, 406–407
and dry chemical systems, 203
electrical risks and, 88
fire hazard properties of
autoignition temperature, 418
flammable limits, 418–419
flash point, 418
fire protection, 428
foam extinguishing systems and, 169
gravity systems for, 426
identification of, 419–420
in industrial occupancies, 340
leaks of, 420
loss control guidelines for, 427–428
outside aboveground tanks for waste storage and, 27
physical properties of
boiling point, 416
specific gravity, 417
temperature and pressure effects, 417
vapor density, 417
vapor pressure, 416
viscosity, 417
water solubility, 417
process hazards of, 61
pumping systems for, 426
removal of, 64
smoking and, 26
spills of, 22
spray applications of, 428
storage of, 405, 570
in one- and two-family dwellings, 318
storing and handling of
drums and other portable containers, 425
hazards of, 420
intermediate bulk containers, 423–425
outside aboveground tanks, 421–423
portable tanks, 423–425
safety cans, 425–426
tanks inside buildings, 423
underground tanks, 420–421
tank vehicles for, 428
transformer fluid, 86
in underground operations, 65
ventilation of, 427
vs. combustible liquids, 416
water spray sprinkler systems and, 165
Flammable vapors
in aerosol manufacturing, 567
in storage occupancies, 352
Flash arrestors, 543–544
Flash point
of flammable and combustible liquids, 418
of oil, 547
testing, 420
Flexible buildings, in educational occupancies, 253
Flexible cords, common faults in, 81
Floor coverings
in hotels, 285
Floor furnaces, 98–99
Floor/ceiling assemblies, 67
construction of, 44
for hotels, 286
for industrial occupancies, 337
inspection checklist for, 595
integrity of, 10
penetrations in fire barriers, protection for, 68–69
Flooring Radiant Panel Test Method, 235
Floors
cleaning and treatment of, 24
interior finishes of, 235
Flow testing
of water main, 153–154
of water spray sprinkler systems, 169
of yard systems, 154–155
Flue pipes
inspection checklist for, 600
for small and residential buildings, 96
Flues
defects in, 103–104
for storage occupancies, 351
Fluid-filled transformers, 85–86
Fluoroiodocarbons (FIC), 207
Fluoroproteins, for foaming agents, 169
Flushing connections
for sprinkler systems, 161
Foam extinguishing systems
description of, 169–171
for business occupancies, 332
for industrial occupancies, 341
testing of
frequency of, 170–171
Foam plastic molding, 481
Foaming agents, 169
Forced-air furnaces, 97–100
Forest products, 396
Formal letters. *See* Inspection reports
Formica, 476
Forward flow tests, for backflow preventers, 145

Foundation systems, 39
Framing members, 39–42
Framing, plank-and-beam, 55
Framing systems
  Type II noncombustible construction of, 52
Freshwater/seawater transfer, for pump driven water mist systems, 192
Friction, in conveying systems, 390
FRP (fiber reinforced plastic), 549
Fuel gas pressure regulators, 508
Fuel gas pressure switches, 508
Fuel gases
  storage of, 436
Fuel oil
  storage of, in inside tanks, 423
Fuel Oil No. 2, flash point of, 418
Fuel pumps, 390
  for storage occupancies, 352
Fuel trains, 506–507
Fuel-dumps, 62
Fuels. *See also* Oil
  for heating systems
    inspection checklist for, 600
  for industrial trucks, 389–390
  storage/handling of, 62
Fume scrubbers, 549
Furnaces
  burner management systems for, 505
  cold work loads and, 516
  definition of, 501, 502–503
  inspection checklist for, 600
  liquid injection for, 516
  location of, 504
  safety shutdown for, 515
  and special atmospheres, 514–516
  types of, 97–100
  warm-air, 96
Furnishings
  hazards of, 235
  inspection checklist for, 615
Furniture, upholstered, 235, 475–476
Fuses, 84–85

G

G-1 powder, for magnesium fire extinguishment, 457
Gas absorption systems, 108
Gas containers
  cylinders
    certification of, 435
    for hot work, 25, 544
  safeguards for, 432–436
  safety devices for
    cryogenic, 434
    piping systems, 434
  storage of, 436–439
Gas meters and snow, 28
Gases
  compressed, uses of, 435
  flammable
    containers for, storage of, 25
    electrical risks and, 88
    smoking and, 26
    waste storage and, 27
  hazards of
    in fire-related entertainment, 436
    in gas-burning appliances, 436
    in industrial heating equipment, 436
    in oxygen-fuel gas cutting and welding, 436
  nonflammable medical gas systems, 439–440
  properties of, 431–432
  safety devices for
    compressed, 433–434
    liquefied, 434
  storage of
    compressed (nonliquefied), 432
    cryogenic liquid, 432
    liquefied, 432
Gasoline
  flashpoint of, 418
  specific gravity of, 417
Gasoline pumps, for storage occupancies, 348
Gauge calibrators, 166
General industrial occupancies, 336
GFCI (ground-fault circuit-interrupters), 85
Girders, 40
Grain elevators. *See also* Storage occupancies
  hazards of, 443–444
Grandstands, in assembly occupancies, 247
Grass, control of, 26–27
Gravity chutes, 370
Gravity furnaces, 97–100
Gravity systems, for flammable and combustible liquids, 426
Gravity tanks, 150–151
Gravity-pneumatic waste systems, 370
Grease filters, 580
Grease removal devices, 332
Ground suction tanks, 151
Ground terminals, 91, 92
Ground-fault circuit-interrupters (GFCI), 85
Grounding
  inspection checklist for, 598
  of lightning down conductors, 91–92
  methods of, 83–84
  vs. bonding, 89–90
Group day-care homes, 255, 256
Group homes. *See* Residential board and care occupancies

H

Hafnium
  characteristics of, 458
  fire extinguishment of, 459
  process hazards of, 458–459
  storage of, 459
Hail resistance, of roofs, 46
Halocarbons, 207
  hazards of, 209
  storage of, 209
Halogenated agent-type stored-pressure fire extinguisher, 220
Halogenated hydrocarbons, ignition of, 420
Halogenated systems
  inspection and maintenance of, 203
  inspection and testing frequency of, 204
  safety considerations of, 203
  toxicity hazards of, 107
  types of storage for, 203
Halogens, 467
Halon
  portable fire extinguishers, for Class C hazards, 218
Halon 1211 (bromochlorodifluoromethane), 199–200, 220
  vs. halon 1301, 203
Halon 1301 (bromotrifluoromethane), 199–200
  for industrial occupancies, 341
  vs. halon 1211, 203
Hand hose line systems, 200–201
  carbon dioxide, 201
Handbook of Compressed Gases, 434
Hanging garments, storage of, 406
Hazard identification, of chemicals, 470
Hazard identification system, NFPA 704, 410, 411–412
Hazard of contents
  inspection checklist for, 590
Hazardous areas
  of ambulatory health care occupancies, 268–269
  of apartment buildings, 295–296
  of business occupancies, 331
  of detention and correctional occupancies, 277
  of educational occupancies, 252
  electrical classification of, 87–89
  fire door assemblies for, 72
  of mercantile occupancies, 324
  of residential board and care occupancies, 310–311
  of vehicles and vessels, 362–363
Hazardous locations
  sprinkler system type exceptions in, 165
Hazardous materials. *See also specific materials*
  in industrial occupancies, 338–339
  labels, radioactive, 382
  maintaining files on, 35
  radioactive, 379–381, 382
  in storage occupancies, 351
  types of, 4
Hazardous processes
  maintaining files on, 35
  in storage occupancies, 352

Hazards
analysis reports
for radioactive materials, 378–379
classification of, 158–160
of contents, 11–13
high, 11–13
level of, 11–13
low, 11, 13
ordinary, 11, 13
Haz-mat teams, 415–416
HCFC (hydrochlorofluorocarbons), 207
Health care facilities
defend-in-place theory of, 260–261
definition of, 259–260
inspection checklist for, 628–631
means of egress for, 263–264
patient protection in, 261–263
protection of openings in, 261–262
safety systems for, 259
and unvented portable kerosene heaters, 96
Health hazard identification, 411
Heat detectors
location of, 132–134
types of, 130–132
Heat distribution systems, 100–101
Heat pumps, 100
Heat tape, 85
Heat transfer fluid systems, 547–548
fire protection for, 548
Heat treating, 533
Heater rooms
inspection checklist for, 601
Heaters
for membrane structures, 364
portable space
for ambulatory health care occupancies, 271
for detention and correctional occupancies, 280–281
hazards of, 315
inspection checklist for, 600
for lodging or rooming houses, 302
for tents, 364–365
Heating appliances
for lodging or rooming houses, 302
Heating systems. *See also specific types*
for assembly occupancies, 245
fuel-fired, 95
inspection checklist for, 600–601
installation of, 101–104
for large buildings, 95
for residential and small buildings, 96
solid fuel, 97
unit heaters, 99–100
Heat-utilization equipment, 501
automatic vs. supervised manual, 509
boilers, 501–502
burner management systems, 504
class A ovens, 512
combustible materials and, 512
combustion control for, 504–509
fans for, 506
flame safeguards for, 508–509
fuel piping for, 506–508
furnaces, 501, 502–503
housekeeping for, 512–513
inspection, testing and maintenance of, 511–512
leak testing of, 511–512
listing of safety equipment for, 510
location of, 503–504
operator intervention for, 510
operator training for, 510
ovens, 501, 502–503
preigntion purge for, 504–505
programs for, 511
pulverized coal systems, 508
purge intervals, 505
safety control testing of, 512
trained operators for, 510–511
written operating instructions for, 510
Heavy timber construction, 55
Helium
in nonflammable medical gas systems, 439
Herbicides, 27
HFC (hydrofluorocarbons), 207
High explosive material, 486
High hazard construction, 49
High hazard industrial occupancies, 336
High-pressure laminating, 481
High-rack storage occupancies, 346
High-rise buildings, 356, 365
hazards in, 355–356
inspection checklist for, 674–677
Hopper/ram assembly, of incinerator charging systems, 373
Horizontal exits
for apartment buildings, 294
for detention and correctional occupancies, 276, 277
inspection checklist for, 594
Horizontal openings, protection of, 71–78
Horizontal sliding doors, for storage occupancies, 349
Hose connections, for dust areas, 449–450
Hose houses
inspection checklist for, 606
Hose stations
for hotels, 287
for storage occupancies, 354
Hoses
coding system for, 543
systems, for flammable and combustible liquids, 428
Hospital occupancies
sprinkler systems for, 123
Hospitals, and radioactive materials, 382
Hot air ducts, autoignition hazards of, 418
Hot process piping, autoignition hazards of, 418
Hot riveting, 533
Hot work
35-foot rule for, 535–536
clothing and other personal protective equipment for, 534–535
for construction, alterations, and demolition operations
inspection checklist for, 593
containers and piping for, 542
cylinders for, 544
designated area location of, 535
fire incident trends in, 544
fire watches for, 541–542
flash arrestors for, 543–544
hoses for, 543
overview of, 533–534
permit system for, 536–540
protective devices for, 543–544
regulators, 542
sprinkler systems for, 542
in storage occupancies, 352
subtle passages to combustibles in, 536
torches and other hot tools for, 543
ventilation for, 535
Hotels
definition of, 283
fire protection systems for, 286–287
front desk and assembly rooms of, 285–286
guest room floors of, 285
hazards in, 283
inspection checklist for, 640–643
inspection observations in, 284–286
interiors of, 284–286
means of egress for, 284
means of escape for, 284
planning and training for, 287
service areas of, 286
top floors of, 284–285
Hot-water systems, 100–101
Housekeeping
during construction, 61
for construction, alterations, and demolition operations
inspection checklist for, 592
for explosives mixing plants, 488
goals of, 19
for industrial occupancies, 340
outdoor, 26–28
principals of, 19–21
problems of, 22–24
requirements of, 19
for storage occupancies, 352
Humidification, 89
Hydrant inspection reports
for industrial occupancies, 341

Hydrants
backflow protection for, 143–144
dry barrel, 147
for flammable and combustible liquids, 428
for industrial occupancies, 341
inspection checklist for, 606
location of, 10
locations of, 143
near apartment buildings, 290
snow and, 28
wall, 147
yard, 146–147
Hydraulic oil operated equipment, 551–552
fire protection for, 552
hazards of, 551
Hydraulic resistance
in pump driven water mist systems, 191
Hydraulic-grade line
plotting of, 155
Hydrocarbon fuels, water solubility of, 417
Hydrocarbon propellants, 563–565
Hydrocarbon surfactants, in foaming agents, 169
Hydrochlorofluorocarbons (HCFC), 207
Hydrofluorocarbons (HFC), 207
Hydrogen
in furnaces, 514
storage of, 436
Hydrogen cyanide, 12, 465
Hydrostatic testing
of clean agent systems, 210
of portable fire extinguishers, 221–223
of water mist systems, 178–179
Hyperbaric chambers, 440
Hypochlorites, 469

I

IBC, in paint finishing systems, 556
Ice build-up, in water storage tanks, 151, 152
Ignitable atmospheres, 527–531
Ignition sources, control of
for flammable and combustible liquids, 427–428
Incinerators, 371–374
auxiliary fuel systems for, 374
charging systems for, 373–374
commercial-industrial, 371–372
equipment design and construction of, 372
fire safety for, 371–372
layout and arrangement of, 372–373
minimum clearances of, for waste-chute and handling systems, 373
sprinkler systems for, 374
types of, 371
waste-handling rooms for, 372
Indoor general storage, for industrial occupancies, 340
Industrial equipment
medium voltage, 87
motors, 87
transformers, 85–86
Industrial Fire Hazards Handbook, 454
Industrial heating equipment, 436
Industrial occupancies
classification of, 336
definition of, 335–336
fire alarm systems for, 342
fire protection systems for, 340–342
fire pumps for, 341
hazardous materials in, 338
housekeeping in, 340
indoor general storage in, 340
inspection checklist for, 666–669
inspection observations in, 340–341
inspection records of, 341
maintenance in, 340
means of egress for, 336–337
misuse of, 231
occupant load for, 336
outdoor storage in, 339–340
protection of openings in, 337–338
special extinguishing systems for, 341
special purpose, 357
sprinkler systems for, 341
water supplies for, 341
Industrial Risk Insurers (IRI), 458
Industrial truck markers, 386–387
Industrial trucks, 351. *See also* Trucks
Inert gases, 207
Infrared scanning equipment, 80
Initiating devices
incorrect wiring of, 141–142
Injection molding, 482
Inns. *See* Hotels
Inorganic acids, 467
Inorganic peroxides, 469
In-rack sprinkler systems
for carpet storage, 406
Inspection(s)
aids for
event matrix as, 121
of apartment buildings, 291–296
of assembly occupancies, 243–246
bar-code readers for, 16
of bulk plants, 438–439
of business occupancies, 329–331, 331–333
cards for, 16
closing interview for, 14
of commercial cooking equipment, 575–583
daily, 16
documentation of, 29
of elevators, 10
equipment for, 7
exterior, 10
of water supplies, 143–144
of fire department connections, 146
of fire detection and alarm systems, 13
of fire doors, 73–76
of fire pumps, 148–150
of fire suppression equipment, 13
forms for
water-based fire protection systems, 143
of heat-utilization equipment, 511–512
of industrial occupancies, 336–340
introductions for, 8–9
of means of egress, 225
of mercantile occupancies, 322–324
note taking during, 32–33
observations during, 9–13
preparation for, 7–8
of pyrotechnic manufacturing sites, 496–498
of pyrotechnic retail sales outlets, 498
scheduling of, 8
sequence of, 9
of smoke alarms, for residential occupancies, 140
of special agent extinguishing systems, 201
of spray process areas, 526–527
of sprinkler systems, 10, 160–162
critical times of, 159
of standpipe and hose systems, 172–174
of storage occupancies, 346–347
summary of, 13–16
of switchboards and panelboards, 81–82
of valves
for water supplies, 145–146
of water mist systems, 181
of water supplies
for sprinklers, 159–160
of wiring
for fire alarm systems, 140–142
of yard hydrants, 147
Inspection forms, 587–588
for air-conditioning and ventilating systems, 602
for ambulatory health care occupancies, 632–635
for apartment buildings, 644–647
for assembly occupancies, 616–619
for automatic sprinkler and other water-based fire protection systems, 607–608
for business occupancies, 662–665
for clean agent extinguishing systems, 612 –613
for construction, alteration, and demolition operations, 591–593
for day-care facilities, 624–627
for detention and correctional occupancies, 636–639

for educational occupancies, 620–623
for electrical systems, 597–599
for fire alarm systems, 604–605
for health care occupancies, 628–631
for heating systems, 600–601
for hotels, 640–643
for industrial occupancies, 666–669
for inspection procedures, 589–590
for interior finish, contents, and furnishings, 615
for lodging or rooming houses, 648–650
for mercantile occupancies, 658–661
for one- and two-family dwellings, 655–657
for portable fire extinguishers, 614
for protection of openings in fire subdivisions, 594–596
for residential board and care occupancies, 651–654
for smoke control systems, 603
for special agent extinguishing systems, 610–611
for special structures and high rise buildings, 674–677
for storage occupancies, 670–673
for water mist systems, 609
for water supplies, 606
Inspection procedures
checklist for, 589–590
Inspection reports, 14–16
final notice, 32
follow-up, 31–32
scheduling of, 31
information in, 14–15
letters
contents of, 34
drafts of, 33
mechanics of, 33–34
maintaining files for, 35
purpose of, 15–16
for water mist systems, 198
written, 29–35, 32
v. checklist, 31
follow-up letter, 31
grammar and, 32
note taking for, 32–33
procedure of, 30–31
purpose of, 29, 30
style of, 30, 32
Inspectors, third-party, 118
Insulation
hot work and, 544
for large buildings, 95
restrictions of, 474
roof, 45
thermoplastic, 46
Interior finishes
for ambulatory health care occupancies
inspection checklist for, 633
for apartment buildings, 295
inspection checklist for, 645
for assembly occupancies
inspection checklist for, 617
and automatic sprinklers, 235
for business occupancies
inspection checklist for, 663
classifications of
and occupancy classifications, 233
for day-care facilities
inspection checklist for, 625
for detention and correctional occupancies, 278
inspection checklist for, 637
for educational occupancies, 252–253
inspection checklist for, 621
fire-retardant coatings on, 234
on floors, 235
for health care occupancies
inspection checklist for, 629
for hotels
inspection checklist for, 641
for industrial occupancies
inspection checklist for, 667
inspection checklist for, 615
for lodging or rooming houses, 300–301
inspection checklist for, 649
for mercantile occupancies, 324
inspection checklist for, 659
for one- and two-family dwellings, 317
inspection checklist for, 656
for residential board and care occupancies
inspection checklist for, 652
for special structures and high rise buildings
inspection checklist for, 675
for storage occupancies
inspection checklist for, 671
trim or incidental, 234
for walls and ceilings, 232–234
flame spread and smoke developed indices of, 232–233
textiles, expanded vinyl and cellular or foamed plastics, 233–234
Intermediate bulk containers (IBC), for storing flammable and combustible liquids, 423–425
International Maritime Organization, 180
Inverse-time circuit breakers, 85
Ionization techniques, 90
IRI (Industrial Risk Insurers), 458
Iron, 460–461
Isocyanate plastics, 405
Isocyanurate plastics, 405
Isotopes, 381–382

J

Jails. *See* Detention and correctional occupancies
Joists, 40–41
heat detectors and, 133
smoke detectors and, 130–131

K

K factor, for sprinkler systems, 401
K-25 sprinkler systems
and rubber tire storage, 404
Kerosene heaters
for lodging or rooming houses, 302
unvented, 96
Kitchens. *See also* Cooking operations
for business occupancies, 332
for detention and correction occupancies, 277
for educational occupancies, 252
exhaust ducts and equipment for, 24–25
for hotels, 286
Knockouts, 81

L

Laboratories
in ambulatory health care occupancies, 269
in educational occupancies, 252
exhaust systems for, 110
fire protection for, 428
hazards of contents classification in, 12–13
Lamps
common faults in, 82–83
inspection checklist for, 598
Large buildings, heating systems for, 95
Large-drop sprinkler systems, 401
and pallet storage, 403
and rubber tire storage, 404
Laundry chutes
inspection checklist for, 595
Leak monitoring systems
for underground tanks, 421
Life Safety Code, 11, 63, 64, 71, 124, 225
and ambulatory health care occupancies, 267
and apartment buildings, 289–290
and assembly occupancies, 243, 248
and business occupancies, 328–329
and construction, of ambulatory health care occupancies, 268
and consumer fireworks retail sales outlets, 498
and day-care facilities, 255, 256
and detention and correctional occupancies, 274, 280
and dwelling units, 294
and educational occupancies, 249
exhibits and trade shows, provisions for, 246
and health care occupancies, 259
and hotels, 284
and industrial occupancies, 336–337
and membrane structures, 364
and mercantile occupancies, 322–323

and one- and two-family dwellings, 316
and patient protection, 261
and piers, 359
and residential board and care occupancies, 307–308
and special amusement buildings, 247
special provisions in, for business occupancies, 334
and special structures and high-rise buildings, 356, 357
and storage occupancies, 346, 349, 354
and tents, 364–365
and towers, 359–360
and underground structures, 361–362
and vehicles and vessels, 363
and wall and ceiling finishes, 474
and water-surrounded structures, 364
Light fixtures
common faults in, 82–83
inspection checklist for, 598
Light hazard occupancies, 158
Lighting
for apartment buildings, 297
Lightning protection, 83, 91–92
inspection checklist for, 599
Limited combustible construction, 49
Limited finishing workstations, 522–524
exhaust systems for, 523
Limited-care facilities, 260
Linen-handling systems, 370
Liquefied gases, 432, 434
Liquid precipitation separators, 456
for zirconium processes, 459
Liquids. *See also specific liquids*
Class II
regulation of, 419
Class IIIA
regulation of, 419
Class IIIB
regulation of, 419
combustible. *See* Combustible liquids
flammable. *See* Flammable liquids
Listed equipment, 510
Lithium. *See also* Alkali metals
burning characteristics of, 453
Local application systems, 200
Lodging or rooming houses
building services for, 302–303
definition of, 299
fire hazards in, 299
fire protection systems for, 303
housekeeping for, 302
inspection checklist for, 648–650
interior finishes for, 300–301
protection of exits of, 301–302
use and code requirements for, 300
Loss control guidelines, for flammable and combustible liquids, 427–428
Low- and intermediate-pressure, single-fluid, compressed gas-driven water mist systems with stored water, 187–188
Low explosive material, 486
Low hazard construction, 49
Lower explosive limit (LEL), 555
of aerosols, 567
of dust concentration, 447
Lower flammable limit (LFL), 528
Low-pressure, twin-fluid, compressed gas-driven water mist systems with stored water, 187
LP-gas, 96–97
in assembly occupancies, 245
bulk plants for, 437
odorized, 438
Lubricants, 22–23
Lucite, 476
Lumber storage, hazards in, 213

M

Magazines, for fireworks, 494
Magnesium
characteristics of, 455
dust, 446
fire extinguishment of, 457
fire hazards of, 464
fires, 218
process hazards of, 455–456
storage of, 456
Magnetic separators, for waste-chute and -handling systems, 375
Mail chutes
inspection checklist for, 595
Main drain test, for sprinkler systems, 166
Maintenance
of fire doors, 73–76
for industrial occupancies, 340
Manometer, 116
Manual activation devices, for commercial cooking equipment, 580
Manual dry system, 172. *See also* Standpipe and hose systems
Manual fire alarm boxes, 127–128
for health care occupancies, 264
Manual wet system, 172. *See also* Standpipe and hose systems
Manufacturing processes
ammonia refrigeration systems, 554–555
hazards of, 547
heat transfer fluid systems, 547–548
hydraulic oil operated equipment, 551–552
metal cleaning and plating operations, 548–550
oil quenching operations, 550–551
paint finishing systems, 555–557
woodworking, 552–553
Marine systems. *See* Water mist systems
Masonry chimneys, 102
Material Safety Data Sheet (MSDS), 339, 420, 463
Materials
noncombustible, in fire resistive construction, 49
noncombustible, in Type IV fire-resistive construction, 55
roofing, 45–46
for roofing, hazards in, 62–63
storage and handling of, 19–21
Materials staging and equipment, 386
Materials-handling systems
commodity classification for, 385
conveying systems for
fire causes in, 390–391
pneumatic, 392
protection of openings for, 391
cranes in, 392–393
equipment for, 386
fire protection for, 78
industrial trucks for, 386–389
fire hazards of, 389
maintenance of, 389–390
refueling and recharging of, 390
materials on receiving dock, 386
materials on shipping dock, 386
storage arrangements in
aisle storage for, 385–386
height of storage for, 385
Mattresses
classification of, 475
hazards of, 235
Maze-type dry collectors, 522
MEC (minimum explosive concentration), 528
Mechanical ventilation, for storage occupancies, 352
Medical gas systems, 439
bulk liquefied oxygen (LOX), 440
hyperbaric chambers, 440
safeguards for
valves and alarm systems, 439
ventilation and construction, 439–440
Medium voltage equipment, 87
Membrane structures, 356, 364. *See also* Special structures and high-rise buildings
Mercantile occupancies
alarm systems for, 324–325
in apartment buildings, 293
characteristics of, 321–322
classification of, 322
construction of, 52
fire protection systems for, 324–325
hazards in, 321
inspection checklist for, 658–661
inspection observations for, 322
interior finishes for, 324
means of egress for, 229, 323
occupant load for, 322–323
portable fire extinguishers for, 324
protection of hazards in, 324

protection of openings in, 324
sprinkler systems for, 157, 325
Metal cable armor, grounding for, 83
Metal chimneys, 102
Metal cleaning operations, 548–550
fire protection for, 549
process hazards of, 549
spill control for, 550
Metal dust, 446
in explosives manufacturing, 489
fire protection for areas with, 450
Metal Guard. *See* G-1 powder
Metal plating operations, 548–550
fire protection for, 549
process hazards of, 549
spill control for, 550
Metals. *See also* Combustible metals *and specific metals*
noncombustible, 460–461
radioactive, 461
Methanamine, 235
Methanol, 514–515
Met-L-X powder
for magnesium fire extinguishment, 457
Mezzanines
for apartment buildings
inspection checklist for, 646
for assembly occupancies
inspection checklist for, 618
for business occupancies
inspection checklist for, 664
for day-care facilities
inspection checklist for, 626
for detention and correctional occupancies
inspection checklist for, 638
for educational occupancies
inspection checklist for, 622
for hotels
inspection checklist for, 642
for industrial occupancies
inspection checklist for, 668
for lodging or rooming houses
inspection checklist for, 649
for mercantile occupancies
inspection checklist for, 660
for residential board and care occupancies
inspection checklist for, 653
for special structures and high rise buildings
inspection checklist for, 676
for storage occupancies
inspection checklist for, 672
Mills, dust hazards of, 449
Mineral buildup, in underground supply mains, 166
Minimum explosive concentration (MEC), 528
Ministorage complexes, 346
Mobile compact shelving storage, of records, 409
Model rocketry, 495
Monitor nozzles, 147–148, 408
Monoammonium phosphate, 203, 219
Montreal Protocol, 220
Motels. *See* Hotels
Motor vehicle terminals, for explosives, 490
Motors, 87
for electrical systems
inspection checklist for, 597
Mounting, of portable fire extinguishers
inspection checklist for, 614
Moveable prefabricated hazardous materials storage buildings, 425
Multiloading pneumatic waste systems, 370
Multiple-chamber incinerators, 371, 372
Multistage incinerators, 369
Municipal fire alarm systems, 125

N

NaK (sodium-potassium alloy), 458
Naphthalene, 465
National Electrical Code (NEC), 85, 88, 110, 140
National Fire Alarm Code, 105, 124
National Fire Codes
and industrial occupancies, 339, 341
and storage occupancies, 351
National Fire Protection Agency (NFPA), 534
hazards of plastics manufacturing and, 481
NFPA *101*, 63
NFPA 221 system, 43
standards for listed equipment, 510
Natural gas, 96–97
hazards of, 501
Negative pressure, smoke hazards of, 115–116
Net positive suction head (NPSH), 191
Neutron monitor, 379
NFPA 704, hazard identification system, 410, 411–412
special hazard identifiers in, 412
NFPA 750
water mist systems and, 178, 180
NFPA Torch Report, 544
Nitrates, 467–468
Nitric acid, 468
Nitrites, 468
Nitroaniline, 464
Nitrocellulose
in spraying applications, 527
Nitrochlorobenzenes, 464
Nitrogen
in furnaces, 514
in nonflammable medical gas systems, 439
Nitromethane, 465
Nitrous oxide
in nonflammable medical gas systems, 439
NOAEL (no observed adverse effect level), of clean agents, 209
Nomex IIIA, 531
Nonbearing walls, 39, 42
fire-resistive ratings of, 50
Noncombustible construction, 49
Noncombustible liquids
in paint finishing systems, 555
Nonflammable medical gas systems, 439–440
Non-power-limited fire alarms (NPLFA), 140
Notification appliances, 135
incorrect wiring of, 141–142
Novelties and trick noisemakers, 495
Nozzles
for clean agent extinguishing systems
inspection checklist for, 612
for water mist systems, testing of, 196–197
NPLFA (non-power-limited fire alarms), 140
NRC (Nuclear Regulatory Commission), 378
Nuclear Regulatory Commission (NRC), 378

O

Occupancies. *See also specific occupancies*
classification of, 9
hazard classification in
extra, 159
light, 158
ordinary, 158–159
and interior finish classifications, 233
Occupant load
for ambulatory health care occupancies
inspection checklist for, 632
for apartment buildings, 293
inspection checklist for, 644
for assembly occupancies, 243
inspection checklist for, 616
for business occupancies
inspection checklist for, 662
for day-care facilities, 255
inspection checklist for, 624
for detention and correctional occupancies, 276
inspection checklist for, 636
for educational occupancies, 250
inspection checklist for, 620
for health care occupancies
inspection checklist for, 628
for hotels
inspection checklist for, 640
for industrial occupancies, 336
inspection checklist for, 666
for lodging or rooming houses
inspection checklist for, 648
for mercantile occupancies, 322–323
inspection checklist for, 658

for one- and two-family dwellings
inspection checklist for, 655
for pyrotechnics, 497
for residential board and care occupancies
inspection checklist for, 651
for special structures and high rise buildings
inspection checklist for, 674
for storage occupancies, 347
inspection checklist for, 670
Occupational Safety and Health Administration (OSHA)
and gas container storage, 436
hot work and, 544
investigations of, 443–444
and safe oxygen levels, 431–432
Office enclosures, small, in storage occupancies, 354
Oil, grades of, 96
Oil quenching operations, 550–551
fire protection for, 550–551
process hazards of, 550
Oil-fired steam boiler, 97
Oily waste, 23
One-family dwellings. *See* Family dwellings
Open array storage, 404
Open flame soldering, 533
Open flames, in assembly occupancies, 247–248
Open structures, 356, 357-358. *See also* Special structures and high-rise buildings
Open-plan buildings, in educational occupancies, 253
Open-plan designs, of business occupancies, 328
Open-shelf storage, of records, 408
Optical flame detectors, 435
Ordinary hazard construction, 49
Ordinary hazard occupancies, 158–159
Organic coatings, 428
Organic peroxides
hazards of, 465
and plastics manufacturing, 477
and spray painting, 518
in spraying applications, 527
storage of, 479–480
Orifice disks, for water mist systems, 195
OS & Y valves, 145
OSHA. *See* Occupational Safety and Health Administration (OSHA)
Outdoor storage, 407–408
for industrial occupancies, 339–340
Outlets, common faults in, 81
Outside aboveground tanks, for storing flammable and combustible liquids, 421–423
Outside screw and yoke (OS & Y) valves, 145
Ovens
burner management systems for, 505
Class A, 512
definition of, 501, 502–503
ductwork for, 513
location of, 504
sprinkler systems for, 514
Overcurrent devices, protection for, 83
Overcurrent protection, 84–85
Overhead lighting, 395
Overspray collectors, 522
Oxides, 466
Oxidizing material, 486
Oxygen
in dust explosions, 445–446
in hot work, 542
liquid, hazards of, 439
in nonflammable medical gas systems, 439
storage of, 436
Oxygen cutting, 533
Ozone treatment, of stored water for mist systems, 194

P

Packing materials, 23–24
hazards of, 395–396
Padded cells, 278
Paint finishing systems, 555–557
fire protection for, 556
process hazards of, 555–556
Palletized storage, 398, 399
Pallets
idle, 402–403
for industrial occupancies, 340
for storage occupancies, 351
types of, 398
Pan conveyors, 391
Panelboards
common faults in, 81–82
inspection checklist for, 598
Parking garages, 346
alarm systems for, 353
exits for, 349
travel distances in, 348
Particle accelerators, 381
Passive smoke-control systems, 119
Patient protection, 261–263
Pensky-Martens closed tester, 107
Perchlorate, 469–470
Perfluorocarbons (PFCs and FCs), 207
Permanganates, 470
Personal care, concept of, 306
Personal protective equipment (PPE), 534–535
inspection checklist for, 589
Persulfates, 470
Phase I elevator recall, 116
Phase II elevator firefighter service, 116
Phenolic plastics, 405
Phosphorus, 466
Photoelectric smoke alarms, for residential occupancies, 140
Photoelectric switches, 391
Photoelectrical cells, for measuring smoke obscuration, 233
Piers, 356, 358-359. *See also* Special structures and high-rise buildings
Pill Test (FF-1-70), 235
Pilot safeties, 101
Piping
for gases, 434
for heating systems
inspection checklist for, 601
penetrations in fire barriers, protection for, 69
for sprinkler systems, 161–162
Pitot tube
assembly, 8
position in flow testing of, 154
pressure readings, 149–150
Plans and reports
inspection checklist for, 589
Plasma cutting, 533
Plastic-media storage, of records, 409
Plastics
in business occupancies, 328–329
cellular and foamed
as wall and ceiling finishes, 233–234
classification of, 477–478
converting, 480
fabricating, 480
fire hazards of, 474–477
fire protection for, 483
hazards of, 321
manufacturing, 480–483
operation hazards of, 480–483
overview of, 474
smoke hazards of, 474, 483
synthesizing, 480
warehousing of, 477–480
Plenums, 97
for limited finishing workstations, 523
Plexiglas, 476
PLFA (power-limited fire alarms), 140
Plug fuses, 84
Plutonium, 379
fire hazards of, 461
hazards of, 377–378
Pneumatic conveyors, 392
Pneumatic jockey pumps, for pump driven water mist systems, 192
Pneumatic waste-handling systems, 370
Point of safety, 307
Pollution abatement devices, 95
Polycarbonates, 476
Polyisocyanurate
and insulation, 474
Polystyrene
and insulation, 474
Polystyrene insulation, 405
Polyurethane
and insulation, 474

Polyurethane foam, 335, 473
Polyurethane insulation, 405
Polyvinyl chloride (PVC), 549
Pool sanitizers, 469
Portable containers, for storing flammable and combustible liquids, 425
Portable fire extinguishers. *See* Fire extinguishers
Positive displacement (PD) pumps, 189
  break tank connections in supply to, 192
  controllers for, 190
  filters for, 191
  unloader valves for, 190–191
Potassium bicarbonate, 203, 219
Potassium chloride, 203, 219
Potassium dichloroisocynurate, 469
Potassium. *See* Alkali metals
Potassium-bicarbonate-based portable fire extinguishers, 218
Powder coating, 518
  apparatus for, 524–526
  booths for, 519–521
  control of ignition sources for, 526–527
  dry filters for, 522–523
  electrostatics and, 519
  explosive prevention, 528–529
  extinguishment methods, 529–530
  ignitable atmospheres in, 527–531
  inspection checklist for, 526–527
  inspection of apparatus and procedures for, 530
  isolation of process for, 528
  limitations of fuel quantity in, 528
  maintenance and, 530
  maze-type dry collectors for, 522
  open-floor, 524
  overspray collectors for, 522
  pressure relief venting, 528–529
  protective clothing for, 531
  rooms, 524
  storage and handling of materials and residues for, 530–531
  training for, 530
  water wash collectors for, 523
Powder guns, 526
Powder-driven fasteners, 533
Power-limited fire alarms (PLFA), 140
PPE (personal protective equipment)
  inspection checklist for, 589
Preaction sprinkler systems, 164
  double interlock, 167
  for refrigerated storage, 405–406
  testing of, 166–167
Prefire planning, for radioactive materials, 379–380
Pressure effects, on flammable and combustible liquids, 417
Pressure regulators
  for stored gases, 433
Pressure relief valves
  for bulk plants, 439
  for stored gases, 433
  for water mist systems, 194
Pressure switches, for proving airflow, 506
Pressure tanks, 151
  for water mist systems
    inspection checklist for, 609
Pressure vessels, for storing gases, 432
Pressurization method, of smoke-control systems, 114, 117–118
Primer, 486
Private fire service mains, 146
Process building, for pyrotechnics, 495
Process hazards, 61
  in construction, alterations, and demolition operations
    inspection checklist for, 591
  fire inspector's knowledge of, 4
Projected beam detectors, 129
Projection rooms, in assembly occupancies, 246
Propane
  hazards of, 431
  storage of, 436
Propellants, 486, 563–565
Protection of openings, for fire barriers
  inspection checklist for, 594–596
Protection systems
  for lodging or rooming houses, 303
  for vehicles and vessels, 363
Public water supplies
  backflow protection for, 143–144
Pulverizers
  dust hazards of, 449
Pump driven water mist systems, 188–192
Pump tests
  for industrial occupancies, 341
Pump-discharge curve, 149
Pumping systems, for flammable and combustible liquids, 426
Purge intervals, for heat-utilization equipment, 505
Pyrolysis products, 68
Pyrophoric materials, 466
Pyrotechnics. *See also* Fireworks
  articles, 495
  in assembly occupancies, 247–248
  civilian products, 494
  composition of, 495, 496
  compressed gas and, 435
  definitions of, 494–496
  emergency planning for, 498
  general safety considerations for, 496
  ignition sources of, 497
  inspection guidelines for, 496–498
  model rocket motors, 495
  novelties and trick noisemakers, 495
  occupancy limits for, 497
  overview of, 493–494
  process building egress requirements, 498
  process buildings for, 495
  quantity of, 497
  salute powder, 496
  separation distances for, 497
  storage requirements for, 497–498
Pyroxilin, 476

Q

Quench tanks, 504
Quick-acting cartridge fuses, 84–85
Quick-response sprinklers
  testing intervals of, 161

R

Raceways
  common faults in, 80
  grounding for, 83
  inspection checklist for, 597
Rack storage, 398–399
Radiant-energy detectors, 134
Radiation protection, 382
Radioactive ionization, 90
Radioactive materials
  containment systems for, 379–380
  external protection from, 383
  hazards analysis reports for, 378–379
  hazards of, 378
  internal protection from, 383
  labeling of, 382
  measuring emissions of, 380
  overview of, 378
  prefire plans for, 379–380
  protection from, 382
  radioactivity machines, 381
  shipping and storage of, 381–382
Radioactive metals, 461. *See also specific radioactive metals*
Radioactivity machines, 381
Rafters, 40
Rags, 23, 24
Rate-of-rise heat detectors, 132
Reciprocating screw injection molding machine, 482
Recirculating systems, for commercial cooking equipment, 583
Recirculation lines, for pump driven water mist systems, 190–191
Record keeping
  cross-referenced filing system, 34–35
  electronic filing system, 35
Record storage, 408–410
  archives and centers for, 410
  bulk, 409
  file rooms for, 410
  mobile compact shelving, 409
  open-shelf, 408
  plastic-media, 409
  risk factors for, 409
  types of, 408–409
  vaults for, 409–410
Records centers, 410
Records, protection of, in business occupancies, 331

Reduced-pressure principle backflow prevention assembly (RPBA) devices, 144
Reentry, for business occupancies, 330
Refrigerants, 107
Refrigerated storage, 405–406
Refuse disposal, 27
Regional distribution centers, 346
Reinforced plastics processing, 482
Remote-controlled release, for detention and correctional occupancies, 274–276
Renovations, 63–64
  inspection checklist for, 592
Repair garages, 428
Report writing, 32
Residential board and care occupancies
  characteristics of, 307–308
  definition of, 305–306
  egress capacity of, 309
  evacuation capability of, 306–307
  facility classification of, 306
  hazards of candles in, 305
  inspection checklist for, 651–654
  large facilities
    compartmentization in, 310
    detection and alarm systems for, 311–312
    fire suppression equipment in, 312–313
    hazardous areas in, 311
    means of egress for, 309
    protection of vertical openings in, 310
  licensing of, 306
  operating features in, 313
  small facilities
    compartmentization in, 310
    detection and alarm systems for, 311
    fire suppression equipment in, 312
    hazardous areas in, 310–311
    means of egress for, 308–309
    protection of vertical openings in, 309
Residential occupancies
  fire alarm systems for, 139–140
  heating systems for, 96
  private fire service mains for, 146
  smoke alarms for, 139–140
Residential sprinkler systems, 157
Restrooms, hazards of contents classification in, 12
Retail displays, for aerosols, 570–571
Retail stores. *See also* Mercantile occupancies
  trash removal at, 21
Retrieval storage systems, 399–400
Rock dusting, 446
Rockefeller Plaza, 355–356
Rocky Flats Plant, 377–378, 380
Role-playing, 229
Roll paper, 396
  classification of, 404
  fire protection for, 405
  storage of, 404–405
Roof vents
  for storage occupancies, 350
Roof/ceiling assemblies
  for industrial occupancies, 337
Roofing mops, 63
Roofs, 44–46
  hail resistance ratings of, 46
  hazards in, 62–63
  insulated steel deck, 45–46
  wind-uplift resistance of, 46
Room spacing
  for visible notification appliances, 137
Rotary kiln incinerators, 371
Rotating machines, 87
RPBA (reduced-pressure principle backflow prevention assembly) devices, 144
Rubber
  classification of, 477–478
  fire hazards of, 474–477
  overview of, 474
  warehousing of, 477–480
Rubbish disposal, 27

S

Safe havens, from explosives, 490–491
Safety cans, for storing flammable and combustible liquids, 425–426
Safety design data forms, 513
Sally ports, 277
Salute powder, 496
Sawdust
  in educational occupancies, 252
Scrap tires, 396
Secondary conductors, 91, 92
Sensitivity, 486
Shafts
  in business occupancies, 333
  fire protection for, 70
Sheet Metal and Air-Conditioning Contractors National Association (SMACNA), 108
Shipping
  of radioactive materials, 381–382
Shipping/receiving areas
  inspection of, 24
Shredders, 375
Sills, 77–78
Site plan, 14
Sleeping areas
  audible notification appliances in, 135
  visible notification appliances in, 138
Sliding doors, for detention and correctional occupancies, 276
Sludge, in water mist systems, 197
Small buildings, heating systems for, 96
Smoke
  and plastics, 474, 483
Smoke alarms. *See also* Alarm systems
  installation of, for residential occupancies, 140
  for one- and two-family dwellings, 317, 318
  for residential board and care occupancies, 311, 312
  for residential occupancies, 139–140
  vs. smoke detectors, 139
Smoke barriers, 43
  for ambulatory health care occupancies, 270
  for detention and correctional occupancies, 280
  for health care occupancies, 261–263
  inspection checklist for, 596
  in means of egress, 227–228
  vision panels in, 76
Smoke concentration decay, 114
Smoke detectors
  around escalators, 71
  ionization, disposal of, 378
  location of, 129–130
  mounting of, 129–130
  types of, 129
  vs. smoke alarms, 139
Smoke partitions
  inspection checklist for, 596
Smoke-control systems
  acceptance testing of, 119–121
  for air-conditioning systems, 109
  airflow method for, 118–119
  exhaust method for, 118
  fire door assemblies and, 73
  history of, 113–114
  inspection checklist for, 603
  mechanical, 117–119
  modern testing of, 114–115
  passive, 119
  pressurized method of, 117–118
Smokeproof enclosures, 115–116
Smoking
  in ambulatory health care occupancies, 271
  in apartment buildings, 289
  in assembly occupancies, 246
  control of, 26
  and dust hazards, 449
  in egress paths, 227
  in residential board and care occupancies, 313
Smoking materials, in business occupancies, 331
Sniff test, 438
Snow removal, 28
Sockets, common faults in, 82
Sodium bicarbonate, 203, 219
Sodium chlorite, 469
Sodium dichloroisocynurate, 469
Sodium hydrosulfite, 467

Sodium. *See also* Alkali metals
burning characteristics of, 453
Sodium-bicarbonate-based portable fire extinguishers, limitations of, 218
Sodium-potassium alloy (NaK), 458. *See also* Alkali metals
Solid fuel appliances
inspection checklist for, 601
Solid fuel central warm-air furnaces, 97
Solid fuel cooking equipment, 584–585
Solid fuel stoves, 97, 100
Solid piling storage, 397
Solvent extraction plants, 428
Space heaters. *See* Heaters
Spark/ember arrestors
for solid fuel cooking equipment, 584
Spark/ember detectors, 134
Spark-producing equipment, 87
Special agent extinguishing systems
carbon dioxide, 201–202
design types of
hand hose line systems, 200–201
local application systems, 200
total flooding systems, 199–200
dry chemical, 203–204
halogenated, 203–204
for industrial occupancies, 341
inspection and maintenance of, 201
inspection checklist for, 610–611
Special amusement buildings, in assembly occupancies, 247
Special atmospheres
air filtration for, 515–516
generated, 514–515
introduction and removal of gases in, 515
liquids for, 515
stored, 514
visual indicators of, 516
Special hazard identifiers, 412
Special industrial explosive materials, 486
Special purpose industrial occupancies, 336
Special structures and high-rise buildings
definition of, 356–357
examples of, 357
inspection checklist for, 674–677
inspection observations in, 357
open structures, 357–358
piers, 358–359
towers, 359–360
underground structures, 361–362
windowless buildings, 360–361
Specific gravity, of flammable and combustible liquids, 417
Spot-type heat detectors, arrangement of, 133
Spot-type smoke detectors, 129
Spray application finishing systems, 556–557
fire protection for, 557
process hazards of, 556–557
Spray guns, 524–526
Spray painting
apparatus for, 524–526
automatic vs. manual, 518
booths for, 519–521
control of ignition sources for, 526–527
dry filters for, 522–523
electrostatics and, 519
explosive prevention and, 528–529
extinguishment methods and, 529–530
filters for, 22
hazards of, 517
ignitable atmospheres in, 527–531
inspection checklist for, 526–527
inspection of apparatus and procedures for, 22–23, 530
residues, 23
isolation of process for, 528
limitations of fuel quantity in, 528
maintenance and, 530
maze-type dry collectors for, 522
open-floor, 524
overspray collectors for, 522
pressure relief venting, 528–529
protective clothing for, 531
rooms, 524
solventborne coatings and, 517–518
storage and handling of materials and residues for, 530–531
training for, 530
water wash collectors for, 523
waterborne coatings, 518
Spray systems
for business occupancies, 332
Sprinkler systems
for aerosol storage, 570
antifreeze, 162–163
for apartment buildings, 296
around escalators, 70
automatic, types of, 401–402
for baled cotton, 407
for business occupancies, 333
for construction, 63
for covered malls, 325
critical inspection times for, 159
for day-care facilities, 256
for detention and correctional occupancies, 273, 279
for dipping and coating finishing systems, 557
dry-pipe, 164
for dust areas, 449–450
early suppression fast response (ESFR), 401
extra large orifice (ELO), 401
fire alarm systems and, 134
grounding and, 83
for health care occupancies, 260–261, 265
for heat transfer fluid systems, 548
for hot work, 542
for hotels, 287
for hydraulic oil operated equipment, 552
for incinerators, 374
for industrial occupancies, 340, 341
inspection checklist for, 607–608
inspection of, 10, 160–162
and interior finishes, 235
K factor, 401
large-drop, 401
for lodging or rooming houses, 303
for mercantile occupancies, 157
for metal cleaning and plating operations, 549
for new apartment buildings, 290
for oil quenching operations, 551
for one- and two-family dwellings, 319
for ovens, 514
preaction and deluge, 164
for residential board and care occupancies, 312–313
retrofitting of, 123
for solid fuel cooking equipment, 584–585
steam smothering, 166
for storage occupancies, 351, 353–354, 400–401
supply piping and backflow device arrangement for, 163
supply piping and valve arrangement for, 163
testing intervals of, 161
testing of, 166–169
frequency of, 167, 168–169
types of, 162–166
very extra large orifice (VELO), 401–402
for waste-processing equipment, including shredders, 375
water spray, 165
water supply of, 159–160
wet-pipe, 162–163
Sprinkler-draft curtain method, 71
Stages, in assembly occupancies, 246
Stair enclosures
fire protection for, 70
inspection checklist for, 594
in means of egress, 227
Stairs
smoke protection for, 115–116
Stairwell doors, for hotels, 286
Stairwells
for business occupancies, 330
for residential board and care occupancies, 308
Standard array storage, 404
Standpipe and hose systems, 171–174
backflow protection for, 144–145

for construction, 63
for construction, alterations, and demolition operations
inspection checklist for, 591
frequency of testing of, 173
for industrial occupancies, 341–342
inspecting and testing of, 172–174
protection for, for detention and correctional occupancies, 279
for storage occupancies, 354
types of, 171–172
Static combs, 90
Static electricity, 89–92, 428
in conveying systems, 391
in electrical systems
inspection checklist for, 598
and plastics manufacturing, 483
in spraying applications, 527
Stationary combustion engines, 428
Statue of Liberty, 360
Steam smothering sprinkler systems, 166
Steam systems, 100–101
Steel
fire hazards of, 460–461
melting point of, 435
Steel Design Manual, 41
Steiner Tunnel Test (ASTM E84), 233, 476
Storage
of aerosols, 568–571
of ammonium nitrate, 489–490
arrangements for, 397–400
arson and, 400
automatic sprinkler systems for, 401–402
of baled cotton, 407
bulk, 397
carousel and retrieval systems, 399–400
of carpets, 406
of chemicals, 470
closed array, 404
commodity classification of, 395–397
during construction, 61
in construction, alterations, and demolition operations
inspection checklist for, 592
in covered malls, 325
of distilled spirits
barrel warehouses, 406–407
finished goods, or case goods, 406
encapsulation, 400
fire causes, 400
fire protection systems, 400–402
for flammable and combustible liquids, 420–426
of fuel, during construction, 62
of hanging garments, 406
high, 395
identification of materials for, 411–412
of idle pallets, 402–403
in industrial occupancies, 339–340
materials, 19–21
in materials-handling systems, 385–386
in one-and two-family dwellings, 318
open array, 404
outdoor, 407–408
outdoor areas, 27
packing materials in, 395–396
palletized, 398, 399
for pyrotechnics, 497–498
rack, 398–399
of radioactive materials, 381–382
of records, 408–410
archives and centers for, 410
file rooms for, 410
risk factors for, 409
types of, 408–409
vaults for, 409–410
refrigerated, 405–406
of roll paper, 404–405
of rubber tires, 27, 403–404, 478–481
solid piling, 397
special occupancy hazards, 407
standard array, 404
theft risks, 400
of wood pallets, 27
Storage occupancies
alarm systems for, 353
building services for, 352
characteristics of, 346
contents of, 346–347
definition of, 345–346
electrical hazards in, 345
exit access in, 347–348
exit discharges for, 349
exits in, 348–349
fire protection systems for, 352–353
fire pumps for, 353
general practices in, 350–351
hazardous processes in, 352
housekeeping for, 352
identification of exits for, 349–350
inspection checklist for, 670–673
inspection of, 346–347
locations of, 348–349
means of egress for, 347–350
occupant load for, 347
portable fire extinguishers for, 353
protection of openings in, 350
sprinkler systems for, 352–353
standpipe hose systems for, 353
travel distances in, 347–348
Storage practices
indoor, 350–351
outdoor, 351
Storage rooms
in detention and correction occupancies, 277
Stored-pressure water extinguisher, 219
Storerooms
for packing materials, 23–24
Stoves, coal- and wood-burning, 317–318
Styrene, 466
Suction booster pumps, for pump driven water mist systems, 191
Suction switches, for proving airflow, 506
Sulfides, 465
Sulfur, 465
Supermarkets. *See* Mercantile occupancies
Supervising stations
for fire alarm systems, 125
Supervisory signals, 125
Suppression systems
actuation, 138
control units for, 138
Surge arrestors, 91
Surveying, during initial inspection, 13–14
Sweat shops, 336
Switch boxes, common faults in, 81
Switchboards
common faults in, 81–82
inspection checklist for, 598
System pressure gauges
for clean agent extinguishing systems
inspection checklist for, 612
for special agent extinguishing systems
inspection checklist for, 610

T

Tank vehicles, for flammable and combustible liquids, 428
Tanks
inside buildings, for storing flammable and combustible liquids, 423
portable, for storing flammable and combustible liquids, 423–425
Teflon, for spray hoses, 528
Temperature effects, on flammable and combustible liquids, 417
Temporary buildings, in educational occupancies, 253
Temporary structures, 60–61
inspection checklist for, 592
Tents, 356, 364–365. *See also* Special structures and high-rise buildings
Test gauges
inspection checklist for, 589
Test Performance of Air Filter Units, 107
Textiles
inspection checklist for, 615
as wall and ceiling finishes, 233–234
Thawing pipe, 533
Theatrical special effects, 495
Thermal decomposition, in clean agent systems, 209
Thermal overload devices, 85

Thermal spraying, 533
Thermal-magnetic circuit breakers, 85
Thermoplastic ABS, 480
Thermoplastic compounding, 481
Thorium, 461
Time-delay cartridge fuses, 84–85
Tires
  storage of, 27, 396, 403–404, 478–481
Titanium
  burning characteristics of, 453
  characteristics of, 457
  fire extinguishment of, 458
  process hazards of, 457–458
Tools
  inspection checklist for, 589
Torch applied roof systems, 45
  safeguarding, 63
Torches, for hot work, 543
Total flooding systems, 199–200
Towers, 356, 359–360. *See also* Special structures and high-rise buildings
Tracer gas analysis, 119
Trade shows, in assembly occupancies, 246–247
Transfer molding, 482
Transformer fluid, hazards of, 415–416
Transformers, 85–86
  fire protection for, 169
  oil-filled
    inspection checklist for, 598–599
  for underground operations, 65
Transoms, for hotels, 285
Transport Canada (TC)
  gas container requirements of, 432
Trash cans, availability of, 21
Trash chutes, during construction, 61
Trash rooms, in detention and correction occupancies, 277
Travel distance
  for ambulatory health care occupancies
    inspection checklist for, 633
  for apartment buildings
    inspection checklist for, 645
  for assembly occupancies
    inspection checklist for, 617
  for business occupancies
    inspection checklist for, 663
  for day-care facilities
    inspection checklist for, 625
  for detention and correctional occupancies
    inspection checklist for, 637
  for educational occupancies
    inspection checklist for, 621
  for health care occupancies
    inspection checklist for, 629
  for hotels
    inspection checklist for, 641
  for industrial occupancies
    inspection checklist for, 667
  for mercantile occupancies
    inspection checklist for, 659
  for residential board and care occupancies
    inspection checklist for, 652
  for special structures and high rise buildings
    inspection checklist for, 675
  for storage occupancies, 347–348
    inspection checklist for, 671
Trichloroisocyanuric acid, 469
Trouble signals, 125
Trucks
  CNG-powered, 387
  diesel-powered, 387
  dual-fuel powered, 387
  gasoline-powered, 387
  hazards of, 389
  industrial, 386–389
  LP-Gas-powered, 387
  maintenance of, 389–390
  markers for, 386–387
  refueling and recharging of, 390
Trusses, 40
Two-family dwellings. *See* Family dwellings
Type B vent systems, 99
Type L vent systems, 99
Type S plug fuses, 84

U

UL (Underwriters Laboratories). *See* Underwriters Laboratories (UL)
UL 181, 108
UL 555, 109
UL 555C, 109
UL 864, 118
Underground operations, 64–65
  inspection checklist for, 592
Underground structures, 356, 361–362. *See also* Special structures and high-rise buildings
Underground tanks, for storing flammable and combustible liquids, 420–421
Underwriters Laboratories (UL), 99
  and gas-burning appliances and industrial heating equipment, 436
  hazards of plastics manufacturing and, 481
  and pallet storage, 403
Unit heaters, 99–100
  for storage occupancies, 351
Unlined metal chutes, 371
Unloader valves, for positive displacement pumps, 190–191
Unstable materials, identification of, 411
Upholstered furniture. *See* Furniture
Uranium, 379
  fire hazards of, 461
Urea potassium bicarbonate, 203, 219
U.S. Coast Guard, 362
Utilities
  for ambulatory health care occupancies
    inspection checklist for, 635
  for apartment buildings
    inspection checklist for, 647
  for business occupancies
    inspection checklist for, 665
  for day-care facilities
    inspection checklist for, 627
  for detention and correctional occupancies
    inspection checklist for, 639
  for hotels
    inspection checklist for, 643
  for industrial occupancies
    inspection checklist for, 669
  for lodging or rooming houses
    inspection checklist for, 650
  for mercantile occupancies
    inspection checklist for, 661
  for one- and two-family dwellings, 317
    inspection checklist for, 657
  for residential board and care occupancies
    inspection checklist for, 654
  for special structures and high rise buildings
    inspection checklist for, 677
  for storage occupancies
    inspection checklist for, 673
Utility shafts
  inspection checklist for, 594
UUKL, 118

V

Valve operating records, for industrial occupancies, 341
Valves, for water supplies
  inspection of, 145–146, 160
Vandalism, 408
Vapor density, of flammable and combustible liquids, 417
Vapor pressure, of flammable and combustible liquids, 416
Vaults, 409–410
Vehicles. *See also* Special structures and high-rise buildings
  classification of, 362
  definition of, 356
  hazard areas in, 362–363
  means of egress for, 362–363
  permanent foundation and mooring of, 362
  unusual nature and character of, 362–363
Vent connectors, 99, 101–102
Ventilation
  of flammable and combustible liquids, 427
  of medical gas systems, 439–440

Ventilation systems, 109–110
  inspection checklist for, 602
Vents, 99, 102
Vertical exits
  for apartment buildings, 294
Vertical openings
  for ambulatory health care occupancies
    inspection checklist for, 633
  for apartment buildings
    inspection checklist for, 645
  for business occupancies
    inspection checklist for, 663
  for day-care facilities
    inspection checklist for, 625
  for detention and correctional occupancies, 278
    inspection checklist for, 637
  for health care occupancies
    inspection checklist for, 629
  for hotels
    inspection checklist for, 641
  for industrial occupancies, 338
    inspection checklist for, 667
  for lodging or rooming houses, 301
    inspection checklist for, 649
  for mercantile occupancies
    inspection checklist for, 659
  protection of, 68–71
  for residential board and care occupancies, 309–310
    inspection checklist for, 652
  for special structures and high rise buildings
    inspection checklist for, 675
  for storage occupancies
    inspection checklist for, 671
Vertical turbine pumps, 148
Very extra large orifice automatic sprinkler systems, 401–402
Vessels. *See also* Special structures and high-rise buildings
  classification of, 362
  definition of, 356
  hazard areas in, 362–363
  means of egress for, 362–363
  permanent foundation and mooring of, 362
  unusual nature and character of, 362–363
Vinyl, expanded
  as wall and ceiling finishes, 233–234
Violations
  notices, 3, 14
  written notices of, 30–32
Viscosity, of flammable and combustible liquids, 417
Visible notification appliances, 136–138
Vision panels, 76–77
  for health care occupancies, 261
Volatility, of flammable and combustible liquids, 416
Voluntary inspections, of family dwellings, 317–319

W

Wall aisle space
  for storage occupancies, 351
Wall finishes, 232–234
  classification of, 474
Wall furnaces, 98
Walls
  bearing, 39, 42
  common, 42, 43
  fire, 42, 43
  fire barrier, 43
  fire-resistive ratings of, 50
  nonbearing, 39, 42
  shear, 42, 43
  for Type IV fire-resistive construction, 55
Warehouses
  aerosols, 570
  general purpose, 569
  for plastics and rubber, 477–480
Warm-air systems, 100
Washington Monument, 360
Waste anesthesia gas disposal (WAGD), 439
Waste chutes, 370–371
  in apartment buildings, 295
  in business occupancies, 332
Waste compactors
  chute termination bins for, 374
  commercial-industrial, 374
  storage rooms for, 374
  types of, 374
Waste disposal, 27
  for business occupancies, 331–332
Waste-chute and -handling systems, 370–371
  chute service opening rooms of, 371
  chute terminal enclosures of, 371
  incineration fire safety, 371–372
  incinerators for
    auxiliary fuel systems for, 374
    charging systems for, 373–374
    commercial-industrial, 371–372
    equipment design and construction of, 372
    layout and arrangement of, 372–373
    waste-handling rooms for, 372
  magnetic separators in, 375
  other waste-processing equipment, 375
  shredders, 375
  types of, 370
  waste chutes, 370–371
  waste compactors for, 374
    chute termination bins for, 374
    commercial-industrial, 374
    storage rooms for, 374
Waste-handling and -processing systems
  overview of, 369–370
Water
  radioactive contamination of, 379–380
Water cylinders, for water mist systems, 186–187
Water flow tests, for industrial occupancies, 341
Water gel, 486
Water main, flow testing of, 153–154
Water mist systems
  acceptance testing of
    flushing and cleaning, 178
    hydrostatic test, 178–179
    preliminary functional test, 179
    review of electrical components, 179
    review of mechanical components, 179
    system documentation and test record, 180–181
    systems operational tests, 180
  ancillary equipment, testing of, 197
  centrifugal fire pump, maintenance of, 189
  detection equipment of, testing of, 197
  discharge testing of, 187
    cycling of systems for, 195
    full, 195
    modified, 195
    orifice disks in, 195
    timing in, 195–196
  end of discharge of, 187
  frequency of testing of, 182–184
  general system components, inspection and testing frequency of, 197
  high-pressure, single-fluid, compressed gas-driven systems with stored water, 185–187
  inspection checklist for, 609
  inspection of, 181
  inspection, testing and maintenance records for, 198
  leakage in, 196
  limitations of dedicated storage reservoirs in, 189
  low- and intermediate-pressure, single-fluid, compressed gas-driven with stored water, 187–188
  low-pressure, twin-fluid, compressed gas-driven with stored water, 187
  nozzles of, testing of, 196–197
  positive displacement (PD) pumps in, 189–192
  preengineered, 187
  pump driven
    break tank connections in, 192
    break tanks and recirculation lines for, 190–191
    controls for, 190

corrosion in, 192
filters for, 191–192
freshwater/seawater transfer in, 192
high-pressure, 190
overview of, 188–190
pneumatic jockey pump for, 192
suction booster pumps in, 191
refilling water cylinders of, 186–187
sectional control valves of, testing of, 196
sludge, hazards of, 197
testing of, 181–185
water storage tanks for
algae growth in, 194
ambient temperature of, 193
bacteriological growth in, 194
components of, 193
low-level trouble signals of, 188, 193
pressure relief valves for, 194
sight glass of, 188, 193
sterilization of, 194
strainers in, 194–195
water quality of, 194–195
Water pressure, standards for fire fighting, 153
Water quality, of water mist systems, 194–195
Water solubility, of flammable and combustible liquids, 417
Water spray sprinkler systems, 165
testing of, 167–169
Water storage tanks
cathodic protection in, 152
inspection of, 151–153
types of, 150–151
Water supplies
arrangements for, 143–144
for industrial occupancies, 341
inspection checklist for, 606
minimum requirements for fire fighting of, 153–155
sources of, 143–144
for sprinkler systems, 159–160
valves and metering of, 144–146
valves for
inspection of, 145–146
Water wash collectors, 523
Waterflow detection devices
alarms for, 134
requirements for, 127–128
Water-surrounded structures, 356, 363–364. *See also* Special structures and high-rise buildings
Weeds, control of, 26–27
Welding. *See also* Hot work
and dust hazards, 449
fire incident trends in, 544
hazards of, 62
near conveying systems, 390
overview of, 533–534
oxygen and, 435
oxygen-fuel gas in, 436
Wet barrel hydrants
inspection checklist for, 606
Wet chemical fire suppression systems
fire alarm systems and, 134
Wet-pipe sprinkler systems
for flammable and combustible liquids, 428
for industrial occupancies, 335
and radioactive materials, 379
testing of, 162, 166
vs. antifreeze sprinkler systems, 162–163
vs. dry-pipe sprinkler systems, 164
Windowless and underground structures
in business occupancies, 334
Windowless buildings, 356, 360–361. *See also* Special structures and high-rise buildings
Windows
for educational occupancies, 251
for one- and two-family dwellings, 315
for residential board and care occupancies, 308–309
Wind-uplift resistance, of roofs, 46
Wired glass, 76–77
Wire-glass windows, for storage occupancies, 350
Wiring, electrical, 80–85
of fire alarms systems, 140–142
pigtail connections in, 141
T-tapping, 141–142
Wood chips
storage of, 397
Wood, for solid fuel cooking equipment, 584
Wood pallets, storage of, 27
Wood-burning stoves, 97, 100, 317–318
Woodworking
fire protection for, 552–553
hazards of, 552
housekeeping for, 553

X

X-ray machines, 381

Y

Yard systems, flow testing of, 154–155

Z

Zinc, hazards of, 460
Zirconium
burning characteristics of, 453
characteristics of, 458
fire extinguishment of, 459
process hazards of, 458–459
storage of, 459

About the Editor

Robert Solomon is the Assistant Vice President for Building and Life Safety Codes at NFPA. He oversees the operations of the department whose projects include NFPA *101*®, *Life Safety Code*®, the NFPA fire test standards, NFPA and HUD Manufactured Housing projects, and *NFPA 5000*™, *Building Construction and Safety Code*™.

Mr. Solomon's previous responsibilities have included serving as the staff liaison to the various NFPA Technical Committees that deal with water-based extinguishing systems. He has authored three editions of NFPA's *Automatic Sprinkler Systems Handbook* and written several chapters in NFPA's *Fire Protection Handbook*.

Prior to starting at NFPA in 1986, Mr. Solomon worked for the Naval Facilities Engineering Command in Charleston, SC. He worked in the Design Division where he was involved in the fire protection design and layout of various buildings and structures.